Magnetospheric Substorms

Geophysical Monograph Series

Including

IUGG Volumes
Maurice Ewing Volumes
Mineral Physics Volumes

GEOPHYSICAL MONOGRAPH SERIES

Geophysical Monograph Volumes

1. Antarctica in the International Geophysical Year   *A. P. Crary, L. M. Gould, E. O. Hulburt, Hugh Odishaw, and Waldo E. Smith (Eds.)*
2. Geophysics and the IGY   *Hugh Odishaw and Stanley Ruttenberg (Eds.)*
3. Atmospheric Chemistry of Chlorine and Sulfur Compounds   *James P. Lodge, Jr. (Ed.)*
4. Contemporary Geodesy   *Charles A. Whitten and Kenneth H. Drummond (Eds.)*
5. Physics of Precipitation   *Helmut Weickmann (Ed.)*
6. The Crust of the Pacific Basin   *Gordon A. Macdonald and Hisahi Kuno (Eds.)*
7. Antarctica Research: The Matthew Fontaine Maury Memorial Symposium   *H. Wexler, M. J. Rubin, and J. E. Caskey, Jr. (Eds.)*
8. Terrestrial Heat Flow   *William H. K. Lee (Ed.)*
9. Gravity Anomalies: Unsurveyed Areas   *Hyman Orlin (Ed.)*
10. The Earth Beneath the Continents: A Volume of Geophysical Studies in Honor of Merle A. Tuve   *John S. Steinhart and T. Jefferson Smith (Eds.)*
11. Isotope Techniques in the Hydrologic Cycle   *Glenn E. Stout (Ed.)*
12. The Crust and Upper Mantle of the Pacific Area   *Leon Knopoff, Charles L. Drake, and Pembroke J. Hart (Eds.)*
13. The Earth's Crust and Upper Mantle   *Pembroke J. Hart (Ed.)*
14. The Structure and Physical Properties of the Earth's Crust   *John G. Heacock (Ed.)*
15. The Use of Artificial Satellites for Geodesy   *Soren W. Henricksen, Armando Mancini, and Bernard H. Chovitz (Eds.)*
16. Flow and Fracture of Rocks   *H. C. Heard, I. Y. Borg, N. L. Carter, and C. B. Raleigh (Eds.)*
17. Man-Made Lakes: Their Problems and Environmental Effects   *William C. Ackermann, Gilbert F. White, and E. B. Worthington (Eds.)*
18. The Upper Atmosphere in Motion: A Selection of Papers With Annotation   *C. O. Hines and Colleagues*
19. The Geophysics of the Pacific Ocean Basin and Its Margin: A Volume in Honor of George P. Woollard   *George H. Sutton, Murli H. Manghnani, and Ralph Moberly (Eds.)*
20. The Earth's Crust: Its Nature and Physical Properties   *John G. Heacock (Ed.)*
21. Quantitative Modeling of Magnetospheric Processes   *W. P. Olson (Ed.)*
22. Derivation, Meaning, and Use of Geomagnetic Indices   *P. N. Mayaud*
23. The Tectonic and Geologic Evolution of Southeast Asian Seas and Islands   *Dennis E. Hayes (Ed.)*
24. Mechanical Behavior of Crustal Rocks: The Handin Volume   *N. L. Carter, M. Friedman, J. M. Logan, and D. W. Stearns (Eds.)*
25. Physics of Auroral Arc Formation   *S.-I. Akasofu and J. R. Kan (Eds.)*
26. Heterogeneous Atmospheric Chemistry   *David R. Schryer (Ed.)*
27. The Tectonic and Geologic Evolution of Southeast Asian Seas and Islands: Part 2   *Dennis E. Hayes (Ed.)*
28. Magnetospheric Currents   *Thomas A. Potemra (Ed.)*
29. Climate Processes and Climate Sensitivity (Maurice Ewing Volume 5)   *James E. Hansen and Taro Takahashi (Eds.)*
30. Magnetic Reconnection in Space and Laboratory Plasmas   *Edward W. Hones, Jr. (Ed.)*
31. Point Defects in Minerals (Mineral Physics Volume 1)   *Robert N. Schock (Ed.)*
32. The Carbon Cycle and Atmospheric CO_2: Natural Variations Archean to Present   *E. T. Sundquist and W. S. Broecker (Eds.)*
33. Greenland Ice Core: Geophysics, Geochemistry, and the Environment   *C. C. Langway, Jr., H. Oeschger, and W. Dansgaard (Eds.).*
34. Collisionless Shocks in the Heliosphere: A Tutorial Review   *Robert G. Stone and Bruce T. Tsurutani (Eds.)*
35. Collisionless Shocks in the Heliosphere: Reviews of Current Research   *Bruce T. Tsurutani and Robert G. Stone (Eds.)*
36. Mineral and Rock Deformation: Laboratory Studies—The Paterson Volume   *B. E. Hobbs and H. C. Heard (Eds.)*
37. Earthquake Source Mechanics (Maurice Ewing Volume 6)   *Shamita Das, John Boatwright, and Christopher H. Scholz (Eds.)*
38. Ion Acceleration in the Magnetosphere and Ionosphere   *Tom Chang (Ed.)*
39. High Pressure Research in Mineral Physics (Mineral Physics Volume 2)   *Murli H. Manghnani and Yasuhiko Syono (Eds.)*
40. Gondwana Six: Structure, Tectonics, and Geophysics   *Garry D. McKenzie (Ed.)*

41 Gondwana Six: Stratigraphy, Sedimentology, and Paleontology *Garry D. McKenzie (Ed.)*

42 Flow and Transport Through Unsaturated Fractured Rock *Daniel D. Evans and Thomas J. Nicholson (Eds.)*

43 Seamounts, Islands, and Atolls *Barbara H. Keating, Patricia Fryer, Rodey Batiza, and George W. Boehlert (Eds.)*

44 Modeling Magnetospheric Plasma *T. E. Moore and J. H. Waite, Jr. (Eds.)*

45 Perovskite: A Structure of Great Interest to Geophysics and Materials Science *Alexandra Navrotsky and Donald J. Weidner (Eds.)*

46 Structure and Dynamics of Earth's Deep Interior (IUGG Volume 1) *D. E. Smylie and Raymond Hide (Eds.)*

47 Hydrological Regimes and Their Subsurface Thermal Effects (IUGG Volume 2) *Alan E. Beck, Grant Garven, and Lajos Stegena (Eds.)*

48 Origin and Evolution of Sedimentary Basins and Their Energy and Mineral Resources (IUGG Volume 3) *Raymond A. Price (Ed.)*

49 Slow Deformation and Transmission of Stress in the Earth (IUGG Volume 4) *Steven C. Cohen and Petr Vaníček (Eds.)*

50 Deep Structure and Past Kinematics of Accreted Terranes (IUGG Volume 5) *John W. Hillhouse (Ed.)*

51 Properties and Processes of Earth's Lower Crust (IUGG Volume 6) *Robert F. Mereu, Stephan Mueller, and David M. Fountain (Eds.)*

52 Understanding Climate Change (IUGG Volume 7) *Andre L. Berger, Robert E. Dickinson, and J. Kidson (Eds.)*

53 Plasma Waves and Istabilities at Comets and in Magnetospheres *Bruce T. Tsurutani and Hiroshi Oya (Eds.)*

54 Solar System Plasma Physics *J. H. Waite, Jr., J. L. Burch, and R. L. Moore (Eds.)*

55 Aspects of Climate Variability in the Pacific and Western Americas *David H. Peterson (Ed.)*

56 The Brittle-Ductile Transition in Rocks *A. G. Duba, W. B. Durham, J. W. Handin, and H. F. Wang (Eds.)*

57 Evolution of Mid Ocean Ridges (IUGG Volume 8) *John M. Sinton (Ed.)*

58 Physics of Magnetic Flux Ropes *C. T. Russell, E. R. Priest, and L. C. Lee (Eds.)*

59 Variations in Earth Rotation (IUGG Volume 6) *Dennis D. McCarthy and William E. Carter (Eds.)*

60 Quo Vadimus *Geophysics for the Next Generation* (IUGG Volume 10) *George D. Garland and John R. Apel (Eds.)*

61 Cometary Plasma Processes *Alan D. Johnstone (Ed.)*

62 Modeling Magnetospheric Plasma Processes *Gordon R. Wilson (Ed.)*

63 Marine Particles: Analysis and Characterization *David C. Hurd and Derek W. Spencer (Eds.)*

Maurice Ewing Volumes

1 Island Arcs, Deep Sea Trenches, and Back-Arc Basins *Manik Talwani and Walter C. Pitman III (Eds.)*

2 Deep Drilling Results in the Atlantic Ocean: Ocean Crust *Manik Talwani, Christopher G. Harrison, and Dennis E. Hayes (Eds.)*

3 Deep Drilling Results in the Atlantic Ocean: Continental Margins and Paleoenvironment *Manik Talwani, William Hay, and William B. F. Ryan (Eds.)*

4 Earthquake Prediction—An International Review *David W. Simpson and Paul G. Richards (Eds.)*

5 Climate Processes and Climate Sensitivity *James E. Hansen and Taro Takahashi (Eds.)*

6 Earthquake Source Mechanics *Shamita Das, John Boatwright, and Christopher H. Scholz (Eds.)*

IUGG Volumes

1 Structure and Dynamics of Earth's Deep Interior *D. E. Smylie and Raymond Hide (Eds.)*

2 Hydrological Regimes and Their Subsurface Thermal Effects *Alan E. Beck, Grant Garven, and Lajos Stegena (Eds.)*

3 Origin and Evolution of Sedimentary Basins and Their Energy and Mineral Resources *Raymond A. Price (Ed.)*

4 Slow Deformation and Transmission of Stress in the Earth *Steven C. Cohen and Petr Vaníček (Eds.)*

5 Deep Structure and Past Kinematics of Accreted Terranes *John W. Hillhouse (Ed.)*

6 Properties and Processes of Earth's Lower Crust *Robert F. Mereu, Stephan Mueller, and David M. Fountain (Eds.)*

7 Understanding Climate Change *Andre L. Berger, Robert E. Dickinson, and J. Kidson (Eds.)*

8 Evolution of Mid Ocean Ridges *John M. Sinton (Ed.)*

9 Variations in Earth Rotation *Dennis D. McCarthy and William E. Carter (Eds.)*

10 Quo Vadimus *Geophysics for the Next Generation* *George D. Garland and John R. Apel (Eds.)*

Mineral Physics Volumes

1 Point Defects in Minerals *Robert N. Schock (Ed.)*

2 High Pressure Research in Mineral Physics *Murli H. Manghnani and Yasuhiko Syono (Eds.)*

Geophysical Monograph 64

Magnetospheric Substorms

Joseph R. Kan
Thomas A. Potemra
Susumu Kokubun
Takesi Iijima
Editors

American Geophysical Union

Published under the aegis of the AGU Books Board.

Library of Congress Cataloging-in-Publication Data

Magnetospheric substorms / Joseph R. Kan . . . [et al.], editors.
    p.   cm. — (Geophysical monograph : 64)
  ISBN 0-87590-030-5
  1. Magnetospheric substorms—Congresses.   I. Kan, Joseph R.
II. Series.
QC809.M35M333   1991
538'.766—dc20   91-34806
                                                                                                                              CIP

ISBN 0-87590-030-5

Copyright 1991 by the American Geophysical Union, 2000 Florida Avenue, NW, Washington, DC 20009, U.S.A.

Figures, tables, and short excerpts may be reprinted in scientific books and journals if the source is properly cited.

   Authorization to photocopy items for internal or personal use, or the internal or personal use of specific clients, is granted by the American Geophysical Union for libraries and other users registered with the Copyright Clearance Center (CCC) Transactional Reporting Service, provided that the base fee of $1.00 per copy plus $0.10 per page is paid directly to CCC, 21 Congress Street, Salem, MA 10970. 0065-8448/89/$01. + .10.
   This consent does not extend to other kinds of copying, such as copying for creating new collective works or for resale. The reproduction of multiple copies and the use of full articles or the use of extracts, including figures and tables, for commercial purposes requires permission from AGU.

Printed in the United States of America.

CONTENTS

PREFACE
   J. Kan, T. A. Potemra, S. Kokubun, and T. Iijima   xiii

1. INTRODUCTION

Development of Magnetospheric Physics
   S. I. Akasofu   3

The Beginning of Substorm Research
   D. P. Stern   11

2. OVERVIEWS

Physics of Magnetospheric Substorms: A Review
   M. I. Pudovkin   17

Substorm Current Systems and Auroral Dynamics
   Y. I. Feldstein   29

A Synthesis Model for Magnetospheric Substorms
   A. T. Y. Lui   43

Observational Constraints for Substorm Models
   G. Rostoker   61

Synthesizing A Global Model of Substorms
   J. R. Kan   73

3. SUBSTORM CURRENTS

An Empirical Model of Substorm-Related Magnetic Field Variations at Synchronous Orbit
   T. Nagai   91

The Relationship Between Ion and Electron Precipitation Patterns and Field-Aligned Current Systems During a Substorm
   T. Iijima, M. Watanabe, T. A. Potemra, L. J. Zanetti, and F. J. Rich   97

Birkeland-Ionosphere Currents of the Magnetospheric Storm Circuit
   L. J. Zanetti, T. A. Potemra, T. Iijima, and W. Baumjohann   111

Simultaneous Observations of the Westward Electrojet and the Cross-Tail Current Sheet During Substorms
   R. E. Lopez, H. Spence, and C. I. Meng   123

Tail Current Disruption in the Geosynchronous Region
   S. Ohtani, K. Takahashi, L. J. Zanetti, T. A. Potemra, R. W. McEntire, and T. Iijima   131

CONTENTS

4. PLASMA SHEET DYNAMICS

Heating and Fast Flows in the Near-Earth Tail
W. Baumjohann 141

The Earthward Edge of the Plasma Sheet in Magnetospheric Substroms
D. N. Baker and T. I. Pulkkinen 147

Association Between Tail Substorm Phenomena and Magnetic Separatrix Distortion
L. R. Lyons 161

Three Dimensional Numerical Simulations of Magnetotail Reconnection
M. Scholer, A. Otto, and G. Gedbois 171

Substorm Features in MHD Simulations of Magnetotail Dynamics
J. Birn and M. Hesse 177

A Magnetosphere Wags the Tail Model of Substorms
G. Atkinson 191

Role of the Near Earth Plasma Sheet at Substorms
A. Roux 201

Properties of the Geotail Plasma Sheet—Theory and Observations
C. J. Owen, S. W. H. Cowley, and I. G. Richardson 215

Tracing and Acceleration of Mid-Tail Ions During Substorms
D. C. Delcourt 225

5. AURORAL SUBSTORM MORPHOLOGY

Auroral Substorm Observed by UV-Imager on Akebono
E. Kaneda and T. Yamamoto 235

Viking Optical Substorm Signatures
J. S. Murphree, R. D. Elphinstone, L. L. Cogger, and D. Hearn 241

Observations of Changes to the Auroral Distribution Prior to Substorm Onset
R. D. Elphinstone, J. S. Murphree, L L. Cogger, D. Hearn, M. G. Henderson, and R. Lundin 257

EXOS-D Observations of Charged Particle Precipitation and Acceleration Process
T. Mukai, N. Kaya, and W. Miyake 277

Auroral and Energetic Particle Signatures During A Substorm With Multiple Expansion
R. Nakamura, T. Oguti, T. Yamamoto, S. Kokubun, D. N. Baker, and R. D. Belian 285

Auroral Substorms
W. J. Heikkila 295

Toward a Better Understanding of the Global Auroral Electrodynamics Through Numerical Modeling Studies
G. Marklund and L. Blomberg 305

6. CHARACTERISTICS OF SUBSTORM PHASES

Statistical Features of the Substorm Expansion Phase as Observed by the AMPTE/CCE Spacecraft
I. A. Daglis, E. T. Sarris, W. I. Axford, G. Kremser, B. Wilken, and G. Gloeckler 323

Auroral Signatures of Substorm Recovery Phase: A Case Study
T. I. Pulkkinen, R. J. Pellinen, H. E. J. Koskinen, H. J. Opgenoorth, J. S. Murphree, V. Petrov, A. Zaitzev, and E. Friis-Christensen 333

A Statistical Study of Substorm Onset Conditions at Geostationary Orbit
A. Korth, Z. Y. Pu, G. Kremser, and A. Roux 343

CONTENTS

7. SUBSTORM ELECTRODYNAMICS

The Contribution of the Boundary Layer EMF to Magnetospheric Substorms
R. Lundin, I. Sandahl, J. Woch, and R. Elphinstone   355

Polar Cap Convection: Steady State and Dynamic Effects
J. J. Moses and P. Reiff   375

Polar Hiss
T. Ondoh   387

Driven and Unloading Electrojets During the Main Phase of a Magnetic Storm
K. Lassen and E. Friis-Christensen   399

Occurrence of Magnetic Flux Transfer Events During Substorm
H. Kawano, S. Kokubun, and K. Takahashi   409

Substorm Electrodynamics
D. Stern   421

Characteristics of the Fields and Particle Acceleration During Rapidly Induced Tail Thinning and Reconnection
R. M. Winglee   425

The Development of Field-Aligned Currents and Auroral Particle Acceleration During Active Times
P. B. Dusenbery, R. M. Winglee, and G. A. Dulk   437

A Nonlinear Dynamic Analogue Model of Geomagnetic Activity
A. J. Klimas, D. N. Baker, D. A. Roberts, and D. H. Fairfield   449

Synergetic Approach to Substorm Phenomenon
Z. Vörös   461

Ring Current Ion With Micropulsations During the Recovery Phase of Geomagnetic Storms
X. Li, A. Chan, and M. Hudson   469

PREFACE

This volume on Magnetospheric Substorms is a compilation of papers invited and contributed to the Chapman Conference on Magnetospheric Substorms held September 3-7, 1990, in Hakone, Japan. The Conference was attended by 149 researchers from 13 countries: Japan, the United States, the U.S.S.R., the United Kingdom, China, Sweden, Norway, Netherlands, France, Finland, Germany, Denmark, and Canada.

Each chapter is organized with an intentional mixture of observational and theoretical papers on similar topics to emphasize the importance of the observation-simulation-theory closure. Of equal importance is the global connection in substorm research. In the past, substorm research has focused on understanding the substorm signatures in the inosphere and in the plasma sheet separately. Future progress may well depend on paying greater attention to the global connection between the ionosphere and the plasma sheet.

In this volume the reader will find diametrically opposed views on the cause-and-effect relationships for substorm phenomena. For example, one of the key issues in substorm research today is the cause-and-effect relationship in the onset of substorm expansion. The long-standing controversy surrounding this issue was brought to light during the conference by calling attention to the following questions: (1) Is the formation of a near-Earth X line uniquely associated with the immediate cause or should it be considered an effect of the onset of substorm expansion? (2) Is the disruption of the cross-tail current the immediate cause or a consequence of the onset of substorm expansion? (3) Is the ionospheric response to an enhanced magnetospheric convection the immediate cause or a consequence of the onset of substorm expansion? An important step toward resolving these questions would be to determine the timing and relative locations of the three key events of a substorm: (1) the substorm expansion onset, as defined by the sudden brightening of the most equatorward arc; (2) the disruption of the cross-tail current, as measured by the dipolarization of the magnetic configuration in the plasma sheet of the same substorm; and (3) the formation of a near-Earth X line associated with the same substorm. Development of the substorm current wedge is another key issue that deserves greater attention.

We believe that investigators in substorm research are now ready to face the challenge of establishing the timing and relative locations of these interrelated events of a substorm—undoubtedly one of the most challenging tasks in space research today. Significant progress depends upon close interaction between observational analysis-synthesis studies and theoretical modeling-simulation efforts. NASA's International Solar Terrestrial Physics Program (ISTP), NSF's Geospace Environment Modeling Program (GEM), and the International Council of Scientific Unions Solar-Terrestrial Energy Program (STEP) will provide powerful tools for solving the controversial issues of substorm physics.

Substorm research is a fast-developing field. It is our hope that this volume will serve as the leading reference in pointing the direction for future research. We believe that significant progress will be made in future substorm research by emphasizing the global connection between the ionosphere and the plasma sheet and the observation-simulation-theory closure. This conference has certainly served its purpose by bringing all of the magnetospheric substorm expertise in the world together.

Joseph R. Kan
Geophysical Institute
University of Alaska Fairbanks

Thomas A. Potemra
Applied Physics Laboratory
The Johns Hopkins University

Susumu Kokubun
Takesi Iijima
Department of Earth and Planetary Physics
Faculty of Science
The University of Tokyo

1. INTRODUCTION

Development of Magnetospheric Physics

S.-I. Akasofu

Geophysical Institute, University of Alaska Fairbanks, Fairbanks, Alaska 99775-0800

Abstract. Magnetospheric physics has grown out of classical disciplines, such as geomagnetism, auroral physics, ionospheric physics and cosmic ray physics after 1958 and has had a fascinating history of development. The purpose of this paper is to furnish the general background for those that follow by describing briefly the historical development. It is emphasized that magnetospheric physics and a study of magnetospheric substorms are still in the developmental stage. A new instrumental development, a new analytical finding or a new theoretical progress could lead this discipline to a new era of development far beyond the present generation can conceive of at this time. Many basic processes we reveal in this discipline may be applicable to solar, stellar and interstellar phenomena.

Magnetospheric physics is a science in which astonishing advances have been made after the advent of space age in 1958, although it had begun first as classic disciplines, such as geomagnetism, ionospheric physics, auroral physics and cosmic ray physics. Until 1958, practically the whole of our knowledge of space physics had been derived from ground-based observations and theoretical research. K. Birkeland [1908] was the first to make an extensive geomagnetic observation in the polar region and to deduce the current systems which are responsible for geomagnetic disturbances. Further, with C. Störmer, he examined motions of individual charged particles in the vicinity of a dipole field. At that time, Birkeland [1908] was fascinated by electrons which were newly discovered by J. J. Thomson in 1897; in fact, J. J. and G. P. Thomson [1903] suggested that the aurora was caused by the passage of electrons through the upper atmosphere after being ejected from the sun. It was generally believed at that time that the aurora was distributed along the auroral zone, a narrow belt which lies parallel to the geomagnetic latitude of 67°. Figures 1a, b and c show a pictorial representation of the progress of our understanding of the magnetosphere from the beginning of this century.

The next major development was the recognition by Chapman and Ferraro [1931] that the solar "corpuscular stream" is what we now call plasma. In fact, Chapman and Ferraro devised a formula similar to that of the Debye length and argued that the solar gas flow in the vicinity of the earth should be treated as a plasma, not motions of individual charged particles. This effort led them to the first theory on the formation of the magnetosphere. They correctly estimated the distance between the earth and the dayside magnetopause.

Chapman [1935] also made an extensive study of geomagnetic records and deduced the so-called 'SD' current system for geomagnetic disturbances. The term 'SD' stands for the solar daily magnetic disturbance [Chapman and Bartels, 1940]. Chapman's concept was that geomagnetic time variations observed at a point on the earth result from the fact that the earth and the observation point rotated once a day under the current system which was fixed with respect to the sun, rather than intrinsic time variations of the current system. This concept was an extension of the Sq variations (the solar quiet day daily variations) which arises from the fact that an observation point on the earth rotates once a day under the ionospheric current system which is driven by a dynamo process in the ionosphere and is fixed with respect to the sun. Such a concept is a reasonable first approximation for the Sq current system, but has later been found to be inadequate for the rapidly changing substorm current systems. Another important point of Chapman's current system is that it is an equivalent current system. Chapman argued that since there is an infinite number of possible current systems for a given distribution of magnetic disturbance vectors on the earth's surface, it is not worthwhile to consider just one possible three-dimensional current system. Thus, he devised the two-dimensional equivalent current system which is mathematically unique. In later years, a number of researchers assumed such an equivalent current system to be the true current system, another prevailing paradigm.

Magnetospheric Substorms
Geophysical Monograph 64
Copyright 1991 American Geophysical Union

4 DEVELOPMENT OF MAGNETOSPHERIC PHYSICS

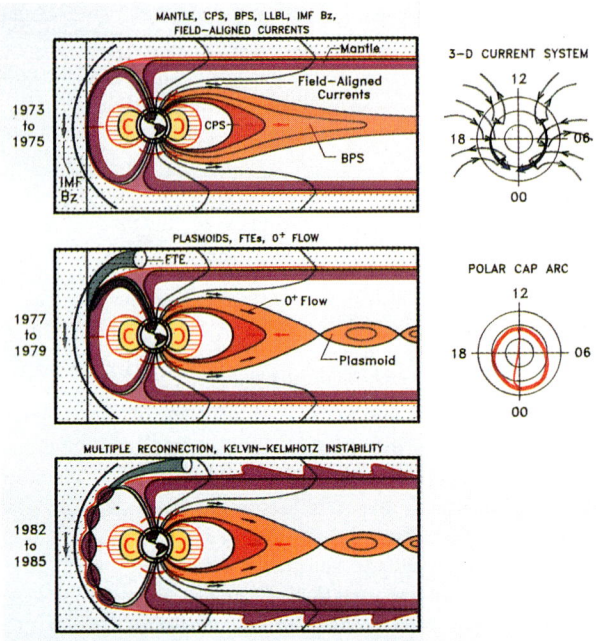

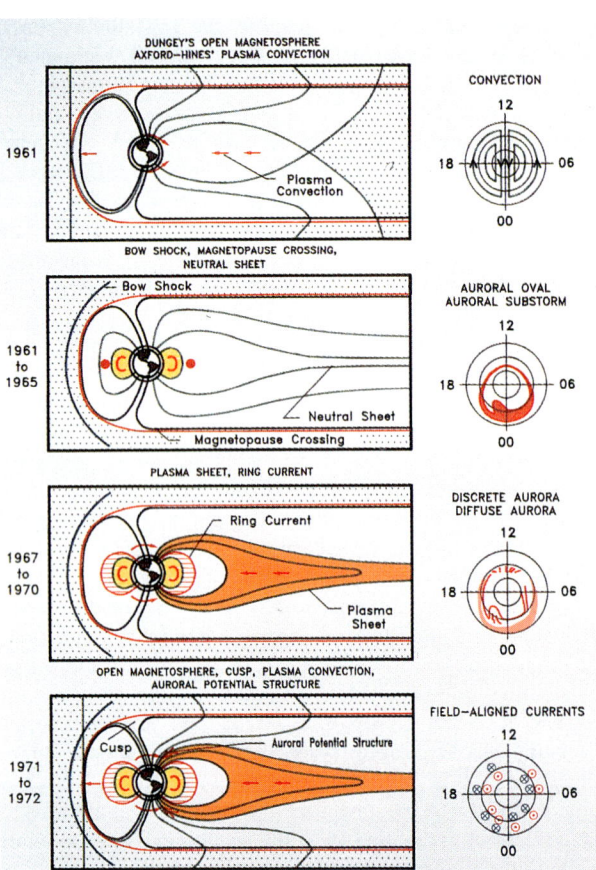

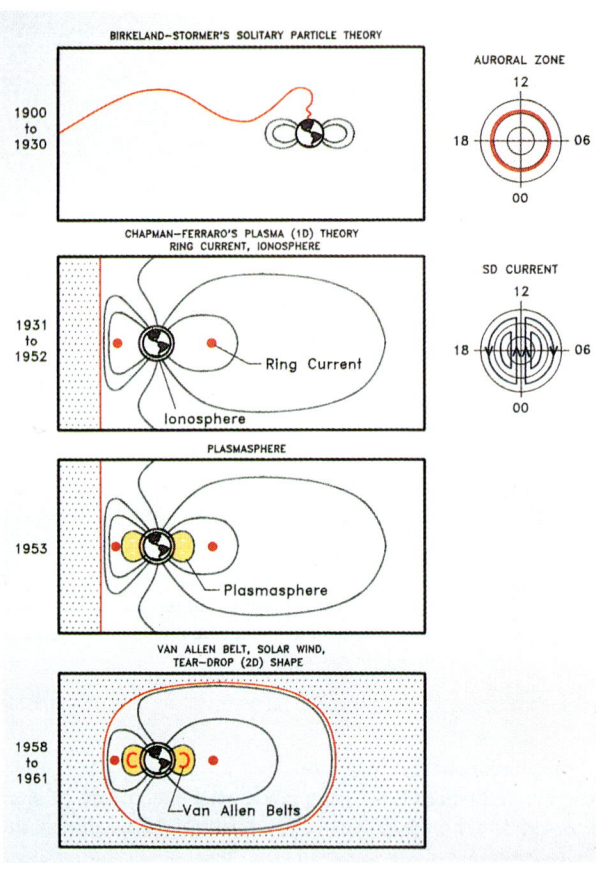

Fig. 1. Pictorial representation of the progress of our understanding of the magnetosphere from the beginning of this century.

In spite of a fascinating initial progress in this discipline, however, there was no significant development until 1958, except for the discovery of the plasmasphere by Storey [1953]. The discovery of the Van Allen belts in 1958 inaugurated the era of space exploration. The motions of trapped particles in the belts are found to carry a westward current. Later, it was found that an abnormal growth of a belt of protons is the long-sought ring current belt which is responsible for the main phase of geomagnetic storms.

The year 1961 was an important milestone in magnetospheric physics. Axford and Hines [1961] proposed that Chapman's SD current system arises from a convective motion of plasma in the magnetosphere. They suggested that only electrons can participate in the ExB convective motion in the lower ionosphere (since the motion of positive ions is inhibited by frequent collisions with neutral particles there). Thus, on the basis of the convection pattern at the ionospheric level, they inferred the convection pattern in the equatorial plane of the magnetosphere.

In the same year, the first major revision to the Chapman-Ferraro theory was suggested by Dungey [1961] who considered the importance of the interplanetary magnetic field (IMF) in the process of the magnetosphere formation. His paper was very important in shaping the course of magnetospheric research. The solar plasma envisaged by Chapman and Ferraro was a diamagnetic plasma. It streams around the earth by confining the earth's magnetic field into a comet-shaped cavity. One of the most important consequences of the diamagnetic plasma flow is that it prevents the penetration of the earth's magnetic field into the plasma. This is a sort of skin effect or shielding. Thus, the earth's magnetic field is completely confined in the cavity. The magnetosphere produced by a diamagnetic plasma may be called the 'closed' magnetosphere. Because of this nature, the closed magnetosphere does not allow for the transfer of the solar wind energy into the magnetosphere, except for a very small amount funneled through the cusp region.

Dungey suggested that magnetic field lines in the solar wind plasma 'reconnect' with those of the earth's magnetic field across the boundary of the magnetosphere. Thus, the magnetosphere becomes 'open', in which magnetic fluxes from the polar region are connected to the IMF fluxes. The reconnection process allows the solar wind to flow across the reconnected magnetic field lines. It is this flow of the solar wind across the reconnected magnetic field lines, which allows a dynamo process to take place. Indeed, it is this dynamo process, the solar wind-magnetospheric dynamo, which supplies the power for most of magnetospheric processes, including magnetospheric substorms. This important point was, however, not explicit in Dungey's paper. The importance of the IMF B_z component in substorm processes had gradually been recognized from satellite observations during the following years. The first important paper in this regard was published by Arnoldy [1971]. However, most researchers at that time were concerned with an idea as to how the reconnection process in the magnetotail might provide the necessary energy for magnetospheric substorms. A study of magnetic reconnection in terms of the MHD formulation was another prevailing paradigm for more than two decades. The concept of magnetic reconnection was originally introduced to explain solar flares. Thus, it was thought a priori at that time that the process was supposed to be spontaneous and explosive with a single X-line. However, an extensive MHD simulation study in recent years appears to show that the magnetic reconnection in the magnetosphere is a driven process, often contains multiple X-lines [Lee and Fu, 1986; Lee et al., 1985].

It was as late as 1974 when the importance of the solar wind-magnetosphere dynamo process was recognized by Gonzalez and Mozer [1974] and verified by Perrault and Akasofu [1978] and Pudovkin et al. [1986]. At present, most researchers agree that the solar wind speed V, the IMF magnitude B and its polar angle θ are three major parameters which determine the power of the dynamo process. However, the exact expression for the power is still controversial.

As mentioned earlier, one of the prevailing concepts in geomagnetism and auroral physics until about 1963 was that of the distribution pattern, either the current system or auroral displays, which is fixed with respect to the sun. Thus, any *time* variations of auroral displays at an observation point on the earth were considered to be due to the motion of the observation point as the earth rotates once a day. The fact that auroral displays change from quiet arcs in the evening sector, active forms in the midnight sector to patchy forms in the morning sector were thought to be due to the fact that an auroral zone observation point rotated under the such a pattern of auroras once a day. All-sky films collected during the IGY provided us with the first opportunity to compare auroral displays in the different sectors. First of all, it was found by Feldstein [1963] that at a particular instant, the aurora is distributed along the auroral oval, not along the auroral zone. In fact, the auroral zone is found to be the locus of the midnight part of the oval on a geographic map, as the earth rotates under the oval once a day. Later, it has become clear that the auroral oval delineates approximately the boundary of the polar cap; the magnetic field lines which 'originate' in the area surrounded by the auroral oval, the polar cap, are reconnected with the interplanetary magnetic field lines.

The all-sky films revealed also that the aurora undergoes global changes. At a time, quiet arcs lie all along the different time sectors, instead of only in the evening sector. Then, auroras undergo sudden changes all along the darkside of the oval in a matter of a few minutes. This change is followed by active displays for 1-3 hours, of which characteristics are different in the different sectors. Then, auroras all along the oval become gradually quiet. This particular type of auroral activity is called the auroral substorm [Akasofu, 1964]. In a typical day (a moderately disturbed day), there occur several auroral substorms. An observation point on the earth rotates under such a continuously changing pattern of the aurora, rather than the fixed pattern. This drastic change of the concept from the fixed pattern had not readily been accepted for at least a decade or so. The author's first paper on the auroral substorm was rejected by the Journal of Geophysical Research. Global images from satellites are needed to clarify the controversy. After the revolutionary development of the auroral imagers aboard the ISIS-2, DMSP, DE, Viking and Akebono satellites since the early 1970's, the global activity of the aurora has been confirmed and studied in detail.

The auroral substorm is now understood to be one of the manifestations of the magnetospheric substorms, in fact the only visible one. During the same period of a few hours, an intense electric current system in the polar ionosphere grows and subsides, causing polar magnetic substorms, another important manifestation of the magnetospheric substorm. There are a large number of geophysical phenomena which occur in association with the auroral and polar magnetic substorms, including micropulsations, x-rays and ionospheric disturbances.

During the first ten years of the space age, magnetospheric physics had made a considerable progress, as a number of satellites surveyed space around the earth. A series of discoveries, both observational and theoretical, have combined to increase our knowledge of the magnetosphere and to stimulate great interest in exploring the magnetosphere. The basic geometry of the magnetosphere, the bow shock, the magnetic field configuration in and around the magnetosphere had been revealed during this period. The author himself was fortunate enough to witness the first passage of the Explorer 10 satellite, a man-made object, across the magnetopause, although this interpretation was not necessarily obvious to everyone at that time.

Our present understanding of the distribution of different plasma regimes, the cusp, CPS, BPS, mantle, LLBL, etc., had, however, to rely on the development of new plasma detectors after 1970. Until about the end of the 1970's, it was generally considered that the source of plasma particles in the magnetosphere was the solar wind. This was another example of the prevailing concept in our discipline. The discovery of singly ionized oxygen atoms in the ring current belt and in the plasma sheet was a great surprise, since it implies that they are of ionospheric origin [Shelley et al. 1972]. Thus, the ionosphere is found to be an important source of plasmas in the magnetosphere.

One of the most important events at the beginning of the 1970s is the discovery of the field-aligned currents and their distribution in the polar ionosphere. Obviously, this finding has demonstrated that the magnetospheric current system is a three-dimensional one. Furthermore, it was found that the field-aligned currents appear in pairs. Each pair consists of field aligned currents, into and out from the ionosphere, in each local time sector. The higher latitude part is called Region 1 current and the lower latitude part is called Region 2 current [Iijima and Potemra, 1976]. It is generally accepted that Region 1 current is directly connected to the solar-wind magnetosphere dynamo and is thus the primary current system.

In the 1970's, ground-based observations of geomagnetic disturbances became systematic. In particular, meridian chains of magnetic observatories, together with an efficient computer code to determine the responsible current systems, have become a powerful tool in a study of the current system [Kamide et al. 1982; Kamide, 1988].

One of the fascinating aspects of the field-aligned currents is that the currents are related to a significant potential drop along the geomagnetic field lines [Gurnett, 1972; Mozer et al., 1977]. The presence of such a potential drop was suggested originally by Alfven [1950]. It is now generally accepted that it is this potential drop which can accelerate current-carrying electrons, so that they can cause ionizations and dissociations of atoms and molecules in the ionosphere. In magnetospheric physics, unlike in some other similar fields, experimenters, analysts and theorists have worked very closely together. We have learned that a close cooperation among the three groups is the only way to make a good progress in this field. In theoretical research, the development of a number of particle codes offered rich rewards to understand complicated behaviors of space plasma. A study of the potential drop along the geomagnetic field lines is a good example, in which the particle codes are extensively used to understand the satellite observations. In spite of such a great effort, however, main features field-aligned of the potential drop still defies explanation.

During the 1970's, disturbance characteristics of the magnetic and electric fields and plasmas in the magnetosphere during magnetospheric substorms were also systematically examined by satellite-borne instruments. As a wealth of satellite information began to accumulate, it was found by the efforts of a large number of analysts that the magnetosphere undergoes a systematic change in association with the auroral and polar magnetic substorms, although it has not necessarily been successful to identify causes of individual substorm features in the polar upper atmosphere with individual aspects of disturbances in the magnetosphere, such as the poleward expanding bulge of the aurora during the expansive phase, the westward traveling surge, pulsating patches. One of the basic problems in this regard is the fact that we are, at present, unable to map accurately such polar phenomena to the magnetotail, and vice versa.

One of the most dramatic features in the magnetosphere during the magnetospheric substorm is thinning and subsequent recovery of the plasma sheet and the formation of plasmoids [Hones 1979, 1984]. At the present time, however, there is no generally accepted explanation of the plasma sheet variations. Indeed, this behavior of the plasma sheet continues to constitute one of the major unsolved problems of magnetospheric physics. On the other hand, it is generally believed that a diversion of the cross-tail current is the cause of the phenomenon called the "dipolarization" and many other phenomena in the polar ionosphere [McPherson and Russell, 1973]. The cause of the diversion of the cross-tail current is, however, still a matter of great controversy. The prevailing idea is that there is some current instability responsible for the interruption of the cross-tail current and the subsequent diversion. On the other hand, there is a minority group which suggests that the ionosphere, activated by the auroral precipitation, is directly responsible for the diversion of the cross-tail current in the plasma sheet.

It is generally believed that the magnetospheric substorm consists of two components: one is directly driven by an enhanced dynamo process and the other is contributed by magnetic reconnection which is thought to convert magnetic energy in the magnetotail to the substorm energy [Rostoker et al., 1987]. This particular disturbance of the magnetosphere occurs after the so-called southward turning of the IMF vector or more accurately after an increase of either one, two or all of V, B and θ, namely an increase of the power generated by the solar wind-magnetosphere dynamo. It is becoming clear that the magnetosphere is a low-pass filter and that the magnetospheric substorm is the most basic mode of magnetospheric disturbances associated with an increase of dynamo power above $\sim 10^{18}$ erg/sec for about one hour or longer. The magnetosphere does not respond in such a way to an increase of the power of much shorter durations (say, 10 min). Much of the power thus generated by the solar wind-magnetosphere dynamo is eventually dissipated as the Joule heat in the polar ionosphere.

Obviously, future systematic studies are needed to settle many crucial issues in understanding the magnetospheric substorm. There are at least several suggested future research projects in this regard.

Imaging the Magnetospheric Substorm

Until about 1975, auroral imaging by spacecraft was not necessarily the top priority project of the scientific community, compared with observations of *invisible* physical quantities, such as electric fields, magnetic fields and particle fluxes. The great success of auroral imaging by Frank et al. [1982] and others reminded us of the important of auroral images and also of imaging in general. There is no doubt that this success will shape the course of magnetospheric research for many years to come.

The next obvious step is to take global images of the entire magnetosphere from a satellite or a lunar base, which can provide the frame of reference for both single point measurements and numerical simulation studies [Swift et al., 1989]. In fact, a new breakthrough in magnetospheric physics and aeronomy might arise from our ability to remote sense the magnetosphere. For example, it would be highly instructive to observe complicated three-dimensional phenomena, such as flux transfer events [Russell and Elphic, 1979], the formation and detachment of plasmoids during substorms, and the response of the near-Earth plasma sheet to the growth and expansive phases of substorms.

Magnetospheric Substorms and Solar Flares

It is quite likely that basic processes associated with magnetospheric substorms and solar flares are similar. This is because:

- Magnetospheric substorms and solar flares are primarily various manifestations of electromagnetic energy dissipation processes. The dissipation is manifested by atmospheric emissions (auroras; flares), x-rays, radio emissions (auroral kilometric radiations; microwave emissions), and a variety of plasma motions (plasmoids; coronal mass ejections).

- There must be dynamo processes that supply the power for magnetospheric substorms and solar flares, respectively. The solar wind-magnetosphere dynamo has been identified for substorms, while a photospheric dynamo is likely to be the power supply process for solar flares.

- Field-aligned currents play a major role in substorm processes, while force-free fields are considered to be the basic field configuration for solar flares. A force-free field arises from field-aligned currents J x B = 0 or (∇ x B) x B = 0. Thus, physics of field-aligned currents must be crucial in understanding both substorms and solar flares. How are field-aligned currents generated in the magnetosphere and in the solar atmosphere?

- The auroral potential structure is essential for accelerating auroral electrons, while there is some indication of a monoenergetic peak of the electron spectrum for solar flares.

Magnetospheric physicists are encouraged to study solar flares and learn the electromagnetic processes in different plasma environments. Likewise, solar physicists are also encouraged to study magnetospheric substorms.

IMF Variations and the Driver Gas

In spite of great progress in magnetospheric physics and solar physics in recent years, our understanding of some of the most crucial aspects of solar-terrestrial physics has not been very much improved. Most magnetospheric physicists are willing to take up processes only after the southward turning of the IMF vector at the front of the magnetosphere and do not venture out to study the nature and causes of IMF vector variations. Note also that we are still far from a full understanding of magnetosphere responses to different types of solar wind/IMF disturbances, different types of Alfven waves, different periods (and frequency of Alfven waves) of disturbances, shock waves, and different interplanetary discontinuities. The so-called *magnetic cloud* proposed by Klein and Burlaga [1982] may also be added as another proposed signature. Since magnetic fields in the driver gas play a major role for the development of geomagnetic storms, it is essential to understand the driver gas in detail.

In conclusion, in any scientific discipline, there may arise a strong tendency to believe that little is left to explore anymore, particularly when a paradigm (not a discipline) is maturing. Today, there are some who believe that magnetospheric physics has matured to the point where only quantitative research is left, verifying, elaborating, consolidating a complete set of facts, resolving minor details (residual ambiguities) and reconciling anomalies. On the contrary, we are still far from understanding many fundamental aspects of the magnetospheric substorm. The cause-effect relationships of many discovered phenomena have not been clearly established. There is no doubt that there are also many undiscovered phenomena and processes.

In the past, the ionosphere was thought to be only a passive medium. Thus, substorm phenomena at the ionospheric level were thought to be simply manifestations of processes which take place in the magnetotail, another prevailing paradigm. It is only recently that new attempts have been made to develop a global model of magnetospheric substorms, which consider the magnetosphere and the ionosphere as two important components of a single system. However, these models are still far from complete. It is a great challenge to uncover many hidden processes by modeling the nonlinear interactions between the magnetosphere and the ionosphere as a coupled system [Kan et al., 1988]. Entirely new directions of research on magnetospheric substorms have also just begun [cf. Baker et al., 1989; Goertz and Smith, 1989].

Acknowledgement. The work reported here is supported by a grant from the National Science Foundation ATM 90-22819.

References

Akasofu, S.-I., 1964. The development of the auroral substorm, *Planet. Space Sci.*, 12, 273.

Alfven, H., 1950. *Cosmical Electrodynamics*, Oxford University Press.

Arnoldy, R. L., 1971. Signature in the interplanetary medium for substorms. *J. Geophys. Res.*, 76, 5189.

Axford, W. I. and Hines, C. O., 1961. A unifying theory of high-latitude geophysical phenomena and geomagnetic storms, *Can. J. Phys.*, 39, 1433.

Baker, D. N., A. J. Klimas and R. L. McPherron, Evolution from moderate to strong geomagnetic activity: An interpretation in terms of chaos theory, paper presented in the International Association of Geomagnetism and Aeronomy meeting, Exeter, England, July 1989.

Birkeland, K. K., 1908. *The Norwegian Aurora Polaric Expedition*, 1902-03. Vols. I and II, H. Ascheboug, Christiania.

Chapman, S., 1935. The electric current-systems of magnetic storms, *Terr. Magn. Atmos. Elect.*, 40, 349.

Chapman, S. and Ferraro, V. C. A., 1931. A new theory of magnetic storms, Part I. The initial phase, *Terr. Magn. Atmos. Elect.*, 36, 77.

Dungey, J. W., 1961. Interplanetary magnetic field and the auroral zone, *Phys. Rev. Lett.*, 6, 47.

Feldstein, Y. I., 1963. Some problems concerning the morphology of auroras and magnetic disturbances at high latitudes, *Geomagn. Aeron.*, 3, 183.

Frank, L. A., J. D. Craven, J. L. Burch and J. D. Winningham, 1982. Polar views of the Earth's aurora with Dynamics Explorer, *Geophys. Res. Lett.*, 9, 1001.

Goertz, C. K. and R. A. Smith, 1989. Thermal catastrophe model of substorms, *J. Geophys. Res.*, 94, 6581.

Gonzalez, W. D. and Mozer, F. S., 1974. A quantitative model for the potential resulting from reconnection with an arbitrary interplanetary magnetic field, *J. Geophys. Res.*, 79, 4186.

Gurnett, D. A., 1972. Electric field and plasma observations in the magnetosphere, in *Critical Problems of Magnetospheric Physics*, p. 123, ed. Dyer, E. R., National Academy of Sciences, Washington, DC.

Hones, E. W., Jr., 1984. Plasma sheet behavior during substorms, in *Magnetic Reconnection in Space and Laboratory Plasma*, Geophys. Monogr. Ser., vol. 30, edited by E. W. Hones.

Hones, E. W., Jr., 1979. Transient phenomena in the magnetotail and their relation to substorms, *Space Sci. Rev.*, 23, 393.

Iijima, T. and Potemra, T. A., 1976. Large-scale characteristics of field-aligned currents associated with substorms, *J. Geophys. Res.*, 81, 3999.

Kamide, Y., 1988. *Electrodynamic Processes in the Earth's Ionosphere and Magnetosphere*, Kyoto Sangry University Press, Kyoto.

Kamide, Y., Ahn, B.-H., Akasofu, S.-I., Baumjohann, W., Friss-Christensen, E., Kroehl, H. W., Mauer, H., Richmond, A. D., Rostoker, G., Spiro, R. W., Walker, J. K and Zaitzev, A. N., 1982. Global distribution of ionospheric and field-aligned currents during substorms determined using magnetic data from six IMS meridian chain: Initial results, *J. Geophys. Res.*, 87, 8228.

Kan, J. R., L. Zhu and S.-I. Akasofu, 1988. A theory of substorms: onset and subsidence, *J. Geophys. Res.*, 93, 5624.

Kleun, L. W. and L. F. Burlaga, 1982. Interplanetary magnetic clouds at 1 a.n., *J. Geophys. Res.*, 87, 613.

Lee, L. C. and Z. F. Fu, 1986. Multiple X-line reconnection, 1, A criterion for for the transition from a single X-line reconnection, *J. Geophys. Res.*, 91, 6807.

Lee, L. C. and Z. F. Fu and S.-I. Akasofu, 1985. A simulation of forced reconnection processes and magnetosphere substorms and storms, *J. Geophys. Res.*, 90, 10,886.

McPherron, R. L., C. T. Russell and M. P. Aubry, 1973. Satellite studies of magnetospheric substorms on August 15, 1988, 9, Phenomenological model for substorms, *J. Geophys. Res.*, 78, 3131.

Mozer, F. S., Carlson, C. W., Hudson, M. K., Torbert, R. B., Parady, B., Yatteau, J. and Kelley, M. C., 1977. Observations of paired electrostatic shocks in the polar magnetosphere, *Phys. Rev. Lett.*, 38, 292.

Perreault, P. and Akasofu, S.-I., 1978. A study of geomagnetic storms, *Geophys. J. R. Astr. Soc.*, 54, 547.

Pudovkin, M. I., V. S. Semenov, M. F. Heyn and H. K. Biernat, 1986. Implications of the stagnation line model for energy input through the dayside magnetopause, *Geophys. Res. Lett.*, 13, 213.

Reiff, P. H., R. W. Spiro and T. W. Hill, 1981. Dependence of polar cap potential on interplanetary parameters, *J. Geophys. Res.*, 86, 7639.

Rostoker, G., S.-I. Akasofu, W. Baumjohann, Y. Kamide and R. C. McPherron, 1987. The roles of direct input of energy from the solar wind and unloading of stored magnetotail energy in driving magnetospheric substorms, *Space Sci. Rev., 46*, 93.

Shelley, E. G., Johnson, R. G. and Sharp, R. D., 1972. Satellite observations of energetic heavy ions during a geomagnetic storm, *J. Geophys. Res., 77*, 6104.

Storey, L. R. O., 1953. An investigation of whistling atmospherics, *Phil. Trans. R. Soc. A., 246*, 113.

Swift, D. W., R. W. Smith and S.-I. Akasofu, 1989. Imaging the earth's magnetosphere, Planet. Space Sci., 37, 379.

Thomson, J. J. and G. P., 1903. *Conduction of Electricity Through Gases*, p. 30, General Publishers, Toronto.

S.-I. Akasofu, Geophysical Institute, University of Alaska Fairbanks, Fairbanks, AK 99775-0800.

THE BEGINNING OF SUBSTORM RESEARCH

David P. Stern

*Laboratory for Extraterrestrial Physics,
Goddard Space Flight Center, Greenbelt, MD 20771*

After observing a great auroral display in Connecticut, on July 1, 1837, E.C. Herrick [1838, cited by *Siscoe*, 1980] wrote (italics in the original):

> "It is worthy of notice, that on this occasion there were two well marked and distinct *seasons of greatest brilliance* or *fits of maximum intensity*, at intervals of about four hours. It will be found on examination of former accounts, that this is a common feature of Auroral exhibitions of unusual brilliance."

The magnetic signature of such "fits of maximum intensity" was first studied in 1899-1900 and 1902-3 by Kristian Birkeland; in his second study he used four subpolar stations in Norway, Spitzbergen, Iceland and Novaya Zemlya [*Birkeland*, 1908; *Bostrom*, 1968; *Chapman*, 1968, sect. 10]. From the data he deduced a new kind of magnetic disturbance, the "elementary polar magnetic storm" with a typical time scale of half an hour. The observed magnetic disturbance field of such events tended to be perpendicular to the auroral arcs, apparently indicating electric currents flowing along such arcs, arriving from space at one end and returning at the other.

In hindsight it is clear that Birkeland observed the effects of the auroral electrojet. If "regions 1 and 2" of field aligned currents are well developed but ionospheric conductivity is relatively undisturbed, the ground signature of the primary circuit is weak [*Fukushima*, 1969, 1976] and the only significant ground disturbance comes from the electrojet, which is a secondary Hall current. In substorms, however, the concentrated westward electrojet flowing in the nightside auroral oval as part of the "substorm wedge" circuit resembles the current deduced by Birkeland. In honor of those early observation, the term "Birkeland currents" was introduced for field-aligned currents flowing into the polar cap and out of it [*Schield et al.*, 1969].

Birkeland claimed to have distinguished five classes of elementary polar magnetic storms but that division was not sustained by later research, and after his tragic death in 1918 his findings were often discounted outside Scandinavia. His most serious critic was Sidney Chapman, who felt that because of their short time scale Birkeland's events could not possibly be regular magnetic storms. He proposed that they were merely phases of such storms, isolated from the context of the entire storm sequence [*Chapman and Bartels*, 1940, sect. 9.21]:

> "Birkeland's typical disturbances often extend over only a few hours, and rarely for as long as a day, and sometimes even longer [*these must have been sequences of substorms*]. This tends to confirm the view that he split up the perturbation artificially into small portions that ought really to be considered together."

He also opposed Birkeland's idea of currents linking the Earth to space (loc cit, sct. 25.10, p. 881). In part because of Chapman's opposition, when polar disturbances were studied [e.g. *Silsbee and Vestine*, 1942], they were generally termed "auroral bays," since they showed up on magnetograms as indentations similar to bays on a coastline.

In his later life Chapman moved (with intermissions) to Alaska [*Akasofu et al.*, 1969] where he was joined by S.-I. Akasofu. Together the two continued studying magnetic storms, incorporating new information from artificial satellites. In "The Ring Current, Geomagnetic Disturbance, and the Van Allen Radiation Belts" [*Akasofu and Chapman*, 1961] they resolved the storm's disturbance field D into three parts, one of them the polar disturbance field DP (p. 1339):

> "DP field: produced by currents flowing in the ionosphere. They are driven by electromotive

forces in the auroral zones. Thence they spread over the main region of the earth between the zones and over the enclosed polar caps. This DP field has a different time scale from that of the magnetic storm. Each such event, which Birkeland [1908] called a polar elementary storm, is here called a DP substorm."

This was one of the earliest usages of the term "substorm" which has now displaced Birkeland's, probably because it is much shorter. Chapman later introduced "auroral substorm" (see below) and Akasofu [1968, 1977] proposed "magnetospheric substorm" to stress the involvement of the entire magnetosphere. Another term used was "polar magnetic substorm" [Akasofu et al., 1966b, Rostoker, 1972] and one may note that Obayashi [1975] has used "auroral flare", to underscore the analogy between substorms and solar flares, both manifestations of impulsive acceleration of particles in space by the release of magnetic energy.

Chapman and Akasofu later studied the relation between magnetic storms and substorms in more detail [Akasofu and Chapman, 1963]. Comparing magnetic storm records near the geomagnetic equator (Honolulu) with those obtained in the auroral zone (College, Alaska), they found that the latter were indeed punctuated by many brief "substorms" not observed near the equator (Figure 1). This was taken as a confirmation of Chapman's view.

However, in the following year Akasofu noted that similar events, though less pronounced, often occured outside magnetic storms and were correlated with outbursts of auroral activity. He proposed to call them "auroral activations," but Chapman disagreed [Akasofu, 1970]:

> "The term "auroral substorm" was introduced by Chapman. A draft of my first auroral substorm paper carried the title 'The Auroral Activation'; he refused to read it until I changed it to 'The Auroral Substorm'."

Following that paper [Akasofu, 1964] Akasofu and his colleagues published a series of detailed studies of the substorm's morphology [Akasofu et al., 1965 a,b, 1966 a,b,c,d,e]. The phases which they identified are still accepted, as is their terminology: onset, expansion, recovery, westward travelling surge.

At about the same time spacecraft began to map the geomagnetic tail and to find evidence for the tail's involvement in substorms. IMP 1, launched in November 1963, observed the structure of the Earth's magnetotail [Ness, 1965, Speiser and Ness, 1967; see also Bame et al., 1967] and allowed its main features to be deduced [Axford et al., 1965]. Later satellites in the IMP series continued such observations and IMP 4 in particular provided information on the profound effect of substorms on the tail's structure [Fairfield and Ness, 1970]. Figures 12 and 13 of that study supported the "slingshot analogy" by which the tail's field lines became stretched before substorms and relaxed rapidly during substorms. At the same time Davis and Sugiura [1966] introduced the AE (Auroral Electrojet) index, still used to gauge substorm activity.

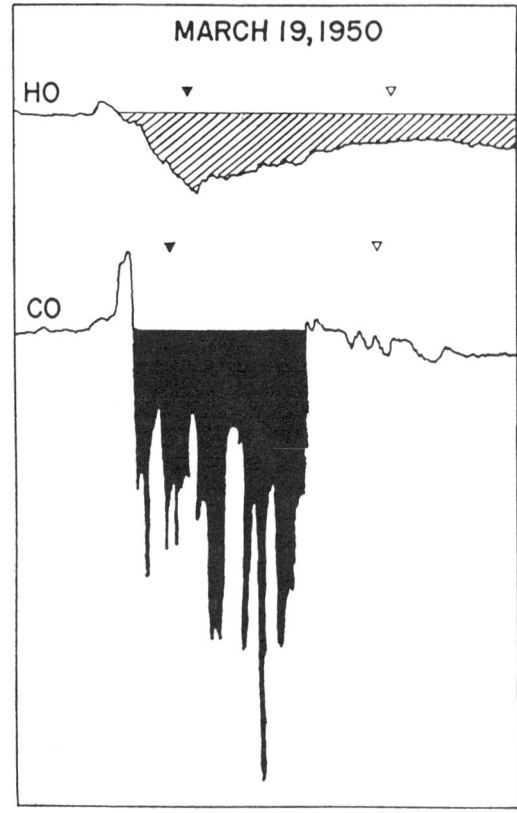

Figure 1

The role of reconnection and of the existence of an "open" nature of the magnetosphere in promoting substorm activity was realized after Fairfield and Cahill [1966; Fairfield 1967, Arnoldy 1971] correlated increased activity with a southward B_z in the interplanetary magnetic field. This led to theories of magnetic energy storage in the tail and of its release in substorms [Siscoe and Cummings, 1969; Aubrey and McPherron, 1971; Coroniti and Kennel, 1972, 1973].

The magnetic observations of the IMPs also supported the notion that particle "dropouts" around the onset time of substorms were caused by a drastic thinning of the plasma sheet. Those events were initially observed by satellites of the Vela series (launched to help enforce the nuclear test ban and carrying no magnetometers) which detected sudden disappearances of the ion flux while the satellite was in the plasma sheet [*Hones et al.*, 1967, 1970]. Also observed were sudden increases of energetic electron fluxes in and near synchronous orbit, noted by Parks and Winckler [1968] and in more detail (with simultaneous ion injections) by DeForest and McIlwain [1971; see also *Konradi*, 1967]. Synchronous spacecraft also detected large magnetic distrubances during substorms [*Cummings et al.*, 1968].

By about 1972 much of our current knowledge on substorms was on hand [*Rostoker*, 1972, *Aubry*, 1972]. A conference on substorms was held in October 1972 [*Vasyliunas and Wolf*, 1973] and an initial coordinated study on the substorms of August 15, 1968 was carried out. The results of that study appeared in 1973 in 9 consecutive articles (J. Geophys. Res., 78, 3044-3149), the last of which [*McPherron et al.*, 1973] presented an interpretation of which many details still persist, e.g., the diversion of part of the cross-tail current into a "substorm wedge." At the same time Hones proposed that substorms were associated with magnetic reconnection around x=-15 R_E [*Hones et al.*, 1973; see p. 130].

Since that time many spacecraft have probed the geomagnetic tail, auroral imagers were added to the arsenal and several interesting explanations were proposed for the substorm; yet overall progress remains rather slow. And although much more is known about the plasma sheet, its storm-time variation and different regions, the sheet's origin, mean flow, stability and dynamics are still poorly understood.

Chapman believed that magnetic storms were the fundamental feature while substorms were just an associated detail; in contrast, substorms are nowadays viewed as fundamental and are studied in great detail, with relatively little attention given to magnetic storms [but see *Kamide*, 1980]. Yet Chapman's 1963 observations (Figure 1) still stand: magnetic storms indeed tend to contain a series of intense substorms, differing from the "garden variety" by injecting appreciable numbers of ions into the inner magnetosphere. How this happens is not clear. Studies of injection events [*DeForest and McIlwain*, 1971; *McIlwain and Whipple*, 1986] suggests that hot plasma suddenly and simultaneously appears on a wide front, energized by electric fields which are not too well understood.

The history of substorm research, it seems, is far from over.

REFERENCES

Akasofu, Syun-Ichi, The development of the auroral substorm, Planet. Space Sci., 12, 273-282, 1964.

Akasofu, Syun-Ichi, The magnetosphere and magnetospheric substorm, p. 33-38 in The Birkeland Symposium on Aurora and Magnetic Storms, A. Egeland and J. Holtet, editors, Centre de la Recherche Scientifique, Paris 1968.

Akasofu, Syun-Ichi, In memoriam Sydney Chapman, Space Science Reviews, 11, 599-606, 1970.

Akasofu, Syun-Ichi, Physics of Magnetospheric Substorms, xviii+599 pp., D. Reidel, Dordrecht 1977.

Akasofu, Syun-Ichi and Sydney Chapman, The ring current, geomagnetic disturbance and the Van Allen radiation belts, J. Geophys. Res., 66, 1321-1350, 1961

Akasofu, Syun-Ichi and Sidney Chapman, The development of the main phase of magnetic storms, J. Geophys. Res., 68, 125-129, 1963.

Akasofu, Syun-Ichi, Sidney Chapman and Ching.-I. Meng, The polar electrojet, J. Atmos. Terr. Phys., 27, 1275-1305, 1965a.

Akasofu, Syun-Ichi, D.S. Kimball and Ching-I. Meng, Dynamics of the Aurora, II, Westward traveling surges, J. Atmos. Terr. Phys., 27, 173-187 1965b.

Akasofu, Syun-Ichi, D.S. Kimball and Ching-I. Meng, Dynamics of the Aurora, III, Westward drifting loops, J. Atmos. Terr. Phys., 28, 189-196, 1966a.

Akasofu, Syun-Ichi, D.S. Kimball and Ching-I. Meng, Dynamics of the Aurora, IV, Polar magnetic substorms and westward traveling surges, J. Atmos. Terr. Phys., 28, 489-496, 1966b.

Akasofu, Syun-Ichi, D.S. Kimball and Ching-I. Meng, Dynamics of the Aurora, V, Poleward Motions, J. Atmos. Terr. Phys., 28, 497-503, 1966c.

Akasofu, Syun-Ichi, D.S. Kimball and Ching-I. Meng, Dynamics of the Aurora, VI, Formation of patches and their eastward motion, J. Atmos. Terr. Phys., 28, 505-511, 1966d.

Akasofu, Syun-Ichi, D.S. Kimball and Ching-I. Meng, Dynamics of the Aurora, VII, Equatorward motions and the multiplicity of auroral arcs, J. Atmos. Terr. Phys., 28, 627-635, 1966e.

Akasofu, Syun-Ichi, Benson Fogle and Bernard Haurwitz, eds., Sidney Chapman, Eighty, University of Colorado Press, 1969.

Arnoldy, Roger L., Signature in the Interplanetary Medium for Substorms, J. Geophys. Res., 76, 5189-5201, 1971.

Aubry, Michel P., A short review of magnetospheric substorms, p. 357-364 in Earth Magnetospheric Processes, Billy McCormack, ed., 1972

Aubry, Michel P. and Robert L. McPherron, Magnetotail changes in relation to the solar wind magnetic field and magnetospheric substorms, J. Geophys. Res., 76, 4381-4401, 1971.

Axford, W. Ian, Harry E. Petschek and George L. Siscoe, Tail of the magnetosphere, J. Geophys. Res., 70, 1231-1236, 1965.

Bame, Samuel J., J.R. Asbridge, H.E. Felthauser, E.W. Hones and I.B. Strong, Characteristics of the plasma sheet in the earth's magnetotail, J. Geophys. Res., 72, 113-129, 1967.

Birkeland, Kristian, The Norwegian Aurora Polaris Expedition 1902-3, A. Aschehoug, Christiania, Norway, 1908, 1913.

Boström, Rolf, Currents in the ionosphere and magnetosphere, p. 445-458 in The Birkeland Symposium on Aurora and Magnetic Storms, A. Egeland and J. Holtet, editors, Centre de la Recherche Scientifique, Paris 1968.

Caan, Michael N, Robert L. McPherron and Christopher T. Russell, Substorm and interplanetary magnetic field effects on the geomagnetic tail lobes, J. Geophys. Res., 80, 191-194, 1975.

Chapman, Sydney, Historical introduction to aurora and magnetic storms, p. 21-29 in The Birkeland Symposium on Aurora and Magnetic Storms, A. Egeland and J. Holtet, editors, Centre de la Recherche Scientifique, Paris 1968.

Chapman, Sidney and Julian Bartels, Geomagnetism, Oxford Univ. Press, New York, 1940.

Coroniti, Frederick V. and Charles F. Kennel, Changes in magnetospheric configuration during substorm growth phase, J. Geophys. Res., 77, 3361-3370, 1972.

Coroniti, Frederick V. and Charles F. Kennel, Can the ionosphere regulate magnetospheric convection?, J. Geophys. Res., 78, 2837-2851, 1973

Cummings, W.D., J.N. Barfield and P.J. Coleman, Magnetospheric substorms observed at the synchronous orbit, J. Geophys. Res., 73, 6687-6698, 1968.

Davis, T. Neil and Masahisa Sugiura, Auroral electrojet activity index AE and its universal time variations, J. Geophys. Res., 71, 785-801, 1966.

DeForest, Sherman E. and Carl E. McIlwain, Plasma clouds in the magnetosphere, J. Geophys. Res., 76, 3587-3611, 1971.

Fairfield, Donald H., Polar magnetic disturbances and the interplanetary magnetic field, Space Research VIII, 107-119, 1967.

Fairfield, Donald H. and Lawrence G. Cahill, Jr., Transition region magnetic field and polar magnetic disturbances, J. Geophys. Res., 71, 155-169, 1966.

Fairfield, Donald H. and Norman F. Ness, Configuration of the geomagnetic tail during substorms, J. Geophys. Res., 75, 7032-7047, 1970.

Fukushima, Naoshi, Equivalence in ground geomagnetic effect of Chapman-Vestine's and Birkeland-Alfven's current systems for polar magnetic storms, Rep. Ionos. Space Res.Jap., 23, 219-227, 1969.

Fukushima, Naoshi, Generalized theorem for no ground magnetic effect of vertical currents connected with Pedersen currents in the uniform-conductivity ionosphere, Rep. Ionos.Space Res.Jap., 30, 35-40, 1976.

Herrick, E.C., Amer. J. of Science, 33, 146, 1838 (cited by Siscoe, 1980).

Hones, Edward W., Jr., J.R. Asbridge, S.J. Bame and I.B. Strong, Outward flow of plasma in the magnetotail following geomagnetic bays, J. Geophys. Res., 72, 5879-5892, 1967.

Hones, Edward W., Jr., S.I. Akasofu, P. Perreault, S.J. Bame and Sidney Singer, Poleward expansion of the auroral oval and associated phenomena in the magnetotail during auroral substorms, J. Geophys. Res., 75, 7060-7074, 1970.

Kamide, Y. Relationship between substorms and storms, p. 425-443 in Dynamics of the Magnetosphere, S.-I. Akasofu, ed., D. Reidel Publ. Co., Dordrecht, Holland 1980.

Konradi, Andrei, Proton events in the magnetosphere associated with magnetic bays, J. Geophys. Res., 72, 3829-3841, 1967.

McIlwain, Carl E. and Elden C. Whipple, The dynamic behavior of plasmas observed near geosynchronous orbit, IEEE Trans. Plasma Sci., PS-14, 874-890, 1986.

McPherron, Robert L., Growth phase of magnetospheric substorms, J. Geophys. Res., 75, 5592-5599, 1970

McPherron, Robert L., Substorm related changes in the geomagnetic tail: the growth phase, Planet. Space Sci., 20, 1521-1539, 1972.

McPherron, Robert L., Christopher T. Russell and Michel P. Aubry, Satellite studies of magnetospheric substorms on August 15, 1968, 9. Phenomenological model for substorms, J. Geophys. Res., 78, 3131-3149, 1973.

Ness, Norman F., The earth's magnetic tail, J. Geophys. Res., 70, 2989-3005, 1965.

Obayashi, Tatsuzo, Energy build-up and release mechanisms in solar and auroral flares, Space Sci. Rev., 17, 195-203, 1975.

Parks, George K. and John R. Winckler, Acceleration of energetic electrons observed at the synchronous altitude during magnetospheric substorms, J. Geophys. Res., 73, 5786-5791, 1968

Rostoker, Gordon, Polar magnetic substorms, Rev. Geophys., 10, 157-211, 1972.

Schield, Milo A., John W. Freeman and Alex J. Dessler, A source for field-aligned currents at auroral latitudes, J. Geophys. Res., 74, 247-256, 1969.

Silsbee, H.C. and E.H. Vestine, Geomagnetic bays, their frequency and current systems, J. Geophys. Res., 47, 195-208, 1942.

Siscoe, George L. and W.D. Cummings, On the cause of geomagnetic bays, Planet. Space Sci., 17, 1795-1802, 1969.

Siscoe, George L., What's in the name "Magnetospheric substorm"?, J. Geophys. Res., 85, 1643-1644, 1980.

Speiser, Theodore W. and Norman F. Ness, The neutral sheet in the geomagnetic tail: its motion, equivalent currents and field line connection through it, J. Geophys. Res., 72, 131-141, 1967.

Vasyliunas, Vytenis M. and Richard A. Wolf, Magnetospheric substorms: Some problems and controversies, Rev. Geophys., 11, 181-189, 1973.

2. OVERVIEWS

PHYSICS OF MAGNETOSPHERIC SUBSTORMS: A REVIEW

M.I. Pudovkin

Institute of Physics, Leningrad University, Leningrad, Stary Peterhoff 198904, USSR

Abstract. A three phase model (initial phase + break-up one + recovery one) of substorms is discussed, and the physical processes which develop in the magnetotail during these phases are considered. These are developed from a single point of view on the substorm as a result of magnetic field reconnection in the magnetotail. Development of spontaneous and triggered break-ups is studied, and characteristics of the behavior of the magnetic field in the magnetosphere during break-ups of the two types are discussed. A method for separating the potential and vortex parts of the magnetospheric electric field is proposed. The variation of those fields and the energetics of the substorm at various phases are also studied.

1. Introduction

Magnetospheric substorms are elementary displays of magnetospheric activity which produce a variety of disturbances. Numerous investigations of these disturbances, carried out for the last three decades, have allowed us to determine the basic physical processes which cause the magnetospheric substorms. These processes include magnetic field reconnection at the magnetopause and in the magnetotail, convection and energization of the magnetospheric plasma, generation of global systems of field-aligned and ionospheric electric currents.

Many details of those processes, as well as the relationship between them, are unclear as yet. In particular, we do not exactly know the mechanism of the transition from one substorm phase to another, and cannot explain the motion of auroral forms during break-ups. This uncertainty manifests itself in the various models of magnetospheric substorms. This makes it crucial to consider additional experimental data to allow us to check theoretical models and to distinguish between them.

First of all we have to agree on what we mean by the term "substorm". The fact is that this term pertains to a wide variety of phenomena, which may differ from each other not only in their intensity, but also in the sequence of the physical processes producing them and by their relationship to the solar wind parameters [Rostoker et al., 1987]. For example, an isolated substorm generated by a short duration variation of the solar wind, develops in a quite a different manner than substorms forming a large, world-wide geomagnetic storm. In addition, apparent display of the substorm depends on the time and space resolution of the instruments used to measure them. Because of these problems, we shall not go into the details of the definition of "substorm", and will study the development of an isolated magnetospheric disturbance which occurs against the background of a relatively quiet magnetosphere. These disturbances may be generated, for example, by a short duration southward excursion of the interplanetary magnetic field, or take place at the very beginning of a more prolonged magnetospheric storm.

As was shown by Pudovkin et al. [1970] and by McPherron [1970], the substorm occurs in three stages: initial (or preliminary), expansion and recovery phases. Characteristic dynamics of auroral forms during these phases is shown in Figure 1 [after Zverev et al. 1991]. One can see in this figure that in the initial phase, a fast and regular equatorward motion of auroral arcs is evident [Belyakova et al., 1968; Pudovkin et al., 1970]. During the expansion phase, there is a rapid jump of the aurora toward the pole [Akasofu, 1964]. And during the recovery phase, the aurora return to their initial position. As is shown by Pudovkin et al. [1990], these motions provide considerable information on the plasma convection, and hence on the electric fields in the plasma sheet, and on the changes in the magnetic field configuration in the magnetotail. This will allow us to determine quantitatively some energetic aspects of substorms.

Our analysis we will begin with the consideration of the influence of the potential and rotational (vortex) electric fields on the motion of auroral forms.

2. Auroral Arc Dynamics and Magnetospheric Electric Fields

This problem was studied in detail by Pudovkin et al. [1991], and we will summarize some results of their analysis. These results are based upon the following suppositions:

a) The motion of auroral arcs is determined by the motion of their sources in the magnetosphere, which not always coinsides with the motion of auroral plasma.

b) The magnetospheric electric field consists of two components: potential $\left(\vec{E}_m^p = -\nabla\varphi\right)$ and vortex $\left(\operatorname{rot}\vec{E}_m^v = -\dfrac{1}{c}\dfrac{\partial \vec{B}}{\partial t}\right)$; the

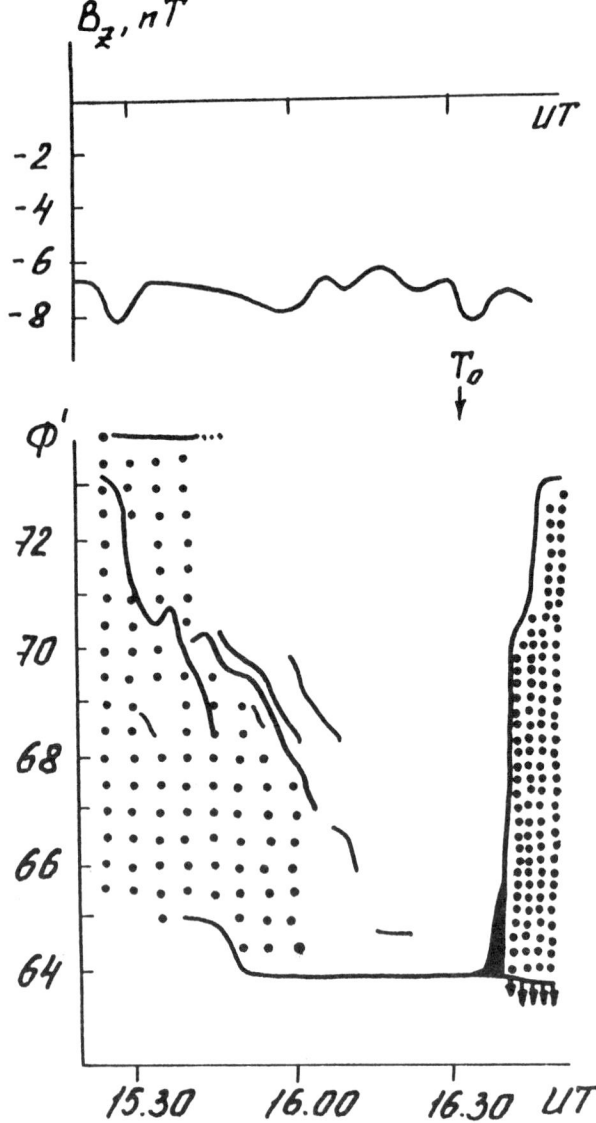

Fig. 1. Dynamics of aurorae on Dec. 18, 1965 at Dixon [after Zverev et al., 1991]. To denotes expansion phase onset. (Upper) IMF B_z component. (Lower) location of aurorae. Dotted area shows location of diffuse aurorae.

latter of which exists in the magnetosphere in the regions where $\left(\dfrac{\partial \vec{B}}{\partial t}\right) \neq 0$ and does not penetrate into the ionosphere. In the ionosphere, there may exist a locally induced electric field, but it may be shown that intensity of these in the auroral zone and in the polar cap $\left(E_i^v \leq 1 \text{mV/m}\right)$ is small with respect to the potential electric fields ($E_i^p \approx 10$ mV/m and more), so that they may be neglected.

c) In the ionosphere and in the magnetosphere, on the closed magnetic field lines, plasma moves with the frozen-in condition. In the magnetosphere in the vicinity of the reconnection line, the frozen-in condition may be violated.

Bearing in mind these assumptions, let us construct a circuit (Figure 2) consisting of a segment (Δl_i) of an auroral arc in the ionosphere; a segment of the geomagnetic field line from the ionosphere to the magnetospheric source of the arc; then of a segment (Δl_m) along the magnetospheric source of the arc; and, finally, of a segment of another magnetic field line backwards to the ionosphere. During geomagnetic substorms, the location and configuration of all these elements of the circuit may significantly change, so that the circuit or some parts of it may move. However, for the circuit under consideration, the magnetic flux (F) across it is always zero and $\left(\dfrac{dF}{dt}\right) = 0$. Hence, the circulation of the electric field along this circuit is zero. When the potential drop along the magnetic field lines is neglected, we have:

$$\vec{E}_m^1 \cdot \vec{\Delta l}_m + \vec{E}_i^1 \cdot \vec{\Delta l}_i = 0 \qquad (1)$$

Here $\vec{E}_m^1$ and $\vec{E}_i^1$ are measured in reference frame moving with the elements $\vec{\Delta l}_m$ and $\vec{\Delta l}_i$, so that

$$\begin{aligned}\vec{E}_m^1 &= \vec{E}_m^0 + \frac{1}{c}\vec{U}_m \times \vec{B}_m \ ; \\ \vec{E}_i^1 &= \vec{E}_i^0 + \frac{1}{c}\vec{U}_i \times \vec{B}_i \end{aligned} \qquad (2)$$

where $\vec{E}_m^0$ and $\vec{E}_i^0$ are measured in a motionless reference frame.

The magnetospheric electric field may consist of two components, potential and vortex: $\vec{E}_m^0 = \vec{E}_m^p + \vec{E}_m^v$, whereas the ionospheric field has only a potential component: $\vec{E}_i^0 = \vec{E}_i^p$. The latter is related to the magnetospheric potential field E_m^p as

$$E_i^p = E_m^p \frac{\Delta l_m}{\Delta l_i} \qquad (3)$$

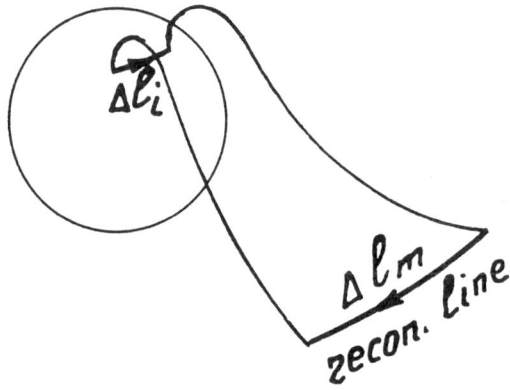

Fig. 2. Electric circuit under consideration.

Inserting (2) into equation (1) and taken into account that the elements $\vec{\Delta l}_m$ and $\vec{\Delta l}_i$ are pointed along the circuit that is in opposite direction, one can rewrite (1) in a form:

$$\left(\vec{E}_m^p + \vec{E}_m^v + \frac{1}{c}\vec{U}_m \times \vec{B}_m\right)_1 \cdot \vec{\Delta l}_m - \left(\vec{E}_i^p + \frac{1}{c}\vec{U}_i \times \vec{B}_i\right)_1 \cdot \vec{\Delta l}_i = 0 \quad (4)$$

The index "1" at the brackets marks the electric field component along the elements $\vec{\Delta l}_m$ and $\vec{\Delta l}_i$ correspondingly.

Now we consider two special cases:

(a) The source of the auroral arc is located on the closed field lines: $B_m \neq 0$, and for the frozen-in condition being fulfilled in the magnetosphere $\left(E_m^1 = 0\right)$, equation (4) reduces to:

$$\vec{E}_{i1}^p = \frac{1}{c}\left(\vec{U}_i \times \vec{B}_i\right)_1 = 0 \quad (5)$$

One can see from equations (5) that, in spite of the existence of the both components of the electric field in the magnetosphere ($\vec{E}_m^p$ and $\vec{E}_m^v$), the auroral arc is driven in the ionosphere by the potential electric field $\vec{E}_i^p$, or $\vec{E}_m^p$ only, and the intensity of the latter may be found from (5) and (3) as:

$$E_{m1}^p = -\frac{1}{c}\left[\vec{U}_i \times \vec{B}_i\right]_1 \cdot \frac{\Delta l_i}{\Delta l_m} \quad (6)$$

(b) If the source of the arc projects to the magnetosphere in the vicinity of the reconnection (neutral) line, $B_m = 0$, and the frozen-in condition $\left(E_m^1 = 0\right)$ is violated. In this case, equation (4) with (3) reduces to:

$$E_{m1}^v = \frac{1}{c}\left[\vec{U}_i \times \vec{B}_i\right]_1 \cdot \frac{\Delta l_i}{\Delta l_m} \quad (7)$$

Thus, the motion of the auroral arc in the ionosphere is determined in this case, by the vortex part of the magnetospheric electric field $\left(\vec{E}_m^v\right)$. When $\vec{E}_m^v$ is directed from dawn to dusk, the arc moves polewards.

It is worth noting that, as was assumed above, the vortex electric field E_m^v does not penetrate into the ionosphere and hence cannot cause any drift of the ionospheric plasma, and similarity of equations (5) and (7) is exact.

Now we turn to the analysis of experimental data on the development of magnetospheric substorms at their various phases.

3. Preliminary Phase of Substorms

The most characteristic feature of auroral dynamics during the initial phase of substorms, is the equatorward drift of auroral arcs (see Fig. 1). According to Pudovkin et al. [1972], this drift proceeds on closed field lines and is caused by the potential component of the magnetospheric electric field (see Section 2 above). This supposition is supported by the close relation between the aurora arc velocity and the ionospheric electric field intensity. Figure 3 [after Kelley et al., 1971] shows variations of the ionospheric electric field measured by means of balloon-born devices and calculated from the auroral arc velocity. One can see a close agreement between the two quantities. At the same time, the poleward jump of the aurora during break-ups is not associated with the change of the sign of the ionospheric electric field [see e.g. Fig. 3 in the paper by Kelley et al., 1971]. Thus, the motion of auroral arcs at the preliminary phase of substorms may be interpreted in terms of the ionospheric and magnetospheric potential electric field.

Figure 4 shows the mean meridional profile of the auroral arc velocity at the preliminary phase in the midnight sector of the auroral zone [Pudovkin et al., 1990]. One can see that the mean arc velocity increases with latitude from 130 m/sec at the equatorward boundary of the auroral zone to 470 m/sec at latitude $\phi = 75°$. This corresponds to the variation of the ionospheric electric field from approximately 5 mV/m at $\phi = 64°$ to ~25 mV/m at the poleward edge of the auroral oval. However, when projected along the geomagnetic field lines onto the equatorial plane of the magnetotail, the electric field proves to be close to uniform ($E \approx 0.4$ mV/m) all along the plasma sheet at $-15 \text{ R}_E \geq x \geq -60 \text{ R}_E$. Only in the nearest parts of the plasma sheet does this field decrease to 0.3 mV/m [Zverev et al., 1988].

Figure 5 [after Pudovkin et al., 1985] shows the velocity of the auroral arc motion in the southernmost region of the auroral zone (observatories Loparskaya, Dixon and Tixi; $\phi \approx 65°$) and its dependence on the Y-component of the solar wind electric field ($E_y = \frac{1}{c}VB_s$, where B_s is the southward component of the IMF). One can see from the data presented in this figure that the auroral arc velocity and hence, the electric field intensity in the ionosphere and in the magnetosphere are proportional to the solar wind electric field. This confirms the suggestion that the former is generated by the solar wind interaction with the geomagnetic field. On the average, the electric field in the magnetosphere E_m^p is approximately 0.3 E_y.

4. Transition to the Active Phase

Turning to the phenomena developing just before break-ups, and for some minutes after them, all break-ups may be divided onto two groups: spontaneous and triggered (or stimulated) ones [Pudovkin et al., 1970]. Examples of these two types of break-ups are shown in Figure 6 [after Pudovkin et al., 1970]. This figure presents the variation of the B_z IMF and development of the geomagnetic disturbances at the mid-latitude observatory, Petropavlovsk for a spontaneous (Fig. 6 left), and for a triggered (Fig. 6, right) break-up. It is seen that, in the both cases the break-ups are preceded by a sufficiently long periods of negative B_z IMF. In addition, one can see that the spontaneous one is not associated with any significant variation of Bz IMF at the moment of the break-up. The triggered one seems to be caused by a rapid turn of the B_z from southwards to the northwards.

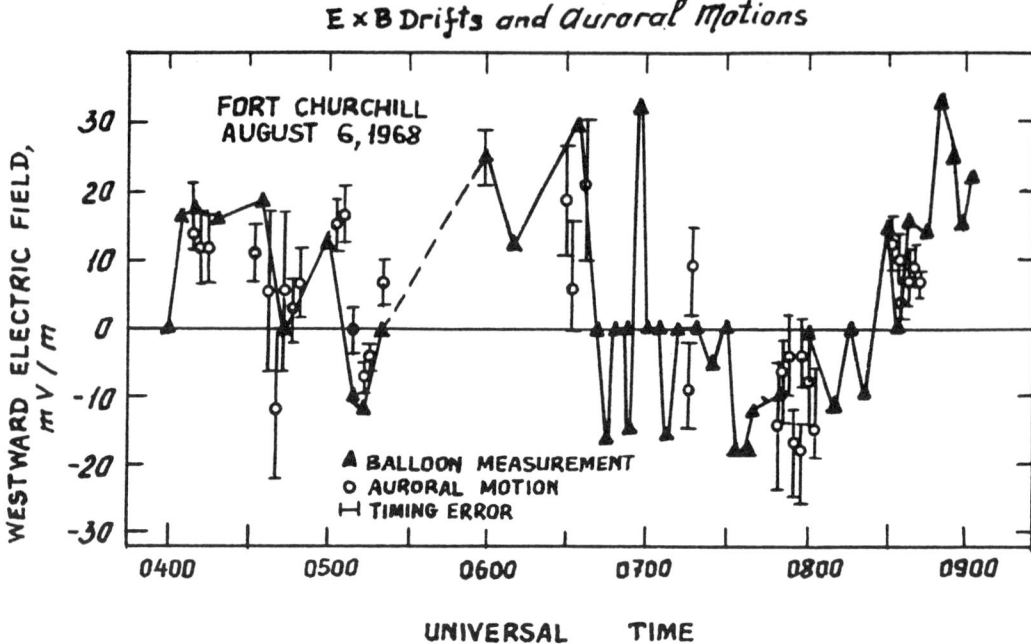

Fig. 3. Electric field in the ionosphere measured on the balloons (triangles) and from the auroral arc motion velocity [after Kelley et al., 1971].

What conditions are favourable for the development of an auroral break-up? Sugiura et al. [1970] and Coleman and McPherron [1970] have shown that, during the initial phase of substorms, a significant stretching of the geomagnetic field lines into the magnetotail and hence intensification of the electric currents in the plasma sheet takes place. The intensity of the currents may be characterized by the value of the angle χ of the inclination of the magnetic field B at the synchronous orbit with respect to the magnetic equatorial plane. Variations of the χ-angle in the course of several break-ups are shown in Figure 7 [after Kozelova et al., 1989]. In this figure, open circles mark the times of the beginning of spontaneous break-ups. The closed circles and triangles mark the onsets of stimulated substorms of various types, and the black squares correspond to triggered break-ups which took place against the background of substorms which began as spontaneous ones. The vertical bars in Figure 7b show the range of variations of the B_z IMF at time interval (-20; +10 min) around the break-up under consideration. One can see that:

a) Spontaneous break-ups begin when $\chi \leq 30°$, when the geomagnetic field lines are greatly stretched into the tail and hence a great amount of the magnetic energy is stored in the magnetotail lobes.

b) Stimulated break-ups may begin at significantly larger (up to 50°) χ-angles, which corresponds to less intensive electric currents in the plasma sheet.

In addition, one can see that the value of the jump of the B_z IMF necessary for triggering a break-up depends on the value of the χ-angle that is on the state of the magnetosphere: $\delta B_z \approx 10\gamma$ when $\chi = 50°$; $\delta B_z \approx 2\gamma$ when $\chi = 40°$ and $\delta B_z \approx 1\gamma$ at $\chi = 30°$. This means that in case when the magnetosphere is ripe for the development of substorms, the difference between spontaneous and triggered break-ups is diminished and a burst-like release of the magnetic energy stored in the tail lobes may result from any small disturbance of external or internal (with respect to the magnetosphere) nature.

Thus, data presented in Figure 7 provide evidence that a spontaneous break-up may begin only when the magnetosphere is ripe for it, and the critical state of the magnetosphere is determined by the intensity of the plasma sheet current. This circumstance allows one to estimate the duration of the initial phase of the substorm [Pudovkin et al., 1990]. If the

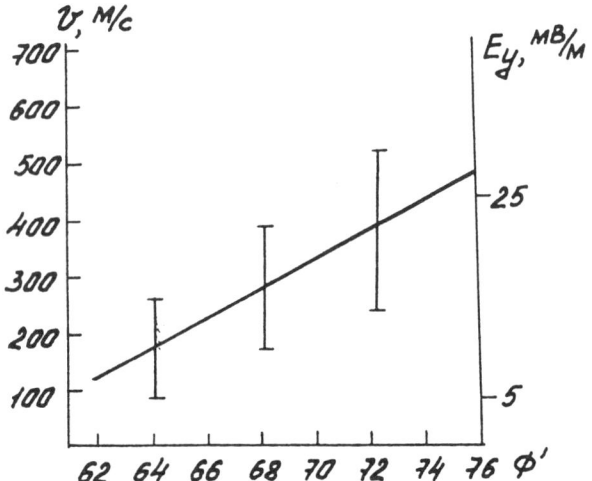

Fig. 4. Meridional profile of auroral arc velocity [after Pudovkin et al., 1990].

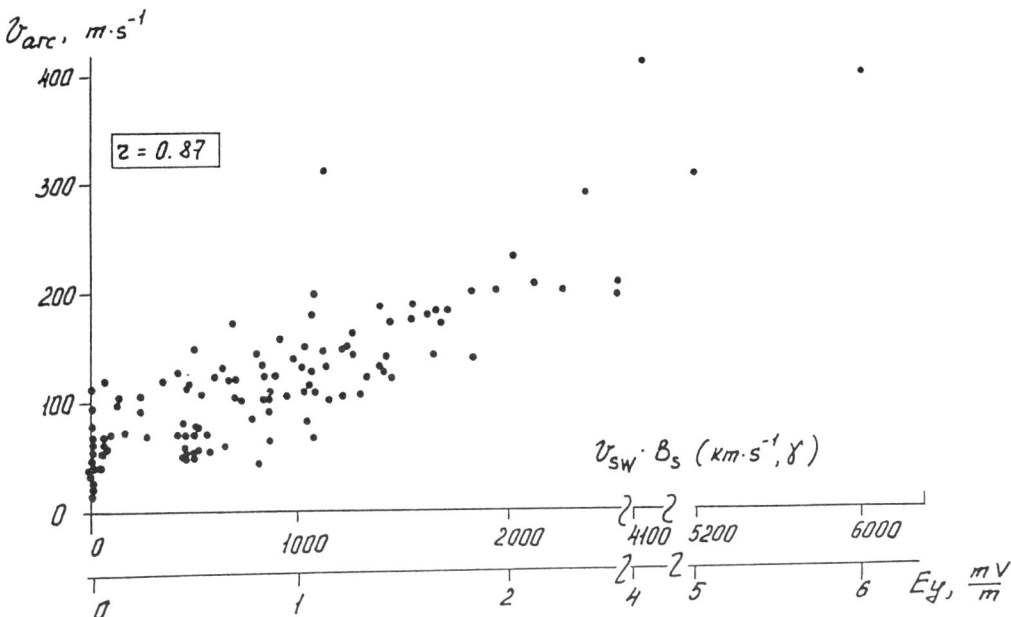

Fig. 5. Auroral arc velocity in dependence on the Y-component of the solar wind electric field [after Pudovkin et al., 1985].

intensity of the magnetotail magnetic field increases during this phase by ΔB_T, then the magnetic flux in the tail lobes increases by the value $\Delta F = \frac{1}{2} \pi R_m^2 \Delta B_T$, where Rm is the radius of the magnetosphere. The rate of the magnetic flux entrance into the lobe equals $(dF/dt) = c\Delta\phi$, and the duration of the initial phase is equal to

$$t = \frac{\Delta F}{dF/dt} = \frac{\pi R_m^2 \Delta B_T}{2c\Delta\phi} \quad (8)$$

Where $\Delta\phi$ is the potential drop across the magnetosphere and equals to the data given in Figure 5 to $\Delta\phi = \frac{0.3}{c} V B_z D_m$; here all the quantities are measured in the Gaussian system of units. The value of ΔB_T is uncertain. However, judging from the data produced by Sergeev and Tsyganenko [1980], at the distance X = $-15\ R_E$, the value of B_T increased during the initial phase of a substorm from 18 γ to 28 γ that is by 10 γ, and when $D_m = 40\ R_E$

$$\tau = \frac{5.6 \cdot 10^4}{vB_s} \quad (9)$$

Here τ is measured in minutes, v in km/sec and B_s in gammas.

It is interesting to compare this result with experimental data. In this connection, Figure 8 [after Dmitrieva and Sergeev, 1983] shows the duration of the preliminary phase of spontaneous (marked by the solid circles) and triggered (marked by the open circles) substorms related to the value of vB_s. As is seen from the figure, the expected value for the spontaneous substorms equality τvB_s = constant really takes place. At the same time, the constant on the right side of the equality equals $9 \cdot 10^4$ from the experiment, instead of $5.6 \cdot 10^4$ from the model. However, taking into account that the value of ΔB_T in our calculation was obtained from only one plasma sheet crossing, and hence may be considered as a rather arbitrary, the agreement between the experimental and model data seems to be reasonable.

Concerning the preliminary phase of triggered substorms, one can see that it is systematically shorter than that of the spontaneous substorms.

5. Expansion and Recovery Phases

It is known that the development of auroral break-ups is associated with a rapid dipolization of the geomagnetic field in the magnetotail [Walker et al., 1976]. This feature of the expansion phase is seen also in Figure 7. One can see that degree of the dipolization depends on the type of the break-up under consideration. In the course of six (of the total amount of seven) spontaneous break-ups, the angle χ increased up to 40°–50°, which means dipolization of the geomagnetic field is far from being complete. At the same time, stimulated break-ups result in increase of the angle up to 65°–70°.

In the ionosphere, the expansion phase of the substorms manifests itself in a rapid motion of auroral arcs toward the pole [Akasofu, 1964]. At the same time, as was shown by Belyakova et al. [1968] and by Pudovkin et al. [1970], there is equatorward motion of auroral arcs located inside the auroral

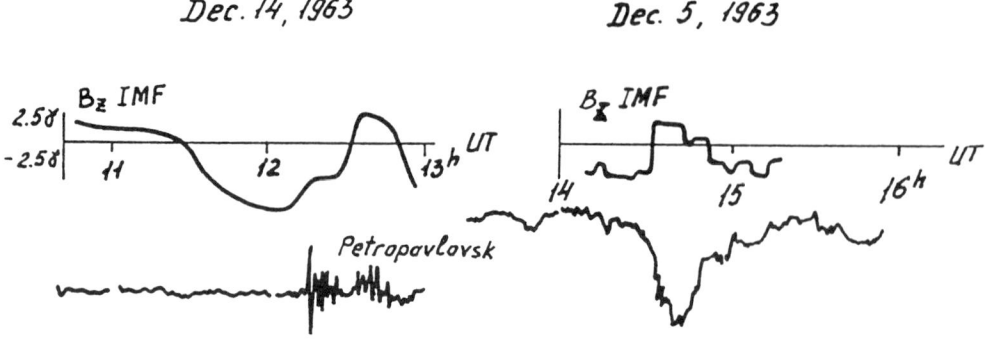

Fig. 6. Two types of auroral break-ups [after Pudovkin et al., 1970]. (Left) spontaneous break-up. (Right) triggered break-up.

bulge. This motion may be seen in Figure 9 [after Pudovkin et al., 1990], where the development of a rather intensive auroral disturbance at Heiss Island ($\phi \approx 74°5$) on February 12, 1985 is evident.

One can see from this figure, that in accordance with the results by Sergeev and Yakhnin [1978], the expansion phase of the substorm consists of a series of pulses, every one of which is associated with the appearance of a new auroral arc, often north of the previous one. And the appearance of arcs coincides (with accuracy of 1-2 minutes) with the onsets of the pulses in the geomagnetic field (δH). If we suppose, after Semenov and Sergeev [1981], Pudovkin and Semenov [1985] that every pulse of δH and associated to it microsubstorm (or micro-break-up) are caused by an elementary burst of the magnetic field reconnection in the magnetotail, we may identify the arcs located at the poleward edge of the auroral bulge with the ionospheric projections of magnetic neutral lines. The arcs and arc fragments located inside the bulge may be connected to some magnetospheric plasma irregularities moving in closed geomagnetic field lines. Consequently, the velocity of the most northern arcs in the vicinity of the poleward boundary of the bulge is determined by the vortex part of the azimuthal magnetospheric electric field, while the equatorward motion of auroral arcs located inside the bulge characterizes the intensity of the potential part of the magnetospheric electric field.

As one can see in the figure, in the first stage of the expansion phase (15.02 UT - 15.05 UT), a very rapid drift of auroral arcs was observed. The mean velocity of the arc motion is 1.6 km/sec at this period. The velocity of the northward arc motion decreased to 0.3 km/sec at 15.05 - 15.13 UT and to 0.44 km/sec at 15.13 - 15.19 UT, and finally at the recovery

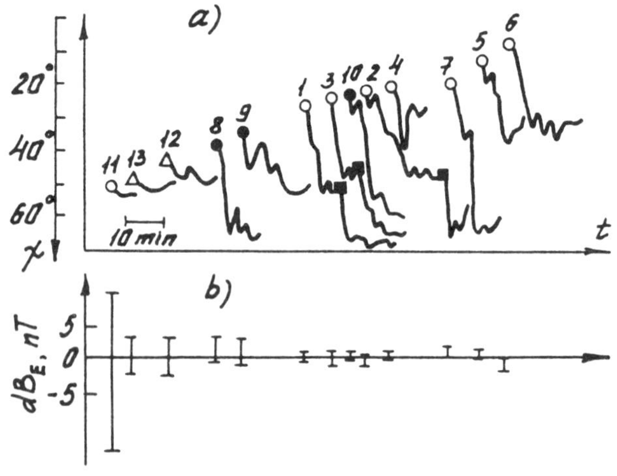

Fig. 7. (Upper) variation of the angle c of inclination of the magnetic field at the geostationary orbit during auroral break-ups of various types [after Kozelova et al., 1989]; o : spontaneous break-ups; • : triggered break-ups; ■ : triggered break-ups developing after spontaneous break-ups; Δ : break-ups stimulated by short pulses of B_z. (Lower) jumps of the IMF B_z component at the time interval between -20 and 10 min around the break-up.

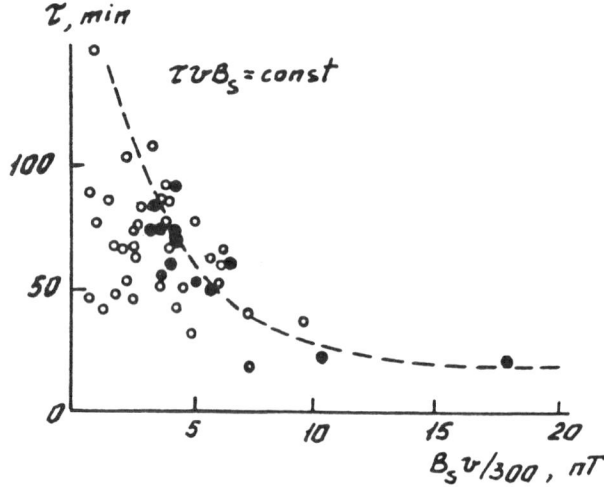

Fig. 8. Duration of the preliminary phase of substorms in dependence on the solar wind electric field [after Dmitrieva and Sergeev, 1983], • : spontaneous beak-ups; o : triggered break-ups.

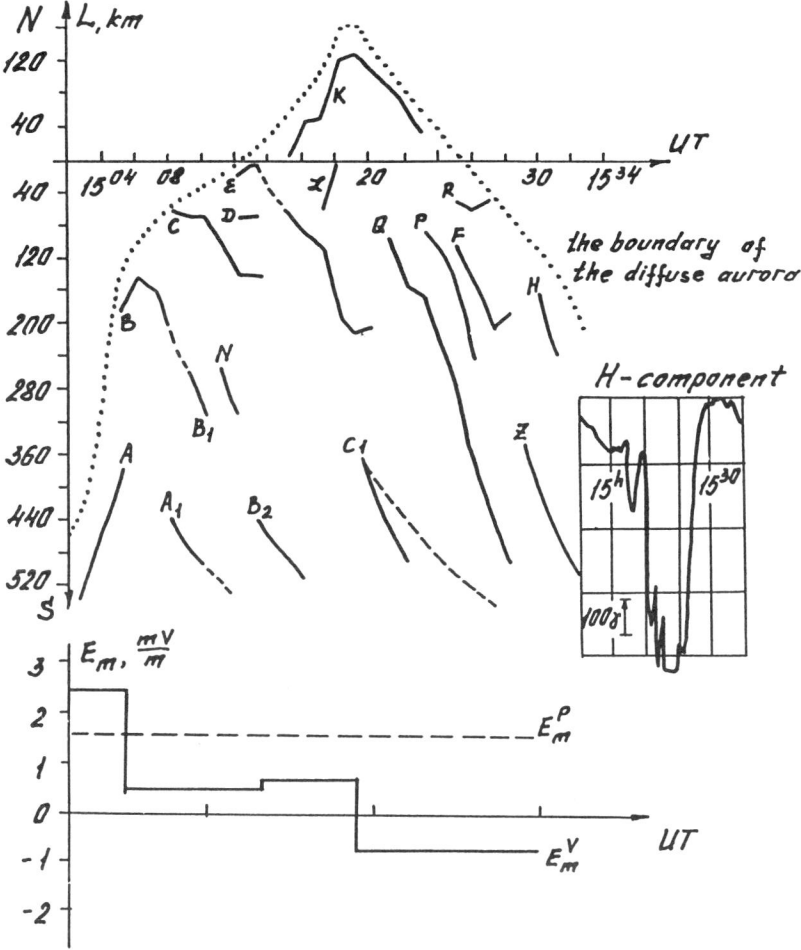

Fig. 9. Development of an auroral substorm on Feb. 12, 1985 (Heiss Island, $\phi \approx 74°$) [after Pudovkin et al., 1990]. (Upper) location of auroral arcs with respect to the zenith of the observatory. (Lower) variations of the calculated intensity of the E_m^v (solid line) and E_m^p (dotted line) electric fields.

phase of the substorm (15.20 - 15.34 UT) the northernmost arc moved equatorwards at the mean speed of about 0.5 km/sec. And as is said above, this arc motion is determined by the vortex component (E_m^v) of the magnetospheric electric field (see formula (7)). The variation of the E_m^v field intensity during the substorm calculated from the arc motion velocity for $\Delta l_m / \Delta l_i = 40$ is shown in Fig.9 (bottom panel) as a solid line. One can see that, at the beginning of the expansion phase, when the amount of the magnetic field energy stored in the magnetotail's lobes is large, the intensity of the vortex electric field is high. Later, when the storage of the free energy is exhausted, the intensity of the E_m^v field decreases, and, finally, at the recovery phase of the substorm, the E_m^v field changes its sign.

At the same time, the velocity of the equatorward motion of auroral arcs within the auroral bulge is rather stable all through the substorm and equals 1 km/sec on the average. The potential component of the magnetospheric electric field E_m^p, calculated according to that velocity (see formula (6)), equals approximately 1.5 mV/m and is directed from dawn to dusk. The intensity of this field is shown in the bottom panel in Fig. 9 as the dotted line.

Having estimated intensity of the potential and vortex components of the magnetospheric electric field, and their variation in the course of a magnetic disturbance, we may consider the energy balance of magnetospheric substorms.

6. Magnetospheric Substorm Energetics

Magnetospheric substorms are known to be caused by the electric fields generated in the magnetosphere by the solar wind. The rate of the energy input into the magnetosphere is determined by the energy function ε of Perreault and Akasofu [1978]. However, this energy does not immediately dissipate within the ionosphere and experiences a series of transformation. Let us attempt to follow the variations of the energy transport into the plasma sheet.

The amount of energy entering the reconnection region from both magnetotail lobes for some time interval may be estimated as

$$\Delta W = \frac{c}{2\pi} \Delta l_{mx} \Delta l_{my} \int (E_m^p + E_m^v) B_T \, dt \quad (10)$$

where B_T is the magnetotail's magnetic field intensity, Δl_{mx} and Δl_{my} are the extents of the reconnection region along and across the plasma sheet, respectively, (see Fig. 10). Inserting expressions (6) and (7) into (10), one obtains:

$$\Delta W = \frac{1}{2\pi} \Delta l_{mx} \Delta l_{iy} B_i B_T \int (U_i^S - U_i^N) dt \quad (11)$$

Here U_i^S is the velocity of the equatorward motion of auroral arcs and arc fragments inside the auroral bulge, and U_i^N is the velocity of the drift of the northernmost arc associated with the reconnection line. The velocity is positive when it is directed southwards. Supposing $B_T = 15\,\gamma$, $B_i = 0.6$ G, $\Delta l_{mx} = 10\, R_E$ and $\Delta l_{iy} = 10^3$ km we can calculate the energy transported into the plasma sheet by the E_m^p and E_m^v fields at various phases of the substorm. The results are given in Table I. One can see from the table that, during the substorm under consideration, the potential component of the magnetospheric electric field transfers the energy into the plasma sheet at approximately constant rate all through the substorm and contributes a total of $2.6 \cdot 10^{21}$ ergs. The vortex part of the E_m field is positive during the expansion phase of the substorm and transports into the plasma sheet about $1 \cdot 10^{21}$ ergs for this period. However, during the recovery phase of the substorm, it becomes negative and takes away from the plasma sheet $0.5 \cdot 10^{21}$ ergs. Therefore, the total energy input associated with the vortex electric field is small in the total energy budget of the substorm.

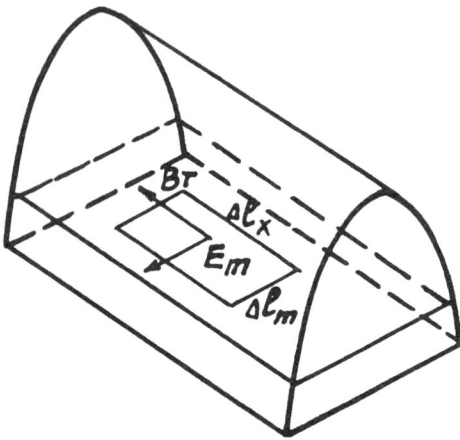

Fig. 10. The magnetotail reconnection region.

TABLE I. Energy budget of the substorm on Feb. 12, 1985, 15.00 – 15.30 UT

	Expansion phase	Recovery phase	Total
ΔW_v (ergs)	$0.9 \cdot 10^{21}$	$-0.5 \cdot 10^{21}$	$0.4 \cdot 10^{21}$
ΔW_p (ergs)	$1.4 \cdot 10^{21}$	$1.2 \cdot 10^{21}$	$2.6 \cdot 10^{21}$
$\Delta W = W_v + W_p$	$2.3 \cdot 10^{21}$	$0.7 \cdot 10^{21}$	$3.0 \cdot 10^{21}$

The variation of the electric fields in the magnetosphere and dynamics of aurora in the midnight sector of the auroral zone in the course of spontaneous and triggered substorms are shown schematically in Figure 11 [after Pudovkin et al., 1991]. The main features of the magnetospheric electric field behaviour are as follows.

a) *Preliminary phase*. In both cases (of spontaneous and triggered substorms), the development of magnetospheric substorms is associated with the appearance, in the magnetosphere, of a dawn-to-dusk electric field with drift of auroral forms toward the equator (see Figures 4 and 5). At the same time, the process of magnetic field reconnection responsible for the generation of the electric field, results in the transfer of the magnetic flux from the day side magnetosphere into the magnetotail. Hence, the magnetic field increases in the tail lobes. The intensification of the magnetic field generates a vortex electric field $\vec{E}_m^v$, and, as was shown by Semenov and Sergeev [1981], within the plasma sheet the $\vec{E}_m^v$ field effectively cancels the potential electric field so that at the distance $x < -10\, R_E$, the total field equals zero. Consequently, acceleration of auroral plasma and other substorm phenomena in the plasma sheet and in the ionosphere are rather weak at this phase of the substorm.

b) *Expansion phase*. During this state of substorms, a rapid dipolization of the tail's magnetic field takes place (see Fig.7). Thus, the vortex electric field E_m^v changes its sign and is now directed from dawn to dusk so that the total electric field in the plasma sheet significantly increases and the energy stored in the magnetic lobes is effectively released. Consequently, the rate of energy dissipation in the plasma sheet, and in the ionosphere exceeds the rate of energy transfer from the solar wind at this phase of substorms.

These features are unique for substorms of both types. At the same time, some details of their development can differ.

As is seen in Figure 6, spontaneous substorms proceed at periods when B_z IMF is negative. Consequently, a rather intensive equatorward motion of auroral forms should be observed. When the surplus magnetic energy in the magnetotail's lobes is expanded, the values of $\partial B/\partial t$ and hence of $\vec{E}_m^v$ decrease. At this stage, the substorm is developing in a quasi-steady-state regime. Also, the energy input into the plasma sheet is determined by the potential electric field $\vec{E}_m^p$ which penetrates into the magnetosphere from the solar wind.

In contrast, development of the active phase of triggered substorms takes place during the period when B_z IMF is

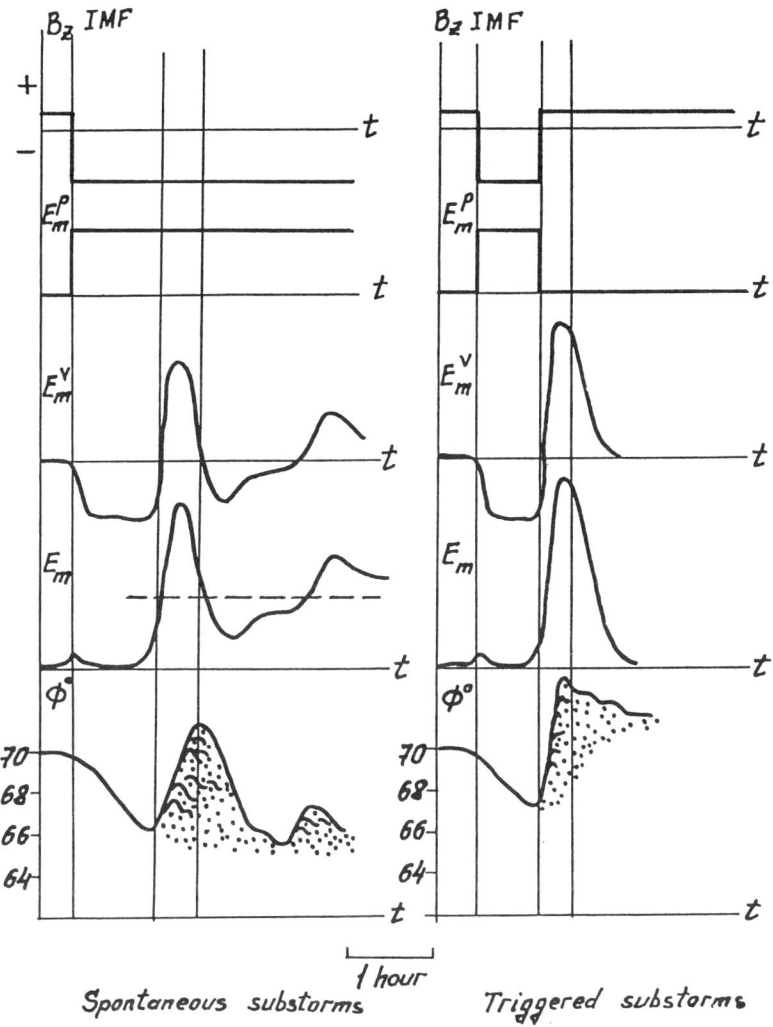

Fig. 11. Scheme of the development of the auroral substorm. (Left) spontaneous break-ups. (Right) triggered break-ups. The lowest panel shows the latitude of location of the northern boundary of the diffuse aurorae, location of discrete auroral arcs moving equatorwards, and location of the diffuse auroral belt on the whole.

positive. Hence, the magnetic field reconnection at the dayside magnetopause has subsided, and the energy needed to drive the substorm is provided by the vortex electric field. Thus, a slow equatorward motion of the aurora may be expected at this time.

c) *Recovery phase.* Dynamics of aurora and variations of the electric field intensity in the magnetotail during the last stage of substorms have not been studied sufficiently yet, and what is discussed below is based only in part on experimental data and has therefore is preliminary.

The final stage of substorms is usually called the "recovery phase". This term suggests that the magnetosphere returns to some initial "quiet" state. However, these "quiet" states are essentially different for spontaneous and triggered substorms.

In the case of spontaneous substorms, the B_z component of the IMF changes insignificantly during the expansion phase. And after it the magnetosphere returns to the pre-break-up state, which may greatly differ from the quiet state. Thus, the recovery phase of a spontaneous substorm proceeds at $E_m^p \neq 0$ and $E_m^v < 0$, and being indistinguishable from the preliminary phase of the next substorm, is associated with the equatorward motion of auroral forms all over the midnight sector of the auroral belt.

Contrary to this, during a triggered substorm, the magnetosphere transits to a really "quiet" state with B_z IMF > 0. Correspondingly, the equatorward motion of auroral forms is relatively slow at this time. In addition, the entire auroral oval contracts during substorms of this type.

7. Conclusions

The data presented above allows us to derive the following conclusions.

1) Development of magnetospheric substorms is controlled by the variations and relative intensity of two components of the magnetospheric electric field: a potential field $\vec{E}_m^p$ and a vortex field $\vec{E}_m^v$. The first component of the electric field is generated by the process of the solar wind - magnetosphere interaction and is determined by solar wind electric field: Es = $-\frac{1}{c}$ Vs Bs. In terms of directly driven and loading-unloading processes, this component of the $\vec{E}_m$ field seems to be responsible for the former. The $\vec{E}_m^v$ component is induced by the variations of the magnetotail's magnetic field and does not correlate directly with the solar wind parameters. It may be suggested that it is responsible for the loading-unloading processes in the magnetosphere.

2) During the initial phase of substorms, the vortex electric field $\vec{E}_m^v$ effectively screens the plasma sheet of the penetration of the $\vec{E}_m^p$ field. Therefore the Joule heat $(\vec{E}_m^p \cdot \vec{j}_m) = 0$, and the Poynting vector entering the magnetosphere transforms mostly into the energy of the magnetotail's magnetic field.

3) When the intensity of the plasma sheet electric currents increases to some critical value, a plasma instability may develop in the plasma sheet. Then the process of the magnetic field reconnection and the auroral break-up begin. This plasma instability may be provoked by both the external (triggered or stimulated break-ups) or internal (spontaneous break-ups) causes. In the case of spontaneous substorms, the duration of the preliminary phase of the substorm (τ) depends on the intensity of the solar wind electric field, $V_s B_s \tau$ = const. In the case of stimulated break-ups, the intensity of the B_z IMF jump needed for the triggering the substorm, depends on the state of the magnetosphere.

4) During the expansion phase of substorms, the vortex electric field $\vec{E}_m^v$ and the $\vec{E}_m^p$ field are directed from dawn to dusk, providing a rapid release of the magnetic energy stored in the magnetotail's lobes. Then, when the supply of the free magnetic energy reduces, $\vec{E}_m^v$ decreases and the substorm develops into a quasi-steady-state regime. During this stage of the substorm, the energy input into the plasma sheet is determined mainly by the potential electric field $\vec{E}_m^p$, hence by the solar wind parameters.

5) The character of the recovery phase of substorms depends on the type of the break-up under consideration.

a) In the case of spontaneous break-ups, the recovery phase of a previous substorm merges with the preliminary phase of the next substorm and these two phases are indistinguishable.

b) In the case of triggered break-ups, the magnetosphere passes from the state corresponding to B_z IMF < 0 to the state with B_z IMF > 0, which is associated with a global reconstruction of the magnetotail and with contraction of the auroral oval.

References

Akasofu, S.-I., The development of the auroral substorm, *Planet. Space Sci.*, *12*, 273, 1964.

Belyakova, S.I., S.A. Zaitseva, and M.I. Pudovkin, Development of polar storm, *Geomagn. Aeron.*, 8, 712-718, 1968.

Coleman, P.J., and R.L. McPherron, Fluctuations in the distant geomagnetic field during substorms: ATS-1, in *Particles and Fields in the Magnetosphere*, D. Reidel Publ. Co., Dordrecht, Holland, 171-194, 1970.

Dmitrieva, N.P., and V.A. Sergeev, Spontaneous and stimulated onsets of the expansion phase of the substorms and duration of the preliminary phase, *Geomagn. Aeron.*, 23, 470-474, 1983.

Kelley, M.C., J.A. Starr, and F.S. Mozer, Relationship between magnetospheric electric fields and the motion of auroral forms, *J. Geophys. Res.*, 76, 5269-5277, 1971.

Kozelova, T.V., M.I. Pudovkin, and L.L. Lazutin, Some peculiarities of the development of stimulated and spontaneous magnetospheric substorms according to satellite and ground-based data, *Geomagn. Aeron.*, 29, 910-915, 1989.

McPherron, R.L., Growth phase of magnetospheric substorms, *J. Geophys. Res.*, 75, 5592-5599, 1970.

Perreault, P., and S.-I. Akasofu, A study of geomagnetic storms, *Geophys. J. Roy. Astron. Soc.*, 54, 547-573, 1978.

Pudovkin, M.I., S.I. Isaev, and S.A. Zaitseva, Development of magnetic storms and the state of the magnetosphere according to the data of ground-based observations, *Ann. Geoph.*, 26, 761, 1970.

Pudovkin, M.I., O.M. Raspopov, L.A. Dmitrieva, V.A. Troitskaya, and R.V. Shepetnov, The interrelation between parameters of the solar wind and the state of the geomagnetic field, *Ann. Geophys.*, 26, 389-396, 1970.

Pudovkin, M.I., S.A. Zaitseva, and S.I. Isaev, Development of magnetospheric substorms, in *Geophysical Studies in the Auroral Zone*, Apatity, 50-56, 1972.

Pudovkin, M.I., and V.S. Semenov, Magnetic field reconnection theory and the solar wind - magnetosphere interaction: a review, *Space Sci. Rev.*, 41, 1-89, 1985.

Pudovkin, M.I., S.A. Zaitseva, T.A. Bazhenova, and V.G. Andresen, Electric fields and currents in the Earth's polar caps, *Planet. Space Sci.*, 33, 407-414, 1985.

Pudovkin, M.I., V.S. Semenov, T.A. Kornilova, and T.V. Kozolova, *Development of Magnetospheric Substorms: Theory and Experiment - Part II.*, 1990.

Pudovkin, M.I., V.S. Semenov, G.V. Starkov, and T.A. Kornilova, On separation of the potential and vortex parts of the magnetotail electric field, *Planet. Space Sci.*, in press, 1991.

Rostoker, G., S.-I. Akasofu, W. Baumjohann, Y. Kamide, and R.L. McPherron, The roles of direct input of energy from the solar wind and unloading of stored magnetotail energy in driving magnetospheric substorms, *Space Sci. Rev.*, 46, 93-111, 1987.

Semenov, V.S., and V.A. Sergeev, A simple semi-empirical model for the magnetospheric substorm, *Planet. Space Sci.*, 24, 271-281, 1981.

Sergeev, V.A., and N.A. Tsyganenko, *The Earth's Magnetosphere*, Nauka, Moscow, 174p., 1980.

Sergeev, V.A., and A.G. Yakhnin, The peculiarities of the auroral bulge development during the expansive substorm phase and the magnetospheric microsubstorm model, *Phys. Solariterr.*, Potsdam, N7, 23-46, 1978.

Sugiura, M., T.L. Skillman, B.G. Ledley, and J.P. Heppner, Magnetic field observations in high regions of the magnetosphere, in *Particles and Fields in the Magnetosphere*, D. Reidel Publ. Co., Dordrecht, 165-170, 1970.

Walker, R.J., K.N. Erikson, R.L. Swanson, and J.P. Winkler, Substorm-associated particle boundary motion at synchronous orbit, *J. Geophys. Res.*, *81*, 5541-5550, 1976.

Zverev, V.L., T.A. Kornilova, M.I. Pudovkin, and G.V. Starkov, A meridional profile of the auroral arc motion velocity, *Geomagn. Aeron.*, *28*, 1031-1033, 1988.

Zverev, V.L., M.I. Pudovkin, and G.V. Starkov, Auroral arc motion and electric field at the preliminary phase of substorms, *Geomagn. Aeron.*, *31*, in press, 1991.

M.I. Pudovkin, Institute of Physics, Leningrad University, Leningrad, Stary Peterhoff 198904, USSR.

SUBSTORM CURRENT SYSTEMS AND AURORAL DYNAMICS

Y.I. Feldstein

IZMIRAN, Troitsk 142092, Moscow Region, USSR

Abstract. The implementation of the auroral oval concept lead to a change in the global interpretation of the space-time distribution of geomagnetic disturbances (1963-1965). It was found that along the auroral oval a westward electrojet stretches from midnight to evening sector poleward of eastward electrojet. A slot is produced between the electrojets in dusk sector. The auroral electrojets overlap in the dusk sector, which was supported by other geophysical phenomena accompanying of electrojets. The slot later was named the Harang discontinuity.

During substorms both driven and unloading processes take place. The equivalent current system during substorm growth phase is similar to DP2 current system, which is directly related to the enhancement of the convective electric field and which is controlled by the solar wind parameters. The driven process in the course of substorm is described by a DP2 current system.

The development of the ring current is accompanied by the equatorward shift of the auroral electrojets. During the magnetic storm main phase the auroral electrojet intensity is closely related to the energy flux supplied to ring current.

An interpretation of the patterns for auroral electron precipitations in the high-latitude upper atmosphere in near-midnight sector during quiet and substorm intervals is presented. The diffuse aurora is mapped to the outer radiation belt, the auroral oval of the discrete forms maps to the central plasma sheet, and a soft precipitation band lying just poleward of the oval maps to the boundary plasma sheet in the tail. A summarizing schematic of the polar precipitation regions and their mapping to the magnetospheric plasma domains is presented.

1. Introduction

Almost 30 years ago, during International Conference on Cosmic Rays and Earth Storms in Kyoto [September 1962], S. Chapman in his lecture "Earth storms: Retrospect and Prospect" traced the history of the evolution of ideas about the magnetic storms and substorm current systems (see Chapman [1962]). He focussed on the statistical laws governing the development of the magnetic storm and the equivalent current system, Dst and DS, of the magnetic storm field. He also

Magnetospheric Substorms
Geophysical Monograph 64
Copyright 1991 American Geophysical Union

discussed the possible relationship of magnetic storms to polar magnetic substorms.

This paper is devoted to the description of the further evolution of both statistical and temporal-spatial regularities of the equivalent current systems of geomagnetic disturbances at high latitudes. The interpretation is presented in the context of our understanding of the most intensive particle precipitations along auroral oval. Equivalent and three-dimensional current systems are discussed at different phases of magnetospheric substorms.

The location of the auroral electrojets is controlled by the ring current in inner magnetosphere. For a strong magnetic storm, as an example, there were obtained quantitative relations between westward electrojet shift as subauroral latitudes and the ring current intensity increase, and between auroral electrojets intensity and interplanetary medium parameters. For the main phases of 10 magnetic storms, the relation between the energy injected into the ring current and the energy entered inside the magnetosphere is presented.

The relationship of various auroral luminosity types to the parameters of precipitating auroral electron fluxes are discussed in numerous papers. Consideration was given to the two concepts concerning the conjugacy of the diffuse and discrete auroral features produced by the auroral electron precipitation with different characteristic structures to the magnetospheric plasma domains in the nightside magnetosphere. Possible changes in the large-scale structure of plasma domains from magneto-quiet intervals, through the beginning of substorm, and to the maximum of the expansive phase are described.

2. Electrojets and Auroral Oval

According to Chapman's [1935] idealized equivalent current system for the total disturbance vector, the concentration of currents occur at auroral zone latitudes (the westward and eastward electrojets in the dawn and dusk LT sector). Harang [1946] obtained essential results for the current system morphology using data from a meridional chain of observatories. He found a discontinuity between the westward and eastward electrojets at pre-midnight hours at auroral zone latitudes. The auroral zone is separated into two segments characterizing the eastward and westward electrojets, respectively. In the dusk sectors intensive currents overlap each other, and the westward current is located polewards from the eastward current. However, the same concentration of equivalent current exists also in the late dusk sector and in the Chapman current system. So Harang's study failed initially to produce any drastic revision of the concepts concerning the high-latitude structure of equivalent currents. The electrojets, as the regions of the highest-intensity currents, continued to be associated with the Fritz-Vestine auroral zone. The earlier concepts were maintained to a great extent, because Harang himself assumed the 67° geomagnetic parallel to be the region of the highest magnetic disturbance along which the auroral electrojets are located.

Burdo [1960] obtained the space-time distribution of high-latitude geomagnetic disturbance field vectors in geomagnetic latitude-local geomagnetic time coordinates. The peaks in the diurnal pattern are located along the spirals extending from the auroral latitudes to the deep inside of the polar cap. However, the Burdo equivalent current system does not, in principle, differ from somewhat modified Chapman system. Namely, the electrojets are located at auroral latitudes, and spread through higher latitudes. The electrojets arise from the dynamo-effect at the ionospheric altitudes. The generation of the electrojets and of the polar magnetic substorm current system was most comprehensively explained in terms of the dynamo-theory for geomagnetic disturbances by Nagata and Fukushima [1952], Fukushima[1953], Cole [1960], and many other researches. Based on the dynamo-theory concepts, Burdo also obtained the three spirals of the highest magnetic disturbances.

So the studies by Harang[1964] and Burdo [1960] have substantiated the fundamental fact that the westward electrojet is not only located within the night-dawn sector of auroral zone; it also extends to the dusk sector at higher latitudes. At the same time, the general opinion of the scientific community that the most active geophysical processes occur on the auroral zone latitudes ($\Phi \sim 67°$) demanded that both eastward and westward electrojets be located within the given zone. A certain concentration of the current lines at higher latitudes was successfully interpreted in terms of the ionospheric dynamo model. The current systems calculated by Fukushima and Oguti [1953] on assumption of an increased ionospheric conductivity along the auroral zone proved to resemble the observed systems both in form and in phase.

The analysis of the observations obtained during the IGY resulted in drastic changes of the concepts concerning the high-latitude auroral distribution. The region of the highest occurrence frequency and of the highest intensity of auroras proved to be oval-shaped. Auroras proved to be closely associated with geomagnetic disturbances. Therefore, the new concept of planetary distribution of auroras resulted in a revision in the early sixties of the earlier concepts concerning the space-time distribution of magnetic disturbances.

From the analysis of the IGY observations of geomagnetic disturbances it followed that the westward electrojet was not located within the night-dawn sector at auroral zone latitudes. It extended to the dusk sector at higher latitudes. In this pattern [Feldstein, 1963] the position of the westward electrojet proves to be closely associated with the position of the auroral oval.

Figure 1a shows the respective equivalent current system of magnetic disturbances observed during winter seasons of the IGY interval. The westward current covers all longitudes. The highest values of the current shift to higher latitudes and decrease in intensity from dawn hours at auroral zone latitudes ($\Phi \sim 67°$) to day hours at $\Phi \sim 78°$, with the currents being closed across polar cap or lower latitudes. In the dusk sector, the westward electrojet is located at higher latitudes compared with the eastern electrojet. The disturbance vector in the horizontal component vanishes on a certain latitude between the two electrojets. A gap between the electrojets is formed in the dusk sector: the latitude of the gap shifts polewards at earlier hours. The given characteristic features of the current system of geomagnetic disturbances was called later the Harang discontinuity. The concept that the westward electrojet is located along the auroral oval and extends to the dusk sector, polewards from the eastward electrojets, during the epoch of substorm maximum was further substantiated and developed by Akasofu et al. [1965] and by other researchers. Figure 1b presents the respective model current system after Akasofu et al. [1965]. The westward electrojet is located along the oval in the dusk sector, at higher latitudes compared with the eastward electrojet. The location of the westward electrojet along the auroral oval is also characteristic of particular disturbances.

After the concept of the auroral oval had been established, the fact that the auroral electrojets overlap in the dusk sector found an obvious interpretation in terms of the general pattern for the occurrence of geophysical events in high latitudes. At the Conference dedicated to Harang's seventieth birthday in 1971, Heppner proposed that the splitting of the ionospheric current system should be called the Harang discontinuity, whereupon the term got commonly adopted.

3. Current Systems During Different Phases of Substorms

The transition from the auroral zone concept to the concept of auroral oval entailed some changes in the spatial regular features of the current system distribution. Other substantial changes have arisen from the development of the magnetospheric substorm concept relevant to quite a complex of events observed on the Earth's surface and in the magnetosphere [Akasofu, 1964, 1968].

Figure 2 presents, after Baker et al. [1984], a schematic pattern of three-dimensional current systems connecting the magnetosphere to the ionosphere during a magnetic substorm. The solar wind-magnetosphere interaction results in separation of charges. The field-aligned currents at the polar cap boundary from the S1 system (a 20-min time constant) are closed by the Pedersen ionospheric currents, by the field-aligned currents at the equatorward boundary of the auroral zone, and by the the partial ring current (the S2 current system: a 1-2-hour time constant). The electric field is directed equatorwards at the dawn side, and polewards at the dusk side. As a result the westward and eastward electrojets of the Hall current are generated in the ionosphere. The current wedge with westward ionospheric current, which is extended far to the dusk sector during active phases of intensive substorms, thereby making the westward and eastward electrojets overlap each other at dusk hours, is located at the poleward side from the convective electrojets in the near-midnight sector.

Substorm evolution is usually divided into three phases, namely, (i) the growth or creation phase when the convection and the associated energy dissipation, as well as the energy accumulation in the magnetospheric tail, get enhanced; (ii) the expansion or active phase when the energy stored in the tail is

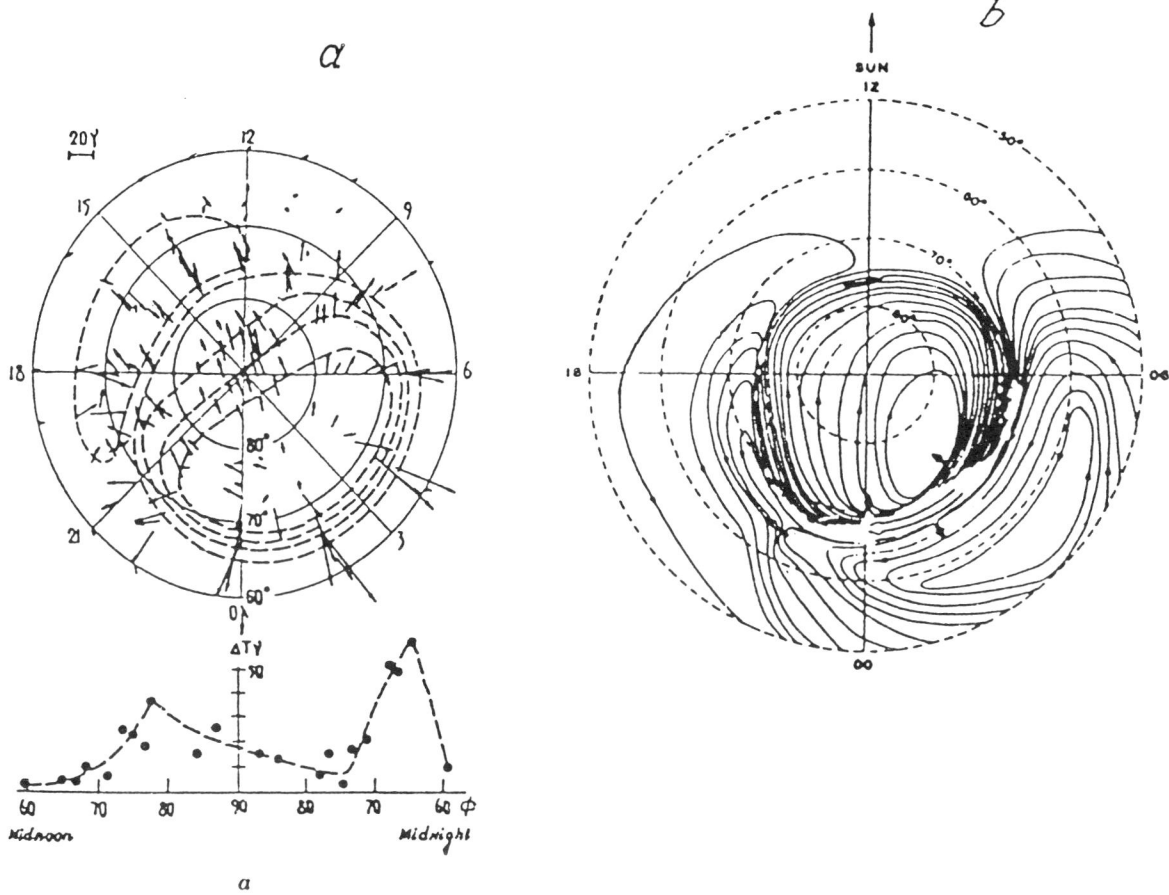

Fig. 1. (a): First depiction of an equivalent current system with a westward electrojet located along auroral oval [Feldstein, 1963]. Arrows indicate direction and magnetude of ΔT intensity of disturbed geomagnetic field vector in horizontal plane is in nT, between the dashed current lines is 20,000 A. (b): Equivalent current system of DP type geomagnetic disturbances for IGY winter season [Akasofu et al., 1965].

suddenly released and dumped to the ionosphere, thereby intensifying auroras and magnetic disturbances; (iii) the recovery phase when the disturbance decays gradually and the magnetic field on the Earth's surface returns to its initial undisturbed level.

The onset of the substorm creation phase is frequently associated with a southward turn of the IMF. Most of the features of the creation phase result from the enhancement of the electric field of convection related directly to the solar wind and IMF parameters. Therefore, the driven process dominates during the creation phase.

Figure 3, I, II shows, after Lyatsky [1978], the current systems during the initial (Fig. 3, I) and final (Fig. 3, II) stages of substorm creation phase. The S1 current system, shown in the upper part of the figure and associated with enhancement of the solar wind electric field, is formed during the early stage. The magnetic disturbance observed at the beginning of the creation phase on the Earth's surface is described by DP-2 system of the equivalent ionospheric current shown at the bottom in the Fig. 3, I and discovered by Nishida [1966,

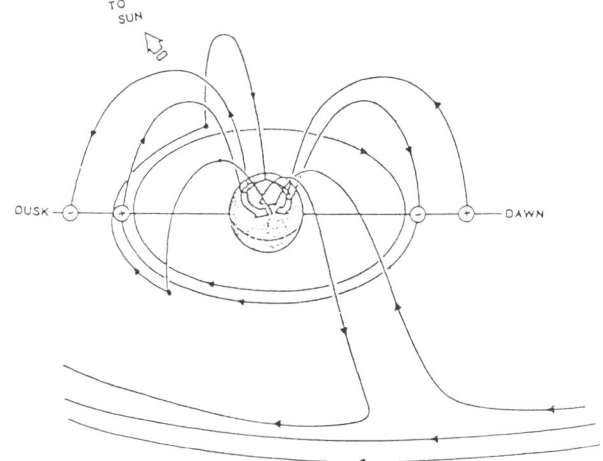

Fig. 2. Three-dimensional current system of magnetospheric substorm active phase [Baker et al., 1984].

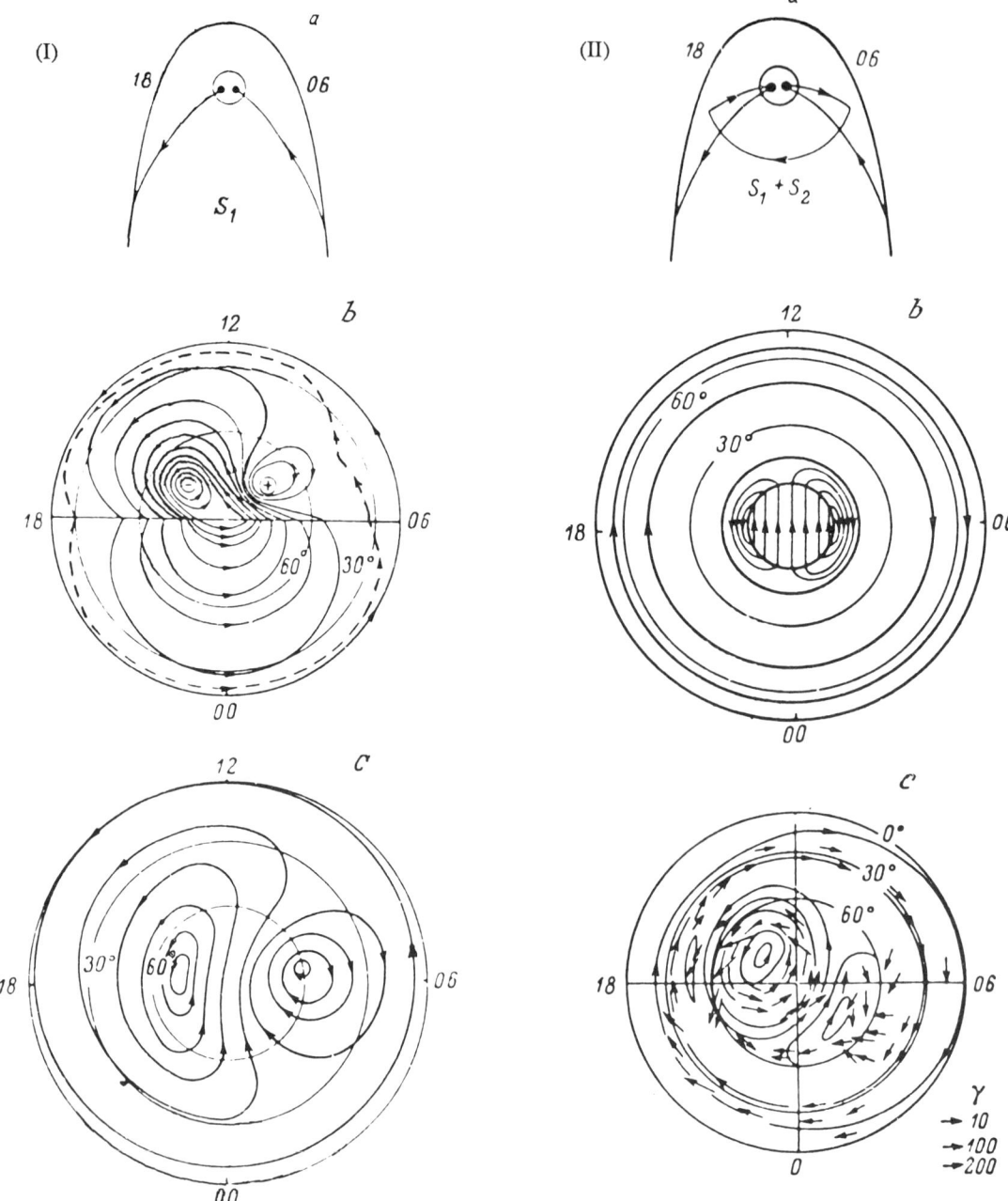

Fig. 3. Equivalent current systems for initial (I) and final (II) stages of substorm creation phase: (a) three-dimensional current system; (b) model equivalent ionosphere current [from Lyatsky, 1978]; (c) equivalent ionosphere current inferred from observations by Nishida et al. [1966] (I) and by Iijima and Nagata [1972] (II).

1968]. The current flows across the polar cap from night to day and is closed through lower latitudes down to the equator without exhibiting any features in the auroral zone.

The current system DP-2 discussed by Nishida is closely connected with southward orientation of IMF(Bz<0). It was first mentioned in the literature in terms of a direct influence of interplanetary medium on fluctuation of geomagnetic field at the Earth's surface. Magnetospheric convection enhancement, which is controlled by IMF Bz<0, is directly connected with driven part of magnetospheric substorm magnetic field.

During the late stage of the creation phase a partial ring current is formed, with field-aligned currents at the equatorward boundary of the auroral zone (the S2 current system in the Fig 3, II). The development of that current system results in middle and low latitudes being shielded from the convection electric field. The magnetic disturbance on the Earth's surface is described by the system of equivalent currents obtained by Iijima and Nagata [1972] and shown at the bottom in the figure. The DP-2 are pushed from middle latitudes to the auroral zone. Equatorward of the latter there are westward azimuthal currents which reflect the magnetic effect of the partial ring current. Analytic model calculations [Lyatsky, 1978] presented in the middle of the Fig. 3, I, II demonstrate that the theoretical and experimental patterns of currents are much alike. The calculations were made assuming that the ionospheric conductivity is uniform at both day and night hours, but suffers a discontinuity at the terminator (in the initial stage of the creation phase) or is uniform (in the final stage of the creation phase).

The substorm expansion phase is associated with a variety of phenomena. These include a rapid decay of the plasma sheet, the restoration of the magnetic field lines extended to the tail to their undisturbed state, rapid plasma motion towards the Earth, intensification of auroras, and the development of a magnetic disturbance on the Earth's surface that is described by a system of equivalent ionospheric currents with a strong concentration of current lines in the auroral zone (the unloading process).

During the expansion phase, the Pedersen and, particularly, the Hall conductivities of the auroral ionosphere increase sharply, thereby giving rise to westward and eastward electrojets. The equivalent ionospheric current system inferred from observations by Iijima and Nagata [1972] is shown at the bottom of Fig. 4c together with the calculation results obtained by Lyatsky [1978] for increased, but uniform (at the top) and nonuniform (at the middle) conductivity in the auroral zone. The main feature of the model current systems consists in the occurrence of two electrojets flowing in the auroral zone towards each other.

During the substorm expansion phase, the poleward region of the westward current, which penetrates deep into the dusk sector, seems to be associated with discontinuity in the currents flowing in the current sheet of the magnetospheric tail and branching to the ionosphere along field lines [Kaufmann, 1987]. A current wedge is formed, with the field-aligned current inflowing to the ionosphere at the eastward end and outflowing from the ionosphere at the westward end of the westward travelling surge. Such an outflowing current is detected during the expansion phase in the vicinity of bright aurora. Using the data from a dense network of magnetometers, of radar measurements of electric fields, and optical observations in the northern Scandinavia, Baumjohann et al. [1981] and Inhester et al. [1981] modelled a three-dimensional current system and obtained a localized, intense field-aligned current flowing out from the ionosphere at the westward edge of active aurora.

4. Auroral Electrojets and the Ring Currents-Dependence on the Interplanetary Plasma Parameters

The magnetic storms is the most global geomagnetic phenomenon. It includes the ring current and a number of other geophysical phenomena such as magnetospheric substorms. That is why magnetic storms were an early object of investigations in solar-terrestrial physics. But recently the

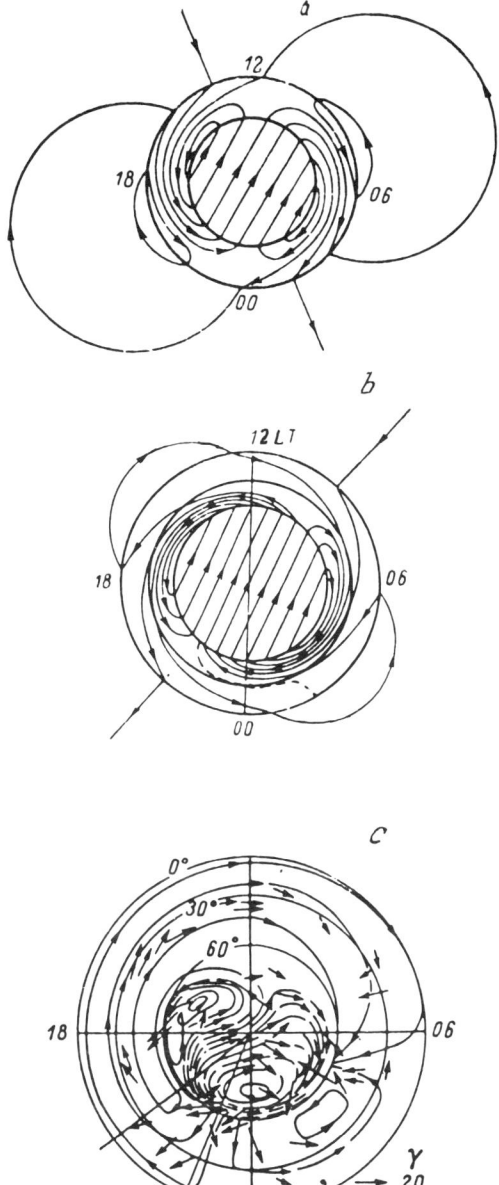

Fig. 4. Equivalent current system of substorm active phase: (a) model equivalent ionosphere current system for enhanced, but homogeneous, conductivity inside the auroral zone; (b) when inhomogeneouty of conductivity inside the auroral zone is taken into account [from Lyatsky, 1978]; (c) equivalent ionosphere currents inferred from observations by Iijima and Nagata [1972].

main attention has turned to magnetic substorms which are assumed to be the elementary events of magnetospheric disturbances. However, a magnetic storm is accompanied by effects indicating that its evolution is different from a simple sum of single substorms. The development of DR in the inner

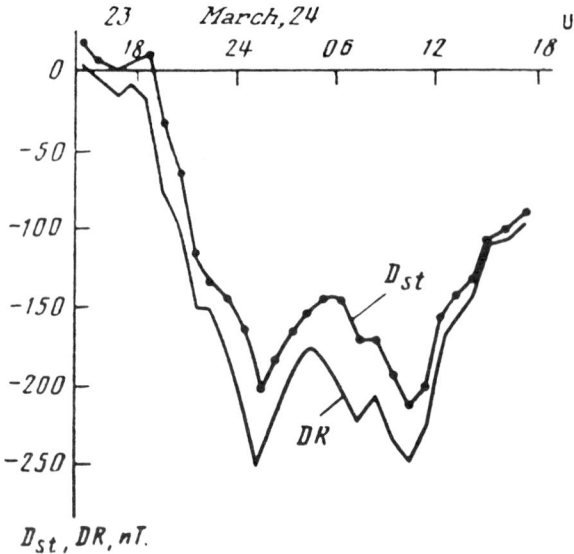

Fig. 5. The variations of Dst and DR magnetic field at geomagnetic equator during the March 23-24, 1969 strong magnetic storm [Sumaruk et al., 1989].

magnetosphere leads to the deformation of geomagnetic field lines, which also affects the location of plasma domains in the magnetosphere [Akasofu and Chapman, 1972]. As a result, auroral phenomena are observed down to middle latitudes. Some aspects of the magnetospheric activity are considered below as exemplified by the magnetic storm of March 23-24, 1969, which attracted the attention of several investigators [Akasofu, 1981a; Tinsley and Akasofu, 1982; Khorosheva, 1986; Sumaruk et al., 1989]. The Dst variation reflects a changing of the intensity of the ring current's symmetric part during the storm. This variation is distinctly traced in the X or H component.

Figure 5 gives the variations of Dst and DR according to hourly values from 11 low-latitude magnetic observatories, spaced rather uniformly with respect to longitude. The DR values were calculated using relation DR = Dst−DCFd+DCFq, where DCF is the magnetic field of the magnetopause current during the disturbed and quiet conditions. The development of DR is accompanied by the intensification of magnetospheric substorm activity, as manifested in auroral electrojet intensifications (the indices AU, AL, AE). The generation of DR and the deformation of geomagnetic field lines shifts the auroral electrojets equatorward. As a result, the AE index, which is based on the geomagnetic field measured at auroral latitude magnetic observatories (the longitudinal chain is at 62 $\leq \Phi \leq 70$), ceases to reflect the electrojet intensity. The use of the ΔX and ΔZ latitude profiles in the near-midnight-early dawn MLT sectors at different UT moments has made it possible to find the position of the westward convective electrojet center as a function of DR intensity. Figure 6 according to Sumaruk et al. [1989] presents the position of the convective electrojet center (the time moments before pronounced substorm activizations were selected before the electrojet becomes stratified and the separated fraction begins moving rapidly polewards) as a function of the DR intensity at the UT indicated by numerals at the points. The crosses are the locations of visual aurorae at that time. The electrojet moves to lower latitudes as DR increases and its position is described by the relation

$$\Phi = 65.2° + 0.035\, DR$$

in the 0>DR>−250 nT interval, where Φ is the corrected geomagnetic latitude in degrees and DR is in nT. Consequently, at DR<−100 nT the western electrojet moves out of the belt of auroral magnetic observatories by which AE indices are defined. Thus to estimate the auroral electrojet intensity during magnetic storms, it is necessary to use either data from subauroral observatories or Z-component variations. This conclusion is in agreement with the analysis by Khorosheva [1986] according to which the electrojets shift out of the belt of auroral observatories at Dst<−40 nT. At values of Dst<−40 nT, AE indices during the magnetic storm interval are defined by using a network of subauroral observatories, and are denoted by AE'.

Let us consider the relationship of magnetospheric activity (estimated in terms of AE' indices) with the energy input function from the solar wind into the ring current. Energy flux into the magnetosphere is controlled by the interplanetary plasma parameters and the IMF. The azimuthal electric field component of solar wind Ey [Burton et al., 1975], the energy flux into the magnetosphere ε [Akasofu, 1981a,b], and the energy flux into the ring current F [Pisarsky et al., 1989] are accepted below as geoeffective characteristics. Pisarsky et al. [1989] assumed the energy flux function $F=F(B_z,\sigma,v)$ and determined its functional form by comparison between the observed and modelled DR variations for some dozen of magnetic storms. The functional form F is given as $F=8.2(B_z-0.67\sigma)\cdot v \cdot 10^{-3} - 14.1(v-300)\cdot 10^{-3}+9.4$, where F in nT·h, B_z and σ (standard deviation of B) in nT, v in km·s^{-1}.

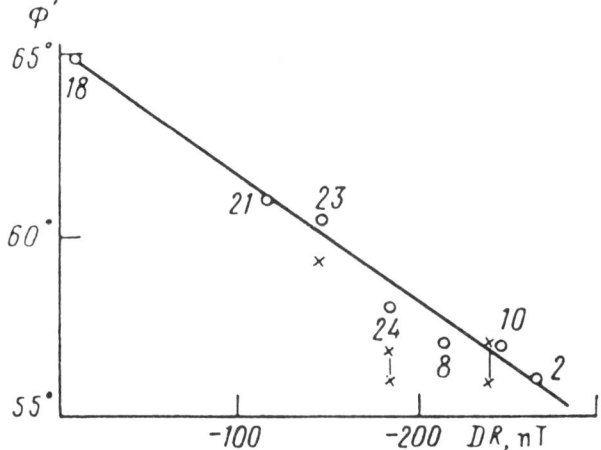

Fig. 6. The position of the westward convective electrojet center in the near-midnight-early dawn MLT sector as a function of DR intensity. The numerals at the dots are hours of UT. The straight line has been obtained by the least squares method [Sumaruk et al., 1989].

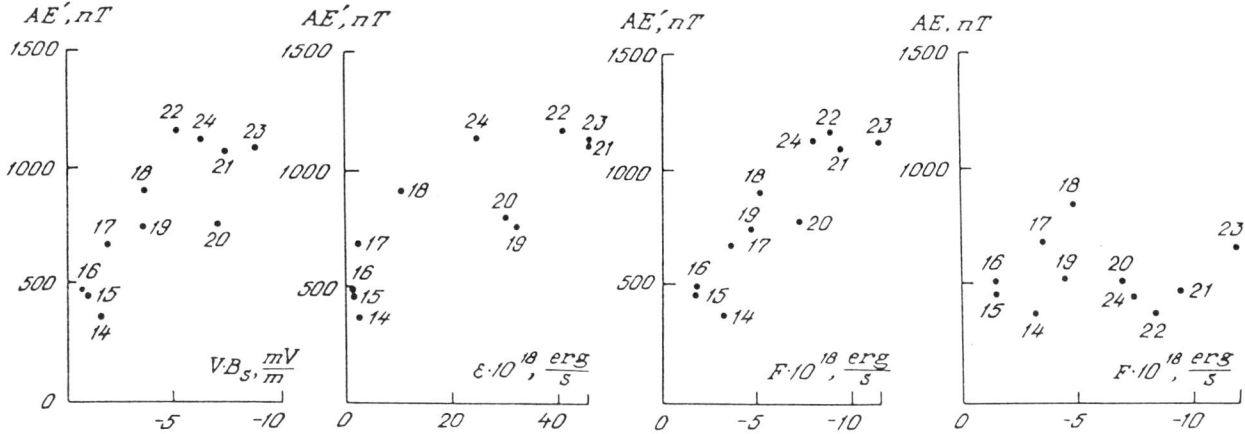

Fig. 7. Intensity of the hourly AE' indices of auroral electrojets versus different combinations of geoeffective parameters in the solar wind which characterize the energy fluxes during the March 23–24, 1969 storm main phase. The solar wind parameters lead the AE' index by 1 hour. The numerals are hours of UT. The right-hand side shows the same for the standard AE index.

Figure 7 gives the dependences of hourly values of AE' on Ey, ε and F also of AE on F. The geoeffective characteristics lead the AE' or AE indices by 1 hour.

All three characteristics are positively correlated with AE' namely with intensifying Ey, ε, or F the AE' index increases. During the storm recovery phase (when all the three characteristics are close to zero) AE' is also small. The linear regression equations and correlation coefficients are as follows:

AE'=86.6·(Bs·v)+443;   r=0.82±0.1 at / Bs·v/≥1mV/m
AE'=13.0ε+527;   r=0.82±0.1 at ε≥2·10^{11}W
AE'=81.9F+326;   r=0.9±0.08 at |F|>10^{11}W

Here AE' is in nT, Bs·v is in mV/m, ε and F are in 10^{11}W.

While the correlation between AE' and all three geoeffective characteristics is high, it is highest between AE' and F. This means that during the storm main phase the auroral electrojet intensity is closely related to the energy flux supplied to ring current. It is no surprise that at low latitudes the field decrease (being a generally accepted indicator of a magnetic storm) occurs during the same time interval when intensive electrojets appear. Due to this fact, the close correlation between substorms and ring current generation takes place at the level of hourly values. The same fact can explain a wide-spread opinion that magnetic storms are nothing else quickly recurring intense substorms.

The relationship of AE' to F makes it possible to estimate the real hourly values of AE' indices during magnetic storm intervals if the interplanetary medium parameters are measured during the intervals. In such cases, the data from the network of subauroral observatories, which contribute much to the AE' values because of the equatorward shift in the electrojet as the ring current develops, must no longer be processed separately.

It should be pointed out that there is approximately a factor of four difference between the absolute values of ε and F characterizing energy fluxes in Fig. 7. Taking into account the fact that the input function F used in the present investigation generates the outer and inner parts of DR field, and that the ratio of the parts is 2, the difference between ε and F increases up to a factor of 6. The relationship between ε and F is defined more accurately according to the hourly values for the main phase intervals of 10 magnetic storms (Fig. 8). The linear relationship between them is obtained by the leaset square method ε = (6.6±–0.3)F+(0.3±0.3), where ε and F are in 10^{11} W, r=0.9±0.05. Almost the same relationship is valid for each individual storm. Thus, only 15% of the value ε comes to the ring current. The difference between ε and F is surely caused by the fact that ε characterizes the total energy flux from the solar wind into the magnetosphere, whereas the value F characterizes the energy flux into the ring current. It is also probable that a difference between ε and F is connected with the fact that ε is only one estimate of the energy into the magnetosphere. F has also been calibrated against the particle energy in the ring current on the basis of Desseler-Parker-Sckopke theorem [Olbert et al., 1968]. Figure 7 also shown the dependence on F of the standard index AE according to Allen et al. [1974] for the previous hour during the storm main phase. The limitation of the observatory network in the auroral zone not only significantly underrates the AE index intensity, but leads to a noticeable spread in the points in the plot. The absence of a correlation between AE and F (r=0.08) can result in a wrong conclusion that the auroral electrojet intensity is not connected with injection into the ring current. When looking for a possible relation between the AE index and the interplanetary medium parameters for a given storm, one can perform an analysis to determine the confidence of the AE index series available. In a number of investigations, no relationship or a weak relationship could reflect nothing more than the need to improve the technique for determining the indices characterizing the auroral electrojet intensity. As an example of an improved technique, the AE' results in a sharply increased of correlation with the interplanetary medium parameters (see Fig. 7). The low AE correlation with the interplanetary medium parameters obtained by Gonzalez et al.

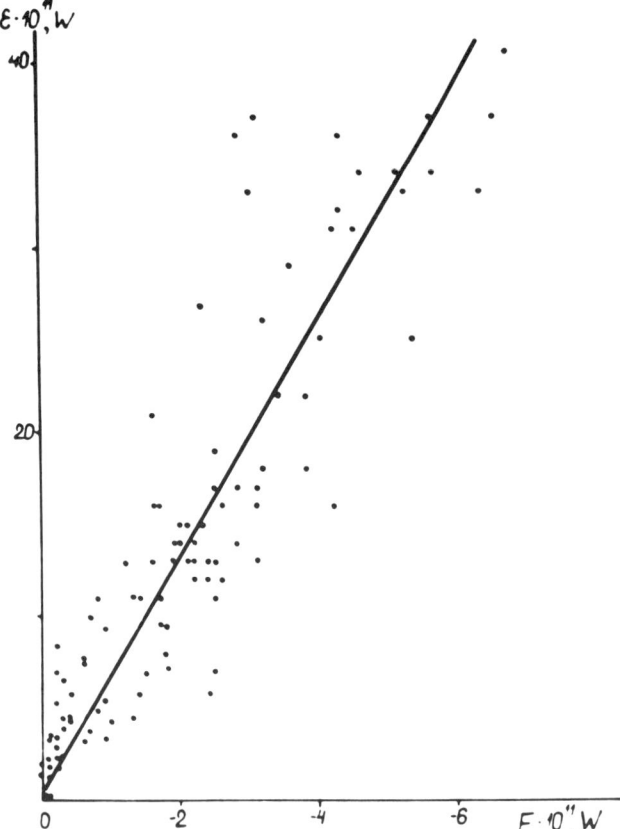

Fig. 8. Dependence of ε on F values during the main phase of 10 magnetic storms. The solid line represents the linear regression equation.

results obtained by Vasyliunas [1970], Feldstein and Starkov [1970], and Lassen [1974] has proved to give the most comprehensive description of the atmospheric luminosity and auroral electron precipitation.

Figure 9 a, b presents the inferred latitudinal morphological patterns of auroral electron precipitations during various substorm phases [Winningham et al., 1975; Feldstein and

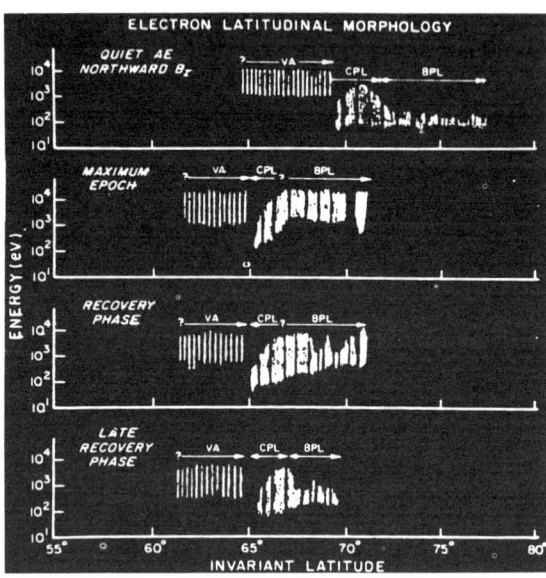

[1989] for the main phase of several magnetic storms can just be explained by the fact that Gonzalez et al. used the auroral observatory data only to calculate the AE index.

5. The Auroral Precipitation and Auroral Luminosity in Connection with the Large Scale Dynamics of Plasma Regions in the Magnetospheric Tail During Substorms

Substorm generation in the magnetosphere is closely associated with the variations of the space-time distribution of charged particle precipitations to the upper atmosphere [Akasofu and Chapman, 1972; Akasofu, 1977; Nishida, 1978; Kamide, 1988].

The relationships of various auroral forms to the parameters of precipitating auroral electron fluxes were discussed by Feldstein and Galperin [1985] and by Galperin and Feldstein [1990]. They examined two patterns discussed elsewhere for the relationships between the diffuse and discrete auroral forms produced by precipitation from plasma regions in the night-time magnetosphere during different phases of a substorm. The pattern proposed by Winningham et al. [1975], by Lui et al. [1977], and by Eastman et al. [1985] was used extensively to interpret the observations, including the latest DE, DMSP, and Viking experiments. The alternative pattern based on the

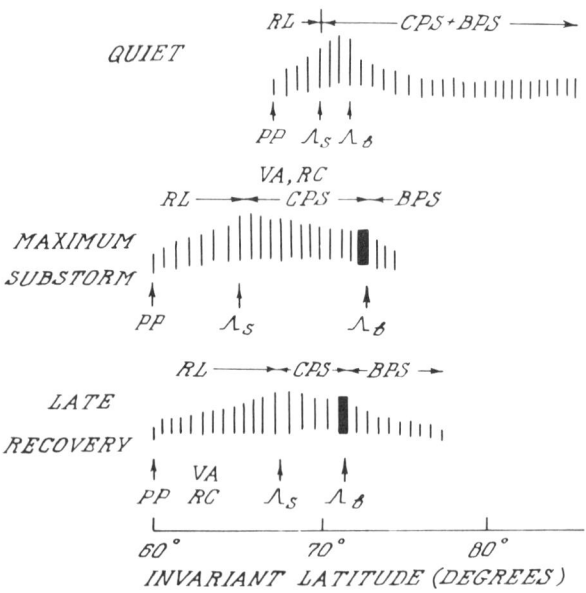

Fig. 9. Schematic representation of the latitude morphology of various types of auroral electron precipitations during different substorm phases and their assignment to generic magnetospheric plasma domains according to Winningham et al. [1975] (a. above) and their reassignment according to Feldstein and Galperin [1985] (b, below).

Galperin, 1985] in the quiet conditions-disturbance peak-recovery phase format. According to Winningham et al. [1975], the precipitations from the central plasma sheet (CPS) give rise to diffuse auroral luminosity, while the precipitations from the boundary plasma sheet (BPS) are associated with discrete auroras (arcs and bands) and cover the poleward region of the auroral electron precipitation zone.

As shown in Fig. 9 a, b, the equatorward region of the precipitation zone contains the electrons whose spectrum gets soft with decreasing latitude (CPS in Fig. 9a and remnant layer RL in Fig. 9b). The diagrams in Fig. 9a and in Fig. 9b differ substantially in that the given precipitations occur polewards from the outer radiation zone (VA in Fig. 9a), but within the radiation zone, equatorwards from the boundary of stable trapping of high-energy electrons, Λs (RL in Fig. 9b). In the pre-midnight sector, this region coincides with zone 2 of field-aligned current (inflowing to the ionosphere in the dusk sector) which is mapped spatially [Roelof, 1989; Iijima et al., 1990] to the zone of ring current (RC) in the inner magnetosphere. This particular type of auroral electron precipitations with a softening spectrum is observed up to the plasmapause (PP) and gives rise to diffuse aurora, while the similarity between the electron spectra in conjugate regions above the ionosphere and in the equatorial plane of the magnetosphere indicate that any potential difference is absent along magnetic field lines. In Fig. 9a, the diffuse aurora is located poleward of Λs, i.e. from the boundary of stable trapping of ≥35 keV electrons, whereas in Fig. 9b the boundary Λs separates the diffuse aurora region located equatorward of Λs and the discrete aurora region located poleward of Λs.

The second substantial difference in the patterns shown in Fig. 9a and in Fig. 9b is that the discrete aurora region is mapped spatially to the CPS (Fig. 9b), rather than to the BPS (Fig. 9a). In the case (i) we avoid the difficulties arising from the necessity to interpret the closing of the major magnetic field flux during active substorm phase from the night-time sector of auroral oval through a sufficiently thin (~0.5–2.0 Re) BPS: (ii) an intensive and long-lived active discrete aurora which form the polar edge of the auroral bulge is mapped on the above mentioned thin layer in the magnetospheric tail [Frank and Craven, 1988] where high-velocity plasma streams are observed [Huang and Frank, 1986]. The given auroral arc is shown in Fig. 9b with black rectangle; (iii) during a quiet interval, the near-midnight sector contains but a single auroral arc on Φ~70° which is difficult to identify with the BPS extending from Φ~71° to Φ~77.5°(Fig. 9a).

The third substantial difference in the examined patterns is that a soft precipitation band with an almost isotropic pitch-angle distribution and a falling spectrum without field-aligned acceleration exists continually in Fig. 9b poleward of the discrete aurora region. The band is broad during the magnetically-quiet interval and narrow during the active substorm phase. In Fig. 9a, such a soft precipitation is absent during the substorm and is associated with discrete auroral forms during quiet interval. According to the pattens shown in Fig. 9a, b, they differ by the character of the relationships of BPS to various luminosity types (precipitations), namely, the BPS is associated with discrete forms (with accelerated electrons) in Fig. 9a and with diffuse aurora (with soft precipitations) is Fig.9b. The occurrence of the BPS-forming auroral electron precipitations poleward from the most equatorward auroral arc in Fig. 9b indicates that the substorms are initiated deep in the plasma sheet, rather than near the plasma sheet boundary.

The auroral electron precipitation patterns shown in Fig. 9a, b at different activity levels reflect the existence of different ideas concerning the localization of the substorm generation onset region, namely, in the remote magnetotail region (a substorm begins in the BPS) or in the inner magnetosphere (a substorm begins at the boundary between the regions of discrete and diffuse auroral forms where an auroral arc is located at the inner boundary of the CPS).

Figure 10a, b presents the scheme of the meridional cross sections of the plasma domain locations in the night-time magnetosphere during the pre-storm quiet period and during the substorm development maximum which correspond to the above mentioned localization ideas.

According to the scheme shown in Fig. 10a [Lyons and Nishida, 1988; Nishida, 1988], the substorm begins along the magnetic field lines threading the BPS and located near the plasma sheet boundary. The substorm onset is due to formation of a new neutral line (NNL) within the magnetotail current sheet which is the source of isotropic precipitation of accelerated ions to the night-latitude atmosphere. Actually, the auroral arcs occur always on the field lines which include the isotropic ion precipitations from the current sheet [Lyons and Evans, 1984; Lyons et al., 1988].

An alternative dynamical process in the magnetospheric tail during a substorm is shown schematically in Fig. 10b. During quiet intervals, the auroral arc occurs at the boundary of stable trapping of high-energy electrons, Λs, which is due to a pronounced curving of magnetic field lines because the current sheet is located near the magnetotail neutral sheet (NS). The CPS located on either sides from the NS is filled with low-energy auroral plasma up to the magnetic field line mapped on the distant neutral line (DNL). The auroral electrons responsible for soft precipitations up to very high latitudes precipitate from throughout the given region to the ionospheric altitudes.

The onset of the auroral substorm expansive phase is marked with brightening or splitting of an existing arc, i.e. deep inside the magnetosphere, near the inner boundary of the plasma sheet, rather than in the periphery of the latter. The substorm expansive phase onset results from violation of the magnetospheric electrodynamic configuration stability inside the closed magnetosphere, thereby giving rise to a discontinuity in the large-scale current flowing across the magnetospheric tail [Kaufmann, 1987]. Such a discontinuity, as well as a branching of a current fraction from the tail to the altitudes of the auroral ionosphere, occur on the earthward side of the plasma sheet. Feasible mechanisms of substorm generation near the inner boundary of the plasma sheet were proposed by Lyatsky [1987], Trakhtengerts and Feldstein [1988], Kan et al. [1987], and Rothwell et al. [1988]. The substorm onset at the inner boundary of the plasma sheet coinciding with the McIlwain injection boundary [Mauk and McIlwain, 1974; Feldstein and Galperin, 1985] removes the contradiction [Siscoe, 1988] between the widely-adopted idea that the discrete auroral forms are mapped on BPS and the fact of substorm onset at the inner boundary of injection.

We assume the following sequence of events in the evolution of a substorm active phase. The expansive phase

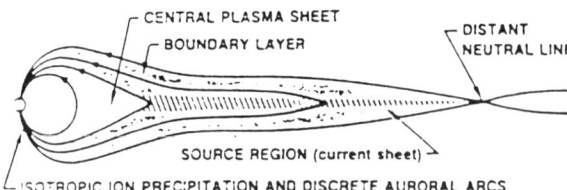
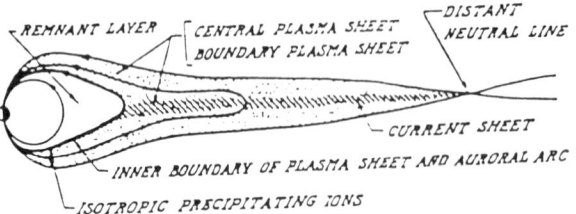
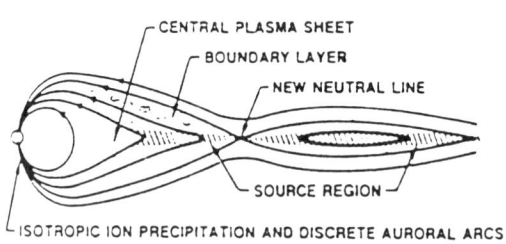
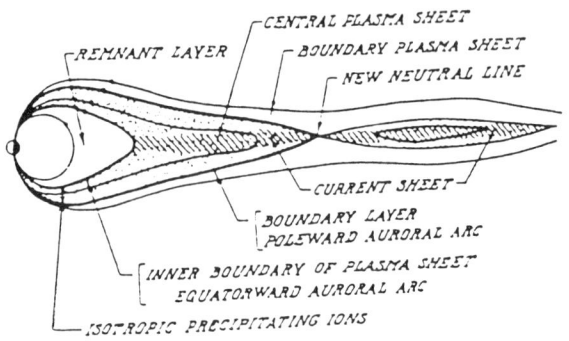

Fig. 10. Schematic representation (not to scale) of the magnetospheric tail plasma domains and some structural features of the magnetotail in the interval of quiet auroras before a substorm onset (at the top) and during substorm expansion phase (at the bottom) according to Lyons and Nishida [1988] on the left (a) and Feldstein and Galperin [1991] at the right (b).

originates deep in the magnetosphere, near the inner boundary of the plasma sheet. The onset is characterized by a brightening of the equatorial arc, or by the occurrence of new nearby arc, which forms the poleward edge of auroral bulge moving to higher latitudes. The movement corresponds to an increase of the geocentric distance of the magnetotail region magnetically connected to the arc. At some time a NNL can be formed across the tail. The NNL is formed at L~18 because of the persisting thinning of the plasma sheet at that distances during the expansive phase [Hones et al., 1984], although the plasma sheet on lower L-shells begins thickening in connection with formation of the auroral bulge. The occurrence of the NNL gives rise to a plasmoid moving rapidly away from the Earth. As a result the plasma sheet gets thicker, thereby giving rise to a pronounced increase of the velocity of the motion of the poleward auroral arc to higher latitudes [Hones, 1985]. Weak substorms may not be accompanied by NNL formation, so they evolve without any rapid jumps polewards [Craven and Frank, 1985; Anger et al., 1987]. Formation of a NNL some time after the onset of substorm expansive phase at a geocentric distance L >18 removes the difficulties [Siscoe, 1988] arising from the NNL formation deep in the plasma sheet at L~8.

During the active period, discrete auroral forms occur in the latitude interval that maps to the magnetospheric tail limited by the equatorial arc projection and by a magnetic field line running through the NNL. In this case the active aurora at the poleward edge of the auroral bulge is mapped on the ionospheric altitudes of the NNL. The mechanism of poleward arc excitation in some substorms differs from electron acceleration in the inverted - V structures for the arcs in the auroral oval. This characteristic feature is responsible for the often persistent conservation of intense auroras at the poleward edge of the auroral bulge.

High-velocity plasma stream during disturbed period were detected not only in the plasma sheet boundary layer, but also near the neutral layer [Baumjohann et al., 1990]. Having been generated near current sheet, the high-velocity plasma streams begin penetrating the entire plasma sheet during substorms, when the active discrete auroral forms are observed throughout the auroral oval. The existence of fast-ion fluxes throughout the plasma sheet at great geocentric distances has been inferred from the analysis [Nishida et al., 1988].

The auroral precipitations from the BPS with its poleward boundary along the line mapped on the DNL will form the soft precipitation region polewards from the auroral oval which will exist during both quiet and disturbed intervals. It cannot be excluded that the BPS will actually disappear during the maximum of strong substorms when the NNL moves away from the Earth to great distance.

The proposed model is one of the versions of the substorm evolution pattern involving formation of a neutral line within the magnetotail current sheet which, in turn, is a modification of the pattern [Lyons and Nishida, 1988] to be brought in agreement with the pattern presented in Fig. 1b. Though Lyons and Nishida [1988] consider the neutral-line formation within the CPS to conflict with the well-known experimental facts, we assume that:

(1) an NNL occurs deep in the central zone of the plasma sheet, rather than in the periphery of latter, within a certain period after the onset of substorm active phase. The CPS covers the magnetotail current sheet on two sides, wherein the substorm neutral line is just formed. The CPS is separated from the tail lobes by the BPS;

(2) the substorms development is not always associated with the NL occurrence. This relates in particular, to weak substorms;

(3) during quiet intervals the discrete auroral arc is mapped by magnetic field lines to the earthward region of the current sheet, rather than to the outer periphery of the latter. The occurrence of such an arc seems to be associated with the extreme values of field-aligned currents near the inner boundary of CPS. The soft diffuse precipitation zone mapped on the tail up to the DNL is located poleward from the arc which may be solitary during quiet interval. During the substorm expansive phase, the discrete arcs are located in the region that maps to the magnetospheric tail between the earthward part of the current sheet and the NNL, i.e. the entire CPS;

(4) the active aurora, which forms the poleward edge of auroral bulge, seems to be mapped to ionospheric altitudes by the magnetic field lines of the NNL in the magnetospheric tail, i.e. is the internal magnetospheric singularity irrelevant to solar wind;

(5) the soft auroral electron fluxes and the faint luminosity polewards from the poleward arc are due to precipitations from the BPS whose external surface is mapped on the DNL. Thus, the magnetotail region where the most active processes occur is located deep in the plasma sheet, rather than in the periphery of the latter. This fact may be treated to indicate indirectly that the energy released during substorm expansion phase is supplied mainly by unloading process of releasing the energy accumulated earlier in the magnetospheric tail, rather than from solar wind (the driven mechanism). Therefore, the BPS is not the channel of direct energy transfer from solar wind to the magnetosphere. The fast ion fluxes in the CPS, the external region included, occur during the substorm active phase because of dissipation of the solar wind energy accumulated earlier in the tail in the form of magnetic field and released during the expansive phase;

(6) the substorm generation processes occur within the region of closed magnetic field lines, for the lines get closed in the night-time magnetosphere throughout the whole plasma sheet, the boundary plasma layer included.

Conclusion

1. The transition from a concept of an auroral zone to a new concept of an auroral oval demanded changes in the planetary scheme of space-time distribution of geomagnetic disturbances. Such a planetary scheme was proposed in early sixties and its peculiar feature was existence of westward electrojet along auroral oval. In dusk sector the westward electrojet is located poleward of eastward one, and the latitude of the westward electrojet increases as it shifts towards earlier hours of local time. A slot is produced between the electrojets in dusk sector. The slot later was named the Harang discontinuity. After the auroral oval concept had been established, the fact that auroral electrojets overlap in the dusk sector found its quite obvious interpretation in terms of the general pattern for the occurrence of geophysical events in high latitudes.

2. Substorm current systems are different during growth and expansion phases. During the growth phase the driven process dominates, which is directly related with the enhancement of the convective electric field, controlled by the solar wind parameters. The equivalent current system during the start of the growth phase is similar to DP-2 current system. It embraces the whole glove from the pole to the equator without current increase in auroral zone. During the active phase unloading processes prevail, manifested by the release of stored in the magnetotail energy, sharp increase of conductivity at high-latitudes and the generation of westward and eastward electrojets.

3. The development of DR is accompanied by the intensification of magnetospheric substorm activity, as manifested in auroral electrojet intensifications. The auroral electrojets shifts equatorward. As a result, the AE index ceases to reflect the electrojet intensity. To estimate the auroral electrojet intensity during magnetic storm it is necessary to use data from subauroral observatories (AE' index).

This new index is closely connected with interplanetary medium parameters. The highest correlation is between AE' and the energy flux F into the ring current. This means that during the magnetic storm main phase, the auroral electrojet intensity is closely related to the energy flux supplied to ring current. The close relationship between AE' and F makes it possible to estimate the real hourly values of AE' indices during magnetic storm interval if the interplanetary medium parameters are measured during those intervals. If energy flux into the mag-netosphere is determined by Akasofu parameter ϵ, then only approximately 15% of the ϵ value is injected into the ring current.

4. An interpretation is presented of the patterns for auroral electron precipitations to the high-latitude upper atmosphere in near-midnight sector as an alternative to the one commonly accepted. The model differs from that of Winningham et al. [1975] that in the near-midnight sector, the diffuse aurora is mapped to the outer radiation belt rather than the CPS, and the oval of the discrete forms maps to the CPS proper, but not to BPS. A diffuse precipitation band lying just poleward of the oval maps to the BPS in the tail.

An active aurora during an expansion phase which forms the expanding poleward border of the auroral bulge maps to the border between CPS and BPS. In this region of the tail Eastmann et al. [1985] and other have observed ion beams and plasma flow irrespective of activity. But during a substorm active phase such high speed flows, sometimes very localized, were observed throughout the CPS by Baumjohann et al. [1990].

A summarizing schematic of the polar precipitation regions and their mapping to the magnetospheric plasma domains is presented. It can be considered as a modification of the Lyons-Nishida [1988] scheme for magnetospheric plasma domains, and it characterizes the relationship of the gross magnetospheric structure to region of the auroral electron precipitations in the upper atmosphere.

Acknowledgements. I would like to thank B.A. Below, Yu, I. Galperin, G.V. Starkov, and P.V. Sumaruk for their participation in investigations which are comprised in this paper. The author acknowledge J. Heppner and K. Snyder for providing data on the solar wind plasma and IMF during the magnetic storm of March 23-24, 1969.

References

Akasofu, S.I., The development of the auroral substorm, *Planet. Space Sci.*, 12, 273, 1964.

Akasofu, S.I., *Polar and magnetospheric substorms*, D. Reidel, Dordrecht, 1968.

Akasofu, S.I., *Physics of magnetospheric substorms*, D. Reidel Publ. Co., Dordrecht, 1977.

Akasofu, S.I, AE and Dst indices during geomagnetic storms, *J. Geophys. Res.*, 86, 4820, 1981a.

Akasofu, S.I. Energy coupling between the solar wind and the magnetosphere, *Space Sci. Res.*, 28, 111, 1981b.

Akasofu, S.I., and S. Chapman, *Solar-terrestrial physics*, Oxford, The Clarendon Press, 1972.

Akasofu, S.I., S. Chapman, and C.-I. Meng, The polar electrojet, *J. Atmosph. Terr. Phys.*, 27, 1275, 1965.

Allen, J.H., C.C. Abston, and L.D. Morris, Auroral electrojet magnetic activity indices AE (11) for 1969, WDCA, Report AGU-31.

Anger, C.D., J.C. Murphree, A. Vallance Jones, Scientific results from the Viking imager: an introduction, *Geophys. Res. Lett.*, 14, 383, 1987.

Baker, D.N., S.I. Akasofu, W. Baumjohann, J.W. Bieber, D.H. Fairfield, B. Mauk, R.L. McPherron, and T.E. Moor, Substorms in the magnetosphere, in *Solar terrestrial physics-present and future*, edited by D.M. Butler, and K. Papadopoulos, Washington, NASA, 1984.

Baumjohann, W., G. Paschmann, and H. Luhr, Characteristics of high-speed ion flows in the plasma sheet, *J. Geophys. Res.*, 95, 3801, 1990.

Baumjohann, W., R.J. Pellinen, H.J. Opgenoorth, and E. Nielsen, Joint two-dimensional observations of ground magnetic auroral zone currents: current systems associated with local auroral breakups, *Planet. Space Sci.*, 29, 431, 1981.

Burdo, O.A., Connection of regular and irregular geomagnetic field variations at high latitudes, *Proc. Arc. Ant. Inst.*, 223, N3, 21, 1960.

Burton, R.K., R.L. McPherron, C.T. Russell, An empirical relationship between interplanetary conditions and Dst, *J. Geophys. Res.*, 80, 4204, 1975.

Chapman, S., The electric current-systems of magnetic storms, *Terrest. Magn.*, 40, 349, 1935.

Chapman, S., Earth storms: Retrospect and prospect, *J. Phys. Soc. Japan*, 17, Suppl. A-1, 6, 1962.

Cole, K., A dynamo theory of the aurora and magnetic disturbance, *Austral. J. Phys.*, 13, 484, 1960.

Craven, J.D., and L.A. Frank, The temporal evolution of a small auroral substorm as viewed from high altitudes with Dynamics Explorer 1, *Geophys. Res. Lett.*, 12, 465, 1985.

Feldstein, Y.I., *The morphology of aurorae and geomagnetism, aurora and airglow*, Publ. House, Academy of Sciences, N10, 121, 1963.

Feldstein, Y.I., and Yu.I. Galperin, The auroral luminosity structure in the high-latitude upper atmosphere: Its dynamics and relationship to the large-scale structure of the Earth's magnetosphere, *Rev. Geophys.*, 23, 217, 1985.

Feldstein, Y.I., and Yu.I. Galperin, A new interpretation of auroral precipitation and luminosity observations from the satellites DMSP, AUREOL, and Viking, *J. Atmosph. Terr. Phys.*, in press, 1991.

Feldstein, Y.I., and G.V. Starkov, The auroral oval and the boundary of closed field lines of geomagnetic field, *Planet. Space Sci.*, 18, 501, 1970.

Frank, L.A., and J.D. Craven, Imaging results from dynamics Explorer 1, *Rev. Geophys.*, 26, 249, 1988.

Fukushima, N., Polar magnetic storms and geomagnetic bays, *J. Fac. Sci., Univ. Tokyo, Sec. II*, 8, 293, 1953.

Fukushima, N., and T. Oguti, II Polar magnetic storms and geomagnetic bays, Appendix I. A theory of Sd-field, *Rep. Ionosph. Res. Japan*, 7, 137, 1953.

Galperin, Yu.I., and Feldstein, Y.I., Auroral luminosity and its relationship to magnetospheric plasma domains, in *Auroral Physics*, edited by C.-I. Meng, 187, 1990.

Gonzalez, W.D., B.T. Tsurutani, A.L.C. Gonzalez, E.J. Smith, F. Tang, and S.I. Akasofu, Solar wind-magnetosphere coupling during intense magnetic storms (1978-1979), *J. Geophys. Res.*, 94, 8835, 1989.

Harang, L., The mean field of disturbance of polar geomagnetic storms, *Terrest. Magn.*, 51, 353, 1946.

Hones, E.W., The poleward leap of the auroral electrojet as seen in auroral images, *J. Geophys. Res.*, 90, 5333, 1985.

Hones, E.W., T. Pytte, and H.I. West, Associations of geomagnetic activity with plasma sheet thinning and expansion: a statistical study, *J. Geophys. Res.*, 89, 5471, 1984.

Huang, C.Y., and L.A. Frank, A statistical study of the central plasma sheet: implications for substorm models, *Geophys. Res. Lett.*, 13, 652, 1986.

Iijima, T., and T. Nagata, Signatures for substorm development of the growth phase and expansion phase, *Planet. Space Sci.*, 20, 1095, 1972.

Iijima, T., and T.A. Potemra, and L.J. Zanetti, Large-scale characteristics of magnetospheric equatorial currents, *J. Geophys. Res.*, 95, 991, 1990.

Inhester, B., W. Baumjohann, R.A. Greenwald, and E. Nielsen, Joint two-dimensional observations of ground magnetic and ionospheric electric fields associated with auroral zone currents, 3. Auroral zone currents during the passage of a westward traveling surge, *J. Geophys. Res.*, 49, 155, 1981.

Kamide, Y., *Electrodynamic processes in the Earth's ionosphere and magnetosphere*, Kyoto Sangyo Univ. Press, Kyoto, 1988.

Kan, J.R., L. Zhu, and S.I. Akasofu, A theory of substorm: onset and subsidence, *J. Geophys. Res.*, 93, 5624, 1988.

Kaufmann, R.L., Substorm currents: growth phase and onset, *J. Geophys. Res.*, 92, 7471, 1987.

Khorosheva, O.V., Relation of magnetospheric disturbances to the parameters of the interplanetary medium, *Geomagn. and Aeronomie*, 26, 447, 1986.

Lassen, K., Relation of the plasma sheet to the nighttime auroral oval, *J. Geophys. Res.*, 79, 3857, 1974.

Lui, A.T.Y., D. Venkatesan, C.D. Anger, S.I. Akasofu, W.J. Heikkila, J.D. Winningham, and J.R. Burrows, Simultaneous observations of particle precipitations and auroral emissions by the ISIS-2 satellite, *J. Geophys. Res.*, 82, 2210, 1977.

Lyatsky, V.B., *The current system of magnetosphere-*

Lyatsky, V.B., The conductivity waves in the magnetosphere-ionosphere system, *Geomagn. and Aeronomie, 27*, 965, 1987.

Lyons, L.R., and D.S. Evans, An association between discrete aurora and energetic particle boundaries, *J. Geophys. Res., 89*, 2395, 1984.

Lyons, L.R., J.F. Fennell, and A.L. Vampola, A general association between discrete auroras and ion precipitation from the tail, *J. Geophys. Res., 93*, 12932, 1988.

Lyons, L.R., and A. Nishida, Description of substorms in the tail incorporating boundary layer and neutral line effects, *Geophys. Res. Lett., 15*, 1337, 1988.

Mauk, B.H., and C.E. McIlwain, Correlation of Kp with the substorm-injected plasma boundary, *J. Geophys. Res., 79*, 3193, 1974.

Nagata, T., and N. Fukushima, III Constitution of polar magnetic storms, *Rep. Ionosph. Res. Japan., 6*, 85, 1952.

Nishida, A., Coherence of geomagnetic DP2 fluctuations with interplanetary magnetic variation, *J. Geophys. Res., 73*, 5549, 1968.

Nishida, A., *Geomagnetic diagnosis of the magnetosphere*, Springer, New York, 1978.

Nishida, A., Critical issues in the solar wind/magnetosphere coupling and the magnetotail dynamics, in Solar-Terrest. Energy Program: major scientific problems, *Proceedings SCOSTEP Symp.*, Helsinky Univ., 31, 1988.

Nishida, A., S.J. Bame, D.N. Baker, G. Gloeckler, M. Scholer, E.J. Smith, T. Terasawa, and B. Tsurutani, Assessment of the boundary layer model of the magnetospheric substorm, *J. Geophys. Res., 93*, 5579, 1988.

Nishida, A., N. Iwasaki, and T. Nagata, The origin of the equatorial electrojet: a new type of geomagnetic variation, *Ann. Geophys., 22*, 478, 1966.

Olbert, S., G.L. Siscoe, and V.M. Vasyliunas, A simple derivation of the Desseler-Parker-Sckopke relation, *J. Geophys. Res., 73*, 1115, 1968.

Pisarsky, V. Yu., Y.I. Feldstein, N.M. Pudneva, and A. Progancova, Ring current and interplanetary medium parameters, *Studia Geophys. et Geod., 33*, 61, 1989.

Roelof, E.C., Remote sensing of the ring current using energetic neutral atoms, *Adv. Space Res., 9*, 195, 1989.

Rothwell, P.L., L.P. Block, M.B. Silevitch, and C.-G., Falthammar, A new model for substorm onsets: the pre-breakup and triggering regimes, *Geophys. Res. Lett., 15*, 1279, 1988.

Siscoe, G.L., *EOS, 69*, 1586, 1988.

Sumaruk, P.V., Y.I. Feldstein, and B.A. Belov, The dynamics of magnetospheric activity during an intensive magnetic storm, *Geomagn. and Aeron., 29*, 110, 1989.

Tinsley, B.A., and S.I. Akasofu, A note of the lifetime of the ring current particles, *Planet. Space Sci., 30*, 733, 1982.

Trakhtengertz, V.Yu., and Y.I. Feldstein, The substorm expansive phase as a result of the turbulent regim of magnetospheric convection, *Geomagn. and Aeron., 28*, 743, 1988.

Vasyliunas, V.M., Low energy particle fluxes in the geomagnetic tail, in *The polar ionosphere and magnetospheric processes*, edited by G. Skovli, Gordon and Breach, N.Y., 25, 1970.

Winningham, J.D., F. Yasuhara, S.-I. Akasofu, and W.J. Heikkila, The latitudinal morphology of 10-eV to 10-keV electron fluxes during magnetically quiet and disturbed times in the 2100-2300 MLT sector, *J. Geophys. Res., 80*, 3148, 1975.

Extended Consideration of a Synthesis Model for Magnetospheric Substorms

A. T. Y. LUI

Applied Physics Laboratory, The Johns Hopkins University, Laurel, Maryland

Investigations in magnetospheric substorm phenomena have led to proposal of several rather diverging views on the phenomenological description of a substorm. These different models, however, have important compatible features which can be synthesized to a single substorm model with capability to account for observed substorm features superior to any individual model. We discuss here such a synthesis model after identifying the principal substorm phenomena and recapitulating several substorm models. The proposed scenario includes current disruptions leading to convection surges and tailward propagation of rarefaction waves, the ionospheric control on the global establishment of a current wedge, wave-induced precipitation, local time expansion of the disturbance region via ballooning and velocity-shear instabilities, plasma sheet heating by resonant absorption of plasma waves, and plasmoid formation. This model represents a plausible coherent picture from integrating the presently existing ones. If magnetic reconnection is defined as any process leading to the breakdown of ideal MHD behavior, then it is manifested in several forms within the substorm episode in this model. In the early phase of substorm expansion, it occurs in current disruptions and convection surges. These activities then instigate the traditional form of magnetic reconnection with formation of separatrices and plasmoids in the later phases of substorm expansion and recovery.

INTRODUCTION

Magnetospheric substorm phenomena encompass a myriad of dynamical disturbances in the magnetosphere, the ionosphere, and even the thermosphere [e.g., *Akasofu*, 1977; *Rostoker et al.*, 1980; *Huang*, 1987; *Cogger and Murphree*, 1990]. The importance of understanding magnetospheric substorm can hardly be overstated, as evident by the assiduous effort by the magnetospheric community in unfolding the substorm mystery. Decades of modern space research employing in situ measurements from spacecraft has yet to consolidate the description of substorm phenomena to one consensus model. In spite of the prevalence of the near-Earth neutral line model as the conventional paradigm for interpretation of substorm features [*Hones*, 1979], alternative substorm models are continually propounded in the literature. The list of contending substorm models includes wave-induced precipitation model [*Parks et al.*, 1972; *Kropotkin*, 1972], current disruption model [*Chao et al.*, 1977; *Lui*, 1979], the interchange or ballooning instability model [*Liu*, 1970; *Roux*, 1985], the boundary layer model [*Rostoker and Eastman*, 1987], the magnetosphere-ionosphere coupling model [*Kan et al.*, 1988; *Rothwell et al.*, 1988], and the thermal catastrophe model [*Smith et al.*, 1986; *Goertz and Smith*, 1989]. Most models account for only part of the entire substorm sequence. Summaries of four of these substorm models were given by Baumjohann [1988, 1989] who noted that these different models are motivated by attempts to explain substorm features observed at different regions in space.

Despite striking differences between these competing models, there are features among them which are compatible and can be unified. According to one critical assessment of the present state of the field, "The need to explore synthesis models becomes evident in view of the fact that there are over ten distinct substorm models, each one addressing some restricted aspect of substorm phenomenology. To assert that only one of these is right and all the rest are wrong is indefensible in terms of a priori probability alone. To assert that none is right, at least conceptually, unjustifiably insults our field." [G. L. Siscoe, private communication, 1990]. Indeed, suggestion has been made, first by Siscoe [1986] and later by others [*Lui*, 1987; *Akasofu*, 1989; *Kan*, 1990], that different substorm models may be combined to produce a unified substorm model superior in consistency with major observed substorm features than any individual model. Along this line of effort, Lyons and Nishida [1988] have incorporated a near-Earth neutral line in the boundary layer, combining two of the substorm models. Recently, Lui [1991] attempted another synthesis by constructing a coherent substorm picture from most of the existing models. The resulted model is primarily a phenomenological model which carries a description of sequential substorm phenomena rather than an identification of specific physical processes. Nevertheless, it provides the framework upon which substorm phenomena may be amenable to quantitative analysis.

In this paper, we explore further the scenario propounded earlier and discuss at length the two elements of the "substorm puzzle." These are the buildup of intense current in the near-Earth region during the growth phase and the rarefaction wave generation during the early substorm expansion phase. To provide a proper perspective of this development, we provide a detailed account on the rationale motivating this work. The intent here is not to assert that this synthesis model is complete and requires no further improvement. On the contrary, we view future modifications of this model to be inevitable as we gain further understanding of the magnetospheric substorm and assimilate more features into the picture.

PRINCIPAL SUBSTORM PHENOMENA

Rostoker et al. [1980] have defined a magnetospheric substorm to be "a transient process initiated on the nightside of the Earth in which a significant amount of energy derived from the solar wind-magnetosphere interaction is deposited in the auroral ionosphere and in the magnetosphere." This broad definition encompasses a large number of phenomena occurring in a wide variety of spatial and temporal scales. The purpose of a substorm model, however, is not to include every detail phenomenon which takes place during a substorm but to extract the principal phenomena so that the basic physical processes involved can be readily revealed.

Three phases of a substorm—growth, expansion and recovery—are often distinguished and phenomena have been identified in various regions of the near-Earth space during each of these substorm phases. Listed in Table 1 are some of these key phenomena in the solar wind, the ionosphere, the near-Earth tail ($-5 R_E \geq X \geq -15 R_E$), the mid-tail ($-15 R_E \geq X \geq -80 R_E$), and the far-tail ($X \leq -80 R_E$). The division of the magnetotail into three spatial regions is found to order substorm phenomena in different parts of the tail well in terms of their characterization and their activation relative to substorm expansion onsets [*Huang*, 1987]. In particular, an important but often unaware substorm feature is that plasma sheet thinning does not occur at all radial distances during the substorm growth phase. Detailed timing analysis suggests that plasma sheet thinning tailward of $X \approx -15 R_E$ occurs only after substorm expansion onset [*Hones et al.*, 1973].

That significant growth phase features occur primarily in the near-Earth tail suggests that substorm expansion disturbance is initiated there and propagates to further downstream distances as the substorm progresses. As will be discussed in a later section, this view is indeed held and with good reasons by many substorm researchers. Later in the expansion and recovery phases, most of the substorm activities in the tail, such as plasmoids and travelling compression regions, shift to the more distant tail of $X \leq -80 R_E$ [*Baker et al.*, 1984; *Slavin et al.*, 1984a].

TABLE 1: Several Key Observed Substorm Phenomena

Regions	Growth (0.5-1 hr)	Expansion (0.5-1 hr)	Recovery (1-2 hr)
Solar wind	Southward turning of IMF	Substorm expansion onset occasionally corresponds to northward turning of IMF or sudden solar wind pressure increase	Leakage of magnetospheric particles upstream
Ionosphere	Enlargement of polar cap size, enhancement of ionospheric electrojets	Sudden auroral arc brightening at expansion onset Formation of WTS and auroral bulge Appearance of bright structures in diffuse aurora Further enhancement of electrojets Outflow of ionospheric ions	Reduction in auroral bulge size and luminosity
Near-Earth tail $5 \leq |X| \leq 15 R_E$	Increase in tail cross-section Gradual plasma sheet thinning Stretched magnetic field configuration Development of field-aligned anistropy of energetic electons	Decrease in tail cross-section, particle injection Relaxed magnetic field configuration spreading in local time Impulsive E-fields	Drift echoes
Mid-tail $15 \leq |X| \leq 80 R_E$		Plasma sheet thinning Transient fast plasma flow Occasionally large fluctuating E-fields	Plasma sheet thickening with fast field-aligned flow at plasma sheet boundary layers
Far-tail $|X| \geq 80 R_E$			B_z perturbation first northward then southward Tailward streaming electrons at plasma sheet entry, changing to isotropic electrons deeper in the plasma sheet

(a) NENL Model

Near-earth neutral line formation

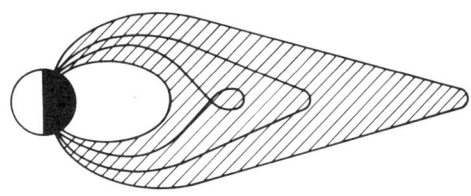

(b) Boundary layer model

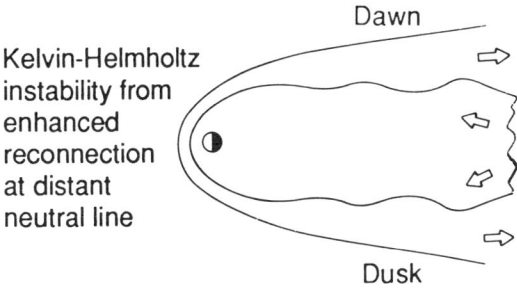

Kelvin-Helmholtz instability from enhanced reconnection at distant neutral line

(c) Thermal catastrophe

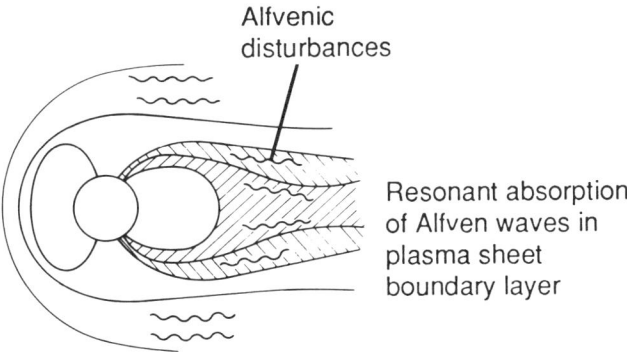

Alfvenic disturbances

Resonant absorption of Alfven waves in plasma sheet boundary layer

(d) M - I coupling

Intense field aligned current from divergence of both Hall and Pederson currents in the ionosphere

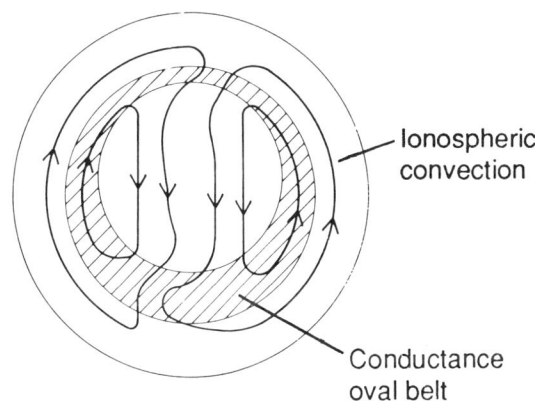

Ionospheric convection

Conductance oval belt

(e) Current disruption model

Deflation of plasma sheet from rarefaction wave launched by reduction in cross-tail current

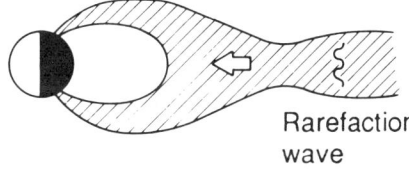

Rarefaction wave

(f) Ballooning instability model

Unstable surface wave between dipolar-like and tail-like field region

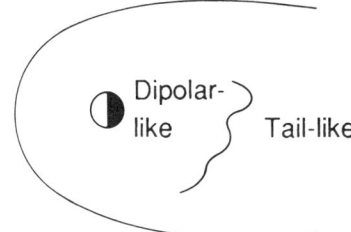

Dipolar-like Tail-like

Fig. 1. The scenario for substorm expansion phase in six substorm models: (a) near-Earth neutral line model—the formation of a near-Earth neutral line, (b) boundary layer model—Kelvin-Helmholtz instability from enhanced reconnection at the distant neutral line, (c) thermal catastrophe model—resonant absorption of Alfvén waves in the plasma sheet boundary layer, (d) magnetosphere-ionosphere coupling model—intense FAC from divergence of Hall and Pederson currents in the ionosphere, (e) current disruption model—deflation of plasma sheet from rarefaction waves launched by current disruption, and (f) ballooning instability model—unstable surface wave between dipolar-like and tail-like field region.

SUBSTORM MODELS

The near-Earth neutral line model, as depicted in Fig. 1a, is the conventional scenario for substorms [e.g., *Hones*, 1979]. It is also the most complete model since all three substorm phases are described. In essence, southward turning of the interplanetary magnetic field (IMF) enhances dayside reconnection and initiates the substorm growth phase. As a result, the tail size and tail lobe field strength increase, leading to plasma sheet thinning and stretched field configuration in the near-Earth tail. A near-Earth X-type neutral line is formed at the downstream distance between 10 to 20 R_E at substorm expansion onset, leading to the severance of the plasma sheet downstream of the neutral line and plasmoid formation. Earthward of the neutral line, the magnetic field becomes dipolar and the energized plasma creates an earthward propagating compressional wave front which later forms the substorm injection boundary near the geosynchronous altitude [*Moore*, 1986]. The neutral line stays in the near-Earth region until substorm recovery, at which time it moves downstream, causing the plasma sheet earthward of the neutral line to thicken.

There are several slight variances of this model. For example, it has been suggested that multiple, small-scale X-type and O-type neutral lines are formed instead of one single large-scale X-type neutral line [*Heikkila and Pellinen*, 1977]. The neutral line has been abstracted to be an extended neutral surface for the consideration of current wedge formation [*Stern*, 1990]. Lyons and Nishida [1988] proposed the substorm neutral line to be located at the source region of the plasma sheet boundary layer which is generally regarded to be in the mid-tail region and beyond. Similarly, Scholer [1987] and Baumjohann [1989] suggested the merging region to be typically tailward of ~ 20 R_E and the near-Earth substorm effects are results of earthward flows derived from this more remote location.

Figure 1b shows the boundary layer model which describes the growth phase similar to the previous model [*Rostoker and Eastman*, 1987]. However, the substorm expansion onset is visualized as due to enhanced reconnection occurring in the distant neutral line, which in turn generates fast earthward flowing plasma sunward of the reconnection site. This region of fast earthward plasma flows lies adjacent to the tailward flowing low latitude boundary layer plasma and consequently leads to the Kelvin-Helmholtz instability (KHI). This instability generates a current wedge along the plasma sheet boundary layer with the evening edge corresponding to the westward travelling surge in the ionosphere. Spatial movement of the plasma sheet boundary layer and central plasma sheet relative to an observing spacecraft in the magnetotail is invoked to explain the field and particle measurements interpreted by others as neutral line signatures. Recently, Lundin et al. [1991] proposed a model based on the dynamo residing in the boundary layer. Current disruption is a consequence of solar wind pressure pulses driving current across an enhanced conductivity channel.

Thermal catastrophe model (Fig. 1c) [*Smith et al.*, 1986; *Goertz and Smith*, 1989] concentrates on the heating of the plasma sheet at substorm expansion as reported by Huang and Frank [1986]. The heating mechanism proposed is resonant absorption of Alfvén waves. During the substorm growth phase, the plasma sheet evolves slowly through a succession of equilibrium states parameterized by a quantity dependent on the incident power flux of Alfvén waves, the local density and the convective velocity. Onset occurs when this parameter reaches a critical value, rendering the plasma sheet opaque to the Alfvénic disturbances and changing its temperature discontinuously. This model can accommodate three different ways in which substorm expansion can be triggered and the preference for substorm expansion onset in the premidnight sector.

Figure 1d depicts the magnetosphere-ionosphere coupling model [*Kan et al.*, 1988] which considers the establishment of the substorm current wedge. Southward turning of the IMF leads to enhanced magnetospheric convection during which the magnetosphere and the ionosphere both adjust to the new externally imposed condition through Alfvén waves bouncing between these two regions. Two criteria for expansion onset are established, namely, the polar cap potential must exceed a threshold value of about 70 kV and the convection reversal region must overlap with the poleward gradient of the diffuse auroral conductance belt in the midnight sector. Substorm recovery begins when either of the conditions is violated. Similarly, Rothwell et al. [1988] considered the electrodynamics of the auroral arcs in terms of two coupled electrical circuits between the ionosphere and the magnetosphere. A scheme is set up to determine the constraints imposed by self-consistent solution of the two circuits. This model can account for the occurrence of substorm current wedge at fairly low L-shells (i.e., 5 to 6) where the magnetic field is too strong to expect an X-type neutral line to form [*Tanskanen et al.*, 1987].

The current disruption model [*Chao et al.*, 1977] was introduced based on the observational result that the expected signatures of a large-scale X-type neutral line (e.g., tailward flows and a drastic change in magnetic field topology) are typically not found within 20 R_E during substorm expansion in spite of a clear indication that current disruption occurs in the near-Earth region of the midnight sector at onset (Fig. 1e). Therefore, this model emphasizes current disruption taking place without the creation of a large-scale X-line. Disruption/diversion of the cross-tail current can account quantitatively for the observed magnetic field reconfiguration in the near-Earth tail during substorm expansion onsets [*Lui*, 1978] and particle injection as well as energization by convection surges [*Mauk*, 1989]. Furthermore, the large-scale plasma sheet thinning and southward dipping of the magnetic field in the mid-tail region can be accomplished with the convection surge coupled with subsequent generation of a rarefaction wave propagating downstream. Launching of a rarefaction wave front deep in the plasma sheet at substorm onset has also been proposed by Kropotkin [1972] in his model.

The ballooning instability proposed by Roux [1985] is based on detailed analysis of geosynchronous measurements in relation to ground observation of westward travelling surges (Fig. 1f). This model focuses on the boundary dividing dipolar-like and tail-like field lines. The pressure gradients of energetic ions and magnetic field are both pointing earthward in the dipolar region and are oppositely directed in the tail-like region. This situation is analogous to a heavy fluid resting on a light fluid and the pressure gradient plays the role of the gravity force. Such a surface is unstable and polarization electric field develops. The unstable surface wave is carried westward by the westward drifting ions. Positive and negative charges accumulate at the edges of the propagating perturbation and generate field-aligned currents. This perturbed surface in the magnetospheric equatorial region maps to the ionosphere as westward travelling surges which are observed to correlate with the trailing negative charge excess of the surface. A kinetic treatment of this instability in the electrostatic limit was studied earlier by Liu [1970] and referred to as the interchange instability.

Parks et al. [1972] advanced an early substorm model without invoking a neutral line. They proposed that the earthward convection in the tail leads to a buildup of temperature anisotropy (with higher perpendicular temperature) as a consequence of the first and second adiabatic invariants being conserved in convection. This temperature anisotropy excites electrostatic waves, which in turn heat the electron population perpendicular to the magnetic field. This allows further development of the temperature anisotropy and promotes whistler wave growth to cause enhanced precipitation. Removal of high beta plasma results in recovery of stretched magnetic field, maintaining or even enhancing the temperature anisotropy via betatron acceleration associated with magnetic field recovery. This sets up a positive feedback loop for substorm onset.

A Critique of Substorm Models

A brief assessment of the strengths and weaknesses of the various substorm models is pertinent before unifying them. As indicated previously, the near-Earth neutral line model is the conventional one and, understandably, is well developed in comparison with the others. The three substorm phases, the thinning and thickening of the plasma sheet, particle energization, and the creation of plasmoids are included.

Controversies about this model are mostly tied to the asserted location of the substorm neutral line near which reconnection signatures are anticipated. If the substorm neutral line is typically formed earthward of 20 R_E, then the model prediction is incompatible with the observation that strong tailward plasma flow is rarely detected in the central plasma sheet within a downstream distance of about 20 R_E [*Lui*, 1979; *Huang and Frank*, 1986], even with a temporal resolution of 4.5 sec from the fastest plasma analyzer flown in the tail region today [*Baumjohann et al.*, 1989, 1990]. On the other hand, southward magnetic field in the neutral sheet has been detected at a downstream distance less than 10 R_E in the midnight sector, but high time resolution analysis reveals that the current disruption region cannot be adequately described by an X-type neutral line [*Lui et al.*, 1988]. Relocating the substorm neutral line to beyond 20 R_E does not remove all the observational conflicts of this model. Many observations suggest that most substorms initiate in the near-Earth region within 20 R_E, as will be discussed in the next section.

A need for improvement for other substorm models exists also. The boundary layer model is proposed to reconcile the lack of reconnection signatures in the near-Earth region and to emphasize the importance of the field-aligned current (FAC) at the western end of the substorm current wedge. The substorm activity takes place mainly in the distant tail region and near-Earth activity is not yet incorporated. The thermal catastrophe model can account successfully the plasma heating associated with substorms but other substorm features such as the current wedge and the plasma flow pattern have not yet been included in this model. Similar criticism can be made to the current disruption model, the ballooning instability model, and the wave-induced precipitation model since they do not describe the substorm recovery phase. The magnetosphere-ionosphere coupling model specifies the condition and evolution of the substorm current system. However, the configurational change and heating of the plasma sheet, the synoptic plasma flow pattern, and the temporal change of the north-south component of the tail magnetic field are not part of this model. Furthermore, since this model portrays the magnetotail as only a scalar quantity in the reflection of Alfvén waves initiated by enhanced magnetospheric convection during the growth phase, the magnetotail in this model is thus incapable of launching a new Alfvén wave from diverting suddenly a portion of the cross-tail current into the ionosphere.

Substorm Initiation Region

One irreconcilable conflict among these models is on the location for substorm onset. The substorm onset location is suggested to be in the mid-tail for the revised near-Earth neutral line model, in the mid or distant tail for the boundary layer model, in the ionosphere for the magnetosphere-ionosphere coupling model, and in the near-Earth tail region for all other models. There is in fact a host of evidences for substorm onset location residing close to the Earth, most of which were discussed in some details by Lui [1991]. Here, we extend the list and use Table 2 to summarize these findings grouped under the three categories of Particles, Fields, and Modelling. Seven particle features are listed in the first category, six in the second, and two in the third. The list contains direct and indirect evidences. The direct evidences are rather strong, which includes the injection boundary location, remote sensing of in situ particle acceleration, time delay in particle energization at radially different sites in the magnetosphere, FAC structures at geosynchronous altitude, in situ field reconfiguration, and radial spreading of magnetic perturbations from substorm current wedge. The others are indirect but are also difficult to dispute. Overall, these evidences are rather overwhelmingly in favor of substorm initiation in the near-Earth region and thus is adopted in our synthesis model.

Growth Phase of the Synthesis Model

The proposed substorm model is intended to build upon all the above models by extracting important features from them so that a single coherent picture emerges. The substorm growth phase feature provided by the near-Earth neutral line model appears to be generally accepted, as shown in the top row of Fig. 2. The magnetosphere undergoes a reconfiguration with southward IMF, with both the dayside magnetopause and the inner edge of the cross-tail current approach earthward. The asymptotic cross-sectional size of the tail increases and is manifested in the ionosphere as an equatorward movement of the polar cap boundary. The near-Earth plasma sheet thins and a more stressed magnetic field configuration in that region is seen. No significant plasma sheet thinning occurs in the mid-tail region. These changes reflect the buildup of a strong

TABLE 2: Evidence for Near-Earth ($|X| \leq 15\ R_E$) Substorm Initiation

Features or Techniques	Results	Ref.
Particles		
Injection boundary	Dispersionless injection of energetic particles at geosynchronous altitude implies near-Earth energization site	1–3
Remote sensing of in-situ particle energization	Ion sounding during current disruption at neutral sheet indicates particle energization occurring earthward of ~8.8 R_E	4
Low-altitude plasma characteristics	Initial brightening arc location at the trapping boundary of energetic electrons implies substorm initiation at the transition region between dipolar and tail-like field	5, 8
Oblique ion inverted V structure	Signature of upward ion beams shows ions generated at the conjugate electron inverted V structure on field lines closed within ~10 R_E	6–7
Plasma characteristics at equatorward border of discrete auroral oval	Plasma properties above the most equatorward arc where initial brightening typically occurs are consistent with the near-Earth central plasma sheet	8
Mapping of discrete arcs to geosynchronous region	Discrete arcs mapped magnetically to geosynchronous region, mapping verified by conjugate spectral characteristics	9
Time delay in particle energization at radially different sites in the magnetosphere	Radially aligned satellite observations in the near-tail show tailward spreading of particle energization from within ~10 R_E	10
Fields		
Field aligned current structures at geosynchronous altitude	Localized current structures in association with auroral breakups at the footprint of geosynchronous satellites	11–12
Remote sensing of current disruption	Field changes in the tail lobe are deduced to originate from current disruption at the near-Earth tail of ~7 R_E	13
Inference from in situ field reconfiguration	Field changes in the plasma sheet are deduced to originate from current disruption in the plasma sheet at ~10 R_E	14
Implication from field line resonance	Field lines of auroral arcs showing initial brightening are closed within ~10 R_E	15
Radial spreading of magnetic perturbations from substorm current wedge	Multi-satellite observations show relaxation of tail-like field starting from within ~10 R_E	10, 16–17
Development of field aligned current system of region 1 sense	Magnetic field perturbations within ~9 R_E indicate region 1 field aligned current	18–19
Modelling		
Tail-like field configuration at geosynchronous altitude during substorm growth phase	Tail-like field geometry at geosynchronous orbit requires large cross-tail current buildup at 7–9 R_E	20–21
Projection of auroral region to magnetospheric equatorial region	Auroral arc brightening locations are mapped to the region of the most intense current at the cross-tail current inner edge	22

1. McIlwain, 1974; 2. Sauvaud and Winckler, 1980; 3. Goertz and Smith, 1989; 4. Lui et al., 1988; 5. Lui and Burrows, 1978; 6. Bosqued et al., 1986; 7. Bosqued, 1991; 8. Galperin and Feldstein, 1991; 9. Mauk and Meng, 1991; 10. Lopez and Lui, 1990; 11. Robert et al., 1984; 12. Roux et al., 1991; 13. Jacquey et al., 1991; 14. Lui, 1978; 15. Samson, 1991; 16. Ohtani et al., 1988; 17. Lopez et al., 1990; 18. Ohtani et al. 1990; 19. Zanetti et al., 1991; 20. Kaufmann, 1987; 21. Tsyganenko,1989; 22. Elphinstone et al., 1991.

cross-tail current mostly in the midnight sector of the near-Earth region at ~6–15 R_E downstream.

The earthward development of cross-tail current is a natural consequence of stress balance. The inner edge of the current sheet must approach earthward to counter-balance the increase in tailward tangential stress at the magnetopause as the IMF becomes southward [*Siscoe*, 1966; *Siscoe and Cummings*, 1969]. Coroniti and Kennel [1972] developed this idea further as a flaring tail

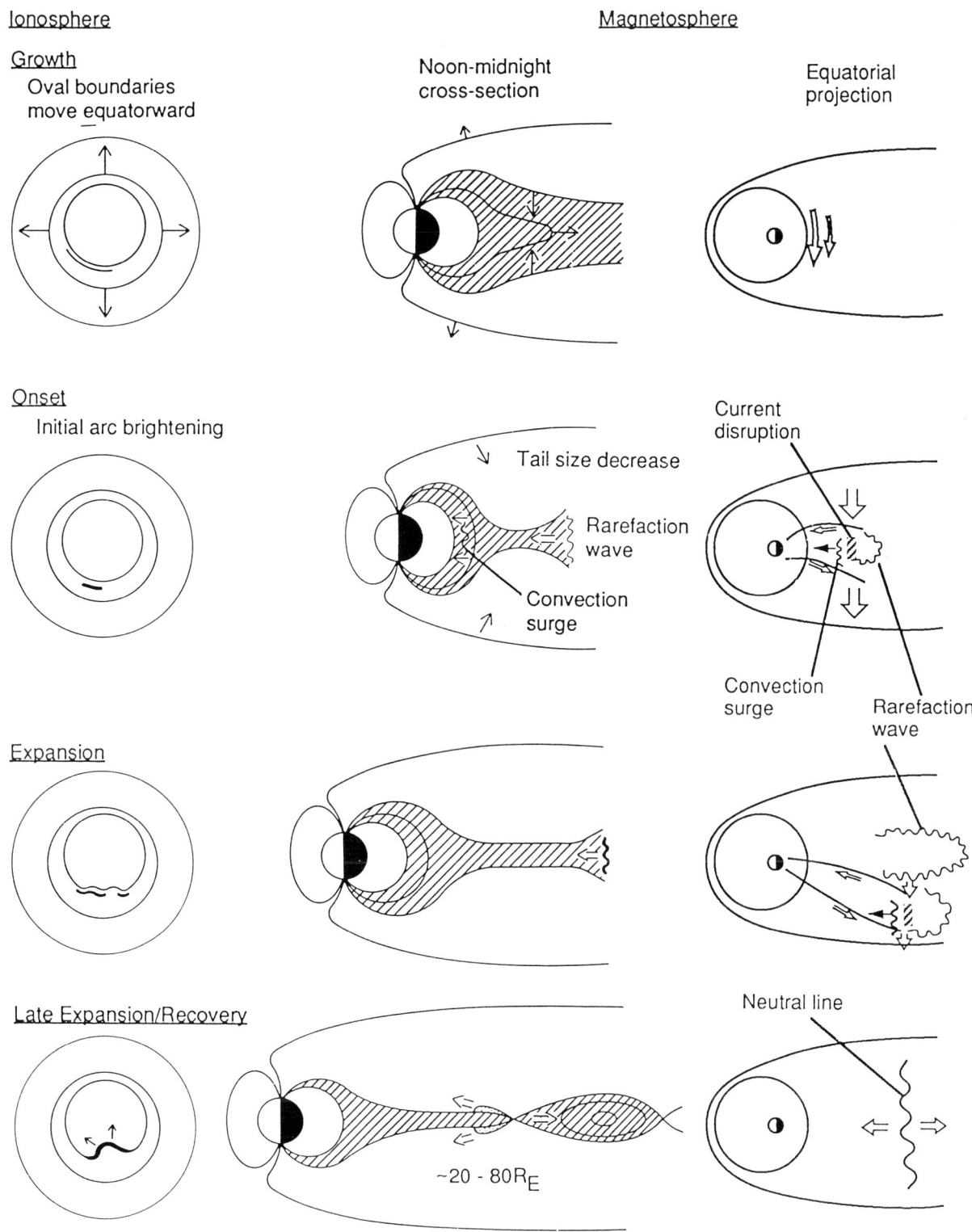

Fig. 2. Sketch of a synthesis substorm model with four stages of substorm development. The first, second and third columns illustrate substorm features in the ionosphere, noon-midnight cross section and neutral sheet region of the magnetosphere, respectively.

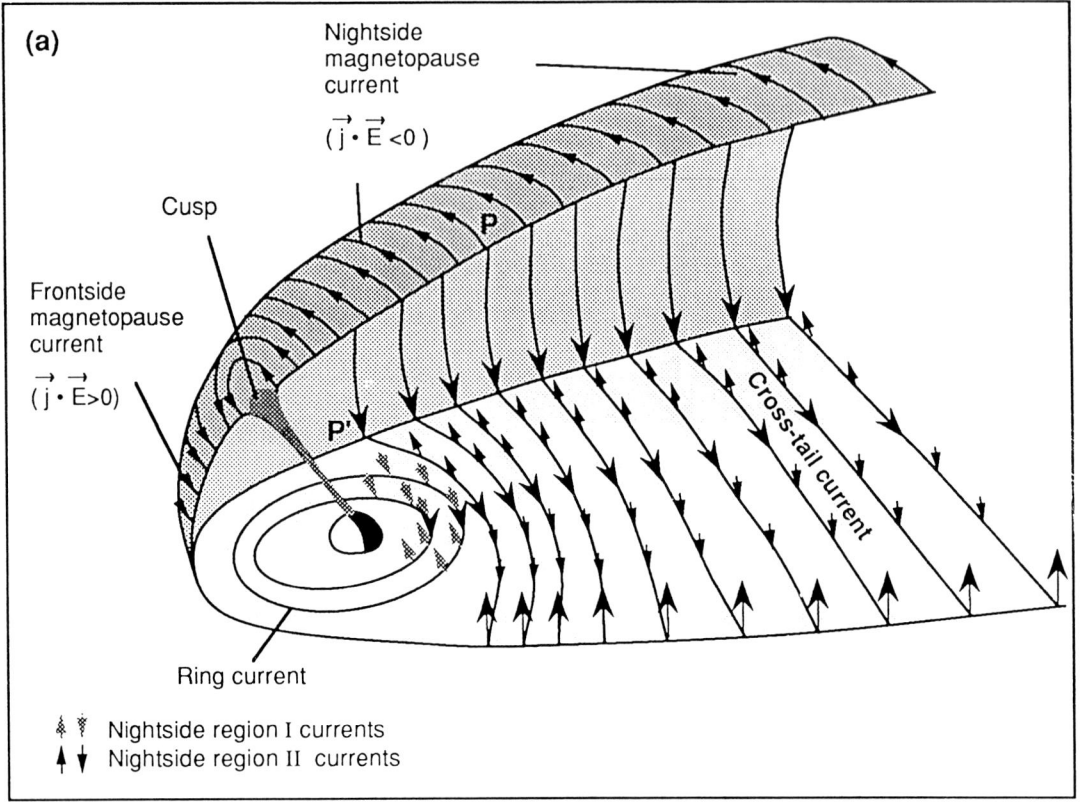

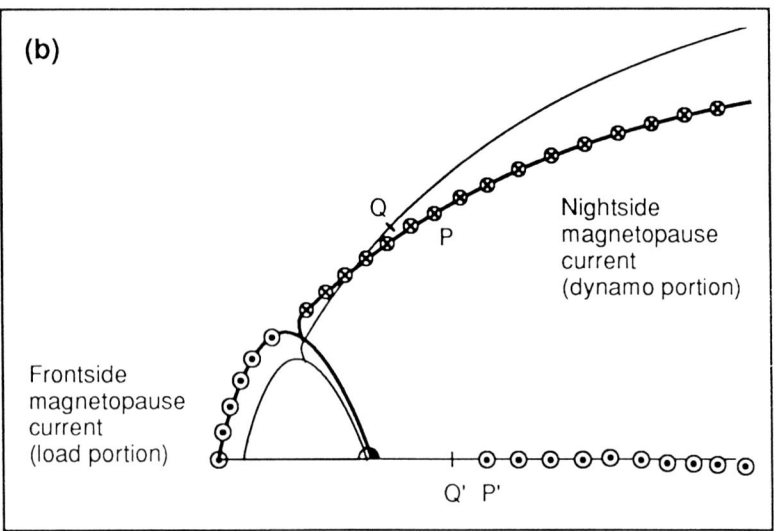

Fig. 3. (Top) A schematic diagram to illustrate the two current closures of the nightside magnetopause current. From the cusp to the location P, the nightside magnetopause current is closed through the frontside magnetopause where $\mathbf{j} \cdot \mathbf{E} > 0$ (load). Tailward of location P, the nightside magnetopause current closes within the magnetosphere (the load) consisting of the cross-tail current, the ring current, the regions I and II FAC, and the ionosphere. (Bottom) The magnetospheric reconfigures during southward IMF period, changing the flaring angle of the magnetopause from the shape denoted by the solid-line to the shape denoted by the dashed-line. This moves the current closure location from P to Q which corresponds to moving the inner edge of the cross-tail current from P' to Q'. Note that the change in the flaring angle is larger in the near-Earth region than at further downstream. Since the magnetopause current enhancement at a given downstream distance depends on the change in the flaring angle, one expects that the enhancement of cross-tail current is larger near the inner edge than in the mid-tail region.

concept for the substorm growth phase. They provided an approximate expression

$$\frac{\Delta x_t}{x_t} \approx \frac{3(x_t/R_t)^2 \Delta R_t/R_t - [1 + (x_t/R_t)^2] \Delta F_t/F_t}{1 + (2 x_t/R_t)^2}$$

which relates the variation of the inner edge location of the cross-tail current x_t with the changes in the amount of magnetic flux in the polar cap F_t and the radius R_t of the magnetotail at the inner edge of the cross-tail current. Let us take a typical situation in which the polar cap boundary expands from 70° latitude to 65° latitude at the end of the substorm growth phase. Concurrently, the tail flaring enlarges the tail radius (at the downstream distance of the inner edge of the cross-tail current sheet) from 20 R_E to 22 R_E. For an initial position of the inner edge of the cross-tail current of $x_t = 10\ R_E$, we obtain $\Delta F_t/F_t \approx 0.55$, and $\Delta x_t/x_t \approx -0.3$. This simple calculation shows the cross-tail current being near the geosynchronous altitude at the end of the growth phase.

A possible scenario of the near-Earth current buildup is illustrated in Figure 3. The top panel emphasizes that the entire nightside magnetopause current downstream of the cusp is the dynamo portion of the current system. The magnetopause current from the cusp to the downstream distance marked by P is closed through the frontside magnetopause which acts as the load for the current circuit whereby energy is consumed in deflecting and accelerating the oncoming solar wind flow around the frontside magnetosphere. The magnetopause current downstream of position P is closed inside the magnetosphere. This portion within the magnetosphere, which is the load for the circuit, includes the nightside plasma sheet and its boundary layer, the ring current, the Regions I and II FAC [*Iijima and Potemra*, 1976, 1978], and the ionosphere (see also Figure 4). Here, the ring current denotes the region where the current perpendicular to the magnetic field forms a closed path within the magnetosphere whereas that for the cross-tail current joins the magnetopause or the low latitude boundary layer. In the bottom panel, we illustrate by the dashed-line the shape of the magnetopause during the growth phase in comparison with the solid-line representing the magnetopause prior to the growth phase. The dayside magnetopause moves in, the cusp moves to lower latitude, and the tail size just behind the cusp decreases to give rise to a larger flare angle. This reconfiguration is consistent with observations and theory [*Unti and Atkinson*, 1968; *Sibeck et al.*, 1991]. It is likely that this configurational change will move the transition point between these two current closures from location P to Q closer to the cusp at the magnetopause.

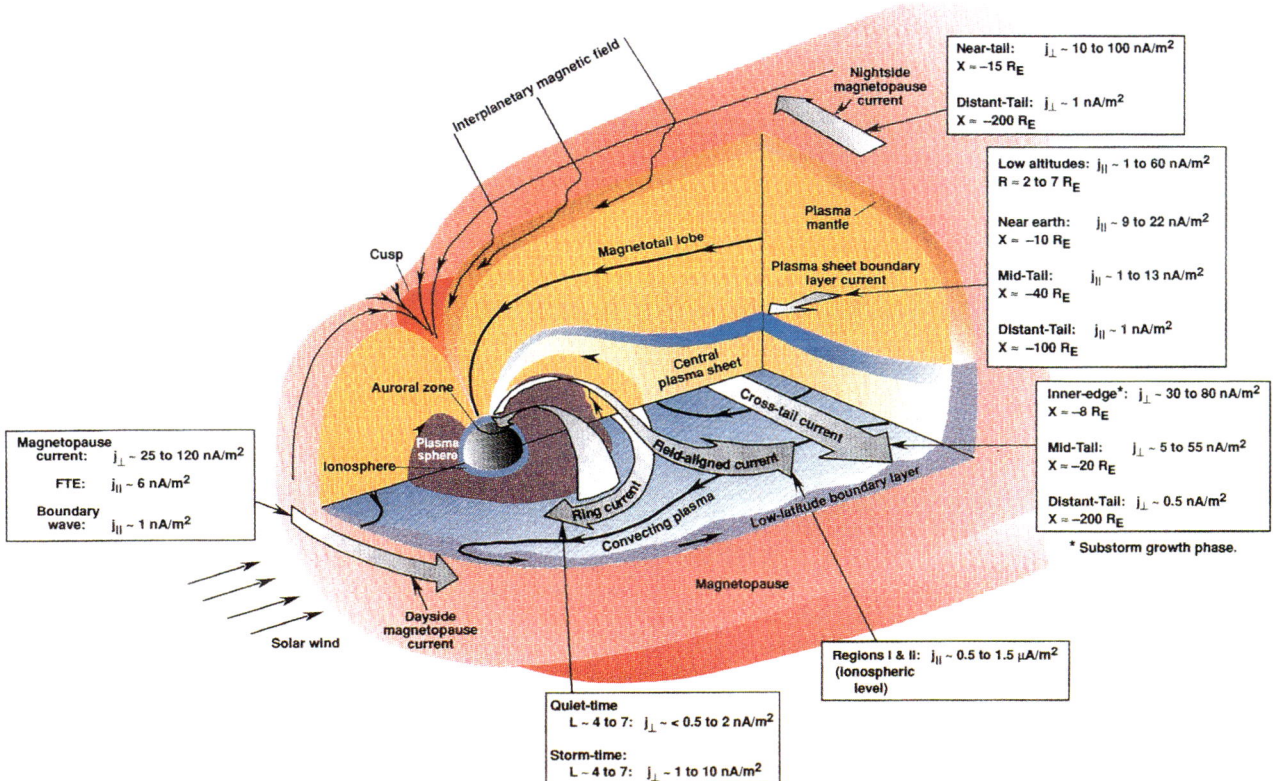

Fig. 4. A diagram illustrating the complex current system at the magnetopause and within the magnetosphere. The range of current density for each current is also shown.

The total magnetopause current at a given downstream distance depends on the flaring angle by which the solar wind dynamic pressure exerts on the magnetopause. The larger flare angle at the nightside magnetopause tailward of location Q results in a larger driver current at the magnetopause. It is reasonable to expect that an increase in the driving current will result in an increase in all the closure currents within the magnetosphere including the cross-tail current. Furthermore, since point Q moves closer to the cusp, a closer inner edge of the cross-tail current is then established, consistent with the stress balance requirement for the tail as discussed earlier. This trend is illustrated in Figure 3 where Q' corresponds to the inner edge of the cross-tail current when the current closure transition is at Q in the nightside magnetopause (and similarly, P' corresponds to P). The enhanced cross-tail current increases the $\mathbf{j} \times \mathbf{B}$ force on the plasma sheet which then pinches the plasma sheet until the pressure gradient force is built up to counter-balance the enhanced $\mathbf{j} \times \mathbf{B}$ force. This leads to plasma sheet thinning occurring primarily in the near-Earth region since the enhanced current is mostly in the inner portion of the cross-tail current corresponding to the larger change in the flaring angle of the near-Earth region. The combined effect of cross-tail current enhancement and the plasma sheet thinning gives rise to a dramatic increase in current density there.

The complexity of the current system arising from the solar wind-magnetosphere dynamo is further illustrated in Figure 4 with observed ranges of current density displayed. Table 3 provides also references from which the values are extracted. The near-Earth cross-tail current during the growth phase represents the strongest perpendicular current within the magnetosphere. Therefore, it can stretch the magnetic field significantly to allow for current sheet acceleration close to the Earth. The FAC associated with the pre-breakup arc may also link to the intense cross-tail current there. This is compatible with the observational constraint of auroral breakup occurring often on the most equatorward pre-existing arc [Akasofu, 1964]. It is also consistent with the field line mapping result of Elphinstone et al. [1991] in which the auroral brightening location maps to the innermost part of the cross-tail current where the current density is intense.

EXPANSION PHASE OF THE SYNTHESIS MODEL

The substorm expansion onset occurs when the intense cross-tail current built up in the near-Earth region suddenly becomes reduced drastically within a narrow longitudinal sector, as illustrated in the second row of Figure 2. The disruption may arise from sudden heating of the plasma sheet in the manner suggested by the thermal catastrophe model [G. L. Siscoe, private communication, 1990], by a cross-field current instability [Lui et al., 1990], by ion tearing via the onset of chaotic electron orbits in the thinned plasma sheet [Coroniti, 1980; Büchner and Zelenyi, 1989], or by electron tearing due to a significant B_y component [Wang et al., 1990]. The ionosphere plays a significant role in restraining or promoting further current diversion. Current disruption creates a complex three-dimensional magnetic field geometry, as schematically illustrated in Figure 5a and behaves more like a dynamo process than a dissipative process, similar to the situation at the dayside magnetopause as discussed by Song and Lysak [1989]. Current disruption forces most of the cross-tail current in the near-Earth region to continue through the ionosphere due to the large inductance of the current system. Although the cross-tail current is directly linked with the magnetopause current driven by the MHD dynamo of solar wind flow across the open magnetic field lines of the tail lobe, as illustrated in Figure 5b, the evolution of the current disruption and the subsequent formation of the substorm current wedge are also dependent on the large inductance of the tail current system and on the response of the ionosphere. Therefore, the substorm process has elements of both a directly driven system and a loading-unloading system [Lee et al., 1985; Liu et al., 1988]. The effect of the ionosphere manifests in pseudo-breakups. Current diversion from the magnetosphere would be prevented if the ionospheric condition is not appropriate for the imposed current. This can be accomplished by the modification of magnetospheric electric field through Alfvén waves bouncing between the two regions. The result is a pseudo-breakup with no subsequent poleward auroral expansion or further enhancement of activities.

The significant role of the ionosphere provides a possible explanation for the characteristics of pseudo-breakups. Murphree [1991] reported from a study of Viking images that pseudo-breakups can occur over a broad region. In our scenario, each pseudo-breakup location corresponds to the magnetic projection of a localized region in the extended thin near-Earth plasma sheet where current disruption criterion is met but the ionospheric condition is unfavorable for the establishment of a current wedge. Murphree [1991] further noted that the site of the eventual auroral brightening that sets off the substorm expansion phase may not necessarily be the same as the previous pseudo-breakup location. This is perceivable as the initial brightening location corresponds to current disruption in the tail mapping to an ionospheric region where a current wedge can be established. This onset location is therefore not necessary the same as the previous pseudo-breakup site.

For a favorable onset condition, the diverted current will flow preferentially along pre-existing auroral arcs because they provide better conductivity channels [Inhester et al., 1981]. Auroral breakup therefore occurs on one of the pre-existing arcs. In this breakup longitudinal sector, the removal of the intense cross-tail current and the creation of a current loop through the ionosphere cause a sudden relaxation of the stretched magnetic field, producing an earthward convection surge whereby the stretched magnetic field line becomes dipolar-like [Mauk, 1989; Delcourt and Sauvaud, 1991]. Electron precipitation can be enhanced further by whistler wave growth as described by Parks et al. [1972]. Substorm injection front is formed earthward of the current disruption region from this collapse of field lines. Plasma tailward of the current disruption region is partially evacuated by the convection surge which results in a rarefaction wave propagating mainly down the tail [Kropotkin, 1972; Chao et al., 1977].

Chao et al. [1977] have modeled the rarefaction wave propagation in a simplified tail field geometry with no magnetic field component normal to the neutral sheet. They noted that the characteristic rarefaction wave speed is the fast mode speed $v_s = (\gamma P_0/\rho_0)^{1/2}$, where γ is the ratio of specific heats, P_0 and ρ_0 are the pressure and density, respectively, in the unperturbed plasma sheet. Behind the rarefaction wave, an earthward plasma flow is

TABLE 3: A Quantitative Summary of Magnetospheric Current System

Region	∥/⊥	Cur. Den. (nA/m^2)	Spatial Scale	Integrated Current	Ambient Field (nT)	References
Magnetopause						
Frontside	$j_\perp$	25 to 120	400 to 2000 km	~50 mA/m	30–50	Berchem & Russell (1982)
FTE	$j_\parallel$	6	1 R_E diameter[1]	0.2 MA	20–30	Saunders et al. (1984)
Boundary wave	$j_\parallel$	1	2 R_E (y-dimension)[1]	~18 mA/m	~60	Sibeck (1990)
Near-Earth X ≈ −15 R_E	$j_\perp$	10 to 100	500 km to 1 R_E[1]	50 mA/m	40–80	Rosenbauer et al. (1975)
Distant-tail X ≈ −200 R_E	$j_\perp$	1	1 R_E[1]	8 mA/m	5–15	Slavin et al. (1984b)
Plasma Sheet and its Boundary Layer						
Low altitudes R ≈ 2.4 to 7 R_E	$j_\parallel$	1 to 60	0.1 to 3 R_E	13 to 150 mA/m	10–150	Kelly et al. (1986)
Inner region X ≈ −10 R_E	$j_\parallel$	9 to 22	1 R_E[1]	60 to 140 mA/m	40–80	Aubry et al. (1972)
Near-Earth X ≈ −20 R_E	$j_\parallel$	3 to 13	0.4 to 0.7 R_E	10 to 30 mA/m	20–30	Frank et al. (1981)
Mid-tail X ≈ −40 R_E	$j_\parallel$	1 to 12	0.1 to 0.4 R_E	1 to 9 mA/m	15–20	Lui & Krimigis (1984)
Distant-tail X ≈ −100 R_E	$j_\parallel$	1	3 R_E[1]	0.5 MA	10–20	Sibeck et al.[2] (1984)
Cross-Tail Current Sheet						
Inner edge X ≈ −8 R_E	$j_\perp$	30 to 80 100	0.2 R_E 0.5 R_E[1]	80 to 110 mA/m 100 to 300 mA/m	10–50	Lui et al. (1991)[3] Kaufmann (1987)[4]
X ≈ −11 R_E	$j_\perp$	~100	400 to 800 km	40 to 80 mA/m	10–50	Mitchell et al. (1990)
Near-Earth X ≈ −20 R_E	$j_\perp$ $j_\parallel$	5 to 55 1 to 4	1.5 to 4 R_E 1.5 to 5 R_E	80 mA/m 0.3 MA	10–20	McComas et al. (1986) Elphic et al. (1986)[2]
Distant-tail X ≈ −200 R_E	$j_\perp$	~0.5	3 R_E[1]	8 mA/m	5–15	Slavin et al. (1984b)
Ring Current and Auroral Region						
Quiet time	$j_\perp$	<1 to 2	4 R_E × 5 R_E	0.4 MA	50–1000	Lui et al. (1987)
Storm time	$j_\perp$	1 to 10	5 R_E × 5 R_E	1.5 MA	50–1000	Lui et al. (1987)
Storm time	$j_\perp$	<1 to 3	~5 R_E	0.29 MA/R_E	50–1000	Iijima et al. (1990)[5]
Regions 1 & 2	$j_\parallel$	500–1500	100 to 500 km	100 to 450 mA/m	50000	Potemra et al. (1979)

[1]assumed; [2]flux rope; [3]estimated; [4]modelling; [5]radial current

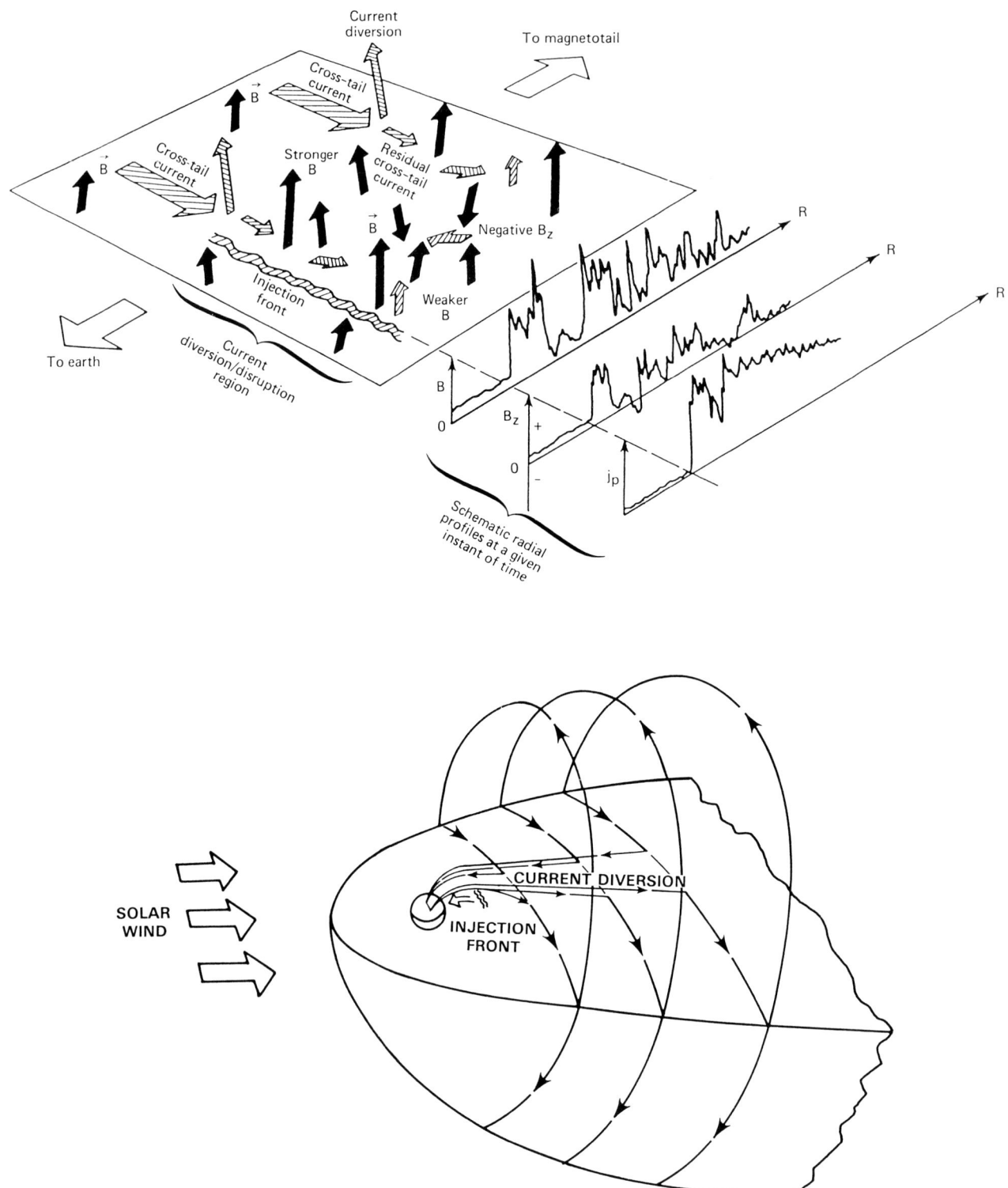

Fig. 5. (a) A schematic diagram to illustrate the complex magnetic field geometry and the turbulent nature of the current disruption (after *Lui et al.*, 1988), (b) current disruption in the cross-tail current forces current diversion into the ionosphere. This leads to the formation of the current wedge and injection front earthward of the disruption region.

induced, with speeds up to the wave mode speed. The pressure P_1 and density ρ_1 behind the rarefaction wave will be lowered and are given by

$$\rho_1 = \rho_0[1 - (\gamma - 1)u/(2v_s)]^{2/(\gamma - 1)},$$

$$P_1 = P_0[1 - (\gamma - 1)u/(2v_s)]^{2\gamma/(\gamma - 1)},$$

where u is the plasma flow speed. Since the pressure in the north-south direction is determined by the external solar wind condition, compression of the rarefied plasma sheet will take place to maintain the pressure balance. The net effect is thinning of the mid-tail plasma sheet and occurrence of transient earthward plasma flow as plasma tailward of the rarefaction region is displaced earthward to fill the partial void created by the convection surge. Mauk [1989] has shown that a convection surge can generate electric fields along the magnetic field self-consistently. In addition, the imposed FAC from the magnetotail can also lead to potential drops along the magnetic field line when it exceeds the threshold governed by the loss cone. This sequential development is advocated by the current disruption model. Current reduction in the disturbance region can lead to current enhancement in the adjacent regions as shown in Figure 6 and indicated by MHD simulation [*Birn and Hesse*, 1991], which may lead to spreading of current disruption radially if the disruption mechanism depends on the local current density. The regions outside the current wedge will also experience further tailward stretching of the field lines which may provoke further disruption, thus allowing the disturbance to spread in local time as well. Akasofu [1972] made the first suggestion in linking the local time widening of the substorm auroral electrojet and substorm current wedge with the local time spreading of current disruption in the tail.

The third row of Figure 2 shows the idea that during the substorm expansion, several locations in the current sheet may be disrupted in a rather irregular fashion and not necessarily be simultaneous. This is consistent with the observation that a substorm expansion phase usually encompasses several substorm intensifications [*Pytte et al.*, 1976; *Semenov and Sergeev*, 1981; *Rostoker et al.*, 1980]. Each of these disruptions represents a substorm intensification as observed on the ground and the intensity of a substorm is related to both the strength and the frequency of substorm intensifications [*Sergeev et al.*, 1986; *Lui*, 1988]. The brief duration of each intensification is governed by the time scale of current disruption and the duration of its associated disturbances (convection surge and rarefaction wave). The sequence of auroral substorm development shown by Rostoker et al. [1987] from the Viking

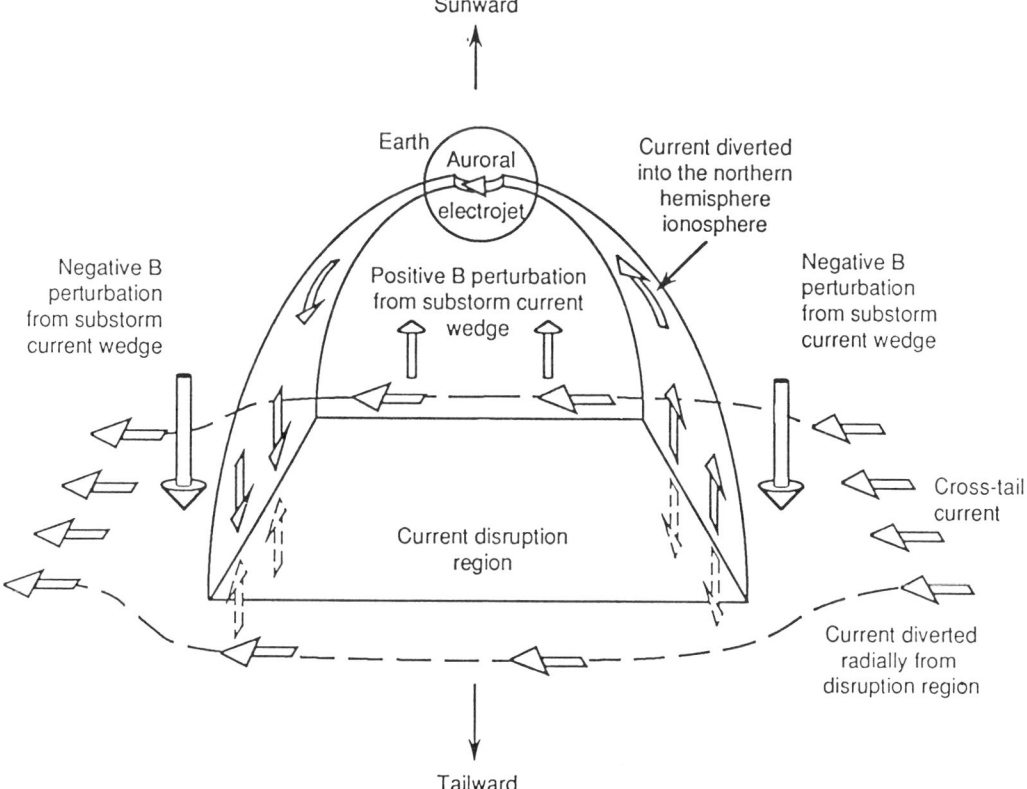

Fig. 6. A diagram to illustrate the magnetic field perturbations and re-routing of currents associated with a current disruption region in the magnetotail. These associated changes can lead to spreading of the disturbance both radially and longitudinally.

imager clearly demonstrates that multiple substorm disturbance regions can occur during the expansion phase of a single substorm. One can draw a number of analogies between a magnetospheric substorm and a thunderstorm as illustrated in Figure 7. Both are natural phenomena involving electrical discharges and covering large areas. However, each intensification of these natural phenomena is extremely transient and localized in nature (intensification in the case of a thunderstorm is represented by lightning).

Current disruption sets up an initial perturbation to the boundary between the tail-like and the dipolar-like field regions. The local time expansion of this perturbed boundary can be in the form of a surface wave as described by the interchange or ballooning instability model. Another feature which may emerge is the establishment of a new velocity shear zone at the western and eastern edges of the convection surge and rarefaction waves which can lead to the onset of KHI and auroral vortices (surges and omega bands). Large magnetic disturbances covering a broad frequency range accompany current disruptions and the propagation of rarefaction waves, allowing thermalization of the plasma sheet through particle resonance with these waves. When the rarefaction wave reaches far downstream, it may encounter the low latitude boundary layer and a strong velocity shear can be set up to yield a larger scale development of KHI and multiple surge forms or vortices, as advocated by the boundary layer model.

RECOVERY PHASE OF THE SYNTHESIS MODEL

After the passage of rarefaction waves, the plasma sheet becomes thin with weak magnetic field normal to the neutral sheet. The extended thin plasma sheet may then be unstable to tearing and subsequent formation of one or multiple large-scale X-lines. The

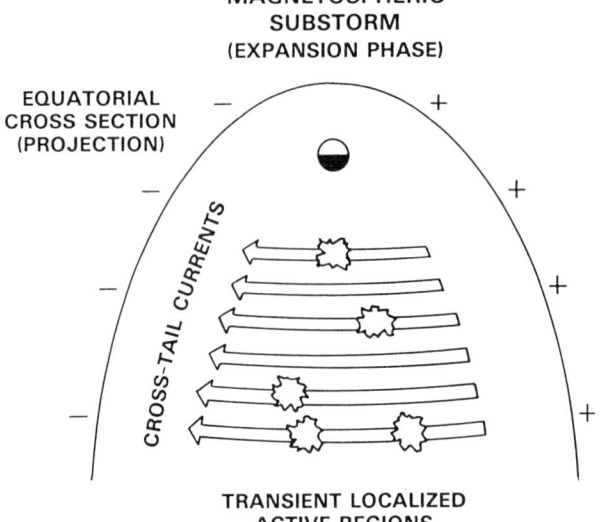

Fig. 7. A comparison between the two natural phenomena of thunderstorm and magnetospheric substorm.

weakening of the B_z component can reduce the stabilizing effect of the electrons on tearing instability [*Lembége and Pellat*, 1982]. Wang et al. [1990] recently showed that collisionless tearing growth can be enhanced by the presence of a B_y component in the neutral sheet region. It is, therefore, anticipated that the reduction of normal field will ultimately lead to separatrix formation at 20–80 R_E downstream in the way similar to that depicted by the neutral line model. Earthward of the reconnection site, plasma sheet begins to thicken with the newly added plasma jetting at the plasma sheet boundary layer while one or more plasmoids are formed and become ejected down the tail. This development is associated with a major poleward expansion of auroras. The substorm activity subsides when the plasma sheet becomes sufficiently thick to hinder significantly the magnetic reconnection across the separatrix.

FURTHER IMPLICATION OF THE SYNTHESIS MODEL

The two stage aspect of a substorm, i.e. first current disruption (accompanied by a convection surge and rarefaction wave propagation) and then reconnection at a neutral line, allows an explanation for the different nature of fast plasma flows between the central plasma sheet and the plasma sheet boundary noted by Baumjohann et al. [1988, 1989, 1990]. They found from comprehensive statistical analyses of IRM plasma data that while the fast flows tend to be field-aligned in the plasma sheet boundary, those in the central plasma sheet have a significant component perpendicular to the magnetic field. Furthermore, the occurrence frequency of fast flows in the central plasma sheet increases with the AE index while that of boundary layer stays relatively constant.

In this synthesis model, earthward flow in the central plasma sheet are related to the transient passing of a rarefaction wave front leading to field line collapse (i.e., the flows have a significant perpendicular component) while flows in the plasma sheet boundary are associated with reconnection at a more distant location, giving rise to flows which are more field-aligned at the near-Earth location. The plasma sheet boundary, during both quiet and substorm times, is always connected with the region near the distant reconnection site and thus the occurrence frequency of flows there is not much affected by substorm activities. In addition, one factor governing the intensity of a substorm in this synthesis model is the occurrence frequency of current disruptions and thus the occurrence frequency of fast flows in the central plasma sheet is related to substorm intensity.

SUMMARY AND CONCLUDING REMARKS

The proposed synthesis model can be summarized as follows. Southward interplanetary magnetic field initiates a reconfiguration of the magnetosphere resulting in an increased tail flaring and an inward intrusion of the cross-tail current in the near-Earth region. This leads to plasma sheet thinning and a drastic increase in current density in the near-Earth region. Expansion phase onset begins when the intense cross-tail current is suddenly reduced in a localized region by a number of possible mechanisms such as (ion or electron) tearing instability, cross-field current instability, interchange or ballooning instability, and thermal catastrophe. Current continuity requirement due to the large inductance of the tail current system forces current to be diverted into the ionosphere. Multiple current disruptions can occur within the expansion phase. Each current disruption leads to the development of a convection surge propagating earthward, forming the injection boundary close in, and a rarefaction wave front launching tailward, giving rise to plasma sheet thinning in the mid-tail region after expansion onset. The substorm disturbance therefore propagates radially in both the earthward and tailward directions. The local time widening of the disturbance region can be achieved by the cross-field current instability, the ballooning instability and the Kelvin-Helmholtz instability. The passing of the rarefaction wave leaves a thin current sheet with reduced magnetic field normal to the neutral sheet which may then be susceptible to the formation of large-scale neutral lines or the initiation of tearing instability at a downstream distance of 20–80 R_E. One or more plasmoids may be formed and eventually ejected further downstream. The substorm activity decreases when this reconnection process in the mid-tail subsides as the plasma sheet becomes thick.

The above substorm scenario bears elements from most previous substorm models. In particular, it bears considerable resemblance to the near-Earth neutral line model. There are also major differences which are (1) current disruption in the near-Earth tail does not sever the plasma sheet nor create large-scale plasmoids; these features are incorporated as later development in the mid-tail when the plasma sheet there is thinned substantially by rarefaction waves, (2) multiple substorm intensifications are emphasized, (3) local time expansion of the disturbance is specifically described, and (4) pseudo-breakups are instances when ionospheric condition is not favorable for current diversion from the magnetosphere. In addition, it resolves the conflict of the near-Earth neutral line model with observations that indicate a lack of both tailward flows and a large-scale magnetic field topology change in the near-Earth region in spite of evidence for substorm initiation there.

It is regarded by many substorm researchers that magnetic reconnection plays a crucial role in substorms. Recently, with the consideration of three-dimensional geometry, magnetic reconnection is generalized to include any process capable of breaking down the ideal MHD behavior, i.e., whenever the frozen-in condition is violated [*Schindler et al.*, 1988]. A more precise criterion for magnetic reconnection in this case is then the condition $\mathbf{B} \times (\nabla \times \mathbf{E}_\parallel) \neq 0$ [*Longmire*, 1963]. Under this definition, magnetic reconnection is constantly occurring in the inner magnetosphere since particles with different energy, mass and charge state take on different drift paths in this region when their drift is no longer dominated by the simple $\mathbf{E} \times \mathbf{B}$ motion. If this definition is adopted, then this synthesis model suggests that magnetic reconnection occurs in several forms within a substorm episode. In the early phase of substorm expansion, it is manifested as current disruption and convection surges. These activities later lead to formation of separatrices and plasmoids, which are the more traditional form of magnetic reconnection, during the late substorm expansion phase or the recovery phase.

There is no doubt that other ways of synthesizing the various substorm models exist. The intention here is to construct a plausible coherent framework having more substorm features accounted for than any individual model. Although we are aware that the inclusion of features from many substorm models does not guarantee the emergence of the correct description of a substorm, we find that most existing models have one or more crucial features required for understanding substorm phenomena comprehensively. We look forward to future analysis and missions to elucidate the true picture of the magnetospheric substorm and modify the proposed synthesis.

Acknowledgments. This work has been supported by the Atmospheric Sciences Section of the National Science Foundation, grant ATM-9000052 to The Johns Hopkins University.

REFERENCES

Akasofu, S.-I., The development of the auroral substorm, *Planet. Space Sci., 12,* 273, 1964.

Akasofu, S.-I., Magnetospheric substorm, a model, in Solar Terrestrial Physics, Part III, ed. by D. Dyer, D. Reidel Publ. Co., Hingham, Md, U. S. A., p. 131, 1972.

Akasofu, S.-I., Physics of Magnetospheric Substorms, D. Reidel Publ. Co., 1977.

Akasofu, S.-I., Substorms, *EOS, 70,* 529, April, 1989.

Aubry, M. P., M. G. Kivelson, R. L. McPherron, C. T. Russell, and D. S. Colburn, Outer magnetosphere near midnight at quiet and disturbed times, *J. Geophys. Res., 77,* 5487, 1972.

Baker, D. N., S. J. Bame, J. Birn, W. C. Feldman, J. T. Gosling, E. W. Hones, Jr., R. D. Zwickl, J. A. Slavin, E. J. Smith, B. T. Tsurutani, and D. G. Sibeck, Direct observation of passages of the distant neutral line (80–140 R_E) following substorm onsets: ISEE-3, *Geophys. Res. Lett., 11,* 1042, 1984.

Baumjohann, W., The plasma sheet boundary layer and magnetospheric substorms, *J. Geomag. Geoelectr., 40,* 157–175, 1988.

Baumjohann, W., The substorm as a global phenomenon, IAGA Assembly, Exeter, July, 1989.

Baumjohann, W., G. Paschmann, N. Sckopke, C. A. Cattell, and C. W. Carlson, Average ion moments in the plasma sheet boundary layer, *J. Geophys. Res., 93,* 11507–11520, 1988.

Baumjohann, W., G. Paschmann, C. A. Cattell, Average plasma properties in the central plasma sheet, *J. Geophys. Res., 94,* 6597–6606, 1989.

Baumjohann, W., G. Paschmann, and H. Luhr, Characteristics of high-speed ion flows in the plasma sheet, *J. Geophys. Res., 95,* 3801–3809, 1990.

Berchem, J., and C. T. Russell, The thickness of the magnetopause current layer: ISEE 1 and 2 observations, *J. Geophys. Res., 87,* 2108, 1982.

Birn, J. and M. Hesse, The substorm current wedge and field-aligned currents in MHD simulations of magnetotail reconnection, *J. Geophys. Res., 96,* 1611, 1991.

Bosqued, J. M., Ion precipitation and the transport of ions accelerated by auroral process, in *Auroral Physics,* ed. by C.-I. Meng, M. J. Rycroft, L. A. Frank, Cambridge University Press, England, p. 143, 1991.

Bosqued, J. M., J. A. Sauvaud, D. Delcourt and R. A. Kovrazhkin, Precipitation of suprathermal ionospheric ions accelerated in the conjugate hemisphere, *J. Geophys. Res., 91,* 7006–7018, 1986.

Büchner, J., L. M. Zelenyi, Regular and chaotic charged particle motion in magnetotaillike field reversals, 1. basic theory of trapped motion, *J. Geophys. Res., 94,* 11821–11842, 1989.

Chao, J. K., J. R. Kan, A. T. Y. Lui, S.-I. Akasofu, A model for thinning of the plasma sheet, *Planet. Space Sci., 25,* 703–710, 1977.

Cogger, L. L., and J. S. Murphree, The UV auroral distribution: its impulsive nature, *Adv. Space Res., 10,* 167, 1990.

Coroniti, F. V., On the tearing mode in quasi-neutral sheets, *J. Geophys. Res., 85,* 6719, 1980.

Coroniti, F. V., and C. F. Kennel, Changes in magnetospheric configuration during the substorm growth phase, *J. Geophys. Res., 77,* 3361, 1972.

Delcourt, D. C. and J. A. Sauvaud, Generation of energetic proton shells during substorms, *J. Geophys. Res., 96,* 1585, 1991.

Elphic, R. C., C. A. Cattell, K. Takahashi, S. J. Bame, and C. T. Russell, ISEE 1 and 2 observations of magnetic flux ropes in the magnetotail: FTE's in the plasma sheet?, *Geophys. Res. Lett., 13,* 648, 1986.

Elphinstone, R. D., D. Hearn, J. S. Murphree and L. L. Cogger, Mapping using Tsyganenko long magnetospheric model and its relationship to Viking auroral images, *J. Geophys. Res., 96,* 1467, 1991.

Frank, L. A., R. L. McPherron, R. J. DeCoster, B. G. Burek, K. L. Ackerson, and C. T. Russell, Field-aligned currents in the Earth's magnetotail, *Geophys. Res. Lett., 86,* 687, 1981.

Galperin, Yu. I. and Ya. I. Feldstein, Auroral luminosity and its relationship to magnetospheric plasma domains, in *Auroral Physics,* ed. by C.-I. Meng, M. J. Rycroft, L. A. Frank, Cambridge Univ. Press, England, p. 207–219, 1991.

Goertz, C. K., and R. A. Smith, Thermal catastrophe model of substorms, *J. Geophys. Res., 94,* 6581–6596, 1989.

Heikkila, W. J., and R. J. Pellinen, Localized induced electric field within the magnetotail, *J. Geophys. Res., 82,* 1610–1614, 1977.

Hones, E. W., Jr., J. R. Asbridge, S. J. Bame, and S. Singer, Substorm variations of the magnetotail plasma sheet from $X_{sm} = -6$ R_E to $X_{sm} = -60$ R_E, *J. Geophys. Res., 78,* 109, 1973.

Hones, E. W., Jr., Transient phenomena in the magnetotail and their relation to substorms, *Space Sci. Rev., 23,* 393, 1979.

Huang, C. Y., Quadrennial review of the magnetotail, *Rev. Geophys., 25,* 529, 1987.

Huang, C. Y. and L. A. Frank, A statistical study of the central plasma sheet: implications for substorm models, *Geophys. Res. Lett., 13,* 652655, 1986.

Iijima, T., and T. A. Potemra, The amplitude distribution of field-aligned currents at northern high latitudes, *J. Geophys. Res., 81,* 2165, 1976.

Iijima, T., and T. A. Potemra, Large-scale characteristics of field-aligned currents associated with substorms, *J. Geophys. Res., 83,* 599, 1978.

Iijima, T., T. A. Potemra, and L. J. Zanetti, Large-scale characteristics of magnetospheric equatorial currents, *J. Geophys. Res., 95,* 991, 1990.

Inhester, B., W. Baumjohann, R. A. Greenwald, and E. Nielsen, Joint two-dimensional observations of ground magnetic and ionospheric electric field associated with auroral zone currents. 3. Three-dimensional currents associated with a westward travelling surge, *J. Geophys. Res., 49,* 155, 1981.

Jacquey, C., J. A. Sauvaud, and J. Dandouras, Location and propagation of the magnetotail current disruption during substorm expansion: analysis and simulation of an ISEE multi-onset event, *Geophys. Res. Lett., 18,* 389, 1991.

Kan, J. R., Developing a global model of magnetospheric substorm, *EOS, 71,* 1083, 1990.

Kan, J. R., L. Zhu, S.-I. Akasofu, A theory of substorms: onset and subsidence, *J. Geophys. Res., 93,* 5624–5640, 1988.

Kaufmann, R. L., Substorm currents: growth phase and onset, *J. Geophys. Res., 92,* 7471–7486, 1987.

Kelly, T. J., C. T. Russell, R. J. Walker, G. K. Parks, and J. T. Gosling, ISEE 1 and 2 observations of Birkeland currents in the Earth's inner magnetosphere, *J. Geophys. Res., 91,* 6945, 1986.

Kropotkin, A. P., On the physical mechanism of the magnetospheric substorm development, *Planet. Space Sci., 20,* 1245–1257, 1972.

Lee, L. C., Z. F. Fu, and S.-I. Akasofu, A simulation study of forced reconnection processes and magnetospheric storms and substorms, *J. Geophys. Res., 90,* 10896, 1985.

Lembége, B., and R. Pellat, Stability of a thick two-dimensional quasineutral sheet, *Phys. Fluids, 25,* 1495, 19820.

Liu, C. S., Low-frequency drift instabilities of the ring current belt, *J. Geophys. Res., 75,* 3789, 1970.

Liu, Z. X., L. C. Lee, C. Q. Wei, and S.-I. Akasofu, Magnetospheric substorms: an equivalent circuit approach, *J. Geophys. Res., 93,* 7366, 1988.

Longmire, C. L., Elementary Plasma Physics, Interscience Publ., New York, 1963.

Lopez, R. E., and A. T. Y. Lui, A multi-satellite case study of the expansion of a substorm current wedge in the near-earth magnetotail, *J. Geophys. Res., 95,* 8009, 1990.

Lopez, R. E., H. Löhr, B. J. Anderson, P. T. Newell, and R. W. McEntire, Multipoint observations of a small substorm, *J. Geophys. Res., 95,* 18897, 1990.

Lui, A. T. Y., Estimates of current changes in the geomagnetotail associated with a substorm, *Geophys. Res. Lett.*, 5, 853, 1978.

Lui, A. T. Y., Observations on plasma sheet dynamics during magnetospheric substorms, in *Dynamics of the Magnetosphere*, ed. by S.-I. Akasofu, Reidel Publ. Co., 563, 1979.

Lui, A. T. Y., Observations on the fluid aspects of magnetotail dynamics, in *Magnetotail Physics*, ed. by A. T. Y. Lui, Johns Hopkins Univ. Press, Baltimore, Maryland, p. 101–118, 1987.

Lui, A. T. Y., What is a magnetospheric substorm expansion made of?, *EOS*, 69, 435, 1988.

Lui, A. T. Y., A synthesis of magnetospheric substorm models, *J. Geophys. Res.*, 96, 1849, 1991.

Lui, A. T. Y., and J. R. Burrows, On the location of auroral arcs near substorm onsets, *J. Geophys. Res.*, 83, 3342–3348, 1978.

Lui, A. T. Y., and S. M. Krimigis, Association between energetic particle bursts and Birkeland currents in the geomagnetic tail, *J. Geophys. Res.*, 89, 10741, 1984.

Lui, A. T. Y., R. W. McEntire, and S. M. Krimigis, Evolution of the ring current during two geomagnetic storms, *J. Geophys. Res.*, 92, 7459, 1987.

Lui, A. T. Y., R. E. Lopez, S. M. Krimigis, R. W. McEntire, L. J. Zanetti, and T. A. Potemra, A case study of magnetotail current sheet disruption and diversion, *Geophys. Res. Lett.*, 15, 721–724, 1988.

Lui, A. T. Y., A. Mankofsky, C.-L. Chang, K. Papadopoulos, and C. S. Wu, A current disruption mechanism in the neutral sheet: a possible trigger for substorm expansions, *Geophys. Res. Lett.*, 17, 745–748, 1990.

Lui, A. T. Y., R. E. Lopez, B. Anderson, K. Takahashi, L. J. Zanetti, R. W. McEntire, T. A. Potemra, D. M. Klumpar, E. M. Greene, R. Strangeway, Current disruptions in the near-Earth neutral sheet region, submitted to *J. Geophys. Res.*, 1991.

Lundin, R., I. Sandah, J. Woch, R. Elphinstone, The contribution of the boundary layer EMF to magnetospheric substorms, this Monograph, 1991.

Lyons, L. R., and A. Nishida, Description of substorms in the tail incorporating boundary layer and neutral line effects, *Geophys. Res. Lett.*, 15, 1337–1340, 1988.

Mauk, B. H., Generation of macroscopic magnetic-field-aligned electric fields by the convection surge ion acceleration mechanism, *J. Geophys. Res.*, 94, 8911–8920, 1989.

Mauk, B. H., and C.-I. Meng, The aurora and middle magnetospheric processes, in *Auroral Physics*, ed. by C.-I. Meng, M. J. Rycroft, L. A. Frank, Cambridge Univ. Press, England, p. 223, 1991.

McComas, D. J., C. T. Russell, R. C. Elphic, and S. J. Bame, The near-Earth cross-tail current sheet: ISEE 1 and 2 case studies, *J. Geophys. Res.*, 91, 4287, 1986.

McIlwain, C. E., Substorm injection boundaries, in *Magnetospheric Physics*, ed. by B. M. McCormac, p. 143, D. Reidel, Hingham, Mass, 1974.

Mitchell, D. G., D. J. Williams, C. Y. Huang, L. A. Frank, C. T. Russell, Current carriers in the near-Earth cross-tail current sheet during substorm growth phase, *Geophys. Res. Lett.*, 17, 583, 1990.

Moore, T. E., Acceleration of low-energy magnetospheric plasma, *Adv. Space Res.*, 6, 103–112, 1986.

Murphree, J. S., Viking optical substorm signatures, this volume, 1991.

Ohtani, S., S. Kokubun, R. C. Elphic, and C. T. Russell, Field-aligned current signature in the near-tail region, 1. ISEE observations in the plasma sheet boundary layer, *J. Geophys. Res.*, 93, 9709, 1988.

Ohtani, S., S. Kokubun, R. Nakamura, R. C. Elphic, C. T. Russell, and D. N. Baker, Field-aligned current signatures in the near-tail region, 2. Coupling between region 1 and region 2 systems, *J. Geophys. Res.*, 95, 18913, 1990.

Parks, G. K., G. Laval, and R. Pellat, Behavior of outer radiation zone and a new model of magnetospheric substorm, *Planet. Space Sci.*, 20, 1391–1408, 1972.

Potemra, T. A., T. Iijima, and N. A. Saflekos, Large-scale characteristics of Birkeland currents, in *Dynamics of the Magnetosphere*, ed. by S.-I. Akasofu, D. Reidel Publ. Co., 165, 1979.

Pytte, T., R. L. McPherron, S. Kokubun, The ground signatures of the expansion phase during multiple onset substorms, *Planet. Space Sci.*, 24, 1115, 1976.

Robert, P., R. Gendrin, S. Perraut, and A. Roux, GEOS 2 identification of rapidly moving current structures in the equatorial outer magnetosphere during substorms, *J. Geophys. Res.*, 89, 819–840, 1984.

Rosenbauer, H., H. Grunwaldt, M. D. Montgomery, G. Paschmann, and N. Sckopke, Heos 2 plasma observations in the distant polar magnetosphere: The plasma mantle, *J. Geophys. Res.*, 80, 2723, 1975.

Rostoker, G., and T. E. Eastman, A boundary layer model for magnetospheric substorms, *J. Geophys. Res.*, 92, 12187–12202, 1987.

Rostoker, G., S.-I. Akasofu, J. Foster, R. A. Greenwald, Y. Kamide, K. Kawasaki, A. T. Y. Lui, R. L. McPherron, and C. T. Russell, Magnetospheric substorms—definition and signatures, *J. Geophys. Res.*, 85, 1663, 1980.

Rostoker, G., A. Vallance Jones, R. L. Gattinger, C. D. Anger, and J. S. Murphree, The development of the substorm expansive phase: The "eye" of the substorm, *Geophys. Res. Lett*, 14, 399–402, 1987.

Rothwell, P. L., L. P. Block, M. B. Silevitch, C.-G. Falthammar, A new model for substorm onsets: the pre-breakup and triggering regimes, *Geophys. Res. Lett*, 15, 1279–1282, 1988.

Roux, A., Generation of field-aligned current structures at substorm onsets, *Proc. ESA Workshop on Future Missions in Solar, Heliospheric and Space Plasma Physics*, ESA SP-235, Garmisch-Partenkirchen, Germany, pp. 151–159, 1985.

Roux, A., S. Perraut, P. Robert, A. Morane, A. Pedersen, A. Korth, G. Kremser, B. Aparicio, D. Rodgers, R. Pellinen, Plasma sheet instability related to the westward travelling surge, *J. Geophys. Res.*, in press, 1991.

Samson, J. C., Transient field-aligned currents associated with the substorm expansive phase, this Monograph, 1991.

Saunders, M. A., C. T. Russell, and N. Sckopke, Flux transfer events: Scale size and interior structure, *Geophys. Res. Lett.*, 11, 131, 1984.

Sauvaud, J.-A., and J. R. Winckler, Dynamics of plasma, energetic particles, and fields near synchronous orbit in the nighttime sector during magnetospheric substorms, *J. Geophys. Res.*, 85, 2043–2056, 1980.

Schindler, M. Hesses, and J. Birn, General magnetic reconnection, parallel electric fields, and helicity, *J. Geophys. Res.*, 93, 5547, 1988.

Scholer, M., Earthward plasma flow during near-Earth magnetotail reconnection: numerical simulations, *J. Geophys. Res.*, 92, 12181–12186, 1987.

Semenov, V. S. and Sergeev, V. A., A simple semi-empirical model for the magnetospheric substorm, *Planet. Space Sci.*, 29, 271, 1981.

Sergeev, V. A., R. J. Pellinen, T. Bosinger, W. Baumjohann, P. Stauning, A. T. Y. Lui, Spatial and temporal characteristics of impulsive structure of magnetospheric substorm, *J. Geophys.*, 60, 186–198, 1986.

Sibeck, D. G., Solar wind dynamic pressure variations: quantifying the statistical magnetospheric response, *Proceedings of Workshop on Plasma Astrophysics (ESA SP- 311)*, 63, 1990.

Sibeck, D. G., G. L. Siscoe, J. A. Slavin, E. J. Smith, S. J. Bame, and F. L. Scarf, Magnetotail flux ropes, *Geophys. Res. Lett.*, 11, 1090, 1984.

Sibeck, D. G., R. E. Lopez, and E. C. Roelof, Solar wind control of the magnetopause shape, location, and motion, *J. Geophys. Res.*, 96, 5489, 1991.

Siscoe, G. L., A unified treatment of magnetospheric dynamics with applications to magnetic storms, *Planet. Space Sci.*, 14, 947, 1966.

Siscoe, G. L., Coupling between the solar wind and the Earth's magnetosphere: Summary comments, in *Solar Wind/ Magnetosphere Coupling*, ed. by Y. Kamide and J. A. Slavin, Terra Scientific Publ. Co., p. 793, 1986.

Siscoe, G. L., and W. D. Cummings, On the cause of geomagnetic bays, *Planet. Space Sci.*, 17, 1795–1802, 1969.

Slavin, J. A., E. J. Smith, B. T. Tsurutani, D. G. Sibeck, H. J. Singer, D. N. Baker, J. T. Gosling, E. W. Hones, Jr., and F. L. Scarf, Substorm associated travelling compression regions in the distant tail: ISEE-3 geotail observations, *Geophys. Res. Lett.*, 11, 657, 1984a.

Slavin, J. A., E. J. Smith, and D. S. Intriligator, A comparative study of distant magnetotail structure at Venus and Earth, *Geophys. Res. Lett.*, 11, 1074, 1984b.

Smith, R. A., C. K. Goertz, and W. Grossmann, Thermal catastrophe in the plasma sheet boundary layer, *Geophys. Res. Lett.*, 13, 1380–1383, 1986.

Song, Y. and R. L. Lysak, Current dynamo effect of 3-D time-dependent reconnection in the dayside magnetopause, *Geophys. Res. Lett., 16*, 911–914, 1989.

Stern, D. P., Substorm electrodynamics, *J. Geophys. Res., 95*, 12057–12068, 1990.

Tanskanen, P., J. Kangas, L. Block, G. Kremser, A. Korth, J. Woch, I. B. Iversen, K. M. Torkar, W. Riedler, S. Ullaland, J. Stadnes, and K.-H. Glassmeier, *J. Geophys. Res., 92*, 7443–7457, 1987.

Tsyganenko, N. A., On the re-distribution of the magnetic field and plasma in the near nightside magnetosphere during a substorm growth phase, *Planet. Space Sci., 37*, 183–192, 1989.

Unti, T., and G. Atkinson, Two-dimensional Chapman–Ferraro problem with neutral sheet, 1, The boundary, *J. Geophys. Res., 73*, 7319, 1968.

Wang, X, A. Bhattacharjee, and A. T. Y. Lui, Collisionless tearing instability in magnetotail plasmas, *J. Geophys. Res., 95*, 15047, 1990.

Zanetti, L. J. T. A. Potemra, T. Iijima, and W. Baumjohann, Birkeland-ionospheric currents of the magnetospheric storm circuit, this Monograph, 1991.

Some Observational Constraints for Substorm Models

GORDON ROSTOKER*

*Canadian Network for Space Research and Department of Physics
University of Alberta, Edmonton, Alberta, Canada, T6G 2J1*

Any model for magnetospheric substorms must attempt to explain the large ensemble of observations of particles and fields acquired directly through *in situ* measurements in the ionosphere and above and indirectly through remote sensing using ground based instrumentation and global imaging by instrumentation sensitive to UV radiation. In this paper I shall emphasize the nature of the ground based observational data base on which the original definition of a substorm was based and the recent observations of auroral dynamics using data from the UV imager aboard the Viking satellite. We shall use the observations to define constraints which any substorm model must satisfy or, at least address. We shall examine the multiplicity of substorm surge forms along the poleward edge of the auroral oval in the evening sector and ask what mechanism might be responsible for this characteristic of substorm activity. We shall examine the directly driven component of substorm activity and ask how this is understood in the framework of present substorm models. Finally we shall critically examine the evidence for the near-Earth neutral line model of substorms in the light of recent *in situ* observations of particles and fields in the magnetotail.

INTRODUCTION

The transfer of energy from the solar wind to the terrestrial magnetosphere-ionosphere system has been a subject of considerable interest to space scientists for most of this century. While a considerable understanding of the solar-terrestrial interaction had evolved in the first part of the twentieth century due to the pioneering studies of *Birkeland* [1908], *Chapman and Ferraro* [1931] and *Alfvén* [1939] amongst others, it was *Akasofu* [1964] who galvanized the community with his description of the phenomenology of the auroral substorm. From the time that high latitude observations were assembled within the framework of the magnetospheric substorm by *Akasofu* [1968], a great deal of effort has been expended trying to understand the physical nature of the substorm process. This ambitious task was made all the more difficult by the introduction of conflicting definitions of the "substorm" over the ensuing twenty years.

---

* Also at Institute of Earth & Planetary Physics

Magnetospheric Substorms
Geophysical Monograph 64
Copyright 1991 American Geophysical Union

By 1980, it was recognized that substorm definition was more than a problem in semantics (Siscoe, 1980) and an initial attempt was made to try to provide a universally acceptable definition of the substorm phenomenon (Rostoker et al., 1980). Unfortunately, this effort did not meet with immediate success and the term substorm still implies different things to different researchers. Therefore, my first objective is to provide a definition of a substorm which will be assumed throughout this paper.

It is now generally recognized that a substorm features two distinctly different components of activity. The first component has been termed directly driven activity and reflects the deposition of energy from the solar wind which enters the magnetosphere the deposition process taking place within a time characteristic of the propagation of Alfvén waves from the magnetospheric boundary regions to the high latitude ionosphere. Directly driven activity is characterized by an eastward electrojet flowing in the auroral oval across the dusk meridian and a corresponding westward electrojet flowing across the dawn meridian. The time scale on which these large scale electrojets varies is commensurate with the impulse response time of the magnetosphere as determined by Clauer et al. [1983], viz. approximately 2 hours. The large scale directly driven system also includes the Birkeland currents and their corresponding north-south ionospheric closures currents first identified by *Zmuda and Armstrong* [1974] and studied in detail by *Iijima and Potemra* [1976]. The geometry of the Hall currents of the directly driven system and their impulse response function are shown in Figure 1a. The second component of substorm activity is often referred to as loading-unloading or the storage-release process. The signature of this process is the substorm expansive phase whose most noted signature is the auroral substorm first identified by *Akasofu* [1964]. The building block of the expansive phase is a longitudinally localized regime of upward field-aligned current emanating from a region of intense auroral luminosity termed the surge. The upward field-aligned current is fed from all sides by ionospheric current which in turn is connected to the magnetosphere by diffusely distributed downward field-aligned current. The ionospheric current is not uniformly distributed and in fact, tends to be concentrated in a latitudinally concentrated westward electrojet. While there is also eastward closure current entering the surge region, it is generally far weaker than the westward electrojet. The lifetime of these building block currents is far shorter than the response time of the directly driven currents, being of the order of 12 min. (Korotova et al., 1991). The geometry of the field-aligned and ionospheric currents associated with the elementary expansive phase system are shown in Figure 1b. Most substorm expansive phases are not restricted to the development of one surge and its associated current system. More typically, expansive phase activity involves the sporadic appearance of surges along the poleward edge of an oval which carries simultaneously varying directly driven currents. The most spectacular outbursts of expansive phase activity often develop after a period of sustained southward interplanetary magnetic field (IMF). The onset of the expansive phase activity often coincides with a reduction of energy flow into the magnetosphere, usually induced by a turning towards the north of the IMF. The ensuing expansive phase activity involves the rapid evolution of an ensemble of building block current elements, each one

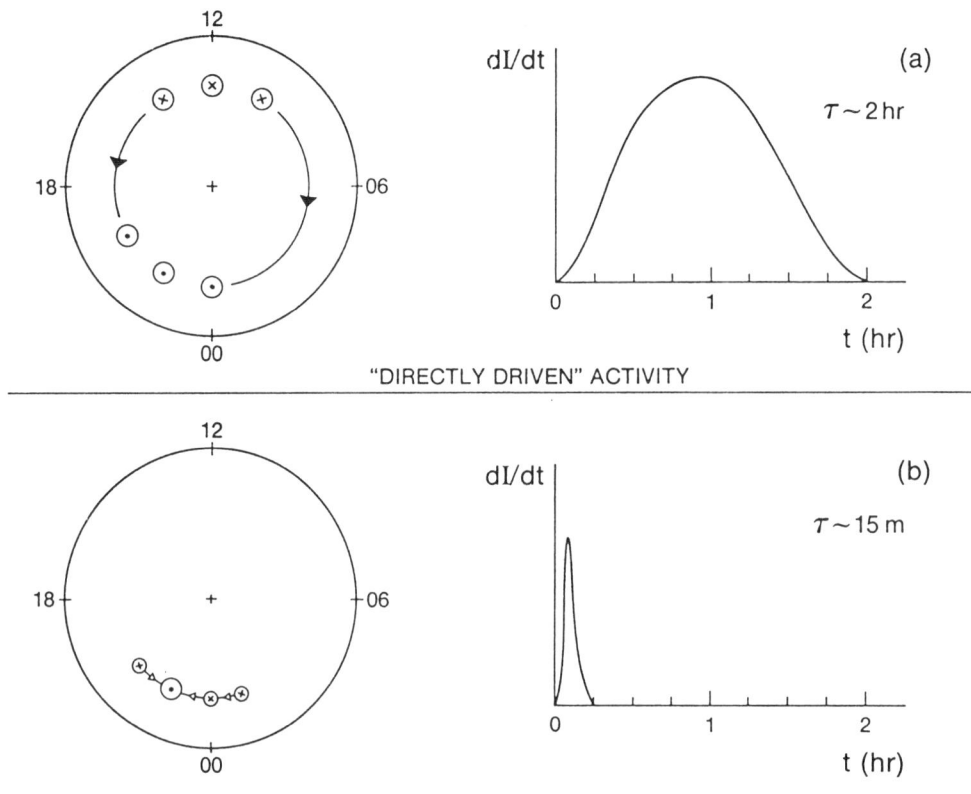

Fig. 1 Electric current systems associated with the two components of substorm activity together with the impulse responses in the current systems. Panel a shows the two large scale electrojets characterizing directly driven activity. A sudden increase in energy input to the magnetosphere results in directly driven currents increasing to a new higher value over a two hour time scale as shown in the figure. Panel b shows a substorm wedgelet which is the current system associated with a single expansive phase onset or intensification. A major substorm involves the sequential activation of an ensemble of such wedgelets which, together constitute the substorm current wedge (after Rostoker, 1991a).

featuring a westward jet element poleward of the previous one (Kisabeth and Rostoker, 1974). Often each newly created surge appears to the west of the previous one, leading to a westward expansion of the substorm disturbed ionosphere (Wiens and Rostoker, 1975). The overall pattern of development of a substorm expansive phase is shown in Figure 2. In this Figure, it is apparent that the recovery of the driven system activity accompanied by the poleward contraction of the poleward edge of the oval is accompanied by a multiplicity of expansive phase intensifications. The substorm current wedge (cf. Baumjohann, 1983) represents the combined effect of the ensemble of small scale current structures which develop during the course of expansive phase activity. The physics of the substorm current wedge lies, not in the behaviour of a large three dimensional current system of the type shown in Figure 3, but rather in the evolution of the spatially localized small scale component parts whose combined effect produces the signature of the wedge.

It is important to realize that both directly driven activity and expansive phase activity contribute to the overall substorm disturbance. In particular, the auroral electrojet index AL is influenced by both the directly driven westward electrojet and expansive phase westward electrojet filaments Often it is difficult to decouple these two contributions in examining variations in AL. However, Figure 4 shows a case where the two contributions are clearly separable, with the disturbance from 0730 - 0818 UT being due purely to the directly driven system and the disturbance after 0818 UT being heavily influenced by expansive phase activity. Figure 5 shows the midnight sector magnetograms for this event as well as the times of IMF Bz polarity reversals. Clearly no expansive phase activity is evident in Figure 5 during the interval 0730-

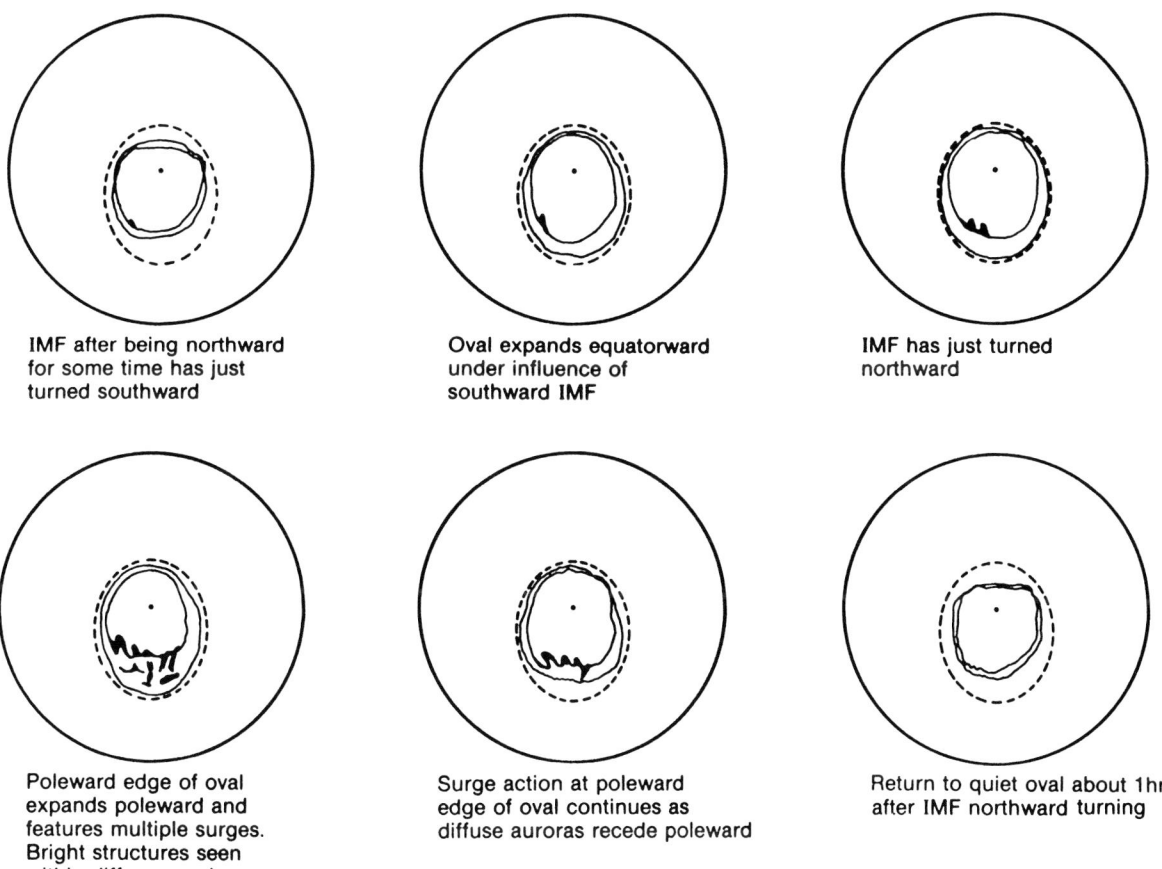

Fig. 2 The response of auroral activity to a southward turning of the IMF and a subsequent northward turning. The southward turning of the IMF signals the growth of the polar cap area and of the strength of the driven system electrojets. The northward turning of the IMF after a period of southward orientation often leads to significant expansive phase activity with multiple surges in the poleward portion of the evening sector oval. Each surge has associated with it a substorm wedgelet. The recovery of the substorm features a poleward retreat of the poleward border of the oval and a reduction in the area of the polar cap (after Rostoker, 1991a).

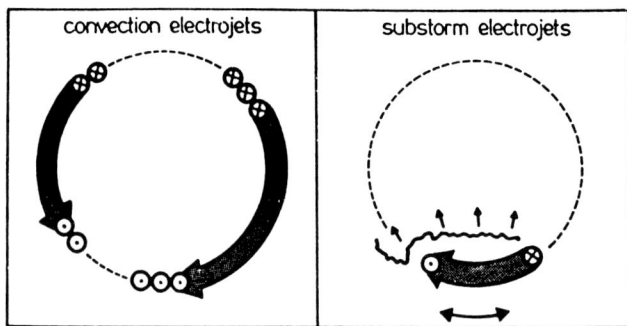

Fig. 3 The component current systems of the substorm (after Baumjohann, 1983). Note that the current wedge across midnight is not a single monolithic current system but rather is an ensemble of wedgelets. The physics of the substorm expansive phase lies not in the properties of the large scale wedge but rather in the properties of the individual wedgelets of which the substorm current wedge is composed.

0818 UT (during the initial increase in AE) and *De Groot-Hedlin and Rostoker* (1987) showed a complete absence of Pi 2 activity in this interval as well.

Before concluding this introduction to the substorm phenomenon, it is useful to describe the framework of particle precipitation and electromagnetic signatures which describe the evening sector auroral oval. Figure 6 shows the profiles of the auroral electrojets, field-aligned currents, the ionospheric electric field, the characteristics of energetic electron precipitation, auroral luminosity and whistler mode noise in a cut across the evening sector oval. The most important point to note here is that the boundary plasma sheet (bps) electrons are found across the regime of upward field-aligned current. This is consistent with the fact that the equatorward portion of the bps is co-located with the poleward portion of the eastward electrojet. This, in turn, is in accord with the contention of *Winningham et al.* [1979] and *Heelis* [1979] that the polarity transition of the meridional component of the ionospheric electric field (viz. the Harang discontinuity) occurs in the heart of the bps. Discrete auroral arcs can be found anywhere in the bps, viz. anywhere in the region of the gradient in the meridional component of the electric field in which the Harang discontinuity is immersed. Any one of these auroral arcs can brighten, signaling the onset of a substorm expansive phase. However, it is found statistically that it is the equatorwardmost arc that brightens at substorm onset (Akasofu, 1964) and this arc is often found in the poleward portion of the eastward electrojet (Baumjohann et al., 1981). The purpose of introducing this discussion is to emphasize one important point. The breakup arc of an auroral substorm is found in the bps which is the site of upward Region 1 field-aligned currents associated with the directly driven system. The breakup does not occur at the equatorward edge of the auroral oval as defined

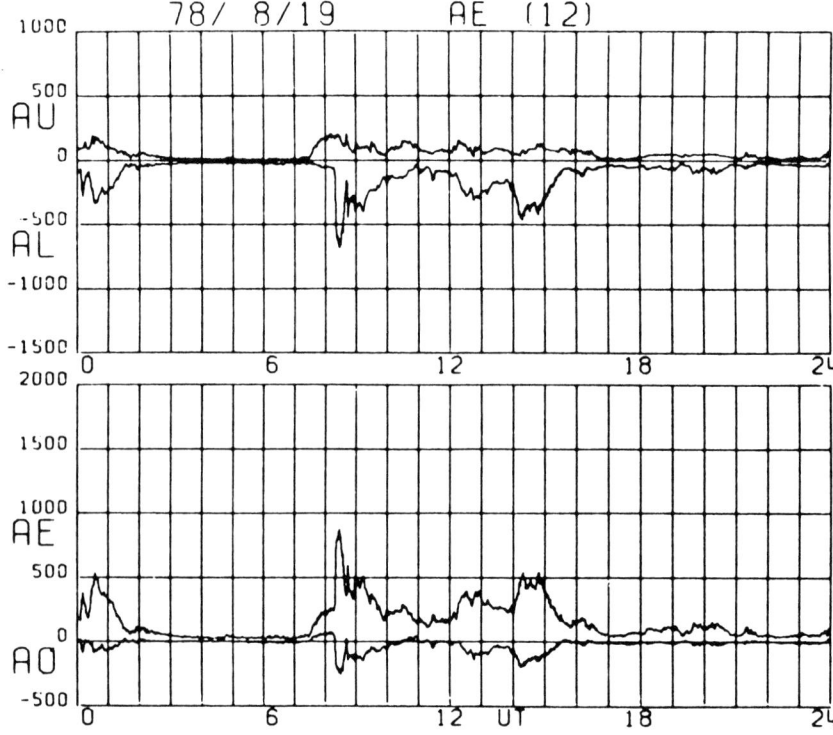

Fig. 4 AE index for August 19, 1978 showing an episode of almost pure directly driven activity from 0730 - 0818 UT followed by an outburst of expansive phase activity.

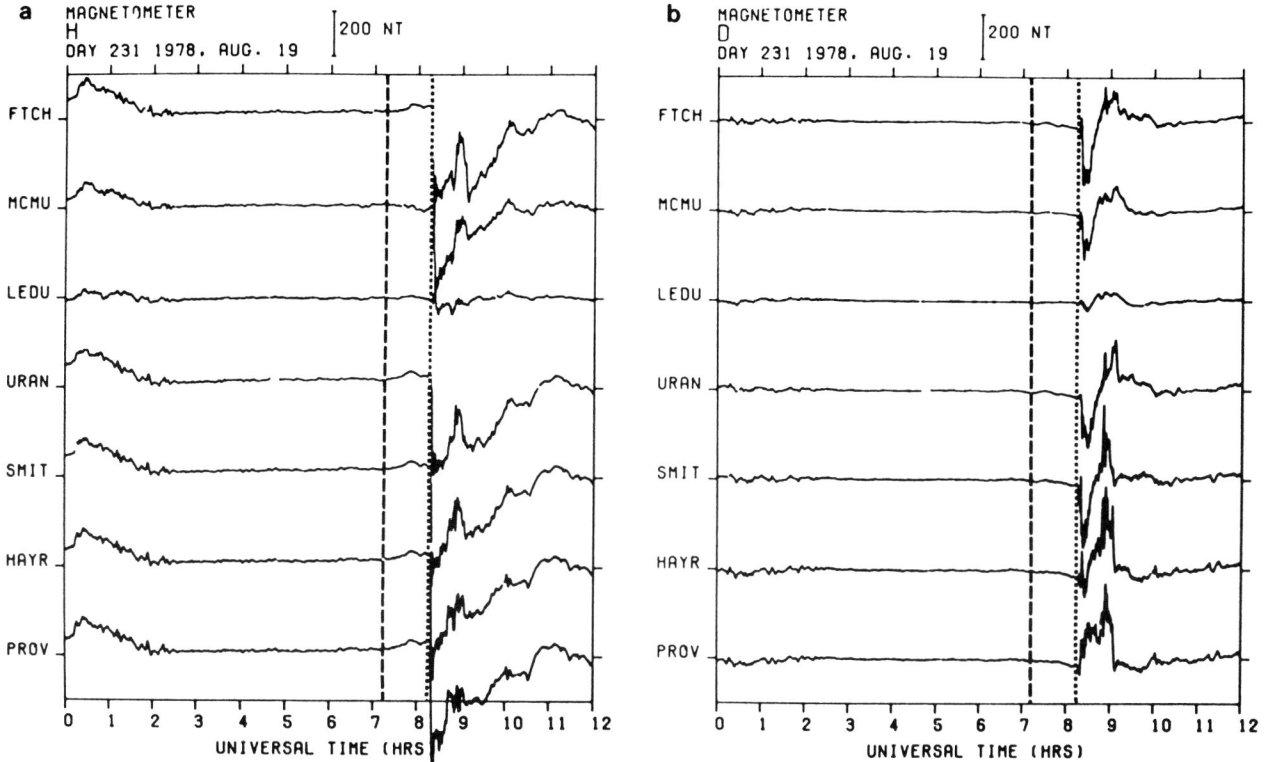

Fig. 5 Midnight sector magnetograms for August 19, 1979 showing the lack of expansive phase activity prior to 0818 UT (after Rostoker, 1983). The southward and subsequent northward turnings of the IMF are marked by vertical dashed lines.

by the so-called diffuse auroras. Furthermore, the co-location of the auroral surge at the western edge of the breakup arc with the Region 1 currents of the driven system has an important implication. It indicates that the upward field-aligned current of the surge is immersed in the Region 1 currents of the driven system and suggests that the localized current of the surge may simply be a redistribution of the Region 1 currents in response to changes in the ionospheric conductivity structure. Above all, it suggests that the source region for the auroral breakup upward field-aligned current is in the same region of the magnetosphere as the source of the driven system upward field-aligned currents. This is a vitally important constraint that any substorm model must be governed by.

Auroral Signatures of Substorm Expansive Phases and the Constraints they Pose

The original definition of a substorm was founded on auroral data obtained by allsky cameras distributed in remote location in high latitude regions. With their limited field of view (i.e., a maximum of 1000 km from horizon to horizon), the distortion of images at the edge of the field of view due to mirror geometry and irregular spacings between sites, allsky cameras had severe limitations in being able to accurately track the evolution of discrete auroral forms. For this reason, the westward expansion of the western edge of the region of auroral disturbance associated with an expansive phase was originally interpreted in terms of the westward drift of an auroral surge (viz. the westward traveling surge). While there were suspicions voiced earlier regarding the question of whether or not surges actually propagated westward over significant distances (cf. Tighe and Rostoker, 1980), it wasn't until the Viking satellite began taking images with exposure times of 1 s and image separations of as little as 20 s that it became clear that most individual surge forms did not propagate westward with velocities large enough to detect. It is true that during the initial moments of surge development, the western edge of the form expands westward as the scale size of the structure grows. However, even this motion ceases after two or three minutes and the resultant form then remains in the same longitudinal sector until it dies away. Figure 7 shows the evolution of surge features along an active oval during a period of intense southward IMF. The surge at the left side of the image stays stationary during the several minutes covered by the images shown here, while new surge forms appear to the west. This event is typical of surge development, and it should be recognized that each surge is the signature of the elementary substorm expansive phase "wedgelet" (cf. Figure 1b).

A second aspect of surge evolution seen in Figure 7 is the multiplicity of surge forms. Evidently the upward field-aligned

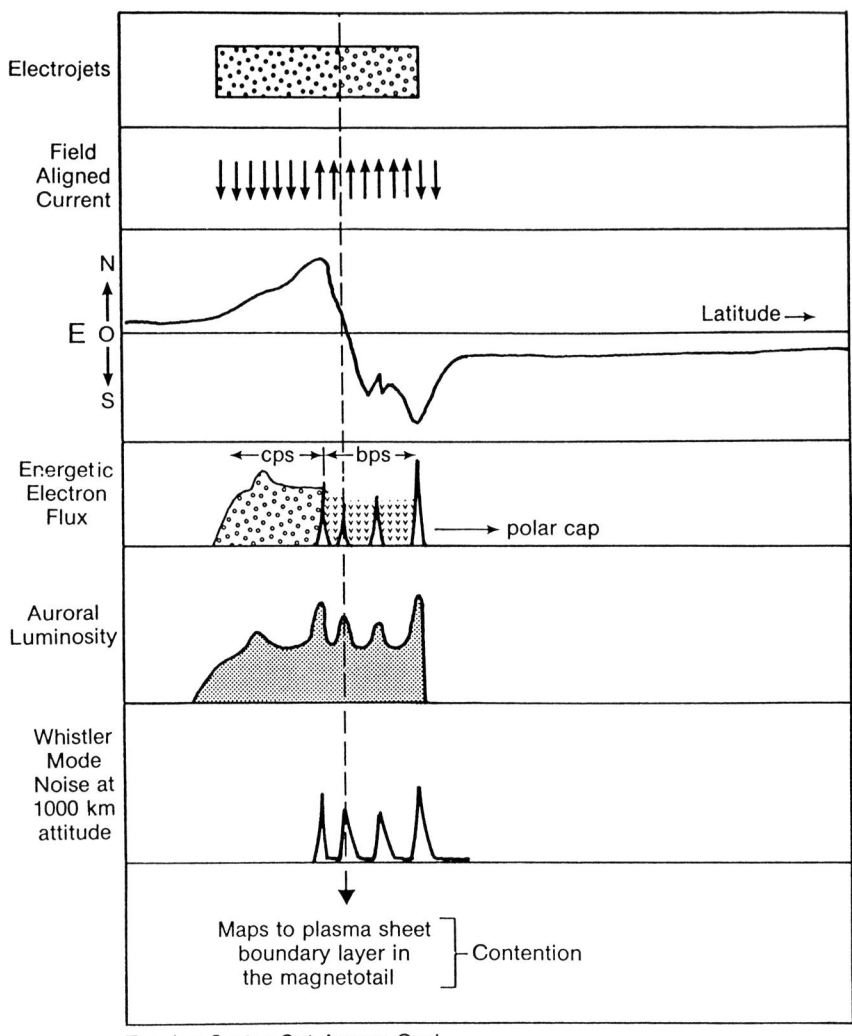

Fig. 6 Variation of current system directionality, electric field, particle precipitation, auroral luminosity and whistler mode noise across the evening sector auroral oval. The breakup arc can occur anywhere within the region of electric field gradient across the Harang discontinuity, even in the poleward portion of the eastward electrojet.

current in the region normally designated as Region 1 exhibits spatially periodic structure during expansive phase activity. Figure 8 shows a second example of surge multiplicity for an instant during an episode of strong expansive phase activity. These surge forms are approximately 400 km in scale size and feature a spatial separation of approximately 400 km as well. The easternmost surge developed in the short space of one minute (or less) and one minute later that characteristic surgelike structure was not evident. Whatever the physics of surge formation may be, it must satisfy the constraint that the growth time be of the order of a minute or less. The tendency for two or more auroral surges to co-exist is not atypical for substorm expansive phase activity. Figure 9 shows the occurrence frequency for surges in terms of whether they occur as isolated features or whether there are two, three, four etc.

surges co-existing. While the most likely scenario during expansive phase activity is for a single surge to be visible, it is clear that if one surge exists at any instant, there is a greater than 50% probability that there will be additional surge forms arrayed along the poleward edge of the evening sector auroral oval. Any substorm model must address the question of the origin of these periodically arrayed surge forms.

Constraints on Substorm Models from Magnetotail Observations

In the final section of this paper, I should like to comment briefly on two of the frameworks in which substorm magnetotail observations are presently ordered. These are the near-Earth neutral line (NENL) model (cf. Hones, 1984) and the

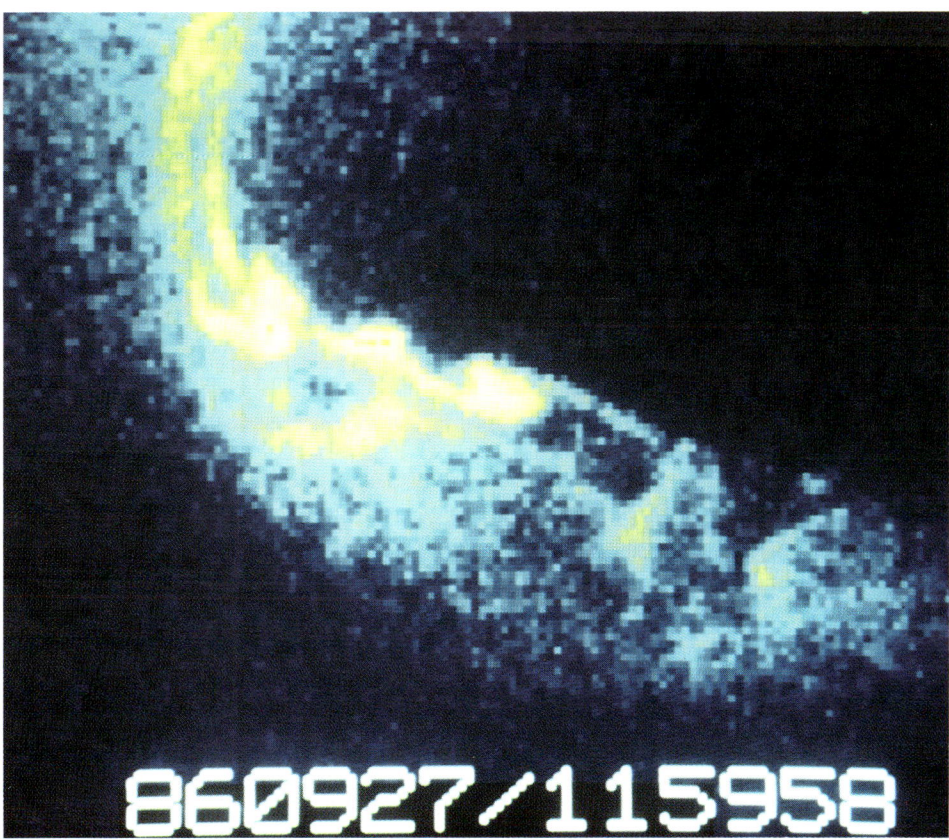

Fig. 7 Viking image taken during a period of strong substorm activity which took place over an interval of strong southward IMF. Three separate surge forms are evident in the center of the frame. The one on the far right had existed for some minutes before this frame and was relatively stationary. The one in the center and the one on the far left developed approximately at the same time. Each surge form represents the western portion of a substorm wedgelet as described in Figure 1.

boundary layer dynamics model (cf. Rostoker and Eastman, 1987). While it is not possible within this short review to discuss the two frameworks in detail, I would like to briefly outline their salient features before commenting on the nature of the magnetotail observations these frameworks are constrained by.

The NENL model is normally shown in the noon-midnight meridian plane as seen in Figure 10. One imagines the sequence of activity to be initiated by an increase in energy flow into the magnetosphere through enhanced frontside merging. The magnetosphere initially has one neutral line in the distant tail as shown in the top left panel. As enhanced frontside merging proceeds, tail reconnection commences at a new neutral line formed close to the Earth. Initially reconnection at the new neutral line is on closed field lines; however, after some time lobe field lines begin to reconnect. This latter time marks the onset of the substorm expansive phase. As reconnection of lobe field lines proceeds, the plasma sheet tailward of the NENL is ejected from the tail as a consequence of the tension in the freshly reconnected lobe field lines. The body of ejected plasma is called a plasmoid. For some time after the plasmoid begins to travel downtail, the NENL remains close to the Earth with reconnection of lobe field lines continuing. Then, for reasons as yet not understood, the NENL begins to move downtail and the near Earth plasma sheet thickens as the substorm enters the recovery phase. Because the NENL model was originally designed to explain only the large monolithic current wedge thought to represent the expansive phase current system (cf. Figure 3), the standard model of the substorm involves only a single plasmoid. The piecewise evolution of the wedge described earlier in this paper demands that there be more than one plasmoid and that all plasmoids be azimuthally localized. Another consequence of substorm expansive phase evolution as detected in the auroral ionosphere is that there be a sequential release of the plasmoid segments as each new expansive phase intensification is initiated. Perhaps the most puzzling thing about the NENL model in terms of meeting observational constraints is how it

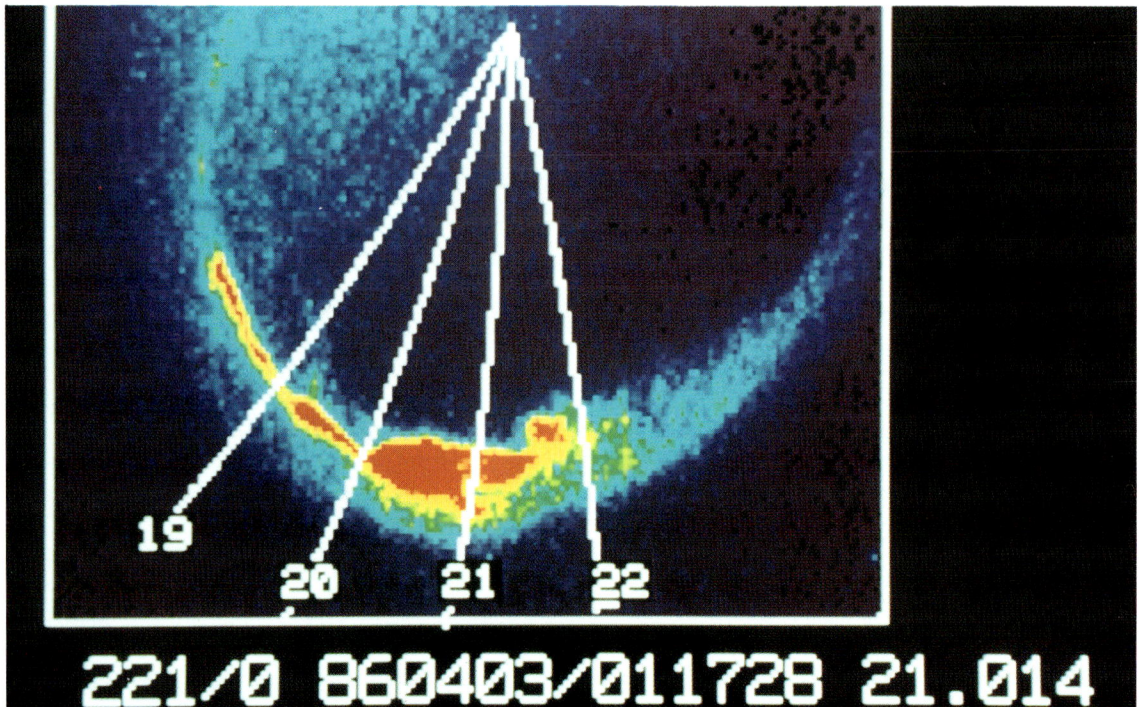

Fig. 8   Viking image showing surge periodicity during an episode of strong expansive phase activity on September 27, 1986 at 1159:58 UT. The three surges seen on this frame typify the wavelike character which can be attributed to the processes which lead to surge formation. The most recently formed surge is the one at the latest local time (~ 19 MLT), showing that each new surge does not necessarily form to the west of the previous one.

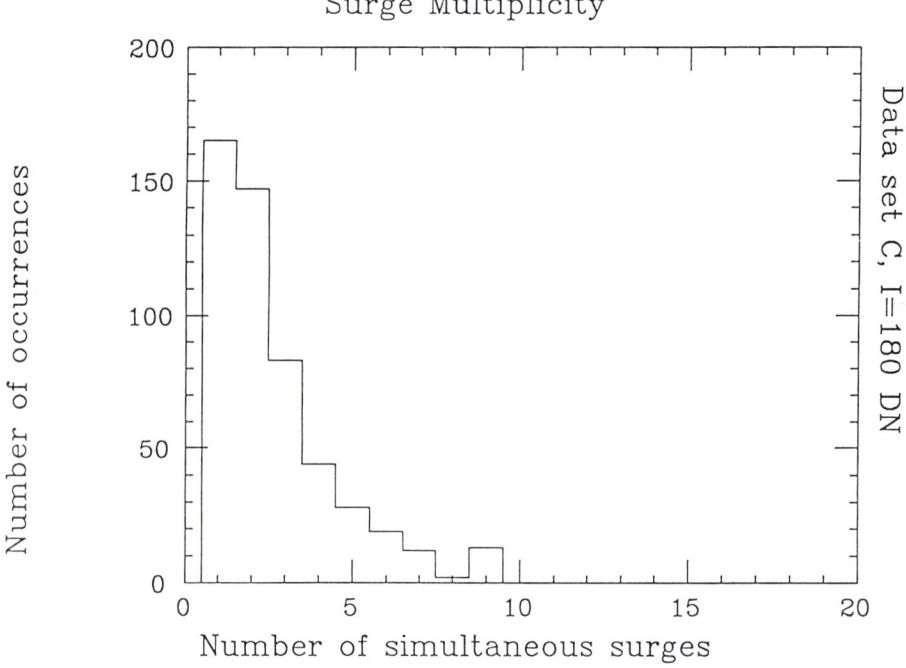

Fig. 9   Occurrence frequency for auroral surges in terms of the number of times that they are seen singly, in pairs, three at a time etc. While the most probable situation is for a single surge to be seen compared to the other possibilities, if one surge exists at all there is a greater than 50% chance that more than one surge will exist at any instant along the evening sector auroral oval (after Kidd and Rostoker, 1991).

relates to the directly driven component of substorm activity. As was mentioned in the introduction to this paper, substorm surge activity originates in a region of the ionosphere co-located with the Region 1 upward field-aligned current of the directly driven system close to or at the Harang discontinuity. Since the directly driven system can become rather large well before the onset of significant expansive phase activity (cf. Figure 4), this suggests that the source of energy for Region 1 currents in the magnetotail is providing that energy prior to the onset of lobe field line reconnection associated with substorm (expansive phase) onset in the NENL model. This question accentuates the need to find some scheme of mapping the Harang discontinuity into the magnetotail, because it is across the latitudinal extent of gradient in the meridional component of the ionospheric electric field that Region 1 currents are found and within which the expansive phase breakup arc is located.

The boundary layer dynamics (BLD) framework for substorms introduced by *Rostoker and Eastman* [1987] is best shown in a projection on the plane of the neutral sheet in Figure 11. The stimulus for this model came from the hydrodynamic analogue for flow around an obstacle (cf. Rostoker, 1984) and draws its strength from explaining the detailed character of substorm activity defined by the projections on the ionosphere and topside ionosphere of the substorm process in the magnetosphere. Like the NENL model, enhancement of energy flow into the magnetosphere (perhaps through frontside merging) leads to a burst of magnetic field line reconnection in the magnetotail at some later time. However, even as the plasma sheet thins during the storage of energy in the magnetotail (a process common to both the NENL and BLD frameworks), the BLD calls on increased reconnection at the distant neutral line to produce enhanced earthward convective flow consistent with the buildup of space charge at the velocity shear zones between the low latitude boundary layer (LLBL) and central plasma sheet (CPS). The increase in that space charge also leads to enhanced Region 1 current flow as can be envisioned on inspection of the three dimensional currents in Figure 11 which couple the magnetosphere to the ionosphere. While not all Region 1 currents necessarily close in those circuits, at least part of those currents can be driven through the extraction of the kinetic energy of the (primarily Earthward) bulk flow of plasma in the CPS through MHD generator processes (cf. Rostoker and Boström, 1976). A sudden enhancement in reconnection at the distant neutral line can lead to enhanced velocity shear at the

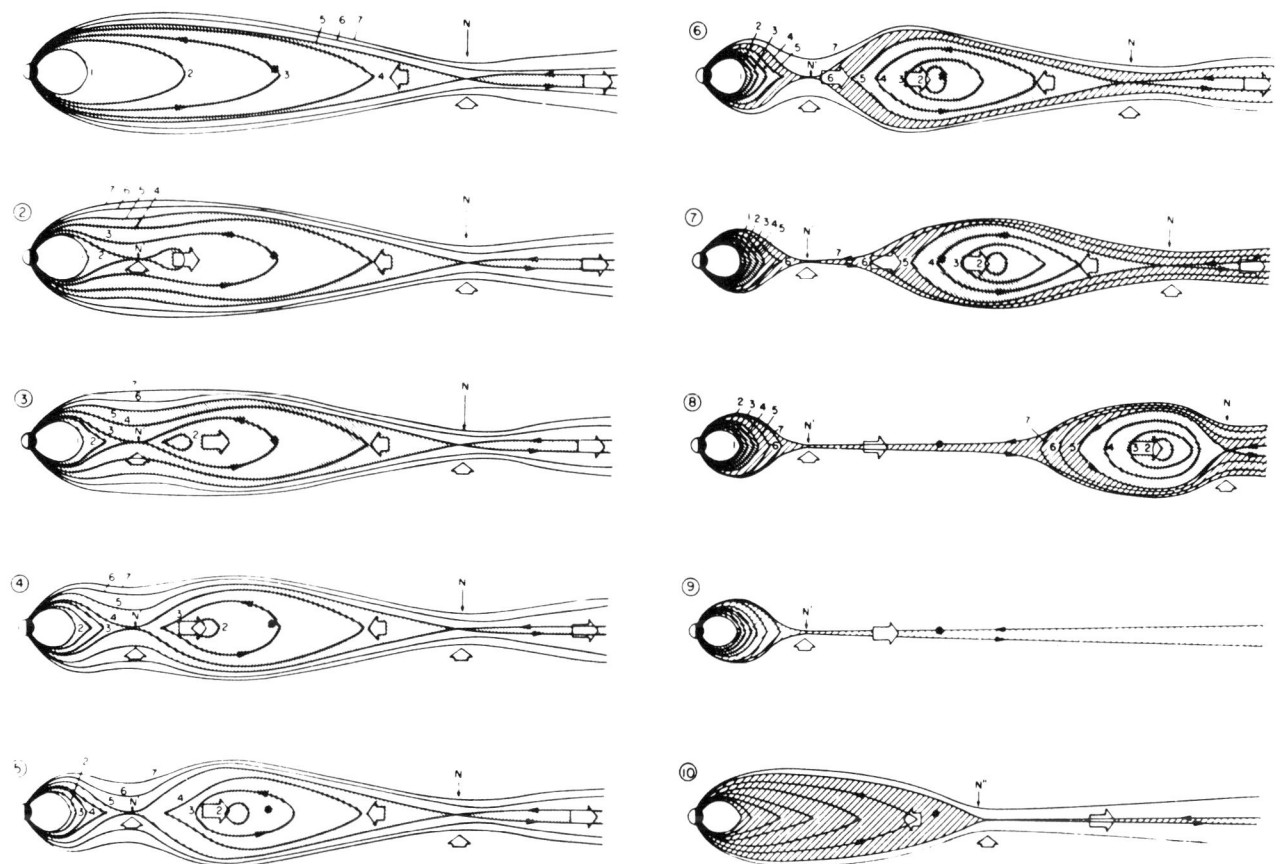

Fig. 10 Development of a plasmoid in the near-Earth neutral line model of substorms (after Hones, 1984).

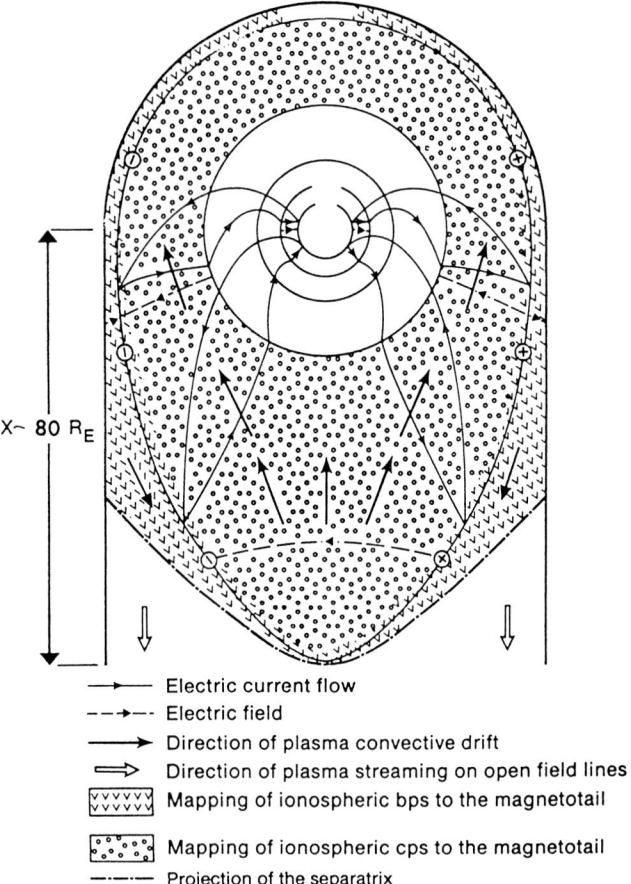

Fig. 11 The configuration of particles and fields in the magnetotail in the framework of the boundary layer dynamics model (after Rostoker [1991b, *Auroral Physics*, p. 266, reprinted with permission from Cambridge UP]). This cartoon shows the situation for purely driven activity.

LLBL/CPS interface and the growth of a Kelvin-Helmholtz instability as shown in Figure 12. The field-aligned potential drop generated in the wave growth regions (cf. Thompson, 1983) leads to the growth of spatially localized upward field-aligned current involving the precipitation of energetic electrons into the ionosphere (leading to the appearance of auroral surges). The BLD model therefore has the potential of explaining the multiplicity of auroral surges as well as pointing to a specific physical mechanism in the magnetotail for provision of energy for surge formation. A primary question surrounding the BLD model centers on the question of whether the Earthward CPS flows are sufficiently strong to permit the Kelvin-Helmholtz instability to grow at a rate commensurate with surge evolution (viz. a growth time of < 5 min.). The required Earthward flow velocities of > 150 km/s at the LLBL/CPS interface have yet to be measured at the relevant distances behind the Earth (X > 20 Re) and enhancements in those flows must yet be shown to be associated with surge growth. However, this is because properly instrumented satellites have not yet monitored those regions of space with those goals in mind. Nonetheless, the lack of information about this important demand which the BLD model must meet leaves an unanswered question insofar as the validity of the model is concerned.

I will conclude this commentary on the two substorm frameworks by asking the reader to consider the following points regarding the constraints which those frameworks must satisfy. First of all, I would like to point out that the stimulus for the NENL model of substorms came from concurrent observations of anti-sunward motion of a measurable portion of the magnetotail particle population and the appearance of a southward directed component of the magnetic field at some position in the magnetotail. Many of these measurements involved only high energy particles and failed to examine the full plasma population to see if it was the net flow which was tailward or only some component of the energy spectrum which exhibited that characteristic. It has only been since the ISEE

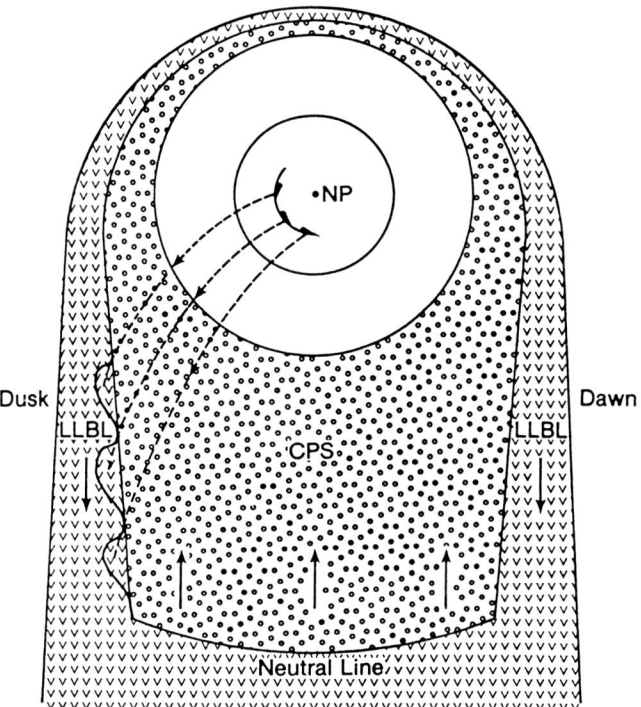

Fig. 12 Development of substorm expansive phase surges in the framework of the boundary layer dynamics model (after Kidd and Rostoker, 1991). Enhanced Earthward CPS convective flow leads to the growth of a Kelvin-Helmholtz instability at the interface between the LLBL and CPS resulting in field-aligned current flow out of the ionosphere into the region of wave growth. The surges may develop much closer to the Earth than the neutral line from which the enhanced flow emanates, depending on the balance between the growth and damping terms for the instability.

mission (cf. Frank et al., 1979) that it has been possible to measure the distribution functions of the plasma sheet plasma and therefore to be able to infer the magnitude and direction of the bulk plasma flow normal to the background magnetic field, viz. to separate field-aligned flows from convective flows. Now that this important capability exists, along with the ability to measure electric fields directly in tenuous plasmas, it is useful to read what is said about the location of neutral lines in the modern literature.

"The Bz (north-south) field component in dipole coordinates was always positive within 0.5 Re of the equatorial current sheet, indicating that neutral lines were never seen inside of 8.8 Re." (after Fairfield et al., 1987)

"The distribution of flow directions strongly favors sunward flow for velocities above 300 km/s indicating that a near-earth neutral line is rarely, if ever, located inside of X_{GSM} = -19 Re." (after Baumjohann et al., 1989)

"A survey has been made of all the electric field data from the University of California, Berkeley, double probe experiments on ISEE-1 (apogee ~ 22 Re) during 1980 when the satellite was in the magnetotail. ...These data suggest that the near earth x-line usually forms tailward of X_{GSM} ≈ -20 Re". (after Cattell and Mozer, 1984)

It is probably necessary to re-examine the interpretations of the satellite data acquired prior to the ISEE mission which formed the basis for the postulation of a near-Earth neutral line earthward of ~ - 20 Re (cf. Hones, 1976, 1977, 1979). One must recognize that, although modern satellite magnetotail data have been interpreted in the framework of the NENL model, the observational basis for the NENL model comes from data acquired prior to 1976 when the hypothesis was first put forward in its modern form by *Hones* [1976]. Yet some modern data, as cited above, put into doubt the interpretation of the older data in terms of supporting the existence of a near-Earth neutral line. One possible basis for resolving the dilemma comes with the proposition that there is more than one way in which a combination of southward tail field and anti-sunward flows can be interpreted. For example, in the BLD framework one can explain this combination of observations in terms of the edge effect of spatially localized field-aligned currents in the plasma sheet boundary layer (where both sunward and anti-sunward plasma flows are regularly observed). Another possible source of negative Bz in the tail are transient pulsation events which accompany substorm expansive phase onset and intensifications (viz. Pi 2 pulsations as detected on the ground). One thing is certain. The contentions cited above are completely inconsistent with the NENL hypothesis if one places the NENL typically inside ~ -20 Re and this represents a constraint which cannot be ignored.

CONCLUSIONS

In the latter part of this paper, I have tried to outline the nature of two frameworks which are presently being considered as candidates in which two substorm processes in the magnetotail may be described. I should emphasize that these are not the only two frameworks in which substorm observations are being interpreted. For example, *Lui et al.* [1990] consider the substorm expansive wedge to originate from the diversion into and out of the ionosphere of crosstail current at the inner edge of the plasma sheet when accentuated plasma sheet thinning leads to the disruption of that crosstail current. In their concept, the region of current diversion is not, however, associated with a near-Earth neutral line. That there is more than one candidate framework should come as no surprise. The paucity of observations in the magnetotail where two or more satellites simultaneously provide relevant data makes the development of a unique framework extremely difficult if not impossible. All one can do at the present time is to try to constrain the candidate frameworks by appealing to all relevant observations, be they *in situ* or through remote sensing of the ionospheric end of the volume of space in which the substorm processes are going on. The NENL model, once it was introduced, was explored primarily through simulating the magnetotail plasma and field environment and trying to establish whether or not reconnection, once initiated, would produce plasma and field signatures which were detected in the data. Little attention was paid to the wealth of signatures found through remote sensing of the ionosphere and topside ionosphere in terms of posing constraints which had to be met. The BLD model, in contrast, draws on the ionospheric and topside ionospheric observations to develop constraints which must be met and does not explore comprehensively the details of the tail observations which, it is contended, cannot in their present state yield a set of constraints which favor one model over another.

The most serious problem faced by substorm researchers who wish to utilize ionospheric observations to study substorm physics is how one can map from the ionosphere into the regions of the magnetosphere where the important energy transport and conversion processes are taking place. At the present time, the only magnetic field model available to achieve this mapping is the one designed by *Tsyganenko* [1987, 1989]. However, this model is not yet fully developed, lacking both realistic magnetopause currents and field-aligned currents. Nonetheless, before the ground observations of substorm phenomena will be able to play a deciding role in determining which, if any, of the existing frameworks for describing substorm phenomena are correct, the magnetic field model will have to be refined to the level where researchers are fully confident in its capabilities. Until then there will continue to be disagreements about the physical processes which describe the phenomenon of the magnetospheric substorm.

Acknowledgements. Thanks are due to World Data Center C2 in Kyoto for supplying the AE indices used in Figure 3. This research was supported by the Natural Sciences and Engineering Research Council of Canada.

REFERENCES

Akasofu, S.-I. The development of the auroral substorm, *Planet. Space Sci.*, **12**, 273, 1964.

Alfvén, H. A Theory of Magnetic Storms and of the Aurorae, Kungl. Sv. Vetensk.-Akad. Handlingar III, 18, **3**, Stockholm, 1939.

Baumjohann, W., R.J. Pellinen, H.J. Opgenoorth and E. Nielsen, Joint two-dimensional observations of ground magnetic and ionospheric electric fields associated with auroral zone currents: current systems associated with the local auroral break-ups, 431-447, **29**, 1981.

Baumjohann, W. Ionospheric and field-aligned current systems in the auroral zone: a concise review, in Advances in Space Research, **2**, 55, 1983.

Baumjohann, W., G. Paschmann and C.A. Cattell. Average plasma properties in the central plasma sheet, *J. Geophys. Res.*, **94**, 6597, 1989.

Birkeland, K. The Norwegian Aurora Polaris Expedition 1902-1903, **1**, 1st section, Aschhoug and Co., Christiania, 1908.

Cattell, C.A. and F.S. Mozer. Substorm electric fields in the earth's magnetotail, in Magnetic Reconnection in Space and Laboratory Plasmas, ed. by E.W. Hones, Jr., 208, *Amer. Geophys. Union*, Washington, DC, 1984.

Chapman, S. and V.C.A. Ferraro. A new theory of magnetic storms, *Terr. Mag. Atmos. Elect.* **36**, 77, 1931.

Clauer, C.R., R.L. McPherron and C. Searls. Solar wind control of the low-latitude asymmetric magnetic disturbance field, *J. Geophys. Res.*, **88**, 2123, 1983.

De Groot-Hedlin, C. and G. Rostoker, Magnetic signatures of precursors to substorm expansive phase onset, *J. Geophys. Res.*, **92**, 5845, 1987.

Fairfield, D.H., M.H. Acuna, L.J. Zanetti and T.A. Potemra. The magnetic field of the equatorial magnetotail: AMPTE/CCE observations at R 8.8 RE, *J. Geophys. Res.*, **92**, 7432, 1987.

Frank, L.A., K.L. Ackerson, R.J. DeCoster and B.G. Burek. Three dimensional plasma measurements within the Earth's magnetosphere, in Advances in Magnetospheric Physics with GEOS-1 and ISEE, ed. by K. Knott, A. Durney and K. Ogilvie, 419, D. Reidel Publ. Co., Hingham, MA, 1979.

Iijima, T. and T.A. Potemra. The amplitude distribution of field aligned currents at northern high latitudes observed by Triad, *J. Geophys. Res.*, **81**, 2165, 1976.

Hones, E.W., Jr. The magnetotail: its generation and dissipation, in Physics of Solar Planetary Environments, ed. by D.J. Williams, 558, *Amer. Geophys. Union*, 1976.

Hones, E.W., Jr., Substorm processes in the magnetotail: Comments on 'On Hot Tenuous Plasmas, Fireballs, and Boundary Layers in the Earth's Magnetotail', by L.A. Frank, K.L. Ackerson and R.P. Lepping, *J. Geophys. Res.*, **82**, 5633, 1977.

Hones, E.W., Jr. Transient phenomena in the magnetotail and their relation to substorms, *Space Sci. Rev.*, **23**, 393, 1979.

Hones, E.W., Jr. Plasma sheet behaviour during substorms, in Magnetic Reconnection in Space and Laboratory Plasmas, ed. by E.W. Hones, Jr., 178, *Amer. Geophys. Union*, Monograph 30, Washington, D.C., 1984.

Kidd, S.R., and G. Rostoker. Distribution of auroral surges in the evening sector, *J. Geophys. Res.*, **96**, 5697, 1991.

Kisabeth, J.L. and G. Rostoker. The expansive phase of magnetospheric substorms 1. Development of the auroral electrojets and auroral arc configuration during a substorm. *J. Geophys. Res.*, **79**, 972, 1974.

Korotova, G., G. Rostoker and M. Connors. Evolution of the component current systems of the expansive phase of a magnetospheric substorm, manuscript in preparation, 1991.

Lui, A.T.Y., A. Mankofsky, C.-L. Chang, K. Papadopoulos and C.S. Wu. A current disruption mechanism of the neutral sheet: a possible trigger for substorm expansions, *Geophys. Res. Lett.*, **17**, 745, 1990.

Rostoker, G. Triggering of expansive phase intensifications of magnetospheric substorms by northward turnings of the interplanetary magnetic field, *J. Geophys. Res.*, **88**, 6981, 1983.

Rostoker, G. Implications of the hydrodynamic analogue for the solar-terrestrial interaction and the mapping of high latitude convection patterns into the magnetotail, *Geophys. Res. Lett.*, **11**, 251, 1984.

Rostoker, G. Auroral signatures of magnetospheric substorms and constraints which they provide for substorm theories, *J. Geomag. Geoelect.*, (in press) 1991a.

Rostoker, G. Overview of observations and models of auroral substorms, in Auroral Physics, ed. by C.-I. Meng, M.J. Rycroft and L.A. Frank, 257, Cambridge, UP, 1991b.

Rostoker, G. and R. Boström. A mechanism for driving the gross Birkeland current configuration in the auroral oval, *J. Geophys. Res.*, **81**, 235, 1976.

Rostoker, G., S.-I. Akasofu, J. Foster, R.A. Greenwald, Y. Kamide, K. Kawasaki, A.T.Y. Lui, R.L. McPherron and C.T. Russell. Magnetospheric substorms - definition and signatures, *J. Geophys. Res.*, **85**, 1663, 1980.

Rostoker, G. and T.E. Eastman. A boundary layer model for magnetospheric substorms, *J. Geophys. Res.*, **92**, 12,187, 1987.

Siscoe, G.L. What's in the name 'Magnetospheric Substorm', *J. Geophys. Res.*, **85**, 1643, 1980.

Thompson, N.B. Parallel electric fields and shear instabilities, *J. Geophys. Res.*, **88**, 4805, 1983.

Tighe, W.G. and G. Rostoker. Characteristics of westward travelling surges during magnetospheric substorms, *J. Geophys. Res.*, **50**, 51, 1981.

Tsyganenko, N.A. Global quantitative models of the geomagnetic field in the cislunar magnetosphere for different disturbance levels, *Planet. Space Sci.*, **35**, 1347, 1987.

Tsyganenko, N.A. A magnetospheric magnetic field model with a warped tail current sheet, *Planet. Space Sci.*, **37**, 5, 1989.

Wiens, R.G. and G. Rostoker. Characteristics of the development of the westward electrojet during the expansive phase of magnetospheric substorms, *J. Geophys. Res.*, **80**, 2109, 1975.

Zmuda, A.J. and J.C. Armstrong. The diurnal flow of field-aligned currents, *J. Geophys. Res.*, **79**, 4611, 1974.

Synthesizing a Global Model of Substorms

J. R. KAN

Geophysical Institute, University of Alaska Fairbanks, Fairbanks, AK 99775

A global model of substorms is synthesized by connecting several basic elements of the substorm phenomenon advanced in recent years. The proposed model emphasizes the nonlinear two-way interaction between the magnetosphere and ionosphere. Substorm features in the ionosphere can be explained as a result of the ionospheric response to an enhanced magnetospheric convection driven by an increased solar wind-magnetosphere interaction. On the other hand, most substorm features in the magnetosphere can be understood in terms of magnetotail reconfigurations caused by an enhanced dayside reconnection during the substorm growth phase and the feedback effect from the ionosphere during the substorm expansion phase. Specifically, the tail-like reconfiguration of the near-earth plasma sheet during the growth phase can result when the magnetospheric convection is almost fully enhanced by an increase in the dayside reconnection. The substorm expansion onset can result from the ionosphere respond almost fully to the enhanced magnetospheric convection, which occurs when the upward field-aligned currents exceeds ~1 $\mu A/m^2$ as demonstrated in the M-I coupling model by *Kan et al.* [1988]. Closing the substorm enhanced field-aligned currents in the plasma sheet can disrupt the cross-tail current to result in the dipolarization of the plasma sheet. The electric field induced by the current disruption can launch a shock wave propagating earthward to drive the plasma injection right after the substorm expansion onset. A near-earth X line can form tailward of the dipolarization region after the onset of an intense substorm. A timing sequence of substorm events has been deduced from the proposed model to identify the cause-and-effect relationships of substorm events. Perhaps, it could also stimulate future observations.

INTRODUCTION

Magnetospheric substorm is a global response of the magnetosphere-ionosphere (M-I) system to an increased interaction with the solar wind. The concept of magnetospheric substorm was first proposed a quarter century ago in a classic paper by *Akasofu* [1964]. The magnetospheric substorm is not a phenomenon of the plasma sheet dynamics alone, nor is it exclusively a phenomenon of auroral activities in the ionosphere. In this regard, all the published substorm models are partial models with biased emphasis either on the plasma sheet dynamics or on the ionospheric dynamics as has been discussed by *Kan* [1990a].

In this paper we propose a global model of substorms in which the electrodynamics of the ionosphere and of the plasma sheet are coupled. Substorms can be viewed as a consequence of nonlinear two-way interactions between the magnetosphere and ionosphere. The global model presented in this paper synthesizes several of our ideas on substorms developed in recent years. Emphasis will be placed in connecting several published elements of the M-I coupling model of substorms. The proposed model is still in a conceptual stage of development.

SIGNATURES OF SUBSTORMS

The signatures of substorms in the ionosphere are well established in comparison with the signatures in the magnetosphere. The spatial and temporal connections between these two sets of signatures are among the most-desired missing information of substorms today.

Ionospheric Signatures of Substorms

(1) Substorm expansion phase onset occurs most typically when the equatorward-most auroral arc (near the poleward edge of the diffuse aurora) suddenly brightens up and subsequently undergoes both poleward and westward expansions [*Akasofu*, 1964]. The expansive phase is a period of about 30 minutes or more during which the brightened auroral forms expand poleward and westward; it ends after the auroral activity reaches its maximum epoch.

(2) Bright auroral arcs are located where intense upward field-aligned currents greater than ~1 $\mu A/m^2$ [e.g., *Kamide and Rostoker*, 1977], carried by precipitating electrons accelerated by kilovolt potential drops along magnetic field lines [e.g., *Arnoldy*, 1974].

(3) The auroral oval expands during the growth phase of a substorm. The growth phase [*McPherron*, 1972] is a period of ~ 30 to 60 minutes before the onset of the substorm expansive phase.

(4) The potential drop across the polar cap increases with the auroral activity as the interplanetary magnetic field turns southward [e.g., *Reiff and Luhmann*, 1986; *Doyle and Burke*, 1983].

(5) The region 1 and 2 field-aligned currents [*Iijima and Potemra*, 1976] are permanent features of the magnetospheric current system. During substorms, field-aligned currents are intensified locally and asymmetrically in the evening-midnight sector. The upward field-aligned currents are most intense near the head of the westward traveling surge [*Baumjohann et al.*, 1981; *Opgenoorth et al.*, 1983] in the evening sector near the Harang discontinuity [*Heppner*, 1977].

(6) The westward and eastward auroral electrojet currents in the ionosphere are intensified during substorms as measured by the AE index that correlates well with the polar cap potential drop [*Ahn et al.*, 1984].

Magnetospheric Signatures of Substorms

(7) The magnetic field becomes more tail-like ~ 6.6 R_E radial distance in the plasma sheet [*Kokubun and McPherron*, 1981] during the growth phase of substorms.

(8) Earthward injection of energetic electrons and ions (> 30 keV) occurs near ~6.6 R_E in the plasma sheet. The magnetic field collapses to a more dipole-like configuration near the substorm expansion onset [e.g., *McIlwain*, 1974; *Kokubun and McPherron*, 1981].

(9) Closure of the intense substorm field-aligned current is mostly in the near-earth plasma sheet [*Galperin and Feldstein*, 1991]. The resulting substorm current system is commonly known as the substorm current wedge [*McPherron et al.*, 1973].

(10) A near-earth X line may form earthward of the preexisting X line in the distant tail, leading to the formation of a plasmoid streaming tailward [e.g., *Hones*, 1979]. The radial distance of the near-earth X line is still uncertain, probably within ~20 R_E at times. The timing of the near-earth X line formation is also unresolved.

(11) Ion temperature increases significantly with the AE index in the central plasma sheet, while it increases much less with AE in the plasma sheet boundary layer (PSBL) under disturbed conditions [*Huang and Frank*, 1986; *Baumjohann et al.*, 1988, 1989a].

(12) High-speed bursty earthward plasma flows (~ 500 to 800 km/s) have been observed in the central plasma sheet inside ~20 R_E; the occurrence frequency increases with AE [*Baumjohann et al.*, 1989b]. *Lui et al.* [1977] observed high-speed earthward plasma flow in the PSBL when the plasma sheet recovers during the recovery phase of substorms. The recovery phase begins at the end of the expansion phase.

MAGNETOSPHERE IONOSPHERE COUPLING PROCESS
ON THE ALFVÉN TIME SCALE

Ionospheric response to an enhanced magnetospheric convection evolves on the Alfvén wave bounce time scale as has been described in the M-I coupling model constructed by *Kan and Sun* [1985]. In the M-I coupling model, the ionospheric Hall and Pedersen conductances are enhanced depending on the upward field-aligned current density. This model was further developed [*Kan et al.*, 1988; *Zhu and Kan*, 1990] into an M-I coupling model of substorms to include the effect of the ionospheric recombination time scale. A briefly summary of the M-I coupling model of substorms is given below.

Reflection of Alfvén waves from the ionosphere depends on the distribution of the ionospheric Hall and Pedersen conductances. The nonuniform Hall conductance can produce the characteristic rotation on the convection electric field, which is a prominent feature observed in the Harang discontinuity near the head of the westward traveling surge [*Heppner and Maynard*, 1987]. By matching the field-aligned current carried by the incident and reflected Alfvén waves to the divergence of the ionospheric current, the reflected wave can be calculated, i.e.,

$$\Sigma_A \nabla \cdot (\mathbf{E}^i - \mathbf{E}^r) = \nabla \cdot (\Sigma_P \mathbf{E}_i + \Sigma_H \mathbf{b}_o \times \mathbf{E}_i) \quad (1)$$

where $\mathbf{E}_i$ (= $\mathbf{E}^i + \mathbf{E}^r$) is the ionospheric electric field, Σ_P and Σ_H are the ionospheric Pedersen and Hall conductances, $\Sigma_A = (\mu_o V_A)^{-1}$ is the conductance of the Alfvén wave, the superscripts "r" and "i" denote quantities of the reflected and incident waves, respectively, and $\mathbf{b}_o$ is the unit vector of the magnetic field in the ionosphere. The reflected wave electric field can be determined by solving $\mathbf{E}^r$ from (1). The nonuniform Hall conductance is responsible for rotating the reflected wave electric field by an angle from the incident wave field.

The reflection of Alfvén waves at magnetospheric boundaries depends on whether the field lines are open or closed [*Kan and Sun*, 1985]. On open field lines the solar wind inertia is sufficiently large so that the solar wind flow is approximately unchanged by the loading effect of the Alfvén wave incident on the magnetopause. In other words, the electric field on open field lines at the magnetopause can be maintained by the solar wind. Therefore, the incident wave electric field must be canceled by the reflected wave field which then leads to the magnetospheric reflection coefficient $R_m(E) = E^r / E^i \approx -1$ on open field lines at the magnetopause. This is equivalent to an input conductance of the solar wind dynamo Σ_D (on open field lines) which is much greater than the Alfvén wave conductance Σ_A. Recently, *Cao and Kan* [1990] showed, by means of a theoretical model, that the reflection coefficient of Alfvén wave at the open magnetopause a complex parameter. It has a real part almost equals to -1, and a small imaginary part due to the loss through conversion of the incident Alfvén wave to the fast and slow modes at the magnetopause.

Reflection of Alfvén wave from the plasma sheet on closed field lines is more complicated. The convection in the plasma sheet is maintained by the inertia of the plasma flow and the $\mathbf{J} \times \mathbf{B}$ force of the cross-tail current. The inertia of the $\mathbf{E} \times \mathbf{B}$ convection in the plasma sheet is limited since the total mass of plasma in the plasma sheet is rather limited.

Consider a pair of identical but oppositely propagating Alfvén waves incident on the plasma sheet from opposite hemispheres. Let $_nR_m(E)$ and $_nT_m(E)$ denote the reflection and transmission coefficients of an Alfvén wave incident on the plasma sheet from the northern hemisphere; $_sR_m(E)$ and $_sT_m(E)$ denote those from the southern hemisphere. The reflection and transmission coefficients for each incident wave are related by $_nT_m(E) = 1 + {_nR_m(E)}$. Each encounter between the pair of incident Alfvén waves and the plasma sheet leads to a reflected wave and a transmitted wave in each hemisphere. The resulting reflected and transmitted waves can be viewed collectively as an equivalent "reflected" wave in each hemisphere. The effective "reflection" coefficient in each hemisphere can be written as

$$R_{mN}(E) = {_nR_m(E)} + {_sT_m(E)} \quad \text{in the northern hemisphere}$$
$$R_{mS}(E) = {_sR_m(E)} + {_nT_m(E)} \quad \text{in the southern hemisphere} \quad (2)$$

where $R_{mN}(E)$ and $R_{mS}(E)$ are respectively the effective "reflection" coefficients in the northern and southern hemispheres of the plasma sheet. Note the important difference in the definition between the effective "reflection" coefficient in the northern hemisphere $R_{mN}(E)$, and the actual reflection coefficient of an incident wave from the northern hemisphere $_nR_m(E)$ as discussed earlier. If the plasma sheet is assumed symmetric in the two hemispheres, the reflection and transmission coefficients are also symmetric in the two hemispheres.

The loading effect of an incident Alfvén wave can be expected to modify the electric field in the plasma sheet. In the limiting case where the inertia including the $\mathbf{J} \times \mathbf{B}$ force in the plasma sheet is negligible, such as near the inner-edge of the plasma sheet, the convection can be expected to be fully modified by the incident Alfvén wave. This means the incident Alfvén wave from one hemisphere can propagate across the plasma sheet to the other hemisphere almost without reflection, i.e., $R_m(E) \approx 0$, so that $T_m(E) = 1 + R_m(E) \approx +1$. Thus, the equivalent "reflection" coefficient at a symmetric plasma sheet due to simultaneous incidence of Alfvén waves from opposite hemispheres is $R'_m(E) = R_m(E) + T_m(E) \approx +1$ according to (2). On the other hand, if the inertia and the $\mathbf{J} \times \mathbf{B}$ force in the plasma sheet are sufficiently large, $R_m(E)$ may be negative, say $R_m(E) = -0.6$ and $T_m(E) = +0.4$. The equivalent "reflection" coefficient due to simultaneous incidence of Alfvén waves from opposite hemispheres is $R'_m(E) = R_m(E) + T_m(E) = -0.2$. On the basis of the above discussion, the reflection coefficient for the outward-traveling Alfvén waves at magnetospheric boundaries can be summarized by

$$R_m \geq -1 \quad \text{on open field lines}$$
$$-1 < R_m < +1 \quad \text{on closed field lines} \quad (3)$$

It may be noted that $R_m(E) = -1$ corresponds to an idealized constant voltage source while $R_m(E) = 1$ corresponds to a constant current source as has been shown by *Kan and Sun* [1985]. The value of $R_m(E)$ on closed field lines can be expected to increase from greater than -1 to $+1$ as one moves from the distant tail toward the inner edge of the plasma sheet. Further study of the reflection of Alfvén waves from the plasma sheet is required.

Figure 1 shows the distribution of field-aligned currents calculated from the M-I coupling model near the substorm expansion onset at $T = 28$ minutes after the magnetospheric convection is enhanced [*Kan et al.*, 1988; *Zhu and Kan*, 1990]. Brightening of the auroral arc for substorm onset occurs at the location where the upward field-aligned current density is most intense near the poleward edge of the diffuse auroral conductance belt in the evening sector, as shown in Figure 1. This agrees with the observational definition of

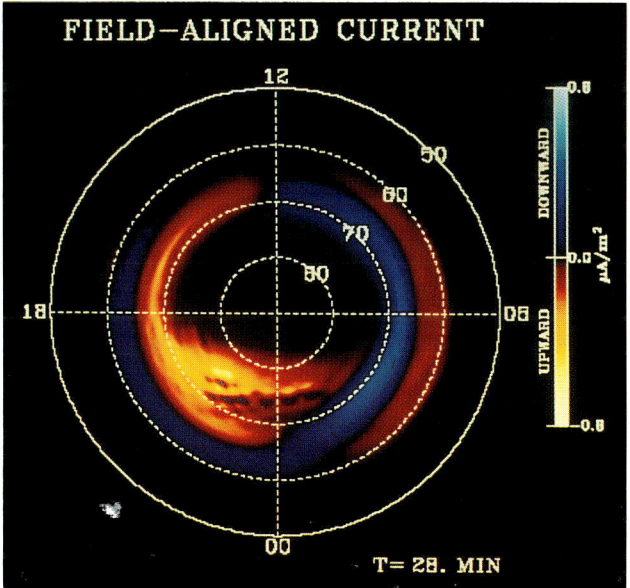

Fig. 1. Distribution of field-aligned currents calcuated from the M-I coupling model of substorms near the substorm expansion onset at $T = 28$ minutes after the magnetospheric convection has been enhanced due to a southward turning of the interplanetary field. Red is for the upward field-aligned current flowing away from the ionosphere; blue is for the downward field-aligned current. The current density increases with the brightness of each color.

substorm onset [*Akasofu*, 1964], keeping in mind that the diffuse auroral conductance is produced by the diffuse auroral precipitation. The field-aligned currents resulting from the M-I coupling process, must be closed in the magnetosphere. The closure of field-aligned currents will be discussed in a later section.

Main Results of the M-I Coupling Model

(a) *Substorm Intensifications on the Alfvén Bounce Time Scale.* The ionospheric convection, the electrojet currents, the field-aligned currents, the Hall and Pedersen conductances are all enhanced on the Alfvén bounce time scale of 2 to 4 minutes. The enhancement of field-aligned currents is in response to an enhanced magnetospheric convection. The temporal behavior of these ionospheric quantities can either increase with irregular oscillations on a 2 to 4-minute time scale or increase in a stepwise manner on a 2 to 4-minute time scale. These temporal behaviors can be associated with Pi 1 and Pi 2 pulsations observed near the onset of substorms [e.g., *Yumoto*, 1986].

The 2 to 4-minute Alfvén bounce time scale of the M-I coupling model can also be related to the 1-3 minute intensification of substorms [*Sergeev, et al.*, 1986b]. Such impulsive intensification of substorms on the ground may have its counterpart in the plasma sheet [*Sergeev et al.*, 1986a; *Baumjohann et al.*, 1989a]. Temporal variation of auroral arcs on a few minutes' time scales have been observed before the substorm expansion onset [*Morse and Romick*, 1982]. Further studies of substorm related phenomena on the Alfvén bounce time scale using ground-based optical and radar observations are underway.

(b) *Divergence of the Hall Current.* The matching between the convection reversal region and the poleward gradient of the diffuse auroral conductance in the evening sector required for the divergence of the Hall current to be upward to reinforce the upward divergence of the Pedersen current [*Kan et al.*, 1988]. This explains why the most intense upward field-aligned current should occur in the evening sector near the Harang discontinuity region where the equatorward-most arc brightens to mark the onset of the substorm expansion phase. This result is consistent with the STARE radar results on the observation of substorm expansion onset near the Harang discontinuity [E. Nielsen, private communication, July 1988].

(c) *Substorm Field-Aligned Currents Initiated From the Ionosphere.* Field-aligned currents are enhanced as a consequence of ionospheric response to an enhanced magnetospheric convection. In other words, the magnetosphere imposes a convection electric field on the ionosphere without providing adequate field-aligned currents to drive the ionosphere. This is evident by examine the field-aligned current density and the electric field associated with the incoming Alfvén wave toward the ionosphere. The ionosphere react to the imposed electric field by demanding more field-aligned current. This is evident by examine the field-aligned current and electric field associated with the reflected Alfvén wave from the ionosphere [*Kan and Sun*, 1985]. This leads to the result that enhancement of field-aligned current in the M-I coupling model always starts from the ionosphere at each reflection of the incident Alfvén wave [*Kan et al.*, 1988]. To demonstrate this result analytically, let us consider the limiting case of a uniform ionospheric conductance. It can be shown that the Hall conductance term drops out of (1) when the conductance is uniform [e.g., *Kan and Sun*, 1985]. The electric field of the reflected Alfvén wave at the ionosphere in this case is given by

$$E^r = [(\Sigma_A - \Sigma_P) / (\Sigma_A + \Sigma_P)] E^i \quad (4)$$

The field-aligned current associated with an Alfvén wave is given by

$$\mathbf{J}_{||} = \pm \Sigma_A (\nabla \cdot \mathbf{E}) \mathbf{b}_o \quad (5)$$

where $\mathbf{b}_o$ is the unit vector of the ambient magnetic field, the "+" and "–" signs are respectively for waves propagating parallel and antiparallel to $\mathbf{b}_o$. The field-aligned current after each reflection at the ionosphere can be written as

$$J_{||} = \Sigma_A \nabla \cdot (\mathbf{E}^i - \mathbf{E}^r) \quad (6)$$

Because $\Sigma_A = (\mu_o V_A)^{-1} \approx 0.4$ mho in the magnetosphere is much less than $\Sigma_P \approx 5$ to 15 mho in the oval ionosphere, one obtains $E^r \approx -E^i$ from (4) and therefore $J_{||}$ after the reflection is increased almost by a factor of 2 as can be seen from (6). Thus, the field-aligned current is enhanced at each reflection of Alfvén wave from the ionosphere. The reason why enhancements of field-aligned currents are initiated in the ionosphere in response to an enhanced magnetospheric convection is the disparity between the Alfvén wave conductance and the ionospheric conductance, i.e., $\Sigma_A \ll \Sigma_P$. This characteristic property ($\Sigma_A \ll \Sigma_P$) of the M-I coupling system is responsible for the behavior that enhancement of field-aligned currents always starts from the ionosphere, although the enhanced convection starts from the magnetosphere. The importance of this result of the M-I coupling system cannot be over emphasized. Unfortunately, it has not been fully appreciated by the substorm community. This result has led me to the idea that the substorm expansion onset should start first in the ionosphere when the field-aligned current exceeds ~ 1μA/m². The closure of this intense substorm field-aligned current through the plasma sheet could be the cause for the dipolarization of the field in the plasma sheet and the formation of a near-Earth X line. The feasibility of the above scenario has been demonstrated by *Sun et al.* [1991] in a "kinematic" model based on the substorm event of March 17, 1978.

(d) *The Minimum Polar Cap Potential for Substorm Expansion Onset.* The potential drop across the polar cap must reach a certain value (~ 70 keV in the model) for the substorm onset to occur about ~30 to 40 minutes after the southward turning of the IMF [*Kan et al.*, 1988]. Recently, *Weimer et al.* [1990a] showed that the *AE* index tends to remain low when the polar cap potential is below ~60 keV, which is consistent with the model prediction of *Kan et al.* [1988].

(e) Saturation of the Polar Cap Potential. Kan et al. [1988] showed that the polar cap potential is reduced to ~79 keV for an input potential of 100 keV imposed on the polar cap of the M-I coupling model. This result can be understood as a consequence of the nonlinear ionospheric loading due to the nonlinear enhancement of the ionospheric conductance by the upward field-aligned currents in the M-I coupling model. The observed saturation of AE index with large southward IMF [*Weimer et al.*, 1990b] is consistent with the model prediction.

TAIL-LIKE RECONFIGURATION OF THE NEAR-EARTH PLASMA SHEET DURING THE SUBSTORM GROWTH PHASE

Tail-like reconfiguration of the plasma sheet have been observed in the near-earth plasma sheet within ~10 R_E radial distance from Earth during the growth phase of substorms [e.g., *Kokubun and McPherron*, 1981; *Baker et al.*, 1987]. The growth phase can last for ~30 to ~60 minutes prior to the substorm expansion onset.

Coroniti and Kennel [1973] proposed that the plasma-sheet thinning [*Hones et al.*, 1967] can result from the fast rarefaction wave launched by the enhanced dayside reconnection following a southward turning of the IMF. *Coroniti* [1985] further proposed that a fast compression wave, launched by the enhanced dayside reconnection, can accelerate the convection on open field lines and increase the lobe field to enhance flaring of the magnetotail.

Recently, *Kan* [1990b] showed that the tail-like reconfiguration region in the near-earth plasma sheet during the substorm growth phase is surrounded by enhanced convection except inside and tailward of the reconfiguration region itself. This result is based on a crude estimate of the electric field distribution surrounding the tail-like reconfiguration region using the observed magnetic field variation and Faraday's law. The simplicity of the procedure ensure the qualitative correctness of the results. Since these results are extremely important in understanding the dynamics of the tail-like reconfiguration process, they should be verified by a quantitative model.

The deduced distribution of enhanced magnetospheric convection during the growth phase can be driven by an enhanced reconnection on the dayside magnetopause. The enhanced dayside convection (due to an enhanced dayside reconnection) can transmit across open field lines at the enhanced convection speed to reach the last open field line in the tail lobe in about 20 minutes. This can cause B_x to increase in the plasma sheet. On the other hand, the enhanced dayside convection can also transmit across closed field lines by the fast rarefaction wave to arrive at the near-earth plasma sheet within ~10 R_E in about 15 minutes, causing B_z to decrease in the plasma sheet. Thus, a tail-like reconfiguration can start within about 15 to 20 minutes after a southward turning of the IMF (interplanetary magnetic field). The duration of the growth phase can be identified with the M-I coupling time scale of ~30 to ~40 minutes. The region-2 field-aligned current is shown to increase to 0.2 $\mu A/m^2$ at the ionospheric level, resulting from the tail-like reconfiguration during the growth phase of substorms. The duration of the substorm growth phase can be identified with the M-I coupling time scale of 8 to 10 Alfvén bounce periods which is about 30 to 40 minutes [*Kan et al.*, 1988]. Specifics of the tail-like reconfiguration process during the growth phase as proposed by *Kan* [1990b] are given below.

Enhancement of Magnetospheric Convection During the Growth Phase

The M-I coupling model [*Kan et al.*, 1988] starts from an enhanced magnetospheric convection. In this subsection we discuss how the magnetospheric convection is enhanced following an enhanced dayside reconnection.

Faraday's law $\nabla \times \mathbf{E} = -\partial \mathbf{B}/\partial t \neq 0$ must be satisfied in the tail-like reconfiguration region, since the magnetic field must be time-dependent during the reconfiguration. The enhanced dayside reconnection electric field will be assumed to be along the y axis of the geocentric solar-magnetospheric coordinate system, i.e., $\mathbf{E} \approx (0, E_y, 0)$. Thus, conditions for the tail-like reconfiguration to occur can be written, from Faraday's law, as

$$\partial E_y/\partial z = \partial B_x/\partial t > 0 \text{ for } z > 0,$$
(inequality sign reverses for $z < 0$)
(7)

$$\partial E_y/\partial x = -\partial B_z/\partial t > 0 \qquad (8)$$

The inequality in (7) states that for B_x to increase with time during tail-like reconfiguration, E_y on open field lines in the tail lobe must be greater than E_y on closed field lines in the plasma sheet so that $\partial E_y/\partial z > 0$. The inequality in (8) states that for B_z to decrease with time during the reconfiguration, E_y must be greater on the earthward side than on the tailward side of the tail-like reconfiguration region $\partial E_y/\partial x > 0$. Thus, E_y must be enhanced in the northern and southern tail lobes and on the earthward side of the tail-like reconfiguration region during the growth phase of substorms.

Figure 2 is a schematic illustration of the electric field profiles required for the tail-like reconfiguration to occur in the plasma sheet. Panels (*a*) and (*b*) show the E_y profiles obtained from (7) and (8), augmented by symmetry and differentiability. Magnitudes of the electric field associated with the tail-like reconfiguration can be estimated from the magnetic field observed during the reconfiguration. The derivatives of E_y shown in panels (*a'*), (*a"*), (*b'*) and (*b"*) are obtained by eye-balling differentiation. The profiles shown in panels (*a*) and (*b*) of Figure 2 requires that during the growth phase the tail-like reconfiguration region must be surrounded by enhanced convection except inside and tailward of the reconfiguration region itself. The key issue is

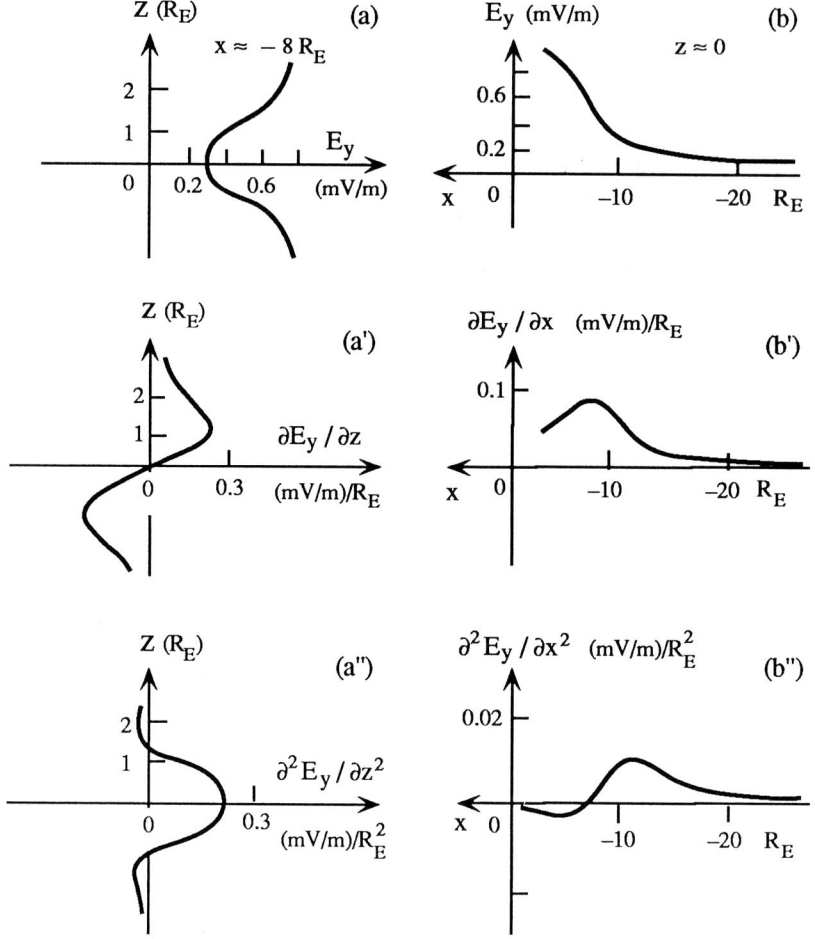

Fig. 2. Profiles of the convection electric field required for thinning in the near-earth plasma sheet based on the observed changes in the magnetic field during the growth phase of substorms.

what causes the magnetospheric convection to be enhanced in such a particular pattern during the growth phase. This issue will be addressed in the next section.

Observations [*Kokubun and McPherron*, 1981; *Baker et al.*, 1987] show that at synchronous radial distance, B_x increases while B_z decreases during the growth phase of substorms. As an example, at $x \approx -7\ R_E$, B_x can increase from ~80 nT to ~120 nT while B_z decreases from ~70 nT to ~40 nT in ~50 minutes during the growth phase of an intense substorm [*Kokubun and McPherron*, 1981]. Using these values, one obtains $\partial E_y/\partial z = \partial B_x/\partial t \approx 40$ nT/3×10^3 s ≈ 0.08 (mV/m)/R_E and $\partial E_y/\partial x \approx 30$ nT/3×10^3 s ≈ 0.06 (mV/m)/R_E which are required in the tail-like reconfiguration region near synchronous orbit.

The dayside reconnection electric field can be expected to increase from $E_y \approx 20$ keV/30 $R_E \approx 0.1$ mV/m to $E_y \approx 100$ keV/30 $R_E \approx 0.6$ mV/m due to a southward turning of the IMF. The scale lengths of the tail-like reconfiguration region in the near-earth plasma sheet can be assumed to be: $L_z \approx 2$ to $4\ R_E$ and $L_x \approx 6$ to $10\ R_E$. Thus, $\partial E_y/\partial z \approx 0.2$ (mV/m)/R_E and $\partial E_y/\partial x \approx 0.1$ (mV/m)/R_E. These values are in reasonable agreement with those required for the tail-like reconfiguration as estimated in the previous paragraph.

The electric field profile required for the tail-like reconfiguration, as given in (4) and (5), should lead to enhanced cross-tail current in the reconfiguration region. Combining Faraday's law with Ampere's law, one obtains

$$\mu_0\ \partial J_y/\partial t = \partial^2 E_y/\partial x^2 + \partial^2 E_y/\partial z^2 > 0 \qquad (9)$$

The inequality is the condition imposed on the electric field by the requirement that the cross-tail current J_y must increase with time in the tail-like reconfiguration region. This is confirmed by the second derivatives of E_y shown in panels (a") and (b") of Figure 2. The magnitude of $\partial^2 E_y/\partial z^2$ is about one order of magnitude greater than that of $\partial^2 E_y/\partial x^2$. Thus, the main contribution to the enhanced cross-tail current comes from the increase in B_x. The sum of the two second derivative terms in (9) is guaranteed to be positive in the tail-like reconfiguration region by the large difference between the two terms as can be seen in Figure 2. The current density enhancement can be estimated by $\Delta J_y \approx \Delta t \, [\Delta E_y/\Delta z^2 + \Delta E_y/\Delta x^2] / \mu_o$. For $\Delta t = 40$ minutes, $\Delta E_y/\Delta z^2 \approx 0.2$ (mV/m)/R_E, $\Delta E_y/\Delta x^2 \approx 0.01$ (mV/m)/R_E (see, Figure 2), $\Delta z \approx 3 \, R_E$ and $\mu_o = 4\pi \times 10^{-7}$ weber/A/m, the cross-tail current density can be enhanced by $\Delta J_y \leq 0.02 \, \mu A/m^2$. The corresponding increase in the cross-tail current intensity is $\Delta \kappa_y = \Delta J_y \, \Delta z \approx 200$ mA/m. This is in agreement with the estimate made by *Kaufmann* [1987] that the cross-tail current must increase from ~30 mA/m to ≥ 100 mA/m within 7 to 10 R_E radial distance to produce the observed the tail-like reconfiguration at synchronous altitude.

Region-2 Field-Aligned Currents Enhanced During the Growth Phase

Region-2 field-aligned currents can be expected to increase during the tail-like reconfiguration as long as (i) the cross-tail current increases in the reconfiguration region and (ii) the tail-like reconfiguration region is most pronounced in the midnight sector of the near-earth plasma sheet. The y-dependence of the tail-like reconfiguration region required in (ii) can be traced to the y-dependence of the dayside reconnection electric field. For a strongly southward IMF, the reconnection electric field should be maximum around the noon meridian and falls off on the two sides.

The conditions for the enhancement of the region-2 field-aligned current in (i) and (ii) can be shown to be the same conditions for confining the tail-like reconfiguration region in the midnight sector. Combining Faraday's law with Ampere's law, one obtains

$$\mu_o \, \partial J_z/\partial t \approx -(\partial/\partial y)\partial B_x/\partial t \approx -(\partial/\partial y)\partial E_y/\partial z$$
$$< 0 \quad \text{Dawnside;}$$
$$> 0 \quad \text{Duskside} \qquad (10)$$

where the inequalities are conditions required for the field-aligned current to have the region-2 current sense. This requires $\partial E_y/\partial z$ to be maximum centered around the midnight meridian, i.e., the tail-like reconfiguration region is confined in the midnight sector of the near-earth plasma sheet.

Figure 3 shows the y-dependence of the electric field required for the enhancement of the region-2 field-aligned current associated with tail-like reconfiguration in the near-earth plasma sheet. The convection electric field E_y on the dawn and dusk sides of the tail lobe is less enhanced than around the noon-midnight meridian, which is necessary for the tail-like reconfiguration to be most pronounced in the midnight sector and for the enhancement of the region-2 field-aligned current.

Toward the end of a tail-like reconfiguration period, the enhanced region-2 field-aligned current can be estimated by $\Delta J_z \approx \Delta t \, [\Delta E_y / \Delta z \, \Delta y] / \mu_o$. For a 40-minute tail-like reconfiguration period with an average $\Delta E_y/\Delta z \approx 0.1$ (mV/m)/R_E (see Figure 2) and $\Delta y \approx 10 \, R_E$, one obtains $\Delta J_z \approx 0.5 \times 10^{-9}$ A/m² near the equatorial plasma sheet which maps to $\Delta J_\parallel \approx 0.2 \, \mu A/m^2$ at the ionospheric altitude. Thus, the region-2 field-aligned current is predicted to increase as a consequence of tail-like reconfiguration in the near-earth plasma sheet during the growth phase of substorms. This is consistent with the observation, based on ground magnetometer data, presented at the fall AGU meeting by *Akasofu et al.* [1988] that the region-2 current increases during the growth phase of substorms.

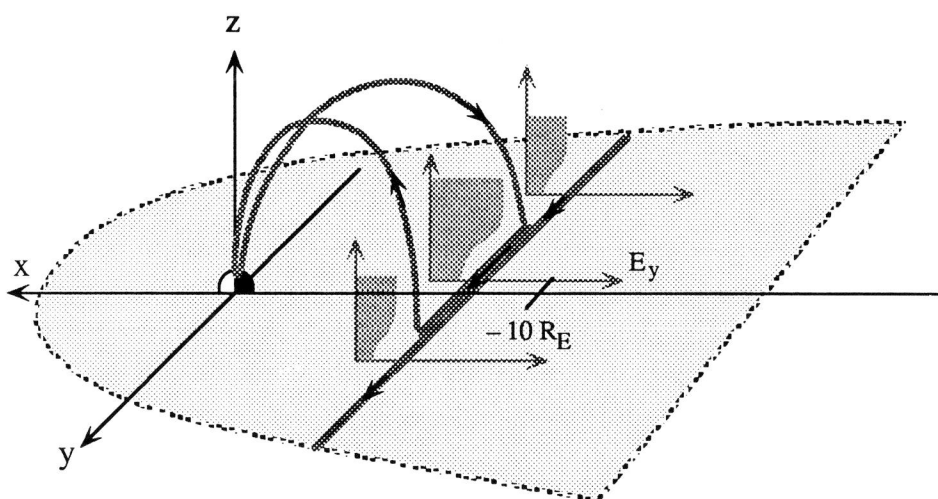

Fig. 3. Enhancement of the region-2 field-aligned current by the tail-like reconfiguration during the growth phase of substorms.

DIPOLARIZATION OF THE PLASMA-SHEET FIELD CONFIGURATION DURING THE SUBSTORM EXPANSION PHASE

Dipolarization of the magnetic field in the near-earth plasma sheet is a well-known signature of the substorm expansion phase [e.g., *Kokubum and McPherron*, 1981]. Recently, *Kan* [1991] proposed that the dipolarization is a consequence of closing the substorm enhanced field-aligned currents in the plasma sheet. The proposed scenario depends critically on the intense substorm field-aligned currents are enhanced first from the ionosphere and subsequently closed in the plasma sheet. The physics behind this behavior of the substorm field-aligned current has been summarized earlier in subsection (c) under the M-I coupling section. The closure current is antiparallel to the cross-tail current to result in disruption (or reduction) of the cross-tail current. Disrupting the cross-tail current leads to induction electric field in the dawn to dusk direction. We are still in the process of developing a model to determine the induction electric field self-consistently from the closure of the substorm field-aligned currents.

Electric Field Enhanced in the Dipolarization Region

The electric field required for a dipolarization in the near-earth plasma sheet can be deduced from the magnetic field observed during the dipolarization event. From the examples given by *Kokubum and McPherron* [1981], B_x can decrease say from ~150 nT to ~80 nT while B_z can increase from ~50 nT to ~100 nT in ~20 minutes during a dipolarization event. Assuming the convection electric field is predominantly in the y direction for simplicity, electric field required for dipolarization can be written from Faraday's law as

$$\partial E_y/\partial z = \partial B_x/\partial t < 0 \quad \text{for } z > 0,$$
(inequality sign reverses for $z < 0$) (11)

$$-\partial E_y/\partial x = \partial B_z/\partial t > 0 \quad (12)$$

where the inequalities ensure that the magnetic field is dipolarizing in the plasma sheet. For $\Delta B_x = 70$ nT, $\Delta B_z = 50$ nT and $\Delta t = 20$ minutes, one obtains: $|\partial E_y/\partial z| = |\partial B_x/\partial t| \approx 0.35$ (mV/m)/R_E and $|\partial E_y/\partial x| \approx |\partial B_z/\partial t| \approx 0.25$ (mV/m)/R_E. Assuming the dimensions of the dipolarization region are: $L_z \approx 2$ to $4 R_E$, $L_y = 3$ to $6 R_E$ and $L_x \approx 6$ to $10 R_E$, the electric field inside the dipolarization region can be estimated to be $\Delta E_y \approx 0.35$ (mV/m)/$R_E \times \Delta z \approx 1.05$ mV/m. The electric field of an enhanced dayside reconnection can be estimated to be ~0.6 mV/m which has been transmitted to the plasma sheet at the end of the growth phase [*Kan*, 1990b]. Therefore, the electric field in the dipolarization region is estimated to be $E_y(z = 0, x \approx -10 R_E) \approx 1.05 + 0.6 \approx 1.6$ mV/m. Likewise, $\Delta E_y \approx 0.25$ (mV/m)/$R_E \times \Delta x \approx 1.5$ mV/m and $E_y(z = 0, x = x_o) \approx 1.5 + 0.6 \approx 2.1$ mV/m, where x_o is at the center of the current disruption region to be discussed later.

The above discussion showed that the electric field (~2.1 mV/m) associated with the dipolarization event is ~3 times the field (~0.6 mV/m) produced by the enhanced dayside reconnection. Such a large electric field cannot come from the enhanced dayside reconnection alone. It could be induced internally by the current disruption process as discussed in the next section.

Electric Field Induced by the Disruption of Cross-Tail Currents

Electric field induced by disrupting the cross-tail current can be written, from Ampere's law and Faraday's law, as

$$\partial^2 E_y/\partial x^2 + \partial^2 E_y/\partial z^2 = \mu_o \, \partial J_y/\partial t < 0 \quad (13)$$

where the inequality is the condition for current disruption.

The magnitude of the current disruption can be estimated from (13) by $\Delta J_y = [\Delta E_y/\Delta z^2 + \Delta E_y/\Delta x^2] \Delta t / \mu_o$ where $\mu_o = 4\pi \times 10^{-7}$ weber/A/m is the permeability of free space. For a substorm expansion phase with duration $\Delta t \approx 20$ min, with field enhancement $\Delta E_y \approx 1.5$ mV/m, in a localized region $\Delta x \approx 6$ to $10 R_E$ and $\Delta z \approx 2$ to $4 R_E$, the cross-tail current density can be reduced by $\Delta J_y \approx 5.5 \times 10^{-3}$ μA/m^2. The corresponding reduction in the cross-tail current intensity is $\Delta \kappa_y = \Delta J_y \, \Delta z \approx 100$ mA/m.

Figure 4 is a schematic illustration of the electric field profiles required for dipolarization in the plasma sheet. Panels (a) and (b) show the E_y profiles deduced from (11) and (12), augmented by symmetry and differentiability, under the assumption that the disruption of the cross-tail current is localized so that the induced electric field falls off from the center of the disruption region, say, at $x \approx -12 R_E$. The electric field in the dipolarization region can be estimated from the magnetic field observed in the near-earth plasma sheet. Derivatives of E_y shown in panels (a'), (a"), (b') and (b") are obtained by eye-balling differentiation.

The region denoted by D is the dipolarization region specified by (12) and can be visualized in panel (b'). The region denoted by CD is the current disruption region where the second derivatives of E_y are negative as shown in panel (b"). Note that the dipolarization region is confined within the current disruption region. The region denoted by T shown in panel (b") could be a thinning region located tailward of the dipolarization region in the plasma sheet. For thinning to occur, it is necessary that $\partial^2 E_y/\partial x^2 > 0$ over power $\partial^2 E_y/\partial z^2 < 0$ as required by (13). This is consistent with the observations [*Hones et al.*, 1967] that thinning occurs around $x \approx -18 R_E$ tailward of the dipolarization region after the substorm expansion onset.

Closure of Substorm Field-Aligned Currents in the Plasma Sheet

The mechanism responsible for the disruption of cross-tail currents is still an open issue at the present time. Popular ideas on current disruption mechanisms include the formation

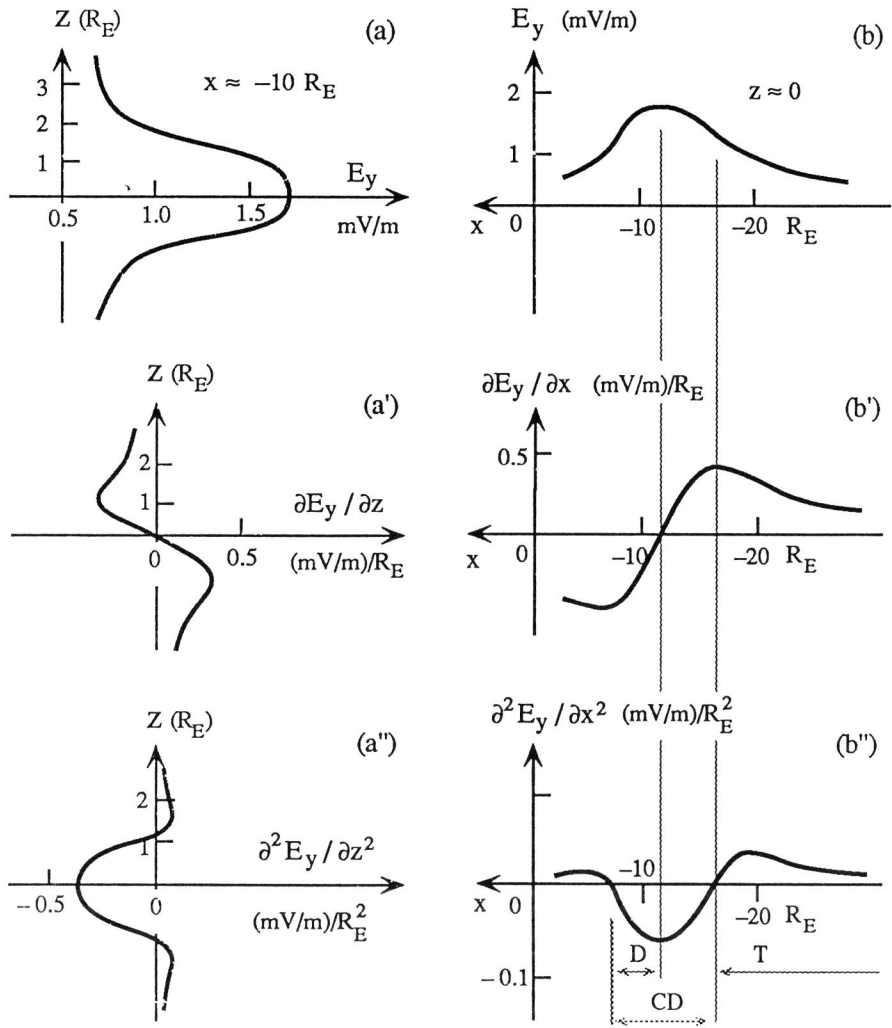

Fig. 4. Profiles of the convection electric field required for dipolarization of the near-earth plasma sheet based on the observed changes in the magnetic field during the expansion phase of substorms.

of a near-earth X line [e.g., *McPherron et al.*, 1973] and a current instability [*Lui et al.*, 1988, 1990] to trigger the substorm expansion onset. *Mitchell et al.* [1990] presented a case study in which the disruption of cross-tail currents in the thinning region appears to occur after the expansion onset. This example [*Mitchell et al.*, 1990] indicates, in our opinion, that current disruption could be a consequence, rather than the cause of the substorm expansion onset.

Kan [1991] proposed that the cross-tail current can be disrupted by the closure of substorm enhanced field-aligned currents in the near-earth plasma sheet. Enhancements of field-aligned currents are initiated from the ionosphere in response to an enhanced magnetospheric convection due to the ionospheric Pedersen conductance is much greater than the Alfvén wave conductance as discussed earlier in this paper. A sudden brightening of the equatorward-most auroral arc at the substorm expansion onset should occur when the upward field-aligned current exceeds ~1 $\mu A/m^2$ near the Harang discontinuity region in the evening sector [*Kan et al.*, 1988]. The substorm enhanced field-aligned currents must be closed in the plasma sheet. The closure of the substorm enhanced field-aligned currents should lead to disruption of the cross-tail current, resulting in an intense induction electric field discussed in the previous section. The intense induction electric field can propagate into the ionosphere to cause further enhancement of field-aligned currents. This leads to a positive-feedback loop which can result in rapid intensification of the substorm current system during the expansion phase.

Closure of the substorm enhanced field-aligned currents in the plasma sheet imposes an additional constraint on the induced electric field as given by

$$\mu_o \partial J_z/\partial t \approx -(\partial/\partial y)\partial B_x/\partial t \approx -(\partial/\partial y)\partial E_y/\partial z$$
$$> 0 \quad \text{Dawnside};$$
$$< 0 \quad \text{Duskside} \qquad (14)$$

where the inequalities are the conditions for the region-1 field-aligned current [Iijima and Potemra, 1976] to increase with time due to the disruption of cross-tail currents during substorms. Thus, $\partial E_y/\partial z$ is required to be maximum in the evening-midnight sector ($y \geq 0$) and falls off towards both the dawn and dusk sectors. The field-aligned current density flowing to and from the plasma sheet during an intense substorm can be estimated from (14) to be $\Delta J_z \approx \Delta t (\Delta E_y/\Delta y\, \Delta z)/\mu_o \approx 2.2 \times 10^{-3}$ μA/m².

Figure 5 shows the y-dependence of the induction electric field required for the closure of the substorm enhanced field-aligned currents originating from the ionosphere in response to an enhanced magnetospheric convection. The upward field-aligned current (out of the ionosphere) is localized around the head of the westward traveling surge in the evening sector while the downward field-aligned current (into the ionosphere) are distributed over a relatively wide region. The resulting substorm current wedge is asymmetric centered in the evening sector. To close the substorm enhanced field-aligned currents in the plasma sheet, the induction electric field E_y must be maximum inside the current disruption region of finite extent centered around $x_o \approx -10$ to $-15\, R_E$, $y_o \approx 2\, R_E$ and $z_o \approx 0$ in the geocentric solar-magnetospheric coordinates as shown in Figure 5. It is clear that the electric field induced by the disruption of the cross-tail current is intrinsically a three-dimensional field. The y-dimension of the current disruption region $L_y \approx 3$ to $6\, R_E$ can be estimated by mapping the westward traveling surge region to the equatorial plane. This is demonstrated recently by [Sun et al., 1991] in a model based on the Tsyganenko [1987] magnetic field model. In the current disruption region, $\mathbf{E} \cdot \mathbf{J} > 0$ before and after the current disruption. The dynamo region for the cross-tail current and the substorm enhanced field-aligned current should be located primarily in the high-latitude magnetopause-boundary-layer region where $\mathbf{E} \cdot \mathbf{J} < 0$.

The substorm enhanced field-aligned current density observed at the ionospheric altitude is $J_{\parallel} \geq 1$ μA/m². When mapped from the ionosphere ($B_i \approx 3 \times 10^4$ nT) into the equatorial plasma sheet ($B_z \approx 50$ nT), the field-aligned current density in the plasma sheet is reduced to $J_z = J_{\parallel} (B_z/B_i) \geq 1.6 \times 10^{-3}$ μA/m². This current density can be closed in the plasma sheet, because it is comparable to the current density $\Delta J_z \approx 2.2 \times 10^{-3}$ μA/m² estimated from the observed dipolarization discussed earlier.

Plasma Injection Driven by the Shock Launched by the Substorm Induced Electric Field

Energetic particle fluxes (~30 to ~300 keV) in the near-earth plasma sheet are enhanced during dipolarization, known as the plasma injection event [e.g., McIlwain, 1974]. The electric field induced by the disruption of the cross-tail current due to the closure of the substorm field-aligned currents has been estimated to be about 3 to 4 times the dayside reconnection electric field as shown in Figure 4. The intense induction electric field in the current disruption region can launch a fast shock wave propagating earthward and a fast rarefaction wave propagating tailward. The plasma pressure and the magnetic field both increase across the fast shock. In this connection, we propose to identify the plasma injection front with the shock front launched by the induction electric field produced by the dipolarization in the plasma sheet in an intense substorm. Fast shocks or large-

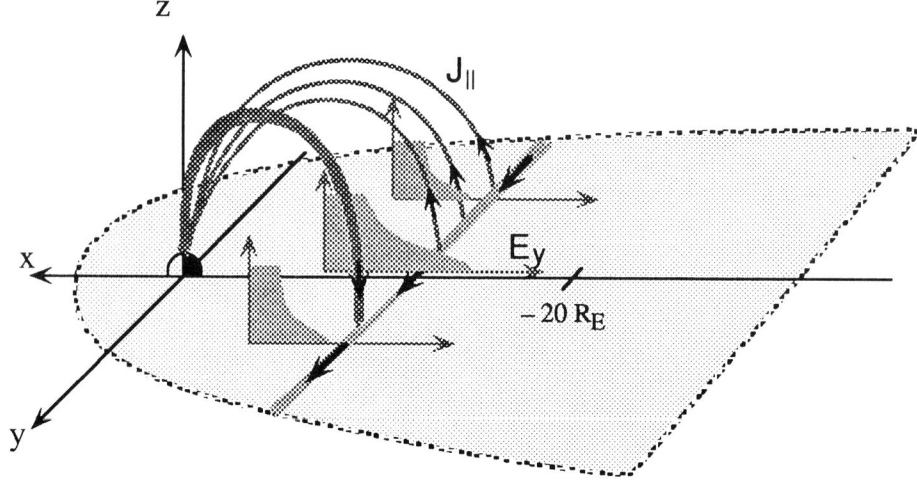

Fig. 5. Enhancement of the substorm field-aligned currents (of the region-1 current sense) initiated in the ionosphere by the substorm expansion onset. Closure of the substorm field-aligned currents in the plasma sheet leads to the disruption of the cross-tail current and the dipolarization of the near-earth plasma sheet after the substorm expansion onset.

amplitude compression waves cannot be expected to be launched during small substorms.

Association of the plasma injection event with a compression wave has been proposed by *Moore et al.* [1981]. This association has not been borne out by observations during a small substorm [e.g., *Lopez et al.*, 1990]. A systematic observational search during intense substorms would be required to determine whether or not a shock is launched by the expansion onset of intense substorms.

The particle flux, enhanced across the shock, can be calculated from the conservation of mass across the shock, i.e., $N_o(V_S - V_o) = N_1(V_S - V_1)$. Thus, the enhanced flux can be written as

$$(N_1V_1 - N_oV_o)/N_oV_o = (N_1 - N_o)V_S/(N_oV_o) \leq 3(V_S/V_o) \quad (15)$$

where V_S is the shock speed, subscripts "o" and "1" indicate upstream and downstream quantities, respectively. The inequality comes from the fact that the density jump N_1/N_o across the fast shock cannot exceed 4 [e.g., *Kantrowitz and Petschek*, 1966]. For an estimate, the convection speed before the injection can be taken to be around $V_o \approx 10$ to 20 km/s (for $E \approx 0.6$ mV/m and $B_z \approx 25$ to 50 nT). The shock speed must be greater than the fast-mode speed, $V_S > V_F \approx 300$ to 500 km/s in the near-earth plasma sheet. Thus, according to (15), particle flux can be estimated to increase by ~50 to ~150 times in an injection event which is consistent with observations [e.g., *Kokubun and McPherron*, 1981]. The thermal energy enhanced during the plasma injection event can be estimated from the temperature jump across the fast perpendicular shock. From the Rankine-Hugoniot condition [*Kantrowitz and Petschek*, 1966], the ratio of the temperature jump across the shock is given by $T_1/T_o \approx 5, 15, 35, 80, ...$ for $M_A \approx 3, 5, 7, 10, ...$, respectively, for $\beta_o = 0.5$ (ratio of plasma pressure to magnetic pressure). The magnetic field jump across the plasma injection front can be estimated by the magnetic field jump across the fast shock. For the perpendicular shock, $B_1/B_o = N_1/N_o$. These predictions of the proposed interpretation of the plasma injection event can be tested by observations.

The minimum time delay between the substorm expansion onset in the ionosphere and the subsequent dipolarization in the plasma sheet is about 2 minutes which is the one-way Alfvén transit time from the ionosphere to the near-earth plasma sheet. The actual time delay can be expected to take a few Alfvén bounce time which can easily add up to several minutes. This time delay is of the same trend as the field-aligned current density enhancement observed in the near-earth plasma sheet which lags behind the substorm onset in the ionosphere as presented by *Iijima* [1989] and *Nagai et al.* [1989]. This time delay may also be related to the 6 minutes time delay between the particle injection at the synchronous orbit and the substorm expansion onset in the auroral zone observed by *Bargatze et al.* [1987]. These authors proposed to explain the observed 6 minutes delay by the difference in travel time between the compression wave traveling from the source region to the synchronous orbit and the auroral electrons traveling to the ionosphere. The fast mode speed is about 500 km/s or more in the near-earth plasma sheet, a 6-minute time delay would place the source region at least 30 R_E from the synchronous orbit, which is too far tailward from the dipolarization region.

Convection Modified by the Substorm Induced Electric Field

The electric field induced by the disruption of the cross-tail current is confined in the dipolarization region as shown in Figure 5. In this subsection we wish to point out that the substorm induced electric field can lead to a convection pattern which is consistent with the vorticity required to close the substorm field-aligned currents in the plasma sheet. However, we have not yet identify the physical process responsible for connecting the closure of the substorm enhanced field-aligned currents to the enhancement of the induction electric field during the expansion phase of substorms.

Figure 6 show the convection pattern produced by the substorm induced electric field. The substorm induced electric field is maximum in the center of the dipolarization which is also the region for the substorm field-aligned currents to close in the plasma sheet. The convection pattern shown in Figure 6 is consistent with the distribution of the substorm induced electric field shown in Figure 5. The convection shown in Figure 6 will be referred to as the substorm induced convection hereafter. The net convection in the plasma sheet after the substorm expansion onset is a superposition of the substorm induced convection on top of the background convection which existed prior to the substorm onset. The vorticity in the substorm induced convection is of the same sense as the vorticity required to close the substorm field-aligned currents by diverting the cross-tail current in the plasma sheet. The vorticity associated with the substorm convection pattern is initiated from the ionosphere, just as the substorm field-aligned currents are initiated from the ionosphere, in response to an enhanced magnetospheric convection.

The discussions in this subsection and in the last subsection show that the plasma pressure in the plasma sheet is spontaneously modified by the shock launched by the substorm induced electric field and the convection electric field is spontaneously modified by the closure of the substorm field-aligned currents. These spontaneous modifications in the plasma sheet are self-consistently produced by the closure of the substorm field-aligned currents in the plasma sheet. The key element is to realize that the substorm field-aligned currents are initiated from the ionosphere in response to enhanced magnetospheric convection due to the Alfvén wave conductance is much less than the ionospheric conductance discussed earlier. This discussion also brings out the nonlinear two-way nature of the M-I coupling process in that an enhanced magnetospheric convection drives the ionosphere while the ionosphere, in response to the enhanced magnetospheric convection, can significantly modify the field configuration and the convection in the plasma sheet.

Formation of a Near-Earth x Line: Consequence of Substorm Expansion Onset

A near-earth x line can be expected to form by (a) disrupting the cross-tail current on the earthward side of the anticipated x line if the amount of disrupted current is large enough, (b) enhancing the cross-tail current on the tailward side of the anticipated x line or (c) a combination of (a) and (b).

As discussed earlier, field-aligned currents must be drastically enhanced first in the ionosphere at the substorm expansion onset. The closure of the substorm enhanced field-aligned current in the plasma sheet can lead to disruption of the cross-tail current and the dipolarization of the near-earth plasma sheet. As shown in panel (b'') of Figure 4, disrupting the cross-tail current in a localized region can lead to an increase of the cross-tail current on the tailward side of the current disruption region. Thus, case (c) appears to be the most likely scenario for the disruption of the cross-tail current. The near-earth x line should be located tailward of the current disruption region and earthward of the thinning region [*Hones*, 1979] after the substorm expansion onset.

Sun et al. [1991] modeled the evolution of the plasma sheet field configuration during an isolated intense substorm on March 19, 1978. They showed that a near-earth x line of ~5 R_E long can form in the evening sector right after the substorm onset. The location of the x line shifts toward the morning sector as the expanse phase progresses. These results of the model are consistent with the theoretically deduced expectations described above.

RECOVERY PHASE OF SUBSTORMS

Recovery phase of a substorm begins at the end of the substorm expansion phase. Existing ideas on the substorm recovery phase can be grouped according to external or internal causes. An example of substorm recovery due to external cause is a northward turning of the IMF. Examples of substorm recovery due to internal causes are the Alfvén shielding effect during a prolonged southward IMF [*Kan et al.*, 1988], and the plasmoid formation during a prolonged southward IMF [*Hones*, 1984].

A substorm can be expected to recover in ~15 to 20 minutes after the IMF turned northward. The time delay is required for the effect of the northward turning of the IMF to reach the near-earth plasma sheet. This is similar to the southward turning of the IMF leading to the growth phase [*Kan*, 1990b], except the compression and rarefaction waves are reversed between these two cases. The reduction of the convection electric field caused by the northward turning of the IMF will reach the near-earth plasma sheet first and then propagates tailward. Thus, the auroral activity can be expected to fade out starting from the equatorward part of the auroral oval and progressing poleward, if the recovery phase is caused by a northward turning of the IMF.

If the recovery phase is caused by the Alfvén shielding effect during a prolonged southward IMF [*Kan et al.*, 1988], the auroral activity should fade out right in the head of the westward traveling surge and recede equatorward. At the same time, the intense auroral electrojets should shift equatorward as demonstrated by *Kan et al.* [1988].

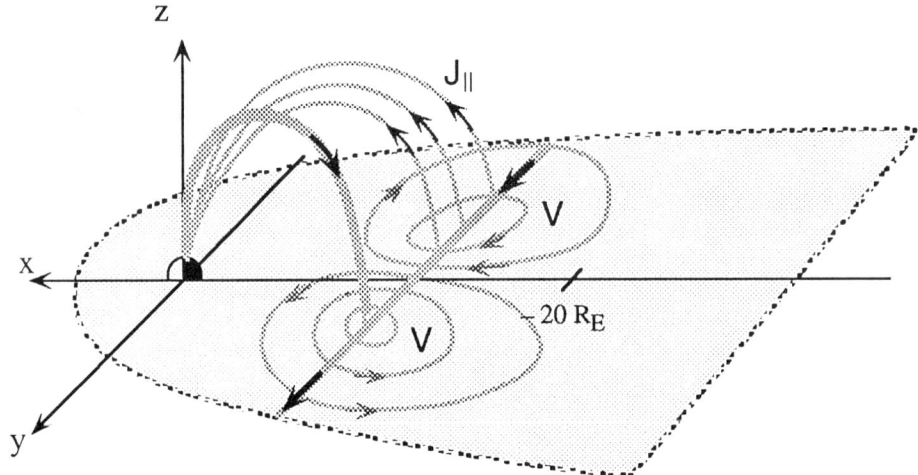

Fig. 6. Convection pattern associated with the substorm induced electric field due to the closure of the substorm enhanced field-aligned currents in the plasma sheet. The resulting substorm induced convection is to be superimposed on the background convection (not shown) existed prior to the substorm expansion onset.

The recovery phase can also be caused by the formation and detachment of the plasmoid from the near-earth plasma sheet. The plasmoid is detached from the near-earth plasma sheet when the last closed field line in the plasma sheet is reconnected at the near-earth X line [*Hones*, 1984]. On the tailward side of the near-earth X line, the field-aligned current would be cut off near the X line first and progressively tailward. On the earthward side, the field-aligned current would shift earthward when the plasmoid is detached. Thus, the equatorward part of the bright auroral forms can be expected to retreat equatorward, while the poleward part fades progressively poleward, leaving a gap of dark region of increasing latitudinal width as the recovery phase progresses. Equatorward retreat of auroral forms is a well established signature of the recovery phase [*Akasofu*, 1964].

The duration of the expansion phase can be identified with the interval between the expansion phase onset and the detachment of the plasmoid. Thus, the expansion phase time scale can be estimated by the time taken for the last closed field line to be reconnected at the near-earth X line after the expansion onset. The electric field in the plasma sheet can be taken as $E_y \approx 1$ mV/m, the magnetic field in the near-earth plasma sheet is around $B_x \approx 80$ nT, so that the convection speed toward the equatorial plasma sheet V_z (km/s) $\approx 10^3 \times [E_y(\text{mV/m}) / B_x(\text{nT})] \sim 12.5$ km/s. The half thickness of the near-earth plasma sheet tailward of the dipolarization region could be about 3 R_E, the time taken for the last closed field line to be reconnected at the near-earth X line can be estimated by $\Delta t \approx 18000$ km / 12.5 km/s ≈ 24 minutes. Thus, the duration of the substorm expansion phase is estimated to be about 20 to 30 minutes.

TIMING OF THE SUBSTORM SEQUENCE OF EVENTS

The cause-and-effect relationship between the sequence of events during a substorm can be identified by the timing of these events. Unfortunately, timing of the substorm sequence of events has not yet been established by observations with enough accuracy to be useful for this purpose. The substorm model described above can lead to a timing sequence of events during a substorm which has been proposed by *Kan* [1991] as summarized below which we hope will be a useful framework for future observational studies.

T_0 = IMF turns southward.
$T_1 \approx T_0 + 2$ min.
 = Growth phase begins.
$T_2 \approx T_1 + 15$ min.
 = Start of tail-like reconfiguration in the near-earth plasma sheet,
 = Region-2 field-aligned current starts to intensify.
$T_3 \approx T_2 + 2$ min.
 = Auroral electrojets start to intensify,
 = Region-1 field-aligned current starts to intensify.
$T_4 \approx T_3 + 30$ min.
 = End of the tail-like reconfiguration in the near-earth plasma sheet,
 = End of the growth phase,
 = Upward field-aligned current exceeds ~1 μA/m^2,
 = Field-aligned potential drop of a few kilovolts develops,
 = Sudden brightening of the equatorward-most auroral arc,
 = Onset of the substorm expansion phase. The 30 minutes in T_4 is the M-I coupling time scale [*Kan et al.*, 1988].
$T_5 \approx T_4 + 2$ min.
 = Disruption of the cross-tail current begins
 = Dipolarization in the near-earth plasma sheet begins,
 = Plasma injection begins driven by the shock launched by the current disruption,
 = A new X line forms tailward of the dipolarization region between T_5 and T_6.
$T_6 \approx T_5 + 20$ min.
 = Substorm enhanced field-aligned current reaches maximum intensity,
 = Electrojets reach maximum intensity,
 = Plasmoid becomes fully detached from the near-earth plasma sheet,
 = Substorm expansion phase reaches maximum epoch, recovery phase begins.

The difference between T_4 and T_5 is the one-way Alfvén travel time from the ionosphere to the plasma sheet. It depends on the Alfvén speed and the length of the field lines at the site of the substorm expansion onset in the ionosphere. The onset site varies from substorm to substorm, so does the time delay of the dipolarization in the plasma sheet from the substorm expansion onset in the ionosphere. Thus, the timing listed above should be taken in perspective with the respective physical processes govening the substorm. These physical processes have been discussed in the paper.

SUMMARY

The global model of substorms proposed in this paper is still in the developing stage. Basic ideas behind this model can be summarized as follows. An enhanced magnetospheric convection is responsible for driving the substorm. However, the substorm expansion onset occurs only when the ionosphere responds fully to the enhanced magnetospheric convection. The tail-like reconfiguration of the near-earth plasma sheet can result from the enhancement of magnetospheric convection due to an increased solar wind-magnetosphere interaction during the growth phase of a substorm. The substorm expansion onset can result from the ionospheric response to the enhanced magnetospheric convection when the upward field-aligned currents exceeds ~1 μA/m^2 as demonstrated in the M-I coupling model by *Kan et al.* [1988]. Closing the substorm enhanced field-aligned currents in the plasma sheet can disrupt the cross-tail current to result in the dipolarization of the plasma-sheet configuration during the expansion phase of substorms. The electric field induced by the current disruption can launch a shock wave propagating earthward to drive the plasma injection right after the substorm expansion onset. A near-earth X line can form tailward of the dipolarization region after the onset of an intense substorm. A timing sequence of events during a substorm has been deduced based on the proposed substorm model. This model is still in the conceptual stage of development. It is our hope that the proposed model will be found useful in future observational and theoretical studies of magnetospheric substorms.

Acknowledgments. This work was supported in part by the NSF grant ATM-8912359 to the University of Alaska Fairbanks.

REFERENCES

Ahn, B.-H., S.-I. Akasofu, Y. Kamide, and J. H. King, Cross polar cap potential drop and the energy coupling function, *J. Geophys. Res., 89*, 11,028, 1984.

Akasofu, S.-I., The development of the auroral substorm, *Planet. Space Sci., 12*, 273, 1964.

Akasofu, S.-I., M. Yamauchi, and J. R. Kan, IMF variations and substorms, SM 21C-07, AGU fall meeting, *EOS, 69*, 1381, 1988.

Arnoldy, R. L., Auroral particle precipitation and Birkeland currents, *Rev. Geophys. Space Phys., 12*, 217, 1974.

Baker, D. N., R. C. Anderson, R. D. Zwickl, and J. A. Slavin, Average plasma and magnetic field variations in the distant magnetotail associated with near-earth substorm effects, *J. Geophys. Res., 92*, 71, 1987.

Bargatze, L. F., D. N. Baker, and R. L. McPherron, Superposed epoch analysis of magnetospheric substorms using solar wind, auroral zone, and geostationary orbit data set, *Magnetotail Physics*, edited by T. Y. Lui, Johns Hopkins University Press, 163, 1987.

Baumjohann, W., R. J. Pellinen, H. J. Opgenoorth, E. Nielsen, Joint two-dimensional observations of ground magnetic and ionospheric electric fields associated with auroral zone currents: current systems associated with local auroral break-ups, *Planet. Space Sci., 29*, 431, 1981.

Baumjohann, W., G. Paschmann, N. Sckopke, C. A. Cattel, and C. W. Carlson, Average ion moments in the plasma sheet boundary layer, *J. Geophys. Res., 93*, 11507, 1988.

Baumjohann, W., G. Paschmann, and C. A. Cattell, Average plasma properties in the central plasma sheet, *J. Geophys. Res., 94*, 6597, 1989*a*.

Baumjohann, W., G. Paschmann, and H. Lühr, Characteristics of high-speed ion flow in the plasma sheet, *J. Geophys. Res., 95*, 3801, 1989*b*.

Cao, F., and J. R. Kan, Reflection of Alfvén waves at an open magnetopause, *J. Geophys. Res., 95*, 4257, 1990.

Coroniti, F. V., and C. F. Kennel, Can the ionosphere regulate magnetospheric convection? *J. Geophys. Res., 78*, 2837, 1973.

Coroniti, F. V., Explosive tail reconnection: the growth and exoansion phases of magnetospheric substorms, *J. Geophys. Res., 90*, 7427, 1985.

Doyle, M. A., and W. J. Burke, S3-2 measurements of the polar cap potential, *J. Geophys. Res., 88*, 9125, 1983.

Galperin, Yu. I., and Ya. I. Feldstein, Auroral luminosity and its relationship to the magnetospheric plasma domains, *Proceedings of the Cambridge Conference on Auroral Physics*, edited by C. I. Meng, Cambridge Press, 207, 1990.

Heppner, J. P., Empirical models of high latitude electric field, *J. Geophys. Res., 82*, 1115, 1977.

Heppner, J. P., and N. C. Maynard, Empirical high latitude electric field models, *J Geophys. Res., 92*, 4467, 1987.

Hones, E. W. Jr., J. R. Asbridge, S. J. Bame, I. B. Strong, Outward flow of plasma in the magnetotail following geomagnetic bays, *J. Geophys. Res., 72*, 5897, 1967.

Hones, E. W. Jr., Transient phenomena in the magnetotail and their relation to substorms, *Space Sci. Rev., 23*, 393, 1979.

Hones, E. W. Jr., Plasma sheet behavior during substorms, in *Magnetic Reconnection in Space and Laboratory Plasmas, Geophys. Monog., 30*, edited by E. W. Hones, Jr., AGU, Washington, D.C., 1984.

Huang, C. Y., and L. A. Frank, A statistical study of the central plasma sheet: implications for substorm models, *Geophys. Res. Lett., 13*, 652, 1986.

Iijima, T., and T. A. Potemra, Large-scale characteristics of field-aligned currents associated with substorms, *J. Geophys. Res., 81*, 3999, 1976.

Iijima, T., Magnetic equatorial currents and field-aligned currents, SM 11B-02, AGU fall meeting, *EOS, 70*, 1268, 1989.

Kamide, Y., and G. Rostoker, The spatial relationship of field-aligned currents and auroral electrojets to the distribution of nightside auroras, *J. Geophys. Res., 82*, 5589, 1977.

Kan, J. R., and W. Sun, Simulation of the westward traveling surge and Pi2 pulsations during substorms, *J. Geophys. Res., 90*, 10,911, 1985.

Kan, J. R., L. Zhu, and S.-I. Akasofu, A theory of substorms: onset and subsidence, *J. Geophys. Res., 93*, 5624, 1988.

Kan, J. R., Developing a global substorm model, *EOS, Transactions, American Geophysical Union, 71*, pages 1083 and 1086-1087, September 18, 1990*a*.

Kan, J. R., Tail-like reconfiguration of the plasma sheet during the substorm growth phase, *Geophys. Res. Lett., 17*, 2309, 1990*b*.

Kan, J. R., Dipolarization: a consequence of substorm expansion onset, *Geophys. Res. Lett., 18*, 57, 1991.

Kantrowitz, A., and H. E. Petschek, MHD characteristics and shock waves, in *Plasma Physics in Theory and Application*, edited by W. B. Kunkel, p.148, McGraw-Hill, New York, 1966.

Kaufmann, R. L., Substorm currents: growth phase and onset, *J. Geophys. Res., 92*, 7471, 1987.

Kokubun, S., and R. L. McPherron, Substorm signatures at synchronous altitude, *J. Geophys. Res., 86*, 11,265, 1981.

Lopez, R. E., H. Lühr, B. J. Anderson, P. T. Newell, and R. W. McEntire, Multipoint observations of a small substorm, *J. Geophys. Res., 95*, 18,897, 1990.

Lui, A. T. Y., E. W. Hones Jr., F. Yasuhara, S.-I. Akasofu, and S. J. Bame, Magnetotail plasma flow during plasma sheet expansions: Vela 5, 6, and Imp-6 observations, *J. Geophys. Res., 82*, 1235, 1977.

Lui, A. T. Y., R. E. Lopez, S. M. Krimigis, R. W. McEntire, L. J. Zanetti, and T. A. Potemra, A case study of magnetotail current sheet disruption and diversion, *Geophys. Res. Lett., 15*, 721, 1988.

Lui, A. T. Y., A. Mankofsky, C.-L. Chang, K. Papadopoulos, and C. S. Wu, A current disruption mechanism in the neutral sheet: A possible trigger for substorm expansions, *Goephys. Res. Lett., 17*, 745, 1990.

McIlwain, C. E., Substorm injection boundaries, in *Magnetospheric Physics*, Ed. B. M. McCormac, p. 143, D. Reidel Publ. Co., Dordrecht-Holland, 1974.

McPherron, R. L., Substorm related changes in the geomagnetic tail: the growth phase, *Planet. Space Sci., 20*, 1521, 1972.

McPherron, R. L., C. T. Russell, and M. P. Aubry, Satellite studies of magnetospheric substorms on August 15, 1968, 9. Phenomenological model for substorms, *J. Geophys. Res., 78*, 3131, 1973.

Mitchell, D. G., D. J. Williams, C. Y. Huang, L. A. Frank, and C. T. Russell, Current carriers in the near-Earth cross-tail current sheet during substorm growth phase, *Geophys. Res. Lett., 17,* 583, 1990.

Moore, T. E., R. L. Arnoldy, J. Feynman, and D. A. Hardy, Propagating substorm injection fronts, *J. Geophys. Res., 86,* 6713, 1981.

Morse, T. H., and G. J. Romick, The fluctuation and fading of auroral arcs preceding auroral substorm onsets, *Geophys. Res. Letts., 9,* 1065, 1982.

Nagai, T., K. Takahashi, R. E. Lopez, R. W. McEntire, T. A. Potemra, and D. M. Klumpar, Structure of Field-aligned currents in the near-Earth magnetotail, SM 12A-4, AGU fall meeting, *EOS, 70,* 1270, 1989.

Opgenoorth, H. J., R. J. Pellinen, W. Baumjohann, E. Nielsen, G. Marklund, and L. Eliasson, Three-dimensional current flow and particle precipitation in a westward traveling surge (obseved during the barium-GEOS rocket experiment), *J. Geophys. Res., 88,* 3138, 1983.

Reiff, P. H., and J. G. Luhmann, Solar wind control of the polar cap voltage, in *Solar Wind-Magnetosphere Coupling*, edited by Y. Kamide and J. A. Slavin, 453, Terra Scientific Publishing Co., Tokyo, Japan, 1986.

Sergeev. V.A., T. Bosinger, and A.T.Y. Lui, Impulsive processes in the magnetotail during substorm expansion, *J. Geophys., 60,* 175, 1986a.

Sergeev, V.A., R.J. Pellinen, T. Bosinger, W. Baumjohann, P. Stauning, and A.T.Y. Lui, Spatial and temporal characteristics of impulsive structure of magnetospheric substorm, *J. Geophys., 60,* 186, 1986b.

Sun, W., J. R. Kan, and S.-I. Akasofu, Evolution of the magnetic field in the plasma sheet during the substorm event on March 17, 1978, *J. Geophys. Res., 96,* in press, 1991.

Tsyganenko, N. A., Global quantitative models of the geomagnetic field in the cislunar magnetosphere for different disturbance levels, *Planet. Space Sci., 35,* 1347, 1987.

Weimer, D. R., N. C. Maynard, W. J. Burke, and C. Lirbrecht, Polar cap potentials and the auroral electrojet indices, *Planet. Space Sci., 38,* 1207, 1990a.

Weimer, D. R., L. A. Reinleitner, J. R. Kan, and S.-I. Akasofu, Saturation of the auroral electrojet current and the polar cap potential, J. Geophys. Res., 95, 18,981, 1990b.

Yumoto, K., Generation and propagation mechanisms of low-latitude magnetic pulsations, *J. Geophys., 60,* 79, 1986.

Zhu, L., and J. R. Kan, Effects of ionospheric recombination time scale on the auroral signature of substorms, *J. Geophys. Res., 95,* 10,389, 1990.

3. SUBSTORM CURRENTS

An Empirical Model of Substorm-Related Magnetic Field Variations at Synchronous Orbit

TSUGUNOBU NAGAI

Meteorological Research Institute
Tsukuba, Ibaraki 305, Japan

The average time sequence of magnetic field variations at geosynchronous orbit has been constructed based on 194 substorm events. The following time sequence emerged from the analysis. First, a change in the magnetic field towards a more taillike configuration starts approximately 45 minutes prior to the expansion phase onset. At the expansion phase onset, field dipolarization occurs in a narrow local time sector centered at 2330 MLT. However, at local times more than 1 hour away from the dipolarization sector, the field becomes more taillike. The region of dipolarization eventually expands longitudinally both eastward and westward, reaching 0300 MLT in 11 minutes and 2000 MLT in 15 minutes. This analysis provides a quantitative picture of the magnetic substorm signatures in each MLT at synchronous orbit.

Introduction

Since the magnetic fields threading geosynchronous orbit intersect the Earth's surface near the auroral oval, magnetic field measurements by synchronous spacecraft offer a great advantage for studying magnetospheric substorms. Early studies using ATS 1 and ATS 6 have demonstrated that the magnetic field at synchronous orbit shows similar variations during different substorms [e.g., Cummings et al., 1968; Walker et al., 1976; Kokubun and McPherron, 1981]. The magnetic field becomes more taillike on the nightside prior to the onset of the substorm expansion phase. The onset of the expansion phase produces a reconfiguration of the magnetic field towards a more dipolar configuration. This process is called dipolarization. Studies using multiple GOES spacecraft have demonstrated the temporal and spatial development of the magnetic substorm signatures [Nagai, 1982a; Nagai et al., 1983]. In association with an onset, the dipolarization occurs only in a longitudinally narrow sector near midnight. The dipolarization onset west of this sector is significantly delayed. The dipolarization onset east of the sector is also significantly delayed. Therefore, the dipolarization region appears to expand both westward and eastward. Subsequent studies have confirmed this qualitative picture [Arnoldy and Moore, 1983; Singer et al., 1985; Nagai, 1987; Nagai et al., 1987].

Recently, it has been recognized that synchronous spacecraft have the capability to monitor substorm activity. This capability makes it possible to relate the field configuration changes in the near-Earth magnetotail with various magnetic field and plasma processes in the more distant magnetotail. It also makes it possible to find detailed relationships between substorm activity in the inner magnetosphere and that on the ground. Furthermore, there has been a growing interest in modeling substorm processes in the near-Earth magnetotail [e.g., Harel et al.,

1981; Kaufmann, 1987]. In this paper an empirical model of the magnetic field variations during substorms at synchronous orbit is constructed in order to give a quantitative basis for utilizing the magnetic field signatures at synchronous orbit.

Data

The magnetic field data are those from GOES 5 and GOES 6 during the Polar Region Outer Magnetosphere International Study interval (March 10-June 16, 1986). During this interval GOES 5 was located within a few degrees of 75° West (geographic) and GOES 6 was located within a few degrees of 108° W. The magnetic latitude was approximately 11° for GOES 5 and 9° for GOES 6, respectively. One-minute average data were produced from the original GOES data sampled at 3.06 sec intervals. The magnetic field data are presented in dipole VDH coordinates. In this system, H is along the dipole axis (northward positive), D is azimuthally eastward, and V is outward. The total field (BT) and the inclination (INC), which is defined as $INC=\tan^{-1}(-H/V)$, are also presented.

One hundred and ninety-four substorm events were selected from the PROMIS interval. The substorms were defined using particle injections and/or particle flux recoveries as observed by the Los Alamos Charged Particle Analyzer on board s/c 1982-019 (37°W), s/c 1984-129 (155°W), and s/c 1984-037 (70°E) and the Space Environment Monitor on board GMS-3 (140°E). Using the network of these four spacecraft, one can easily detect substorms at any UT time [e.g., Nagai, 1982b]. The substorm time, which is defined as the start time of flux change, may not be the precise onset time of the substorm expansion phase because of inadequate spacecraft location relative to the substorm activity. However, this will not effect the results in the present study because we only use the substorm time as a rough measure for the onset time.

Analyses

For each substorm event, the onset time of local dipolarization is determined for GOES 5 and GOES 6. When the GOES spacecraft were located in the 1800-0300 MLT, the local dipolarization was detected for more than 80 % of the events. The dipolarization onset time was defined as the time when the field inclination (INC) reaches its local minimum near the substorm time. The difference of the GOES 5 dipolarization onset time from the GOES 6 dipolarization onset time was examined for events occurring in the 2200-1000 UT window. The results are presented in Figure 1. Each delay/advance time is plotted as a function of the substorm time in UT. Although the data points are scattered, a definite pattern can be found. When both GOES 5 and GOES 6 are in the pre-midnight sector (before 0500 UT), the local dipolarization starts first at GOES 5 and then at GOES 6. In this case GOES 5 is near midnight. Both GOES 6 and GOES 5 are in the post-midnight sector, the local dipolarization starts first at GOES 6 and then at GOES 5. In this case GOES 6 is near midnight. This analysis clearly demonstrates that the dipolarized region expands both eastward and westward.

This azimuthal expansion can be modeled by a simple function. Let T(x) be the dipolarization onset time at local time x, with respect to the universal onset time of a substorm. Then, the time difference ΔT between GOES 5 and GOES 6 can be formally written as

$$\Delta T = T(x) - T(x - \Delta x) \approx \Delta x \, dT/dx \qquad (1)$$

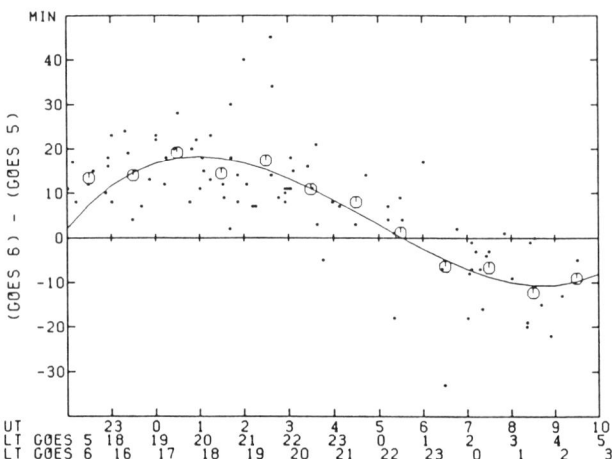

Fig. 1 Time lag of the local dipolarization onset between GOES 6 and GOES 5 as a function of substorm onset time. Large circle is the average in each local time bin. The 3-rd order polynomial equation fitting the bin averages is also plotted.

where Δx is the difference (=2.2 hrs) in local time. From least squares fitting of a polynomial to the data in Figure 1, we get

$$\Delta T = 0.06x^3 + 0.19x^2 - 2.31x - 1.13 \quad (2)$$

By integrating equation (2), we can get the expression for T(x). The integrated function includes an integral constant. It is known that in some cases a local dipolarization at synchronous orbit is observed exactly at the onset of the expansion phase, which is determined by well-known ground magnetic field signatures; Pi 2 onset and mid-latitude positive bay onset. A good example is the August 19, 1978 event presented in Nagai [1982a]. Hence, we can determine the integral constant in order to make the integrated function having zero as minimum. The integrated function can be represented by

$$T = -0.01x^4 - 0.06x^3 + 1.16x^2 + 1.13x + 0.27 \quad (3)$$

In this formula, T is in minutes and x is MLT hours. This function has its minimum at 2332 MLT, as seen in Figure 2.

The average magnetic field variations at each local time in the 1800-0300 MLT sector are obtained by superposed epoch analysis. For example, the variations at 00 MLT are derived from the events for which the substorm times are in the 2300-0100 MLT period. The magnetic field data for each spacecraft are superposed by choosing the local dipolarization onset as zero epoch. In these data, the quiet day diurnal variation (the averages of three near-by quiet days) is subtracted. Although the average variations at GOES 5 are slightly different from those at GOES 6 because of magnetic latitude effect, we make averages of these two variations in order to have good statistics. And then the average variations are plotted taking account of the averaged delay time determined by the equation (2).

Figure 3 shows the average AU and AL variations which were obtained by using the events in the 2300-0100 MLT interval. It is important to note that the substorm events used in this analysis are not isolated. Prior to the zero epoch, the AL is almost constant around -150 nT and the AU is above 90 nT. The averaged AU shows an increase after T=-40 minutes. This might be a 'growth phase' signature. The averaged AL develops after zero epoch and it reaches its minimum at T=+33 minutes. The amplitude of the AL change is 150 nT. The AL does not return to the pre-substorm value even after T=+120 minutes.

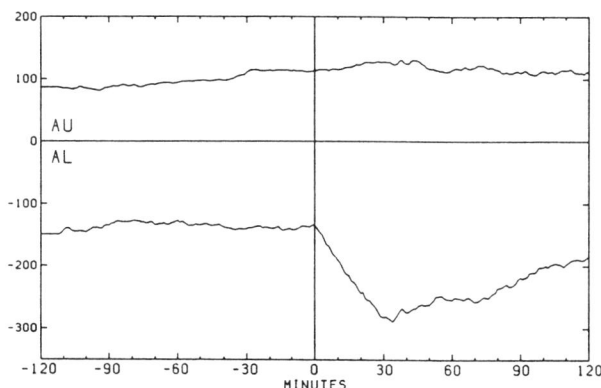

Fig. 3 Average variations of AU and AL in the substorms used in this analysis.

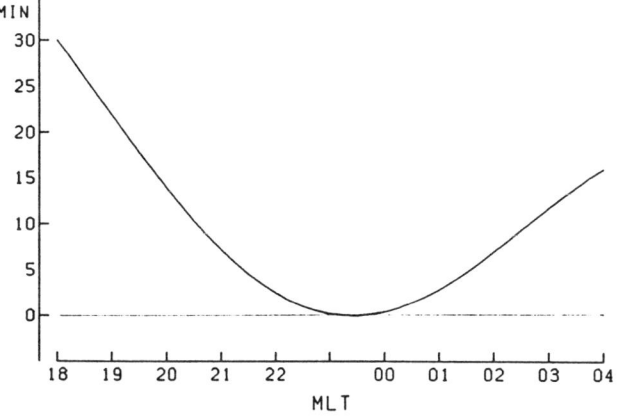

Fig. 2 Delay time of local dipolarization onset at synchronous orbit in each local time with respect to the onset of the substorm expansion phase.

The average magnetic field variations are presented in Figure 4. The zero minute in the figure is the onset time of the substorm expansion phase. Although only perturbations are derived in each component, there is still a trend corresponding to the diurnal variation. This is because the diurnal variation in disturbed conditions is different from that in quiet conditions. This is also because geomagnetic conditions at T=+120 minutes are different from those at T=-120 minutes, as seen in Figure 3. The

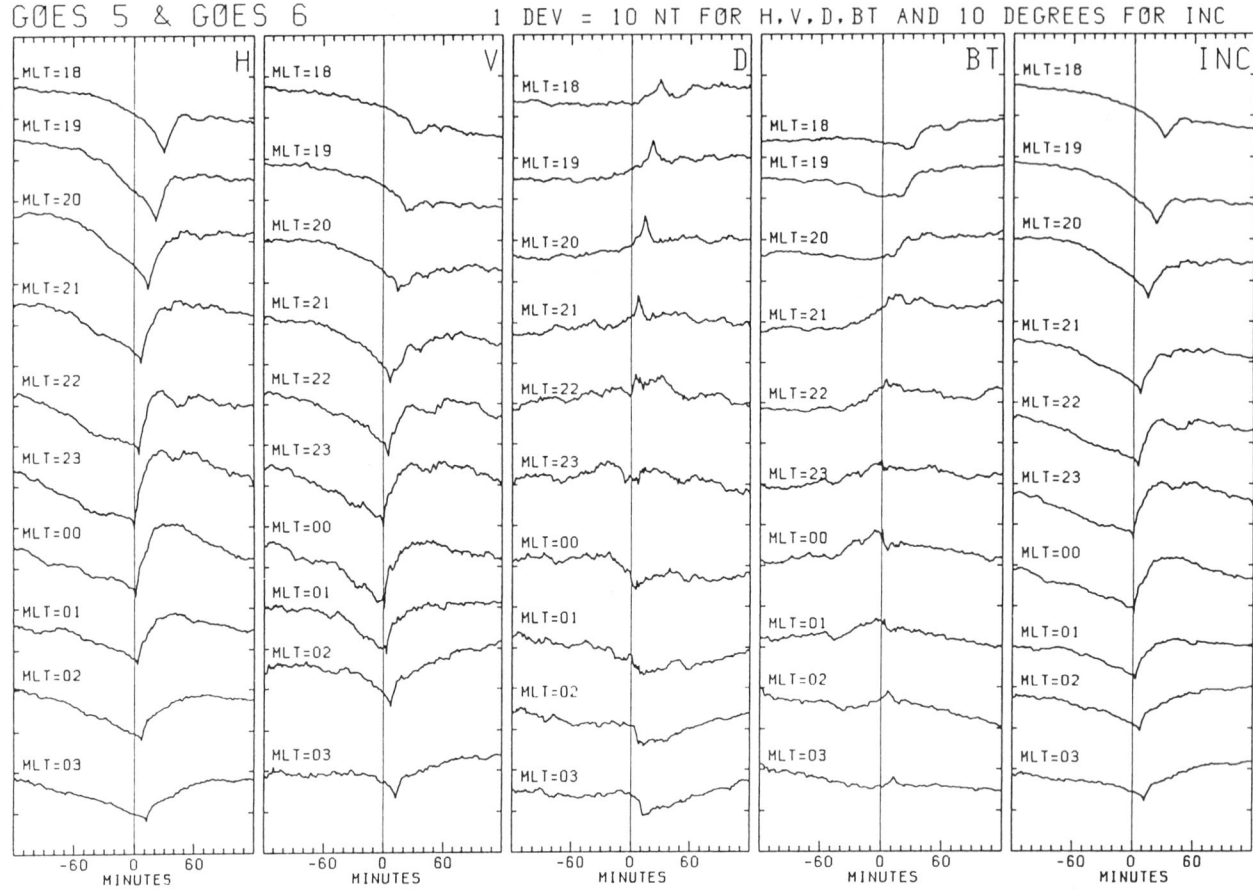

Fig. 4 Average magnetic field variations during substorms derived from GOES 5 and GOES 6 data. MLT=18 means that the zero time corresponds to the 1800 MLT for the trace.

total field increases prior to the onset time and it returns to the pre-substorm level in association with the dipolarization in the 2100-0300 MLT interval. This is characteristic at high latitudes. At low latitudes the total field decreases prior to the onset time because the perturbations in the H component are dominant in comparison with the perturbations in the V component.

The D component shows effect due to substorm-associated field-aligned currents. A sharp change in the D component, positive in the 1800-2200 MLT sector and negative in the 0000-0300 MLT sector, respectively, starts at T=0, except at 2300 LT. Gradual variations appear to start prior to T=0. Some of the events have several onsets prior to the major onset. These onsets produces these variations prior to T=0. Irregular variations at 2300 MLT may be due to the fact that the substorm center differs in MLT for each event. A positive or negative D peak occurs at the local dipolarization onset except 2300 MLT. The effect of the field-aligned currents almost disappears around T=+60 minutes.

The salient features of the field configuration changes are as follows:

T=-45 minutes

It is difficult to determine unambiguously the start time when the magnetic field becomes more taillike. However, after T=-45 minutes the total field increases evidently in the 2100-0100 MLT sector and the increase in the total field is the most evident at 0000 MLT. Although the total field shows an increase in the 0200-0300 MLT sector, the start time of the increase is significantly delayed. In the 1800-2000 MLT sector, the total field shows a decrease prior to the onset time.

T= 0 minute

A dipolarization in the field configuration starts near 2330 MLT. The longitudinal extent of the dipolarization region cannot be determined in this analysis. The data in which GOES 5 and GOES 6 are near midnight are examined for the period October 1984 - December 1986. There is no simultaneous (within 1 minute) local dipolarization onset at GOES 5 and GOES 6 for 30 well-isolated substorm events. This result indicates that the longitudinal extent of the initial dipolarization region is less than 2 hours. Outside the dipolarization region the field becomes further taillike. At T=+5 minutes the dipolarization starts near 2000 MLT and near 0100 MLT.

T=+25 minutes

The local dipolarization is in progress near 0300 MLT and the local dipolarization starts around 1900 MLT. In the 2100-0100 MLT sector the dipolarization is completed by this time. By T=+60 minutes the dipolarization is completed.

Concluding Remarks

The magnetic field at synchronous orbit shows significant changes prior to the onset of the expansion phase. This fact indicates that the 'growth phase signature' is more evident in the near-Earth magnetotail. Although the magnetotail shows the 'growth phase signature' in a wide longitudinal extent, the initial onset starts only in a longitudinally narrow sector of the magnetotail. Hence, the 'growth phase signature' and the 'expansion phase signature' coexist in the near-Earth magnetotail just after the onset time. The magnetic field returns to the pre-substorm state during the expansion phase and any signature is not evident in the near-Earth magnetotail for the recovery phase. These characteristics are important for using the magnetic field data at synchronous orbit.

Acknowledgments. The author thanks S. Kokubun and K. Takahashi for informative discussion. The GOES data were obtained from the NOAA National Geophysical Data Center. The Los Alamos particle data are those in "Electron and ion data from geosynchronous satellites 1982-019, 1984-037, and 1984-129" PROMIS SERIES Volume 6 published by Los Alamos National Laboratory.

References

Arnoldy, R. L., and T. E. Moore, Longitudinal structure of substorm injections at synchronous orbit, J. Geophys. Res., 88, 6213, 1983.

Cummings, W. D., J. N. Barfield, and P. J. Coleman, Jr., Magnetospheric substorms observed at the synchronous orbit, J. Geophys. Res., 73, 6687, 1968.

Harel, M., R. A. Wolf, P. H. Reiff, R. W. Spiro, W. J. Burke, F. J. Rich, and M. Smiddy, Quantitative simulation of a magnetospheric substorm, 1, Model logic and overview, J. Geophys. Res., 86, 2217, 1981.

Kaufmann, R. L., Substorm currents: Growth phase and onset, J. Geophys. Res., 92, 7471, 1987.

Kokubun, S., and R. L. McPherron, Substorm signatures at synchronous altitude, J. Geophys. Res., 86, 11265, 1981.

Nagai, T., Observed magnetic substorm signatures at synchronous altitude, J. Geophys. Res., 87, 4405, 1982a.

Nagai, T., Local time dependence of electron flux changes during substorms derived from multi-satellite observations at synchronous orbit, J. Geophys. Res., 87, 3456, 1982b.

Nagai, T., Field-aligned currents associated with substorms in the vicinity of synchronous orbit: 2. GOES 2 and GOES 3 observations, J. Geophys. Res., 92, 2432, 1987.

Nagai, T., D. N. Baker, and P. R. Higbie, Development of substorm activity in multiple-onset substorms at synchronous orbit, J. Geophys. Res., 88, 6994, 1983.

Nagai, T., H. J. Singer, B. G. Ledley, and R. C. Olsen, Field-aligned currents associated with substorms in the vicinity of synchronous orbit: 1. The July 5, 1979 substorm observed by SCATHA, GOES 3 and GOES 2, J. Geophys. Res., 92, 2425, 1987.

Singer, H. J., W. J. Hughes, C. Gelpi, and B. G. Ledley, Magnetic disturbances in the vicinity of the synchronous orbit and the substorm current wedge: A case study, J. Geophys. Res., 90, 9583, 1985.

Walker, R. J., and K. N. Erickson, R. L. Swanson, and J. R. Winckler, Substorm-associated particle boundary motion at synchronous orbit, J. Geophys. Res., 81, 5541, 1976.

THE RELATIONSHIP BETWEEN ION AND ELECTRON PRECIPITATION PATTERNS AND FIELD-ALIGNED CURRENT SYSTEMS DURING A SUBSTORM

T. Iijima[1], M. Watanabe[1], T.A. Potemra[2], L.J. Zanetti[2], and F.J. Rich[3]

Abstract. We show in this study the persistency of large-scale characteristics in auroral particle precipitation patterns during a prolonged disturbed period which included a growth phase and multiple expansion onsets. The north-south conjugacy of these patterns and their relationship with the field-aligned currents are also discussed. We have investigated these characteristics by using the magnetic field and the plasma measurements acquired with the DMSP F7 satellite at an altitude of ~840 km, and in the premidnight MLT sector. Principal characteristics determined here include the following: (1) The plasma precipitation characteristics during substorms consist of three distinctive patterns, or parts that are denoted here as "C", "B", and "A" from the lowest latitude, respectively, for both the ions and electrons; (2) The C plasma precipitation pattern is characterized by a double-energy composite structure of the ions that comprises the high-energy component (energy>a few keV) and the low-energy component (<a few keV). Its high-latitude limit is indicative of the outer boundary in the magnetosphere of the ion low- energy component domain, and the outer boundary of the earthward injecting plasma. Both of these moved toward the earth throughout the course of a prolonged disturbed period. The C pattern is thought to be the quasi-persistent core part of the plasma precipitation; (3) The B plasma precipitation pattern occurs poleward of C. The B pattern is primarily characterized by a highly structured intensity enhancement in the E-t spectrum of both the ions and electrons, which included the features suggestive of an existence of field-aligned electric field directed toward and away from the earth. The B pattern expanded drastically over a much wider latitudinal span, especially after the onset of substorm expansion phase through the recovery phase, and is thought to be the explosive part of the plasma precipitation; (4) At the polewardmost latitude of the plasma precipitation (adjacent to and poleward of B), there exists a distinctive part, A that is principally characterized by its association of a field-aligned flow of the ions. The A pattern was not recognizable during the growth phase, but occurred exclusively after the onset of substorm expansion phase; (5) A general relationship between the field-aligned currents and the plasma precipitation was evident for both the northern and southern hemisphere. Namely, the traditional Region 2 current system nearly was associated with the C part (quasi-persistent core part) and the traditional Region 1 current system was associated mostly with the B part (explosive part) of the plasma precipitation; (6) In the same premidnight MLT sector, the observed field-aligned currents above the northern (winter) polar ionosphere had the normal pattern for the evening-side traditional Region 2 and Region 1 current systems. Whereas, above the southern (summer) polar ionosphere, the observed field-aligned currents had the morning-side pattern of the traditional Region 2 and Region 1 current systems (with the morning-side pattern overlapping with the evening-side pattern at the equatorwardmost latitudes).

Introduction

Over more than two decades, a number of studies have investigated the various characteristics of magnetospheric substorms. These included the electric current systems in the ionosphere and the magnetosphere, and the plasma populations filling the magnetic flux tubes over the wide area of the magnetosphere down to the ionospheric altitudes. For example, from measurement of charged particles or the magnetic field, the spatial distribution pattern and its changes during substorms were determined for the electron precipitation (in the energy range from 10 eV to 10 keV and in the 2100-0300 MLT sector) at the ISIS 1,2 satellite altitudes (which consists of the CPS and BPS precipitations) by Winningham et al. [1975]; and for the large-scale field-aligned currents at ~800-900 km altitude (which consists of the Region 1 and Region 2 systems with the complicated, multiple currents in the ~2000-2400 MLT sector) by Iijima and Potemra [1978]. Later, with simultaneous magnetic field and plasma and/or electric field measurements, various implications for the source regions and the source mechanisms of the field-aligned currents were presented in relation to the magnetospheric boundary layer, the plasma sheet

---

[1]Department of Earth and Planetary Physics, Faculty of Science, The University of Tokyo, Bunkyo-ku, Tokyo 113, Japan.

[2]Applied Physics Laboratory, The Johns Hopkins University, Laurel, Maryland 20723, U.S.A.

[3]Geophysics Laboratory, Hanscom Air Force Base, Bedford, Massachusetts 017311, U.S.A.

Magnetospheric Substorms
Geophysical Monograph 64
Copyright 1991 American Geophysical Union

98 PARTICLES AND CURRENTS DURING A SUBSTORM

and the ring current (radiation belt) domain. Toward the nightside MLTs, the field-aligned currents were seen to be collocated exclusively with the charged particles that are usually used to identify the field lines threading the plasma sheet [e.g. Klumpar, 1979; Frank et al., 1981; Sugiura et al., 1984; Kelley et al., 1986; Heinemann et al., 1989; Fujii et al., 1990].

In the past studies, however, the concurrent characteristics of the field-aligned currents, the ion population, and the electron population of the plasma sheet domain, have not been clarified throughout the course of substorm. The main purpose of this paper is to determine the substorm-associated, concurrent characteristics of the field-aligned currents and the precipitating ions and electrons at an altitude of ~840 km in the premidnight MLT sector, and their implications for the magnetotail dynamics. We will report a case study of a prolonged disturbed period which included the growth phase and multiple expansion onsets.

Data

We have investigated the characteristics of the field-aligned currents and the earthward fluxes carried by the ions and electrons within the 30 eV to 30 keV energy range, by using the magnetic field and the plasma measurements acquired with the DMSP F7 satellite which encircled the earth at an altitude of ~840 km in the prenoon-premidnight MLT (magnetic local time) sector. For the details of these experiments, see Rich et al. [1985]. As in the past studies, the vector residual ΔB (the measured magnetic field minus the model main field [Langel et al., 1980]) was separated into $\Delta B_{\parallel}$ and $\Delta B_{\perp}$ (magnetic perturbation parallel and transverse to the model main field). The $\Delta B_{\perp}$ was further separated into $\Delta B_{\perp 1}$ (transverse to the satellite track) and $\Delta B_{\perp 2}$ (parallel to the satellite track). Here, the presence of the field-aligned current has been determined by the $\Delta B_{\perp 1}$ component, assuming an infinite current sheet that is transverse to the satellite track. Starting from the differential number flux, we have calculated the total number flux (particles/cm^2 · sr · sec), the total energy flux (eV/cm^2 · sr · sec) and the average energy (eV) for both the earthward flux of the ions and the electrons by integrating the flux over the energy range.

In this paper, we will focus on the features of the field-aligned currents and the earthward plasma flux (plasma precipitation at an altitude of ~840 km) that were observed during a prolonged disturbed period occurring from ~0600 to ~1700 UT, December 13, 1985. The onset time of an (first) expansion phase was determined as ~0740 UT, by examining the ground-based world-wide geomagnetic data obtained both in the auroral zone (the AE index is given below) and at middle and low latitudes (not shown here) as in the past study [e.g. Iijima and Nagata, 1972]. During this prolonged disturbed period, the magnetic field data and the plasma flux data were acquired with the DMSP F7 during seven consecutive orbits over the northern polar ionosphere and six consecutive orbits over the southern polar ionosphere, as shown in Figure 1.

Principal Parts of Plasma Precipitation

First, we will demonstrate the features of the precipitating (at an altitude ~840 km) ions and electrons. We determined them

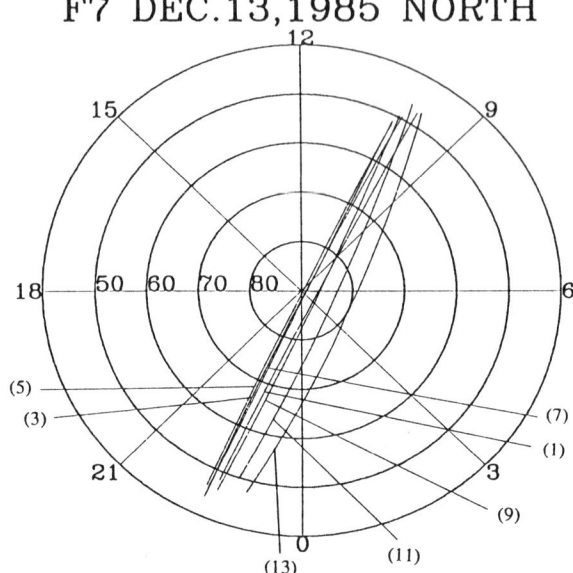

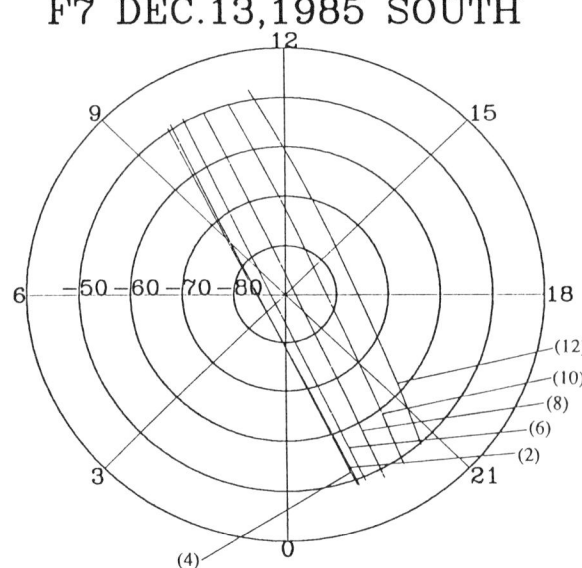

Fig. 1. Polar plot in ML (magnetic latitude) - MLT (magnetic local time) dial of the orbital tracks with the DMSP F7 satellite for our present data from ~0600 to ~1700 UT, December 13, 1985. Top: seven consecutive orbits above the northern polar ionosphere. Bottom: six consecutive orbits above the southern polar ionosphere.

by using primarily the energy flux vs. time (≈magnetic latitude) spectrogram (i.e. the so-called E-t spectrum) (Plates 1 and 2), the velocity distribution function of precipitating particles (Figure 3) and by the profiles in magnetic latitude of the total number flux, the total energy flux and the average energy (Figure 2). Plate 1 shows the E-t spectrum for six consecutive orbits above the southern polar ionosphere, arranged in time sequence from the top, corresponding to the growth phase, to the bottom corresponding to the recovery phase for both the ions (left) and electrons (right). Plate 2 corresponds to the E-t spectrum for seven consecutive orbits above the northern polar ionosphere, in time sequence from the top corresponding to the growth phase, to the bottom two corresponding to the recovery phase for both the ions (left) and electrons (right). We have ascribed the characteristics of plasma precipitation to as conforming to three distinctive patterns, or parts that are denoted here as "C", "B" and "A" from the lowest latitude side, respectively, for both the ions and electrons.

Part C. This corresponds to the equatorwardmost part of the plasma precipitation, and exists almost persistently throughout the growth phase, the expansion phase and the recovery phase of substorm. The vertical white lines in Plates 1 and 2 indicate the high-latitude limit of this C part. Electron signature: The C part corresponds to an arch-shaped structure that was observed at the lowest latitudes in every E-t spectrum in Plates 1 (right) and 2 (right). This part is bounded usually on the high-latitude limit by a sharp, almost vertically enhanced spectrum (centered around $\sim$10 keV and with its trend extending down to $\sim$100 eV, and/or less than this) that is indicative of a dispersionless-like intensity augumentation, independent of the earthward velocity of the particles. The right and the left example in Figure 2 (thin line for electrons) corresponded to the third top panel in Plate 1 (right) and the second top panel in Plate 2 (right), respectively. In both examples, the C electron precipitation part that was primarily identified with the E-t spectrum, is signified in the profiles of the total number flux and the total energy flux as that its low-latitude limit is initiated approximately by a sharp increase in these profiles and its high- latitude limit is bounded nearly by a plateau in these profiles. These characteristics in the profiles of the total (number and energy) fluxes that signified the C electron precipitation part were observed commonly in all examples of 13 orbits in our present study. Ion signature: The C part corresponds to a double-energy composite structure that was observed at the lowest latitudes in every E-t spectrum in Plates 1 (left) and 2 (left). One exhibits the high-energy component, which prevails from 30 keV (high-energy limit in the DMSP F7 experiment) down to a range of keV, and the other exhibits the low-energy component which extends from 30 eV (low-energy limit in the experiment) upto $\sim$keV range. For the examples observed during the growth phase (the top panel in both Plate 1 (left) and 2 (left)), C part is bounded on the high-latitude limit by a sharp cut off of the high-energy component. In the other examples observed during the expansion phase and the recovery phase in Plates 1 (left) and 2 (left), the low-energy component of C part exhibits a dispersion-like trend in the intensity enhancement which begins at the lowest energy at the lowest latitude, extends toward the higher energy ($\sim$keV) at the higher latitude and merges into the high-energy component. The high-latitude limit of the C part is then identified as the end of this dispersion-like trend. The right and the left example in Figure 2 (thick line for ions) corresponded to the third top panel in Plate 1 (left) and the second top panel in Plate 2 (left), respectively. Both examples show that the low-latitude limit of the C ion precipitation part (identified with the E-t spectrum) is initiated by a sharp increase in the profiles of the total number flux (and nearly also in the total energy flux) and the high-latitude limit of the C part corresponds approximately to a plateau in these profiles. Furthermore, this C ion precipitation part corresponds to a trough in the profile of the average energy. These characteristics in the profiles of the total (number and energy) fluxes and the average energy were commonly observed in all examples except for the growth phase examples.

Examples of the velocity distribution function, f(v), that corresponded to the C plasma precipitation part are shown in Figure 3 (the bottom left). We have plotted f(v) with a linear energy scale and a logarithmic phase space density scale. The straight line fit of f(v) implies that the distribution is Maxwellian, with its slope inversely proportional to the temperature of particle population. The f(v) of the C electron precipitation part is close to a straight line, suggesting a Maxwellian having a low average energy. The f(v) of the C ion precipitation part was approximately represented by two contiguous broken lines, one with a sharp slope (suggesting low energy ions), and the other with a gradual slope (high energy ions). These characterized the double energy composite structure, as already ascribed to a primary signature of the C ion precipitation part with the E-t spectrum. The demarcation energy that divides the low energy ions and the high energy ions resides around a few keV.

Part B. This part occurs poleward of C, and is bounded on its high-latitude by the low-latitude limit of the A (mentioned below). In Plates 1 (right) and 2 (right), the E-t spectrum of the precipitating electrons exhibited a relatively discrete, intermittent intensity enhancement in multiple, poleward of the vertical white line (poleward of C). In Plates 1 (left) and 2 (left), the E-t spectrum of the precipitating ions showed a curtain-shaped intensity enhancement mostly in the high-energy component (a range of keV), poleward of C (poleward of the vertical line).

The f(v) characterizing this B precipitation part includes three types, which are shown by three upper panels in Figure 3. Namely, part B-1 type (top left), part B-2 type (top middle) and part B-3 type (top right). The f(v) of both the ions and electrons in the part B-1 are approximated by the straight lines, suggesting that the distributions resemble a theoretical Maxwellian. We duplicated the f(v) of part B-1 by a dotted line in the f(v) of part B-2 and part B-3. It is seen that the ion f(v) of part B-2 and electron f(v) of part B-3 depart from a theoretical Maxwellian. For part B-2, the ion f(v) seems to result from the shifting of a dotted line ($\sim$theoretical Maxwellian) by $\sim$10 keV toward the higher energy, and the electron f(v) diminished, especially for the energies larger than $\sim$10 keV, which suggests an existence of field- aligned electric field directed toward the earth. For part B-3, the electron f(v) seems to be a shifting of a dotted line ($\sim$theoretical Maxwellian) by $\sim$15-20 keV toward higher energy and ion f(v) became totally vanishingly small, suggesting an existence of field-aligned electric field directed away from the earth.

The right and left examples in Figure 2 correspond to the third top panel in Plate 1 and the second top panel in Plate 2. Both examples show that the B part of both the precipitating ions and electrons includes an irregular variation in the profiles of the total (number and energy) fluxes and the average energy. As seen

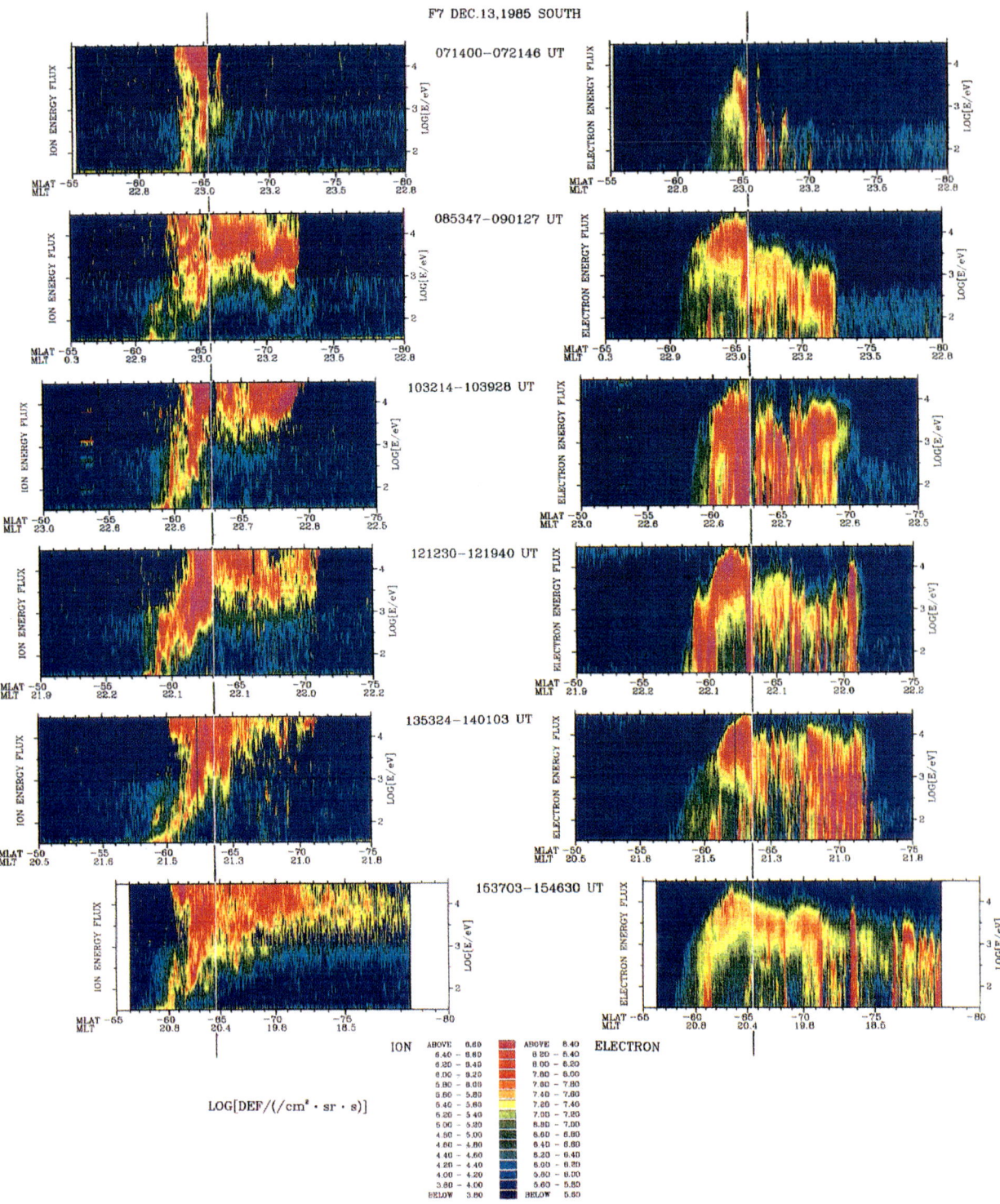

Plate 1. Energy flux vs. time (≈magnetic latitude) spectrogram for six consecutive orbits above the southern polar ionosphere that are arranged in time sequence from the top corresponding to the growth phase to the bottom corresponding to the recovery phase for precipitating ions (left) and electrons (right). The vertical white lines indicate the high-latitude limit of the C part of the plasma precipitation (see the text).

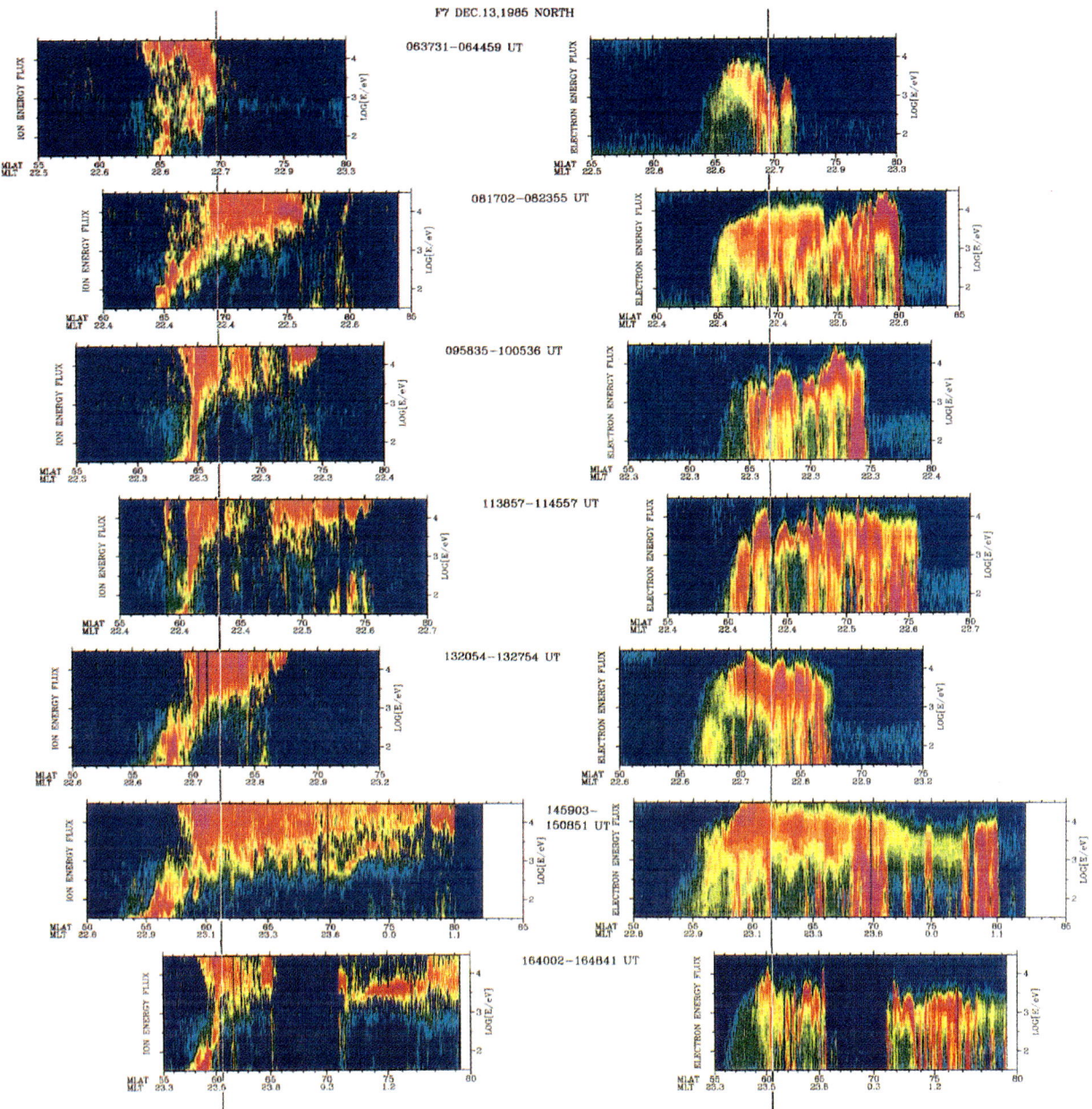

Plate 2. Energy flux vs. time spectrogram for seven consecutive orbits above the northern polar ionosphere, arranged in time sequence from the top corresponding to the growth phase to the bottom two corresponding to the recovery phase for precipitating (left) and electrons (right). The vertical white lines show the high-latitude limit of the C part plasma precipitation.

in Plates 1 and 2, although the B part is discernible (but very weakly), even during the growth phase (the top panel) for the precipitating electrons and ions, it is strongly augumentated in intensity and expands drastically in latitude (mostly poleward) when the substorm has progressed into and through the expansion phase and the recovery phase.

Part A. This corresponds to the polewardmost part of the plasma precipitation and was observed exclusively during the expansion phase and the recovery phase, except for the growth phase. We have identified this part primarily by the following two facts: The profiles in the total number flux and the total energy flux show a bump that is followed by a sharp drop to the

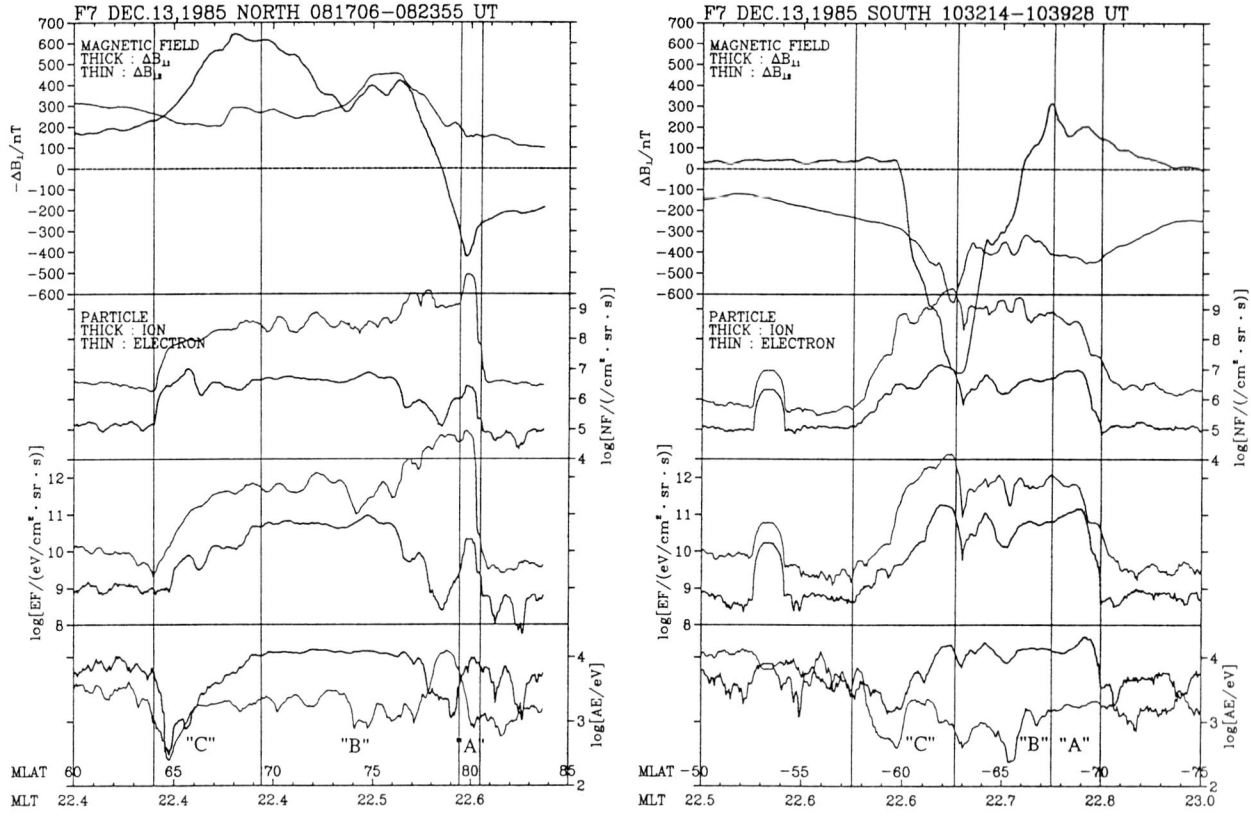

Fig.2. Examples of concurrent characteristics of field-aligned current (top panel) and plasma precipitation (second top: total number flux, third top: total energy flux, bottom: average energy) of ions (thick line) and electrons (thin line) observed during the substorm expansion phase above the northern polar ionosphere (left block) and the southern polar ionosphere (right block) in the premidnight MLT sector. The "C", "B", "A" bounded by vertical lines denote three distinctive parts of the plasma precipitation (see the text).

adjacent background level for both the ions and the electrons, as seen in two examples, labelled as "A", given in Figure 2; The f(v) with which we have identified the A precipitation part are shown at the bottom right in Figure 3. The f(v) of A ion precipitation part seems to result from the shifting of a dotted line ($\sim$theoretical Maxwellian) along the velocity axis. The concurrent f(v) of the precipitating electrons shows nearly a Maxwellian distribution. This different feature of the f(v) between the ions and the electrons suggests that the ion f(v) exhibits a drifting Maxwellian, not a field-aligned energization. We have used the measurements confined to a single look direction and therefore, a drifting Maxwellian, if this were recognizable, does not prove the existence of a net earthward bulk flow of the ions but implies the existence of a field-aligned flow of the ions. This feature could distinguish the A precipitation part from the B precipitation part, which was hardly recognizable in the E-t spectrum of the ions and the electrons.

In summary, both for the precipitating electrons and ions, the C part is the quasi-persistent core part residing at the equatorwardmost of the precipitation, the B part is the explosive part that expands drastically after the onset of substorm expansion phase, and the A part is the polewardmost layer associated with the existence of a field-aligned flow of the ions.

Relationship with Field-aligned Currents.

In Figure 2, the top panel shows the transverse magnetic disturbance whose period is longer than 9 seconds (equivalent to a spatial wave length of $\sim$68 km at an altitude of $\sim$840 km) by a numerical filtering technique. Here, we have determined the field-aligned currents exclusively by the $\Delta B_{\perp 1}$ component (transverse to satellite track, directed to dawn-to-dusk direction and shown by thick line in the figure) by assuming an infinite current sheet that is elongated transverse to the satellite track, approximately in the magnetic east-west direction. In the right block example (above the southern polar ionosphere), we can identify the $\Delta B_{\perp 1}$ disturbance (positive $\Delta B_{\perp 1}$ corresponding to westward component) with three large-scale field-aligned currents: (1) from the lowest latitude, current flowing away from the ionosphere (59.5°-62.7° in magnetic latitude (ML)) which is nearly collocated with the C part (the quasi-persistent core part) of the plasma precipitation (which was marked by the ion precipitation in this figure), (2)

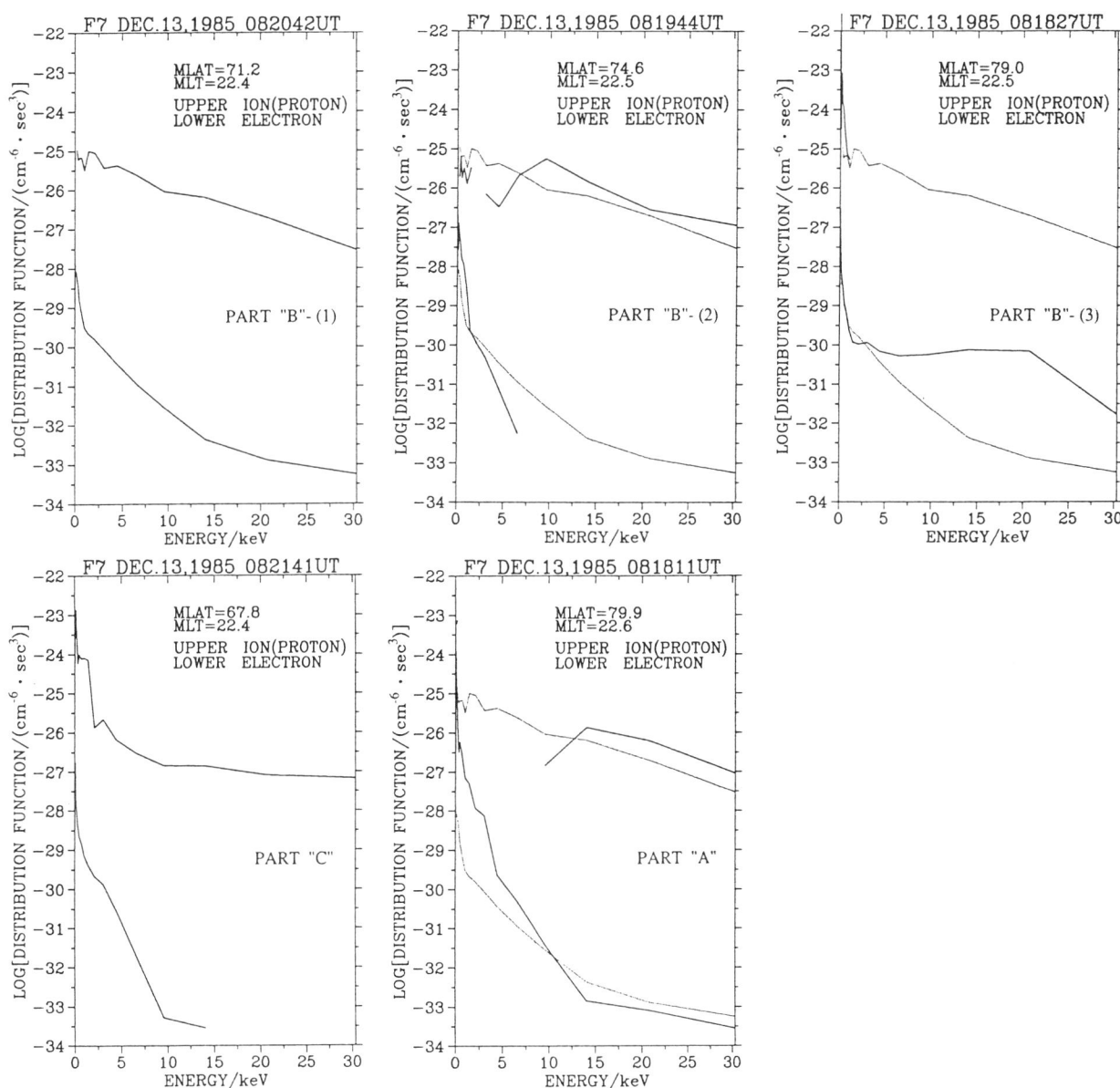

Fig. 3. Velocity distribution function of precipitating ions (upper line) and electrons (lower line) that are used in this study to characterize three distinctive parts ("C", "B" and "A") of the plasma precipitation. This example corresponds to the second top panel in Plate 2 and the left block in Figure 2.

current flowing into the ionosphere (62.7°-67.3° ML) which approximately corresponds to the B part (the explosive part) of the plasma precipitation, and (3) much weaker current flowing away from the ionosphere at the polewardmost part (67.3°-73.5°) which spanned the A part (the layer of field-aligned flow of the ions) of the plasma precipitation and further poleward. If we refer to a classic phenomenological model by Iijima and Potemra [1978], these field-aligned currents seem to conform to the morning-type Region 2 and Region 1 current systems, associating a reversed Region 1-sense current system at the polewardmost latitudes. In the

left block example (above the northern polar ionosphere, negative $\Delta B_{\perp 1}$ corresponding to eastward component), we can identify the presence of three large-scale field-aligned currents: (1) from the lowest latitude, current flowing into the ionosphere (62.5°-68.0° ML) which nearly corresponded to the C part (the core part) plasma precipitation, (2) current flowing away from the ionosphere (68.0°-79.7° ML) that was approximately collocated with the B part (the explosive part) plasma precipitation, and (3) much weaker current flowing into the ionosphere at the polewardmost

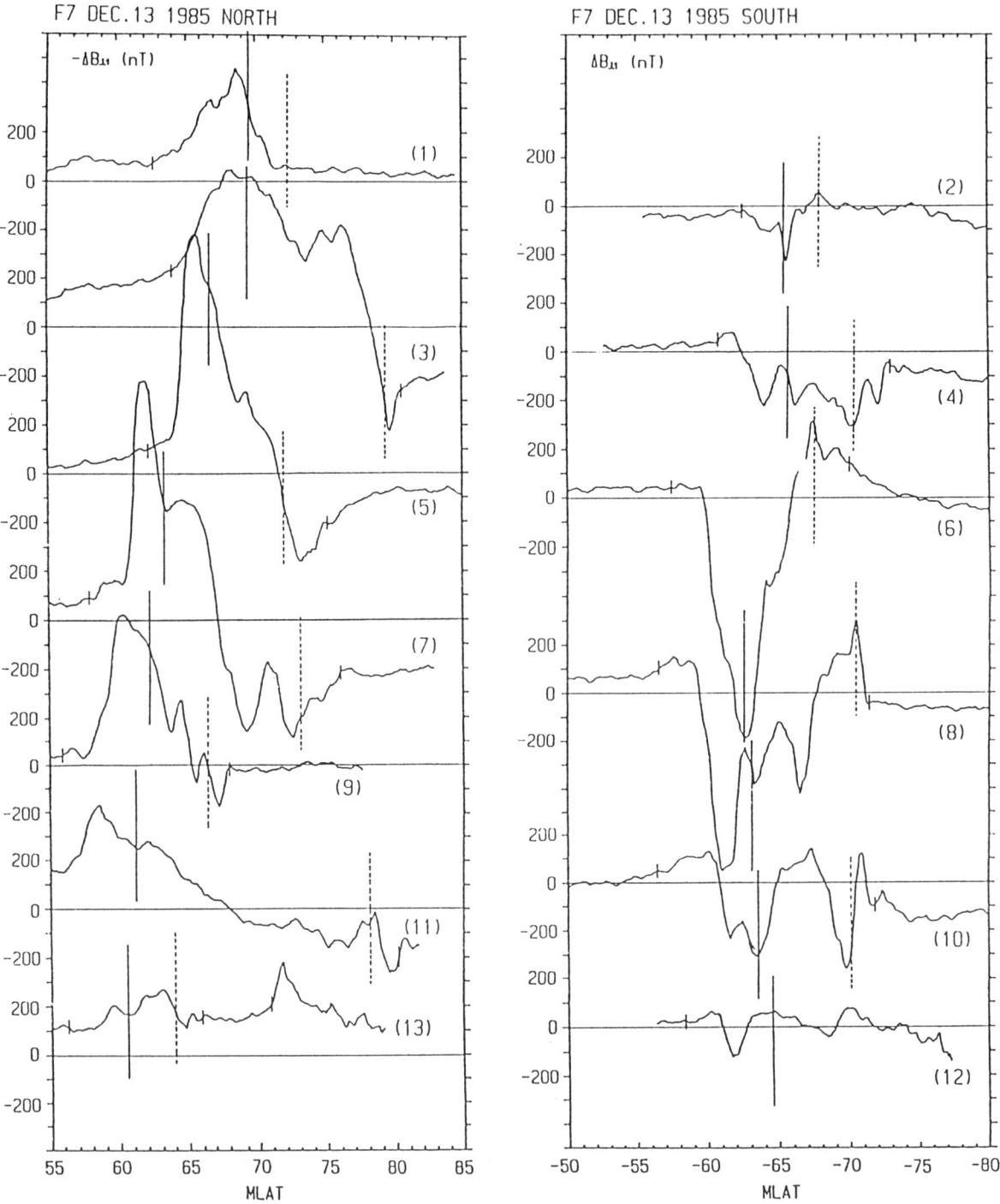

Fig.4. Transverse magnetic disturbance component, $\Delta B_{\perp 1}$ that is used here to identify a field-aligned current sheet that is elongated transverse to the satellite track. Left block: for seven consecutive orbits above the northern polar ionosphere. Right block: for six consecutive orbits above the southern polar ionosphere. Both of them are arranged in time sequence from the top corresponding to the growth phase to the bottom corresponding to the recovery phase. The solid and dotted vertical line denotes, respectively, the C part / B part boundary and the B part / A part boundary of the ion precipitation. Low-latitude limit of the C part and the high-latitude limit of the A part are marked by short lines.

part (79.7°-82.1° ML) which corresponded nearly to the A part (the layer of field-aligned flow of the ions) plasma precipitation. These field-aligned currents seem to conform to the evening-type Region 2 and Region 1 current systems, associating a reversed Region 1-sense current system at the polewardmost latitudes.

Figure 4 shows the $\Delta B_{\perp 1}$ disturbances that identify the field-aligned currents for seven consecutive orbits above the northern polar ionosphere (left block) and six consecutive orbits above the southern polar ionosphere (right block). These are arranged in time sequence from the top (corresponding to the growth phase) to the bottom (corresponding to the recovery phase). All these orbits were included in the premidnight MLT sector shown in Figure 1. In each panel, the solid vertical line and the dotted vertical denotes, respectively, the boundary between the C part and the B part and the boundary between the B part and the A part of the ion precipitation. The low-latitude limit of the C part and the high-latitude limit of the A part are also marked by the short lines. The following facts are noteworthy in this figure. In the nearly same pre-midnight MLT sector, the field-aligned current systems exhibits a set of reversed flow directions between the northern (winter) and the southern (summer) hemisphere. Above the northern polar ionosphere (left block of Figure 4), the observed field-aligned currents comprise a large current flowing into the ionosphere at the lowest latitude, a large current flowing away from the ionosphere adjacent to and poleward of the former, and except for the growth phase (top panel), a small current flowing into the ionosphere at the most poleward latitudes. These conform to the Region 2 and Region 1 field-aligned currents which had the normal pattern for the evening-side belt of the traditional model [Iijima and Potemra, 1978], associating a reversed Region 1-sense current at the most poleward latitudes. Above the southern polar ionosphere (right block of Figure 4), the observed field-aligned currents has the morning-side pattern of the traditional Region 2 (flowing away from the ionosphere) and Region 1 (flowing into the ionosphere) systems, associating a reversed Region 1-sense current at the most poleward latitudes. Furthermore, in the bottom, second and third panels for the southern hemisphere, a small (amplitude ≤ 100 nT), but definite field-aligned current flowing into the ionosphere is observed at the lowest latitudes, with flow direction the same as that of the evening-type Region 2 system (opposite to the morning-type Region 2 system). This is indicative of a possible collocation of the evening-type Region 2 system equatorward of the morning-type Region 2 system in the midnight sector. For both the northern and the southern hemispheres, a large-scale field-aligned current system at the lowest latitudes (corresponding to the traditional Region 2 system) occurrs equatorward of the vertical solid line, within the region of the C part (the quasi-persistent core part) plasma precipitation. A large-scale field-aligned current system, that is observed adjacent to and poleward of the former, corresponding to the traditional Region 1 system, spans nearly the B part (the explosive part) plasma precipitation that is confined between the vertical solid and dotted line. A weaker field-aligned current, observed at the most poleward latitudes, and showing a reserved Region 1-sense flow direction, occurs poleward of the vertical dotted line and spans the A plasma precipitation part.

Discussion

The principal objective of our paper is to show the persistency of large-scale characteristics of auroral particle precipitation patterns during a prolonged disturbed period which included a growth phase and multiple expansion onsets, and the north-south conjugacy and relationship of these patterns with the field-aligned currents. Figure 5 reviews various characteristics of the electric current and the plasma precipitation that were determined here by a case study on December 13, 1985, $\sim$0600-1700 UT. This summary includes electric current intensity (upper panel) of the ground-based AE activity and the northern hemispheric Region 1 and Region 2 field-aligned current intensity, and spatial relationship among the C, B and A parts of the precipitating ions and electrons. Also included are the downward field-aligned current (flowing into the ionosphere) and upward field-aligned current (flowing away from the ionosphere) above the southern polar ionosphere (middle panel) and the northern polar ionosphere (bottom panel).

Plasma Precipitation Characteristics

We have identified in this study the characteristics of the ion precipitation and the electron precipitation (both in the energy range from 30 eV to 30 keV) at an altitude of $\sim$840 km in the premidnight MLT sector with three distinctive parts, C, B and A from the lowest latitude toward the pole. The C part is the core part of the plasma precipitation that exists quasi-persistently throughout the growth phase, the expansion phase and the recovery phase of substorm. This part was characterized by an arch-shaped structure in the E-t spectrum of the precipitating electrons with its high-latitude being bounded by a sharp, almost vertically enhanced (dispersionless-like) spectrum, suggestive of the outer boundary of the earthward injecting electron plasma. Frank [1971] demonstrated the existence of plasma sheet by the energy density vs. L profiles of the low-energy ions and electrons (both $\sim$0.1 keV-50 keV) that were observed with the Ogo 3 satellite near local magnetic midnight and the magnetic equatorial plane. The energy density profile of the electron plasma sheet shows persistently during geomagnetically disturbed periods that a gradual increasing of the energy density followed by a decreasing of it at lower L values (corresponding to the equatorward part of the auroral belt), beyond which more erratic variations occur at higher L values (see Figures 1, 2 and 4 in Frank [1971]). Therefore, we believe that the C electron precipitation part, which we have identified by an arch-shaped structure in the E-t diagram, (in turn, a gradual increasing of the precipitating energy flux followed by a decreasing of it), is related to the inner part of electron plasma sheet actually existing near the magnetospheric equatorial region.

The C part is also characterized by the double-energy composite structure of the precipitating ions that comprises the high-energy component (energy>a few keV) and the low- energy component (energy<a few keV) throughout the course of the substorm. During the expansion phase and the recovery phase, the ion low-energy component exhibits a dispersion in the E-t spectrum, suggestive of the filter effect of the earthward convective flow of the

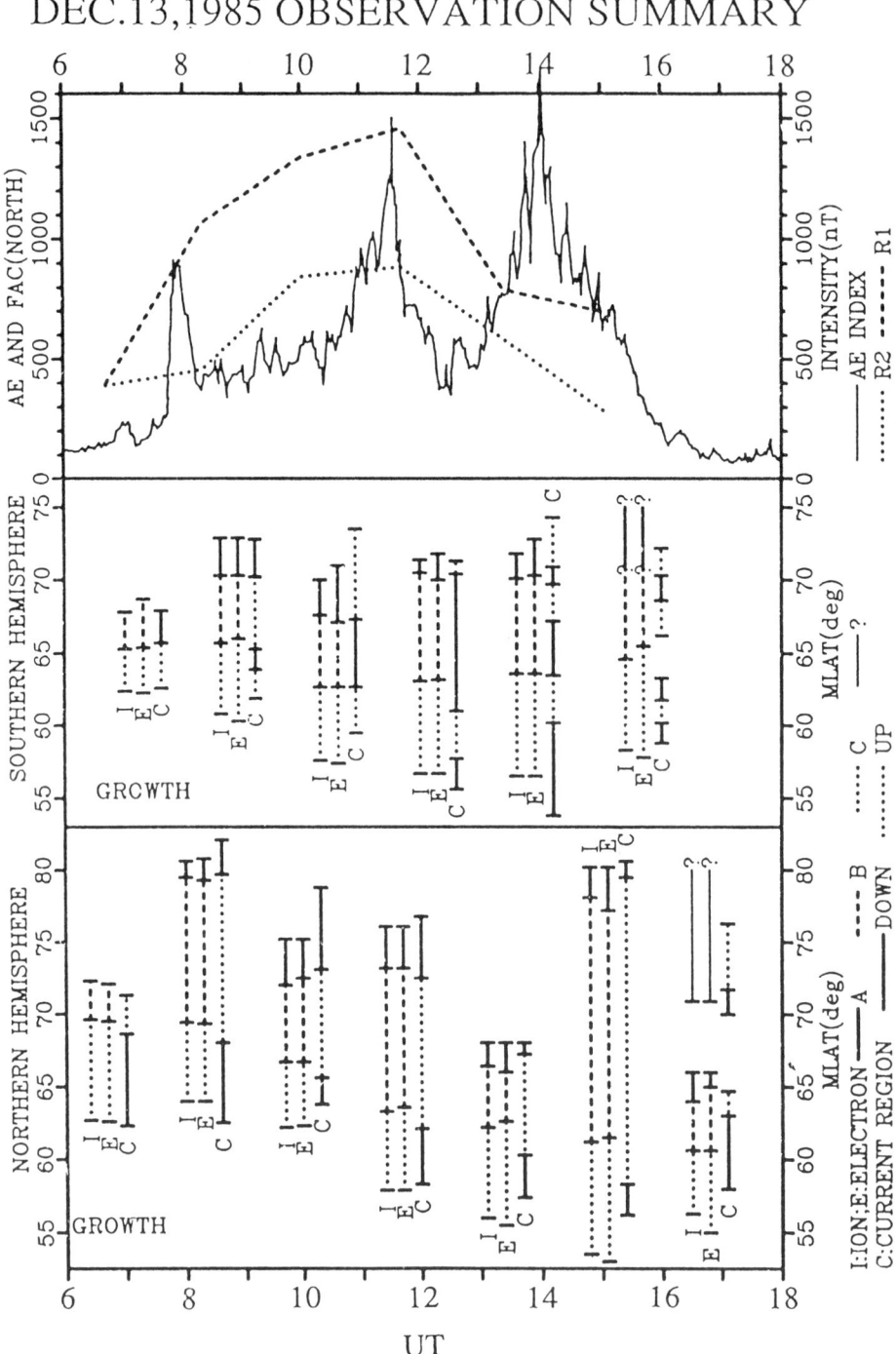

Fig.5. Summary diagram that reviews dynamic changes of various characteristics throughout the course of a prolonged disturbed period from 0600 to 1700 UT, December 13, 1985. Upper panel: electric current intensity (ground-based electrojet (AE) activity (solid), northern hemispheric Region 1 field-aligned current (broken) and northern hemispheric Region 2 current (dotted)). Middle (southern hemispheric) and bottom (northern hemispheric): spatial relationship among C part (dotted), B part (broken) and A part (solid) of precipitating ions and electrons, and downward field-aligned current flowing into the ionosphere (solid) and upward field-aligned current flowing away from the ionosphere (dotted).

magnetospheric plasma on the precipitating ions and its enhancement after the onset of the substorm expansion phase.

The high-latitude boundary of the C part of both the ion precipitation and the electron precipitation exhibits an equatorward shift as the substorm progresses from the growth phase through the expansion phase and the recovery phase. This suggests that both the outer boundary (in the magnetosphere) of the low-energy ion domain, and the outer boundary of the earthward injecting plasma never resided in the same position, but moved dynamically toward the earth throughout the course of substorm.

As seen in figure 5, the C ion precipitation part and the C electron precipitation part spatially coincided well with each other. The overall characteristics associated with the C part of the plasma precipitation are quite similar between the northern and the southern hemisphere, and are close to being geomagnetic conjugate.

The B part of the plasma precipitation occurs poleward of the C precipitation part. The B ion precipitation part and the B electron precipitation part spatially coincided well with each other. This B part of the plasma precipitation appeared even during the growth phase and diminished within a relatively narrow latitudinal region, but developed and expanded drastically over a much wider latitudinal span after the onset of substorm expansion phase. Therefore, this B part can reasonably be denoted as the explosive part of the plasma precipitation. As noticed in the top third and fourth panels in Plate 1, and the top fourth panel in Plate 2, the interface between the C part and the B part of the plasma precipitation showed a discontinuity in the E-t spectra, which is indicative of the different sources and/or the different generation mechanisms of these two parts.

Winningham et al. [1975] first presented the spatial distribution of the electron precipitation patterns (in the energy range from 10 eV to 10 keV) in the 2100-0300 MLT sector at the ISIS 1, 2 satellite altitudes. This consists of the CPS precipitation (the central region of the plasma sheet that is a relatively stable in response to substorms) and the BPS (the outer boundary layer of the plasma sheet that is most dynamic in response to substorms). Although their CPS precipitation resembles the C electron precipitation part in our present study, we have used the concurrent characteristics of both the precipitating ions and electrons to strongly suggest that the BPS precipitation could be resolved further into the B part (the explosive part) and the A part (the layer of field-aligned flow of the ions). The A part of the plasma precipitation in our study may correspond to the energetic particles boundaries which associate the earthward streaming ions along the outermost boundary of the plasma sheet studied by Lyons and Evans [1984].

The important question, however, still remains as to how these characteristics of the plasma precipitation could be modified during the extremely quiet period with a vanishing small IMF (interplanetary magnetic field) and the strongly northward IMF period, which seems to be important to fully understand the source mechanisms of the plasma precipitation.

Implication for Sources of Field-aligned Currents

As evident in Figure 4, the traditional Region 2 field-aligned current nearly corresponds to the C part plasma precipitation (the quasi-persistent core part). Whereas, the traditional Region 1 field-aligned current corresponds mostly to the B part plasma precipitation (the explosive part). In the top second, and the bottom second and third panels of the right block of Figure 4, the field-aligned currents exhibit the highly structured currents poleward of the Region 2 system (i.e. the large-scale current at the lowest latitudes). Both in the northern and southern hemisphere, the interface between the C part and the B part of the plasma precipitation did not coincide with the boundary between the Region 2 and the Region 1 current system, but the low-latitude part of the Region 1 system was involved within the C part. This is consistent with the earlier study by Klumpar [1979], who confirmed poor correlation between the Region 2/Region 1 current interface crossing the premidnight auroral oval and the boundary signatures in the particle precipitation. Sugiura et al. [1984] and Fujii et al. [1990] have, however, associated the Region 2/Region 1 interface with the CPS/BPS boundary of the electron precipitation in the pre-midnight MLT sector. The reason may be mainly due to the fact that our identification of the C, B and A parts of the plasma precipitation was carried out by using the concurrent characteristics of both the precipitating ions and electrons. Consequently our present identification of the plasma precipitation is basically different from the identification of the electron precipitation with the CPS and the BPS.

Our present results have shown that the Region 2 current system corresponds to, and occurrs within the C ion precipitation part which is generally initiated by a sharp increase in the profile of the total energy flux in magnetic latitude, and is bounded on the high-latitude limit approximately by a plateau in this profile. This feature has also been noticed by Heinemann et al. [1989]. By using the discussion mentioned previously, the magnetospheric magnetic flux tubes of the Region 2 field-aligned current system in the premidnight MLT sector are thought to correspond to the quasi-persistent core part of the plasma precipitation whose high-latitude limit is bounded by the outer boundary in the magnetosphere of the ion low-energy component domain and the magnetospheric outer boundary of the earthward plasma injection.

On the other hand, the magnetospheric magnetic flux tubes of the Region 1 field-aligned current system in the premidnight sector correspond mostly to the explosive part of the plasma precipitation which varied and expanded drastically poleward (tailward in the high-altitude magnetosphere) after the onset of the substorm expansion phase. Furthermore, even when the Region 1 current system exhibited exclusively a large-scale structure as seen in Figure 4 (the top third panel both on the left and right), the corresponding B part of the plasma precipitation exhibited a complicated structure (comprising a discrete, intermittent finer multiple structures) of the intensity enhancement in the E-t spectra (see the top third panels both in Plates 1 and 2). This suggests that the generator of the Region 1 current system is related basically, not to the fine structures of the quantities that are arranged in magnetic latitude (in radial direction in the magnetotail) but rather to the large-scale structures occurring in the azimuthal direction in the magnetotail.

We have found here that the three region pattern of the field-aligned current system comprising the traditional Region 2, the Region 1 and the reversed Region 1-sense current system (at the polewardmost latitudes) occurs not only for the evening-type pattern of the current systems, (with the Region 2 system flowing into the ionosphere and the Region 1 system flowing away from the ionosphere) as already discussed by Iijima and Potemra [1978] (see their Figure 13), but also for the morning-type pattern of the current systems (with the Region 2 system flowing away from the ionosphere and the Region 1 system flowing into the ionosphere and the reversed Region 1-sense current system associating at the polewardmost latitudes). Furthermore, the field-aligned currents occurring at the polewardmost latitudes (which included partly the highest latitude portion of the Region 1 system and mostly the reversed Region 1-sense system poleward of the Region 1) should be addressed to the distinctive magnetospheric magnetic flux tubes that correspond to the A part of the plasma precipitation.

In the high-altitude magnetosphere, a spacecraft takes a long time to traverse a distance that corresponds to a spatial scale of the auroral belt at the ionospheric altitude. Therefore, the possible presence of the field-aligned current, and its flow direction have been identified at a spatially special boundaries, say, at the interface between the tail lobe and the plasma sheet, which included the Region 1-sense flow direction pattern of the field-aligned currents [Fairfield, 1973; Sugiura, 1975; Ohtani et al., 1988] and the multiple flow direction pattern [e.g. Aubry et al., 1972; Elphic et al., 1985; Frank et al., 1981]. These observations, however, do not mean that the source region of the field-aligned currents exists exclusively at the high-latitude boundary of the plasma sheet. In fact, in the middle-altitude magnetosphere, by using the dual satellite experiments (with the ISEE 1 and 2), the Region 2- and Region 1-sense field-aligned current systems were observed to correspond fully to the charged particles that are usually used to identify the magnetic field lines threading the magnetospheric plasma sheet [Kelley et al., 1986]. As already mentioned in our present results, the substorm-associated field-aligned current systems corresponded entirely to the three parts of the plasma precipitation, which implies that the magnetic flux tubes of the field-aligned current systems involve a huge volume, presumably almost the whole volume of the magnetospheric plasma sheet/ring current domain.

In this study, in the nearly same pre-midnight MLT sector, the traditional Region 2/ Region 1 field-aligned currents in the northern (winter) hemisphere show the normal pattern for the evening-side oval. Whereas, the field-aligned currents in the southern (summer) hemisphere have the morning-side pattern of the traditional Region 2 / Region 1 currents, or showed signs of the morning-side pattern overlapping with the evening-side pattern. This fact implies that the Harang discontinuity (demarcating the ionospheric eastward Hall current and the westward Hall current) seems to be shifted more duskward in the southern (summer) auroral oval. Vasyliunas [1970] demonstrated that when the ionospheric conductivities (Σ_P and Σ_H) are enhanced by a factor 10 (by keeping the Σ_P/Σ_H ratio constant) in the auroral belt, the Harang discontinuity is shifted by $37°$ (≈ 2.5 hrs in MLT) more duskward, in comparison to the case of non- enhancement of auroral-belt conductivities. Therefore, even if the Region 1/2 currents have a common source region in the magnetosphere, they can map to different MLTs over the higher- conductivity oval and the lower-conductivity oval. In our present data, the earthward energy flux carried by electrons (30 eV-30 keV) had nearly the same maximum value of the order of the 10^{12} eV/cm^2 · sr · sec both in the northern and southern auroral oval. The ionospheric conductivities (Σ_P and Σ_H) due to such precipitating energy flux were estimated to be, at least, more than 10 mho [Kamide and Matsushita, 1979]. The ionospheric background conductivities (Σ_P and Σ_H) due to solar EUV were estimated to be at most, 1 mho in the nightside auroral belt during the equinox. Even if, the background conductivities were enhanced by factor 1.5 during the summer and diminished by factor 0.5 during the winter, the effective conductivities are due primarily to auroral particle precipitation and are thought to be nearly the same factor of magnitude over the northern and southern auroral oval. The seasonal change of the ionospheric conductivities may not be a primary reason of the Harang discontinuity over the northern and southern hemisphere. The alternative reason may be the accuracy in MLT calculation. In this study, we have used the eccentric geomagnetic dipole coordinates and calculated the MLT of the point of interest in space. We did not, however, take into account of the effects of the seasonal variation in the sun's declination and the variation in the sun's apparent position due to the eccentricity of the earth's orbit, which can cause a variation of ~ 1 hour of MLT for a fixed UT over the course of a year [Baker and Wing, 1989]. We think that this reason should be carefully considered for our data in near future. Finally, we will leave a possibility that Region 1/2 field-aligned current systems have source regions basically dislocated above the north and the south of the effective geomagnetic equator of the magnetospheric plasma sheet/ring current domain.

Summary

We have investigated the characteristics of the field-aligned currents and the the precipitating ions and electrons (within the 30 eV to 30 keV energy range) at the altitude of ~ 840 km in the premidnight MLT sector, by using the magnetic field and the plasma measurements acquired with the DMSP F7 satellite. We report here a case study of a prolonged disturbed period from ~ 0600 to ~ 1700 UT, December 13, 1985, which included a growth phase and multiple expansion onsets. Principal characteristics determined here include the following: (1) The plasma precipitation characteristics during substorms conform to three distinctive parts that are denoted here as "C", "B", and "A" from the lowest latitude, respectively, for both the ions and electrons; (2) The C electron precipitation part is characterized by an arch-shaped structure in the E-t (energy flux vs. time $\approx$magnetic latitude) spectrum with its high-latitude limit being bounded by a sharp, almost vertically enhanced (dispersion less-like) spectrum (centered around ~ 10 keV and with its trend extending down to ~ 100 eV and/or less than this), suggestive of the outer boundary of the earthward injecting electron plasma. The C ion precipitation part is characterized by a double-energy composite structure in the E-t spectrum that comprises the high- energy component (energy>a few keV) and the low-energy component (<a few keV). This low-energy component exhibits a dispersion-like trend in the intensity enhancement in the E-t spectrum that is indicative of the filter effect

of the earthward convective flow of the magnetospheric plasma on the precipitating ions, especially after the onset of the substorm expansion phase. The high-latitude limit of the C plasma precipitation part, in turn, is the outer boundary in the magnetosphere of the ion low-energy component and the outer boundary of the earthward injecting plasma. Both of these regions move dynamically toward the earth throughout the course of substorm. The C part is thought to be the quasi-persistent core part of the plasma precipitation; (3) The B plasma precipitation part occurs poleward of the C precipitation part. The B ion precipitation part is primarily characterized by a curtain-shaped structure in the E-t spectrum that consists exclusively of the high-energy component (energy>1 keV), which is very different from the double-energy composite structure of the C part. The B electron precipitation part is characterized by a highly structured intensity enhancement in the E-t spectrum that includes the so-called inverted-V shaped spectrum and also the vertically enhanced discrete, intermittent spectrum. The velocity distribution functions, f(v) of the precipitating ions and electrons, corresponding to the B part, include nearly theoretical Maxwellian for both the ions and electrons, a shifted Maxwellian f(v) for ions with the energy shifting by ~ 10 keV toward the higher energy (suggestive of an existence of field-aligned electric field $E_{\parallel}$ directed toward the earth), and a shifted Maxwellian f(v) for electrons with the shifting energy being ~ 15-20 keV toward the higher energy (suggestive of the existence of $E_{\parallel}$ directed away from the earth). The B part is thought to be the explosive part of the plasma precipitation that expands drastically over a much wider latitudinal span especially after the onset of substorm expansion phase through the recovery phase; (4) At the polewardmost latitude of the plasma precipitation (adjacent to and poleward of the B part), there exists a distinctive part, the A part that is principally characterized by its association of the field-aligned flow of the ions. This A part is hard to distinguish unambiguously from the B part in the E-t spectra of both the precipitating ions and electrons, but is definitely identified by the contrastive features of f(v)'s between the ion precipitation and the electron precipitation. The A part was not recognizable during the growth phase but occurred exclusively after the onset of substorm expansion phase; (5) A relatively general relationship between the field-aligned currents and the plasma precipitation holds for both the northern and southern hemisphere. Namely, a large-scale field-aligned current system that occurs at the lowest latitudes (i.e. traditional Region 2 system) approximately corresponds to the quasi-persistent core part of the plasma precipitation (the C part). Whereas, traditional Region 1 field-aligned current system that occurs adjacent to and poleward of the Region 2 system, corresponds mostly to the explosive part of the plasma precipitation (the B part). The boundary between the Region 2 and Region 1 current system does not coincide with the interface between the C and B plasma precipitation parts. The field-aligned currents occurring at the most poleward latitudes corresponds to the A plasma precipitation part, which includes the highest-latitude portion of the Region 1 current system and the reversed Region 1-sense current system, poleward of the Region 1; (6) The generators of the field-aligned current systems are believed to be ascribable, not to the fine structures of the quantities that are arranged in magnetic latitude as seen in the plasma precipitation, but rather to the large-scale structures occurring in the azimuthal direction in the magnetosphere; (7) In the nearly same pre-midnight MLT sector, the observed field-aligned currents above the northern (winter) polar ionosphere have the normal pattern for the evening-side traditional Region 2 and Region 1 current systems. Whereas, above the southern (summer) polar ionosphere, the observed field-aligned currents have the morning-side pattern of the traditional Region 2 and Region 1 current systems (with the morning-side pattern overlapping with the evening-side pattern at the equatorwardmost latitudes). This is suggestive of a number of causes: an effect of the seasonal variation in the ionospheric conductivity, a variation of one hour of MLT determination for a fixed UT over the course of a year due to the seasonal variation in the sun's declination and the eccentricity of the earth's orbit, and a possible existence of dislocated source region of field-aligned current systems above the north and the south of the effective geomagnetic equatorial plane in the magnetosphere.

Acknowledgements. M.Watanabe and T.Iijima are grateful to T.Ono and H.Miyaoka at Aurora Center Division, National Institute of Polar Research for their help in the acquisition of the DMSP F7 particle data. The work of T. Iijima was supported by the Ministry of Education, Japan, grant 01540348. The analysis of DMSP data is supported by the Air Force Office of Scientic Research. The work of T.A. Potemra and L.J. Zanetti was supported by the National Science Foundation and the Office of Naval Research.

References

Aubry, M.P., M.G. Kivelson, R.L. McPherron, C.T. Russell, and D.S. Colburn, Outer magnetosphere near midnight at quiet and disturbed times, *J. Geophys. Res.* ,*28* , 5487- 5502, 1972.

Baker, K.B., and S. Wing, A new magnetic coordinate system for conjugate studies at high latitude, *J. Geophys. Res.* ,*94* , 9139-9143, 1989.

Elphic, R.C. P.A. Mutch, and C.T. Russell, Observations of field-aligned currents at the plasma sheet boundary: An ISEE-1 and 2 survey, *Geophys. Res. Lett.* ,*12* , 631-634, 1985.

Fairfield, D.H., Magnetic field signatures of substorms on high-latitude field lines in the nighttime magnetosphere, *J. Geophys. Res.* ,*78* , 1553-1562, 1973.

Frank, L.A., Relationship of the plasma sheet, ring current, trapping boundary, and plasmapause near the magnetic equator at local midnight, *J. Geophys. Res.* ,*76* , 2265- 2275, 1971.

Frank, L.A., R.L. McPherron, R.J. DeCoster, B.G. Burek, K.L. Ackerson, and C.T. Russell, Field-aligned currents in the earth's magnetotail, *J. Geophys. Res.* ,*86* , 687-700, 1981.

Fujii, R., R.A. Hoffman, and M. Sugiura, Spatial relationship between Region 2 field-aligned currents and electron and ion precipitation in the evening sector, *J. Geophys. Res.* ,*95* , 18939-18947, 1990.

Heinemann, N.C., M.S. Gussenhovern, D.A. Hardy, F.J. Rich, and H.-C. Yeh, Electron/ion precipitation differences in relation to Region 2 field-aligned currents, *J. Geophys. Res.* ,*94* , 13593-13600, 1989.

Iijima, T., and T. Nagata, Signatures for substorm development of the growth phase and expansion phase, *Planet. Space Sci.* ,*20* , 1095-1112, 1972.

Iijima, T., and T.A. Potemra, Large-scale characteristics of field-aligned currents associated with substorms, *J. Geophys. Res.* ,*83* , 599-615, 1978.

Kamide, Y., and S. Matsushita, Simulation studies of ionospheric electric fields and currents in relation to field-aligned currents 2. Substorms, *J. Geophys. Res.* ,*84* , 4099-4115, 1979.

Kelley, T.J., C.T. Russell, R.J. Walker, G.K. Parks, and J.T. Gosling, ISEE 1 and 2 observations of Birkeland currents in the earth's inner magnetosphere, *J. Geophys. Res.* ,*91* , 6945-6958, 1986.

Klumpar, D.M. Relationships between auroral particle distributions and magnetic field perturbations associated with field-aligned currents, *J. Geophys. Res.* ,*84* , 6524-6532, 1979.

Langel, R.A., R.H. Estes, G.D. Mead, E.B. Fabiano, and E.R. Lancaster, Initial geomagnetic field model from MAGSAT vector data, *Geophys. Res. Lett.* ,*7* , 793-796, 1980.

Lyons, L.R., and D.S. Evans, An association between discrete aurora and energetic particle boundaries, *J. Geophys. Res.* ,*89* , 2395-2400, 1984.

Ohtani, S., S. Kokubun, R.C. Elphic, and C.T. Russell, Field-aligned current signatures in the near-tail region 1. ISEE observations in the plasma sheet boundary layer, *J. Geophys. Res.* ,*93* , 9709-9720, 1988.

Rich, F.J., D.A. Hardy, and M.S. Gussenhoven, Enhanced ionosphere-magnetosphere data from the DMSP satellites, *EOS Trans. AGU* ,*66* , 513-514, 1985.

Sugiura, M., Identifications of the polar cap boundary and the auroral belt in the high-latitude magnetosphere: A model for field-aligned currents, *J. Geophys. Res.* ,*80* , 2057-2068, 1975.

Sugiura, M., T. Iyemori, R.A. Hoffman, N.C. Maynard, J.L. Burch, and J.D. Winningham, Relationships between field-aligned currents, electric fields, and particle precipitation as observed by Dynamic Explorer-2, in *Magnetospheric Currents,* edited by T.A. Potemra, pp.96-103, Geophysical Monograph 28, AGU, Washington D.C., 1984.

Vasyliunas, V.M., Mathematical models of magnetospheric convection and its coupling to the ionosphere, in *Particles and Fields in the Magnetosphere,* edited by B.M. McCormac, pp.60-71, D. Reidal, Hingham, Mass., 1970.

Winningham, J.D., F. Yasuhara, S.-I. Akasofu, and W.J. Heikkila, The latitudinal morphology of 10-eV and 10-keV electron fluxes during magnetically quiet and disturbed times in the 2100-0300 MLT sector, *J. Geophys. Res.* ,*80* , 3148-3171, 1975.

Equatorial, Birkeland, and Ionospheric Currents of the Magnetospheric Storm Circuit

L. J. ZANETTI,[1] T. A. POTEMRA,[1] T. IIJIMA,[2] AND W. BAUMJOHANN[3]

Magnetospheric-ionospheric current systems involved in the geomagnetic storm process are presented with both Birkeland as well as cross field currents evaluated from Magnetic Field Experiment (MFE) data obtained from the AMPTE/CCE and MAGSAT spacecraft. Current systems developed from the low altitude MAGSAT measurements during a day-long event are compared to the recent CCE statistical results of equatorial (4-9 R_E) azimuthal (ring) and radial currents, in both cases derived from MFE data taken during periods when Kp was 3 to 7 and DsT was −40 to −70nT. These data are examined from different altitudes in an effort to distinguish paired Region 1 and Region 2 sense Birkeland current sheets and currents transverse to the magnetic field to models of line circuits, for example the current wedge system. A three-dimensional image of the Birkeland and Hall current system near the ionosphere has been developed from the MAGSAT MFE data taken during a steadily active storm period on March 21, 1980. The results of the AMPTE/CCE and MAGSAT MFE data analyses can be summarized as follows: 1) near midnight, the divergence of the equatorial azimuthal and radial currents as a function of L-shell gives pairs of Region 1 and Region 2 sense Birkeland currents; 2) the radially integrated divergence of the equatorial current system is in the Region 2 sense for all local times; 3) this high altitude storm-time circuit is within the 9 R_E CCE apogee; 4) the low altitude system reveals Region 1 and Region 2 Birkeland current pairs collocated with the storm-enhanced electrojet current, even when the electrojet is split; 5) the dayside electrojet current is completed with a sunward Hall return current across the center of the polar ionosphere; and 6) consistent three-dimensional current systems are derived independently from high and low altitude analyses.

INTRODUCTION

To assess the large-scale magnetospheric current system, Magnetic Field Experiment (MFE) data from high and low altitude spacecraft have been analyzed at each location during similar and steady geomagnetic storm conditions. This study does not involve particular substorm events, but rather the time-averaged storm circuits presented in this study may integrate many expansion phase periods [*Akasofu*, 1964]. Given the global nature and the dynamics involved in the storm and substorm process, the current circuit involved is difficult to observe and, to date, impossible to determine uniquely from single or multiple event studies. There is ambiguity of model interpretation of magnetic field signatures during storm events [*Nagai*, 1982, 1987; *Lopez, et al.*, 1989; *Lopez and Lui*, 1990]; nevertheless the current wedge model continues to survive intact [*McPherron, et al.*, 1973; *Baumjohann*, 1983]. Among other issues, we consider the location of the nightside current disruption and diversion, appearing to be within 15 R_E as cited by the previous references as well as other studies [*Heppner*, 1967; *Siscoe and Cummings*, 1969; *Lui, et al.*, 1978, 1991; *Kelly, et al.*, 1986; *Ohtani, et al.*, 1988,1990]. Of particular interest here is the question of where the Birkeland currents of the current wedge map to in the magnetosphere. Many models identify only the outer edges of the plasma sheet populations as the regions where these currents flow. This concept is supported by recent MHD simulations of reconnection within the magnetotail [*Sato, et al.*, 1983, 1984; *Walker and Sato*, 1984] which show standard current wedge currents forming on the newly created plasmasheet boundary layer. *Nagai* [1982] suggested, however, that because the geosynchronous current wedge signatures are so sensitive to the local time position of the satellite, the current wedge currents must close relatively close to the geostationary regions. More recent studies [*Barfield, et al.*, 1985, 1986] have repeated this claim as well as those mentioned above. There is clearly a great deal of uncertainty as to the magnetospheric closure of the current wedge currents [*Mauk and Meng*, 1986; *Stern*, 1990], as well as the closure of the large scale equatorial currents [*Mauk and Zanetti*, 1987, and references therein].

Some recent efforts in the low altitude regions include advancing the techniques for inverting ground magnetometer measure-

---
[1]The Johns Hopkins University, Applied Physics Laboratory, Johns Hopkins Road, Laurel, MD 20723-6099.
[2]Geophysics Research Laboratory, The University of Tokyo, Bunkyo-ku, Tokyo 113 Japan.
[3]Max-Planck- Institute für Extraterrestrische Physik, 8046 Garching, Germany.

ments in order to reconstruct substorm current systems, systems which include Birkeland currents. *Kamide and Baumjohann* [1985] invert ground magnetometer measurements and use the Rice conductivity model to assess the ionospheric currents and potential contours, and then proceed to calculate the Birkeland current distribution and auroral indices. The substorm phases manifest themselves as the DP-2 [*Bostrom*, 1975] two-cell current system during the expansion and the DP-1 enhanced westward electrojet, substorm current wedge during the explosive "dipolarization." To produce realistic Birkeland current regions, a reasonable conductivity was used and its location was modulated to coincide with the ionospheric current distributions.

Convection patterns were further checked with STARE data by *Baker and Kamide* [1985] who suggested caution when one applies a statistical conductivity model to dynamic substorm situations. More recently, optical images have been used to infer the conductivity structure [*Kamide, et al.*, 1986; *Marklund, et al.*, 1987], resulting in the electrodynamics of the auroral zone. *Clauer and Kamide* [1985] further analyze the DP-2 current system with regard to the IMF and conclude it to be driven by the solar wind, whereas the DP-1 is clearly associated with the substorm onset and merely triggered by the southward turning of the IMF. It is not clear that conductivity gradients are necessary for proper ionospheric Birkeland, Hall, and Pedersen currents if the magnetospheric potential and current generators are properly considered, at least for the quasi-steady-state process. *Ahn, et al.*, [1986] inverted ground magnetic perturbations and used a conductivity model to produce ionospheric current patterns for magnetically disturbed times. These patterns were similar to quiet patterns, except for higher Joule heating at the westward extent of the westward electrojet and except for major particle energy flux in the morning westward electrojet. The Birkeland currents calculated from the divergence of the horizontal currents do not retain the sharp features seen in the observations because of the integration or averaging (essentially smoothing) of small spatial scale features in both ground magnetometer data and conductivity models. The three-dimensional Hall, Pedersen, and Birkeland current system associated with the enhanced westward electrojet as deduced by ground observations has been thoroughly reviewed [*Baumjohann*, 1983] as well as detailed [*Baumjohann, et al.*, 1980, *Inhester, et al.*, 1981].

We present here details of results from the high-altitude analysis of MFE data from the 4 to 9 R_E equatorial region coverage of the AMPTE/CCE spacecraft [*Iijima, et al.*, 1990]. This analysis was statistical, using 180 days worth of data in 1984 and 1985 when Kp was 3 to 7 and DsT was −40 to −70 nT, in an attempt to define steady-state magnetospheric storm conditions. The results of this present study involve the details of the Birkeland current distribution in the nightside region resulting from the divergence of the equatorial current system. Thus the prime objective of this effort is to establish the form of the connection of the equatorial current system to the ionospheric auroral zone current system as well as the distribution of the auroral zone system itself.

Having chosen similar geomagnetic conditions, steady state (> 1 day) storm activity with Kp = 3 to 4 and DsT = −40 nT, the three-dimensional ionospheric and Birkeland current system of the auroral zone and polar cap has been determined from MFE data taken on March 21, 1980 by the MAGSAT spacecraft. The expanded auroral zone current systems include the electrojets and collocated paired Birkeland current systems as well as split electrojet circuits on the nightside; these split currents are also accompanied by paired Birkeland currents. Widespread, polar-cap, sunward Hall current also exists such that the electrojet circuit is balanced. At the obvious risk of long time-averages of a dynamic system, we present remarkably consistent three-dimensional current systems derived independently from high and low altitude measurements taken during similar geomagnetic storm conditions.

OBSERVATIONS AND ANALYSIS

Figure 1 is a reproduction of Figure 9 of *Iijima, et al.*, [1990] which shows the calculated divergence of the horizontal equatorial current distribution determined from magnetic field measurements from the AMPTE/CCE mission. The results were statistical compi-

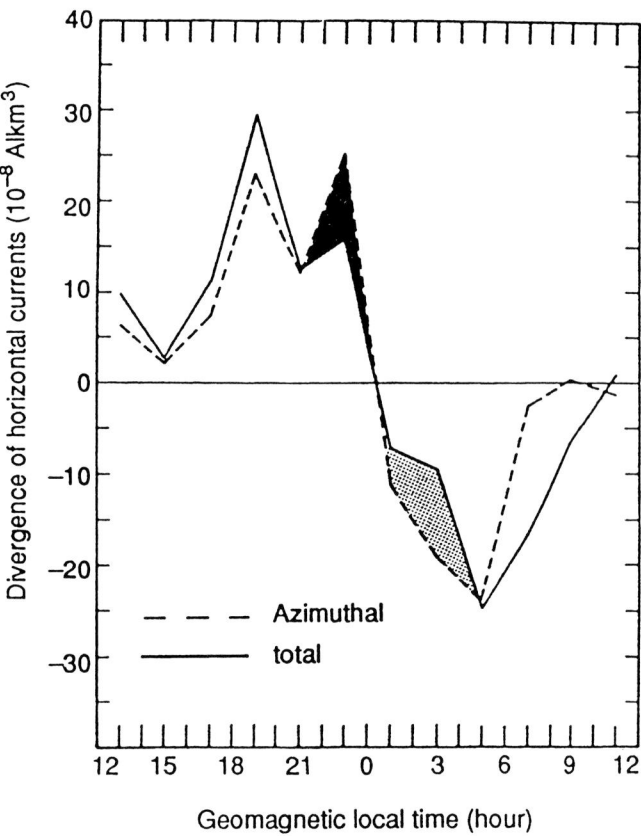

Fig. 1. The meridionally integrated divergence of the total and azimuthal equatorial current as a function of magnetic local time, from Figure 9, *Iijima, et al.*, 1990.

lations during moderately active geomagnetic storm conditions. The graph shows the divergence of the horizontal current system (i.e., diverted out of the equatorial plane presumably along magnetic field lines) in units of 10^{-8} A/km^3 as a function of magnetic local time. The current divergence is integrated along a local time meridian. The purpose here in duplicating these results is to emphasize the contributions to the divergence of the different components as well as the local time meridian detail. The two traces indicate the total integrated current divergence by a solid line and the contribution of the azimuthal (ring) current by a dashed line. These divergences are positive, or away from the equatorial plane before local midnight, and thus they are in the same sense as the Region 2 Birkeland current system, which is toward the ionosphere in the dusk sector. Likewise, the postmidnight divergence of the equatorial horizontal current is toward the equator, consistent with the dawn sector Region 2 Birkeland current system. A difference to note is that the total divergence near midnight is actually less than the contribution from the divergence of the azimuthal ring current (dashed line). Thus the contribution to the divergence of the equatorial radial current is opposite to the total and to the azimuthal current divergence (i.e., in the Region 1 Birkeland current sense).

It is consistent then that the radial equatorial currents participate in the storm circuit and may contribute to the pairs of Birkeland currents existing in the equatorial plane. It is furthermore possible, but not conclusive from these studies, that these radial currents may participate in the connection of these oppositely directed pairs of Birkeland currents in the local midnight ±3 hours region as suggested by some models [e.g., *Kan, et al.*, 1990]. Nevertheless, the sum divergence of horizontal equatorial current is in the Region 2 sense for L shell values less than 9 R_E. The circuit involved in the storm process is observable at the outer section of these near Earth L shells, being of opposite polarity to the nominal azimuthally westward (ring) current. And finally, the divergence of the radial current is in the Region 1 Birkeland current sense near local midnight, indicating the existence of pairs of Birkeland currents at these local times and within 9 R_E at the magnetic equator.

Figure 2 is a summary of the equatorial current results from the AMPTE/CCE MFE data analysis. Shown are the vector representations of the equatorial current from 4.0 to 8.8 R_E and for all local times; noon is at the top. The length of the vector indicates current intensity (A/km). Approximately half a year's data were used in 1984 and 1985, such that the distribution was uniform in local time, and the selection criteria was for magnetic conditions of Kp 3 to 7 and Dst -40 to -70 nT. These data results were inferred from steady state disturbances on the order of the 5 to 50 nT which remained after the removal of the 1980 IGRF reference field. The CCE MFE data have been calibrated in-flight to an absolute accuracy of 1 nT per axis, and the drift of the various axes has been less than 1 nT over the five-year data life span.

The generally westward azimuthal equatorial current is evident and recognizable as the traditional ring current. This azimuthal current has a peak in amplitude near midnight and a minimum near noon, consistent with an Earthward plasma pressure gradient throughout this L shell region; this pressure gradient has been observed [*Lui, et al.*, 1987; *Roelof*, 1987; *Hamilton, et al.*, 1988]. In addition, however, there is a local minimum at midnight at the outer L shell region which is the integrated effect of the storm or substorm discharge circuit, the current wedge circuit which reduces the extended magnetotail to a more dipolar configuration. This local minima is exaggerated by the actual maximum in the azimuthal current being actually closer to 22 MLT, indicating that the plasma pressure may be actually greatest nearer these local times. These local times are consistent with previous observations of the median of the substorm current wedge existing premidnight [*Lopez, et al.*, 1988; *Arnoldy and Moore*, 1983; *Nagai, et al.*, 1982]. The new results shown in Figure 2 include a radial component of the equatorial current distribution shifting the vectors outward, away from the Earth at dawn and inward, toward the Earth, at dusk. As discussed above, there is a slight clockwise shift in the symmetry of the radial currents to a few hours before midnight. This distribution of radial equatorial current is also consistent with the currents necessary to balance an azimuthal pressure gradient resulting from a peak in the plasma pressure near midnight. These equatorial current results are discussed in greater detail by *Iijima, et al.*, 1990.

The purpose of Figure 2 is not just to summarize the equatorial current results from the CCE data analysis but also to study the detail of the current divergence along local time meridians. The discussion of the data in Figure 1 concluded that divergences of the azimuthal versus radial currents contributed in the same sense for the dayside sector and in the opposite sense near midnight. These divergences are coded with circled dots and crosses indicating currents away from and toward the equatorial region, respectively. For the dayside sector, the divergence is always in the sense of the Region 2 Birkeland current system. As mentioned, the net, meridionally integrated, equatorial current diversion is also in the Region 2 sense from approximately 21 MLT to 5 MLT. Throughout this nightside region the divergence of the radial equatorial current is in the Region 1 sense but is not very strong. The divergence of the azimuthal current dominates with overall net current out of the volume premidnight and into the volume postmidnight at the inner L shells. Thus this nightside divergence is in the Region 2 sense at the inner L shells but is in the Region 1 sense at the outer L shells; the Region 1 current is particularly evident premidnight. Thus from the divergence of a statistical distribution of equatorial current determined during moderately active times, Region 1 and Region 2 Birkeland current pairs are present in the 21 MLT to 5 MLT sectors in the equatorial plane.

Figure 3 is a 64 × 64 pixel grid of the disturbance to the low altitude magnetic field (B_z in a direction parallel to the model field (1980 IGRF)). The positive and negative signatures align themselves with the edges of the distant horizontal current distribution that causes the disturbances and that distribution is presumed to be in the ionosphere. These data have been assembled from northern and southern passes of the MAGSAT spacecraft during moderately active (Kp = 3 to 4, DsT = -40 nT) conditions on March 21, 1980. These conditions correspond closely to the mean values of magnetic activity encountered during the CCE statistical analysis discussed in the first set of figures. The orbital coverage is in the geographic dawn-dusk meridian with the movement of the magnetic pole expanding the noon-midnight coverage to 74° and 67° MLAT, respectively. Thus complete coverage of a magnetically

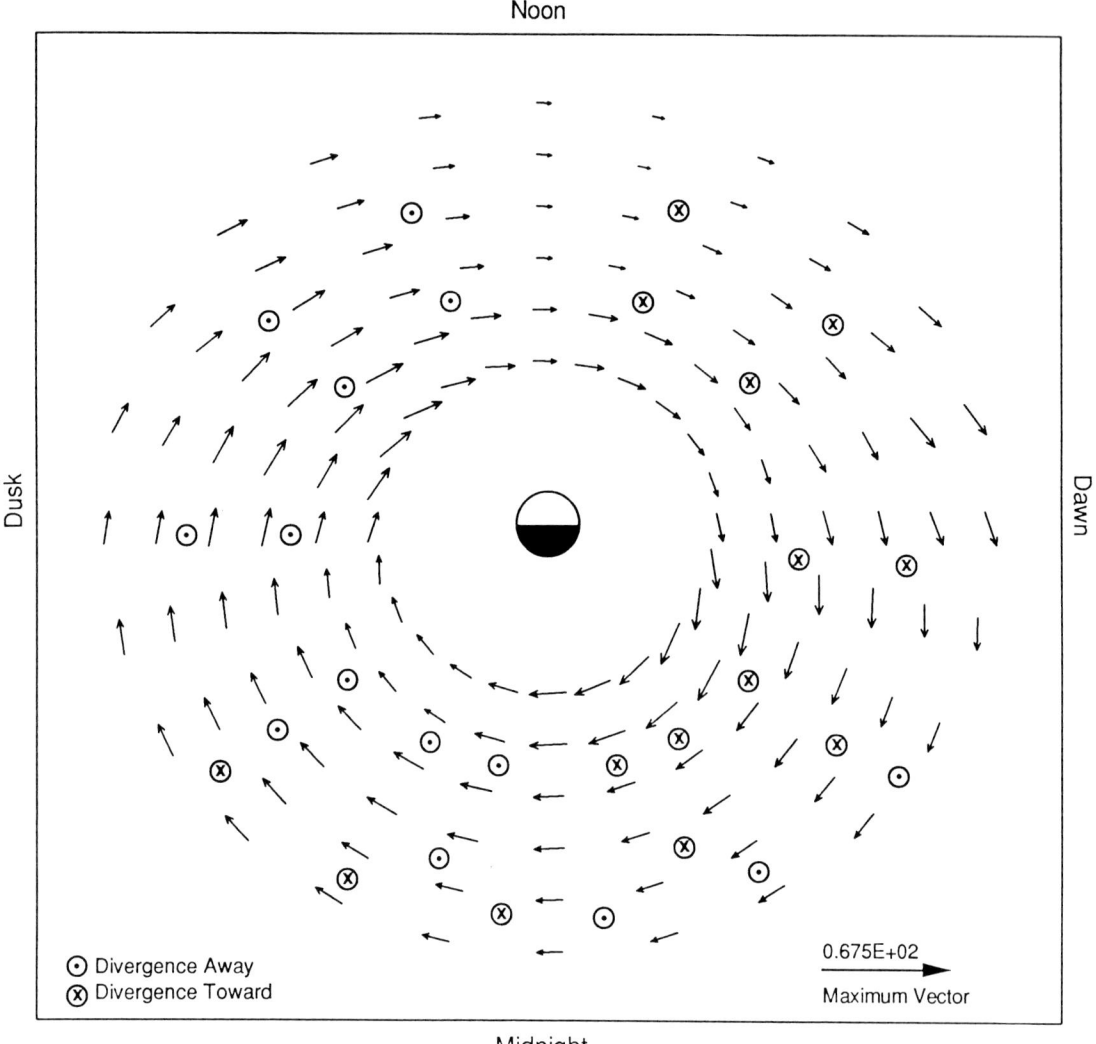

Fig. 2. Vector representation of the equatorial current as a function of L shell and local time as determined by AMPTE/CCE magnetic field experiment data analysis.

active auroral zone current system is, unfortunately not available in the noon and midnight regions, and thus there is a lack of magnetic perturbation in those areas of Figure 3 although, no doubt, currents exist there. These are the same data used in a composite study of auroral ionospheric current systems [Zanetti, et al., 1984] where one dimension (the dawn-dusk orbital transit direction) of B_z was directly inverted [Mersman, et al., 1979; Kuppers, et al., 1979] and the perpendicular ionospheric Hall-horizontal current distribution was determined. The incompleteness in such a composite is that the noon-midnight direction is ignored as are the cross terms of a proper two dimensional inversion analysis and thus ionospheric currents parallel to the dawn-dusk direction are not determined.

A remote sensing, two-dimensional, horizontal current analysis has been performed for the first time on the MAGSAT data of March 21, 1980. These results are preliminary at this point and the emphasis in this study will be on the information that had been lost by doing a composite of one-dimensional dawn-dusk analyses [Plate 1, Zanetti, et al., 1984]. In the former study, the ionospheric currents in the noon-midnight direction were well analyzed, maximizing at dawn and dusk local times, whereas this analysis will accent the other direction, i.e., currents in the dawn-dusk direction and in particular the structure in the current system observed in the postmidnight hours. These analyzed ionospheric current distributions are the toroidal components and will be compared to the disturbances (B_x, B_y) transverse to the 1980 IGRF model field. These toroidal currents are the Hall currents in the absence of conductivity gradients perpendicular to the electric field and toroidal and Hall are used synonymously. The poloidal Birkeland–

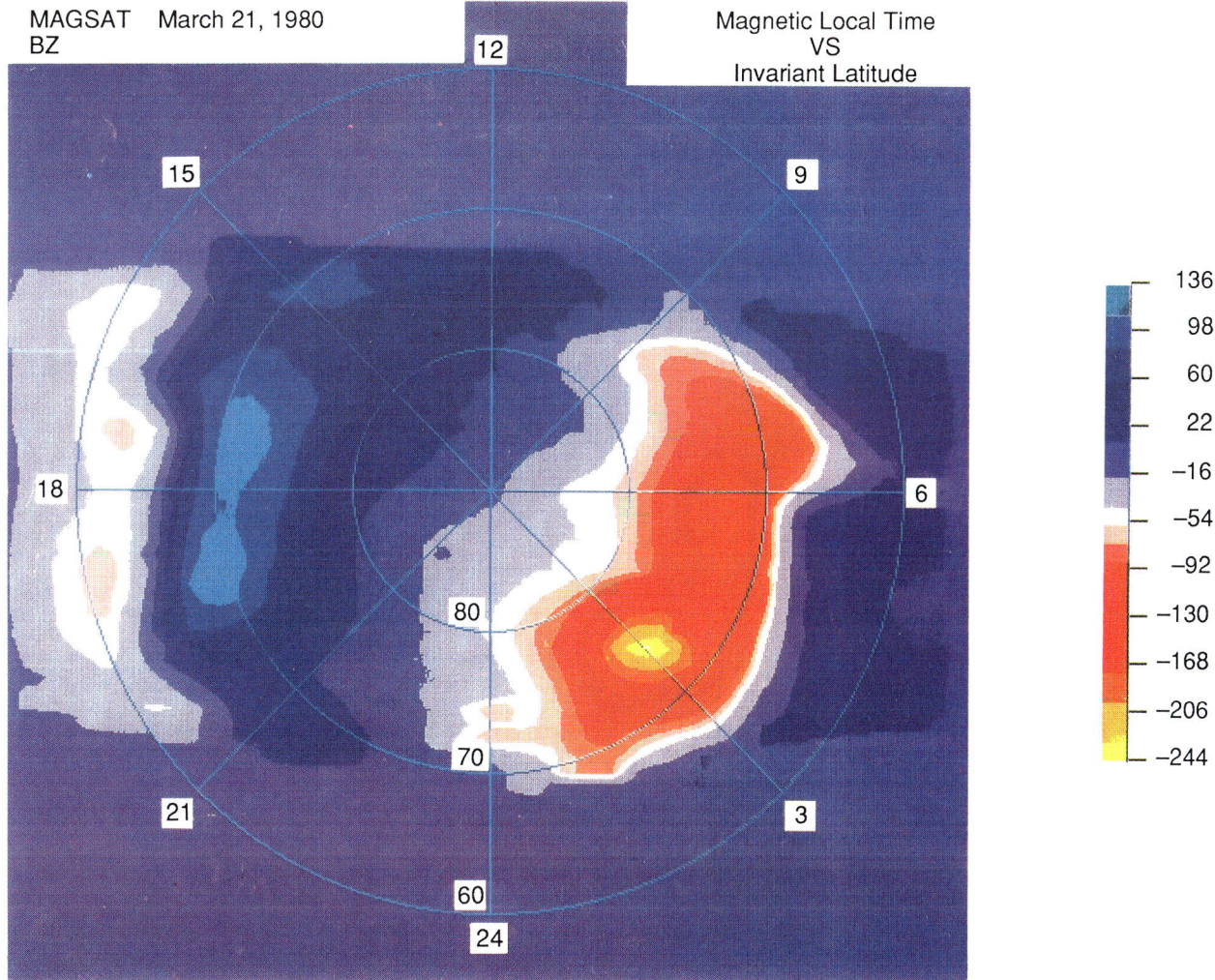

Fig. 3. MAGSAT high latitude disturbance magnetic field data parallel to the model field as a function of magnetic latitude and local time.

Pedersen circuit is not registered in the B-parallel disturbance signature except near the edges of the various circuits or given conductivity gradients, which would necessitate electric field measurements in order to completely assess the three-dimensional circuit. These transverse disturbances are indications of local Birkeland currents and the poloidal circuit.

Table 1 is a summary of the Hall/toroidal current analysis method as implemented on the MAGSAT parallel disturbance data (B_z) in a two-dimensional ionosphere. This method is a direct inversion technique of the magnetic field to the kernel source distribution as opposed to an iterative modeling technique intended to reproduce or fit various magnetic signatures to proposed current distributions. By separating out effects in the parallel direction, a direction free of the local Birkeland current disturbances, that component is curl free and the problem is reduced to solving Laplace's equation (Equation 3, Table 1) in the volume bounded by ± infinity in both horizontal directions, and the ionosphere and observation plane in the vertical direction. The derivative of the potential is defined in the observation plane (i.e., B_z). Furthermore the problem is conformally mapped to Cartesian coordinates (parallel is now vertical) to simplify the solution for the magnetic potential (and to simplify the derivative) which will be oscillatory in two dimensions and exponential in the third. The vertical dependence is physically imposed to be exponential. Thus Equation 4 (Table 1) represents the solution to the vertical (B_z) magnetic field disturbance vector. Fourier transforms of the B_z component's horizontal dependence are used primarily for convenience because they are oscillatory. The coefficients are a function of wavelength or scalelength and the vertical (z) dependence of Φ (or $\overline{B}$) is exponentially increasing with decreasing z. Equations 5 and 6 of Table 1 define the relationship of the B_x and B_y components (dawn and sun directions, respectively) to the B_z solution, and recall that

TABLE 1. Ionospheric Hall Current Analysis (2-D)

Direct method of inverting the magnetic effects of a distant current distribution
Analyzing $B_{\parallel}$ (B_z), free of Birkeland current disturbances
Assumptions: current in thin layer in the ionosphere
NO conductivity assumptions or input

$$\nabla \times B_z = 0 \quad (1)$$

$$B_z = -\nabla\Phi \quad (2)$$

$$\nabla^2\Phi = 0 \quad (3)$$

Transfer problem to Cartesian coordinates, solutions to Φ are oscillatory and exponential, physically impose the exponential z dependence of Φ

$$B_z(x,y) = \sum_{m,n} C_z(k_m,k_n)\ e^{ik_m x}\ e^{ik_n y}\ e^{-\sqrt{k_m^2 + k_n^2}\,z} \quad (4)$$

$$B_x(x,y) = \frac{ik_m}{\sqrt{k_m^2 + k_n^2}} B_z(x,y) \quad (5)$$

$$B_y(x,y) = \frac{ik_n}{\sqrt{k_m^2 + k_n^2}} B_z(x,y) \quad (6)$$

Calculate B in the ionosphere immediately adjacent to the current distribution, then using Ampere's law

$$J_x(x,y) = \frac{2}{\mu_0} B_y(x,y) \quad (7)$$

$$J_y(x,y) = \frac{2}{\mu_0} B_x(x,y) \quad (8)$$

---

the measured horizontal counterparts will contain disturbance effects from both the ionospheric currents as well as the local Birkeland currents. The current distribution $(\vec{J}(x, y))$ is calculated directly from the B_x and B_y components using Ampere's law. Considering each measurement as a segment of the current distribution, the integrated B around the closed path will have vertical components cancelled by adjacent segments and half the field above and half the field below the distribution. Thus Equations 7 and 8 indicate the relationship between the horizontal field (e.g., B_y) measured immediately above the ionospheric current distribution (e.g., J_x).

The results of Equation 7, B_y (x,y) (equivalent through Equation 7 to $J_x(x, y)$), are displayed as the colored background of Figure 4. The color bar indicates intensity and is scaled plus (red) and minus (blue) based on the sign of the B_y (sun-directed) disturbance in order to identify dawn-directed versus dusk-directed current.

These disturbances are minimized at dawn and dusk as expected and increase in amplitude away from those local times. The signs of the disturbances are negative at 9 MLT corresponding to a plus x or dawn-directed ionospheric current. A similar disturbance and current exists at 20 MLT also. A dusk-directed disturbance reaches a maximum at 16 MLT and 4 MLT indicating current in the negative x direction (duskward) with the primary interest of this figure being the enhanced westward electrojet in the post-midnight section. This is the section of current that appeared to decrease in the one-dimensional composite of March 21, 1980 [Plate 1, Zanetti, et al., 1984], although the split electrojet structure was actually more apparent in that composite. In addition, there is a strong polar cap return current evident in the one-dimensional composite and in the two-dimensional B_x (J_y) results that are not reproduced here. Individual orbit analyses will be shown in subsequent figures such that certain details of these events can be appreciated, that is, the accurate magnitude of the polar cap return current and the double westward electrojet structure with associated Birkeland current pairs.

In order to assess the three-dimensional ionospheric current system for March 21, 1980, the transverse to $\vec{B}$ ($\vec{B}_\perp$, $(\overrightarrow{B_x + B_y})$) disturbance vectors for each of the composite orbits are calculated and superimposed upon the Hall current pattern. The most obvious, and certainly not unimportant, feature is the collocation of the Hall current electrojets and the transverse disturbances ($\vec{B}_\perp$), the gradient of which indicates of Birkeland currents. Furthermore, even the signatures of both the $\vec{B}_\perp$ and the distribution of horizontal current match, which indicates oppositely directed, constant intensity Birkeland current pairs and center-peaked ionospheric Hall electrojet distributions. The multiple pairs of Birkeland currents are obvious at about 3 MLT associated with the structure in the horizontal currents. Slight mismatches are evident near 15 MLT that result from the proximity of the geographic pole for those orbits. Misalignments with an oval at about 19 to 20 MLT hint at possible movements and spatial dynamics of the current systems. The continuous alignment of the vectors indicates the presence of sheets of Birkeland current, and these sheets follow the contours of ionospheric Hall currents. The general vector direction of the $B_\perp$ disturbances in the auroral zone, those collocated with the Hall currents, is generally in the antisunward or away from local noon direction. These disturbances are consistent with the large-scale Region 1 and Region 2 Birkeland current system of the auroral zone. Note that there are $B_\perp$ disturbance vectors across the entire polar cap, generally in the opposite, sunward direction. This points to an imbalance in the poleward Region 1 Birkeland current both at dusk and at dawn. Consistent with the observed sunward polar cap ionospheric Hall current, we suggest that this imbalance of the Region 1 Birkeland system is completed with Pedersen current across the polar cap. The distribution of vertical Birkeland currents as indicated by the $\vec{B}_\perp$ disturbances combined with the two-dimensional distribution of ionospheric Hall current constitutes the three-dimensional ionospheric current system in the polar and auroral regions during moderately active times.

Given the coarse resolution of the two-dimensional ionospheric current analysis, the detail of the structure and the absolute values of the current intensities are not well determined. Two major points

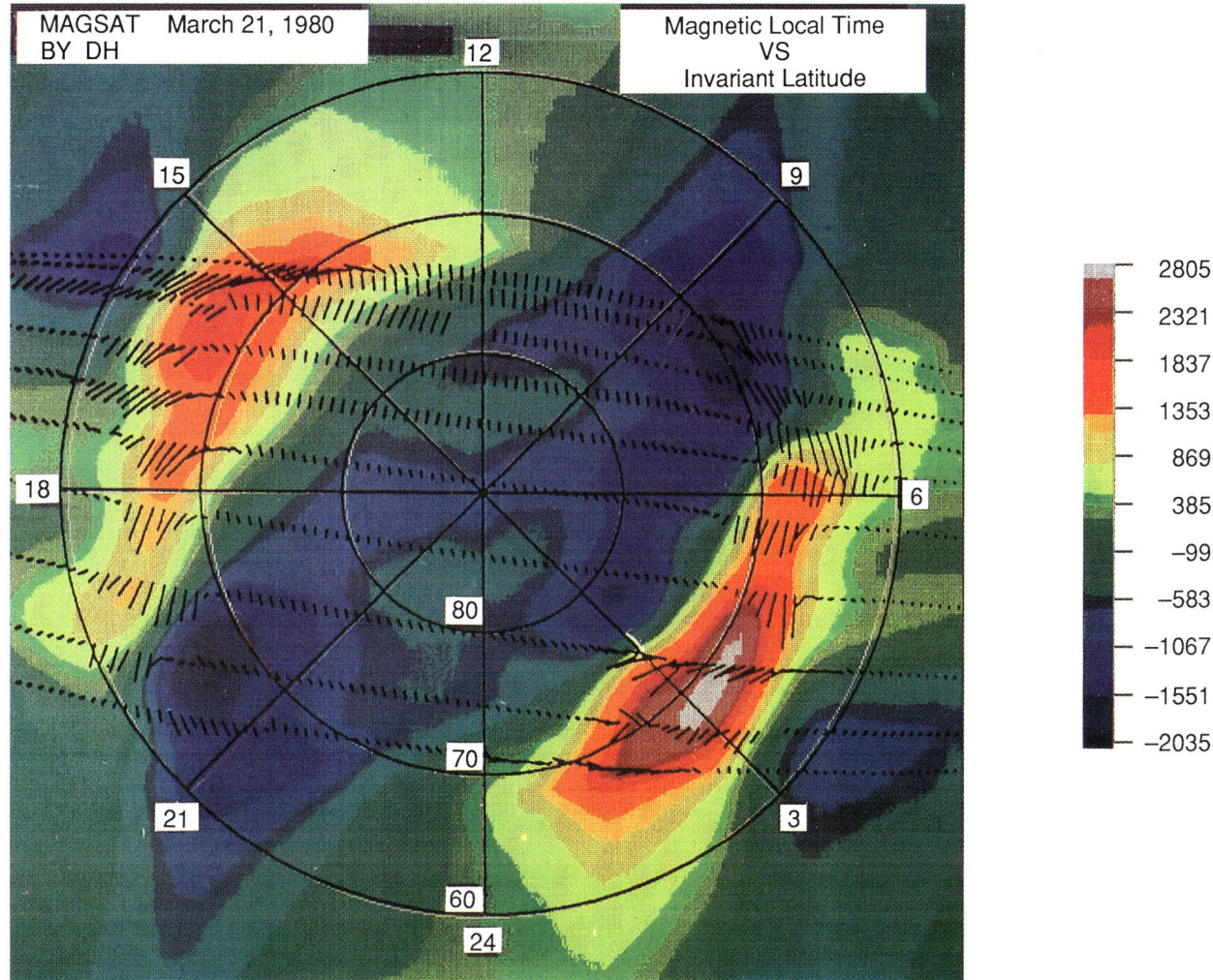

Fig. 4. Two-dimensional inversion analysis results of the B_y component (sunward) in the ionosphere; indicative of Hall current in the dawn-dusk direction, as a function of magnetic latitude and local time.

of this study are thus studied in greater detail: (1) the relationship of the Birkeland and Hall-Pedersen currents as the electrojet is split on three successive orbits from 5:00 UT to 8:30 UT (midnight to 4 MLT region) and, (2) the integrated ionospheric Hall current in the auroral zone compared with the integrated polar cap sunward ionospheric current. To address these questions, two orbits were chosen to represent the dayside and the nightside Birkeland and ionospheric current distributions.

Figure 5 shows the results of the analysis of disturbances to the ambient magnetic field of the ionosphere as recorded at 14:25 UT on the dayside (Figure 5a) and at 8:25 UT on the nightside (Figure 5b) on March 21, 1980. The dawn-dusk orbits of the MAGSAT spacecraft crossed the noon-midnight meridian at approximately ± 10° MLAT. Both Figures 5a and 5b have zero abscissa located at the noon-midnight meridian and have dusk represented by negative km and dawn by positive km. The two panels show the transverse disturbance and the ionospheric Hall current distribution flowing perpendicular to the orbital track. The tracks are shown to the right of each panel against a magnetic latitude and local time grid with the statistical Region 1 and Region 2 Birkeland currents [*Iijima and Potemra*, 1976] shown for reference. In each example, the top panel is the absolute value of the transverse disturbance field plotted from dusk to dawn; this is essentially a reproduction of a $\vec{B}_\perp$ vector trace in Figure 4 but without regard for the direction, and scaled from 0 to 700 nT amplitude. Thus the perturbation is non-zero across the polar cap and is actually sunward, as noted in the discussion of the data in Figure 4. In the space between the top and bottom panels are groups of arrows determined from the gradients in $\vec{B}_\perp$ that point up or down to indicate Birkeland current away from or toward the ionosphere, respectively, labelled R1, R2, and PC for Region 1, Region 2, and polar cap. These are the large-scale Birkeland current regions in an attempt to categorize this current

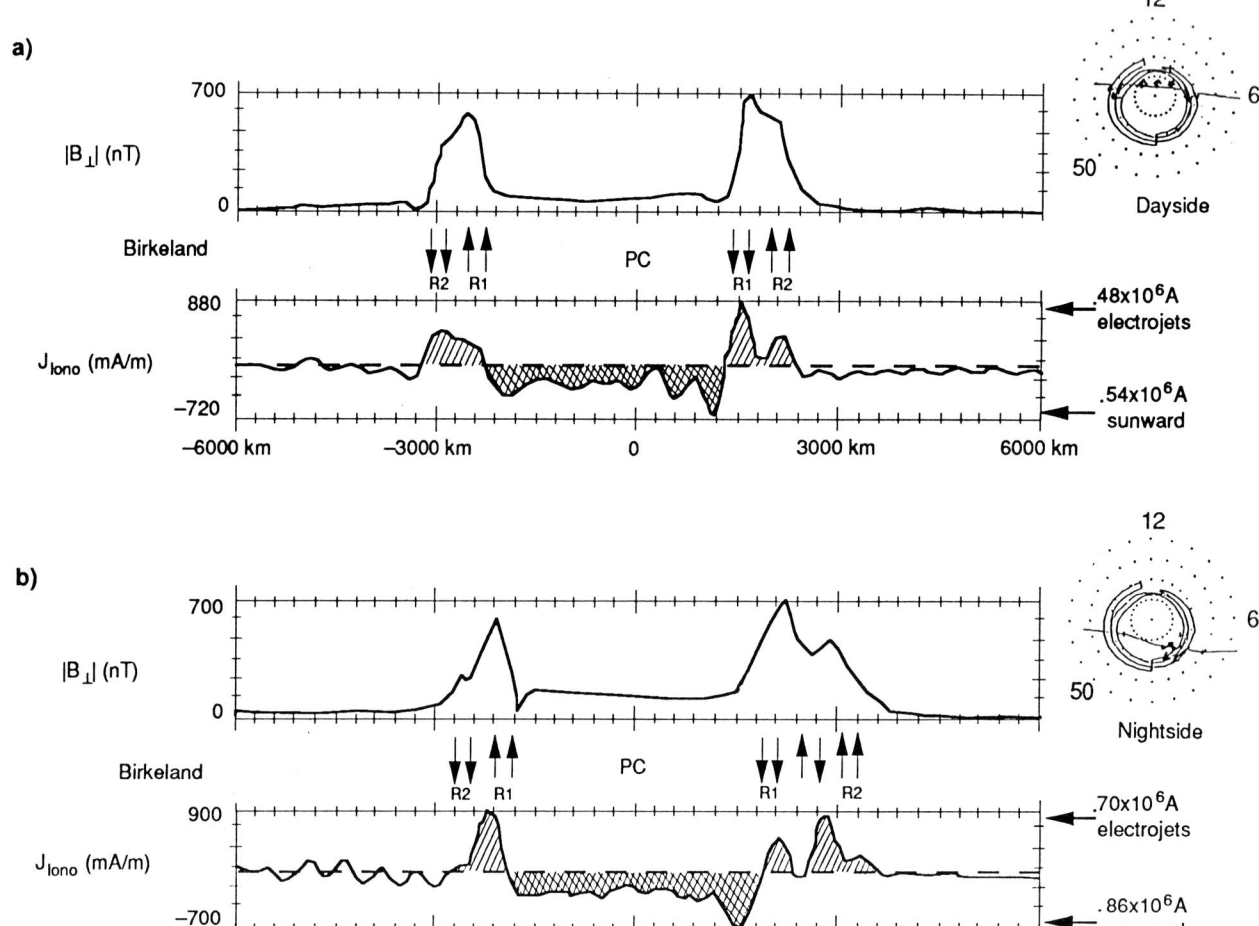

Fig. 5. Two individual MAGSAT overpasses of the auroral zone shown as detailed examples of (a) dayside and (b) nightside Birkeland and ionospheric currents. Data are plotted from dusk to dawn in projected distance (km) with zero at the noon-midnight meridian.

system into the statistical Birkeland current patterns [*Iijima and Potemra*, 1978]. As noted by *Iijima and Potemra* [1978], multiple pairs of large-scale Birkeland currents exist in the night sector during magnetic storm periods; this is also observed in the nightside passes on the dawn side for this study. One purpose of Figure 5 is to show the good correspondence between the $\vec{B}_\perp$ signature and the distribution of Hall current. A one-dimensional (along the orbital track) analysis, performed to directly solve Laplace's equation for the magnetic potential and the ionospheric current distribution, is plotted in the bottom panel of both the dayside and nightside examples. The $\vec{B}_\perp$ signature and the distribution of Hall current are collocated and remarkably similar; the electrojet current increases toward the interface of an R1 and R2 sense Birkeland current pair. This association holds true even for the multiple pair of Birkeland currents in the dawn nightside region, each with a Hall current (electrojet) maxima. There is an interesting exception; the mismatch of the $\vec{B}_\perp$ and the ionospheric current distribution at the polar cap edge of the nightside dawn system may perhaps indicate an enhanced ionospheric current loop of the storm circuit, which may not have associated Birkeland currents. Nevertheless, as one traces from the pole toward the 3 and 4 MLT areas of the auroral zone in the nightside 8:25 UT example (Figure 5b), a positive $\vec{B}_\perp$ slope indicating a downward, Region 1 sense Birkeland current is collocated with increasing ionospheric current. The subsequent reversed negative gradient in $\vec{B}_\perp$, interpreted as an upward-directed Birkeland current, is accompanied by decreasing ionospheric current. The same pattern is then repeated for the Birkeland current pair equatorward of this poleward pair just discussed. The Birkeland current densities involved are nominal for large-scale features, of the order of 1 to 10 $\mu A/m^2$.

The ionospheric current density (J_{Iono}) distribution is shown from dusk to dawn, is scaled in millamperes per meter (mA/m) approaching 900 mA/m in the electrojet regions (generally antisunward or away from local noon), and also a few hundred millamperes per meter across the polar cap toward the sun. It is necessary to comment here on the uniqueness of the MAGSAT data in relation to the accumulation of magnetic field measurements, either space or ground recorded. The MAGSAT magnetic field data had on the order of 6 nT rms absolute vector accuracy from a combination of .5 nT per axis magnetometer accuracy [Acuna, 1980] and laser-ranging, star camera attitude accuracy of 15 arc-seconds [Fountain, 1980], with of course the careful integration of these accuracies during data analysis [Langel, et al., 1982]. The perturbations used for the ionospheric current analysis were on the order of hundreds of nanoTesla, clearly above these unparalleled accuracy floors. Given these facts, no attempt was made to remove baselines; only model fields were subtracted. The shaded areas above zero in Figure 5 are the generally antisunward electrojet currents, the integration of which is 0.48×10^6 amps for the dayside pass. The integration of the sunward polar cap current is 0.54×10^6 amps; thus the electrojet circuit is more than balanced by the sunward Hall current. Likewise, the sunward polar cap current of 0.86×10^6 amps on the nightside overbalances the 0.70×10^6 amp electrojet current. The nightside pass was not directly dusk to dawn, which may have skewed it toward the nightside and into an enhanced storm-related current circulation at around 3 MLT. Nevertheless, there is no need, in this case, for additional or unbalanced Birkeland current into the ionosphere on the dayside to feed the electrojet circuits; the circuit on the dayside at least can circulate wholly within the ionosphere.

DISCUSSION AND SUMMARY

In an effort to construct an averaged magnetospheric current system during magnetically active periods, the equatorial AMPTE/CCE magnetic field analysis results have been compared with the ionospheric MAGSAT results during similar geomagnetic conditions. Both sets of results are either statistical or time averaged with Kp between 3 and 7 and DsT between −40 and −70 nT for the CCE analysis and Kp 3 to 4 and DsT = −40 nT for the MAGSAT case on March 21, 1980. The connection, of course, between the two systems is the Birkeland current system, and these results have confirmed the Region 1 and Region 2 system [Iijima and Potemra, 1978] as well as its variability and structure around midnight local times. Primary results of this study indicate that the current systems involved in the magnetospheric storm circuit appear within the apogee of the AMPTE/CCE orbit (<9 R_E) substantiating near-Earth current disruption models quoted in the Introduction. Also, in this averaged sense, balanced pairs of oppositely directed Birkeland currents making up the large-scale auroral current system are inferred from both the CCE and MAGSAT data near local midnight times. For this observation, the complicated ionospheric current system a few hours near midnight may require meridionally unbalanced or "net" Birkeland currents that may be completed in longitude; for example, net inward and outward Birkeland currents complete the enhanced westward electrojet of the current wedge circuit. Net Birkeland currents that would accommodate horizontal currents in the dayside are not necessary for this March 21, 1980, MAGSAT case since the integrated current in the dawn and dusk electrojets are equalled by sunward Hall current over the entire polar cap.

The equatorial current system from $L = 4$ to 9 R_E as determined from a statistical CCE magnetic field analysis [Iijima et al., 1990] was reproduced in order to examine the details of the divergence of that system. The divergence of the azimuthal (ring) current indicates Birkeland current of the Region 2 sense at all local times on the inner L shells. At the outer L shell region near local midnight there is a depression in the nominal maximum of the azimuthal current. This maximum should be expected from the 3 to 4 R_E midnight plasma pressure peak observed during storm periods [Lui, et al., 1987; Roelof, 1987; Mauk and Zanetti, 1987]. This depression in the azimuthal current is the result of the addition of the storm circuit, which flows opposite to the azimuthal ring current and diminishes or exhausts the tail current, equivalent to the diversion of current into the ionospheric load and the return of the nightside magnetosphere to a more dipolar configuration. This depression of the azimuthal current also gives a diversion that is in the same sense as the Region 1 Birkeland current, that is, toward the ionosphere postmidnight and away from the ionosphere premidnight.

The radial equatorial current [Iijima, et al., 1990] is outward near dawn and inward near dusk, with maxima in both cases. This distribution of radial currents is also consistent with and necessary to balance a plasma pressure maximum at local midnight, as referred to above. As shown in the dayside sector of Figure 1, the divergence of such a radial current distribution will add to the Region 2 sense Birkeland current that must exist from the azimuthal current divergence. In the nightside sector, however, the effect is to diminish the total Region 2 Birkeland current; i.e., the divergence of the radial equatorial current is in the Region 1 sense. If these current systems persist during times of a dipolar field, for example during the collapse of the tail system, than the two distributions (CCE and MAGSAT) map to each other quite well; including the R1/R2 interface at 68° MLAT (~7 R_E); it is thus reasonable to expect partial observation of Region 1 at midnight and within the CCE apogee. In summary, in the outer half of the CCE equatorial L shell coverage at local midnight ± 4 hours, all gradients add to a net Region 1 sense divergence. Conversely, the L shell integrated total divergence of current (Figure 1) is in the Region 2 sense at all local times. Thus from the equatorial current system as determined from a statistical analysis of the AMPTE/CCE magnetic field measurements, pairs of Region 1 and Region 2 Birkeland currents coexist near local midnight and participate in the magnetic storm circuit.

This first application of a complete two-dimensional ionospheric current analysis to spacecraft data has revealed both the auroral electrojet currents and the sunward Hall return current over the polar cap. These results were derived from MAGSAT data taken on March 21, 1980, during moderately active conditions. The second dimension of the ionospheric plane was employed to compensate for the direction in which the spacecraft did not travel. These MAGSAT results show that the electrojet appears to split into two branches from about 4 MLT toward midnight. The possibility that this may be caused by a time effect was disregarded because three successive orbits over the southern hemisphere showed similar electrojet structure. This Hall current is presumed to be the eastward beginnings of the storm-enhanced, westward electrojet. Consistent with the above, dominant auroral enhancements in two branches surrounding the expanding auroral activity of the nightside storm region have been observed, e.g., the March 1989 giant storm [*Allen, et al.*, 1989] and *Murphree* [1990] with implications of dual current channels. These auroral enhancements would indicate increased conductivity channels in the nightside oval.

Field disturbances perpendicular to the local magnetic field indicate the presence of Birkeland current systems and are shown to be collocated with the ionospheric electrojet system. This colocation holds even for the split in the electrojet at 2 to 4 MLT, that is, pairs of Region 1, Region 2 sense Birkeland currents are associated with each branch of the electrojet circuit, although the correlation between the Hall current distribution and the Birkeland currents is not as strict as it is at other local times. Triangular-shaped transverse magnetic field disturbance signatures, typical for this case as well as for most auroral oval situations, indicate adjacent pairs of constant density Birkeland current sheets whose sign depends on the sign of the disturbance gradient. Presumably, these constant density Birkeland currents are connected meridionally via Pedersen currents. If that is true, then the distribution of Pedersen current will have the same signature as that of the $\vec{B}_\perp$ disturbance, a triangular shape peaked at the center (Birkeland current pair interface) where the entire flux of Pedersen current must pass. Furthermore, if the ratio of Hall to Pedersen conductivity, not the absolute values but only the ratios, is constant then the Hall current distribution must also have the same triangular signature and that is what is generally observed in this as well as other cases studied. Proceeding through the logic in reverse, since similar signatures are observed for the disturbance fields and the ionospheric Hall current distribution (away from the current wedge extremities), it is safe to assume that conductivity ratios are constant and that Birkeland currents are completed by Pedersen currents in the ionosphere which therefore exist with commensurate and collocated Hall currents.

An exception to the above exists at the edges of the expanding current wedge [e.g., *Baumjohann and Opgenoorth*, 1984, for the surge head] where there are clear conductivity gradients and mismatched conductivities from dynamic particle precipitation and aurora during the storm process. The mismatch in the poleward pair of Birkeland currents and Hall current distribution (although the equatorward pair and Hall channel match) may indicate such conductivity gradients which could well be expected in this 2-4 MLT region. *Iijima and Nagata* [1972] and *Bostrom* [1975] proposed a cell of current circulation (called DP-1) associated with the substorm initializations that is superimposed on the pre-storm, two-cell current circulation. This extra cell not only enhances the westward electrojet near midnight but also develops a returning eastward current poleward of the auroral zone. This cell has been observed in the nightside orbital pass example (Figure 5) and should be apparent with the higher resolution two-dimensional ionospheric current analysis. In the dawn section of the nightside example of the March 21, 1980 case, Birkeland currents did not coexist along with these polar cap eastward DP-1 return currents, as would be expected since these are on open field lines hooked to the tail lobe. As one moves equatorward into the closed magnetospheric and plasma sheet field lines, the Birkeland currents participating in the storm unloading process appear and are associated with the enhancement of the westward electrojet. The Region 1 sense and Region 2 sense Birkeland currents in this nightside region are balanced, not unbalanced in the Region 1 sense as would be expected for the dawnside of the current wedge. Perhaps an imbalance exists more toward midnight than could be observed by the MAGSAT orbital coverage, the compliment of the surge head net current often observed premidnight.

Finally, we wish to extrapolate these "balanced" conditions to the dayside current system where even more uniform conductivity is to be expected given the addition of solar illumination as well as less erratic particle precipitation. Recall that for these "balanced" conditions to exist it is sufficient to have only constant Pedersen-to-Hall conductivity ratios; it is not necessary to have totally uniform conductivity; i.e., auroral zone conductivity enhancements are certainly permissible, even necessary to determine the locations of the major currents. The first point is that, from the individual and collective MAGSAT auroral zone transverse magnetic field disturbance data, the Region 1 and Region 2 Birkeland currents are nearly balanced except for slightly larger Region 1 currents, primarily on the dayside. Given all of the above arguments, these slight increases of Region 1 currents are most likely connected across the polar cap with Pedersen currents. The second point is that the integrated sunward Hall current across the polar cap balances the sum of the dawnside westward and duskside eastward electrojet currents. There have been suggestions [e.g., *Rostoker, et al.*, 1982; *Pellinen, et al.*, 1982] that the electrojet current system should be supplied on the dayside by unbalanced Birkeland current into the ionosphere. At least for the averaged three-dimensional storm current system represented by this March 21, 1980, MAGSAT example, such a net dayside Birkeland current system into the ionosphere is neither observed nor necessary.

The current system from an integrated number of substorms has been determined from magnetic field data from the AMPTE/CCE and MAGSAT spacecraft during similar magnetic conditions and exists on L shells that are within the apogee of the AMPTE/CCE spacecraft (9 R_E) near local midnight. Near the midnight local times, these results indicate that the high and low altitude communication systems consist of a superposition of the steady-state Region 1 and Region 2 Birkeland current pairs, which exist at other local times, and a system that involves the storm circuit current wedge. These two systems would be consistent with systems often referred to as the driven system, which is driven by the imposition of the IMF electric field onto the magnetosphere, and the dissipative system of the storm circuit, which collapses the flux of the enhanced magnetotail created in the growth phase previous to the substorm or storm period. These two systems' names are, of course, not to be taken as misnomers with regard to the ionosphere since both dissipate energy (Joule Heating) in the Pedersen current. The details of the divergence of the radial and azimuthal equatorial current from the CCE analysis indicate pairs of Region 1 and Region 2 sense Birkeland currents at a given local time meridian near local midnight consistent with the pairs of Birkeland currents that are collocated with the ionospheric electrojet current systems as determined from the MAGSAT analysis.

Acknowledgments. We wish to acknowledge the JHU/APL Space Department and M. Acuna (NASA/GSFC) for the successful experiments, spacecraft and mission data for the MAGSAT and AMPTE/CCE projects. The data analysis assistance of D. Holland of JHU/APL is also appreciated. Both missions were supported by NASA; the correlative aspects are supported by the National Science Foundation and the Office of Naval Research. We also wish to acknowledge the constructive criticism obtained during the refereeing process.

REFERENCES

Acuna, M. H., The MAGSAT precision vector magnetometer, *Johns Hopkins APL Tech. Dig., 1,* 210, 1980.

Ahn, B. -H., Y. Kamide and S. -I. Akasofu, Electrical changes of the polar ionosphere during magnetospheric substorms, *J. Geophys. Res., 91,* 5737, 1986.

Akasofu, S. -I., The development of the auroral substorm, *Planet. Space Sci., 12,* 273, 1964.

Allen, J., H. Sauer, L. Frank and P. Reiff, "Effects of the March 1989 Solar Activity," *EOS, Transactions, Am. Geophys. Union., 70,* 1479, 1990.

Arnoldy, R. L. and T. Moore, Longitudinal structure of substorm injections at synchronous orbit, *J. Geophys. Res., 88,* 6213, 1983.

Baker, K. B. and Y. Kamide, "A comparison of ionospheric electric fields inferred from Scandinavian Twin Auroral Radar Experiment drift data and from Global International Magnetospheric Study magnetometer data," *J. Geophys. Res., 90,* 1339, 1985.

Barfield, J. N., C. S. Lin and R. L. McPherron, "Observations of magnetic field perturbations at GOES2 and GOES3 during the March 22, 1979 substorms: CDAW6 analysis," *J. Geophys. Res., 90,* 1289, 1985.

Barfield, J. N., N. A. Saflekos, R. E. Sheenan, R. L. Carovillano, T. A. Potemra and D. Knecht, "Three-dimensional observations of Birkeland currents," *J. Geophys. Res., 91,* 4393, 1986.

Baumjohann, W. and J. Opgenoorth, "Electric fields and currents associated with active aurora, in *Magnetospheric Currents,*" Geophysical Monograph 28, edited by T. A. Potemra, p. 77, *Am. Geophys. Union,* Washington, D.C., 1984.

Baumjohann, W., "Ionospheric and field-aligned current systems in the auroral zone: A concise review," *Adv. Space Res., 2,* pp. 55-62, 1983.

Baumjohann, W., J. Untiedt and R. A. Greenwald, "Joint two-dimensional observations of ground magnetic and ionospheric electric fields associated with auroral zone currents, 1. three-dimensional current flows associated with a substorm-intensified eastward electrojet," *J. Geophys. Res., 85,* 1963, 1980.

Bostrom, R., in: *Physics of the Hot Plasma in the Magnetosphere,* pg. 341, Plenum, New York, 1975.

Clauer, C. R. and Y. Kamide, "DP1 and DP2 current systems for the March 22, 1979 substorms," *J. Geophys. Res., 90,* 1343, 1985.

Fountain, G. H., F. W. Schenkel, T. B. Coughlin and C. A. Wingate, "The MAGSAT attitude determination system," *Johns Hopkins APL Tech. Dig., 1,* 194, 1980.

Hamilton, D. C., G. Gloeckler, F. M. Ipavich, W. Studemann, B. Wilken, and G. Kremser, "Ring current development during the great geomagnetic storm of February 1986," *J. Geophys. Res., 93,* 14343, 1988.

Heppner, J. P., M. Sugiura, T. L. Skillman, B. G. Ledley and M. Cambel, "OGO-A Magnetic field observations," *J. Geophys. Res., 72,* 5417, 1967.

Iijima, T. and T. Nagata, "Signatures for substorm development of the growth phase and expansion phase," *Planet. Space Sci., 20,* 1095, 1972.

Iijima, T. and T. A. Potemra, "Large-scale characteristics of field-aligned currents associated with substorms," *J. Geophys. Res., 83,* 599, 1978.

Iijima, T. and T. A. Potemra, "The amplitude of field-aligned currents at northern high latitudes observed by TRIAD," *J. Geophys. Res., 81,* 2165, 1976.

Iijima, T., T. A. Potemra and L. J. Zanetti, "Large-scale characteristics of magnetospheric equatorial currents," *J. Geophys. Res., 95,* 991, 1990.

Inhester, B., W. Baumjohann, R. A. Greenwald and E. Nielsen, "Joint two-dimensional observations of ground magnetic and ionospheric electric field associated with auroral zone currents." 3. Three-dimensional currents associated with a westward travelling surge, *J. Geophys. Res., 49,* 155, 1981.

Kamide, Y. and W. Baumjohann, "Estimation of electric fields and currents from international magnetospheric study magnetometer data for the CDAW6 intervals: Implications for substorm dynamics," *J. Geophys. Res., 90,* 1305, 1985.

Kamide, Y., J. D. Craven, L. A. Frank, B. -H. Ahn and S. -I. Akasofu, "Modeling substorm current systems using conductivity distributions inferred from DE auroral images," *J. Geophys. Res., 91,* 11235, 1986.

Kan, J. R., T. Iijima and S. -I. Akasofu, "A model of coupled radial and azimuthal current loops associated with substorms," *J. Geophys. Res., 95,* 21291, 1990.

Kelly, T. J., C. T. Russell, R. J. Walker, G. K. Parks and J. T. Gosling, "ISEE 1 and 2 observations of Birkeland currents in the Earth's inner magnetosphere," *J. Geophys. Res., 91,* 6945, 1986.

Kuppers, F., J. Untiedt, W. Baumjohann, K. Lange and A. G. Jones, "A two-dimensional magnetometer array for ground-based observations of auroral zone electric fields during the International Magnetospheric Study (IMS)," *J. Geophys., 46,* 429, 1979.

Langel, R. A., G. Ousley and J. Berbert, "The MAGSAT mission," *Geophys. Res Lett., 9,* 243, 1982.

Lopez, R. E. and A. T. Y. Lui, "A multisatellite case study of the expansion of a substorm current wedge in the near-Earth magnetotail," *J. Geophys. Res., 95,* 18897, 1990.

Lopez, R. E., A. T. Y. Lui, D. G. Sibeck, K. Takahashi, R. W. McEntire, L. J. Zanetti and S. M. Krimigis, "On the relationship between the energetic particle flux morphology and the change in the magnetic field magnitude during substorms," *J. Geophys. Res., 94,* 17105, 1989.

Lui, A. T. Y., "Estimates of current changes in the geomagnetotail associated with a substorm," *Geophys. Res. Lett., 5,* 853, 1978.

Lui, A. T. Y., R. E. Lopez, R. W. McEntire, B. Anderson, K. Takahashi, L. J. Zanetti, T. A. Potemra, D. M. Klumpar, E. Greene, and R. Strangeway, "Detailed study of near-Earth current disruptions," submitted to *J. Geophys. Res.,* 1991.

Lui, A. T. Y., R. W. McEntire and S. M. Krimigis, "Evolution of the ring current during two geomagnetic storms," *J. Geophys. Res., 92,* 7459, 1987.

Marklund, G. T., L. G. Blomberg, T. A. Potemra, J. S. Murphree, F. J. Rich, and K. Stasiewicz, "A new method to derive 'instantaneous' high latitude potential distributions from satellite measurements including auroral imager data," *Geophys. Res. Lett., 14,* 439, 1987.

Mauk, B. H. and C. -I. Meng, "Plasma injection during substorms," *Phys. Scr., 78;* also Proc. Sixth International Symposium on Solar-Terrestrial Physics (Toulouse, France, 30 June—5 July, 1986) , 1987.

Mauk, B. H. and L. J. Zanetti, "Magnetospheric Electric Fields and Currents," *Review of Geophysics, 25,* 541, 1987.

McPherron, R. L., C. T. Russell and M. P. Aubry, "Satellite studies of magnetospheric substorms on August 15, 1968," 9. Phenomenological model for substorms, *J. Geophys. Res., 78,* 3131, 1973.

Mersmann, U., W. Baumjohann, F. Kuppers and K. Lange, "Analysis of an eastward electrojet by means of upward continuation of ground-based magnetometer data," *J. Geophys., 45,* 281, 1979.

Murphree, J. S., "Viking UV imager observations: implications for substorm modelling," (abstract) *European Geophysical Society*, XV General Assembly, Copenhagen, April, 1990.

Nagai, T., "Field-aligned currents associated with substorms in the vicinity of synchronous orbits," 2. GOES2 and GOES3 observations, *J. Geophys. Res.*, *92*, 2432, 1987

Nagai, T., "Observed magnetic substorm signatures at synchronous altitude," *J. Geophys. Res.*, *87*, 4405, 1982.

Ohtani, S., S. Kokubun and C. T. Russell, "Radial expansion of the tail current disruption during substorms: A new approach to the substorm onset region," submitted to *J. Geophys. Res.*, 1990.

Pellinen, R. J., W. Baumjohann, W. J. Heikkila, V. A. Sergeev, A. G. Yahnin, G. Marklund, and A. O. Melnikov, *Planet. Space Sci.*, *30*, 371, 1982.

Roelof, E. C., "Energetic neutral atom image of a storm-time ring current," *Geophys. Res. Lett.*, *14*, 652, 1987.

Rostoker, G., M. Mareschal and J. C. Samson, "Response of dayside net downward field-aligned current to changes in the interplanetary magnetic field and to substorm perturbations," *J. Geophys. Res.*, *87*, 3489, 1982.

Sato, T., R. J. Walker and M. Ashour-Abdalla, "Driven magnetic reconnection in three dimensions, energy conversion and field-aligned current generation," *J. Geophys. Res.*, *89*, 9761, 1984.

Sato, T., T. Hayashi, R. J. Walker and M. Ashour-Abdalla, "Neutral sheet current interruption and field-aligned current generation by three-dimensional driven reconnection," *Geophys. Res. Lett.*, *10*, 221, 1983.

Siscoe, G. L. and W. D. Cummings, "On the cause of geomagnetic bays," *Planet. Space Sci.*, *17*, 1795, 1969.

Stern, D. P., "Substorm Electrodynamics," *J. Geophys. Res.*, *95*, 12057, 1990.

Walker, R. J. and T. Sato, "Externally driven magnetic reconnection, in Magnetic Reconnection in Space and Laboratory Plasma," Geophysical Monograph 30, edited by E. W. Hones, Jr., p. 272, *Am. Geophys. Union*, Washington, D.C., 1984.

Zanetti, L. J., W. Baumjohann, T. A. Potemra and P. F. Bythrow, "Three-dimensional Birkeland-ionospheric current system, determined from MAGSAT, in Magnetospheric Currents," Geophysical Monograph 28, edited by T. A. Potemra, p. 123, *Am. Geophys. Union*, Washington, D.C., 1984.

Simultaneous Observations of the Westward Electrojet and the Cross-Tail Current Sheet During Substorms

R.E. Lopez,[1] H. Spence,[2] and C.-I. Meng[3]

In this paper we present the results of a study using data from AMPTE/CCE, ground stations, and DMSP F7 during substorms. We have examined 12 events during which CCE was located near the neutral sheet and observed the disruption of the cross-tail current. During these events there were simultaneous ground magnetic field data near to the CCE local time sector. DMSP F7 crossed the nightside auroral oval within 40 minutes of substorm onset at CCE during five of the events. We find that the data are consistent with a near-Earth substorm initiation. We also find that the Tsyganenko [1987] magnetic field model requires an additional tail-like stress during substorms, which corresponds to an equatorward displacement of auroral zone field lines. Assuming that the center of the electrojet is drawn from the near-Earth region, the magnitude of this displacement is about 1° to 2°.

1.0 Introduction

One of the most important signatures of a magnetospheric substorm is the reconfiguration of the magnetotail magnetic field towards a more dipolar orientation [*McPherron et al.*, 1973], which is known as a dipolarization. The dipolarization of the magnetic field is due to the reduction of a portion of the cross-tail current [*Lui*, 1978; *Kaufmann*, 1987]. That reduction has been interpreted in terms of the formation of a substorm current wedge [e.g., *McPherron et al.*, 1973] which diverts a portion of the current into the ionosphere via field-aligned currents (FACs), thereby reducing the equatorial current within a longitudinally limited sector. The sector encompassed by the current wedge expands longitudinally with time [e.g., *Nagai*, 1982]. The current wedge closes in the ionosphere through an enhanced westward electrojet, which terminates in a structure known as the westward travelling surge (WTS) [*Baumjohann et al.*, 1981; *Lühr and Buchert*, 1988].

The origin of the FACs which feed the substorm current wedge, and thus the westward electrojet and the WTS, is an important question which bears on the basic nature of substorms. The boundary layer model [e.g., *Rostoker and Eastman*, 1987] explicitly addresses this issue. That model postulates that the WTS is composed of FACs flowing through the plasma sheet boundary layer at the boundary between the lobe and the plasma sheet, which implies that the FACs map to the distant tail. On the other hand, models such as the near-Earth neutral line model [e.g., *McPherron et al.*, 1973] and the current disruption model [e.g., *Akasofu*, 1972] suggest that the FACs which constitute the current wedge have a relatively near-Earth origin.

The near-Earth ($\lesssim 10\ R_E$) region of the magnetotail beyond geosynchronous orbit has been extensively explored by the AMPTE/CCE satellite. Among other results, it has been found that the phenomenological features of the dipolarization, and the associated injection of energetic particles, depend critically upon the position of the observing satellite relative to the neutral sheet [*Lopez et al.*, 1988a; 1989]. A satellite near the neutral sheet will observe an increase in the total field magnitude, whereas a satellite far from the neutral sheet will observed a decrease in the field magnitude [*Lopez et al.*, 1988a]. Several cases have been published in which CCE was very close to the neutral sheet during substorms [*Takahashi et al.*, 1987; *Lui et al.*, 1988; *Lopez et al.*, 1989; 1990a]. In those events CCE observed dramatic magnetic field and energetic particle variations during what has been termed current sheet disruption, and during one of these events the current sheet disruption was apparently linked to a WTS observed on the ground [*Lopez et al.*, 1990a].

The observation of current sheet disruption in the near-Earth magnetotail, along with multisatellite studies [*Lopez et al.*, 1988b; 1990b; *Lopez and Lui*, 1990; *Ohtani et al.*, 1988], extensive analysis of the geosynchronous energetic particle data [e.g., *Baker et al.*, 1984; *Baker and McPherron*, 1990], a thorough review of auroral morphology [*Feldstein and Galperin*, 1985], modeling studies [e.g., *Kaufmann*, 1987], and the fact that the most equatorward discrete arc is the one that brightens at the onset of the substorm [*Akasofu*, 1964] provide considerable support for a near-Earth

---

[1] Applied Research Corporation, Landover, Maryland
[2] Space Sciences Laboratory, The Aerospace Corporation, Los Angeles, California
[3] Applied Physics Laboratory, The Johns Hopkins University, Laurel, Maryland

Magnetospheric Substorms
Geophysical Monograph 64
Copyright 1991 American Geophysical Union

substorm initiation region. If this is the case, in the early phase of a substorm the westward electrojet must be linked exclusively to the near-Earth magnetotail. We have examined a number of events during which simultaneous observations of the westward electrojet (made by ground stations) and of the near-Earth cross-tail current near the neutral sheet (made by CCE) were available. The ionospheric footpoints of the field lines threading CCE were calculated using the Tsyganenko [1987] field model. During some of the events data from the low-altitude DMSP F7 satellite were available and these have allowed us to place those events in a broader global context. We find that the observations are consistent with the near-Earth initiation of substorms, and that during substorm periods the Tsyganenko [1987] model requires an additional tail-like stress corresponding to an equatorward displacement of auroral zone field lines in the ionosphere of about 1° to 2° in magnetic latitude, assuming that the center of the the substorm electrojet is fed by currents drawn from the near-Earth magnetotail.

2.0 Data

The data to be presented consist of magnetic field measurements from AMPTE/CCE, ground magnetograms from a variety of stations, and precipitating particle data from DMSP F7. CCE is in an equatorial, elliptical orbit with apogee at ~ 8.8 R_E. The magnetic field experiment is described in detail by Potemra et al. [1984]. Ground magnetometer data from Tixie Bay, College, Leirvogur, Syowa, Abisko, and the stations of the EISCAT magnetometer cross were used during this study; those data generally have a time resolution of 1 minute. DMSP F7 is a sun-synchronous polar-orbiting satellite with an altitude of about 840 km, a period of 101.5 minutes, and an orbital plane approximately along the 1035-2235 local time meridian. A more detailed description of the DMSP spacecraft and the instrumentation may be found in Hardy et al. [1984] and Gussenhoven et al. [1985].

The CCE magnetometer data were examined to find substorms that occurred when CCE was not too far from the neutral sheet. Some of these are current sheet disruptions that have been described in detail in previous studies [Takahashi et al., 1987; Lui et al., 1988; Lopez et al., 1989; 1990a]. Others are events that occurred when CCE was relatively close to the neutral sheet, but not within the main distribution of the cross-tail current. They resemble the transitional event discussed by Lopez et al. [1989]. Twelve events were selected due to the relatively close longitudinal proximity of an auroral zone ground station, or set of stations. For each event we have calculated the ionospheric footpoints at 100 km altitude of the field line threading CCE using the model of Tsyganenko [1987]. The truncated version of the model was used with the growth phase Kp value, and all ground coordinates are given in PACE corrected geomagnetic coordinate system [Baker and Wing, 1989]. Software to implement this coordinate system is available from K. Baker, who may be contacted at APLSP::BAKER. In the following section two events will be discussed in some detail, and an overview of the remaining 10 events will be presented.

3.0 Observations

The first event to be discussed occurred on 7 June 1985, at 2209 UT. This event is notable for its excellent magnetometer coverage

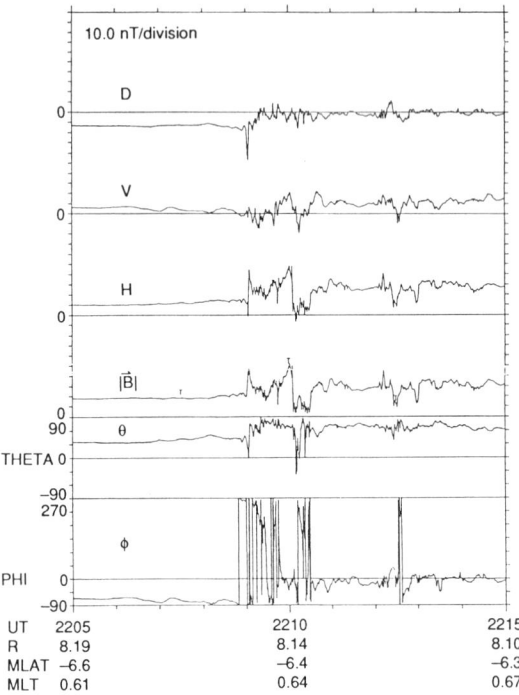

Fig. 1. CCE magnetometer data for 7 June 1985. The data are 0.125s samples and are in VDH coordinates. The onset began at ~ 2209 UT.

and the unusually close longitudinal conjunction between CCE and the ground stations. However, DMSP F7 data are not available for this event. CCE was located at a radial distance of 8.1 R_E, a magnetic local time of 0038 MLT, and a magnetic latitude of -6.4°. Given that K_p was 2, the estimated position of the neutral sheet was 0.2 R_E north of CCE [Lopez, 1990]. This "neutral sheet" position, which is to say the point at which the V component is expected to reverse sign, is 0.3 R_E north of the minimum magnetic field value along the field line threading CCE in the Tsyganenko [1987] model. The model ionospheric footpoint at 100 km of the CCE field line was 66.6° N, 105.9° E in PACE coordinates.

The magnetic field data for the event are presented in Figure 1. These data are 0.125s samples and are in VDH coordinates, where V is positive radially outward, H is positive northward along the dipole axis, and D is positive eastward. The V component of the magnetic field was positive, confirming that CCE was south of the neutral sheet. The onset of the event occurred at 2209 UT, when dramatic variations in the magnetic field began. At this time the flux of energetic particles (not shown) dramatically increased. These signatures are typical of substorms observed by CCE when the satellite was near the neutral sheet [Lopez et al., 1989]. Also, at 2212:30 UT, there was a brief intensification of activity in both the magnetic field and energetic particles. The particle data indicate that the source of the energetic ions at that time was tailward of CCE, suggesting that the activity moved tailward. A similar phenomena was reported by Lopez et al. [1989] during another current disruption event.

Magnetometer data from the EISCAT magnetometer cross (provided courtesy of H. Koskinen and T. Pulkkinen) are presented in Figures 2 and 3, which show the X (positive northward) and Z (positive downward) components, respectively. The vertical order of the stations reflects their respective latitudes (although some stations such as KIL and KAU are almost at the same latitude—their separation is mainly longitudinal). Those data show that the substorm-associated magnetic deflections began at 2209 UT, in excellent agreement with the onset at CCE. On closer inspection, there was also an intensification of activity at 2213 UT, again in accord with the CCE observations. At the 2209 UT onset of activity, KAU and stations poleward of it observed positive Z perturbations, MUO observed essentially no Z variation, and PEL recorded a negative Z deflection. The sign of the Z perturbation allows one to determine the latitudinal position of the center of the westward electrojet relative to the observing station [e.g., *Samson and Rostoker*, 1983]. Therefore we infer that at the onset of the activity the westward electrojet was centered directly over MUO, at 64.7° N, 106.6° E. These coordinates assume that the electrojet was at an altitude of 100 km, the altitude of the field line footpoints, and this same altitude will be used in calculating all ground coordinates. The stations PEL and KAU are 1° equatorward and poleward, respectively, of MUO, and the Z perturbations indicate that the electrojet was poleward and equatorward, respectively, of those stations. Thus in this case it is possible to localize the latitudinal position of the center of the electrojet with a precision of ~ 1°.

We may also use the ground data to determine the poleward boundary of the westward electrojet. At 2212 UT, positive Z perturbations of roughly equal magnitude were observed at KIL and

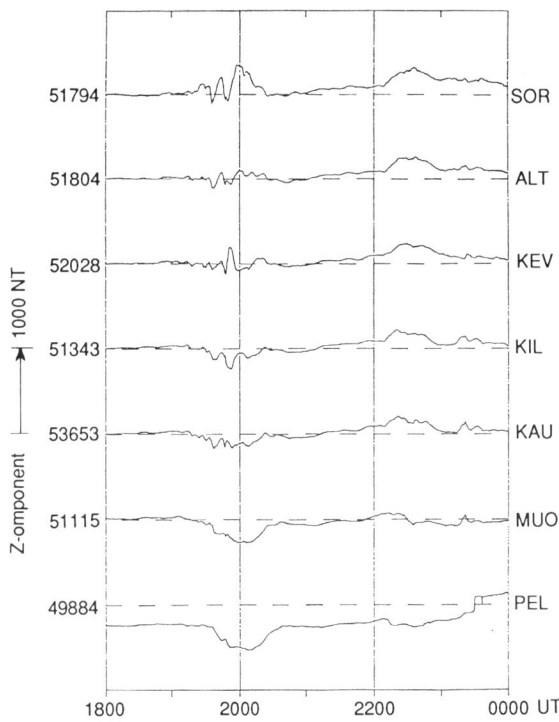

Fig. 3. Z component (positive downward) magnetometer data from the EISCAT magnetometer chain on 7 June 1985. Perturbations associated with the increase in the westward electrojet began at 2209 UT.

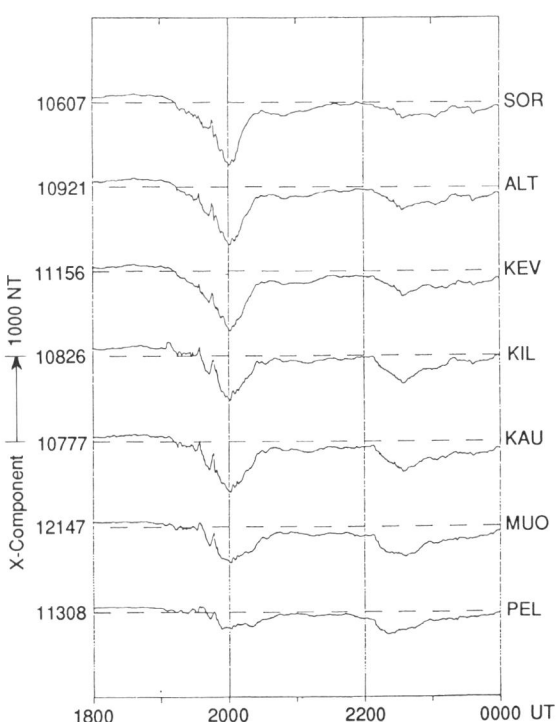

Fig. 2. H component (positive northward) magnetometer data from the EISCAT magnetometer chain on 7 June 1985. The negative bay indicating an increase in the westward electrojet began at 2209 UT.

KAU. Since these were also the largest such perturbations, we infer that the poleward boundary of the westward electrojet at that time was located roughly at 65.7° latitude, midway between KIL and KAU. On the other hand, the negative Z perturbation at PEL was roughly half the value of the positive ones at KIL and KAU, so, assuming a uniform electrojet, the equatorward boundary of the westward electrojet was equatorward of PEL (63.3°, which maps to 6 R_E in the tail). As stated above there was an intensification of activity at 2213 UT, which resulted in an decrease in the Z component at PEL, a very slight decrease in Z at MUO, and an increase in Z at stations poleward of MUO. This suggests that the poleward boundary of the electrojet moved significantly poleward, but that the center of the electrojet moved only slightly poleward. Such a poleward motion would be consistent with the tailward motion of the current disruption region inferred from the CCE data. Later, at 2230 UT, the center of the electrojet moved poleward of MUO to a position equatorward of KIL.

The second event occurred on 28 August 1986 at 1153 UT, and it has been previously examined by Takahashi et al. [1987]. It is one of the most dramatic examples of the turbulent disruption of the current sheet. This event does not have as good a conjunction between the magnetometers and CCE as the previous one, however, DMSP F7 was superbly placed to provide critical information about auroral precipitation morphology. CCE was at a radial distance of 8.1 R_E, a magnetic local time of 2326 MLT, a magnetic latitude of −2.3°, and K_p was 2. The empirical neutral sheet model of Lopez [1990] places CCE precisely at the neutral sheet, while the minimum magnetic field magnitude along the Tsyganenko [1987] field line threading CCE was located 0.3 R_E north of the satellite. The ionospheric foot of that field line was located at 66.7° N, 116.9° W. The CCE magnetic field data are presented in Figure 4; the difference in the time axis between this figure and that in Takahashi et al. [1987] is due to a 20s offset in (all) the CCE data that has been corrected in this figure. The V component was negative indicating that CCE was slightly above the neutral sheet, in contrast to the neutral sheet model prediction. The field magnitude was ~ 8 nT; this is a typical value when CCE is very close to the neutral sheet during the late growth phase. The onset of activity occurred just before 1153 UT, and there was an intensification of activity just before 1200 UT.

Magnetic field data from the ground station at College (65.3° N, 98.4° W) are presented in Figure 5. College was 1.2 hours east of the CCE footpoint, which adds some uncertainty to the determination of the latitudinal position of the westward electrojet at the CCE local time meridian. At 1153 UT, College observed a slowly growing westward electrojet equatorward of the station that was probably associated with the substorm, but there was no signature in the D component. A much more rapid development of the electrojet, this time accompanied by a D perturbation, occurred just after 1200 UT, several minutes after the 1153 UT onset at CCE. We consider this later development to be the expansion of the substorm current wedge into the College sector. The seven minute delay may be a local time effect since on average it takes ~ 7 minutes for the edge of a substorm current wedge to expand one hour of local time [Lopez et al., 1988b]. Also, it may be that the 1200 UT intensifica-

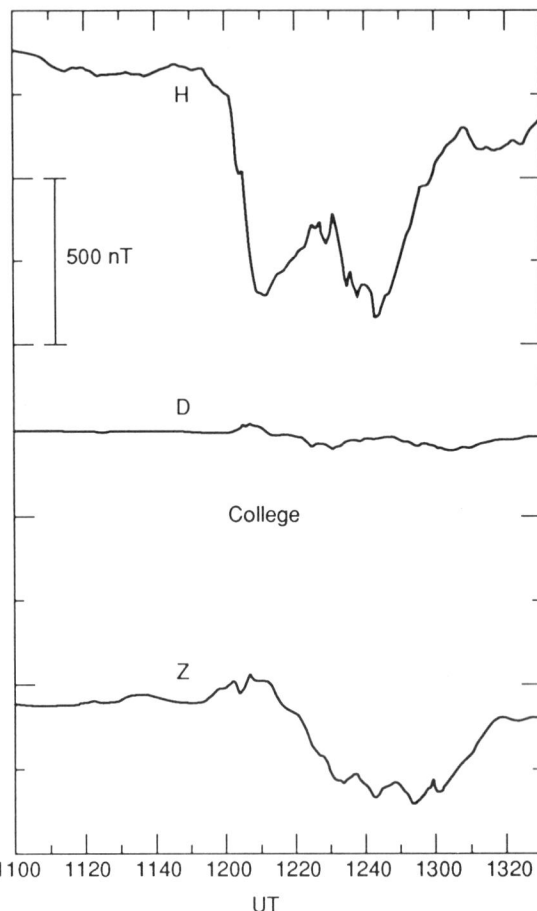

Fig. 5. Ground magnetogram from College on 28 August 1986. The onset began shortly after 1200 UT.

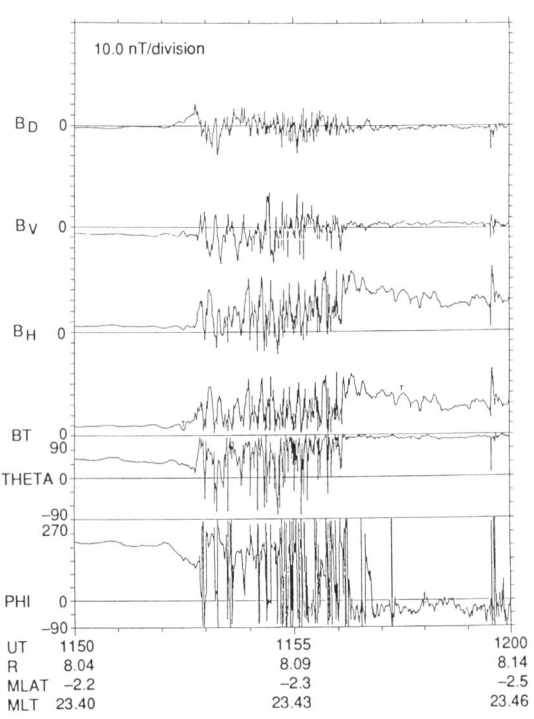

Fig. 4. CCE magnetometer data for 28 August 1986. The data are 0.125s samples and are in VDH coordinates. The onset began at ~ 1153 UT.

tion at CCE was responsible for the 1200 UT electrojet enhancement recorded at College, and that the expansion of the longitudinal sector affected by the substorm expanded in discrete steps as discussed by Wiens and Rostoker [1975].

Since CCE did not move very much from 1153 UT to 1200 UT, the model ionospheric footpoint is essentially the same in both cases. The real footpoints were almost certainly different since there had been a measurable change in the field between 1153 UT and 1200 UT. At the onset of the sharp H decrease, the Z component exhibited small negative and positive perturbations. This suggests that the electrojet center oscillated slightly over the station, or that several small, variable electrojets formed in close latitudinal proximity. Therefore we conclude that at the onset of the event the electrojet center was located at ~ 65.3° N in the College sector, and at ~ 1215 UT the electrojet center moved poleward of College. Just after the 1153 UT onset at CCE, DMSP F7 passed over auroral zone 1.2 hours west of the CCE footpoint. The particle data from this pass are presented in Figure 6. Electrons and ions from 30 eV to 30 keV are plotted with the lowest energies at the center of the plot. The geomagnetic coordinates on the plot are in the PACE coordinate system. The equatorward edge of the auroral zone is an

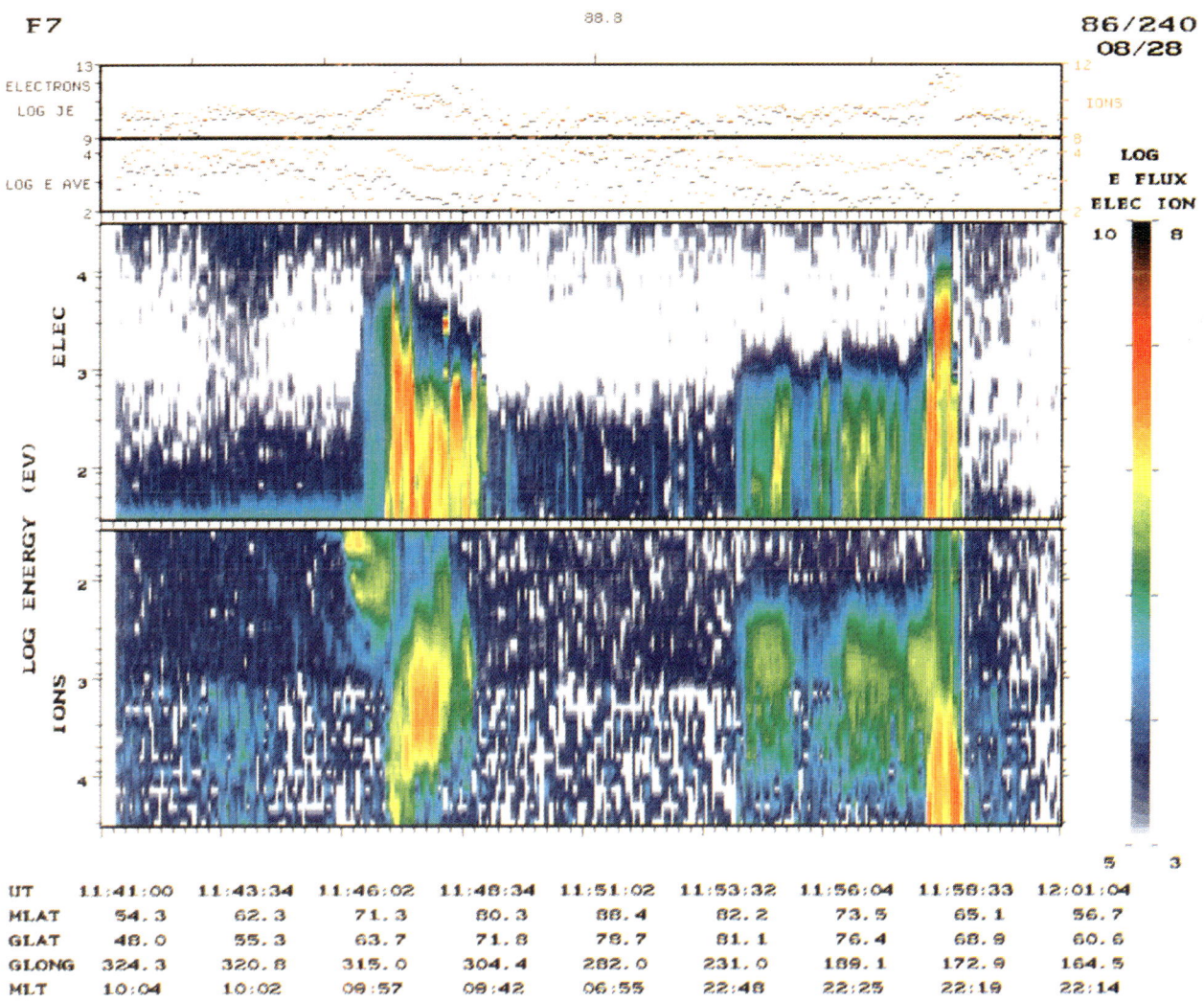

Fig. 6. Precipitating particle data DMSP F7 on 28 August 1986. DMSP F7 passed over the polar cap just ~ 1 hour to the west of CCE in close temporal proximity to the onset of the event. At 1154 UT it encountered the boundary between the lobe and the plasma sheet at a latitude of 80°. Continuing equatorward, the satellite reached the poleward edge of a bright arc at 1158 UT. The arc was located at ~ 66°, and we associate it with the substorm westward electrojet.

example of the low altitude signature of the dispersionless substorm injection boundary discussed by Newell and Meng [1987]. Poleward of the injection boundary a bright discrete feature was observed. It was composed of a monoenergetic (~ 4 keV) electron region, with a lower energy feature at the poleward boundary. The latitude of the poleward edge of the monoenergetic electron part was 65.5°, and the entire discrete arc extended to a latitude of 66.2°. It is important to note that poleward of the arc there was a region filled with weak discrete features, and that the boundary between the lobe and the plasma sheet was located at a geomagnetic latitude of 80° just 11.2° west of the CCE footpoint.

The DMSP F7 image data (not shown) indicates that the bright arc seen in the particle data was associated with the substorm. The event took place in late August and DMSP F7 was in the northern hemisphere; this is not the best situation for auroral photography. However, DMSP F7 crossed the arc before it crossed the terminator, and a limited region just to the east and west of the flight path is visible. The point at which the arc was crossed was the westernmost extension of a bright region which had expanded poleward just to the east of DMSP F7, while to the west the arc was narrow, and it faded out. The satellite apparently crossed the most equatorward discrete arc just as it started to brighten as the substorm was expanding into that sector. Assuming that the poleward edge of the bright arc is near the center of the the auroral electrojet [e.g., *Samsom and Rostoker*, 1983], we find that one hour to the west of CCE the electrojet was centered at ~ 66°, and one hour to the east it was centered at ~ 65.3°. From this information we estimate that the latitude of the electrojet center in the CCE sector was ~ 65.7°.

The remaining ten events have been analyzed in a similar fashion as those discussed above. The results of this investigation are presented in Table 1. Each event, including those discussed above, is listed by the time (within 5 minutes) and date of occurrence. The second column gives the radius (in R_E) and the magnetic local time (in SM coordinates) of CCE at the time indicated for each event. The third column lists the ionospheric footpoints of the field line threading CCE. Northern and southern footpoints are given for those events for which data from both auroral zones were available. The center positions of the westward electrojets were determined exclusively from the ground magnetometer data, with the exception of the 28 August 1986 event (discussed above), and southern electrojet on 7 May 1985. That position was determined from the poleward boundary of a discrete arc (the only one) observed by DMSP F7. Since it was almost certain that this arc was the one that brightened, its position may be used as an indicator of the electrojet position. The arc was observed 20 minutes before the onset of the substorm. During the growth phase discrete arcs drift equatorward [e.g., *Tanskanen et al.*, 1987]. Therefore the position of the arc 20 minutes before the breakup provides us with the maximum possible latitude for the center of the substorm electrojet. The boundary between the plasma sheet and lobe was determined from DMSP F7 particle data. The position of the boundary (magnetic latitude and longitude), and the time at which the boundary was crossed are listed. This position is listed only for those events when F7 reached the boundary within 40 minutes of the onset, which limits us to five out of the twelve events. The final column is labeled "Shift." It will be discussed in the next section when we examine the implications of these results.

4.0 DISCUSSION

The results presented in Table 1 are completely consistent with the near-Earth substorm initiation scenario. The electrojet positions found in this study are also consistent with the results of Craven and Frank [1987], who found that during substorms auroral brightening begins at ~ 66°, then expands poleward. Moreover, the neutral sheet at R ~ 8 to 9 R_E consistently maps to the equatorward edge of the auroral zone, which is additional evidence that the near-Earth region surveyed in detail by CCE is indeed the region of substorm initiation. In those cases where the center of the electrojet is well-placed, that latitudinal position was always equatorward of the CCE footpoint, and there is no case were the center of the electrojet was observed to be initially poleward of the CCE footpoint. Moreover, in the case of the 28 August 1986 event, it is obvious that the most equatorward portion of the auroral oval brightened first.

The available evidence argues strongly against the boundary layer model [e.g., *Rostoker and Eastman*, 1987]. Again, the 28 August 1986 case is unambiguous. The brightening arc and the center of the substorm westward electrojet were located 14°

Table 1: CCE Footprints Relative to Electrojet Centers

Event	R^1, MLT^2	CCE Footprint^3	Electrojet Center	Plasma Sheet–Lobe Boundary	Shift
7 May 1985, 1105 UT	8.7, 0039	67.6 N, 81.7 W 67.8 S, 83.1 W	> 65.3 N, 98.4 W < 67.0 S, 93.5 W	68.9 S, 93.7 W, 1044 UT	< 2.3 > 0.8
17 May 1985, 2350 UT	8.7, 0111	66.8 N, 90.4 E 67.6 S, 90.2 E	< 65.2 N, 103.2 E > 66.8 S, 72.8 E		> 1.6 < 1.2
1 June 1985, 2120 UT	8.6, 2340	66.9 N, 100.1 E	65.7 N, 107.1 E		~ 1.2
1 June 1985, 2315 UT	8.8, 0016	67.9 S, 83.4 E	> 66.8 S, 72.8 E		< 1.1
3 June 1985, 2240 UT	8.8, 0010	67.4 N, 91.8 E 68.2 S, 92.8 E	65.2 N, 103.2 E > 66.8 S, 72.8 E		~ 2.2 < 1.4
7 June 1985, 2210 UT	8.1, 0039	66.6 N, 105.9 E	64.7 N, 106.6 E		~ 1.9
20 June 1985, 2310 UT	8.2, 2355	66.3 N, 78.3 E 67.5 S, 79.7 E	< 65.3 N, 69.1 E > 66.8 S, 72.8 E		> 1.0 < 0.7
9 August 1986, 1450 UT	8.5, 0040	66.9 N, 137.0 W	> 65.2 N, 164.2 W	73.2 N, 77.5 W, 1500 UT	< 1.7
24 August 1986, 1345 UT	8.0, 2332	66.5 N, 142.0 W	> 65.2 N, 164.2 W	68.1 N, 90.8 W, 1319 UT	< 1.3
28 August 1986, 1155 UT	8.1, 2325	66.7 N, 116.9 W	~ 65.7 N, 116.0 W	80 N, 125.6 W, 1154 UT	~ 1.0
29 August 1986, 2050 UT	8.7, 0012	67.0 N, 120.3 E	> 65.2 N, 103.2 E		< 1.8
30 August 1986, 1220 UT	8.7, 2351	67.1 N, 115.5 W	> 65.3 N, 98.4 W	73.7 N, 88.5 W, 1256 UT	< 1.8

^{1}CCE position in R_E
^{2}SM local time
^{3}All latitudes and longitudes in PACE geomagnetic coordinates

equatorward of the boundary between the lobe and the plasma sheet. The same is true for the 7 May 1985 case, when the presumed electrojet center was 1.9° equatorward of the boundary. Since the westward electrojet terminates in the surge head and the westward electrojet center is located near the poleward edge of the surge [e.g., *Baumjohann et al.* 1981; *Samson and Rostoker*, 1983], our results are consistent with the proposition that the surge is connected to the near-Earth tail, at least in the initial phase of a substorm [e.g., *Lopez et al.*, 1990a].

The other major result of this study concerns the stretching of the field during the growth phase of a substorm. As the cross-tail current intensifies, the field becomes more tail-like. This corresponds to an equatorward shift in the ionospheric footpoint of a field line that maps to a fixed equatorial crossing point in the tail. An important question is the magnitude of that shift. If we assume that the electrojet current is being drawn primarily from the near-Earth tail, then the observed difference between the electrojet center and the mapped footpoint can provide an indication of that shift. This difference is listed in the final column of Table 1. On the other hand, if the poleward boundary of the westward electrojet corresponds to the disruption region, then the inferred shift could be considerable smaller. For example, for the 7 June 1985 event, we were able to determine that the poleward boundary of the electrojet was at ~ 65.7°. The CCE footpoint was at 66.6°, which implies a shift of only 0.9°. The other event for which we have good latitudinal magnetogram coverage occurred on 1 June 1985 at 2120 UT. In that case the poleward boundary was at least at 67.2° (the position of the most poleward station), so that the disruption region could map to the poleward portion of the westward electrojet without any modification of the magnetic field model.

Since we do not have adequate ground coverage to determine the full latitudinal extent of the electrojet in every case, we shall simply assume that the near-Earth region connects to the center of the electrojet during the disruptions. This gives an equatorward displacement of auroral field lines in general between 1° and 2°, although we recognize that this is probably an upper limit. Another point of interest is that the latitudinal position of the electrojet varies whether one considers the northern or southern hemispheres. Part of this difference may be due to the seasonal dependence of conjugacy, as discussed by Wu et al. [1991]. However, in some cases it could be a local time effect since the electrojet moves poleward during substorms and different longitudes will have experienced varying degrees of poleward expansion.

5.0 CONCLUSION

We have examined a dozen substorm events during which CCE was located near to, or within, the cross-tail current, and for which there are auroral zone magnetometer data from a nearby local time sector. DMSP F7 observations of the polar cap are available for five of the events. We find that the observations are consistent with a near-Earth substorm initiation region, and inconsistent with the model of Rostoker and Eastman [1987]. Furthermore we find that additional tail-like stress is required in the Tsyganenko [1987] model during substorms. The magnitude of the stress is such to produce an equatorward displacement of auroral zone field lines from 1° to 2° in latitude, assuming that the near-Earth region provides the bulk of the current for the westward electrojet.

Acknowledgments We are pleased to acknowledge T. Potemra and L. Zanetti, who generously supplied the CCE data, K. Takahashi, who supplied the plotting software for CCE data, and K. Baker, who supplied the PACE software. The ground magnetograms were provided by the World Data Center, with the exception of the EISCAT magnetograms, which were provided H. Koskinen, T. Pulkkinen, and the Finnish Meteorological Institute under the aegis of the U.S.–Finnish Auroral Workshop, and the Syowa magnetograms, which were provided by the Japanese Antarctic Research Expedition. We would also like to thank B. Mauk, P. Newell, and K. Takahashi for several helpful conversations. The work performed at APL was supported by NASA under Task 1 of contract N00039-89-C-5301, contract NAS 5-31208, and by NSF grant ATM-8713212. The work performed at The Aerospace Corporation was supported by the US Air Force System Command's Space System Division under contract No. F04701-88-C-0089.

REFERENCES

Akasofu, S.-I., The development of the auroral substorm, *Planet. Space Science*, 12, 273–282, 1964.

Akasofu, S.-I., Magnetospheric substorms: A model, in *Solar Terrestrial Physics/1970: Part III*, ed. by D. Dyer, pp. 131–151, D. Reidel, Dordrecht-Holland, 1972.

Baker, D. N., Particle and field signatures of susbtorms in the near magnetotail, in *Magnetic Reconnection in Space and Laboratory Plasmas*, Geophys. Monogr. Ser., vol. 30, edited by E. W. Hones, Jr., pp. 193–202, AGU, Washington, D. C., 1984.

Baker, D. N:, and R. L. McPherron, Extreme energetic particle decreases near geostationary orbit: A manifestation of current diversion within the inner plasma sheet, *J. Geophys. Res.*, 95, 6591–6599, 1990.

Baker, K. B., and S. Wing, A new magnetic coordinate system for conjugate studies at high altitudes, *J. Geophys. Res.*, 94, 9139–9143, 1989.

Baumjohann, W., R. J. Pellinen, H. J. Opgenoorth, and E. Nielsen, Joint two-dimensional observations of ground magnetic and ionospheric electric fields associated with auroral zone currents: Current systems associated with local auroral breakups, *Planet. Space Sci.*, 29, 431–447, 1981.

Craven, J. D., and L. A. Frank, Latitudinal motions of the aurora during substorms, *J. Geophys. Res.*, 92, 4565–4573, 1987.

Feldstein Y. I., and Yu. I. Galperin, The auroral luminosity structure in the high-latitude upper atmosphere: Its dynamics and relationship to the large-scale structure of the Earth's magnetosphere, *Rev. of Geophys.*, 23, 217–275, 1985.

Gussenhoven, M. S., D. A. Hardy, F. Rich, and W. J. Burke, High-level spacecraft charging in the low-altitude polar auroral environment, *J. Geophys. Res.*, 90, 11009–11023, 1985.

Hardy, D. A., L. K. Schmitt, M. S. Gussenhoven, F. J. Marshall, H. C. Yeh, T. L. Shumaker, A. Hube, and J. Pantazis, Precipitating electron and ion detectors (SSJ/4) for the block 5D/flights 6-10 DMSP satellites: Calibration and data presentation, *Rep. AFGL-TR-84-0317*, Air Force Geophysics Laboratory, Hanscom Air Force Base, Mass., 1984.

Kaufmann, R. L., Substorm currents: growth phase and onset, *J. Geophys. Res.*, 92, 7471–7489, 1987.

Lopez, R. E., The position of the magnetotail neutral sheet in the near-Earth region, *Geophys. Res. Lett.*, 17, 1617–1620, 1990.

Lopez, R. E., and A. T. Y. Lui, A multisatellite case study of the expansion of a substorm current wedge in the near-Earth magnetotail, *J. Geophys. Res.*, 95, 8009–8017, 1990.

Lopez, R. E., D. G. Sibeck, A. T. Y. Lui, K. Takahashi, R. W. McEntire, T. A. Potemra, and D. Klumpar, Substorm variations in the magnitude of the magnetic field, *J. Geophys. Res.*, 93, 14444–14452, 1988a.

Lopez, R. E., D. N. Baker, A. T. Y. Lui, D. G. Sibeck, R. D. Belian, R. W. McEntire, T. A. Potemra, and S. M. Krimigis, The radial and longitudinal propagation characteristics of substorm injections, *Adv. Space Res.*, (9)91–(9)95, 1988b.

Lopez, R. E., A. T. Y. Lui, D. G. Sibeck, K. Takahashi, R. W. McEntire, L. J. Zanetti, and S. M. Krimigis, On the relationship between the energetic particle morphology and the change in the magnetic field magnitude during substorms, *J. Geophys. Res., 94* 17105–17119, 1989.

Lopez, R. E., H. Lühr, B. J. Anderson, and P. T. Newell, Multipoint observations of a small substorm, *J. Geophys. Res., 95*, 18897–18912, 1990a.

Lopez, R. E., D. N. Baker, R. D. Belian, R. W. McEntire, and T. A. Potemra, A case study of a radially antisunward propagating local substorm onset, *Planet. Space Sci.*, 771–784, 1990b.

Lühr, H. and S. Buchert, Observational evidence for a link between currents in the geotail and in the auroral ionosphere, *Ann. Geophys., 6*, 169–176, 1988.

Lui, A. T. Y., Estimates of current changes in the geomagnetic tail associated with a substorm, *Geophys. Res. Lett., 5*, 853–856, 1978.

Lui, A. T. Y., R. E. Lopez, S. M. Krimigis, R. W. McEntire, L. J. Zanetti, and T. A. Potemra, A case study of magnetotail current sheet disruption and diversion, *15, Geophys. Res. Lett.*, 721–724, 1988.

Makita, K., and C.-I. Meng, Average electron precipitation patterns and visual aurora characteristics during geomagnetic quiescence *J. Geophys. Res., 89*, 2861–2872, 1984.

McPherron, R. L., C. T. Russell, and M. P. Aubry, Satellite studies of magnetospheric substorms on August 15, 1968: 9. Phenomenological model for substorms, *J. Geophys. Res., 78*, 3131–3149, 1973.

Nagai, T., Observed magnetic substorm signatures at synchronous altitude, *J. Geophys. Res., 87*, 4406–4417, 1982.

Newell, P. T., and C.-I. Meng, Low altitude observations of dispersionless substorm plasma injections, *J. Geophys. Res., 92*, 10063–10072, 1987.

Ohtani, S., S. Kokobun, R. C. Elphic, and C. T. Russell, Field-aligned current signatures in the near-tail region, 1. ISEE observations in the plasma sheet boundary layer, *J. Geophys. Res., 93*, 9709–9720, 1988.

Rostoker, G., and T. E. Eastman, A boundary layer model for magnetospheric substorms, *J. Geophys. Res., 92*, 12187–12201, 1987.

Samson, J. C., and G. Rostoker, Polarization characteristics of Pi2 pulsations and implications for their source mechanisms: Influence of the westward travelling surge, *Planet. Space Sci., 31*, 435–458, 1983.

Takahashi, K., L. J. Zanetti, R. E. Lopez, R. W. McEntire, T. A. Potemra, and K. Yumoto, Disruption of the magnetotail current sheet observed by AMPTE/CCE, *Geophys. Res. Lett., 14*, 1019–1022, 1987.

Tanskanen, P., J. Kangas, L. Block, G. Kremser, A. Korth, J. Woch, I. B. Iversen, K. M. Torkar, W. Riedler, S. Ullaland, J. Stadnes, and K.-H. Glassmeier, Different phases of a magnetospheric substorm on June 23, 1979, *J. Geophys. Res., 92*, 7443–7457, 1987.

Tsyganenko, N. A., Global quantitative models of the geomagnetic field in the cislunar magnetosphere for different disturbance levels, *Planet. Space Sci., 35*, 1347–1358, 1987.

Wu, Q., T. J. Rosenberg, L. J. Lanzerotti, C. G. Maclennan, and A. Wolfe, Seasonal and diurnal variations of the latitude of the westward auroral electrojet in the nightside polar cap, *J. Geophys. Res., 96*, 1409–1419, 1991.

Tail Current Disruption in the Geosynchronous Region

S. Ohtani[1,2], K. Takahashi[1], L.J. Zanetti[1], T.A. Potemra[1], R.W. McEntire[1], and T. Iijima[2]

The location and the spatial scales of the onset region of tail current disruption remains a significant outstanding problem of the substorm process. We examine two successive substorm events observed in the synchronous region with data from the GOES 5, GOES 6, and AMPTE/CCE spacecraft. The three spacecraft were conveniently located in the premidnight sector during the events and observed significant differences in the magnetic signatures. The multisatellite comparison indicates that: (1) the onsets took place probably in the synchronous region; (2) the radial scale of the onset region was of the order of 1 R_E or less; and (3) the azimuthal scale was also of the order of 1 R_E or less. We infer that AMPTE/CCE was located very close to the onset region during the second event and investigate the sequences of changes in the magnetic field and energetic particle fluxes with a time resolution on the order of seconds. The results indicate that there is a distinctive interval (explosive growth phase) just prior to the local current disruption. The duration of this interval is typically 1 min, much shorter than the so-called growth phase. During this interval: (1) the intensity of the tail current increases significantly; and (2) the energization of ions takes place in association with the enhancement of the tail current intensity. We suggest that a feedback mechanism is responsible for this enhancement.

1. INTRODUCTION

Magnetic field reconfiguration and the increase in energetic particle fluxes are distinctive features associated with magnetospheric substorms in the near-Earth region. The magnetic field in the nightside magnetosphere changes from a stressed (or tail-like) to a more dipolar configuration during substorms [e.g., *Cummings et al.*, 1968]. This change is ascribed to the sudden decrease in tail current intensity [*Fairfield and Ness*, 1970]. The current system, which is often referred to as the current wedge, models the three-dimensional closure of the disrupted current [*McPherron et al.*, 1973]. Changes in energetic particle fluxes are closely correlated with local magnetic field changes in the course of substorms [e.g., *Lezniak and Winckler*, 1970; *Erickson et al.*, 1979; *Sauvaud and Winckler*, 1980]. In particular, the increase in fluxes, which is often called "injection," has been regarded as a result of the current disruption.

It is expected that in the substorm onset region spacecraft would observe a sharp change of magnetic field configuration and a dispersionless injection simultaneous with the commencement of ground activity. Previous statistical studies have reported that the onset region is located at $X \leq -20 R_E$ [*Hones et al.*, 1973; *Nishida and Nagayama*, 1973; *Ohtani et al.*, 1990]. Some recent case studies have shown that the current disruption occurs initially in the near-Earth magnetotail ($X \leq -9 R_E$) [*Takahashi et al.*, 1987; *Lui et al.*, 1988; *Lopez et al.*, 1989, 1990]. *Kaufmann et al.* [1987] have also suggested that the tail current intensity changes most drastically at altitudes between 7 and 9 R_E during substorms.

It is well known that the earliest onset takes place in a relatively localized region. *Arnoldy and Moore* [1983] have reported that the onset sector is about three hours wide at synchronous orbit. *Lui* [1978] deduced, from a model of magnetic field variations during a substorm, that the radial extent of the current disruption region is in the range of 0.4 to 3.5 R_E. The development of substorms would correspond to the expansion of the current disruption region in the tailward [*Ohtani et al.*, 1988, 1990; *Lopez and Lui*, 1990], the earthward [*Russell and McPherron*, 1973], and the azimuthal directions [*Nagai*, 1982; *Arnoldy and Moore*, 1983]. Such a picture would be consistent with the expansive auroral development during the substorm expansion phase [e.g., *Akasofu*, 1964; *Rostoker et al.*, 1987; *Kaneda and Yamamoto*, 1991].

In this paper we use magnetometer data from GOES 5 and GOES 6 and magnetometer data [*Potemra et al.*, 1985] and medium energetic particle analyzer (MEPA) data [*McEntire et al.*,

[1]The Johns Hopkins University Applied Physics Laboratory, Laurel, MD.
[2]Geophysics Research Laboratory, University of Tokyo, Tokyo, Japan.

Magnetospheric Substorms
Geophysical Monograph 64
Copyright 1991 American Geophysical Union

1985] from the Charge Composition Explorer (CCE) of the Active Magnetospheric Particle Tracer Explorers (AMPTE) mission. AMPTE/CCE has an equatorial elliptical orbit with an apogee of 8.8 R_E and a period of about 16 hours. Hence the comparison of substorm-associated signatures between CCE and the geosynchronous satellites would be very useful for examining spatial structures of substorm onsets. First, we examine a geomagnetically disturbed period on July 30, 1985. The three spacecraft were conveniently located during the period of interest and observed two distinctive events. Second, we investigate the time sequence of changes in the magnetic field components and in the energetic particle fluxes during the second event with a time resolution on the order of seconds. Finally, spatial and temporal development of the current disruption is discussed.

2. OBSERVATION

In this section we examine the disturbed period of 0300-0500 UT on July 30, 1985. The AE index continued at very quiet levels from July 29 (the average AE index of July 29 was 110 nT) with no evident disturbance for more than 6 hours until ~0325 UT, the onset time of the first substorm. During this period, the GOES 5 (253.3° W geomagnetic longitude, 11.1° N geomagnetic latitude) and GOES 6 (330.9° W, 9.9° N) geosynchronous spacecraft were located in the premidnight sector; GOES 6 was located at a local time (MLT) meridian approximately 1.5 hours earlier than that of GOES 5. AMPTE/CCE was inbound, also in premidnight local time sector in the northern hemisphere. The latitude of the AMPTE/CCE spacecraft was almost the same as those of the two GOES spacecraft, and the geomagnetic latitude of AMPTE/CCE gradually increased from 8.7° to 9.2° during the period.

Figure 1 shows the H, V, and D components from GOES 5, AMPTE/CCE, and GOES 6 magnetic field data during the period. The location of the spacecraft is indicated at the bottom of the figure (also see Figure 3). Here H is antiparallel to the dipole axis, V points radially outward at the magnetic equator, and D completes the right-hand orthogonal system (positive eastward). The first distinctive signature observed by the satellites was the sudden commencement of disturbances around 0325 UT, followed by irregular variations in all the magnetic components. The tail reconfiguration around 0420 UT was the other substorm-associated event during the period. Figure 2 presents the AMPTE/CCE ion flux measurements by the medium energy particle analyzer (MEPA) over the energy range of 32 keV to 365 keV, plotted on the same scale as the CCE magnetometer data. In association with the commencements of magnetic disturbances around 0325 UT and 0420 UT, the energetic particle fluxes increased significantly. In the following we examine these two events in more detail.

2.1. 0325 UT Event

Figure 3a shows the locations of the three spacecraft at 0325 UT in the local time-radial distance diagram, projected on the equatorial plane. The GOES 5 and GOES 6 satellites were at 22.3 and 20.8 MLT, respectively, at synchronous orbit. The AMPTE/CCE was 0.9 R_E beyond synchronous orbit and located 0.3 hour MLT duskward of GOES 5.

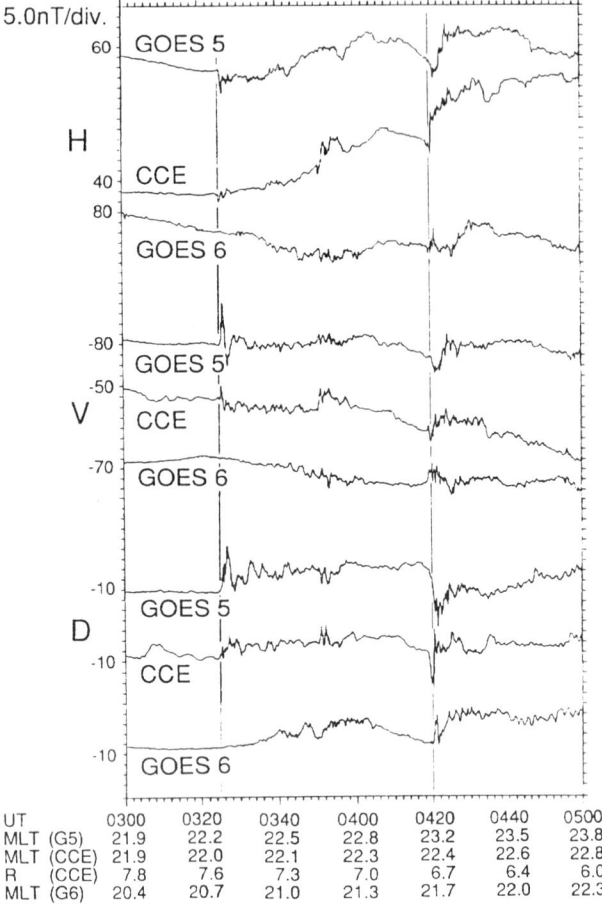

Fig. 1. The H, V, and D component from GOES 5, AMPTE/CCE, and GOES 6 magnetic field data during the disturbed period of 0300-0500 UT on July 30, 1985. The spacecraft locations are indicated at the bottom.

At 0325 UT GOES 5 observed an abrupt commencement of disturbances in every component. AMPTE/CCE observed almost the same signatures; H decreased transiently, V increased suddenly, followed by the large negative deviation, and D increased almost steplike. No evident time lag of magnetic signatures can be found between the two satellites. The commencement of the magnetic disturbance was accompanied by a dispersionless increase in energetic particle fluxes (see Figure 2). At the same time, a positive bay onset was observed at Fredericksburg (256.5° geomagnetic longitude, 64.6° geomagnetic latitude). Hence the disturbances observed at GOES 5 and AMPTE/CCE can be regarded as signatures associated with the substorm onset.

It should be noted that the amplitude of the perturbations at AMPTE/CCE was almost one third of that at GOES 5, despite the small separation between the two spacecraft, 0.3 hours in MLT and 0.9 R_E in radial distance. For example, the magnitude of the initial increase in V was about 20 nT at GOES 5, while it was ~ 7 nT at AMPTE/CCE. This fact suggests that GOES 5 was located closer to the center of the change in current intensity associated with the substorm onset, which would be regarded as the onset region, than

2.2. 0420 UT Event

The other distinctive signature is the tail reconfiguration, with large D deviations around 0420 UT (see Figure 1). A sharp depression in the X (north-south) component started at Churchill (68.2° geomagnetic longitude, 326.4° geomagnetic latitude) at 0420 UT, and an enhancement in disturbance level in the ULF band was observed at Hermanus (81.7°, −33.7°) almost simultaneously. However, it is difficult to determine the exact time of Pi 2 onset due to large background noise. Energetic particle fluxes were enhanced without any evident energy dispersion at AMPTE/CCE, except for the flux of the lowest energy (see Figure 2). Here again, we infer that the magnetic disturbances observed at synchronous altitude were related to the substorm onset. In this event, the three spacecraft were almost at the same radial distance in the premidnight sector. AMPTE/CCE was just beyond synchronous altitude, at $R \sim 6.7\ R_E$, between GOES 5 (23.2 MLT) and GOES 6 (21.7 MLT) in local time (see Figure 3b). The separation between AMPTE/CCE and the GOES spacecraft was ~1.3 R_E. The location of the space-

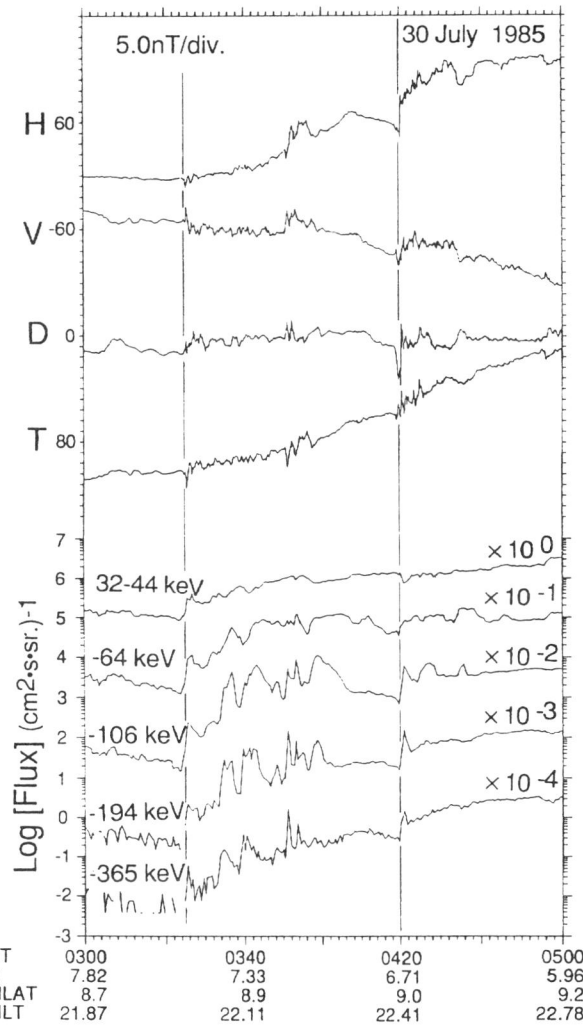

Fig. 2. Magnetic field and energetic particle measurements from AMPTE/CCE for the July 30, 1985, event. The particle data are spin-averaged over 24 s and the magnetic field data are 6.2-s median values.

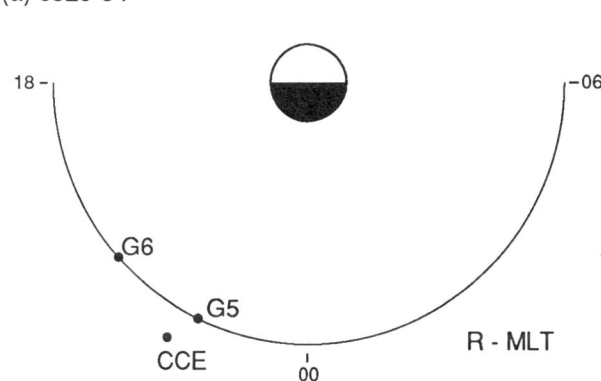

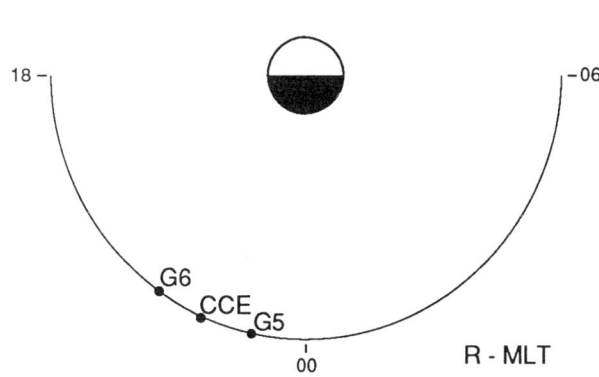

Fig. 3. Locations of AMPTE/CCE, GOES 5 (G5), and GOES 6 (G6) in the local time—radial distance diagram at (a) 0325 UT and (b) 0420 UT on July 30, 1985.

AMPTE/CCE. Furthermore, the signs of the field change were the same in all the magnetic components at the two satellites. This suggests that the two satellites were on the same side of the onset region; otherwise, the sign of deviation would have been opposite at least in one component. Hence it may be expected that the current intensity changed principally very close to, or inside of, synchronous orbit at the 0325 UT substorm onset. Taking into account the closely spaced locations of AMPTE/CCE and GOES 5, we infer that the radial scale of the substorm onset was of the order of 1 R_E.

While GOES 5 and AMPTE/CCE observed the abrupt commencement of the disturbances, GOES 6 observed no evident signatures, despite a local time separation of only 1.5 hours from GOES 5. More importantly, the magnetic field at GOES 6 gradually became more tail-like even after the substorm onset. H tended to decrease and $|V|$ tended to increase, suggesting that the current intensity was enhanced outside of the onset sector.

craft is very convenient for examining the azimuthal structure of the substorm onset in the near-Earth magnetotail. The magnetic signatures were surprisingly different among the three spacecraft. Most importantly, the H component increased by 30 nT within 5 s at AMPTE/CCE (also see Figure 4). GOES 5 observed the tail reconfiguration about 2 min later. The signature at GOES 6 was more complicated and was not as sharp as that observed at AMPTE/CCE. Such differences indicate that the principal current disruption took place probably nearest to AMPTE/CCE. The azimuthal scale of the current disruption would be of the order of 1 R_E in local time. We also infer that the onset took place very close to the synchronous orbit, within 1 R_E, although we have no information on the radial structure of this substorm onset. Otherwise, it would be very difficult to explain the significant difference in magnetic signatures observed with these three closely spaced satellites.

In addition, the time sequence of the D disturbances was complicated among the three spacecraft. In previous studies the substorm-associated D deviation has been explained in terms of the formation of the wedge current system [e.g., *McPherron et al.*, 1973; *Nagai*, 1982]. In this model the D deviation is expected to be observed exclusively after the initial current disruption. AMPTE/CCE observed that the sharp D depression started prior to the abrupt increase in H (0420 UT), followed by the abrupt recovery almost simultaneous with the H increase. GOES 5 also observed the commencement of the D depression before 0420 UT, and the recovery of the D deviation corresponded to the commencement of the local field reconfiguration at the GOES 5 location. The commencement of these D deviations might be related to the slight increase in V at 0418 UT observed at GOES 6. Contrarily, at the GOES 6 location, the positive D deviation started simultaneously with the abrupt H increase at AMPTE/CCE. These sequences seem to be too complicated to ascribe to a single current circuit as was proposed in previous studies. A small current disruption may have taken place locally around GOES 6 at 0418 UT. The complicated signatures may also be ascribed to the multiple-sheet structure of field-aligned currents, which has been pointed out from low-altitude observations in the premidnight sector [e.g., *Rostoker et al.*, 1975; *Iijima and Potemra*, 1976]. The spatial structure of the field-aligned currents in the premidnight magnetosphere would be one of the outstanding problems of the substorm process, although such questions would be beyond the scope of this study.

2.3. Current Disruption at AMPTE/CCE

Here it would be useful to examine in more detail the tail current disruption observed with AMPTE/CCE, which was located nearest to the onset position. Figure 4 shows the magnetic field and energetic particle flux signatures during the 4 min surrounding the sharp increase in H at ~0420 UT (see Figures 1 and 2). Plotted are 0.125-s magnetic field data and particle flux data from the two MEPA channels (31–42 and 62–102 keV) having the highest time resolution (32 samples per spin period, 6 s). MEPA scans the plane perpendicular to the spin axis roughly along the Earth-sun line and divides this plane into 32 sectors. Hence the detector samples

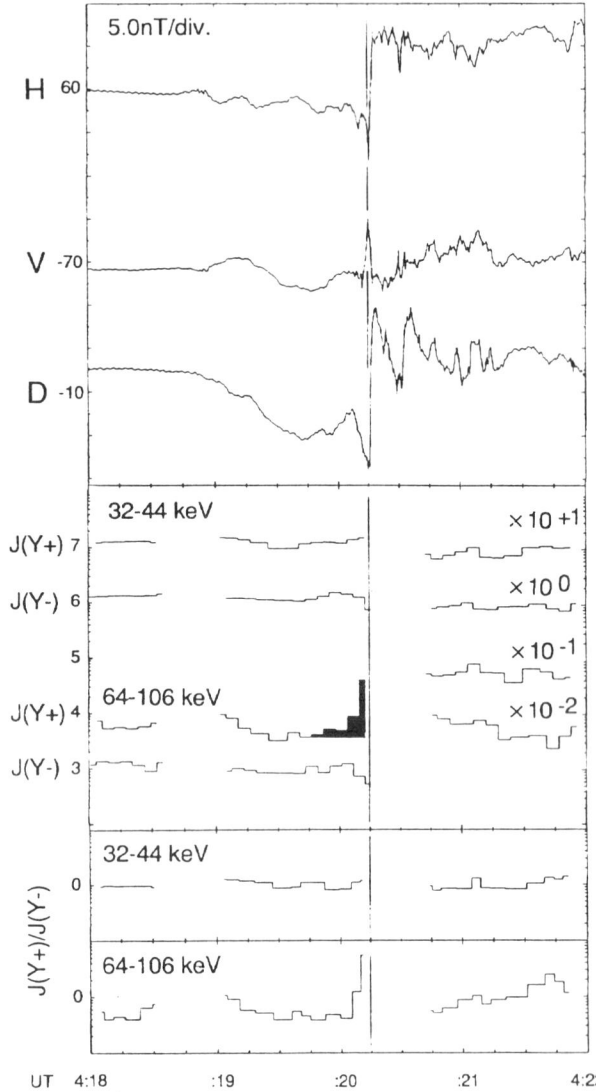

Fig. 4. High time-resolution measurements of the magnetic field and the energetic particle of the two MEPA channel. The particle data are plotted separately with respect to the sign of the GSE Y direction of the detector angle; J(Y+) and J(Y−) represent the flux coming from the positive and negative Y direction, respectively. The ratio between J(Y+) and J(Y−) is plotted in the bottom two panels.

fluxes of particles with $\alpha \sim 90°$ (α: pitch angle) twice every spin. We refer to the flux of particles with $75 < \alpha < 105°$ coming from positive and negative GSE Y directions as J(Y+) and J(Y−), respectively. When the north-south component is positive, as is usually observed in the magnetotail, J(Y+) and J(Y−) represent the flux of particles which have their guiding centers inside (earthward and/or equatorward) and outside (tailward and/or poleward) of the spacecraft, respectively. Therefore, we can infer the radial inhomogeneity of energetic particle population from the difference in J(Y+) and J(Y−). The Larmor radii of 30 keV and 60 keV ions for a magnetic field strength of 100 nT are 250 km and 360 km, respectively. The ratio between J(Y+) and J(Y−) is plotted for both the energy ranges at the bottom of the figure.

The H component increased almost steplike at 0420:15 UT, following the very sharp depression just several seconds before. This depression is ascribed to the enhancement of the tail current, as will be presented below. On the other hand, the sharp increase in H can be regarded as the initial signature of the tail reconfiguration (see Figures 1 and 2), and therefore we should consider it in terms of the current disruption. Hence the tail current disruption is inferred to take place just tailward of the spacecraft. The D component also increased simultaneously. The spikelike deviation in V seems to be related to the sudden increase in H and D. V tended to increase after this spike (also see Figures 1 and 2).

The 24-s averages of the energetic particle fluxes increased without any evident energy dispersion (Figure 2), suggesting that the event was a typical dispersionless injection. However, further careful inspection of the MEPA data (Figure 4) reveals that the time sequence and anisotropy of the energetic particle fluxes were closely related to the magnetic changes mentioned above, although there is an unfortunate data gap just after the H increase. First, the change in the fluxes began before the H increase. Second, both in the lower (32-44 keV) and the higher (64-106 keV) energy channels, J(Y+) tended to increase and J(Y−) tended to decrease. In particular, J(Y+) in the higher-energy range increased more than one order during a few tens of seconds. This flux anisotropy indicates that the population of energetic particles increased tailward of the spacecraft. One may ascribe this increase to the particle injection. However, the injection of energetic particles should be caused by an enhancement in the dawn-to-dusk electric field, which would result from the tail current disruption. In such a case the spacecraft should observe the increase in H, not the sharp depression as is actually observed, almost at the same time with the increase in J(Y+). (Note that we should consider the change in current intensity tailward of the spacecraft, since the particle population changed principally tailward of the spacecraft.) On the contrary, the significant time lag between the increases in H and J(Y+) suggests that the energization and the enhancement in tail current intensity took place just prior to the current disruption. The present result indicates that there is a distinctive interval which precedes the current disruption by a few tens of seconds.

3. DISCUSSION

In the two substorms of July 30, 1985, the magnetic field data revealed strikingly different features among the three spacecraft spaced closely in the synchronous region. We have inferred that the onsets of the substorms took place in the synchronous region. Although it is almost impossible to derive a general conclusion from the two substorms, the present study contributes to the consensus of an onset location closer to the Earth than has been previous acknowledged ($|X| < 20\ R_E$ [Hones et al., 1973; Nishida and Nagayama, 1973; Ohtani et al., 1990]; $|X| < 9\ R_E$ [Takahashi et al., 1987; Lui et al., 1988; Lopez et al., 1989]).

The radial and azimuthal extent of the onset region was estimated to be of the order of $1\ R_E$. This scale size of the region is slightly smaller than values reported previously [Arnoldy and Moore, 1983; Lui, 1978]. And this region actually might be smaller still. Generally, multisatellite observations are very useful for examining the spatial scale of phenomena in the magnetosphere, but the separation between satellites is the limiting factor in measuring the scale. Here the accuracy of the present determination is set by the separation between the spacecraft, $\sim 1\ R_E$. More dense observations in the synchronous region would be necessary for specifying the spatial structure of the onset region more precisely. In particular, the north-south scale of the current disruption remains to be understood.

It is well known that substorm signatures propagate in the magnetosphere both in the azimuthal [Nagai, 1982; Arnoldy and Moore, 1983] and the radial directions: earthward [Russell and McPherron, 1973] and tailward [Ohtani et al., 1988, 1990; Lopez and Lui, 1990]. Hence the expansion of the current disruption could be depicted as in Figure 5, where we took into account the results of previous studies. The speed of the azimuthal expansion at synchronous altitude is between 10 and 100 km/s; the westward expansion is significantly faster than the eastward expansion [Nagai, 1982; Arnoldy and Moore, 1983]; and the tailward expansion speed is a few hundred kilometers per second [Lopez and Lui, 1990; Ohtani et al., 1990]. This figure does not necessarily mean that the current disruption expands continuously. It is possible that

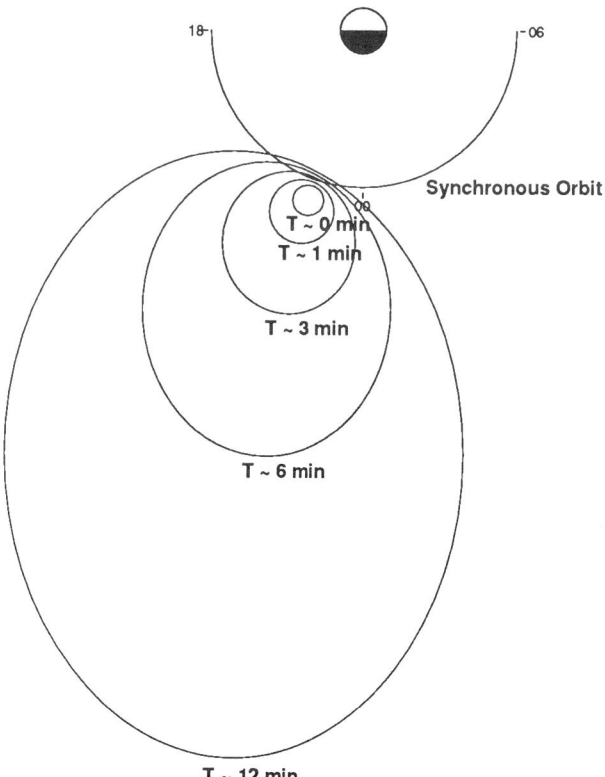

Fig. 5. Schematic illustration of the expansion of the tail current disruption during a substorm. The disruption region, which is depicted as the eclipses, expands both azimuthally and radially from a narrow region in the course of a substorm.

the current disruption develops only in a restricted region. For example, in the 0325 UT event, substorm signatures were observed with GOES 5 and AMPTE/CCE, whereas no evident signature was detected with GOES 6. Magnetospheric conditions for the expansion of the current disruption are one of the most important problems of substorm mechanisms.

The high-time-resolution comparison between magnetic field and energetic particle flux changes associated with the current disruption would be very useful for assessing such conditions. We examined the AMPTE/CCE magnetometer data and MEPA data in detail for the 0420 UT event, when the spacecraft was located very close to the onset region. The most distinctive signature is the sharp depression in the H component and the energization of particles tailward of the spacecraft, which started about a few tens of seconds before the current disruption. Figure 6 schematically shows the sequence of the change in tail current intensity. It is well known that the tail current develops during the so-called growth phase. The present result indicates that the intensity of the tail current is sharply enhanced during a less than 1-min interval prior to the current disruption. This interval may be called "explosive growth phase," although the gross picture would also be valid outside of the onset region; in this case the timing of the reference should be the local disruption of the tail current, instead of the substorm onset. In the 0420 UT event, the magnitude of the H recovery at 0420:15 UT was ~30 nT, while that of the depression during this short interval (0418:55 - 0420:15 UT) was ~15 nT (see Figure 4). This fact suggests that the increase in the tail current intensity in the explosive growth phase is comparable to that in the conventional growth phase.

McPherron et al. [1987] have inferred that the scale height of the plasma sheet thinned to ~400 km at $X \sim -13\,R_E$ before the current disruption in the March 22, 1979 event (CDAW 6 event) [also see *Ohtani et al.*, 1988]. This scale height is comparable to the Larmor radius of ions with an energy of ~3 keV for a magnetic field strength of 20 nT, indicating that non-adiabatic behavior of ions is important in considering the structure of the current sheet. Recently, *Mitchell et al.* [1990] examined the carriers of the tail current in the course of a substorm. They reported that the enhancement of the cross-tail current is carried by non-adiabatic ions just prior to the local disruption of the tail current. The enhancement in tail current intensity and the particle energization could be explained in terms of a positive feedback between the current intensity and the number of non-adiabatic ions. The motion of ions would become non-adiabatic when the thickness of the plasma sheet becomes comparable to their Larmor radius. As the intensity of the tail current increases, the plasma sheet becomes thinner, and consequently more ions become non-adiabatic and contribute further to the tail current. Moreover, the thinning of the plasma sheet may be accompanied by the decrease in magnetic strength in the plasma sheet, which results in the increase in the Larmor radius of ions. The results of *Mitchell et al.* [1990] favor the above scenario. A similar feedback mechanism is also possible for the curvature drift of particles [e.g., *Kaufmann et al.*, 1987]. The energization of ions can also be ascribed to the increase in the mobility of ions in the dawn-to-dusk direction, because the drift is in the same direction as the large-scale electric field, and therefore ions obtain more energy as they drift duskward.

4. SUMMARY

We examined the July 30, 1985 substorm period by using three spacecraft in the geosynchronous region, GOES 5, GOES 6, and AMPTE/CCE. During this period the spacecraft were conveniently located in the premidnight sector. It is found that substorm-associated magnetic signatures are significantly different among the three spacecraft despite small separations. We infer that substorm signatures propagate from a relatively narrow region in the synchronous region. The spatial scale of the initial current disruption would be of the order of 1 R_E or less, both in the radial and the azimuthal directions. We examined the magnetic field and energetic particle flux changes observed with AMPTE/CCE at the 0420 UT substorm onset in detail. In this event, AMPTE/CCE was inferred to be very close to the onset region. Careful inspection of the high resolution data indicates that there is a distinctive short (~ 1 min) interval (explosive growth phase) prior to the current disruption. The tail current intensity is enhanced explosively during this interval. We suggest that a feedback mechanism is responsible for this enhancement in tail current intensity.

Acknowledgments. The authors wish to thank A. T. Y. Lui, B. A. Mauk, and R. E. Lopez for fruitful discussions. The induction magnetometer data of Hermanus was provided by P. Sutcliffe. Ground-based magnetograms were provided through World Data Center C2 for Aurora, National Institute of Polar Research. Work at APL was supported by NASA and NSF. Correlative aspects of this work at the University of Tokyo were supported by the Japanese Society for the Promotion of Science for Japanese Junior Scientists under grant 01790262.

REFERENCES

Akasofu, S. -I., The development of the auroral substorm, *Planet. Space Sci.*, 12, 273, 1964.
Arnoldy, R. L. and T. E. Moore, Longitudinal structure of substorm injections at synchronous orbit, *J. Geophys. Res.*, 88, 6213, 1983.
Cummings, W. D., J. N. Barfield, and P. J. Coleman, Jr., Magnetospheric substorms observed at the synchronous orbit, *J. Geophys. Res.*, 73, 6887, 1968.

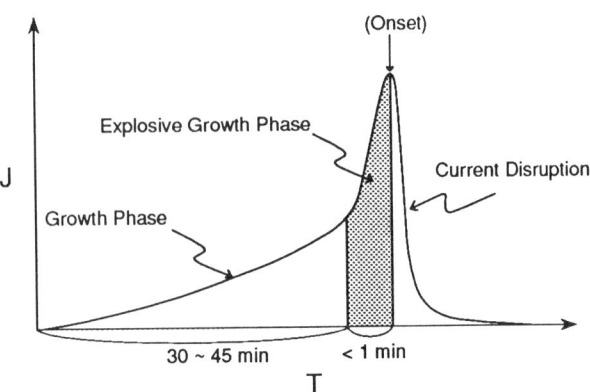

Fig. 6. Schematic illustration of the development of the tail current in the course of substorms. The vertical axis represents the local current intensity. In the onset region the commencement of the current disruption coincides with a substorm onset. The increase in tail current intensity is accelerated transiently prior to the current disruption.

Erickson, K. N., R. L. Swanson, R. J. Walker, and J. R. Winckler, A study of magnetosphere dynamics during auroral electrojet events by observations of energetic electron intensity changes at synchronous orbit, *J. Geophys. Res.*, 84, 931, 1979.

Fairfield, D. H. and N. F. Ness, Configuration of the geomagnetic tail during substorms, *J. Geophys. Res.*, 75, 7032, 1970.

Hones, E. W., J. R. Asbridge, S. J. Bame, and S. Singer, Substorm variations of the magnetotail plasma sheet from $X_{SM} \approx -6\ R_e$ to $X_{SM} \approx -60\ R_e$, *J. Geophys. Res.*, 78, 109, 1973.

Iijima, T. and T. A. Potemra, The amplitude distribution of field-aligned currents at northern high latitudes observed by Triad, *J. Geophys. Res.*, 81, 2165, 1976.

Kaneda, E., and T. Yamamoto, Auroral substorms observed by UV-imager on AKEBONO, *this issue*, 1991.

Kaufmann, R. L., Substorm currents: Growth phase and onset, *J. Geophys. Res.*, 92, 7471, 1987.

Lezniak, T. W. and J. R. Winckler, Experimental study of magnetospheric motions and the acceleration of energetic electrons during substorms, *J. Geophys. Res.*, 75, 7075, 1970.

Lopez, R. E., A. T. Y. Lui, D. G. Sibeck, K. Takahashi, R. W. McEntire, L. J. Zanetti, and S. M. Krimigis, On the relationship between the energetic particle flux morphology and the change in the magnetic field magnitude during substorms, *J. Geophys. Res.*, 94, 17105, 1989.

Lopez, R. E., H. Lühr, B. J. Anderson, P. T. Newell, and R. W. McEntire, Multipoint observations of a small substorm, *J. Geophys. Res.*, 95, 18897, 1990.

Lopez, R. E. and A. T. Y. Lui, A multisatellite case study of the expansion of a substorm current wedge in the near-Earth magnetotail, *J. Geophys. Res.*, 95, 8009, 1990.

Lui, A. T. Y., R. E. Lopez, S. M. Krimigis, R. W. McEntire, L. J. Zanetti, and T. A. Potemra, A case study of magnetotail current sheet disruption and diversion, *Geophys. Res. Lett.*, 15, 721, 1988.

Lui, A. T. Y., Estimates of current changes in the geomagnetotail associated with a substorm, *Geophys. Res. Lett.*, 5, 853, 1978.

McEntire, R. W., E. P. Keath, D. E. Fort, A. T. Y. Lui, and S. M. Krimigis, The medium-energy particle analyzer (MEPA) on the AMPTE/CCE spacecraft, *IEEE Trans. Geosci. Remote Sensing*, GE-23(3), 230-233, 1985.

McPherron, R. L., C. T. Russell, and M. P. Aubry, Satellite studies of magnetospheric substorms on August 15, 1968, 9. Phenomenological model for substorms, *J. Geophys. Res.*, 78, 3131, 1973.

McPherron, R. L., A. Nishida, and C. T. Russell, Is near-Earth current sheet thinning the cause of auroral substorm onset ?, paper presented at the conference on *Quantitative modeling of magnetosphere-ionosphere coupling processes*, Kyoto Sangyo University, Kyoto, Japan, March 9-13, 1987.

Mitchell, D. G., D. J. Williams, C. Y. Huang, L. A. Frank, and C. T. Russell, Current carriers in the near-Earth cross-tail current sheet during substorm growth phase, *Geophys. Res. Lett.*, 17, 583, 1990.

Nagai, T., Observed magnetic substorm signatures at synchronous altitude, *J. Geophys. Res.*, 87, 4405, 1982.

Nishida, A. and N. Nagayama, Synoptic survey for the neutral line in the magnetotail during the substorm expansion phase, *J. Geophys. Res.*, 78, 3782, 1973.

Ohtani, S., S. Kokubun, R. C. Elphic, and C. T. Russell, Field-aligned current signatures in the near-tail region, 1. ISEE observations in the plasma sheet boundary layer, *J. Geophys. Res.*, 93, 9709, 1988.

Ohtani, S., S. Kokubun, and C. T. Russell, Radial expansion of the tail current disruption during substorms: A new approach to the substorm onset region, submitted to *J. Geophys. Res.*, 1990.

Potemra, T. A., L. J. Zanetti, and M. H. Acuna, The AMPTE/CCE magnetic field experiment, *IEEE Trans. Geosci. Remote Sensing*, GE-23(3), 246-249, 1985.

Rostoker, G., A. V. Jones, R. L. Gattinger, C. D. Anger, and J. S. Murphree, The development of the substorm expansive phase: the "eye" of the substorm, *Geophys. Res. Lett.*, 14, 399, 1987.

Rostoker, G., J. C. Armstrong, and A. J. Zmuda, Field-aligned current flow associated with intrusion of the substorm-intensified westward electrojet into the evening sector, *J. Geophys. Res.*, 80, 3571, 1975.

Russell, C. T., and R. L. McPherron, The magnetotail and substorms, *Space Sci. Rev.*, 15, 205, 1973.

Sauvaud, J.-A., and J. R. Winckler, Dynamics of plasma, energetic particles, and fields near synchronous orbit in the nighttime sector during magnetospheric substorms, *J. Geophys. Res.*, 85, 2043, 1980.

Takahashi, K., L. J. Zanetti, R. E. Lopez, R. W. McEntire, T. A. Potemra, and K. Yumoto, Disruption of the magnetotail current sheet observed by AMPTE/CCE, *Geophys. Res. Lett.*, 14, 1019, 1987.

4. PLASMA SHEET DYNAMICS

Heating and Fast Flows in the Near-Earth Tail

WOLFGANG BAUMJOHANN

Max-Planck-Institut für extraterrestrische Physik, Garching, Germany

Two important aspects of magnetospheric substorm activity in the near-Earth tail (10–$20\,R_E$) are the strong heating of ions and electrons in the central plasma sheet and the occurrence of fast plasma transport across magnetic field lines near the neutral sheet. The particles in the plasma sheet boundary layer are much less affected by substorm activity. The latter observation rules out boundary layer substorm models and the thermal catastrophe model cannot explain the fast cross-field flows near the neutral sheet. Only the near-Earth neutral line model seems able to explain the fast flows and eventually, if combined with current sheet acceleration in the presence of a weak northward magnetic field component, the strong heating.

1. INTRODUCTION

A magnetospheric substorm comprises many different phenomena and processes and there are numerous models trying to explain these features or at least certain facets. In the present review I shall focus on the two key phenomena observed in the near-Earth tail, at radial distances of 10–$20\,R_E$, namely strong plasma heating and fast ion flows near the neutral sheet. The description of these features in the next two sections will mainly be based on recent results from the AMPTE/IRM satellite observations, but reference will also be made to studies which have used data from the ISEE spacecraft. In the last section I shall point out the need to explain both features in a single, unified substorm model.

2. PLASMA HEATING

Huang and Frank [1986] were the first to point out that one of the key features to be observed in the near-Earth tail during a substorm was the strong heating of the 'thermal' keV-ions in the central plasma sheet. The study of *Baumjohann et al.* [1989] corroborates this earlier result. In addition, *Baumjohann et al.* [1988] showed that the ions in the plasma sheet boundary layer are heated to a much lesser extent during strong geomagnetic disturbance.

Figure 1 is based on the results of *Baumjohann et al.* [1988, 1989], but covers about twice as much data (more than 260,000 IRM measurements of ions between 20 eV and 40 keV, each covering 4.5 sec). It can clearly be seen that the temperature of the ions (derived from the moments of the distribution function) in the central plasma sheet (CPS) increases linearly with the logarithm of the AE index, from about 1.5×10^7 K during very quiet times to more than 6.5×10^7 K during highly disturbed intervals.

The ion temperature in the plasma sheet boundary layer (PSBL), on the other hand, also increases linearly with the logarithm of AE, but only by a factor of two between very quiet and very active intervals. It seems interesting to note that the central plasma sheet and the plasma sheet boundary layer have the same low temperature (1.5×10^7 K or about 1 keV) during very quiet intervals ($AE < 30\,\mathrm{nT}$).

However, it is not only the ion temperature that increases strongly with geomagnetic activity, the electron temperature varies the same way, too. Figure 2 presents the relation between electron and ion temperature in the central plasma sheet in a more quantitative way (it is similar to Figure 9 of *Baumjohann et al.* [1989] but again includes twice as much data). Apparently these two temperatures show a very high degree of correlation, with a linear correlation coefficient of 0.91 and $T_i = 7.06 \cdot T_e$. More than 80% of the data points have ion-to-electron temperature ratios between 5 and 10. This linear relation holds for temperature variations over nearly two decades, i.e., both in a hot and cold plasma sheet. It corroborates the conclusion of *Christon et al.* [1988] that the plasma and energetic ion and electron populations respond collectively as a single unified particle population during plasma sheet temperature transitions.

The near-earth central plasma sheet ion-to-electron temperature ratios are similar to those found by *Slavin et al.* [1985] at distances of $|X| = 30$–$60\,R_E$ and roughly agree with the T_i/T_e ratios in distant tail plasmoids [*Richardson et al.*, 1987] observed by the ISEE-3 space-

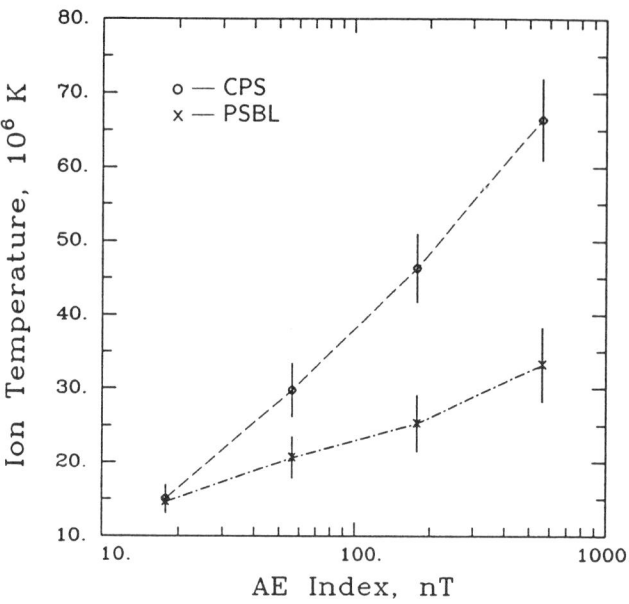

Fig. 1. Average ion temperature versus AE index. Observations from the plasma sheet boundary layer are marked by crosses, those from the central plasma sheet denoted by circles. The error bars are equivalent to one fifth of the variance in the data [after *Baumjohann et al.*, 1988, 1989].

craft. This similarity hints that plasmoids in the deep tail are populated by plasma from the near-earth plasma sheet, which has been severed at a near-earth neutral line.

From the results of three recent studies based on AMPTE IRM and CCE suprathermal ion data [*Baumjo-*

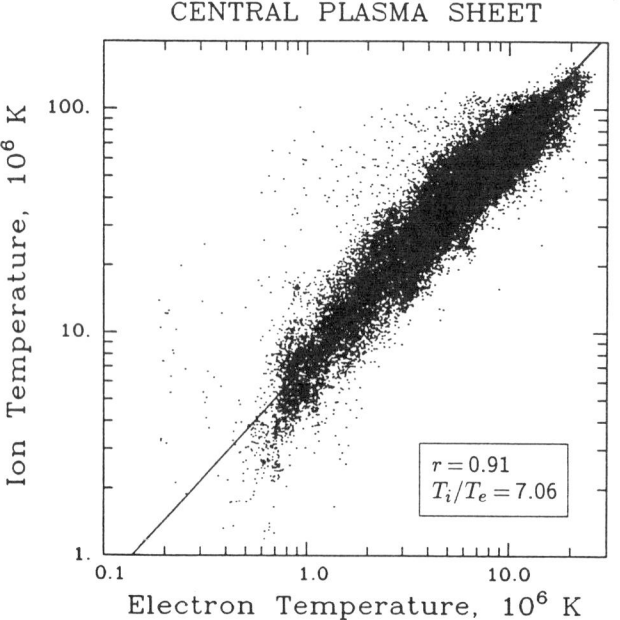

Fig. 2. Scatter diagram of the relationship between ion and electron temperature in the central plasma sheet. The solid lines give the regression lines [after *Baumjohann et al.*, 1989].

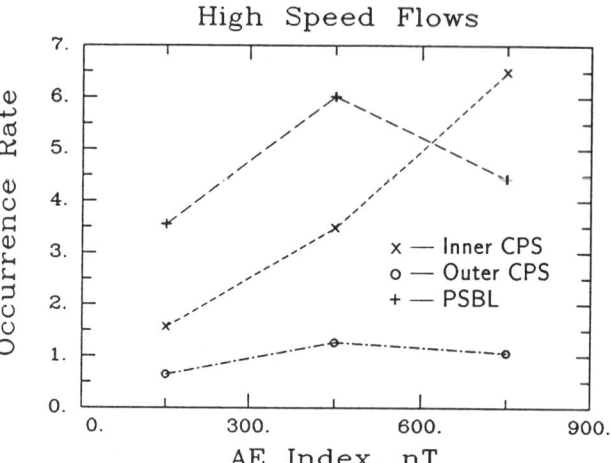

Fig. 3. Occurrence rate (in percent) of high-speed ($V_i > 400$ km/s) ion flow samples in the three plasma sheet regions as a function of the AE index [from *Baumjohann et al.*, 1990b].

hann et al., 1990a; *Kistler et al.*, 1990, 1991] one can even conclude that also all major ion species (H^+, He^+, He^{++}, and O^+) are heated in parallel, according to their energy-per-charge ratio. Furthermore, the heating seems to occur locally everywhere in the near-Earth tail.

The IRM survey studies did not distinguish between substorm phases, but *Huang et al.* [1991], using ISEE Lepedea plasma data, clearly showed that the major heating of the central plasma sheet occurs just after substorm onset. Huang et al. further noted that the heating process must be non-adiabatic, since ion density and temperature are anticorrelated during these events.

3. Fast Ion Bulk Flow

Figure 3 [from *Baumjohann et al.*, 1990b] shows the occurrence rates of high speed flows (bulk flows in excess of 400 km/s) in three different layers of the plasma sheet for different ranges of AE, again using the IRM tail survey data. If one averages over all levels of disturbance, the following picture emerges: From all 4.5-sec ion flow samples taken in in the plasma sheet boundary layer (more than 73,000) about 5% exhibit velocities in excess of 400 km/s. This rate drops to 1% out of 148,000 samples in the outer central plasma sheet, but rises again, to an average of about 3% of the 52,000 samples obtained in the inner central plasma sheet, i.e., the neighborhood of the neutral sheet (defined by $B_{xy} = (B_x^2 + B_y^2)^{1/2} < 15$ nT or $B_z/B_{xy} > 0.5$). Thus there is a 3:1:5 chance to detect ion flows with velocities greater 400 km/s when going from the neutral sheet to the boundary layer.

However, if one considers the behavior of the fast flow occurrence rate during substorm activity, it varies by less than a factor of two in the plasma sheet boundary layer. It is only near the neutral sheet, where the fast flow occurrence rate strongly increases with increasing AE, reaching the same or even slightly higher levels than in the plasma sheet boundary layer during strongly disturbed conditions. Thus the high speed plasma flow in the plasma sheet boundary layer is less affected by substorm activity than that near the neutral sheet, similar to the behavior of the ion temperature described in the previous section.

From all high speed ion flows with a velocity in excess of 400 km/s observed by the IRM satellite during the two 4-month periods it spent in the magnetotail at distances between 9 and 19 R_E (more than 6000 4.5-sec samples), the overwhelming majority was directed earthward (Figure 4). Less than 100 samples had a distinct tailward component. For bulk speeds in excess of 600 km/s, virtually no ion flow sample had a tailward component. These observations corroborate the ISEE results of *Cattell and Mozer* [1984] and indicate that a near-Earth neutral line is rarely located inside 20 R_E.

Figure 5 gives the fraction of those high speed flows which have a dominant component perpendicular to the magnetic field ($V_\perp > V_\parallel$), again separately for the three plasma sheet regions. In the two outer layers, the relative fraction ranges between 20% (for velocities of 400–500 km/s) and 5% (for $V_i > 800$ km/s), thus supporting the findings of, for example, *Lui et al.* [1978] that high speed flows in the plasma sheet boundary layer and the outer central plasma sheet are nearly field-aligned, often resulting from an imbalance of counterstreaming ion beams.

In the neighborhood of the neutral sheet, things look different. Here between 50% and 80% of all high speed flows have a dominant perpendicular component. Except for flows above 1100 km/s, there is even a tendency for an increasing fraction of perpendicular flows with increasing flow velocity. Thus in the inner central plasma sheet most of the high-speed plasma transport occurs across field lines rather than along them.

Figure 6 shows that high speed flows are rather bursty throughout the plasma sheet. The majority of all flows stays uninteruptedly at high speed levels for no more than 10 sec. About 40% of all flow bursts were observed for only one 4.5 sec measurement interval, with the percentage becoming greater for higher velocity levels. It may

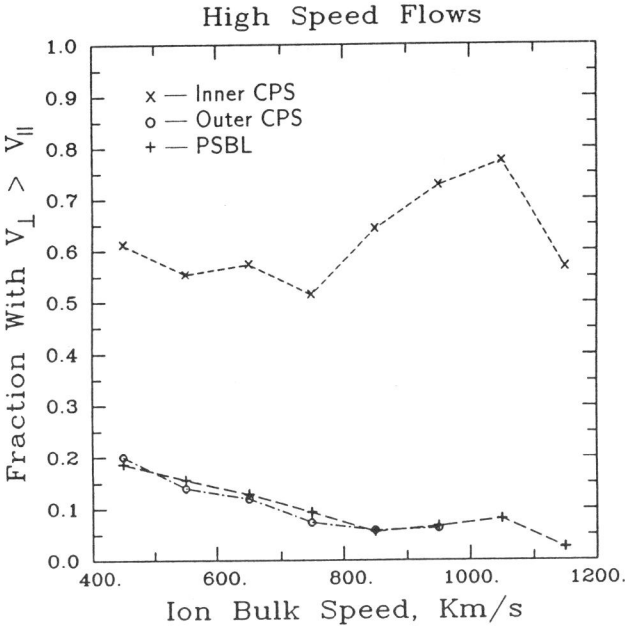

Fig. 5. Relative fraction of high-speed flow samples where the component perpendicular to the magnetic field is stronger than the parallel velocity component ($V_\perp > V_\parallel$) in the three plasma sheet regions as a function of the ion bulk speed [from *Baumjohann et al.*, 1990b].

well be that the duration of many high speed flows is still shorter. Typically several of these high speed flow bursts are grouped into an event lasting for some minutes. Of course, with data from just one satellite one is unable

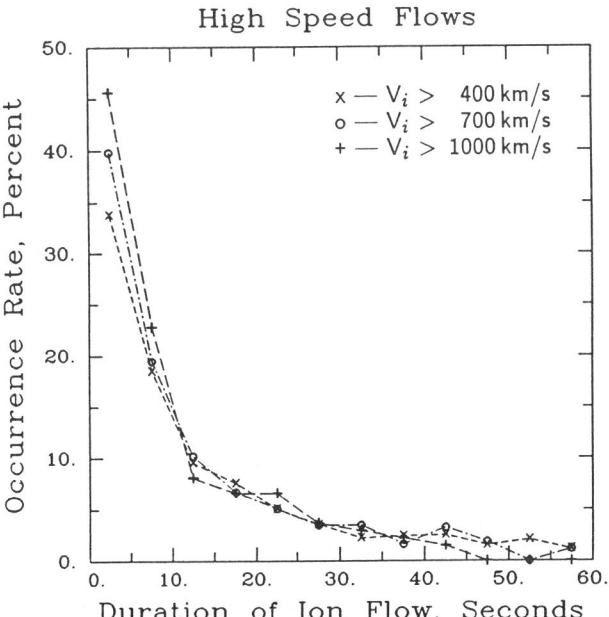

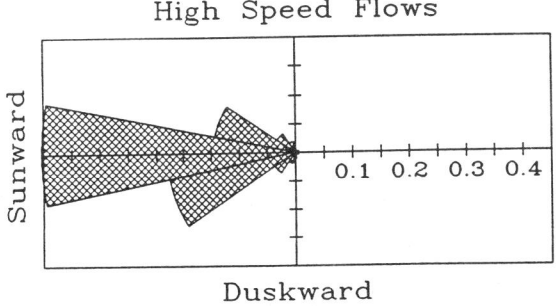

Fig. 4. Angular distribution of high-speed ion velocity vectors ($V_i > 400$ km/s; projected into the X_{GSM}-Y_{GSM} plane) from all data obtained in the plasma sheet [after *Baumjohann et al.*, 1990b].

Fig. 6. Relative occurrence rate of plasma sheet ion flow events with velocities in excess of a certain magnitude versus (continuous) duration of the flow events [from *Baumjohann et al.*, 1990b].

to decide whether the burstiness results from temporal or spatial variations, i.e., whether the ions in the whole plasma sheet move at such high velocities for short periods of time or whether the short durations are caused by rapid traversals of thin layers of plasma streaming at high velocities and intermittent regions of slower plasma motion.

4. Summary

The key features of substorm heating in the near-Earth tail can be summarized as follows:
- major heating occurs in the central plasma sheet;
- major heating occurs at substorm onset;
- heating occurs in a non-adiabatic fashion;
- heating occurs throughout the near-Earth tail;
- electrons and ions are heated in parallel;
- ions with the same E/Q are heated in parallel.

The key features of substorm-associated fast ion bulk flow can be summarized as follows:
- major fast flow activity occurs near the neutral sheet;
- fast neutral sheet flows cross magnetic field lines;
- fast flows are bursty and short-lived;
- fast flows inside 19 R_E are primarily Earthward.

5. Discussion

The observations summarized in the previous section clearly show that most of the substorm-associated changes in the near-Earth tail occur in the central plasma sheet and neutral sheet neighborhood rather than in the plasma sheet boundary layer. Thus it is rather unlikely that substorm models emphasizing the dynamics of the plasma sheet boundary layer, such as those by *Rostoker and Eastman* [1987] or *Lyons and Nishida* [1988] will have much further impact on our understanding of the substorm process.

The thermal catastrophe model [*Smith et al.*, 1986] is successful in explaining the strong heating of the plasma sheet ions in terms of resonant absorption of low frequency waves from the tail lobes. However, it does not explain why electrons and ions with the same energy-per-charge are heated in parallel. More importantly, it seems unlikely that the fast cross-field plasma transport near the neutral sheet can be incorporated into this model.

In the most-accepted substorm model, the so-called near-Earth neutral line model [e.g., *Hones*, 1979], the fast plasma transport across magnetic field lines near the neutral sheet is readily explained as the result of magnetic reconnection at a newly formed near-Earth neutral line (beyond 19 R_E). That simulations done on the basis of this model do not exhibit all fast flow features yet, e.g., the burstiness of the flow, may likely be a sign that the present simulations are still too simplified, and should not be taken as an argument against the near-Earth neutral line model.

The present near-Earth neutral line model does not address nor explain the strong heating of the plasma sheet during substorms. However, the heating might be included by way of current sheet acceleration in a magnetic field geometry as is produced by a near-Earth neutral line. Actually, *Jaeger and Speiser* [1974] followed the trajectory of ions in a current sheet with a weak northward B_z component (as would be expected Earthward of a near-Earth neutral line) and found that the ions are energized roughly according to their energy-per-charge ratio. (Note that the more recent simulations of current sheet acceleration [e.g., *Cowley*, 1980; *Lyons and Speiser*, 1982], which yielded energy-per-mass acceleration, have magnetic field geometries reminiscent of a distant neutral line.)

References

Baumjohann, W., G. Paschmann, N. Sckopke, C. A. Cattell, and C. W. Carlson, Average ion moments in the plasma sheet boundary layer, *J. Geophys. Res.*, 93, 11,507–11,520, 1988.

Baumjohann, W., G. Paschmann, and C. A. Cattell, Average plasma properties in the central plasma sheet, *J. Geophys. Res.*, 94, 6597–6606, 1989.

Baumjohann, W., D. Sachsenweger, and E. Möbius, Suprathermal ion fluxes in the plasma sheet, *Geophys. Res. Lett.*, 17, 275–278, 1990a.

Baumjohann, W., G. Paschmann, and H. Lühr, Characteristics of high-speed ion flows in the plasma sheet, *J. Geophys. Res.*, 95, 3801–3809, 1990b.

Cattell, C. A., and F. S. Mozer, Substorm electric fields in the Earth's magnetotail, in *Magnetic Reconnection in Space and Laboratory Plasmas*, edited by E. W. Hones, Jr., pp. 208–215, AGU, Washington, 1984.

Christon, S. P., D. G. Mitchell, D. J. Williams, L. A. Frank, C. Y. Huang, and T. E. Eastman, Energy spectra of plasma sheet ions and electrons from $\sim 50\,\text{eV}/e$ to $\sim 1\,\text{MeV}$ during plasma temperature transitions, *J. Geophys. Res.*, 93, 2562–2572, 1988.

Cowley, S. W. H., Plasma populations in a simple open model magnetosphere, *Space Sci. Rev.*, 26, 217–275, 1980.

Hones, E. W., Jr., Transient phenomena in the magnetotail and their relation to substorms, *Space Sci. Rev.*, 23, 393–410, 1979.

Huang, C. Y., and L. A. Frank, A statistical study of the central plasma sheet: Implications for substorm models, *Geophys. Res. Lett.*, 13, 652–655, 1986.

Huang, C. Y., L. A. Frank, G. Rostoker, J. Fennel, and D. G. Mitchell, Nonadiabatic heating of the central plasma sheet at substorm onset, *J. Geophys. Res.*, 95, in press, 1991.

Jaeger, E. F., and T. W. Speiser, Energy and pitch angle distributions for auroral ions using the current sheet acceleration model, *Astrophys. Space Sci.*, 28, 129–144, 1974.

Kistler, L. M., E. Möbius, B. Klecker, G. Gloeckler, F. M. Ipavich, and D. C. Hamilton, Spatial variations in the suprathermal ion distributions during substorms in the plasma sheet, *J. Geophys. Res.*, 95, 18,871–18,885, 1990.

Kistler, L. M., E. Möbius, W. Baumjohann, G. Paschmann, and D. C. Hamilton, Pressure changes in the plasma sheet during substorm injections, *J. Geophys. Res.*, 96, in press, 1991.

Lui, A. T. Y., L. A. Frank, K. L. Ackerson, C.-I. Meng, and S.-I. Akasofu, Plasma flows and magnetic field vectors in the plasma sheet during substorms, *J. Geophys. Res.*, 83, 3849–3857, 1978.

Lyons, L. R., and A. Nishida, Description of substorms in the tail incorporating boundary layer and neutral line effects, *Geophys. Res. Lett.*, 15, 1337-1340, 1988.

Lyons, L. R., and T. W. Speiser, Evidence for current sheet acceleration in the geomagnetic tail, *J. Geophys. Res.*, 87, 2276–2286, 1982.

Richardson, I. G., S. W. H. Cowley, E. W. Hones, Jr., and S. J. Bame,

Plasmoid-associated energetic ion bursts in the deep geomagnetic tail: Properties of plasmoids and the postplasmoid plasma sheet, *J. Geophys. Res.*, *92*, 9997–10,0013, 1987.

Rostoker, G., and T. E. Eastman, A boundary layer model for magnetospheric substorms, *J. Geophys. Res.*, *92*, 12,187–12,202, 1987.

Slavin, J. A., E. J. Smith, D. G. Sibeck, D. N. Baker, R. D. Zwickl, and S.-I. Akasofu, An ISEE 3 study of average and substorm conditions in the distant magnetotail, *J. Geophys. Res.*, *90*, 10,875–10,895, 1985.

Smith, R. A., C. K. Goertz, and W. Grossmann, Thermal catastrophe in the plasma sheet boundary layer, *Geophys. Res. Lett.*, *13*, 1380–1383, 1986.

The Earthward Edge of the Plasma Sheet in Magnetospheric Substorms

D.N. Baker[1] and T.I. Pulkkinen[2]

Magnetospheric substorms represent a fundamental global interaction of the magnetosphere-ionosphere system with the solar wind. Continuing observations and established models suggest that the inner portion of the plasma sheet (roughly 6-15 R_E geocentric distance) is a critical region for substorm development. We review the standard model of substorms as it applies to the near Earth region. We discuss the storage of extracted solar wind energy during the substorm growth phase and the effects seen in the inner plasma sheet region during the substorm expansion phase, focusing on the established fact that substorm onset is accompanied by intense hot plasma injection events near geostationary orbit. Among the most intriguing new aspects of the growth phase to be appreciated are the extremely intense cross-tail currents that often develop 10-30 minutes prior to the expansion phase onset. This is normally accompanied by extremely tail-like field and deep energetic particle dropouts are often seen. We address questions attendant to this phenomenon including the basic processes responsible for the current formation, the carriers of the intensified currents, and the implications of these results for various global substorm models. Recent data from multiple-spacecraft constellations in the 6-9 R_E region have also provided new insights into substorm onset mechanisms. Evidence is mounting that intense, localized "disruptions" of the cross-tail current play a key role in the complex processes that occur at substorm onset. The basic breakdown of the currents within the inner plasma sheet ($r < 10$ R_E) may fundamentally constitute the "local" substorm onset. We place such ideas into the broader context of traditional models of substorms such as the neutral line model, as well as more recent dynamical models.

Introduction

The portion of the Earth's magnetic environment extending from ~15 Earth radii (R_E) geocentric distance to hundreds of R_E on the nightside can be termed the distant magnetotail. It is a region comprised of a central slab of hot, diamagnetic plasma (the plasma sheet) which carries strong cross-tail currents (the current sheet). The plasma sheet is overlain by cold, plasma-sparse regions above and below called the tail lobes (see Fig. 1). The distant magnetotail is a region of relatively weak axial (i.e., Earthward-tailward) gradients and it constitutes a large storage reservoir for energy that is extracted from the ambient solar wind flow (Fairfield, 1973). The tapping and conversion of energy in the magnetotail is generally believed to play a fundamental role in magnetospheric dynamics (McPherron et al., 1973).

The inner, or near-Earth, part of the magnetotail may be thought of as the region extending from roughly 6 R_E to a distance of ~15 R_E. This part of the Earth's magnetic envelope is critically important, both structurally and dynamically, in the overall magnetospheric system. It is in this region that the magnetosphere goes from a dipole-like to a tail-like configuration. Hence, this is a region of strong plasma and magnetic field gradients. As we will discuss in this review, it is also a region that is strongly implicated as one searches for the initial location of global magnetotail instabilities related to substorm onsets.

The 'Standard' Substorm Model

Whereas magnetospheric substorms appear to produce a relatively clear sequence of events in the deep magnetotail (e.g., Hones et al., 1984; Baker et al., 1987), it has long been evident that the near-Earth magnetotail is extremely complex

---

[1] Laboratory for Extraterrestrial Physics, NASA/Goddard Space Flight Center, Greenbelt, MD 20771
[2] Finnish Meteorological Institute, Box 503, SF-00101 Helsinke, Finland

Magnetospheric Substorms
Geophysical Monograph 64
This paper is not subject to U.S. copyright. Published in 1991 by the American Geophysical Union.

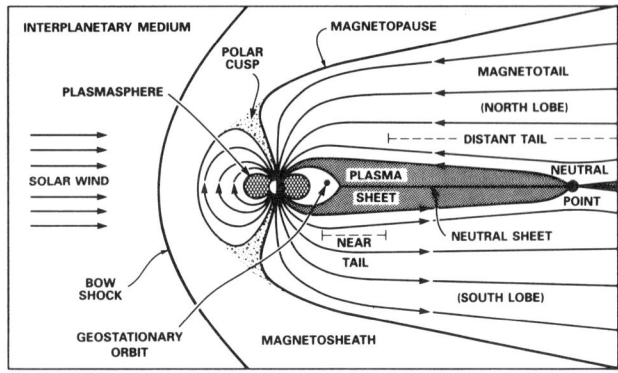

Fig. 1. A cross-sectional diagram of the Earth's magnetosphere and its environs showing the principal plasma regions.

in its behavior during substorms (e.g., Fairfield et al., 1981). Nonetheless, the essential aspects of the physical processes occurring throughout the coupled solar wind-magnetosphere-ionosphere system have been embodied in a basic paradigm concerning the substorm sequence (McPherron et al., 1973). This framework of the fundamental substorm pattern has been embellished over the past two decades (see reviews by Hones, 1979 and Baker et al., 1984), but the near-Earth neutral line model has remained reasonably established as the "standard" model of geomagnetic activity. As long as observations have been limited to ground measurements and one or two high altitude spacecraft, it has been difficult to modify the basic picture in any significant or conclusive way (see, for example, Rostoker and Eastman, 1987; Goertz and Smith, 1989). However, with the advent of multiple, global auroral imaging spacecraft and large constellations of in situ satellite probes, the greater complexity of substorm variations has been demonstrated. Thus, many of the details of the standard model have been questioned in recent work. We review here the essential elements of each phase of the standard model. In a subsequent section of this review, we address several of the recent results that suggest possible modifications that may be required.

Growth Phase

In the idealized situation of a completely isolated substorm, the magnetosphere begins essentially in its "ground" state (somewhat as depicted in Fig. 1). With the southward turning of the interplanetary magnetic field (IMF), dayside magnetic reconnection is greatly enhanced and solar wind energy transfer to the magnetosphere is increased. This initiation of strongly enhanced energy input begins the traditionally defined growth phase of the substorm (McPherron, 1970).

Within roughly the last decade, there developed a controversy over whether or not a substorm growth phase actually exists (e.g., Akasofu, 1981). The view was put forward that substorms were dominantly or even exclusively, driven directly by the solar wind. In this view, it was suggested that energy storage in the magnetotail - and the subsequent release of this stored energy - played no significant role in the substorm process (Akasofu, 1985). However, subsequent statistical approaches and various case studies (e.g., Baker et al., 1986) have shown that substorms clearly have both directly driven and loading-unloading aspects. Roughly speaking, one can view the growth phase of the substorm as being due dominantly to the directly driven processes, while the expansion phase is dominantly due to unloading of magnetotail energy. However, this simplistic separation should not be taken too literally, since solar wind direct-driving of the magnetosphere can continue even as tail unloading is progressing (Rostoker et al., 1987).

In the near-Earth plasma sheet, the growth phase manifests itself by the taillike stretching of magnetic field due to greatly intensified cross-tail currents (McPherron, 1970; Kaufmann, 1987). This typically progresses for about one hour before the expansion phase and it often causes a substantial decrease in energetic particle fluxes at 6.6 R_E (e.g., Baker et al., 1978). These changes in the particle intensities in the near-tail are due to adiabatic particle drifts in the highly distorted magnetic field that develops during the growth phase.

There remain significant issues concerning the growth phase. For example, which portion of the near-Earth plasma sheet particle population provides the current carriers for the newly formed intense cross-tail current? When, exactly, does magnetic reconnection begin during the growth phase? Where and why does reconnection initiate? What are the details of cross-tail current "disruption" that apparently terminates the growth phase and thus lead to the substorm expansion phase?

Expansion Phase

In the standard model, the expansion phase is taken to begin when a near-Earth neutral line forms at ~15 R_E geocentric distance (McPherron et al., 1973; Hones, 1979). The neutral line formation is assumed to disrupt (in some relatively unspecified way) the cross-tail current in the vicinity of the neutral line. The neutral line is envisioned as extending across a finite portion of the tail width. The disrupted cross-tail current is hypothesized to flow along magnetic field lines down into the ionosphere on the morning side of the tail (see, also, Iijima and Potemra, 1976). The diverted current is viewed as then flowing across the auroral ionosphere as the intensified westward (substorm) electrojet. Finally, the electrojet current is assumed to close back into the magnetosphere by flowing outward along field lines on the dusk side of the tail. The complete pattern of diverted cross-tail current flow is known as the current wedge model (McPherron et al., 1973).

Fundamental questions concerning the standard model revolve around why the near-Earth neutral line forms where it does. What causes the formation of the neutral line in the first place? How does the neutral line actually disrupt cross-tail current flow? What are the details of the current wedge formation and how do the plasmas carrying current perpendicular to **B** in the plasma sheet form the parallel (field-aligned) currents of the wedge? What is the 3-D topology of the resulting plasmoid, and how exactly do the X- and O-type neutral lines connect to one another?

An important facet of the neutral line model is shown in Figure 2 (Baker et al., 1979). Fig. 2a shows the late growth phase situation with a very taillike field throughout the inner plasma sheet, including at 6.6 R_E. Fig. 2b shows the situation which might exist at substorm onset with a new neutral line at ~15 R_E in the magnetotail. In this picture, it is assumed that magnetic reconnection has allowed the distended, taillike magnetic field lines observed in the late growth phase to relax to a much more dipolar configuration. It is also suggested that the rapid onset of reconnection (and the concomitant changes in the magnetic fields near the neutral line) would produce strong induced electric fields. Thus, it was hypothesized by Baker et al. (1979) that intense bursts of high-energy (E ~ 1 MeV) particles are the direct consequence of strong acceleration in the vicinity of the substorm neutral line. Pellinen and Heikkila (1984), as an example, calculated accelerations from 1 keV to 1 MeV in 3s at the neutral line by assuming a field variation dB/dt = 2.5 nT/s. As depicted in the diagram of Fig. 2b, Baker et al. suggested that the energetic particles and hot plasma are "injected" into the vicinity of geosynchronous orbit as a consequence of the collapse of reconnected field lines into their more dipolar configuration.

A successful model which describes both qualitatively and quantitatively the dispersive drift of hot substorm plasma particles in the outer magnetosphere was put forward by Mauk and McIlwain (1974). This "injection boundary" model suggests that at the initiation time of the substorm expansion phase, particles are basically energized at all locations so as to fill an extended region tailward of a well-defined, spiral-shaped injection boundary. In order to explain both ion and electron injection and drift characteristics, the boundary must spiral toward Earth as one moves toward local midnight from dusk; the boundary must analogously spiral away from Earth in the post-midnight sector (see Fig. 3). The Mauk and McIlwain injection boundary model invokes specific magnetic and electric field configurations, but it makes no statement on how, specifically, the hot plasmas tailward of the injection boundary are produced in the first place.

Moore et al. (1981) studied the dynamics of plasma injection events using two-satellite techniques. They concluded that hot plasma suddenly appeared in the near-Earth plasma sheet behind an Earthward propagating "injection front". Moore et al. thus suggested that the injection process was of finite duration and was due to a nearly shocklike compression wave that was launched from the more distant tail (10-15 R_E). This

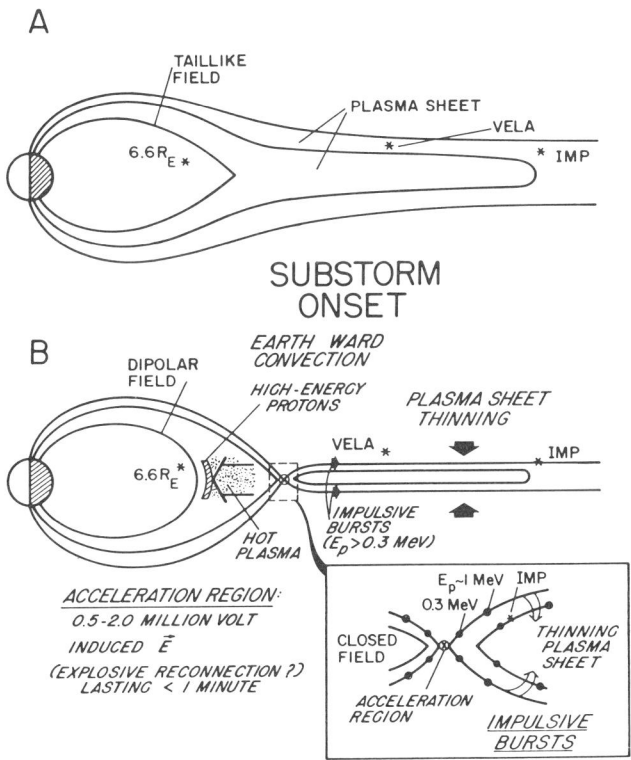

Fig. 2. Illustration depicting the sequence of energetic particle effects in the model of Baker et al. (1979). (a) The inner magnetosphere just prior to substorm onset showing the buildup of stress evidenced by the tail-like field. (b) The magnetosphere just after onset showing a dipolar field configuration and the accelerated proton bunches streaming sunward toward the trapped radiation zones and antisunward along the thinning plasma sheet.

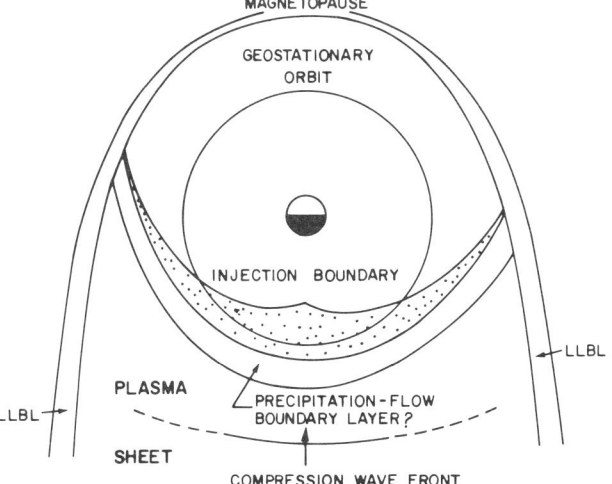

Fig. 3. Equatorial plane schematic of an hypothesis concerning the formation of an injection boundary near synchronous orbit. A compression wave impulse propagates earthward from the magnetotail through the pre-existing inner edge of the plasma sheet, producing an earthward moving injection front with maximum net displacement in a sector near midnight (from Baker et al., 1984).

wave subsequently propagated Earthward at a speed of ≥100 km/s where it eventually formed an injection boundary configuration. In subsequent work, the injection boundary and propagating injection front models (developed, respectively, by B. H. Mauk and T. E. Moore) were combined as shown in Figure 3 into a single phenomenological model (from Baker et al., 1984). If it is assumed that the Earthward propagating compression wavefront in Fig. 3 is initiated by rapid reconnection at the substorm neutral line, then the composite model of Fig. 3 can be considered an elaboration of the substorm injection model illustrated in Fig. 2b.

It must be recognized that the rather complex spatial configurations shown in Figures 2 and 3 were inferred from single (or perhaps double) spacecraft measurements; much of the model development was based upon observation of repetitive features from many case studies. However, without multi-spacecraft observations it has proven difficult to separate radial from azimuthal boundary propagation. It also has been difficult to determine where, when, and how the initial substorm disruptions begins. In particular, the global sequence and timing of expansion phase events in the near tail and in the ionosphere has been very difficult to pin down definitively.

Recovery Phase

The recovery phase is perhaps the least understood aspect of the entire substorm sequence. Certainly, from the standpoint of the observations, this substorm feature is one of the most predictable and one of the most impressive of the substorm phases. The auroral and near-tail data show rather clearly in the recovery phase that: (1) There is a rapid thickening of the plasma sheet; (2) There is an increase in the B_z component of the magnetic field; and (3) There is a decrease in the strength of high-latitude currents and auroral luminosity and a shift of the auroral activity from evening to morning sector (see Pulkkinen et al., this issue). Low-altitude particle measurements also suggest that a narrow region of intense particle fluxes persists at the high-latitude edge of the auroral zone (probably mapping to the plasma sheet boundary layer) while the lower latitude portion of particle flux mapping to the hot, thick plasma sheet decays away (see Baumjohann (1988) for a review).

The increase of B_z in the near-tail implies an increased volume of closed field lines as would result from reconnection. An increase of hot plasma in the plasma sheet would be the result of such reconnection and the newly formed closed field lines provide the means for confining the energized plasma. In the standard model the substorm neutral line, after remaining in its near-Earth location for a period ranging from 0.5 to 2 hours, suddenly moves tailward very rapidly to some unspecified distance. As it does so, the plasma jetting earthward from it, threaded by newly reconnected closed field lines (with northward B_z-component), replenishes the plasma sheet, causing it to increase in thickness as well as length.

Why the plasma sheet remains thin throughout the expansion phase and then recovers so suddenly is not clear on theoretical grounds. It is possible that convection in the inner magnetosphere suddenly ceases in response to changes of the imposed electric field controlled by northward turning of the interplanetary magnetic field. It seems unlikely, however, that solar wind-induced changes could occur so rapidly that they would produce the abrupt thickening and the putative rapid motion of the neutral line down the tail seen in the typical recovery phase. Remarkably, sometimes recovery seems to proceed until tail energy is exhausted and a "ground state" of the magnetosphere is reached. At other times the activity decreases substantially yet for some reason the tail remains in an enhanced energy state. The reason for such differences from event to event is not known, but may be related to the conditions required for onsets. Thus, this may demonstrate how interdependent the various aspects of substorms really are: The recovery of the substorm may be closely related to the very processes responsible for its onset.

New Observational Results

Intense Cross-Tail Currents in the Late Growth Phase

It has been observed for a very long time at 6.6 R_E that relatively rapid and very deep energetic particle "dropouts" occur late in many substorm growth phases (e.g., Walker et al., 1976). These dropouts can persist for 10-30 minutes and they are normally followed by a rapid energetic particle flux recovery and a fresh particle injection event. The dropout is associated with extremely taillike magnetic field configuration, while the injection of particles is identified with a dipolar reconfiguration of the field. The particle injection and field "dipolarization" are identified as the beginning of the substorm expansion phase onset (e.g., Sauvaud and Winckler, 1980).

Baker and McPherron (1990) have ascribed the strong energetic particle dropout events seen at synchronous orbit to the development of extremely intense cross-tail currents quite close to the Earth. Figure 4 shows a well-studied example of the phenomenon. The GOES 3 spacecraft was very near spacecraft (S/C) 1977-007 on March 22, 1979, and the figure shows the GOES magnetic field measurements from 0900 to 1300 UT. The *H* (positive north), *D* (positive east), and *V* (positive radially outward) field components are plotted. At the top of the figure electron data from S/C 1977-007 are shown. A typical growth phase sequence of taillike field development began at ~1015 UT. This development continued until ~1040 UT, at which time the field began to assume a much more taillike configuration, reaching a maximum negative *V* component at 1054 UT. At this time the field began rapidly to reconfigure toward a dipolar state. This reconfiguration brought the first step of the recovery of the energetic particles and hot plasma at geostationary orbit (see the upper panel). The complete dipolar recovery of the magnetic field at GOES 3 was not accomplished until ~1104 UT. This latter time was the period of the apparent large injection of freshly accelerated energetic electrons and ions.

The development of a taillike field was observed farther out in the magnetosphere in the same meridian as the two geosynchronous spacecraft in this case. During the event the ISEE 1 and 2 spacecraft were inbound below the equatorial plane at a radial distance of approximately 13 R_E. Beginning at 1010 UT, B_z in the plasma sheet decreased as B_x increased. By 1050 the northern boundary of the plasma sheet approached the two spacecraft. By equating the total plasma and magnetic pressure at the ISEE 2 spacecraft to magnetic pressure in the lobe, it was inferred that the lobe field increased with time and that both spacecraft were measuring nearly the total lobe field of 60 nT. At 1054 UT the substorm expansion began, and the current sheet rapidly moved upward across the lower spacecraft (ISEE 2) and nine minutes later the current sheet passed over ISEE 1 at higher latitude. The time delays between crossings provided a vertical velocity which allowed an estimate of the current sheet thickness of ~400 km.

Taken together, synchronous orbit and other such near-tail observations imply that a very thin current sheet develops throughout the inner plasma sheet during many substorm growth phases. In 20-30% of substorm cases there is evidence of a nearly complete dropout of energetic particle and hot plasma fluxes along with an extremely taillike magnetic field configuration (as discussed above). Multiple-spacecraft observations show that the flux dropouts and taillike fields are restricted to a localized region within ± ~2 hours of local midnight. The flux dropouts invariably occur in a period of 10-30 min duration in what is identified as the late growth phase just prior to the expansion phase onset of the substorm (Baker and McPherron, 1990).

Kaufmann (1987) has used an idealized two-dimensional sheet current model to estimate the cross-tail current properties necessary to account for geostationary orbit growth phase observations. The model consists of a very thin current sheet extending from $X = -6\ R_E$ to $-11\ R_E$ at the equatorial (Z=0) plane. A centered dipole representing the Earth's main field is also used. In order to obtain the very taillike field at 6.6 R_E that is observed during late substorm growth phases (see Figure 4), it is necessary to substantially increase cross-tail currents in the near-Earth region. If the field is almost parallel to the equatorial plane at 6.6 R_E, the integrated dawn-to-dusk current K_y must be ≥ 300 mA/m = 1.9×10^6 A/R_E, which is a factor of $\geq$ 10 over the normal tail current near 15 R_E. One would expect the total field above such an intensified current sheet to increase greatly. This is the case: for the example shown in Fig. 4 above, the total field more than doubled near 6.6 R_E at the time of the maximum field stretching.

Why are there such intense, relatively localized cross-tail currents late in the growth phase? What is the nature of these currents? How do the currents dissipate at the substorm expansion phase onset? These are key issues in our developing understanding of substorms.

Sources of the Cross-Tail Current

The inner part of the Earth's magnetotail is characterized by strong magnetic field gradients, by substantial plasma pressure gradients, and by significant field line curvature. These nonhomogeneities of the field and plasma distributions must

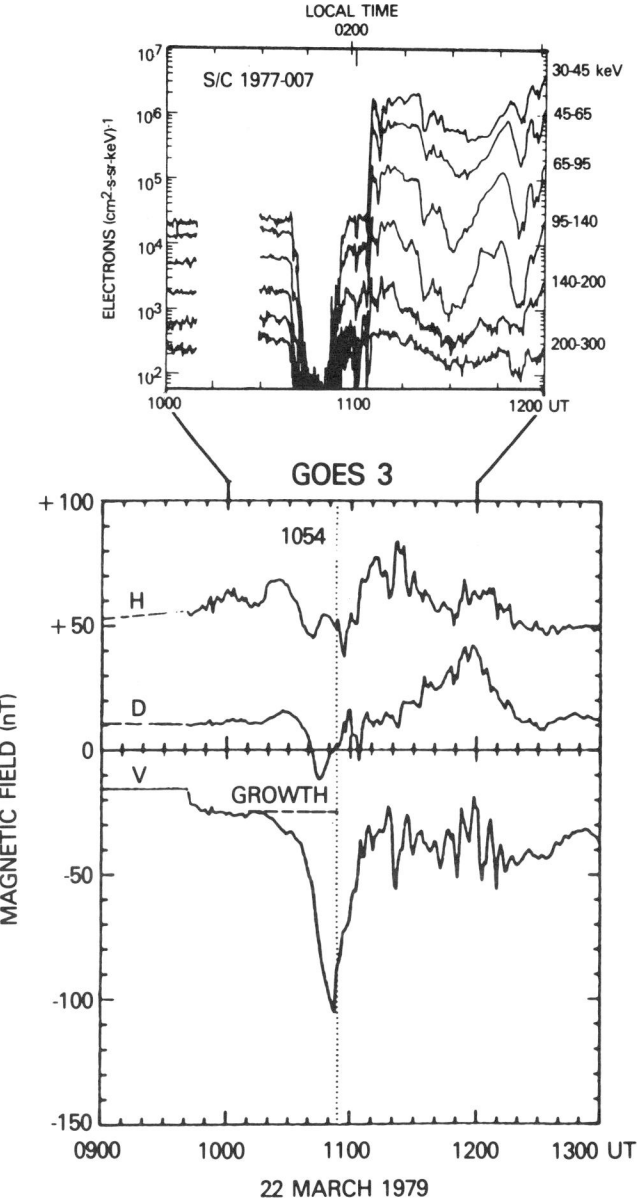

Fig. 4. Combination of (upper panel) energetic electron data from S/C 1977-007 and (lower panel) concurrent data obtained from the magnetometer onboard the GOES 3 spacecraft for March 22, 1979. This event shows an extreme particle dropout and very taillike field development late in a substorm growth phase (from Baker and McPherron, 1990).

be responsible for the distortions that develop late in the substorm growth phase. In general (Parker, 1957), the current density arising from gradient and curvature drifts due to the magnetic field configuration can be written as

$$j_D = \frac{B}{B^2} \times \left(\frac{P_\perp}{B} \nabla B + \frac{P_\parallel}{B^2} (B \cdot \nabla) B \right)$$

In addition, the current density due to the gyration effects of the plasma particles is given by

$$\mathbf{j}_g = \frac{\mathbf{B}}{B^2} \times \left(\nabla P_\perp - \frac{P_\perp}{B} \nabla B - \frac{P_\perp}{B^2}(\mathbf{B} \cdot \nabla)\mathbf{B} \right)$$

The total cross-tail current is the sum of these two components

$$\mathbf{j} = \mathbf{j}_D + \mathbf{j}_g = \frac{\mathbf{B}}{B^2} \times \left(\nabla P_\perp + \frac{P_\parallel - P_\perp}{B^2}(\mathbf{B} \cdot \nabla)\mathbf{B} \right)$$

In the above formulation, $P_\perp = 1/2 \int f(v) m v^2 \sin^2 \alpha \, dv$ and $P_\parallel = \int f(v) m v^2 \cos^2 \alpha \, dv$ are the perpendicular and parallel pressures, respectively.

An isotropic pressure ($P_\parallel = P_\perp$) means that the curvature drift ($P_\parallel(\mathbf{B} \cdot \nabla)\mathbf{B}$) term is canceled by the gyro-orbit crowding effect ($-P_\perp(\mathbf{B} \cdot \nabla)\mathbf{B}$). Thus, in order to have an enhanced cross-tail current due to curvature drifts, there must be a pressure anisotropy ($P_\parallel > P_\perp$). In order for there to be a large change in the currents flowing across the inner plasma sheet late in substorm growth phases, there obviously must be significant changes in the plasma distributions and pressures.

Recently, Mitchell et al. (1990) have studied the detailed properties of the plasma sheet electron and ion distributions during a well-observed substorm growth phase. They found that relatively early in the growth phase sequence, the excess of $P_\parallel$ over $P_\perp$ for ~1 keV electrons locally (at ~10 R_E) can account for much of the cross-tail current required to maintain the observed magnetic field configuration. In their study, Mitchell et al. (1990) noted that these thermal and suprathermal electrons provide a stable sheet structure and allow the progressive development of a very thin current-carrying region. Hence, even though the thermal ion motion is non-adiabatic in this region during the early and mid growth phase, the substantial cross-tail currents will be carried by the electrons.

In the late growth phase, Mitchell et al. (1990) conclude that the current sheet may become so thin that even the thermal electrons can become substantially unmagnetized (see, also Buchner and Zelenyi, 1987). They conclude that late in the growth phase, the required cross-tail currents are carried largely by thermal (1 ~ 10 keV) ions executing serpentine (Speiser-type) drifts across the plasma sheet. Figure 5 shows the proposed sequence put forward by Mitchell et al. In Fig. 5a the inner magnetosphere is in its quiescent, ground-state configuration with the field having a rather dipolar form. As the growth phase commences, Fig. 5b shows that the cross-tail current sheet intensifies and moves closer to the Earth. It is suggested (Fig. 5c) that as magnetic flux is increasingly "added" to the tail lobes, the near-Earth current sheet thins substantially and further intensifies. We note here that, of course, an intensification of the cross-tail current is intimately tied to the increase of the lobe magnetic field during the growth phase. In the solenoidal configuration of the magnetotail, the lobe fields are the result of plasma sheet currents and the thickness of the plasma sheet is a self-consistent balance between magnetic pressure (lobes) and plasma thermal pressure (plasma sheet).

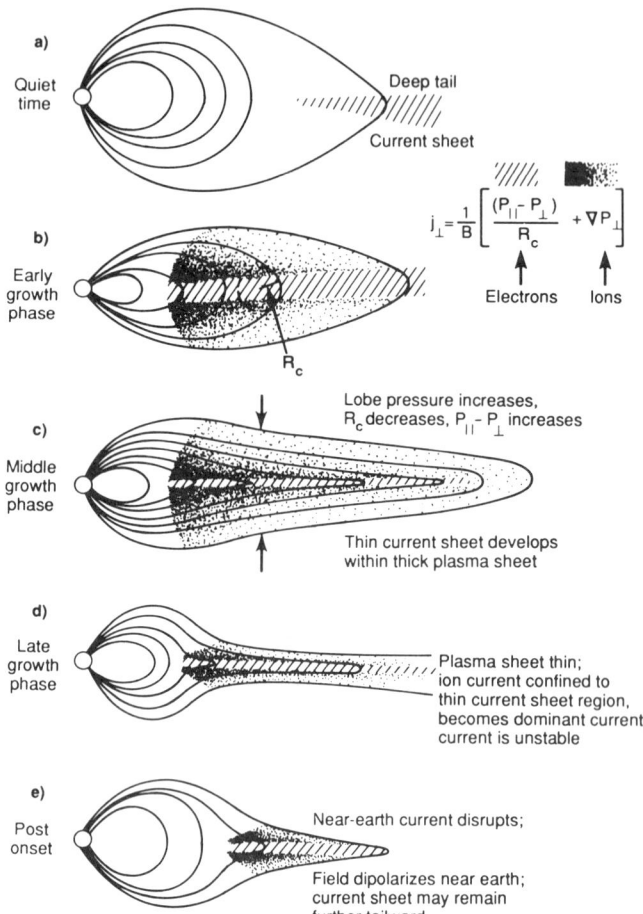

Fig. 5. The sequence of states of the magnetotail in the growth phase as presented by Mitchell et al. (1990). Stipled shading indicates $\nabla P_\perp$ ion current term; hatched shading indicates $P_\parallel > P_\perp$ electron current term.

In panels b and c, Mitchell et al. (1990) emphasize the role of anisotropic ($P_\parallel > P_\perp$) electrons in carrying the current. In panel 5d, they show the kind of extremely thin plasma sheet discussed by Baker and McPherron (1990) late in the growth phase (see Figure 4) and Mitchell et al. suggest the role of both anisotropic electrons and Speiser-type ions at this stage. Finally, in Fig. 5e the near-Earth cross-tail current "disrupts" according to Mitchell et al. and the near-tail field dipolarizes; this is the expansion phase onset.

Current Diversion Within the Near-Tail

In the standard neutral line model it is generally assumed that the onset of the substorm expansive phase is closely associated in time with the formation and pinching off of a plasmoid by an X-type neutral line. However, early descriptions of the neutral line model and theoretical treatments clearly proposed that magnetic reconnection would begin on closed field lines within the central plasma sheet

(McPherron et al., 1973; Coroniti, 1985). The issue then becomes one of time scale between initial neutral line formation on closed central plasma sheet field lines and expansive phase onset as identified by various auroral zone and mid-latitude magnetic indicators. The present X line model (Hones, 1979) would suggest that this time scale is quite short (a few minutes or less).

Recently, Baker and McPherron (1990) have proposed that late in the growth phase (i.e., 10-30 min before the expansion phase onset) the substorm neutral line is formed on closed field lines in the central plasma sheet (see Fig. 6a). They suggest

CURRENT DIVERSION

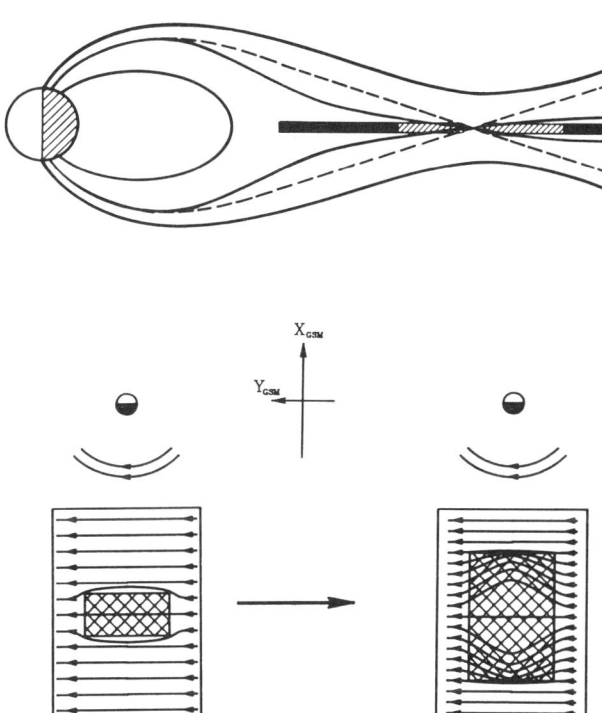

Fig. 6. (Upper panel) Meridional plane cut through the plasma sheet at the time, late in a substorm growth phase, when an X-type neutral line is assumed to form in the center of the plasma sheet. The evolution of the plasma sheet and current sheet region as cross-tail current is partially disrupted and diverted away (earthward and tailward) from the original X line location. The hatched area within the plasma sheet is the current sheet region; the solid region in the lower panel is the area of diverted, intensified cross-tail current.

(Lower panel) Top view of the tail current sheet nightside ring current region during the period illustrated above. The panels show the time development of disrupted and diverted cross-tail curvature drift currents after the initial formation of a near-Earth X-type neutral line (crosshatched region). Eventually, cross-tail currents are strongly concentrated at the earthward edge of the current sheet (adapted from Baker and McPherron, 1990).

that the reconnection begins in a quite localized - and maximally tearing-unstable - region at ~15 to 20 R_E geocentric distance. They assume that this slow initial reconnection tends to affect the cross-tail current locally. Baker and McPherron propose that the current changes within the close vicinity of the neutral line do not lead to immediate diversion of current into the auroral ionosphere. Rather, they suggest that the plasma sheet plasmas redistribute themselves such that part of the cross-tail current near the neutral line diverts slightly earthward and slightly tailward of the initial neutral line region. In contradistinction to the standard picture, current flow into the ionosphere would not occur immediately in this scenario, but rather would be inhibited by the large ionospheric resistivity at this early stage of substorm sequence and by the large inductances within such a global-scale current system.

Thus, in this extention of the standard model, the current from the localized X line region reestablishes itself slightly away from the original disruption region and the currents along the new current pathways intensify. As the cross-tail integrated current increases, the force balance condition demands that the plasma sheet be compressed until the plasma pressure gradients balance the increased $\mathbf{J} \times \mathbf{B}$ force. This thins the plasma sheet, which, in turn, further decreases the radius of curvature of plasma sheet field lines. Such a reduction of the field line radius of curvature increases the curvature drift current in a self-consistent way (see previous section) within the plasma sheet so as to maintain (or reestablish) pressure balance.

As the field lines away from the X line increase in curvature, the drift current in the new drift pathway will also tend to be changed. Baker and McPherron (1990) suggest this as affecting primarily the ion drift current carriers, but the thermal electrons would also be strongly affected (Mitchell et al., 1990). Thus the intensified cross-tail currents some distance from the original neutral line are in turn altered, and an even more intensified current diverts farther earthward (and tailward). The original X line formation and current disruption thus sets up a positive feedback system which progressively diverts and intensifies the cross-tail current in the plasma sheet.

Viewed from above the X-Y plane, the current diversion sequence would look as illustrated in the schematic diagrams in Fig. 6b. This shows two different states of the magnetotail system. The central plasma sheet with cross-tail current vectors is shown, along with a sketch of the outer ring current relative to the Earth. An original region of cross-tail current disruption near the initial neutral line position is shown by the crosshatched rectangle. The curvature drift current that had flowed through this region is envisioned as diverting around the turbulent reconnection region. In the right hand portion of the figure the substantially progressed diversion and intensification of the cross-tail current is illustrated. Eventually, according to this picture, a significant portion of the cross-tail current from the entire inner tail region could be redistributed into the fairly narrow region at the Earthward edge of the original plasma sheet. This strong current sheet would produce very taillike fields surprisingly close to the Earth. Such a model would provide an explanation of the extreme magnetic field distortions seen in many substorms (as illustrated in Fig. 4). Since the magnetic field configuration also controls the local energetic particle population, the thin plasma sheet predicted by this model would also correspond to a similarly thin region of particle confinement.

Current Disruption at Expansion Phase Onset

It is an idea dating back to the earliest models (e.g., Atkinson, 1967) that the wedge current-flow pattern during substorms is due to a reduction of cross-tail current which then is diverted to flow through the auroral ionosphere. The reduction/diversion of cross-tail current has been termed "disruption" in earlier studies (e.g., Lui, 1978). It was argued that neutral line formation at the expansion phase onset would disrupt the cross-tail current and produce the observed large-scale field-aligned current (FAC) pattern (Atkinson, 1967; McPherron et al., 1973). Lui (1978) modeled the current disruption as a reversed (dusk-to-dawn) cross-tail current system. However, the nature of the cross-tail currents in the growth phase has remained relatively unclear until recently (see Mitchell et al., 1990), and little has been known about the details of the current disruption process. This has been changing with improved in situ observations.

High time resolution measurements of magnetic fields and energetic particles with the Active Magnetospheric Particle Tracer Explorer/Charge Composition Explorer (AMPTE/CCE) spacecraft near the satellite apogee (8.8 R_E) have permitted study of several current disruption events (Takahashi et al., 1987; Lui et al., 1988). These events are relatively rare inside 8 R_E geocentric distance (Takahashi et al., 1987) and they appear to have a high degree of spatial and temporal variability. Takahashi et al. observed strong southward magnetic fields at r ~ 8.1 R_E with a quasi-periodic recurrence time of ~ 13 seconds. They concluded that the CCE spacecraft was in the expansion-phase current disruption region and they further concluded that the observations provided "strong evidence ... of magnetic reconnection in the plasma sheet at the onset of this particular substorm."

On the other hand, a similar (but non-identical) set of CCE investigators (Lui et al., 1988) studied a different substorm event for which the spacecraft was at r ~ 8.8 R_E. Lui et al. (1988) concluded that their event was similar in many respects to that of Takahashi et al. (1987). Lui et al. characterized the current disruption as a turbulent plasma process, but argued that their results did not support the local presence of an X-type neutral line. Lui et al. (1988) did invoke a highly localized and time-dependent breakdown of cross-tail current flow and they argued that the field changes ($\partial B/\partial t$) attendant to this event played a strong role in local particle acceleration. Is this (Takahashi et al., 1987), or is this not (Lui et al., 1988), magnetic reconnection in the near-Earth magnetotail at substorm onset?

We would argue that on the macroscale, magnetic reconnection (or what comes to the same) must occur in order to cause the dramatic magnetic reconfigurations that occur in the near-tail at substorm onset. Reconnection is also, in effect, required to tap and to dissipate the stored magnetic energy in the magnetotail during the expansion phase. However, it may be the case that on the microphysical scale the processes appear much more complex than the simple, idealized models of X-line formation would suggest. Thus, one could reconcile the apparently discrepant results from AMPTE that have been published.

More broadly, one could argue that the substorm neutral line is normally located at r ~ 15-20 R_E, as the standard model asserts. Following the line of reasoning of Baker and McPherron (1990), one might suppose that the dominant X-line configuration remains at that position and there is the current diversion within the plasma sheet as shown in Fig. 6. In this model, however, there has been no global change of the topology of the magnetotail, since reconnection has progressed on closed field lines, and currents have been largely redistributed only within the plasma sheet. In this late stage of the growth phase the near tail would have reached a much more unstable state with the cross-tail currents intensified in a narrow sheet at the inner edge of the tail. Reconnection would have progressed to nearly the interface between plasma sheet and lobe field lines.

The expansion phase of the substorm in this scenario may be the nearly explosive change of state of this highly unstable system. Two principal occurrences probably force this change of state. First, the progression of reconnection to the edge of plasma sheet field lines means that the Alfven speed in the inflow region of the X-type neutral line will increase by a factor of 10 or more. Since V_A controls the reconnection rate the transition to reconnection of open (lobe) field lines could mean a large change in the rate of energy dissipation in the tail. Second, the eventual abrupt disruption of the intense cross-tail current at the inner edge of the plasma sheet would produce a dramatic diversion of current into the ionosphere. In combination, these two processes could produce tremendous energy dissipation at expansion phase onset. Moreover, there would be significant disturbance levels both at the original reconnection region (~ 15-20 R_E) and in the very near tail (< 10 R_E).

Thus, we may conclude that current disruption is a breakdown of the orderly cross-tail current represented by ion drifts (curvature drift and Speiser-type) and by electron drifts (dominantly due to $P_{\parallel} > P_{\perp}$ as reported by Mitchell et al (1990)). We would suggest that the current disruption reported by AMPTE investigators at r ~ 8 R_E may not represent the entire substorm onset story. There may be significant global reconnection effects at greater geocentric distances at the same time as the localized disruptions that occur at the inner edge of the plasma sheet. Both features are fundamentally important to the substorm onset during large events.

Formation and Development of the Current Wedge

In a series of recent papers, Lopez et al. (1990a,b,c) and Lopez and Lui (1990) have studied the initiation of substorms using ground-based magnetometers, low-altitude spacecraft, and multiple near-geostationary spacecraft. Lopez et al. have

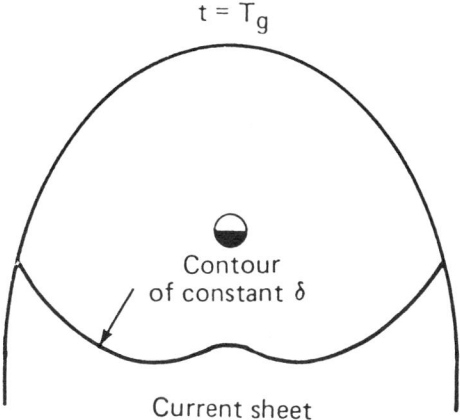
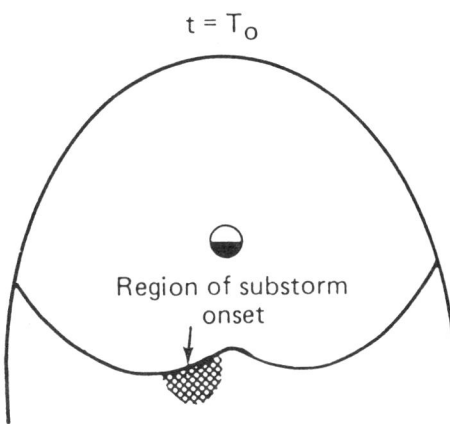
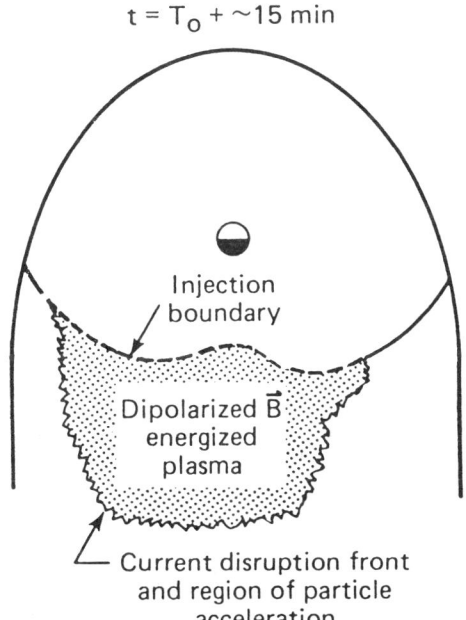

presented compelling evidence that initial substorm onset is very localized in longitude. By careful comparison of data from closely spaced satellites in the equatorial plane, it is concluded that disruption of the cross-tail current begins in a narrow region (often in the premidnight sector) and then spreads rapidly in local time (Lopez and Lui, 1990). In contrast to the earlier work of Moore et al. (1981), it is concluded that current disruption and particle acceleration begin relatively close to the Earth and propagate radially outward into the more distant tail (Lopez et al., 1990a,b,c).

Figure 7 is from Lopez et al. (1990a). In this figure the authors have attempted to determine the effect of a magnetic perturbation (due to the cross-tail current) as compared to the dipole magnetic field that would otherwise be present. The ratio of the current sheet perturbation to the dipole field is defined as δ. Using qualitative arguments about the expected local time and radial variations of δ that would be seen in the magnetotail, Lopez et al. derive a contour of roughly constant δ as shown in Fig. 7a. Note that this contour is similar in shape to the injection boundary seen in Fig. 3 as derived by Mauk and McIlwain (1974).

Lopez et al. (1990a) conclude that regions of larger δ (more sheet-like geometry and smaller B_z) will be more unstable to tearing mode growth. Generally, δ is larger further from the Earth and so one would expect the instability to occur deep in the magnetotail. However, Lopez et al. (1990a) invoke plasma compositional arguments to conclude that current disruption will initiate closer to the Earth, probably along the constant δ contour. Lopez et al. suggest that at a time T_g (the end of the growth phase) there is a contour of δ which is the minimum value required to permit current disruption. This contour has the closest earthward approach in the midnight region. As noted, this is the origin of the double spiral injection boundary. The contour location (i.e., the injection boundary) depends on Kp, with the boundary forming closer to Earth during periods of higher Kp. At time T_0 (expansion onset) a localized sector near the earthward edge of the current sheet (at the contour of minimum stability) becomes unstable due to the presence of O^+, and the current disruption process begins.

Lopez et al. contend that the region affected by the disruption of the cross-tail current sheet will expand both in local time and radius as the region of instability spreads. They suggest

Fig. 7. A qualitative scenario for the generation of the injection boundary. At time T_0 (expansion phase onset), an instability is triggered in a spatially limited region of the cross-tail current sheet near the Earth. The disruption energizes particles and spreads outward away from the point of origin. At $T_0 + 15$ min a substantial portion of the near-Earth cross-tail current has been disrupted, the magnetic field has reconfigured to a more dipolar orientation, and an energetic plasma has been produced whose dispersionless earthward edge is the contour of $\delta_{critical}$. This contour is a double spiral whose closest point to Earth is in the midnight region (from Lopez et al., 1990a).

that within the unstable region, particles are locally energized as the disruption front passes over them. Ions which are accelerated at or beyond the boundary will be energized in a dispersionless fashion, since the acceleration is local. This qualitative model provides an explanation for the dispersionless nature, shape, and Kp dependence of the injection boundary. In direct contradistinction to the model of Fig. 3, the results in Fig. 7 would suggest that there is an outward, not inward, wave of particle "injection" (see, also, Fairfield, 1988).

In related work, Nakamura et al. (1990) have studied detailed auroral morphology data and have compared these results to energetic particle signatures at geosynchronous orbit. The results of this comparison are shown in Figure 8. Nakamura et al. conclude that intense, azimuthally localized north-south aligned auroras (N-S auroras) form the active parts of the auroral bulge at expansion onset. These authors found that abrupt ("transient") dispersionless particle injections at synchronous orbit are well correlated with the development of N-S auroras both spatially and temporally. Nakamura et al. propose that localized, very strong electric fields are responsible for the acceleration and inward transport of the energetic particles during the expansion phase. These electric fields are estimated to be of relatively short duration (< 10 min) and are thought to map directly to the N-S auroral structures.

Away from the narrow longitudinal region of the N-S auroras (and the associated dispersionless injections), Nakamura et al. (1990) identify diffuse auroral regions linked to weakly drifting, dispersive energetic electron populations. Thus, like Lopez et al. (1990a), Nakamura et al. propose a very localized

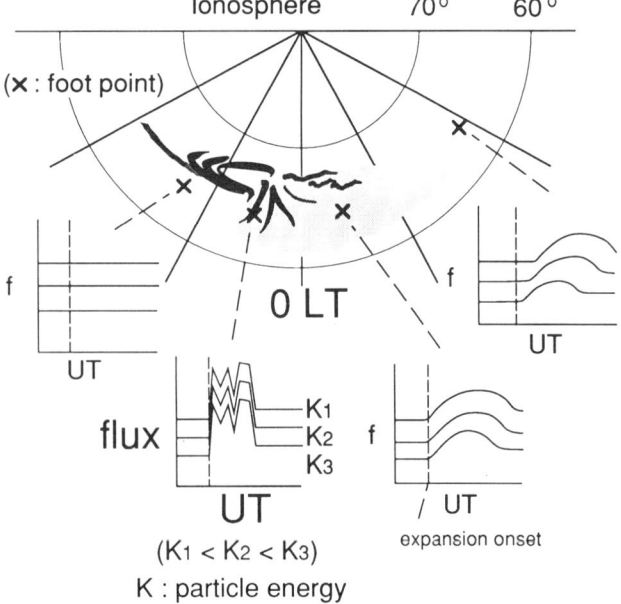

Fig. 8. A representation of the types of energetic elctron flux variations seen at different local times relative to the properties of the expansion aurora (from Nakamura, 1990).

substorm initiation and particle acceleration region. However, the latter authors are much more supportive of an earthward transport of particles rather than a tailward transport.

Key Requirement: Improved Magnetic Modeling

Empirical Models: A Magnetospheric Roadmap

One of the most difficult tasks in substorm studies is to map accurately from given observed auroral features in the ionosphere to the magnetically connected regions in the equatorial magnetotail. Without such a magnetospheric "roadmap", it is essentially impossible to deduce the nature and location of magnetotail disturbances that are causally related to auroral substorm events. Empirical magnetic modeling has provided an extremely valuable framework in which to determine general magnetospheric-ionospheric connectivity (e.g., Tsyganenko (1989)). Such models provide an average, basically time-independent picture of the magnetospheric configuration.

Recently, Pulkkinen et al. (1991) have studied in some detail a very strong substorm event that was well-observed both in the near-tail region and in the auroral zone. Auroral images were available for this event from spacecraft both in the northern hemisphere (Viking) and in the southern hemisphere (DE 1). As a first attempt to map auroral features from ionospheric altitudes to the equatorial plane in the magnetotail, the Tsyganenko (1989) magnetic field model together with the IGRF 1985 coefficients for the internal geomagnetic field were used. The Tsyganenko model gives the magnetic field as a sum of terms representing each of the magnetospheric current systems, two of which--the ring current and the cross-tail current--lie basically within the equatorial plane. The field is given as a function of the dipole tilt angle between the Sun-Earth line and the dipole equator, and the Kp-index indicating the level of geomagnetic activity. The upper two panels of Figure 9 show the current integrated over the thickness of the current sheet (Fig. 9a) and also a set of representative field lines in the meridian plane of the ISEE 1 spacecraft (146° longitude) for this case (Fig. 9c) computed using the appropriate tilt angle and largest level of magnetic activity (Kp ≥ 5-) allowed by the model. In Fig. 9a, the contributions of the tail current and the ring current are shown separately.

Pulkkinen et al. (1991) compared the Tsyganenko model magnetic field line tilts and field strengths with measurements from ISEE 1, GEOS 5, and SCATHA for the substorm event in question. They found that the unmodified model provides a reasonable representation of the near-tail field properties very early in the substorm growth phase, but by late in the growth phase the model representation is completely inadequate. The authors conclude that, quite obviously, a more realistic modeling approach is necessary to give an instantaneous map of the magnetospheric configuration.

Time-dependent Field Modeling

Pulkkinen et al. (1991) demonstrate a method of providing an essentially time-dependent model of the magnetospheric

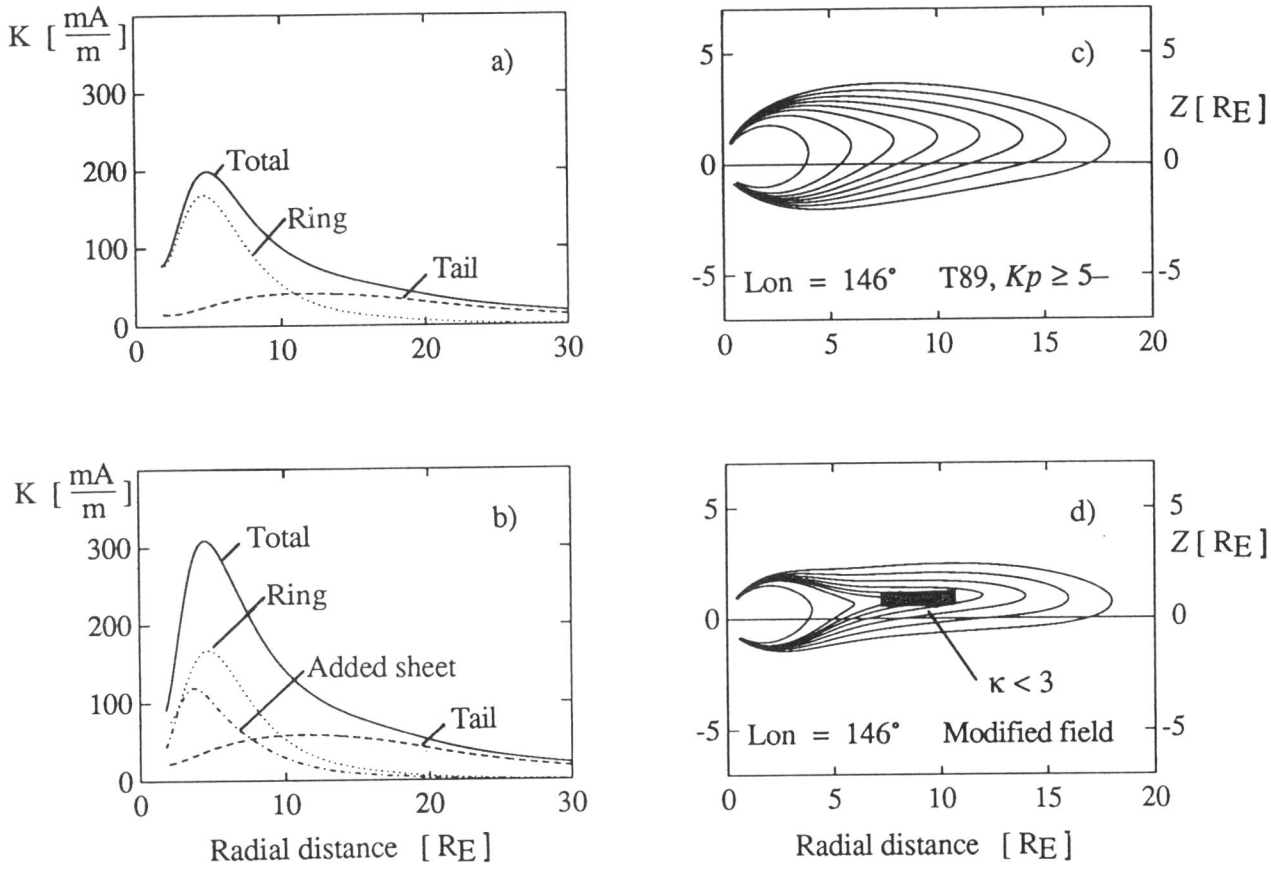

Fig. 9. Left: Current intensities integrated over the current sheet thickness (in mA/m) in the ISEE 1 meridian a) in the basic T89 model, and b) in the modified model of the Pulkkinen et al. (1991). Total current (solid line), tail current (dotted line), and ring current (dashed line) are shown for both cases. The additional current sheet is shown by dash-dotted line. Right: Field line projections to the ISEE 1 meridian computed using c) basic T89 model and d) the modified model. The region of chaotic electron motion for 1 keV electrons (see text) is shown hatched.

magnetic field by gradually increasing the cross-tail current, thinning the current sheet, and adding another localized tail current system with increasing intensity in the basic Tsyganenko model. This gives a global growth phase field model which compares favorably with measurements from several spacecraft in the magnetotail. The modified, time-dependent Tsyganenko model allows a systematic mapping of auroral features recorded by the auroral imagers into the near-tail region. These results, and a mapping study utilizing auroral images from Viking and the Tsyganenko 1987 model (Elphinstone et al., 1990), suggest that the auroral brightenings correspond to regions of thinned and intensified current sheet remarkably close to the Earth.

Specifically, Pulkkinen et al. (1991) include growth phase effects for the event under study by varying the Tsyganenko model parameters and by modification of the functional form of the field components. The current sheet is thinned locally by modification of the X and Y dependent function determining the current sheet thickness in the model (see Pulkkinen 1991) with the minimum thickness, location of the minimum thickness, and size of the thinned region as free parameters. The cross-tail current is intensified by enhancement of the model tail current by a constant factor. However, because the tail current peaks beyond 10 R_E, its enhancement cannot represent stretching of the field in the near-Earth region between 6 and 10 R_E. In order to further enhance stretching close to the Earth, a new thin current sheet is added to the model. The form of the current distribution is similar to the model ring current term, but the peak intensity, location of current maximum, and the current sheet thickness are set according to the observed degree of field stretching.

A time-dependent model for the growth phase was constructed by linearly varying the parameters describing the enhancement of the currents during growth phase until substorm onset. The field at the beginning of the growth phase is represented by the unmodified Tsyganenko model using the highest level of magnetic activity (Kp ≥ 5-). The minimum thickness of the current sheet, the intensification of the tail and ring currents, and the peak intensity of the new current are varied linearly during that period, whereas the remaining parameters assume the constant values used at the beginning of the growth phase.

The event studied by Pulkkinen et al. took place during a strongly disturbed period, which brought the activity very close to the Earth. To obtain best results in comparison with the field measurements from all spacecraft during the growth phase, the current sheet was thinned until it reached a minimum 10% of the original value by the time of the onset, and the tail current was enhanced by 30%. A new thin current sheet was added to the model with an intensity maximum of ~120 mA/m at 4.7 R_E at the ISEE. Figures 9b and 9d contrast the modified model (integrated currents and field line configurations, respectively) with the unmodified model results (Fig. 9a,c). The enhancement of the tail current leads to stretching of the field mainly beyond 10 R_E, whereas enhancement of the ring current or addition of a new current sheet closer to the Earth are needed to obtain tail-like field configurations in the near-Earth region. The additional current sheet introduced by Pulkkinen et al. shifts the current maximum somewhat Earthward. The strongly enhanced current in the near tail region stretches the field lines to much more tail-like configuration, with the field almost parallel to the current sheet at a distance of ~ 10 R_E.

Modeling of the growth phase by current systems linearly enhancing with time gives insight into how the current systems evolve in the near-Earth tail. The strong tail-like development of the field within the geostationary orbit inferred from the available measurements is represented by several changes, most notably the additional thin current sheet in the Tsyganenko model. The resulting cross-tail current is about 300 mA/m peaking at a geocentric distance of 5 R_E (Fig. 9c). This important modification of the standard model is in substantial agreement with the results of Kaufmann (1987).

Chaotic Particle Motion in Realistic Field Models

A global, time dependent magnetic field model can--in addition to providing useful information on the causal relationships and interconnections of the different processes taking place in the magnetosphere-ionosphere system--be used to investigate the stability properties of the current sheet during this development. Several plasma instabilities have been suggested to grow in the magnetotail during substorms, to account for the rapid energy conversion of tail magnetic to plasma kinetic energy. On macroscale, the growth of the instability would then lead to reconnection and the topological changes in the field such as formation of the substorm neutral line and the release of a plasmoid. The growth rates of these instabilities are often dependent on the ratio of the field components parallel and normal to the current sheet, and the thickness and intensity of the current sheet. Thus the global models providing these parameters can be used to infer the location of the unstable regions of the current sheet at each time.

The ion tearing instability provides the required marginal stability during the growth phase and effectively converts energy from magnetic to kinetic form. However, under average conditions in the tail, the ion tearing is stabilized by the adiabatic motion of the electrons (Galeev and Zelenyi, 1974). The ion tearing mode may only be excited at the point where this stabilizing effect is removed by pitch-angle scattering of the electrons, suggested to be due to either plasma turbulence (Coroniti, 1985) or transition of the electron motion from regular to chaotic (Buchner and Zelenyi, 1987). Thus, a stage of current sheet thinning during the substorm growth phase that may be of ultimate interest is the point at which thermal electrons become unstable due to chaotization of their orbits.

Buchner and Zelenyi (1987) considered a two-dimensional field configuration $\mathbf{B} = B_0 \tanh(Z/\lambda)\mathbf{x} + B_n \mathbf{z}$ and concluded that stretching of the tail leads to chaotization of electron motion and consequent onset of ion tearing. The degree of chaotization is described by the parameter $\kappa = (B_n/B_0)\sqrt{\lambda/\rho_{eo}}$ with λ the thickness of the field-reversal region, and ρ_{eo} the electron Larmor radius in the lobe field B_0. The chaotic domain is approximately given by $\kappa<3$, and the instability onset occurs when κ approaches 1. Under quiet conditions in the near-tail region both B_n and λ are large, and thus κ is large. During the growth phase B_n and λ decrease while B_0 increases, decreasing κ until transition to chaos takes place.

Using the obtained global model for the field and the κ-parameter of Buchner and Zelenyi (1987), Pulkkinen et al. (1991) determined the regions of the current sheet that are unstable towards the growth of the tearing mode. As the Tsyganenko model does not include field-aligned currents, the model field configuration of Pulkkinen et al. is nearly two-dimensional locally, and the effects of magnetic shear introduced by a cross-tail field component to the stability of the current sheet cannot be included in the treatment. When computing κ from the model, λ was chosen to be the current sheet half thickness, the lobe field was the field component 5 R_E above the current sheet parallel to the current sheet and the plane defined by the field lines, and B_n was the field component normal to the current sheet. The unstable region determined by Pulkkinen et al. (1991) from the condition $\kappa < 3$ for 1 keV electrons is shown hatched in the field line plot of Fig. 9d. Note that using the basic field model (Fig. 9b), the 1 keV electrons are stable throughout the tail. In the modified model, however, the enhancement of the currents destabilizes 1 keV electrons in the region between 8 and 12 R_E. Thus the results of Pulkkinen et al. suggest that the auroral brightenings correspond to regions of the thin, tearing-unstable current sheet in the near-Earth region.

Summary and Conclusions

We have presented here a brief review of substorms with an emphasis on their effects in the near-Earth magnetotail. We have discussed observations and modeling results in terms of a standard substorm paradigm. Substantial new insights into substorm dynamics have been emerging in the past few years owing to more comprehensive, multi-spacecraft observations and owing to improved instrumentation (e.g., frequent sequential auroral images). The intrinsically better global monitoring of substorm effects within the magnetosphere has revealed the clear need for more complex descriptions of substorm processes. To date there is no complete, entirely self-consistent theory of substorm physics. New ideas concerning onset mechanisms of substorms (e.g., Goertz and Smith, 1989; Korth et al., 1991) may eventually supplant parts of the standard model, or the newer models may prove to be less successful than the near-Earth neutral line description. In either case, such new ideas are very valuable in focusing on the weaker or less clear elements of the standard paradigm, thus forcing a continuing development of our understanding.

By concentrating on the near-Earth part of the magnetotail during substorms, we obviously treat only a portion of the entire substorm picture. However, as the results presented here demonstrate, the inner plasma sheet is a key region with respect to substorm initiation. Evidence continues to mount-- as we have reviewed in this paper--that much of the energy that is stored during the growth phase is suddenly and dramatically converted to thermal and directed plasma energy in the near-tail at expansion phase onset. Any substorm model that does not take cognizance of this feature will surely be an incomplete description of geomagnetic activity.

High time resolution measurements from constellations of three or more spacecraft in the near-tail--when tied together by time-dependent magnetic models--are revealing to us a rich array of plasma physical phenomena. We have yet to clearly understand the full meaning of "current disruption" as it applies to the near-Earth magnetotail. Indeed, we still cannot say for sure where (or, perhaps, whether) an X-type neutral line forms during the substorm. Certainly there are interesting local disturbances seen between 6 and 15 R_E that suggest that intense cross-tail currents are rapidly broken down. Evidently, in this process, a great deal of energy is tapped and converted to hot plasma and energetic particles. We are moving systematically now to integrate these local disturbance features into the global substorm models.

References

Akasofu, S.-I., Energy coupling between the solar wind and the magnetosphere, Space Sci. Rev., 28, 121, 1981.

Akasofu, S.-I., Explosive magnetic reconnection: Puzzle to be solved as the energy supply process for magnetospheric substorms? EOS, 66, 9, 1985.

Atkinson, G., An approximate flow equation for geomagnetic flux tubes and its application to polar substorms, J. Geophysics. Res., 72, 5373, 1967.

Baker, D. N., and R. L. McPherron, Extreme energetic particle decreases near geostationary orbit, J. Geophys. Res., 95, 6591, 1990.

Baker, D. N., P. R. Higbie, E. W. Hones, Jr., and R. D. Belian, High-resolution energetic particle measurements at 6.6 R_E, Low-energy electron anisotropies and short-term substorm predictions, J. Geophys. Res., 83, 4863, 1978.

Baker, D. N., R. D. Belian, P. R. Higbie, and E. W. Hones, Jr., High-energy magnetospheric protons and their dependence on geomagnetic and interplanetary conditions, J. Geophys. Res., 84, 7183, 1979.

Baker, D. N., S.-I. Akasofu, W. Baumjohann, J. W. Bieber, D. H. Fairfield, E. W. Hones, Jr., B. H. Mauk, R. L. McPherron, and T. E. Moore, Substorms in the magnetosphere, Solar Terrestrial Physics--Present and Future, NASA Publ. 1120, chap. 8, p.8-1, 1984.

Baker, D. N., L. F. Bargatze, and R. D. Zwickl, Magnetospheric response to the IMF: Substorms, J. Geomag. Geoelectr., 38, 1047, 1986.

Baker, D. N., R. C. Anderson, R. D. Zwickl, and J. A. Slavin, Average plasma and magnetic field variations in the distant magnetotail associated with near-Earth substorm effects, J. Geophys. Res., 92, 71, 1987.

Buchner, J. and Zelenyi, L. M., Chaotization of the electron motion as the cause of an internal magnetotail instability and substorm onset, J. Geophys. Res., 92, 13456, 1987.

Baumjohann, W., The plasma sheet boundary layer and magnetospheric substorms, J. Geomag. Geoelectr., 40, 157, 1988.

Coroniti, F. V., Explosive tail reconnection: The growth and expansion phases of magnetospheric substorms, J. Geophys. Res., 90, 7424, 1985.

Elphinstone, R. D., K. Jankowska, J. S. Murphree, and L. L. Cogger, The configuration of the auroral distribution for interplanetary magnetic field BZ northward, 1. IMF BX and BY dependencies as observed by the Viking satellite, J. Geophys. Res., 95, 5791, 1990.

Fairfield, D. H., Magnetic field signatures of substorms on high latitude field lines in the night time magnetotail, J. Geophys. Res., 78, 1553, 1973.

Fairfield, D. H., Multipoint measurements of magnetotail dynamics, Adv. Space Res., 8, (9)97, 1988.

Fairfield, D. H., R. P. Lepping, E. W. Hones, Jr., S. J. Bame, and J. R. Asbridge, Simultaneous measurements of magnetotail dynamics by IMP spacecraft, J. Geophys. Res., 86, 1396, 1981.

Galeev, A. A. and L. M. Zelenyi, Tearing instability in plasma configurations, Sov. Phys. JETP, 43, 1113, 1976.

Goertz, C. K., and R. A. Smith, The thermal catastrophe model of substorms, J. Geophys. Res., 94, 6581, 1989.

Hones, E. W., Jr., Transient phenomena in the magnetotail and their relation to substorms, Space Sci. Res., 23, 393, 1979.

Hones, E. W., Jr., D. N. Baker, S. J. Bame, W. C. Feldman, J. T. Gosling, D. J. McComas, R. D. Zwickl, J. A. Slavin, E. J. Smith, and B. T. Tsurutani, Structure of the magnetotail at 220 R_E and its response to geomagnetic activity, Geophys. Res. Lett., 11, 5, 1984.

Iijima, T., and T. A. Potemra, The amplitude distribution of field-aligned currents at northern high latitudes observed by Triad, J. Geophys. Res., 81, 2165, 1976.

Kaufmann, R. L., Substorm currents: Growth phase and onset, J. Geophys. Res., 92, 7471, 1987.

Korth, A., Z. Y. Pu, G. Kremser, and A. Roux, A statistical study of substorm onset conditions at geostationary orbit, this volume, 1991.

Lopez, R. E., and A. T. Y. Lui, A multisatellite case study of the expansion of a substorm current wedge in the near-Earth magnetotail, J. Geophys. Res., 95, 8009, 1990.

Lopez, R. E., D. G. Sibeck, R. W. McEntire and S. M. Krimigis, The energetic ion substorm injection boundary, J. Geophys. Res., 95, 109, 1990a.

Lopez, R. E., D. N. Baker, R. D. Belian, R. W. McEntire, and T. A. Potemra, A possible case of radially antisunward propagating substorm onset in the near-Earth magnetotail, Planet. Space Sci., 38, 771, 1990b.

Lopez, R. E., H. Luhr, B. J. Anderson, P. T. Newell, and R. W. McEntire, Multipoint observations of a small substorm, J. Geophys. Res., 95, 18897, 1990c.

Lui, A. T. Y., Estimates of current changes in the geomagnetic tail associated with a substorm, Geophys. Res. Lett., 5, 853, 1978.

Lui, A. T. Y., R. E. Lopez, S. M. Krimigis, R. W. McEntire, L. J. Zanetti, and T. A. Potemra, A case of magnetotail current disruption and diversion, Geophys. Res. Lett., 15, 721, 1988.

Mauk, B. H., and C. E. McIlwain, Correlation of Kp with the substorm-injected plasma boundary, J. Geophys. Res., 79, 3193, 1974.

McPherron, R. L., Growth phase of magnetospheric substorms, J. Geophys. Res., 75, 5592, 1970.

McPherron, R. L., C. T. Russell, and M. P. Aubry, Satellite studies of magnetospheric substorms on August 15, 1968, o. Phenomenological model for substorms, J. Geophys. Res., 78, 3131, 1973.

Mitchell, D. G., D. J. Williams, C. Y. Huang, L. A. Frank, and C. T. Russell, Current carriers in the near-Earth cross-tail current sheet during substorm growth phase, Geophys. Res. Lett., 17, 583, 1990.

Moore, T. E., R. L. Arnoldy, J. Feymann, and D. A. Hardy, Propagating substorm injection fronts, J. Geophys. Res., 86, 6713, 1981.

Nakamura, R., T. Oguti, T. Yamamoto, S. Kokubun, D. N. Baker, and R. D. Belian, Relationships between energetic particle injection and development of the north-south aligned auroras, J. Geophys. Res., submitted, 1990.

Parker, E. N., Newtonian development of the dynamical properties of ionized gases of low density, Phys. Rev., 107, 924, 1957.

Pellinen, R. J., and W. J. Heikkila, Inductive electric fields in the magnetotail and their reaction to auroral and substorm phenomena, Space Sci. Rev., 37, 1, 1984.

Pulkkinen, T. I., D. N. Baker, D. H. Fairfield, R. J. Pellinen, J. S. Murphree, R. D. Elphinstone, R. L. McPherron, J. F. Fennell, and R. E. Lopez, Modeling of the growth phase of a substorm using the Tsyganenko model and multi-spacecraft observations, Geophys. Res. Lett., in press, 1991.

Pulkkinen, T. I., A study of magnetic field and current configurations at time of a substorm onset, Planet. Space Sci., in press, 1991.

Pulkkinen, T. I., R. J. Pellinen, M. J. J. Korkinen, H. J. Opgenoorth, J. S. Murphree, V. Petrov, A. Zaitzer, and E. Fris-Christensen, Auroral signatures of substorm recovery phase: A case study, this volume, 1991.

Rostoker, G., and T. E. Eastman, A boundary layer model for magnetospheric substorms, J. Geophys. Res., 92, 12187, 1987.

Rostoker, G., S.-I. Akasofu, W. Baumjohann, Y. Kamide, and R. L. McPherron, The roles of direct input of energy from the solar wind and unloading of stored magnetotail energy in driving magnetospheric substorms, Space Sci. Rev., 46, 93, 1987.

Sauvaud, J.-A., and J. R. Winckler, Dynamics of plasma, energetic particles, and fields near synchronous orbit in the nighttime sector during magnetospheric substorms, J. Geophys. Res., 85, 2043, 1980.

Takahashi, K., L. J. Zanetti, R. E. Lopez, R. W. McEntire, T. A. Potemra, and K. Yumoto, Disruption of the magnetotail current sheet observed by AMPTE/CCE, Geophys. Res. Lett., 14, 1019, 1987.

Tsyganenko, N. A., Magnetospheric magnetic field model with a warped tail current sheet, Planet. Space Sci., 37, 5, 1989.

Walker, R. J., K. N. Erickson, R. L. Swanson, and J. R. Winckler, Substorm-associated particle boundary motion at synchronous orbit, J. Geophys. Res., 81, 5541, 1976.

Association Between Tail Substorm Phenomena and Magnetic Separation Distortion

L. R. LYONS

Space Sciences Laboratory, The Aerospace Corporation, Los Angeles, CA 90009

Important geomagnetic-tail phenomena have been observed to occur in association with substorm expansion phases. Often attributed to the systematic development and movement of a substorm neutral line, these phenomena include plasma sheet thinning and expansion, magnetic field stretching and dipolarization, particle injections, and outward and earthward plasma flows. It is proposed here that, out to at least $x \approx -22\,R_e$, these observed phenomena result from a mapping into the tail of the well-known substorm current wedge that is observed at synchronous orbit. Magnetic perturbations of the current wedge ought to be associated with a longitudinal distortion of the separatrix between open and closed magnetic field lines. Estimates based on observational evidence suggest that such a distortion can indeed account for tail phenomena during expansion phases, and that the distortion also maps along magnetic field lines to the auroral bulge that is observed in the ionosphere. The separatrix-distortion hypothesis thus provides a plausible and unifying explanation for expansion-phase phenomena observed in the auroral ionosphere, at synchronous orbit, and in the tail out to $x \approx -22\,R_e$.

I. Introduction

Substorms have different signatures in different regions of the magnetosphere. An important signature of the expansion phase is the plasma sheet thinnings often seen near onset out to $x \approx -22\,R_e$ (the apogee of ISEE) via the transit of satellites from the plasma sheet to the geomagnetic tail lobe [Hones et al., 1967; Dandouras et al., 1986]. Such observations [e.g., Hones et al., 1986] tend to show increased stretching of the magnetic field as the plasma sheet thins. Thinnings are typically followed by plasma sheet expansions later during a substorm [Hones et al., 1984]. In addition to the thinnings, dipolarization of the magnetic field has also been reported near expansion phase onset out to $x \approx -22\,R_e$ [Fairfield et al., 1981; Huang et al., 1991]. Huang et al. reported further that particle injections occur concurrently with the dipolarizations, and that these events have characteristics very similar to those observed at synchronous orbit. Such events were found by Huang et al. to be common in the tail within approximately 3 hours of magnetic midnight.

It seems that a complete description of substorm phenomenology for the tail should account for the approximately equal probabilities of observing either (a) plasma sheet thinning followed by plasma sheet expansion, or (b) dipolarization concurrent with particle injection near substorm onset within a few hours of midnight. Such a description should also account for plasma flows observed in the tail during substorms, which are occasionally outward as the plasma sheet thins but nearly always earthward as the plasma sheet expands [Hones et al., 1973; Lui et al., 1977a, b].

Thinnings, followed by expansions, of the plasma sheet are often interpreted in terms of the temporal evolution of a neutral line that forms earthward of $x \approx -22\,R_e$ at onset [e.g., McPherron, 1979; Hones et al., 1984]. However, this description does not explicitly include dipolarizations and concurrent particle injections as a common occurrence out to $x \approx -22\,R_e$.

In this paper, I consider the possibility that the tail phenomena described above are the direct result of an extension into the tail of the longitudinal structure of the substorm current wedge. This structure has been invoked to account for magnetic observations at synchronous orbit [e.g., Nagai, 1982, 1987] and has been proposed to be associated with the bulge in the auroral oval that forms in the ionosphere [Rostoker and Hughes, 1979; Tighe and Rostoker, 1981]. The interpretation of tail phenomena presented here depends upon the proposal [Lyons et al., 1990] that the longitudinal structure of the auroral bulge reflects a distortion of the separatrix between open and closed magnetic field lines, a distortion which is associated with the magnetic perturbations of the substorm current wedge.

II. Longitudinal Structure

Following the onset of a substorm expansion phase, a region of active aurora spreads poleward such that a portion of the aurora extends into the polar cap region previously devoid of aurora [Akasofu, 1964, 1977]. Images [Akasofu, 1977; Craven and Frank, 1985, 1987; Rostoker et al., 1987] show that the poleward

Magnetospheric Substorms
Geophysical Monograph 64
Copyright 1991 American Geophysical Union

motion of a portion of the poleward boundary of the aurora often leads to a "bulge" in the auroral oval that protrudes into the polar cap. This bulge feature is illustrated by the solid curve in Figure 1, which is based on images presented in the above references. During the substorm growth phase, which precedes the expansion phase onset, the poleward boundary of the aurora is approximately circular, as illustrated by the dashed line in Figure 1.

The distinctive evolution of the poleward boundary, as determined from Viking satellite images [Anger et al., 1987] during the development of an auroral surge on September 24, 1986, is shown in Figure 2 [from Lyons et al., 1990]. The heavy, solid curves in Figure 2 were obtained by drawing smooth curves along the poleward boundary of identifiable aurora in each of a sequence of Viking images obtained approximately once per minute. For spatial reference, each panel in Figure 2 contains geographic coordinates, a bar identifying the magnetic meridian of the Sondrestrom radar in Greenland, a dot along the bar giving the location of the radar, and (for comparison with the later curves) a thin, solid line marking the poleward boundary of the aurora from an image taken 1 min prior to the sequence of heavy curves.

Figure 2 shows that, as the surge developed, the poleward boundary of the aurora moved poleward at longitudes east of the head of the surge and equatorward at longitudes west of the head of the surge. (The head of the surge, where the poleward boundary of the aurora assumes an approximately north-south orientation, was located slightly to the west of the Sondrestrom meridian in this case.) This motion of the poleward boundary gives a significant longitudinal distortion of the boundary, which leads to an auroral bulge of the sort illustrated by the expansion phase

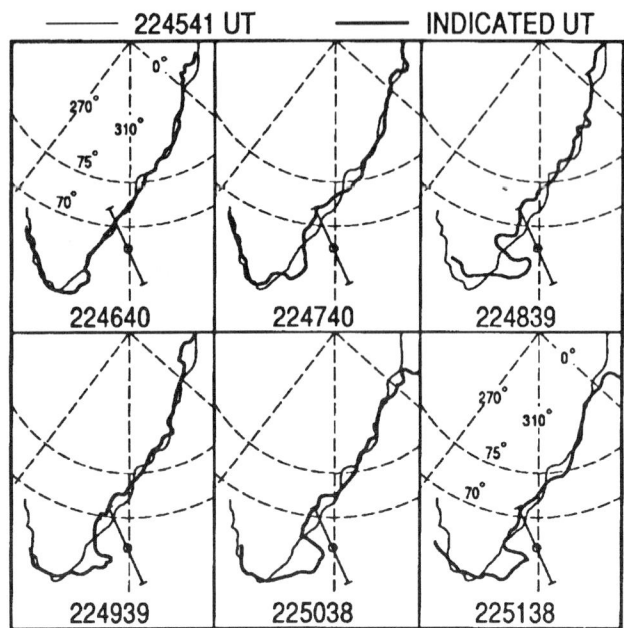

Fig. 2. Poleward boundary of identifiable aurora from a sequence of Viking images for September 24, 1986, obtained during a period of rapid development of an auroral surge near Sondrestrom. Geographic coordinates are used. The lighter curve in each panel is from an initial image for comparison with the boundary for the indicated UT. The bar identifies the field of view of the Sondrestrom radar along its magnetic meridian, and the open circle on the bar gives the radar location.

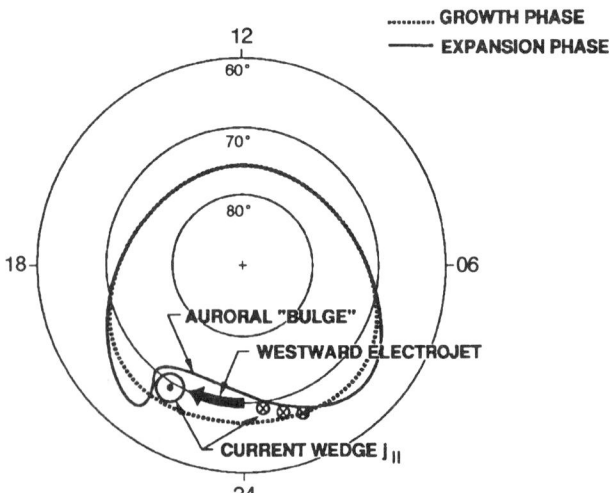

Fig. 1 Sketch of the poleward boundary of the auroral oval at times during the growth phase of a substorm (just prior to expansion phase onset) and during the expansion phase. The spatial relation of the auroral bulge to the field-aligned wedge currents $j_\parallel$ and westward electrojet is also illustrated for the expansion phase

boundary in Figure 1. This distortion usually begins in a relatively narrow longitude sector and then expands both eastward and westward.

If it is assumed that the poleward boundary of the aurora lies along, or adjacent to, the separatrix between open and closed magnetic field lines, then the development of an auroral bulge must be associated with a corresponding distortion of the separatrix. This same assumption was used by Craven and Frank [1987] and Frank and Craven [1988] to estimate variations in the area of open, polar cap field lines during substorms. It is not necessary that the poleward boundary of the aurora lie precisely at the separatrix for the inferred distortion of the separatrix to be valid, but it is necessary that the boundary lie near the separatrix and that the shape of the boundary approximate that of the separatrix.

The western portion of an auroral bulge typically contains at least one azimuthally localized surge having enhanced upward field-aligned currents and auroral intensity. The field-aligned current associated with a surge [Inhester et al., 1981; Opgenoorth et al., 1983] is believed [Rostoker and Hughes, 1979; Tighe and Rostoker, 1981] to be the dusk portion of the substorm current wedge that develops during a substorm expansion phase [McPherron, 1973; Baumjohann, 1983]. The current-wedge includes an enhanced westward electrojet in the ionosphere that connects downward field-aligned currents to the east and upward field-aligned currents to the west. The relation of the auroral bulge to the westward electrojet and the field-aligned currents of the wedge is illustrated in Figure 1 for the idealized case of a single surge near the western edge of the bulge region.

As illustrated in Figure 1, the distortion of the poleward boundary of the aurora at the edges of the auroral bulge occurs in the region of the wedge field-aligned currents. These field-aligned currents map along magnetic field lines into the magnetosphere, and their magnetic effects are readily observable in the vicinity of synchronous orbit [e.g., Nagai, 1982, 1987].

Magnetic field lines in the tail become stretched away from the Earth during a substorm growth phase. This stretching is particularly dramatic at synchronous orbit, where its signature is an increase in the magnitude of B_x and a decrease in the magnitude of B_z (GSE coordinates). At expansion phase onset, the magnetic field at synchronous altitude returns to a more dipolar orientation within the longitude range spanned by the current wedge. This reconfiguration is observed to begin in a relatively narrow longitude sector near midnight and to then expand both eastward and westward [Arnoldy and Moore, 1983], as does the auroral bulge. Particle injections are observed to accompany this "dipolarization" of the magnetic field [Moore et al., 1981; Nagai 1982].

During an expansion phase, the current wedge expands in longitude along with the magnetic field dipolarization. The occurrence of dipolarization at any particular longitude coincides with the passage of the current wedge across that longitude. Such a passage of the current wedge is identifiable by B_y perturbations at the time of dipolarization. The magnetic perturbation is northward at longitudes within the current wedge, which changes the field to an orientation that appears more nearly dipolar. Outside the current wedge, however, the magnetic field perturbation is southward. This causes an increased stretching of the field. Such stretching of the field outside the current wedge after expansion phase onset has been observed by Nagai [1982, 1987] and by Arnoldy and Moore [1983]. Gelpi et al. [1987] compared the post-onset longitudes of stretching and dipolarization at synchronous orbit to the longitudes of surges observed on the ground. They found that the stretching occurred west of the head of the surge and that the dipolarization occurred east of the head of the surge, as is expected if the upward field-aligned currents were located near the head of the surge, as illustrated in Figure 1.

III. EXTENSION TO THE TAIL: SEPARATRIX-DISTORTION HYPOTHESIS

Assuming that wedge currents extend into the tail to distances well beyond synchronous orbit, their magnetic effects ought to be observable at such distances as well as near synchronous orbit. In fact, Rostoker and Eastman [1987] and Eastman et al. [1988] have reported clear magnetic signatures, which they attribute to the current wedge, from ISEE at $x = -15 R_e$ to $-22 R_e$ in the tail. This offers a simple explanation for the magnetic field dipolarizations observed in the tail at substorm onset. Such dipolarizations would be expected to occur at longitudes within the current wedge that forms at onset. Plasma sheet thinnings, which are also observed in tail, would thus be expected at longitudes outside the current wedge. Under this scenario, the "dipolarizations" would not actually be a dipolarization of the entire magnetic field at longitudes within the current wedge unless the northward B_z perturbation reduced the total cross-tail current. However, the field would turn to a more dipolar orientation at an observation location within the plasma sheet.

The changes in the magnetic field component approximately normal to the cross-tail current sheet found by Huang et al. [1991] after onset were $\Delta B_z \sim 5 - 10$ nT. Such perturbations can be attributed to field-aligned currents $\sim 10^5$ A [Lyons, 1991] if the outward and earthward field-aligned components of the current wedge are assumed to be separated by $\sim 2.5 R_e$. The Tsyganenko [1987] model maps $\Delta y = 2.5 R_e$ across the tail at $x = -15$ to $-22 R_e$ to $\Delta y \approx 1.25 R_e$ at $x = -6.6 R_e$ and to $\Delta y \approx 750$ km in longitude in the auroral ionosphere [H. Spence, private communication, 1990]. The reduction in Δy by a factor of ~ 2 between the tail and synchronous orbit suggests that ΔB_z would be $\sim 10 - 20$ nT at synchronous orbit if the field-aligned current magnitude were the same there as in the tail. However, observations of Nagai [1987], which show that ΔB_z is $\sim 30 - 40$ nT at synchronous orbit, suggest that part of the field-aligned wedge current closes at radial distances closer to the Earth than $15 R_e$. Nevertheless, a significant portion of the wedge current appears to extend well out into the tail.

Figure 3 shows a cross section of the geomagnetic tail and includes arrows which illustrate the magnetic perturbations from the field-aligned wedge currents. As illustrated, the ΔB_z perturbations are perpendicular to the cross-tail current sheet. This ΔB_z should lead to an increase in closed magnetic flux at longitudes spanned by the current wedge and to a decrease in closed magnetic flux at longitudes outside the wedge. Thus, as the current wedge forms during the substorm expansion phase, the separatrix between open and closed magnetic field lines should develop a bulge at longitudes within the wedge, as illustrated by the expansion phase separatrix in Figure 3. Such a change in the distribution of closed flux in the tail should develop along with the formation of the current wedge.

A distortion of the separatrix in the tail, as illustrated in Figure 3, will map along field lines to the auroral ionosphere and produce a poleward protruding bulge at longitudes within the current wedge. The amplitude of the bulge can simply be estimated from flux conservation. Let us assume that the $\Delta B_z = 10$ nT across the midplane of the tail within the current wedge, and that this ΔB_z extends over a $50 R_e^2$ area ($\Delta y = 2.5 R_e$ and $\Delta x = 20 R_e$). This gives an additional closed magnetic flux $\Delta \Phi = 500$ nT R_e^2 within the current wedge. Since the vertical component of B in the auroral ionosphere is about 5×10^{-5} T, the area of the bulge there must be about $10^{-2} R_e^2$ ($\approx 4 \times 10^5$ km²). Using the mapping of $\Delta y = 2.5 R_e$ in the tail to 750 km in longitudinal extent in the ionosphere obtained from the Tsyganenko [1987] model, we obtain that the auroral bulge ought to extend about 540 km ($\sim 5°$ in latitude) poleward of the poleward boundary of the growth phase auroral oval. This is reasonable protrusion into the polar cap for an auroral bulge.

The simple calculation above ignores B_y effects and the opening of field lines at longitudes outside the current wedge (which should be responsible for the equatorward motion of the poleward boundary of the aurora outside the current wedge); however, these corrections should be factor of ~ 2 corrections.

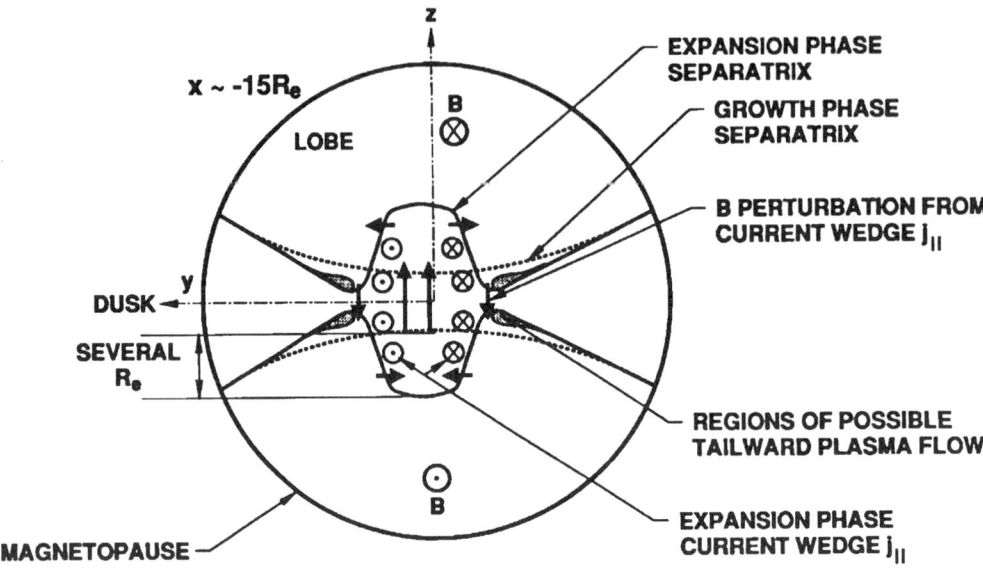

Fig. 3. Illustration of phenomena predicted to occur in the tail in association with the formation of the substorm current wedge.

Also, three-dimensional evaluations of magnetic field distortion by the current wedge have been performed by Vasil'ev et al. [1986], and they obtained ionospheric mappings of the distortion that appear realistic for surges. Thus, it appears that the magnetic perturbations produced by the wedge currents can close sufficient flux to account for the auroral bulge.

A simple calculation can also be performed to estimate the extent to which the distorted separatrix extends into the lobes of the tail. At a distance ($\sim 15\ R_e$) beyond which about one-half of the wedge-produced flux $\Delta \Phi$ closes across the midplane of the tail, the portion of the distorted separatrix that extends into each of the growth phase lobe regions must enclose ~ 250 nT R_e^2 of magnetic flux. By taking B_x to be 20 nT in the lobes and by taking $\Delta y = 2.5\ R_e$ as before, we find that the separatrix must extend $\sim 5\ R_e$ into each of the pre-existing lobes. If the region of newly closed field lines becomes filled with plasma sheet plasma, we should expect an $\sim 10\ R_e$ expansion of the total width of the plasma sheet in the z-direction at longitudes within the current wedge. We should further expect a significant thinning of the plasma sheet outside the current wedge. Both the expected expansion and the expected thinning of the plasma sheet during the substorm expansion phase are illustrated in Figure 3. The current wedge and associated distortion of the separatrix should expand in both the +y and −y directions as the expansion phase progresses.

IV. INTERPRETATION OF TAIL OBSERVATIONS IN TERMS OF SEPARATRIX DISTORTION

The distortion of the separatrix illustrated in Figure 3 represents an extension of the known longitudinal structure of the current wedge and auroral surge into the tail, and it offers a possible description of phenomena observed in the tail during the substorm expansion phase.

At longitudes spanned by the current wedge that forms at expansion phase onset, the magnetic field should appear to dipolarize and the induced electric field associated with the dipolarization should energize trapped particles. Satellites within the plasma sheet and the longitude range of the current wedge at onset should thus observe magnetic field dipolarization and concurrent dispersionless injections. Such events should be similar to those observed at synchronous orbit. The observations of Huang et al. [1991] are consistent with these expectations.

At longitudes outside the current wedge, the plasma sheet should first thin as the amount of closed flux decreases. Thinning should be observed close to the time of onset if the spacecraft longitude is sufficiently near (within a few R_e of) the current wedge. It should be observed later during the expansion phase at more distant longitudes. This latter expectation is consistent with the recently reported ISEE observation [Hones et al., 1990] that a tail plasma sheet thinning observed at 0130 MLT occurred 20 min after the onset of a substorm on May 4, 1986. Thinning should lead occasionally to a very thin plasma sheet outside the current wedge after onset. Indeed, very thin cross-tail current sheets (~ 400 km) have been reported by McPherron et al. [1987] and Mitchell et al. [1990] to occur after onset but prior to dipolarization in the near-Earth tail ($x \approx -12\ R_e$). Particle data obtained during the time of the observations of Mitchell et al. indicate that the thin current sheet had developed to the east of the expansion phase current wedge [Williams et al., 1990].

Following the thinning of the plasma sheet at any particular longitude, the plasma sheet should expand in z as the field-aligned current of the current wedge moves across that longitude. This scenario is consistent with the observation that the tail plasma sheet thins and then expands after expansion phase onset.

The present model may also account for outward flowing plasma at the edge of the rapidly thinning plasma sheet. Outside the current wedge, the ionospheric mapping of the separatrix between open and closed field lines moves equatorward in response to the decrease in B_z across the midplane of the tail. If the separatrix moves equatorward faster than does the convecting plasma, then the separatrix will overtake the convecting plasma, thereby transferring plasma from closed field lines to open field lines [Lyons et al., 1989]. Such a transfer of plasma can be visualized as "reverse reconnection." It would require that the induced electric field exceed the convection electric field at the tail magnetic X-line, so that $E \cdot J < 0$ there [Vasyliunas, 1984]. Even without this, longitudinal drift in the tail could carry energetic particles across the distorted separatrix and lead to their appearance as bursts of flowing particles on open field lines.

The plasma and/or energetic particles transferred to open field lines will be lost from the magnetosphere by flowing outward along open field lines that lie adjacent to the separatrix, as illustrated by the small shaded regions in Figure 4. Such outward plasmas flows should occur only as the plasma sheet in the tail thins and not as it expands. Flows observed at the outer boundary of the plasma sheet [Hones et al., 1973; Lui et al., 1977a, b] are consistent with this prediction. Also, the observations in Figure 2 show that rapid equatorward motion of the poleward boundary of the aurora occurred just to the west of the head of the developing surge, which was located near the Sondrestrom radar. The boundary moved equatorward by about 250 km between 2245:41 UT and 2248:39 UT, which corresponds to a speed of ~1.4 km/s, which was considerably faster than the equatorward plasma drift measured by the radar along the radar magnetic meridian at approximately the same time [Lyons et al., 1990]. We thus presume that plasma was transferred from closed to open field lines over this longitude range at that time, which would have lead to outward plasma flows in the tail.

The above discussion provides an interpretation of general tail phenomena observed during substorm expansion phases in terms of the separatrix-distortion hypothesis. It thus provides a means for re-interpreting detailed sets of data that have been obtained in the tail during specific expansion phase events and interpreted in terms of the temporal formation and movement of a near-Earth neutral line. It would be profitable to reexamine such data to see whether they can be interpreted alternatively (or better) in terms of separatrix distortion. Here, I consider two such data sets.

April 24, 1979

A substorm having an onset between 1111 UT [Eastman et al., 1988] and 1112 UT [Hones et al., 1986] on April 24, 1979, (CDAW 7) has been the subject of several papers. Figure 4, from Hones et al., shows ISEE 2 magnetic field, ion pressure, and ion flow data for a time interval that includes the substorm. Following onset, the ion pressure first went up, and then decreased below the level of detectability by about 1116 UT as the satellite went into the lobe. The satellite reentered the plasma sheet just after 1140 UT. The exit of the satellite from the plasma sheet was presumably the result of plasma sheet thinning, and the return of the plasma sheet to the satellite was presumably the result of plasma sheet expansion. Plasma flow was outward at the outer boundary of the thinning plasma sheet and earthward at the outer boundary of the expanding plasma sheet. According to the interpretation of Hones et al., a neutral line had formed earthward of the satellite at substorm onset so as to produce the plasma sheet thinning and associated outward flows. The later expansion of the plasma sheet was interpreted as the result of a tailward retreat of the neutral line.

However, the observations in Figure 4 are equally consistent with the separatrix distortion hypothesis. The satellite was at $y \approx 5 R_e$, and thus could reasonably have been west of the current wedge that formed at onset. Consistent with this expectation, B_z went negative as the plasma sheet thinned soon after onset. This is what would be expected from the opening of field lines west of the current wedge. The plasma sheet appears to have thinned rapidly and thus could have left previously trapped plasma on freshly opened field lines. This interpretation would account for the outward flows associated with the plasma sheet thinning in Figure 4. (At the end of the period of outward flow,

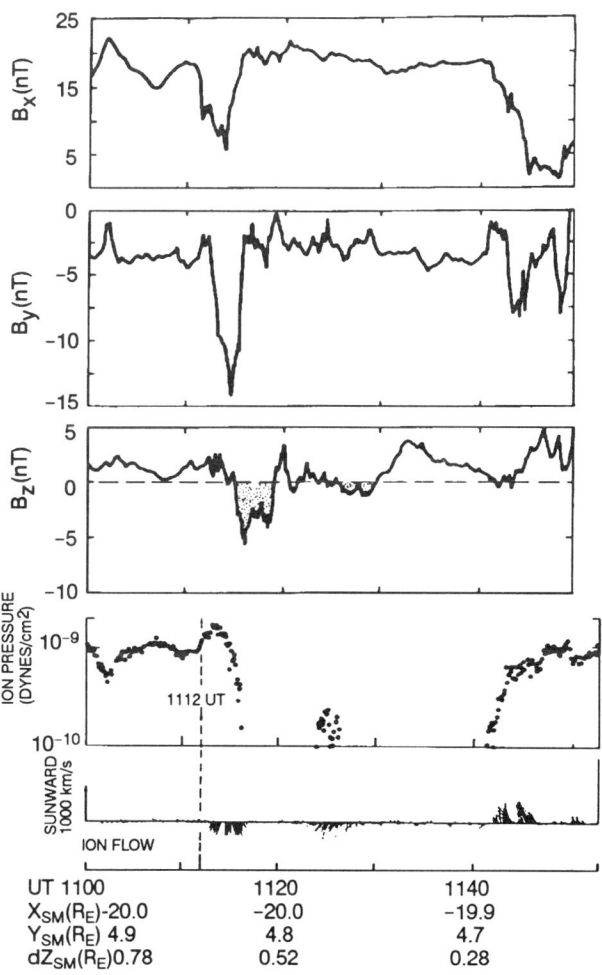

Fig. 4. Magnetic field, ion pressure, and ion flow data obtained from ISEE 2 during a substorm on April 24, 1979. Expansion phase onset (1112 UT) is indicated. The satellite position is given along the bottom of the figure, where dZ is the distance above an estimated position of the tail current sheet (from Hones et al. [1986]).

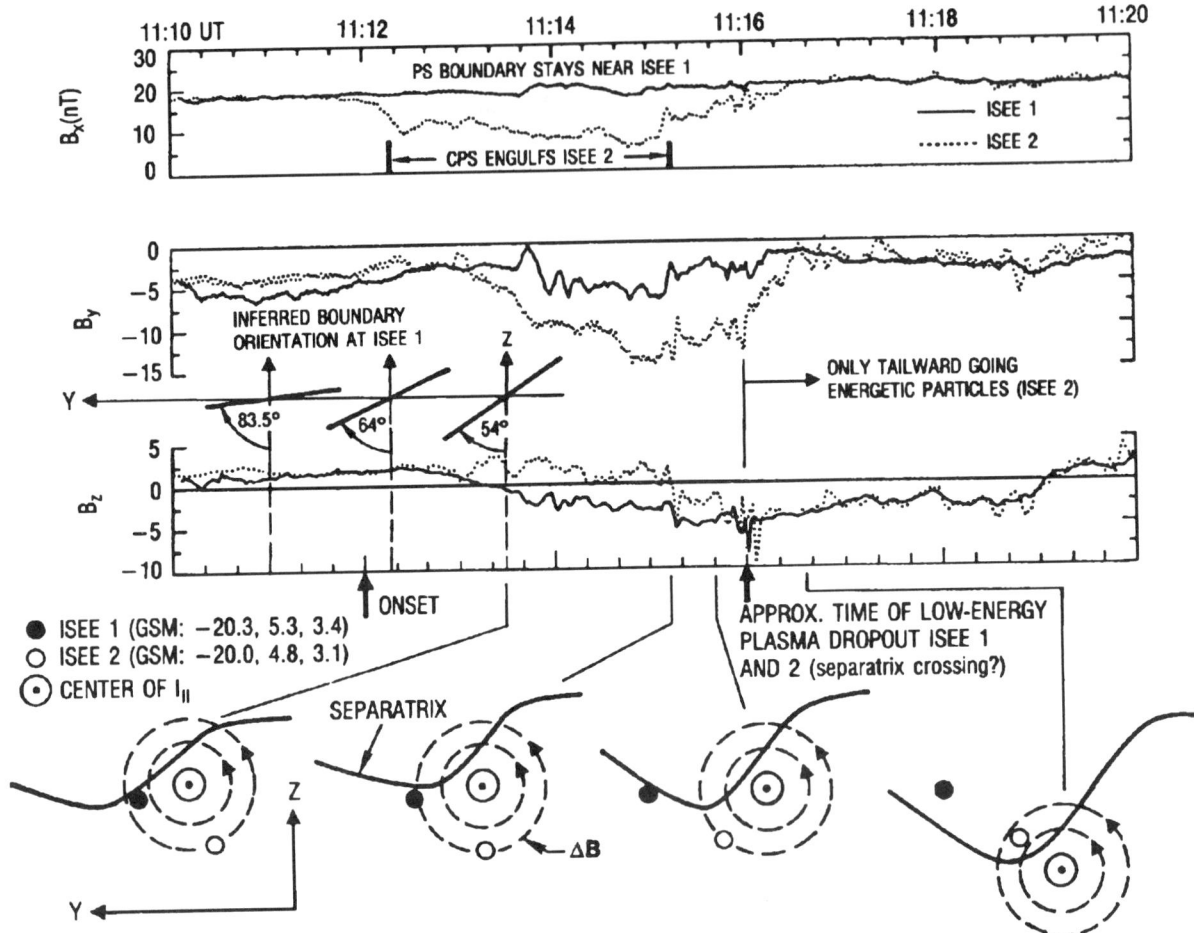

Fig. 5. Observations from a 10 min interval that includes the times of onset and plasma sheet thinning of the April 24, 1979, substorm. Simultaneous ISEE 1 and ISEE 2 magnetometer data, as well as the orientations of the outer boundary of the plasma sheet in the y,z plane, are from Kettmann et al. [1990]. Approximate time of the dropout of plasma sheet particles are from Figures 6 and 8 of Hones et al. [1986]. Both Hones et al. and Kettmann et al. found outward-going energetic particles for a time after 1116 UT on ISEE 2. Locations of the two spacecraft in the y,z-plane, relative to the center of the outward field-aligned current distribution, are based sed on Eastman et al. [1988]. Separatrix configurations and locations are based on the separatrix-distortion hypothesis. In both the boundary orientation and the spacecraft location diagrams, dusk (positive y) is to the left.

during the interval 1116:00 to 1116:30 UT, there were no observable earthward going ions [see Figure 9 of Hones et al., 1986]; this is difficult to explain solely in terms of the boundary layer dynamics model of Rostoker and Eastman [1987]). The B_y perturbation near onset suggests that the satellite was located quite near the initial location of the field-aligned currents.

Expansion of the plasma sheet in z about 25 min later can be seen to have been accompanied by a positive change in B_z and a significant B_y perturbation. This is consistent with the passage of the outward field-aligned wedge current across the longitude of the satellite at that time. Earthward flow is observed along the outer boundary of the expanding plasma sheet, as would be expected for the usual plasma sheet boundary layer [Eastman et al., 1984]. This flow most likely resulted from interactions with the distant cross-tail current sheet [Lyons and Speiser, 1982].

Figure 5 summarizes observations from a 10 min interval that includes the time of plasma sheet thinning. The simultaneous ISEE 1 (solid curve) and ISEE 2 (dashed curve) magnetic field data and the orientations of the outer boundary of the plasma sheet in the y,z plane (as inferred from ISEE 1 energetic particle observations) are from Kettmann et al. [1990]. The approximate time of the dropout of plasma sheet particles on ISEE 1 and 2 was obtained from Figures 6 and 8 of Hones et al. [1986]. Both Hones et al. and Kettmann et al. found outward-going energetic particles after 1116 UT at ISEE 2. The diagrams at the bottom of Figure 5 show the locations of the two spacecraft in the y,z plane relative to the center of the outward field-aligned current distribution. These locations are based on the analysis of Eastman et al. [1988]. The separatrix locations and configurations in the diagrams are based on the separatrix distortion hypothesis discussed here. Dusk (positive y) is to the left in both the boundary orientation and the spacecraft location diagrams.

At ISEE 1, B_x remained approximately at its lobe value throughout the entire time interval included in Figure 5. This implies that ISEE 1 remained near the outer boundary of the plasma sheet until about 1116 UT, when it entered the lobe. Consistent with this interpretation, earthward plasma flows typical of the plasma sheet boundary layer were observed on ISEE 1 during the interval 1111–1113 UT [see Figure 13 of Eastman et al., 1988]. Before the substorm onset, ISEE 2 was also near the plasma sheet boundary; however, after onset, ISEE 2 went further into the plasma sheet for a few minutes, as indicated by the increase in plasma pressure [Hones et al., 1986] and concurrent decrease in B_x.

Figure 5 shows that after onset, B_z remained positive at ISEE 2, but decreased and became negative by about 1113 UT at ISEE 1. Also By became increasingly negative at ISEE 2 but remained approximately constant at ISEE 1. These observations suggest that the outward field-aligned current at onset was centered at a longitude between the two spacecraft and at about the same z value as ISEE 1. This position is shown in the leftmost diagram at the bottom of Figure 5. The B_z perturbation associated with the wedge current is expected to have closed field lines at the longitude of ISEE 2 but to have opened field lines at the longitude of ISEE 1. Thus, the magnetic separatrix should have moved closer to ISEE 1 after onset, but should have moved farther above ISEE 2. These motions would have distorted the magnetic separatrix in the manner shown in the diagram. In particular, the separatrix should have sloped towards positive z with decreasing y in the vicinity of the outward wedge current at this time (~1112 –1115 UT). This is opposite to the slope normally expected on the dusk side of midnight. Kettmann et al. [1990] inferred the orientation of the outer boundary of the plasma sheet for times just after onset. As is shown near the middle of Figure 5, the inferred boundary orientations have the same unexpected slope as predicted by the separatrix distortion hypothesis!

A few minutes after onset (about 1115 UT), B_z also became negative at ISEE 2. Eastman et al. [1988] interpreted this to mean that the center of the outward wedge current moved towards the midnight meridian (opposite to the expected direction), as sketched in the middle two diagrams at the bottom of Figure 5. As this movement occurred, the plasma sheet would be expected to thin at the longitude of ISEE 2. Consistent with this expectation, B_x goes up and the plasma pressure goes down [Hones et al., 1986] at ISEE 2 soon after B_z becomes negative. Thinning of the plasma sheet continued after this time at least until ~1116 UT, when both spacecraft entered the lobe. At ISEE 2, tailward-going energetic particles are seen for a few minutes following 1116 UT, as if the rapid opening of field lines had left some particles on freshly opened field lines.

The above discussion suggests that the detailed observations around the time of onset of the substorm on April 24, 1979, can be understood clearly in terms of the separatrix-distortion hypothesis.

May 3, 1986

Another substorm that has been studied in detail is one with an onset at 0919 UT on May 3, 1986 (CDAW 9). Ground based magnetograms from Alaska, together with magnetic field and 6 keV electron data from ISEE 1 for the period 0900 UT to 1030 UT (from Hones et al. [1987]) are shown in Figure 6. Hones et al. suggested that a new neutral line had formed just tailward of ISEE 1 at onset, and that this neutral line had begun to retreat tailward at 1003 UT. However, the data shown in Figure 6 are also consistent with the separatrix-distortion hypothesis.

The H-component magnetogram data in Figure 6 show that electrojet activity began near the onset time at about 65° geomagnetic latitude. Activity then moved poleward, and multiple particle injections (at 0919, 1936, and 0942 UT) were seen in synchronous orbit data presented by Hones et al. [1987]. While electrojet activity at lower latitudes decreased after about 0953 UT, activity at 71° latitude continued to increase, especially after 1003 UT.

During this time interval, ISEE 1 was located 3.5 R_e west of midnight, and the 6 keV electron data indicate that the plasma sheet gradually thinned at that location from 0920 to 0930 UT. This thinning would be expected if ISEE were well to the west of the expanding substorm current wedge. The measured B_z de-

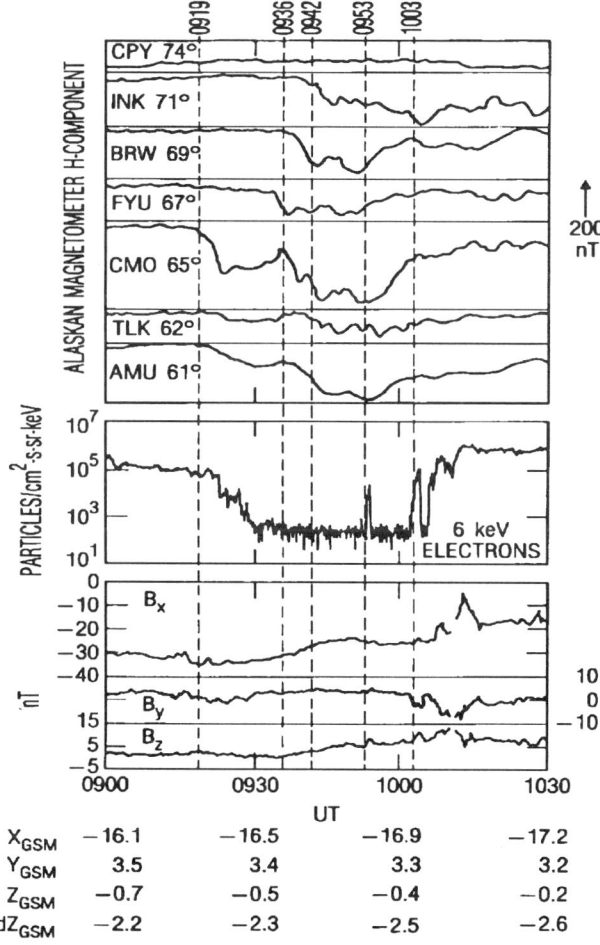

Fig. 6. Ground-based magnetogram data and ISEE 1 magnetic field and 6-keV electron data for the period of the substorm on May 3, 1986 (from Hones et al. [1987]).

creased very slightly during this period of time; however, it is difficult to determine whether this decrease is significant. Later, the plasma sheet expanded in z back over ISEE 1, initially between 1002 UT and 1004 UT, and finally during the period 1006 UT to 1012 UT. During these two time intervals, the magnetic field data at ISEE showed positive B_z and negative B_y perturbations. This is consistent with what would be expected if the current wedge had expanded irregularly across the longitude of ISEE 1 during this time period.

Viking images of the aurora are available for this substorm period, starting at 0930:19 UT (see Figure 3 of Hones et al. [1987]), and Hones et al. [1990] found a very interesting association between the auroral evolution and the expansion of the plasma sheet. They noted that an auroral surge intensified and moved westward from 1003 UT to 1009 UT. This time interval corresponds to the time of enhanced electrojet activity at 71° latitude mentioned above. The surge was located between 70° and 75° invariant latitude, and field line mapping performed by Hones et al. [1990] suggests that the surge moved westward across the field line of ISEE 1 at about the time that the plasma sheet expansion was observed. This is just what is expected if the current wedge, and thus the associated distortion of the separatrix and poleward boundary of the aurora, moved across the longitude of ISEE during this time period.

The above association between an auroral surge observed in the ionosphere and the plasma sheet expansion and B_z changes at ISEE are essentially the same as the association observed at synchronous orbit by Gelpi et al. [1987] between auroral surges and the substorm current wedge. This supports the hypothesis that substorm phenomena observed in the tail out to at least $x \approx -22 R_e$ result from the extension into the tail of the longitudinal structure associated with the current wedge that is observed at synchronous orbit and in the auroral zone.

V. CONCLUSIONS

Tail phenomena that occur out to at least $x \approx -22 R_e$ during the expansion phases of substorms, previously interpreted in terms of a near-Earth neutral line that forms at onset and later moves outward, can be understood alternatively in terms of a distortion of the magnetic separatrix by the substorm wedge current. Moreover, the auroral bulge that develops during a substorm expansion phase can be understood as the ionospheric mapping of the distorted separatrix [Lyons et al., 1990]. Since distortion of the separatrix should result from magnetic perturbations associated with the wedge currents, the distortion should expand in longitude as the current wedge expands. A significant portion of the field-aligned wedge current must extend out to at least $x \approx -22 R_e$ under this hypothesis, though part of the wedge current may well close nearer than this to the Earth.

The separatrix-distortion hypothesis accounts for the approximately equal probabilities of observing magnetic field "dipolarization" accompanied by particle injection and of observing plasma sheet thinning followed by plasma sheet expansion. Either phenomena may occur near substorm expansion phase onset within a few hours of midnight. Field dipolarization and concurrent particle injection are expected to occur at longitudes spanned by the current wedge that forms at onset. Thinnings, which should be accompanied by enhanced stretching of the magnetic field and perhaps by negative B_z, should be observed outside the current wedge. Since the current wedge expands in longitude with time, thinnings should be followed by dipolarizations and concurrent plasma sheet expansions. Thinnings may be observed near the time of onset at longitudes near the initial range spanned by the current wedge. However, both thinnings and expansions should be increasingly delayed relative to onset with increasing longitudinal displacement from the initial location of the current wedge.

Outward plasma flows can also be explained within the context of the separatrix-distortion hypothesis. Such flows would be expected to develop if closed field lines containing trapped plasma become open at a rate faster than the rate at which plasma is convected towards lower latitudes. This would allow previously trapped plasma to escape the tail along freshly open field lines. Even if this condition for outward plasma flow were not met, longitudinal drift in the tail could carry energetic particles across the distorted separatrix and lead to their appearance as bursts of particles flowing outwards along open field lines.

Separatrix distortion offers a unifying explanation for expansion phase phenomena observed in the auroral ionosphere, at synchronous orbit, and in the tail out to at least $x \approx -22 R_e$. The cause of expansion phase onset has not been considered here, but the separatrix-distortion hypothesis may well be compatible with proposed causes, such as enhanced reconnection associated with the formation of a neutral line somewhere in the tail [Coroniti, 1985], instability arising from magnetosphere-ionosphere coupling processes [Kan and Kamide, 1985; Rothwell et al., 1986], and thermal catastrophe [Goertz and Smith, 1989]. To be consistent with the separatrix-distortion hypothesis discussed here, neutral line formation at substorm onset would have to occur either outside the longitude range spanned by the current wedge or beyond $x \approx -22 R_e$.

Acknowledgments. This study has greatly benefited from numerous stimulating discussions with my colleagues D. N. Baker, O. de la Beaujardière, J. D. Craven, E. W. Hones, Jr., C. Y. Huang, A. T. Y. Lui, A. Nishida, G. Rostoker, and M. Schulz. Also, I appreciate the very helpful comments on the manuscript provided by M. Schulz. The work has been supported by NASA grant NAGW-2126 (Space Physics Theory Program), NSF grant ATM-8800602, and the Aerospace Sponsored Research Program.

REFERENCES

Akasofu, S.-I., The development of the auroral substorm, *Planet. Space Sci., 12*, 273, 1964.

Akasofu, S.-I., *Physics of Magnetospheric Substorms*, pp.7-9, D. Reidel Publ. Co., Dordrecht, Holland, 1977.

Anger, C. D., et al., Scientific results from the Viking imager: An introduction, *Geophys. Res. Lett., 14*, 383, 1987.

Arnoldy, R. L., and T. E. Moore, Longitudinal structure of substorm injections at synchronous orbit, *J. Geophys. Res., 88*, 6213, 1983.

Baumjohann, W., Ionospheric and field-aligned current systems in the auroral zone: A concise review, *Adv. Space Res., 2*, 55, 1983.

Coroniti, F. V., Explosive tail reconnection: The growth and expansion phase of magnetospheric substorms, *J. Geophys. Res., 80*, 7427, 1985.

Craven, J. D., and L. A. Frank, The temporal evolution of a small auroral substorm as viewed from high altitudes with Dynamics Explorer 1, *Geophys. Res. Lett.*, *12*, 465, 1985.

Craven, J. D., and L. A. Frank, Latitudinal motion of the aurora during substorms, *J. Geophys. Res.*, *92*, 4565, 1987.

Dandouras, J., H. Rème, A. Saint-marc, J. A. Sauvaud, G. K. Parks, K. A. Anderson, and R. P. Line, A statistical study of plasma sheet dynamics using ISEE 1 and 2 energetic particle flux data, *J. Geophys. Res.*, *91*, 6861, 1986.

Eastman, T. E., L. A. Frank, W. K. Peterson, and W. Lennartsson, The plasma sheet boundary layer, *J. Geophys. Res.*, *89*, 1553, 1984.

Eastman, T. E., G. Rostoker, L. A. Frank, C. Y. Huang, and D. G. Mitchell, Boundary layer dynamics in the description of magnetospheric substorms, *J. Geophys. Res.*, *93*, 14, 431, 1988.

Fairfield, D. H., R. P. Lepping, E. W. Hones, Jr., S. J. Bame, and J. R. Asbridge, Simultaneous measurements of magnetotail dynamics by IMP spacecraft, *J. Geophys. Res.*, *86*, 1396, 1981.

Frank, L. A., and J. D. Craven, Imaging results from Dynamics Explorer 1, *Rev. Geophys.*, *26*, 249, 1988.

Gelpi, C., H. J. Singer, and W. J. Hughes, A comparison of magnetic signatures and DMSP auroral images at substorm onset: Three case studies, *J. Geophys. Res.*, *92*, 2447, 1987.

Goertz, C. K., and R. A. Smith, Thermal catastrophe model of substorms, *J. Geophys. Res.*, *94*, 6581, 1989.

Hones, E. W., Jr., J. R. Asbridge, S. J. Bame, and I. B. Strong, Outward flow of plasma in the magnetotail following geomagnetic bays, *J. Geophys. Res.*, *72*, 5879, 1967.

Hones, E. W., Jr., J. R. Asbridge, S. J. Bame, and S. Singer, Magnetotail plasma flow measured by Vela 4A, *J. Geophys. Res.*, *78*, 5463, 1973.

Hones, E. W., Jr., T. Pytte, and H. I. West, Jr., Associations of geomagnetic activity with plasma sheet thinning and expansion: A statistical study, *J. Geophys. Res.*, *89*, 5471, 1984.

Hones, E. W., Jr., T. A. Fritz, J. Birn, J. Cooney, and S. J. Bame, Detailed observations of the plasma sheet during a substorm on April 24, 1979, *J. Geophys. Res.*, *91*, 6845, 1986.

Hones, E. W., Jr., C. D. Anger, J. Birn, J. S. Murphree, and L. L. Cogger, A study of a magnetospheric substorm recorded by the Viking auroral imager, *Geophys. Res. Lett.*, *14*, 411, 1987.

Hones, E. W., Jr., et al., A tale of two substorms (abstract), *EOS - Trans. Am. Geophys. Union*, *71*, 593, 1990.

Huang, C. Y., L. A. Frank, G. Rostoker, J. Fennell, and D. G. Mitchell, Nonadiabatic heating of the central plasma sheet at substorm onset, *J. Geophys. Res.*, 1991 (in press).

Inhester, B., W. Baumjohann, R. A. Greenwald, and E. Nielsen, Joint two-dimensional observations of ground magnetic and ionospheric electric fields associated with auroral zone currents, 3, Auroral zone currents during the passage of a westward traveling surge, *J. Geophys.*, *49*, 155, 1981.

Kan, J. R., and Y. Kamide, Electrodynamics of the westward traveling surge, *J. Geophys. Res.*, *90*, 7615, 1985.

Kettmann, G., T. A. Fritz, and E. W. Hones, Jr., CDAW 7 revisited: Further evidence for the creation of a near-Earth substorm neutral line, *J. Geophys. Res.*, *95*, 12,045, 1990.

Lui, A. T. Y., E. W. Hones, Jr., F. Yasuhara, S.-I. Akasofu, and S. J. Bame, Magnetotail plasma flow during plasma sheet expansions: Vela 5 and 6 and IMP 6 observations, *J. Geophys. Res.*, *82*, 1235, 1977a.

Lui, A. T. Y., L. A. Frank, K. L. Ackerson, C.-I. Meng, and S.-I. Akasofu, Systematic plasma flow during plasma sheet thinnings, *J. Geophys. Res.*, *87*, 4815, 1977b.

Lyons, L. R., Magnetotail processes associated with auroral surge formation, Geomagnetism and Geoelectricity, 1991 (in press).

Lyons, L. R., and T. W. Speiser, Evidence for current-sheet acceleration in the geomagnetic tail, *J. Geophys. Res.*, *87*, 2276, 1982.

Lyons, L. R., M. Schulz, and J. F. Fennell, Trapped-particle evacuation: Source of magnetotail bursts and tailward flows? *Geophys. Res. Lett.*, *16*, 353, 1989.

Lyons, L. R., O. de la Beaujardiere, G. Rostoker, S. Murphree, and E. Friis-Christensen, Analysis of substorm expansion and surge development, *J. Geophys. Res.*, *95*, 10,575, 1990.

McPherron, R. L., Satellite studies of magnetospheric substorms on August 15, 1988, *J. Geophys. Res.*, *78*, 3044, 1973.

McPherron, R. L., Magnetospheric substorms, *Rev. Geophys. Space Phys.*, *17*, 657, 1979.

McPherron, R. L., A. Nishida, and C. T. Russell, Is near-Earth current sheet thinning the cause of the auroral substorm onset? in *Quantitative Modeling of Magnetosphere-Ionosphere Coupling Processes*, edited by Y. Kamide and R. A. Wolf, p. 252, Kyoto Sangyo Univ., Kyoto, 1987.

Mitchell, D. G., D. J. Williams, C. Y. Huang, L. A. Frank, and C. T. Russell, Current carriers in the near-Earth cross-tail current sheet during substorm growth phase, *Geophys. Res. Lett.*, *17*, 583, 1990.

Moore, T. E., R. L. Arnoldy, J. Feynman, and D. A. Hardy, Propagating substorm injection fronts, *J. Geophys. Res.*, *86*, 6713, 1981.

Nagai, T., Observed Magnetic Substorm signatures at synchronous altitude, *J. Geophys. Res.*, *87*, 4405, 1982.

Nagai, T., Field-aligned currents associated with substorms in the vicinity of synchronous orbit, 2, Geos 2 and Geos 3 observations, *J. Geophys. Res.*, *92*, 2432, 1987.

Opgenoorth, H. J., R. J. Pellinen, W. Baumjohann, E. Nielsen, G. Marklund, and L. Eliasson, Three-dimensional current flow and particle precipitation in a westward travelling surge (observed during the Barium-GEOS rocket experiment), *J. Geophys. Res.*, *88*, 3138, 1983.

Rostoker, G., and T. E. Eastman, A boundary layer model for magnetospheric substorms, *J. Geophys. Res.*, *92*, 12,187, 1987.

Rostoker, G., and T. J. Hughes, A comprehensive model current system for high-latitude magnetic activity - II. The substorm component, *Geophys. J. Royal. Astron. Soc.*, *58*, 571, 1979.

Rostoker, G., A. Vallance Jones, R. L. Gattinger, C. D. Anger, and J. S. Murphree, The development of the substorm expansion phase: The "eye" of the substorm, *Geophys. Res. Lett.*, *14*, 399, 1987.

Rothwell, P. L., M. B. Silevitch, and L. P. Block, Pi2 pulsations and the westward traveling surge, *J. Geophys. Res.*, *91*, 6921, 1986.

Tighe, W. G., and G. Rostoker, Characteristics of westward traveling surges during magnetospheric substorms, *J. Geophys. Res.*, *50*, 51, 1981.

Tsyganenko, N. A., Global quantitative models of the geomagnetic field in the cislunar magnetosphere for different disturbance levels, *Planet. Space Sci.*, *35*, 1347, 1987.

Vasil'ev, E. P., M. V. Mal'kov, and V. A. Sergeyev, Three dimensional effects of a Birkeland current loop, *Geomagnet. and Aeron.*, 26, 88, 1986.

Vasyliunas, V. M., Steady state aspects of magnetic field merging, in *Magnetic Reconnection in Space and Laboratory Plasmas*, edited by E. W. Hones, Jr., p. 25, Amer. Geophys. Union, Washington, D.C., 1984.

Williams, D. J., D. G. Mitchell, C. Y. Huang, L. A. Frank, and C. T. Russell, Particle acceleration during substorm growth phase and onset, *Geophys. Res. Lett.*, 17, 587, 1990

Three-Dimensional Numerical Simulations of Magnetotail Reconnection

M. Scholer[1], A. Otto[2] and G. J. Gadbois[1]

We have performed three-dimensional MHD simulations of tail reconnection. We start from a two-dimensional equilibrium of the tail configuration, with line-tied magnetic field lines at the near-Earth boundary. Reconnection is initiated by placing a localized resistivity at some fixed position in the center of the plasma sheet. The region of finite resistivity is limited in the dawn-dusk direction. This leads to rapid reconnection which exhibits the following features. Reconnection results in fast earthward and tailward flows with large gradients in the y direction. These, in turn, lead to large $\partial B_z/\partial y$ gradients, which contribute to a tailward current. Plasma sheet plasma is drawn from dawn and dusk in the equatorial plane into the reconnection region. This inflow produces tailward of the neutral line a magnetic field B_y component, which is in the northern hemisphere positive on the dawn side and negative on the dusk side. The term $\partial B_y/\partial z$ also contributes to a tailward current. Near the neutral line the equatorial dawn-dusk cross-tail current is diverted on the earthward side first earthward, then tailward. The opposite is true on the tailward side. This diversion leads to a decrease of the total current near the neutral line. Current lines originating at higher latitudes above the neutral line are deflected toward the equatorial plane. This leads to a large current density near the neutral line, although the total current decreases.

Introduction

The ISEE-3 Geotail Mission has provided strong evidence for the existence of large scale plasmoids in the deep geomagnetic tail in the course of geomagnetic substorms. The identification of the plasmoids is based on the typical north-south excursion of the magnetic field [Hones et al., 1984] and on the observations of isotropic energetic electron distributions [Scholer et al., 1984] when the plasmoids move with velocities up to ~ 1200 km/s over the spacecraft.

Two-dimensional magnetohydrodynamic (MHD) simulations of tail reconnection have been successful in producing some of the observed features [e.g., Birn, 1980; Forbes and Priest, 1983; Lee et al., 1985]. In particular, the simulations by Birn [1980] started from a realistic initial tail configuration with a northward B_z in the plasma sheet and a flaring tail lobe, thus naturally producing loop like closed field lines within the plasma sheet due to near-Earth reconnection. In order to follow the evolution of the plasmoid a distant neutral line has to be taken into account in the initial equilibrium. First attempts to do this in a 2-D MHD simulation were performed by Hautz and Scholer [1987]. They showed that plasmoids are indeed pinched off and ejected into the downtail direction. Recently, Otto et al. [1990] have studied in detail the evolution and acceleration of plasmoids in the framework of 2-D MHD simulations with a constant resistivity througout the system. Scholer and Hautz [1991] have extended the analysis of Otto et al. [1990] for cases where the resistivity is localized at some position in the current sheet.

In two-dimensional MHD simulations the electrical current has only a y (dawn-dusk) component; thus, two-dimensional simulations can not describe field-aligned currents. Three-dimensional simulations of tail reconnection have first been performed by Birn and Hones [1981]. Starting from a three-dimensional self-consistent tail model, including flaring in y and z and plasma sheet thickening toward the flanks, Birn and Hones computed the response to a sudden global enhancement of resistivity. They obtained in the course of near-Earth reconnection region 2 currents within the plasma sheet. Sato et al. [1984] analyzed the three-dimensional aspect of driven reconnection. Local reconnection in a one-dimensional current sheet was initiated by prescribing an inflow from the upper and lower boundaries which occurred only over a limited region in y. Sato et al. obtained in their sim-

[1]Max-Planck-Institut fur extraterrestrische Physik, Garching, Germany
[2]Ruhr-Universität Bochum, Bochum, Germany

ulations the region 1 currents. They argued that the cross-tail curent is diverted through the slow shocks which emerge in the course of the reconnection process. Recently, *Birn and Hesse* [1991] and *Scholer and Otto* [1991] have taken up again the problem of region 1 and 2 current generation in MHD models of tail reconnection. However, these papers deal mainly with the currents earthward of the neutral line and are not so much concerned with the current structure in and around the departing plasmoid.

We report here three-dimensional MHD simulations of tail reconnection with emphasis on the three-dimensional magnetic field line and electrical current topology in the region tailward of the near-Earth neutral line. In particular, we start from a two-dimensional equilibrium of the tail configuration, with line-tied magnetic field lines at the near-Earth boundary. Reconnection is initiated by placing a localized resistivity at some fixed position in the center of the plasma sheet.

INITIAL CONFIGURATION AND NUMERICAL PROCEDURE

We have investigated magnetotail reconnection by means of a three-dimensional compressible and resistive MHD code. The resistive MHD equations are used in the form discussed by *Otto et al.* [1990] and allow for arbitrary resistivity as a function of x, y, and z. In the computations the actual variables are normalized to the following quantities: magnetic field to the lobe magnetic field, B_o, density by the density difference between plasma sheet center and lobe ($\rho_o = \rho_c - \rho_l$), both taken at a near-Earth boundary, $x = 0$, length by the half thickness λ of the plasma sheet at $x = 0$, velocity by the Alfvén velocity $V_{Ao} = B_o/(\mu_o\rho_o)^{1/2}$, and time by the Alfvén transit time $\tau_A = \lambda/V_{Ao}$. Here, μ_o is the vacuum permeability. The initial configuration is a two-dimensional magnetotail configuration according to *Birn et al.* [1975] with a northward (positive B_z) component throughout the plasma sheet and a flaring tail magnetic field. The B_z component decreases along the x axis with $1/(x - x_c)$, and the B_x component at $x = 0$ is equal to B_x of a Harris sheet (i.e., $B_x(0, z) = B_o \tanh(z/\lambda)$). The initial temperature is assumed to be uniform and the dimensionless lobe density is initially 0.25. The latter is equal to the plasma beta in the lobe at $x = 0$

The calculations are performed inside a rectangular box surrounded by four planes. The size of the box is $-60 \leq x \leq 0$, $-20 \leq y \leq 20$, and $0 \leq z \leq 10$. A grid of $53 \times 43 \times 33$ grid spacings in x, y, and z was used. The grid spacing is uniform in x and y. In z, i.e., perpendicular to the plasma sheet, a nonuniform spacing is used in order to increase the resolution in the current layer. In the center of the current layer ($z = 0$) the resolution is $\Delta z = 0.05$. The magnetic field is constant at the left hand boundary ($x = 0$). Here, the normal component of the velocity is set equal to zero, but a tangential flow is allowed. At $y = y_{min} = -20$ (dawn side) and $y = y_{max} = 20$ (dusk side) periodic boundary conditions have been imposed. The plasma is allowed to enter freely and exit at the right side ($x = -60$) of the numerical box. Solid wall boundary conditions are imposed at the top $z = z_{max} = 10$. Since intially $B_y = 0$, symmetry can be imposed at the plane $z = 0$. For the numerical integration of the MHD equations we employed a Leapfrog scheme, which is an explicit finite differences method of second order accuracy in space and time.

In order to initiate reconnection at a specified distance x_N we apply a localized resistivity and keep it fixed in space and time. The resistivity is given by

$$\eta = \eta_o \cosh^{-2}(z/z_o) \cosh^{-2}(y/y_0) \cosh^{-2}((x - x_N)/x_o))$$

The parameters used in the simulation are: $\eta_o = 0.01$, $x_c = -30$, $z_o = 2.0$, $y_o = 3.0$, $x_o = 3.0$, and $x_N = -12.5$. The chosen resistivity generates a reconnection region at about x_N with an extent in the dawn-dusk direction of about 6 times the plasma sheet half width.

RESULTS

The localized resistivity leads to fast reconnection which is limited in the dawn-dusk direction. At $t \approx 100$ closed field lines in the center of the tail ($y = 0$) are reconnected and lobe field lines begin to reconnect. Reconnection is weaker away from the central meridian toward dawn and dusk. Figure 1 shows velocity vectors at $t = 100$ originating from 6 planes parallel to the left hand boundary at different x values. The planes in the top part are placed at $x = -12, -32, -50$, the planes in the bottom part are placed at $x = -2, -20, -50$. One can see the tailward flow, which is fastest in the central meridian, the divergence of the tailward flow toward dawn and dusk in the more distant tail, i.e. within the departing plasmoid, the occurrence of an upward flow due to the bulging plasmoid near the right hand (open) boundary, and an inflow from dawn and dusk into the reconnection region close to the neutral line at $x \approx -12$. The inflow of plasma from the sides enhances the reconnection rate as compared to two-dimensional reconnection and has also been reported from simulations of driven magnetic reconnection by *Sato et al.* [1984].

Since the magnetic field is essenially frozen-in the reconnection flow leads to a corresponding change of the magnetic field topology. The magnetic field topology at $t = 100$ is shown in Figures 2 and 3. In both Figures are shown various types of magnetic field lines. Figure 2 shows field lines which originate from $z_o = 1$ (top) and from $z_o = 3$ (bottom), respectively, at $x = -50$. Field lines originating from $z = 1$ at $x = -50$ are field lines at the boundary between lobe and plasma sheet. The field line in the meridian plane coming from the distant tail reverses at $x \approx -40$. Since the magnetic field is essentially frozen into the plasma it can be inferred that near the central meridian lobe plasma fills the region be-

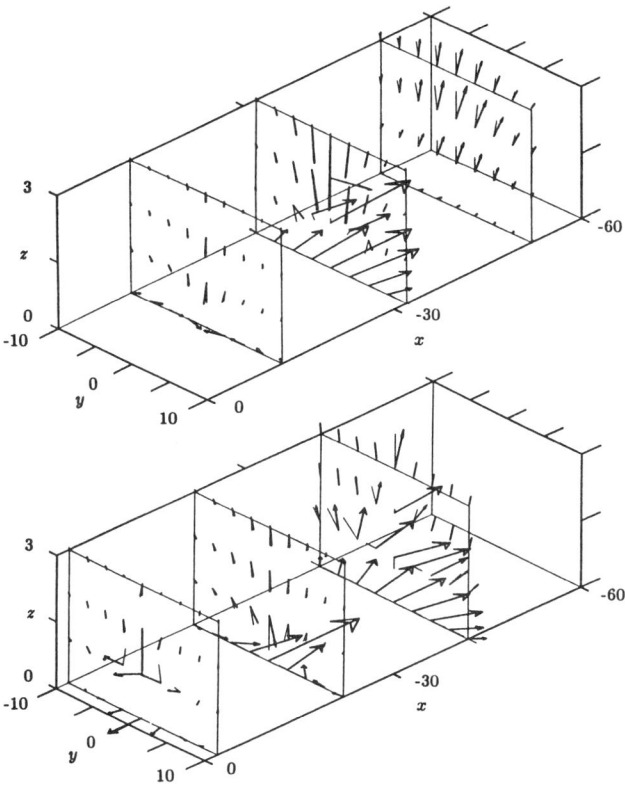

Fig. 1. Flow vectors originating from various planes parallel to the $x = 0$ plane at $t = 100$.

tween the neutral line and $x \approx -40$, i.e., dense plasma sheet plasma can only be found at distances $x \leq -40$. To the dawn and dusk side of the central meridian plasma sheet plasma can still be found closer to the neutral line. This is due to the faster reconnection rate in the center of the tail as compared to the dusk and dawn side, respectively. The plasmoid is free to move tailward when all closed field lines are reconnected and lobe field lines start to reconnect. Since this starts first in the center of the tail and proceeds from there to the sides at a finite speed, plasmoid disconnection occurs at different times for different longitudes. The decrease of the reconnection rate toward dawn and dusk also leads to large magnetic shear. Figure 3 shows in the top part magnetic field lines which come close to the equatorial plane at $x \sim -15$. It can be seen that these field lines diverge around the departing plasmoid like a magnetic bottle. This effect is mainly due to the fact that field lines are drawn into the reconnection region from the sides. There exists a large gradient of the z magnetic field component in the y direction. The x component of the electric current is given by $j_x = \partial B_z/\partial y - \partial B_y/\partial z$. The y dependence of the reconnection rate is via the second term responsible for a x component of the current, which is mostly field aligned. Thus, there exists tailward of the reconnection line a region 1 like field aligned current. The bottom part of Figure 3 shows magnetic field lines which start at $z = 2$ at the left hand boundary. This shows the dipolarization in a limited region across the tail. The strong dependence of the B_z magnetic field component on y at the boundaries of the dipolarization region contributes here to region 1 field aligned currents [Birn and Hesse, 1991; Scholer and Otto, 1991].

We now discuss in some detail the currents which correspond to the magnetic field topology at $t = 100$. We do this by presenting electric current lines which have been computed from the electrical current density, $\mathbf{j}$, in the same manner as the magnetic field lines from the magnetic field vector $\mathbf{B}$. Figure 4 shows in the top part current lines originating from the dawn side ($y = -10$) at $z = 0.3$. This shows the change of the dawn-to-dusk cross tail current system in the course of reconnection. Earthward of the neutral line the current comes from dawn, is first diverted in the earthward direction, and then returns tailward toward the dusk side. A corresponding diversion occurs on the tailward side of the plasmoid. In the equatorial plane this current diversion is, of course, perpendicular to the magnetic field. The diversion of cross tail current lines

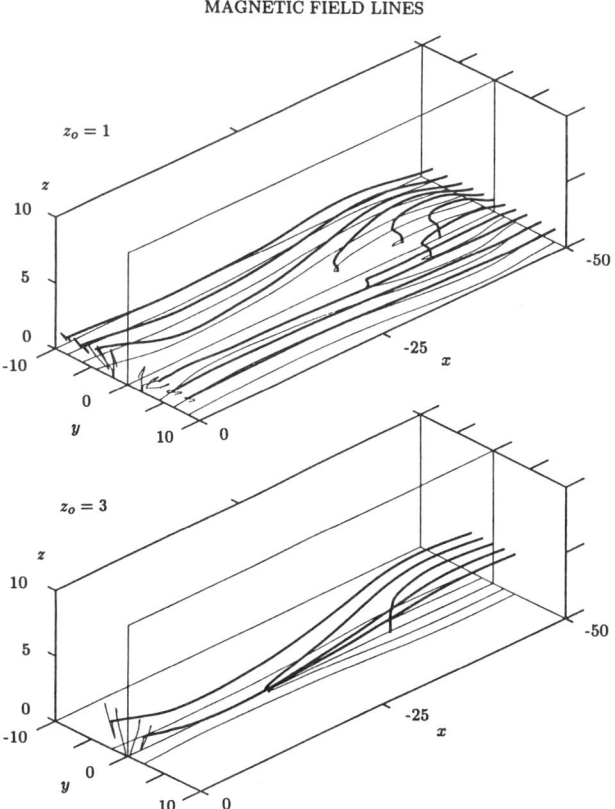

Fig. 2. Magnetic field lines originating from $z_o = 1$ (top) and from $z_o = 3$ (bottom), respectively, at $x = -50$.

MAGNETIC FIELD LINES

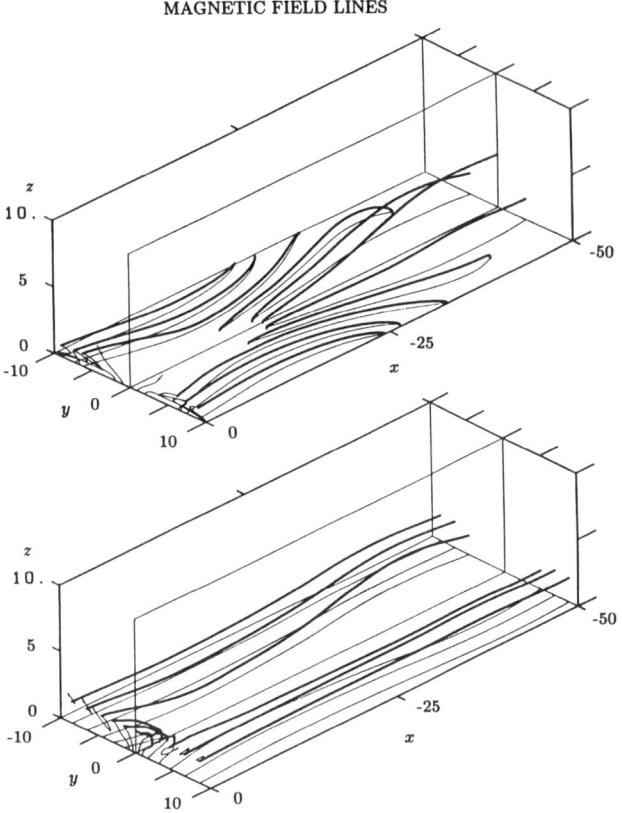

Fig. 3. Magnetic field lines passing through $z = 0.2$ at $x = -15$ (top) and magnetic field lines originating from $z = 2$ at the left hand boundary, $x = 0$.

orignating at higher latitudes leads to a component of the current parallel to the magnetic field, and thus to field aligned currents. The usual schematic showing the diversion of the whole cross tail current along field lines into the ionosphere is misleading: the cross tail current is merely diverted around the reconnection region, which leads to current components parallel to the magnetic field. Tailward of the neutral line the cross tail current is diverted over the departing plasmoid. In Figure 4 are also shown the projections of the current lines onto the $y = -10$ and the $z = 0$ plane. The projections onto the $y = -10$ plane should reveal any diversion of the current in the x direction. There exists a strong diversion in the region close to the neutral line, which becomes weaker in the more distant tail. No current component in the x direction is seen in the current layer around the central part of the plasmoid.

The bottom part of Figure 4 shows current lines which come close to the equatorial plane ($z = 0.2$) at $y = 0$. Near the neutral line the current lines originating from a wide range in latitude converge toward the equatorial plane. This leads to a local current density maximum, although the earthward and tailward diversion of the bulk of the cross tail current causes a reduction of the total current [see *Birn and Hesse*, 1991]. There are closed current loops within the plasmoid. Their direction is opposite to that of the cross-tail current, i.e., the current flow is dusk to dawn at low latitudes and returns from dawn to dusk at high latitudes. This reverse current is responsible for the dent-like magnetic field lines on the earthward side of the plasmoid (Figure 2, top).

Figure 5 shows current lines, which orignate from $y = -10$ at very small z values, i.e. at $z = 0.05$ in the top part and at $z = 0.2$ in the bottom part. These current lines are representative for the bulk of the cross-tail current. In addition to the current diversion near the neutral line Figure 5 exhibits an interesting feature: tailward of the neutral line part of the cross-tail current which comes from dawn is diverted on the dawn side in a loop-like fashion into the distant tail. From there it returns again on the dusk side to the near neutral line region (not shown in Figure 5). This current flows tailward at latitudes below the magnetic current layer generated by the diversion of the cross-tail current around the plasmoid: the top part of Figure 5 shows one current line which diverts around

ELECTRIC CURRENT LINES

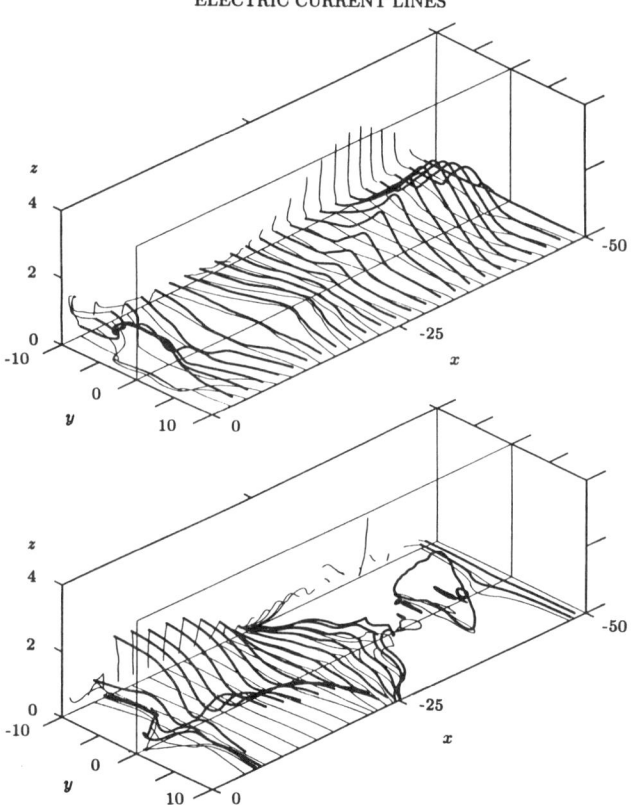

Fig. 4. Electric current lines originating from the dawn side ($y = -10$) at $z = 0.3$ (top) and electric current lines which come close to the equatorial plane at $y = 0$, i.e., at the central meridian (bottom).

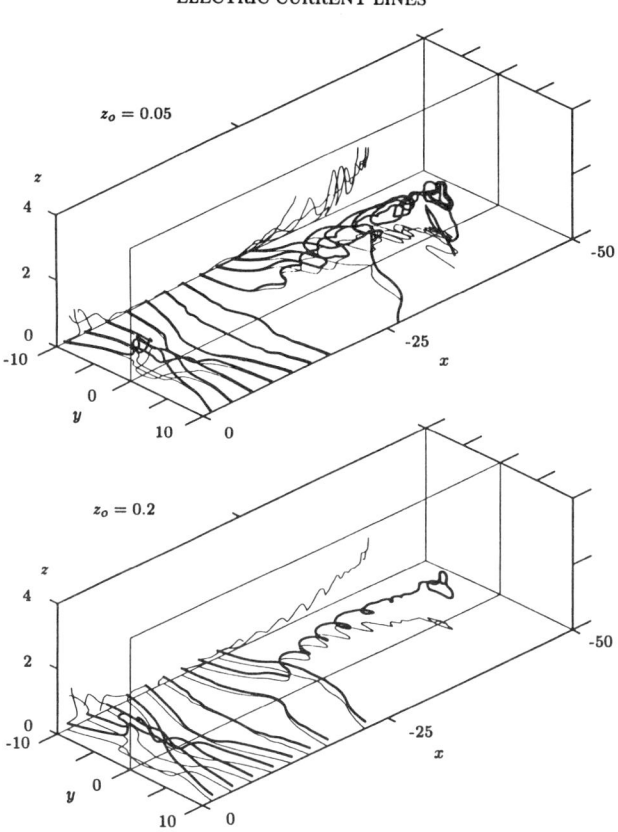

Fig. 5. Electric current lines orignating from $y = -10$ at very small z values, i.e. at $z = 0.05$ (top) and at $z = 0.2$ (bottom).

the plasmoid (the most tailward current line starting on the dawn side). From the projection of this current line onto the $z - x$ plane it can be seen that the tailward (on the dusk side earthward) current loops flow below the the current lines which constitute the current layer around the plasmoid. The tailward current loop on the dawn side produces a B_x magnetic field component which is tailward in the center of the loop and earthward at larger z values outside of the loop. Thus, the current loop leads to a tailward magnetic field in the northern part of the tail near the neutral sheet and to an increase of the earthward component at larger heights. The tailward magnetic field corresponds to the dent-like field lines at the dawn and dusk side earthward of the plasmoid center in Figure 1 (upper part). As can be seen from Figure 1, lower part, the field lines in the meridian plane earthward of the plasmoid do not show the dent-like feature. Between the the dawn side loop and the corresponding dusk side loop the B_x magnetic field component will be enhanced. This leads to the enhancement of the earthward directed tail field seen in the center in Figure 1 (lower part) and is ultimately due to the inflow of plasma into the reconnection region from the dawn and dusk side.

Summary

Placing a finite resistivity limited in the dawn-dusk (y) direction somewhere in a two-dimensional current sheet leads to rapid reconnection with the following features.

1. Reconnection results in fast earthward and tailward flows with large gradients in the y direction. These, in turn, lead to large $\partial B_z/\partial y$ gradients, which contribute to a tailward current.

2. Plasma sheet plasma is sucked from dawn and dusk in the equatorial plane into the reconnection region. This inflow increases the reconnection rate. Since the magnetic field is convected with the flow, this inflow produces a magnetic field B_y component, which is positive on the dawn side and negative on the dusk side in the northern hemisphere. Since B_y decreases with decreasing z, the term $\partial B_y/\partial z$ also contributes to a tailward current.

3. Near the neutral line the equatorial cross-tail current is diverted on the earthward side first earthward, then tailward. The opposite is true on the tailward side. This leads to a decrease of the total current near the neutral line. Current lines originating at higher latitudes all come close to the neutral line. This leads to a large current density near the neutral line, although the total current is smaller.

4. In the distant tail the cross-tail current is diverted around the departing plasmoid. There are closed current lops within the plasmoid which are responsible for the dent-like magnetic field structure. Part of the cross-tail current immediately tailward of the neutral line flows tailward on the dawn side into the plasmoid and returns earthward on the dusk side. These current loops are responsible for the dent-like magnetic field closer to the dawn and dusk side earthward of the plasmoid and the enhancement of the field near the meridian plane caused by the inflow from dawn and dusk into the neutral line.

References

Birn, J., Computer studies of the dynamic evolution of the geomagnetic tail, *J. Geophys. Res.*, *85*, 1214, 1980.

Birn, J., and E.W. Hones, Jr., Three-dimensional computer modeling of dynamic reconnection in the geomagnetic tail, *J. Geophys. Res.*, *86*, 6802, 1981.

Birn, J., and M. Hesse, The substorm current wedge and field-aligned currents in MHD simulations of magnetotail reconnection, *J. Geophys. Res. 96*, 1611, 1991.

Birn, J., R. R. Sommer, and K. Schindler, Open and closed magnetospheric tail configurations and their stability, *Astrophys. Space Sci.*, *35*, 389, 1975.

Forbes, T. G., and E. R. Priest, On reconnection and plasmoids in the geomagnetic tail, *J. Geophys. Res.*, *88*, 863, 1983.

Hautz, R., and M. Scholer, Numerical simulations on the structure of plasmoids in the deep tail, *Geophys. Res. Lett.*, *14*, 969, 1987.

Hones, E. W., Jr., et al., Structure of the magnetotail at 220 R_E and its response to geomagnetic activity, *Geophys. Res. Lett.*, *11*, 5, 1984.

Lee, L. C., Z. F. Fu, and S.-I. Akasofu, A simulation study of forced reconnection processes and magnetospheric storms and substorms, *Geophys. Res. Lett.*, *90*, 10,896, 1985.

Otto, A., K. Schindler, and J. Birn, Quantitative study of the nonlinear formation of plasmoids in the Earth's magnetotail, *J. Geophys. Res.*, *95*, 15,0230, 1990.

Sato, T., R. J. Walker, and M. Ashour-Abdalla, Driven magnetic reconnection in three dimensions: Energy conversion and field-aligned current generation, *J. Geophys. Res.*, *89*, 9761, 1984.

Scholer, M., G. Gloeckler, B. Klecker, F. M. Ipavich, D. Hovestadt, and E. J. Smith, Fast moving plasma structures in the distant magnetotail, *J. Geophys. Res.*, *89*, 6717, 1984.

Scholer, M., and R. Hautz, On acceleration of plasmoids in magnetohydrodynamic simulations of magnetotail reconnection, *J. Geophys. Res.*, *96*, 3581, 1991.

Scholer, M., and A. Otto, Magnetotail reconnection: Current diversion and field-aligned currents, *Geophys. Res. Lett.* *18*, 733, 1991.

SUBSTORM FEATURES IN MHD SIMULATIONS OF MAGNETOTAIL DYNAMICS

Joachim Birn and Michael Hesse

Los Alamos National Laboratory, Los Alamos, New Mexico

Abstract. We present a review and extended analysis of characteristic results from our three-dimensional resistive MHD simulations of an unstable magnetotail evolution, which develops without the necessity of external driving or prescribed localization of nonideal effects (although in reality these effects may be important). These modes involve magnetic reconnection at a near-Earth site in the tail, consistent with the near-Earth neutral line model of substorms. The evolution tailward of the reconnection site is characterized by plasmoid formation and ejection into the far tail, plasma sheet thinning between the near-Earth neutral line (X line) and the departing plasmoid, and fast tailward flow. This tailward flow, however, occupies large sections of the plasma sheet only at larger distances from the X line, while it occurs spatially and temporally much more limited close to the X line. The region earthward of the X line is characterized by a dipolarization, propagating from midnight toward the flank regions and, perhaps, tailward. It is associated with the signatures of the substorm current wedge: reduction and diversion of part of the cross-tail current from a region surrounding the reconnection site and an increase of region 1 type field-aligned currents. A mapping of these currents to the Earth on the basis of an empirical magnetic field model shows good agreement of the mapped current system with the observed region 1 field-aligned current system and its substorm associated changes, including also a nightward and equatorward shift of the peaks of the field-aligned current density. The evolution of the mappings of the boundaries of the closed field line region bears strong resemblance to the formation and expansion of the auroral bulge. The consistency of all of these details with observed substorm features strongly supports the idea that substorm evolution in the tail is that of a large scale nonideal instability. This does not exclude, however, that external forces are responsible for driving the tail into the unstable regime.

1. Introduction

The dynamics of the magnetosphere is strongly influenced by an interplay between microscopic and macroscopic effects. At present no model exists that permits the treatment of both scales simultaneously for realistic two- or three-dimensional configurations. The large-scale dynamics is usually treated by a fluid approach, that is, magnetohydrodynamics (MHD). This approach includes microscopic aspects only in the form of transport coefficients, which are usually assumed in an ad hoc fashion, and restricting assumptions on the pressure tensor and the energy equation. The pressure is usually assumed to be isotropic and the energy equation is typically considered only in the limits of negligible heat flux or constant temperature, equivalent to large heat flux. It can usually be argued [e.g., Vasyliunas, 1975] that for typical magnetospheric parameters plasma behavior should be governed by ideal MHD, except in rather localized regions where microscopic, nonideal processes become important. It thus seems plausible that macroscopic structures and their evolution can be simulated by an MHD model, even if microscopic mechanisms that provide the nonideal effects are not modeled correctly. Obviously, this need not be the case for evolution time scales and stability transitions that are not associated with ideal MHD.

The present paper is an attempt to assess the dynamic evolution of the magnetotail on the basis of numerical simulations using a three-dimensional resistive MHD code. Our intention is not only to demonstrate that these simulations reproduce many features observed in association with magnetospheric substorms or inferred from such observations but also to identify possible limitations and, perhaps, inconsistencies between the MHD results and observations. The results to be presented largely represent an extended analysis of earlier simulations.

Our approach can be understood as a nonlinear stability analysis of the magnetotail configuration within nonideal (here, resistive) MHD. We start out from a realistic (ideal MHD) equilibrium configuration, which is found to be stable in the absence of resistivity. If resistivity is switched on, however, an unstable mode evolves. Imposing an initial perturbation is generally not necessary, because the initial diffusion generates a sufficient perturbation.

The emphasis is on an evolution that is not externally driven, although, more realistically, external forces may be responsible for changing the state of the magnetotail and for "driving" it towards instability. Even in that case, however, it is important to identify the properties of the unstable modes in order to distinguish them from effects that are directly related to the driving forces. Our studies most closely represent a model in which an evolution, satisfying ideal MHD, has reached a critical point where a nonideal process, modeled by an anomalous resistivity, suddenly sets in.

Since deviations from ideal MHD are found to be significant only in a limited spatial region, we believe that the characteristics of the resistive modes studied in this paper will be quite similar to those of collisionless modes, provided these are unstable, too. This view is supported by the similarity of resistive and collisionless tearing modes of a plane current sheet, to which the modes found in our magnetotail simulations still bear resemblence. Since we do not know exactly where nonideal effects may become important, we choose uniform resistivity, although, again, this may not be the most realistic assumption.

2. Numerical Procedure, Initial and Boundary Conditions

Our computer code solves the nonlinear time-dependent MHD equations represented in the following form:

$$\frac{\partial \rho}{\partial t} = -\nabla \cdot (\rho \mathbf{v}) \quad (1)$$

$$\frac{\partial (\rho \mathbf{v})}{\partial t} = -\nabla \left(p + \frac{B^2}{2} \right) - \nabla \cdot (\rho \mathbf{v}\mathbf{v} - \mathbf{B}\mathbf{B}) \quad (2)$$

$$\frac{\partial \mathbf{B}}{\partial t} = \nabla \times (\mathbf{v} \times \mathbf{B}) + \eta \nabla^2 \mathbf{B} \quad (3)$$

$$p = \rho/2 \quad (4)$$

Constant resistivity and constant temperature are assumed in (3) and (4), respectively. All quantities are normalized on the basis of three scaling units, which can be chosen arbitrarily. These are the scale length L_z for variations with z, representing the current sheet half-width L_z, the lobe field strength, and the density difference between the center of the plasma sheet and the lobe, all chosen at $y = 0$ and the near-Earth boundary $x = 0$. A coordinate system is used which is essentially equivalent to the GSM system (disregarding the effects of dipole tilt and aberration of the tail axis from the Earth-sun line) with x pointing earthward along the tail axis, y pointing duskward, and z pointing northward. The dimensionless resistivity η in (3) is the inverse of the Lundtquist number (magnetic Reynoldsnumber) based on the scale L_z and a characteristic Alfvén speed v_A. For simplicity, and because earlier simulations did not show any strong influence of the form of the energy equation on the overall dynamic evolution, an isothermal law (4) is used throughout. The simulations thus do not provide information about possible heating associated with the dynamic evolution.

The equations are solved by using an explicit leapfrog scheme (for more details, see Birn [1980] and Birn and Hones [1981]). The computations are done on a cartesian grid inside a rectangular box $0 \geq x \geq -x_e$ with typical values for x_e of 60 or 120, and $0 \leq y \leq 10$, $0 \leq z \leq 10$. The typical number of grid points is $35 \times 13 \times 19$ in the x, y, and z direction. Uniform grids in x and y, but a nonuniform grid in z are used, in order to obtain higher resolution in the plasma sheet and, in particular, in the singular tearing layer expected to be imbedded in the plasma sheet near $z = 0$. The grid spacing in z thus increases from about 0.06 at $z = 0$ to about 1.5 in the lobes. The time step is 0.025 (in units of a characteristic Alfvén time L_z/v_A), sufficiently smaller than required by numerical stability of our explicit code. For simplicity, we concentrate on configurations with mirror symmetry with respect to both the $y = 0$ (midnight meridian) and $z = 0$ (equatorial) planes. Studies of configurations with asymmetries associated with a net cross-tail magnetic field component B_{yN} [Birn and Hesse, 1990, 1991a; Hesse and Birn, 1990] have demonstrated that for typical cross-tail field strengths of a few percent of the lobe field strength, the dynamic evolution of the fields is not significantly altered when $B_{yN} \neq 0$, although the magnetic topology becomes much more complicated. Restriction to the symmetric case allows us to describe the magnetic evolution in familiar terms, including "neutral lines, separatrices," etc., which are not necessarily well defined in the general case $B_{yN} \neq 0$.

Our box size corresponds to twice that size in R_E, if an equilibrium scale length $L_z = 2R_E$ is assumed, which may be characteristic for a slightly compressed average tail. Using a typical Alfvén speed v_A of 1000 km/sec, our time unit corresponds to about 12 sec. Note that a more local application of the results to the near tail at about 10-15 R_E distance is also possible, where a much more significant compression of the plasma/current sheet is found before the onset of a substorm [e.g., McPherron et al., 1987; Mitchell et al.; 1990]. Using $L_z = 1000$ km and the same Alfvén speed as above, the time unit is only 1 sec.

Symmetry boundary conditions are used at $y = 0$ and $z = 0$. At all other boundaries, except at the distant one, $x = -x_e$, we assume $\mathbf{v} = 0$ and hold the normal magnetic field component fixed, representing solid reflecting walls with an effective field line tying, while vanishing normal derivatives are imposed on the tangential magnetic field components, density and pressure. The distant boundary $x = -x_e$ is assumed to be open with vanishing normal derivatives of density, velocity, and pressure, and a convective condition $d/dt = 0$ for the tangential magnetic field components in the case of outward flow, while the normal magnetic field component and the tangential magnetic field components for inward flow are held fixed. Our near-Earth boundary condition is basically equivalent to an ideally conducting massive ionosphere (if the weak diffusion due to the presence of uniform resistivity is neglected). In addition to this one, we have also used a more realistic boundary condition which includes the convection at the boundary associated with electric fields that result from the closure of field-aligned currents through a resistive ionosphere and map back into the tail [Hesse and Birn, 1990b].

The numerical simulations start out from initial configurations which are derived from an asymptotic tail equilibrium theory [Birn, 1987]. Depending upon the particular focus of an investigation, slightly different initial equilibria are used. The present paper concentrates on results from two simulations previously discussed by Hesse and Birn [1991a] (run A) and Birn and Hesse [1991b] (run B). The initial configuration of run A consists of a long tail ($x_e = 120$) including a far neutral line; it was designed to study the evolution of the plasmoid during its formation, propagation, and severance from Earth. The configuration of run B is shorter ($x_e = 60$) but more realistic in the near-Earth region, containing quiet-time region 1 type [Iijima and Potemra, 1976] field-aligned currents near the plasma sheet/lobe boundary. This configuration was designed for a more realistic study of the near tail region, where the substorm current wedge forms, with one version also including the above-mentioned effects of electric fields associated with the closure of field-aligned currents through a resistive ionosphere [Hesse and Birn, 1991b].

In all cases the unstable evolution is initiated by imposing finite resistivity (corresponding to a Lundtquist number of 200), modeling the possible consequence of a stability transition that releases the ideal MHD constraint. Comparative runs without resistivity have been performed over times similar to the resistive runs. These runs did not show the growth of instabilities nor the rise of a significant wave level, which proves both ideal MHD stability, at least on the time scales of several hundred Alfvén times considered, and sufficient accuracy of the initial equilibrium solutions.

3. Plasmoid Formation and Evolution

The main feature of the most widely used substorm model, as described, e.g., by Hones [1977], is the merging of magnetic field lines in the near-Earth plasma sheet, initially forming closed loops from reconnection of closed plasma sheet field lines (the "plasmoid") and later, when reconnection involves lobe field lines, producing open field lines of IMF type, which envelope the now severed plasmoid. The study of this topological evolution requires the presence of a well defined boundary between plasma sheet and lobes, a separatrix associated with a neutral line in the more distant tail. Our studies of the plasmoid formation and severance process thus start out from an initial state that includes such a distant neutral line, which in the quasi-static tail theory is

associated with a cessation of tail flaring and the lobe magnetic field strength approaching a minimum [e.g., Wiechen and Schindler, 1988; Birn, 1989; Schindler et al., 1989]. The results presented below are based on a simulation (run A) that is described by Hesse and Birn [1991a], however, with emphasis on the asymmetric case $B_{yN} \neq 0$.

ISEE 3 observations in the distant magnetotail [e.g., Siscoe et al., 1984; Tsurutani et al., 1984; Slavin et al., 1985] show that B_z on average becomes very small beyond about 100 R_E distance, thus indicating the transition to a nearly one-dimensional neutral sheet configuration with a "Y type" neutral line at its inner edge [Wiechen and Schindler, 1988]. Accordingly, we have modeled the tail region beyond the distant neutral line initially as a current sheet with very weak southward B_z. We find, however, [see also Hesse and Birn, 1991a] that rather weak perturbations during the initial evolution lead to a slight increase of B_z, making it positive but, perhaps, even more realistic, so that the neutral line moves tailward and approaches the distant simulation box boundary. The magnitude of B_z, however, stays small, so that the distant region retains the character of a nearly plane one-dimensional current sheet, and the location of the plasma sheet/lobe interface closer to the Earth is not significantly changed.

The dynamic evolution of the magnetic field and its topology is demonstrated by Figures 1 and 2. Figure 1 shows, for several different times indicated in the panels, magnetic field lines in the midnight meridian plane (upper part of each panel) and characteristic boundaries in the equatorial plane (lower part). The solid lines in the lower parts of the four panels represent the neutral lines, consisting of X points earthwards and O points tailwards. The dashed lines represent intersections of the separatrix field lines, originating from X points, with the equatorial plane, initially marking the tailward end of the plasmoid region, indicated by shading, and later, when the plasmoid becomes severed, its earthward edge, which is then defined by the separatrix originating from the distant neutral line. The dotted lines indicate the location where B_z becomes equal to 0.1% of the typical lobe field strength, which we use as an indication of the earthward edge of the distant "neutral sheet" region. The dotted lines in the near-Earth region which appear in the last two panels represent additional intersections of the field lines originating from the lines $B_z = 0.001$ (in units of the typical lobe field strength) in the far tail with the equatorial plane, which may alternatively be considered as the earthward plasmoid edge of its severed portion.

The upper parts of the panels in Figure 1 clearly show the picture of plasmoid formation and tailward propagation, visualized in the near-Earth reconnection model of substorms [Hones, 1977] and found in a number of two-dimensional [e.g., Birn, 1980; Forbes and Priest, 1983; Min et al., 1985; Lee et al., 1985; Hau and Wolf, 1987; Hautz and Scholer, 1987; Otto et al., 1990] and three-dimensional [e.g., Birn and Hones, 1981; Brecht et al., 1982; Birn, 1984; Walker and Ogino, 1989] resistive MHD simulations. Lobe reconnection

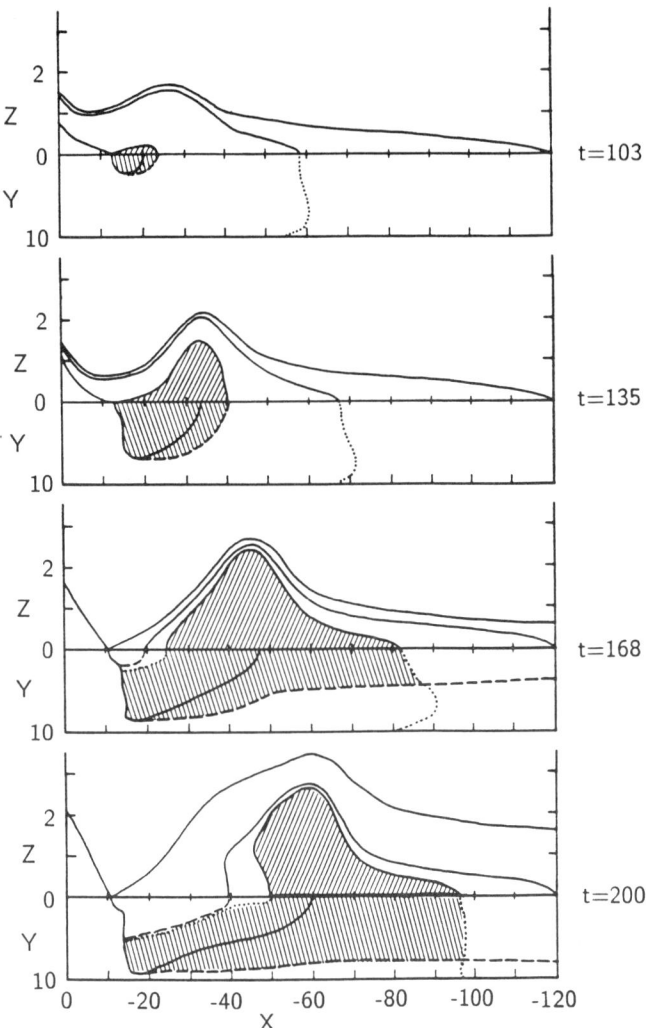

Fig. 1. Magnetic field evolution, run A. The upper part of each panel shows magnetic field lines in the midnight meridian plane and the lower part characteristic boundaries in the equatorial plane with solid lines representing magnetic neutral lines, dashed lines indicating intersections with separatrix field lines, connected with the X type neutral lines, and the dotted lines representing locations where $B_z = 0.001$ in the far tail or, closer to the Earth, where field lines connected with those locations intersect the equatorial plane again. The plasmoid region is indicated by shading.

Figure 2 further emphasizes the cross-tail variations of the plasmoid formation and severance process at a fixed location $x = -26.3$, tailward of the near-Earth X line. It shows the separatrix intersections at that location associated with the near-Earth X line (solid lines) and the distant X line (dashed lines). Initially, as shown in the first panel, the solid line marks the extent of the closed loop plasmoid region (indicated by dark shading) and the dashed line the plasma sheet/lobe interface. Penetration of the two separatrices indicates the partial severance of the plasmoid, associated with the formation of open, IMF type, field lines (lightly shaded region). In the last panel of Figure 2 the central part of the plasmoid has disappeared at $x = -26.3$, leaving only IMF type field lines behind, while the outer parts still show plasmoid regions surrounded at higher latitudes by not yet reconnected plasma sheet field lines. The lateral expansion of the plasmoid boundary, visible in the first two panels of Figure 2, and the form of this boundary at its lateral edges are consistent with a detailed ISEE 1 and 2 data analysis of an April 24, 1979 substorm by Kettmann et al. [1990], which shows the formation of an energetic particle layer inside the plasma sheet with an apparent lateral edge similar to the plasmoid boundary shown in Figure 2.

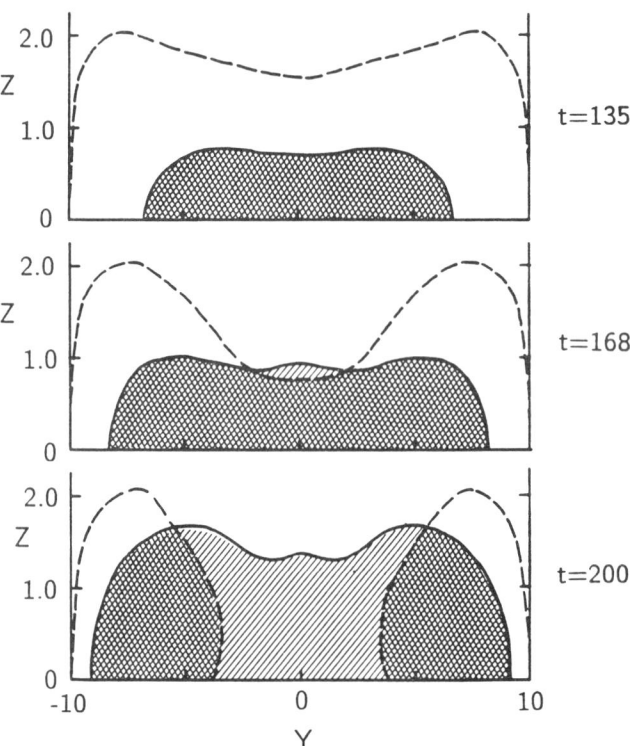

Fig. 2. Separatrix intersections with the plane $x = -26.3$, run A. The solid line connects to the near-Earth X line and the dashed lines to the distant X line. Heavy shading indicates the plasmoid and light shading open, IMF type, field line regions.

and plasmoid severance start between $t = 135$ and $t = 168$. The lower parts of the panels show the y dependence of the process. Both plasmoid formation and severance start out from midnight, where B_z initially is smallest, and propagate outward toward the flanks. This effect has been demonstrated recently through a coordinated data study by Hones et al. [1988]. Note that the plasmoid severance process is not completed yet in the last panel at $t = 200$.

Figure 2 also shows, in its central part, the thinning of the plasma sheet or, rather, of the closed field line region, which is one of the most commonly observed substorm features beyond about 15-20 R_E distance. The associated decrease in density is similar to that demonstrated by Birn and Hones [1981], with isodensity contours resembling the dashed line in the middle panel of Figure 2. We should note, however, that the density drop at a specific location is not quite as drastic as in reality, due to a relatively high lobe density of 25% of the plasma sheet density, which is imposed for numerical reasons (to limit the magnitude of the Alfvén speed). The thinning eventually leads to the complete disappearance of closed field lines in this region, while the properties of the outer regions do not change that drastically.

4. Dipolarization and Substorm Current Wedge

The dominant and most repeatable substorm feature in the near tail region inside of about 15-20 R_E is a "dipolarization" of the magnetic field, following a stretching in the substorm growth phase. This dipolarization, manifested mainly by an increase of B_z, is generally thought to be caused by, or associated with, a diversion of cross-tail current toward the Earth on the dawn side and away on the dusk side of the tail ("region 1" type currents [Iijima and Potemra, 1976]), closing in the ionosphere through the westward electrojet. This current circuit is called the "substorm current wedge" [McPherron et al., 1973].

A simulation that shows the main features of the substorm current wedge in association with the tail instability (run B) has been presented recently [Birn and Hesse, 1991b], starting out from an initial configuration which already includes quiet-time region 1 type field-aligned currents near the plasma sheet boundaries. This simulation shows plasmoid formation and ejection similar to the one discussed in the previous section. Here, we demonstrate the main results associated with the substorm current wedge, which are found again as properties of the unstable mode without necessity of an external driving.

Figures 3 and 4 demonstrate the reduction of the cross-tail current as a function of x and y, respectively. They show that this reduction is found to be localized around the near-Earth neutral line region located at $x \approx -10$, extending about 10-20 R_E in x and 3-5 R_E in y, concentrated near midnight. The reduction of the cross-tail current is found to be associated with its diversion and an increase of field-aligned currents, extending toward the near-Earth boundary. The evolution of these field-aligned currents at $x = 0$ is shown in Figure 5 by contour lines of constant $j_\parallel$, representing multiples of 0.0125 (one unit corresponding to about $3 \times 10^{-9} A/m^2$). Regions of enhanced $j_\parallel$ ($|j_\parallel| > 0.0125$) are indicated by hatching; single hatching corresponds to tailward currents and cross-hatching to earthward currents. The panels clearly show the initial region 1 type current system and, starting at about $t = 140$, the formation and increase of an additional region 1 system at somewhat lower latitudes and closer to midnight. The current system in the lobes with opposite sense to the region 1 system is associated with the increase of B_z toward the flanks in the equilibrium configuration and with the fact that the pressure, and consequently the current density, do not completely vanish in the lobes. They may be an artifact and are of no consequence for the dynamic evolution. The increase of the total region 1 type field-aligned current at $x = 0$ flowing into or away from each hemisphere is shown in Figure 6. One dimensionless

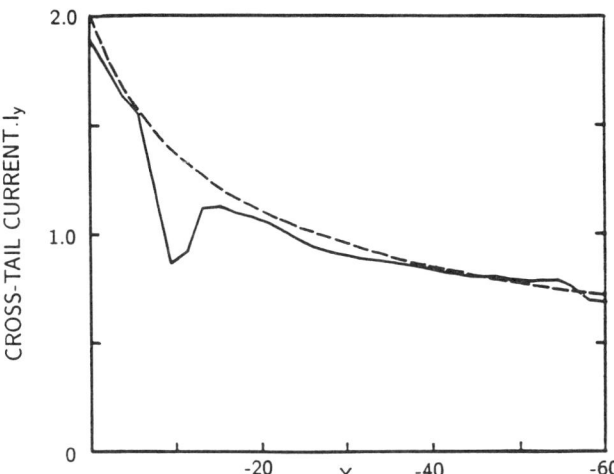

Fig. 3. Cross-tail current, integrated in z, as a function of x at $y = 0$ for $t = 0$ (dashed line) and $t = 180$ (solid line), run B. One dimensionless unit corresponds to about 0.03 A/m.

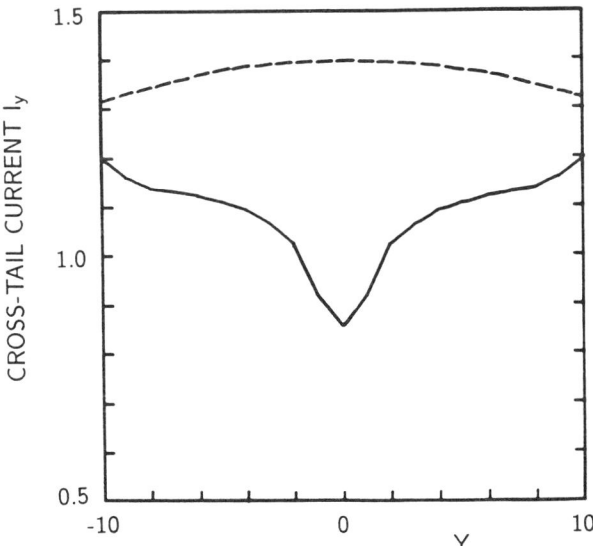

Fig. 4. Cross-tail current, integrated in z, as a function of y at $x = -9.4$, close to the location of the near-Earth X line for run B. The solid line belongs to $t = 170$ and the dashed line to $t = 0$. The units are the same as in Figure 3.

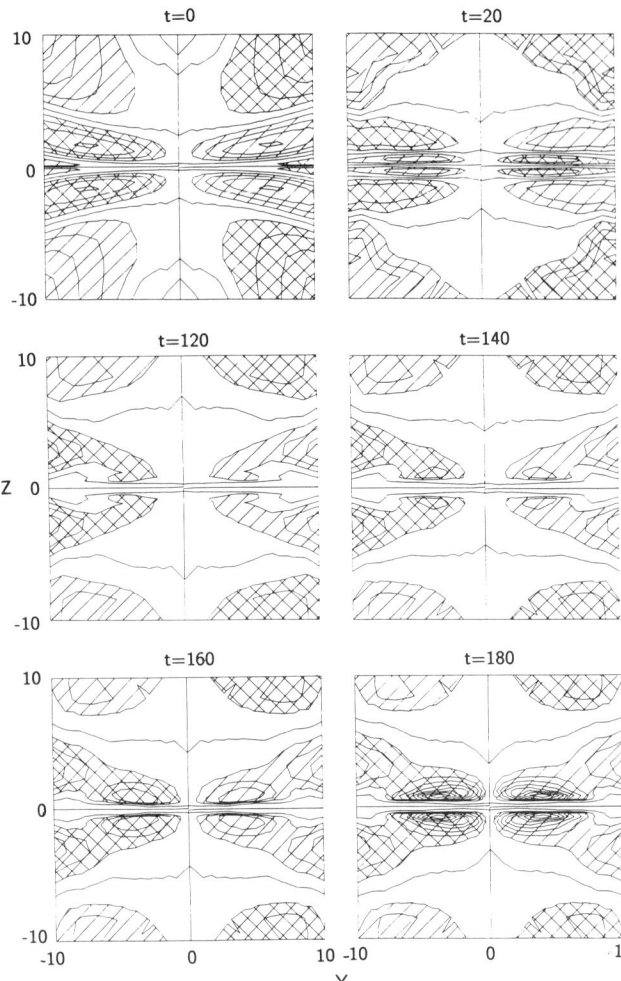

Fig. 5. Contours of constant field-aligned current density $j_\|$ at $x = 0$ for run B at different times as indicated. The increment of $j_\|$ between adjacent contours is 0.0125, one unit corresponding to about 3×10^{-9} A/m^2. Single hatching represents tailward currents and cross-hatching earthward currents exceeding 0.0125 in magnitude.

unit corresponds to about 0.4×10^6 A. The increase starts at about the time of the neutral line formation.

The spatial and temporal variations associated with the dipolarization are demonstrated by Figure 7, showing the cross-tail variation of B_z in the equatorial plane at various times and locations in x as indicated. The two uppermost panels belong to locations earthward of the near-Earth X line and the two bottom panels to locations tailward of it. The two top panels clearly demonstrate the increase of B_z, associated with the dipolarization (Note the compressed scale of the top panel). The top panel also shows how the dipolarization propagates outward longitudinally, consistent with observations [e.g., Lopez et al., 1988a; Lopez and Lui, 1990].

While the lateral expansion of the increase in B_z in the $\pm y$ directions is very pronounced, except close to the X line, the propagation along the x direction is less obvious. This is demonstrated by Figure 8, which shows B_z as a function of time for three different locations on the x axis earthward of the X line. Whereas the temporal ordering between the increases of B_z at $x = -5.6$ and $x = -7.5$ is not clear, the increase of B_z at $x = -9.4$ is obviously delayed. This apparent tailward propagation may in fact also be consistent with observations [Lopez et al., 1988b].

The results of run B, discussed in this section, are obtained with a near-Earth boundary condition $\mathbf{v} = 0$, which implies an effective tying of field lines at a nearly ideal conducting boundary. A more realistic boundary condition must take electric fields into account which result from the closure of field-aligned currents through a resistive ionosphere. A simulation with such a boundary condition but otherwise identical initial and boundary conditions has been performed recently [Hesse and Birn, 1991b]. It is found that the effects of ionospheric resistance on the dynamic tail evolution are small and that the features described above are essentially unchanged, even quantitatively. We will therefore not discuss these results in further detail here.

5. Near-Earth Signatures

Our simulations do not contain the inner region of the magnetosphere surrounding the Earth. However, since we find that tail instabilities develop without being strongly influenced by the near-Earth boundary conditions considered, we might simply use the results of the tail instabilities as input for an inner magnetospheric model to map their effects to the Earth. Let us first demonstrate further results found at the near-Earth boundary.

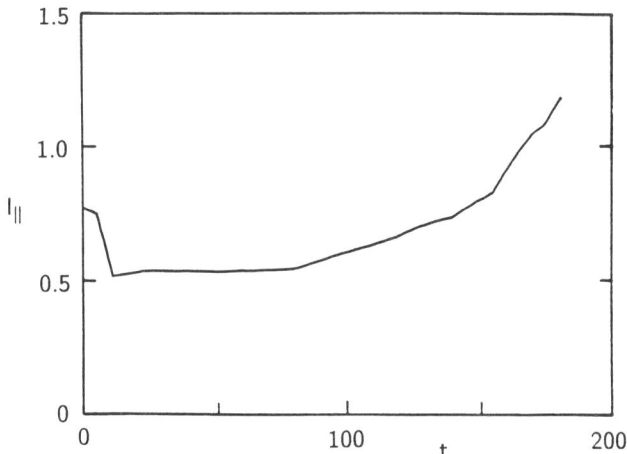

Fig. 6. Temporal evolution of the total region 1 type field-aligned current at $x = 0$ for run B. One dimensionless unit corresponds to about 4×10^5 A.

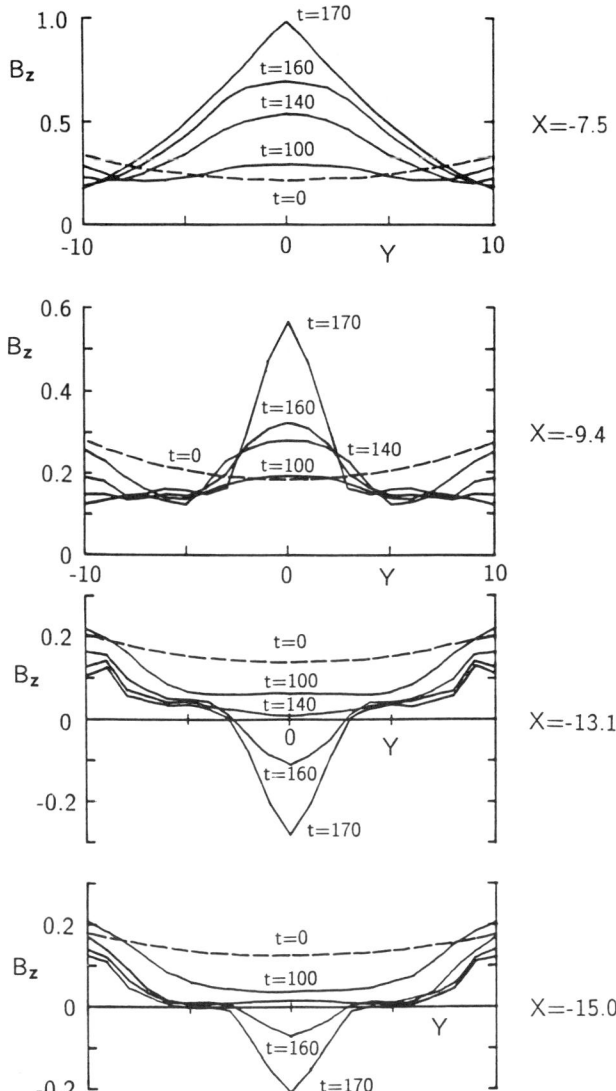

Fig. 7. Cross-tail variation of B_z at $z = 0$ for run B at various times and locations in x as indicated.

Figure 9 shows the signatures of the separatrices of run A at the boundary $x = 0$, representative also for other locations earthward of the near-Earth X line. The solid lines represent the location of the magnetic flux surface connected to the near-Earth X line and the broken lines connect to the distant X line. Different times are indicated in the figure; the short-dashed line belongs to $t = 200$, the dash-dotted lines to $t = 168$, and the long-dashed line to $t = 103$, being close to the one for $t = 135$, which is not shown. The figure demonstrates how field lines connected to the near-Earth reconnection region appear inside of the plasma sheet boundary, expand and propagate poleward, subsequently taking over as the boundary of the closed field line region, so that this boundary also moves to higher latitudes while expand-

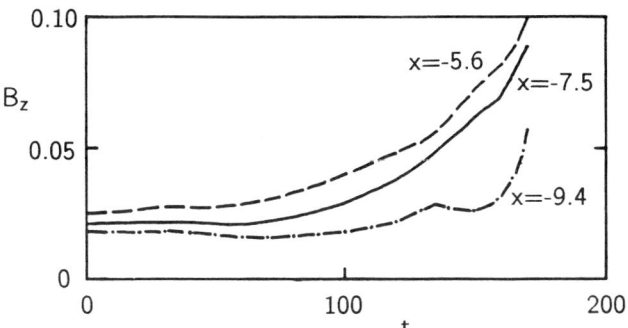

Fig. 8. Temporal evolution of B_z for run B at $y = 0$, $z = 0$ and three locations in x earthward of the near-Earth X line.

ing longitudinally. The corresponding pattern for run B is shown in Figure 10 in the same way as in Figure 9. Since run B did not include a distant X line, we used the boundaries of the simulation box to identify as open all field lines that cross this boundary. The short dashed line in Figure 10 thus connects to the distant boundary $x = -60$, without significant change for different times, and the long-dashed lines to the boundaries $|y| = 10$. These flank boundaries affect

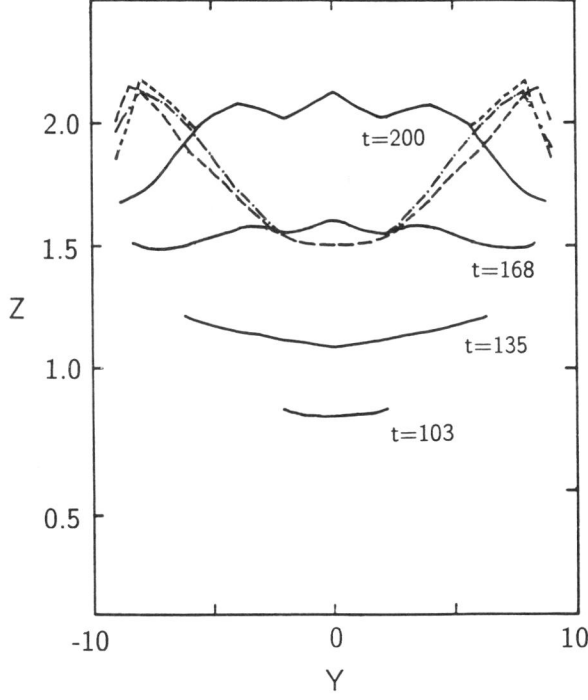

Fig. 9. Separatrix intersections with the near-Earth boundary $x = 0$, run A, for different times as indicated. Solid lines connect to the near-Earth X line and broken lines to the distant X line with the short-dashed lines belonging to $t = 200$, the dash-dotted lines to $t = 168$, and the long-dashed line to $t = 103$. The separatrix belonging to the distant X line for $t = 135$, which is close to the one for $t = 103$, is not shown.

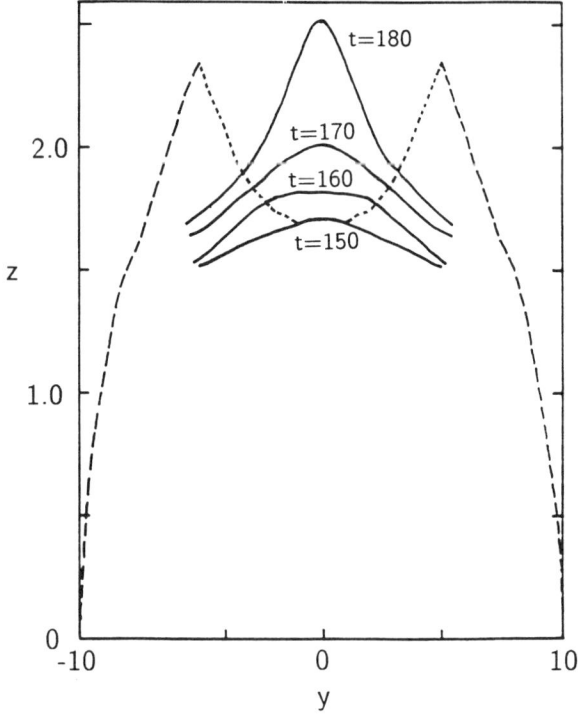

Fig. 10. Same as Figure 9 but for run B. The dashed lines, connecting to the boundaries of the simulation box in the x, y plane, are almost identical for all times shown.

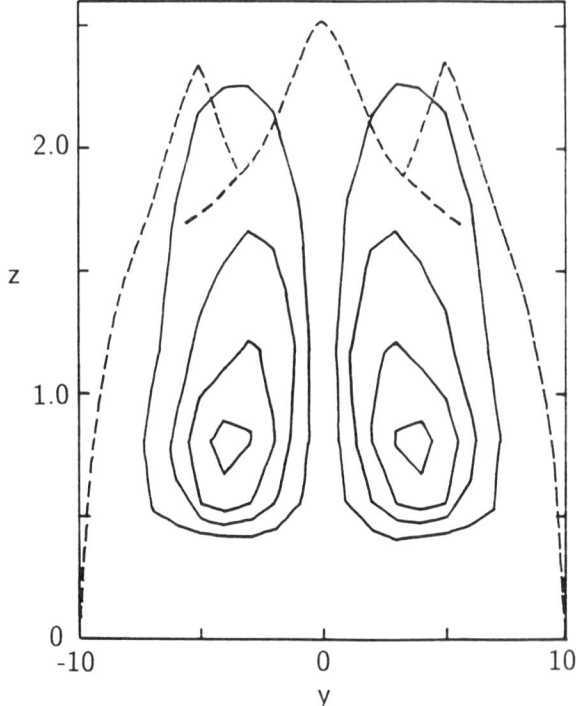

Fig. 11. Contours of constant current density at $x = 0$ for run B (solid lines) and the boundary of the closed field line region, as in Figure 10, for $t = 180$ (dashed line). Contours are shown for $j_\parallel = \pm 0.025, \pm 0.05, \pm 0.075,$ and ± 0.10.

more field lines than in run A, because of a larger flaring in y of the near-tail field, associated with the presence of the quiet-time field-aligned currents.

The distribution of the field-aligned currents in relation to the characteristic boundaries is demonstrated in Figure 11 for run B at $t = 180$, corresponding to the last panel of Figure 5. The solid lines show contours of $j_\parallel$ for $|j_\parallel| \geq 0.025$ with increments in $j_\parallel$ of 0.025, corresponding to earthward currents for $y < 0$ and tailward currents for $y > 0$. The dashed lines represent the separatrix surfaces for $t = 180$ as shown in Figure 10. The enhanced field-aligned currents peak well inside the plasma sheet boundary, while they extend close to that boundary. The fact that they partially extend beyond the closed field line region has again to do with the fact that density, pressure, and electric currents do not vanish abruptly in the open lobe region.

The evolution of the open/closed boundaries in Figures 9 and 10 bears a strong resemblance to the formation and expansion of the auroral bulge [Akasofu, 1977]. To get a somewhat more realistic impression of how these boundaries might look at the surface of the Earth, we have used an empirical magnetic field model for the inner magnetosphere, the Tsyganenko [1987] (long) model for the largest K_p values, to map the patterns of Figures 10 and 11 to the Earth along magnetic field lines. We emphasize that such a mapping cannot be expected to give any kind of absolute accuracy.

The Tsyganenko model represents an average field model, not necessarily self-consistent, without any of the dynamic effects described in the previous sections and no possibility for a precise matching with our tail model. We will therefore simply assume that the patterns as represented in Figures 8–10 are present at some interface between the tail and the inner magnetosphere and that the inner magnetospheric field, represented by the Tsyganenko model, is not greatly disturbed by the currents mapped along its field lines from the tail. For the interface we choose the location $X = -10R_E$ (in GSM coordinates), which is expected to be earthward of the reconnection site but at a location where the Tsyganenko field is already sufficiently tail-like to allow for some rough matching (or, rather, patching) to our tail fields. A scale length of 2 R_E gave reasonable consistency of the two tail widths in y, but a different scale length of 3 R_E in the z direction had to be used to match the two current sheet widths.

Figures 12 and 13b represent the mappings of Figures 10 and 11, respectively, with solid and dashed contours corresponding to those in Figures 9 and 10. In addition, the dash-dotted lines show a mapping of a circle of $5R_E$ radius around the Earth in the equatorial plane. By varying the different patching parameters, we confirmed that the characteristic patterns of Figures 12 and 13 are representative

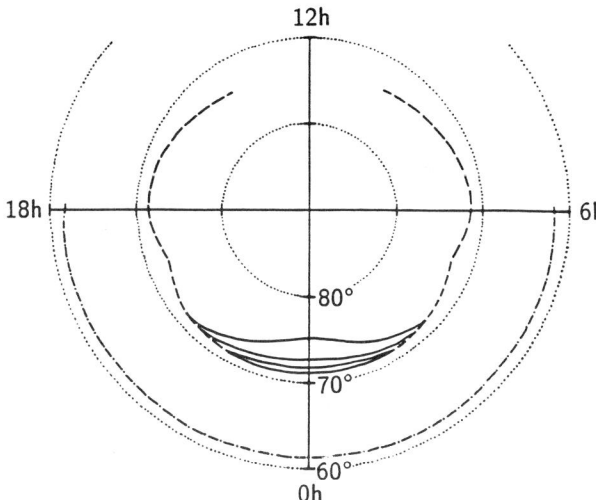

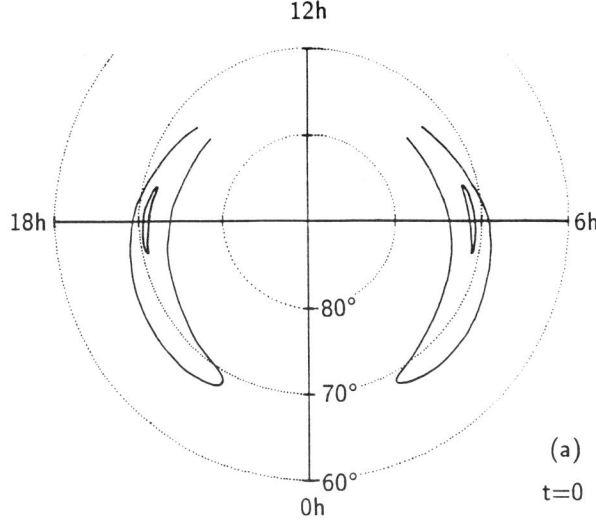

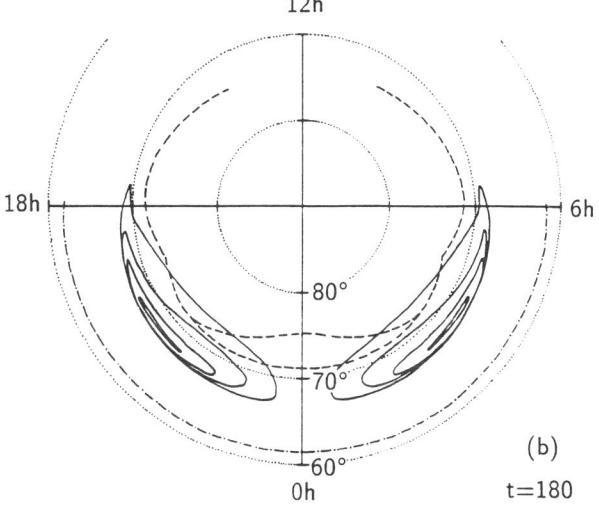

Fig. 12. Mapping of the separatrix surfaces of Figure 10 to the north hemisphere of the Earth's surface, using the long Tsyganenko [1987] model for $K_p \geq 5^-$. The solid lines represent the footpoints of field lines connected with the near-Earth X line for $t = 150, 160, 170,$ and 180 increasing in latitude. The dashed line represents the boundary of field lines closing inside the simulation box; the part sunward of the kinks in this contour connects to the flank boundaries. The dash-dotted line represents a mapping of a circle of 5 R_E radius around the Earth in the equatorial plane.

as far as relative locations and the relative spatial extent of patterns are concerned. The absolute locations, however, are less certain; they were found to vary by as much as about 4° in latitude and several hours in local time, increasing in uncertainty away from midnight. Figure 12 shows that the northward expansion of the boundary of the closed field line region at the near-Earth boundary in our simulation corresponds to a poleward expansion at the Earth of about 4°, roughly consistent with observations [e.g., Akasofu, 1977]. The time scale of this expansion is about 6–9 min, again quite consistent with observations. We should note, however, that the time scale in our simulations is strongly influenced by the arbitrarily chosen resistivity value η (representing a Lundtquist number of 200), so that this consistency seems more the result of an adequate choice of η. The boundaries in Figure 12 do not show any signs of a surge. This is not surprising in view of the limited resolution of our simulation in the y direction, corresponding to about 1 hour local time, and the fact that the twisting of flux tubes, associated with field-aligned currents, is not included in the Tsyganenko field model.

In addition to the mapping of the field-aligned current contours for $t = 180$ in Figure 13b, Figure 13a shows the mapping of the contours of the initial distribution of $j_\parallel$, corresponding to the first panel of Figure 5. The levels of $j_\parallel$ correspond to the same values ± 0.025 and ± 0.05 as the

Fig. 13. Mapping of the contours of constant $J_\parallel$ of Figure 11 to the Earth's surface, using the long Tsyganenko [1987] model with $K_p \geq 5^-$, (a) for $t = 0$, (b) for $t = 180$. The solid contours correspond to current density values $|j_\parallel| \geq 0.025$ with increments of $\Delta j_\parallel = 0.025$ at the near-Earth boundary of our simulation box, increasing in magnitude from the lowest value towards the innermost contour. The dashed lines represent the boundaries of the closed field line region, shown in Figure 12, for $t = 150$ (lower latitude) and $t = 180$ (higher latitude portion). The dash-dotted line maps to a circle in the equatorial plane of 5 R_E radius around the Earth.

two outermost contours of the currents in Figure 13b. The comparison between Figures 13a and 13b not only shows the increase in the peak current density (corresponding to about $0.7 \mu A/m^2$) but also a latitudinal widening and a nightward and equatorward shift of the peak, all signatures being consistent with observations [Iijima and Potemra, 1978]. The reduction of dayside field-aligned currents, which is not observed, is not surprising in view of the fact that these current regions here contain only contributions from the flank regions of the tail. Our model includes these currents as initial conditions but not the presumed driving mechanism, the interaction with the solar wind, which seems necessary to maintain or even increase these currents. The distribution of the field-aligned currents in Figure 13 thus largely reflects the observed distribution of nightside region 1 currents [e.g., Iijima and Potemra, 1976]. Similar currents are found also in global magnetospheric simulations [e.g., Walker et al., 1987; Walker and Ogino, 1989]. Note, however, that the field-aligned currents presented by these authors are integrated along field lines, because of boundary conditions which suppress the currents near the Earth. It is also not quite obvious to what extent these currents were due to the external driving by the solar wind or to an internal instability. Some indications for the latter are reported, such as a filamentary and time-varying character.

6. Plasma Flow

One of the main observations that has originally motivated the development of the near-Earth reconnection model is the onset of fast tailward flow in association with plasma sheet thinning at substorm onset [e.g., Hones et al., 1976; Hones, 1977, 1979] in the tail region beyond about 15-20 R_E, documented in a large number of individual substorm events [e.g., Hones and Schindler, 1979]. More recently, however, critique of the near-Earth reconnection model also is based largely on plasma flow properties or, rather, on the apparent absence of expected flow properties. The two major critique points are (1) the possibility of misinterpreting the encounter of a spatial layer of fast flow, particularly, the plasma sheet boundary layer, as a temporal onset of flow [Eastman et al., 1988] and (2) the fact that statistical analyses of spacecraft data in the region earthward of about 23 R_E typically do not show a prolonged occurrence of fast flows [Huang and Frank, 1986; Baumjohann et al., 1989], that might be expected from standard reconnection models [e.g., Petschek, 1964; Vasyliunas, 1975] for reconnection events in the near tail. However, the standard models are steady state, i.e., time-independent, and thus not necessarily representative of the temporally varying tail configuration. Furthermore, these models are two-dimensional, neglecting variations in the y direction. Detailed studies of the flow properties in two-dimensional simulations, presumably characteristic of the vicinity of the midnight meridian plane only, have been presented, e.g., by Birn et al. [1986]. They found that earthward flows are typically weaker than tailward flows and increase later. The earthward flow also showed a tendency of being stronger in boundary layers adjacent to the magnetic separatrix. This effect was found more pronounced in cases that included a driving external electric field.

Here we focus on three-dimensional aspects, including, particularly, the cross-tail variation of the flow properties. Figures 14 and 15 show the plasma bulk flow velocity v_x for run B at $z = 0$ as a function of x and y, respectively, at various locations in either y or x, as indicated, and for three different times, $t = 100$ (dashed lines), $t = 140$ (dash-dotted lines), and $t = 170$ (solid lines). The two figures show that earthward speeds are typically relatively small, except at the latest time in a very localized region around midnight and next to the near-Earth X line, which is located at $x \approx -10$. This result was found also in the earlier two- and three-dimensional simulations [e.g., Birn, 1980; Birn and Hones, 1981; Birn et al., 1986]. The same holds for the tailward flow within about 10 units ($\sim 20 R_E$) distance tailward of the X line. Further tailward, however, fast tailward flows occupy a much larger region, developing close to the time of neutral line formation ($t \approx 130$). Fast tailward flows in the vicinity of the X line are thus more localized in space and in time than in two-dimensional simulations, while at larger distances the properties become more similar to the two-dimensional case. Figure 15 also shows that significant spatial variations of the flow direction may be present in the vicinity of the X line.

The localization of the flow velocity in the z direction is shown in Figure 16 for $t = 160$ and $t = 170$, $y = 0$, and several locations in x as indicated. The figure demonstrates that also in the z direction fast tailward flow occupies a region that increases in width with larger distances from the X line. Again, we see that fast flows in the vicinity of the X line arise rather late, after $t = 160$. The earthward flow at $x = -9.4$ in panel (b) shows a peak at $z \approx 0.2$, which is located just inside the separatrix, similar to the results reported by Birn et al. [1986].

7. Summary and Discussion

On the basis of two recent three-dimensional resistive MHD simulations of magnetotail dynamics, we have demon-

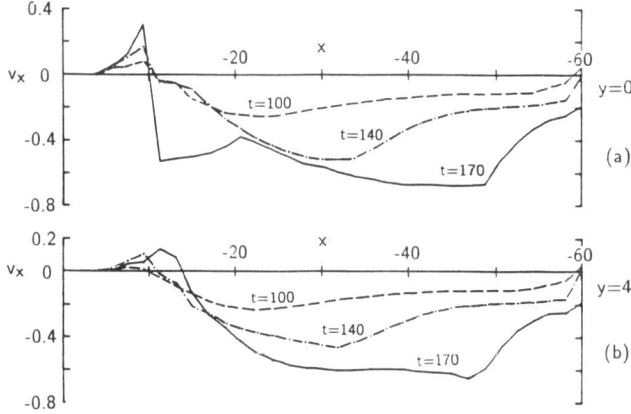

Fig. 14. Flow velocity v_x as a function of x for run B at $z = 0$ and different times, as indicated; (a) for $y = 0$, (b) for $y = 4$.

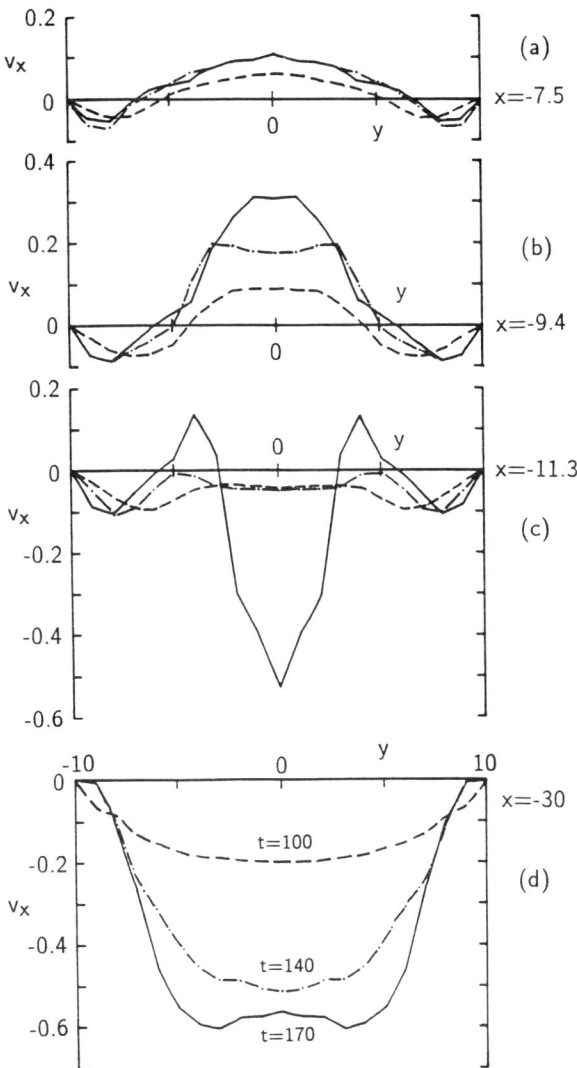

Fig. 15. Flow velocity v_x as a function of y for run B at $z = 0$ and different times, as indicated; (a) for $x = -7.5$, (b) for $x = -9.4$, (c) for $x = -11.3$, (d) for $x = -30$.

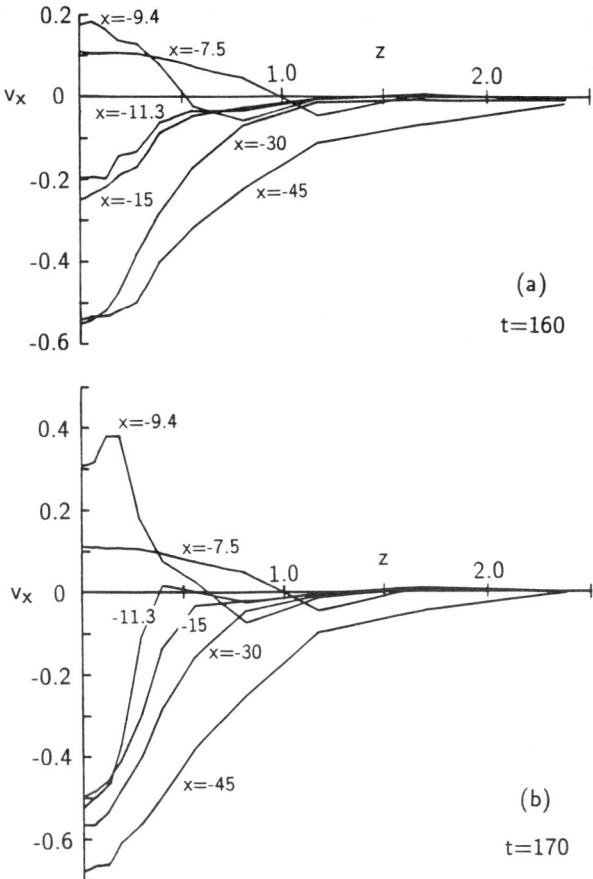

Fig. 16. Flow velocity v_x as a function of z for run B at $y = 0$ and different values of x, as indicated; (a) for $t = 160$, (b) for $t = 170$.

strated characteristic properties of the most unstable, nonideal, tail mode, which develops from initial resistive diffusion without the necessity of external driving or predetermined localization of the nonideal terms. These conditions may not be the most realistic for an actual substorm evolution, because one would expect that in reality a driving process during the substorm growth phase would force the tail into the unstable regime and that nonideal processes occur only localized. The present approach, however, allows us to distinguish properties of the unstable tail modes from driven effects.

Our paper largely represents a summary and extension of results reported earlier. The initial states constitute the most sophisticated and realistic equilibrium configurations of the magnetotail which we were able to derive. They include a flaring of the tail field in both y and z directions, variations of the characteristic thickness of the plasma and current sheet with x and y, associated also with variations of B_z, the decrease of tail flaring with distance down the tail and the transition into a nearly one-dimensional current sheet configuration, whose earthward end is associated with a distant, nearly Y-type neutral line, and the presence of region 1 type field-aligned currents near the plasma sheet boundaries, which are closely related to the decrease of the tail flaring.

Consistent with the neutral line model of substorms, the simulations show the development of a near-Earth reconnection region, which is associated with a characteristic break of plasma and field properties between earthward and tailward of the reconnection site. The tailward properties are dominated by plasmoid formation and ejection into the distant tail. This is accompanied by a thinning of the plasma sheet between the reconnection site and the departing plasmoid and by fast tailward flow, which occupies a large section of the tail, however, only at larger distances down the tail.

Earthward of the reconnection site we find only weak (earthward) flow, except very late in the simulation in a very localized region close to the near-Earth X line. The dominant effect in the near tail earthward of the X line is the dipolarization of the magnetic field, which starts near midnight and propagates laterally toward the flank regions and, perhaps, tailward. The dipolarization is accompanied by features which are attributed to the "substorm current wedge": a reduction and diversion of part of the cross-tail current from a region surrounding the near-Earth X line, leading to an increase of the region 1 type field-aligned current system. The dipolarization is also associated with an expansion of the closed field line region when reconnection proceeds to involve lobe field lines and the plasmoid becomes disconnected. This expansion starts near midnight and propagates outward.

To provide a rough idea of the expected signatures of the magnetic boundaries and field-aligned currents at the Earth, we have used the Tsyganenko [1987] model to map structures found at our near-Earth boundary along field lines from the tail to the Earth's surface. The expansion of the closed field line region maps to a poleward expansion of the open/closed boundary, which bears a strong resemblance to the formation and expansion of the auroral bulge. The field-aligned currents from our tail simulation are found to map to the nightside region 1 equatorward of the open/closed boundary. The peak values of the field-aligned current density and the total field-aligned current are found to increase by a factor of about 2. The peaks of $j_\parallel$ shift nightward and equatorward, also consistent with observations [Iijima and Potemra, 1978]. While the relative locations of these features at the Earth's surface are not very sensitive to the rough patching of our tail model to the (time-independent) Tsyganenko model, absolute locations cannot be derived with confidence. Nevertheless, the locations derived in our fit agree reasonably well with observed auroral distributions.

The overall consistency of the features found in our simulations, which agree largely with features found in global simulations, with observed substorm signatures strongly support the view that these signatures are to be attributed to a tail instability, even if it is externally induced. These features are generally consistent with the near-Earth reconnection model of substorms. Apparent inconsistencies of observations with this substorm model consist basically of two effects: (1) the absence of some of the expected features of near-Earth reconnection in some tail observations during geomagnetic activity and (2) a much more variable character of observed tail properties during such activity. The absence of the expected reconnection flow signatures in the near tail can partially be explained by the fact that a significant portion of geomagnetic activity seemingly cannot be attributed to a tail instability [Rostoker et al., 1987]. Apart from that, however, on the basis of our simulation results, we conclude that the main reason why such signatures may not be found is their spatial localization. This localization is most pronounced in the vicinity of the near-Earth X line, so that it may not surprise that observationally clear signatures are found less often in that region. Note also that the region tailward of the near-Earth X line gets significantly compressed after the onset of the tail instability, while the region earthward of it widens. This makes the detection of tailward flow much less likely than that of earthward flow. The three-dimensional time-dependent simulations also provide hints of a more detailed and variable structure than two-dimensional and steady state reconnection models. Nevertheless, it is still not detailed and variable enough to account for all relevant observations. This clearly shows the limitations of an MHD model and, in particular, of the assumed smoothly distributed resistivity modeling the nonideal mechanism.

Acknowledgments. This work was supported by the U.S. Department of Energy through the Office for Basic Energy Sciences and by NASA. M. H. gratefully acknowledges support from the Los Alamos National Laboratory Director's postdoctoral program.

References

Akasofu, S.-I., *Physics of Magnetospheric Substorms*, D. Reidel Publ. Comp., Dordrecht, Holland, 1977.

Baumjohann, W., G. Paschmann, N. Sckopke, C. A. Cattell, and C. W. Carlson, Average ion moments in the plasma sheet boundary layer, *J. Geophys. Res.*, **93**, 11,507, 1988.

Birn, J., Computer studies of the dynamic evolution of the geomagnetic tail, *J. Geophys. Res.*, *85*, 1214, 1980.

Birn, J., Three-dimensional computer modeling of dynamic reconnection in the magnetotail, in *Magnetic Reconnection in Space and Laboratory Plasmas, Geophys. Monogr. Ser.*, vol. 30, edited by E. W. Hones, Jr., p. 264, AGU, Washington, D.C., 1984.

Birn, J., Magnetotail equilibrium theory: The general three-dimensional solution, *J. Geophys. Res.*, **92**, 11,101, 1987.

Birn, J., Three-dimensional equilibria for the extended magnetotail and the generation of field-aligned current sheets, *J. Geophys. Res.*, **94**, 252, 1989.

Birn, J., and E. W. Hones, Jr., Three-dimensional computer modeling of dynamic reconnection in the geomagnetic tail, *J. Geophys. Res.*, *86*, 6802, 1981.

Birn, J., and M. Hesse, The magnetic topology of the plasmoid flux rope in a MHD simulation of magnetotail reconnection, in "Physics of Magnetic Flux Ropes," edited by C. T. Russell, E. R. Priest, and L. C. Lee, *Geophys. Monograph 58*, American Geophys. Union, Washington DC, p. 655, 1990.

Birn, J., and M. Hesse, MHD simulations of magnetotail reconnection in a skewed three-dimensional tail configuration, *J. Geophys. Res.*, in press, 1991a.

Birn, J., and M. Hesse, The substorm current wedge and

field-aligned currents in MHD simulations of magnetotail reconnection, *J. Geophys. Res.*, in press, 1991b.

Brecht, S. H., J. G. Lyon, J. A. Fedder, and K. Hain, A time dependent three-dimensional simulation of the earth's magnetosphere: Reconnection events, *J. Geophys. Res.*, 87, 6098, 1982.

Forbes, T. G., and E. R. Priest, On reconnection and plasmoids in the geomagnetic tail, *J. Geophys. Res.*, 88, 863, 1983.

Hau, L.-N., and R. A. Wolf, Effect of a localized minimum in equatorial field strength of resistive tearing instability in the geomagnetotail, *J. Geophys. Res.*, 92, 4745, 1987.

Hautz, R., and M. Scholer, Numerical simulations on the structure of plasmoids in the deep tail, *Geophys. Res. Lett.*, 14, 969, 1987.

Hesse, M., and J. Birn, Magnetic reconnection in the magnetotail current sheet for varying cross-tail magnetic field, *Geophys. Res. Lett.*, in press, 1990.

Hesse, M., and J. Birn, Plasmoid evolution in an extended magnetotail, *J. Geophys. Res.*, in press, 1990a.

Hesse, M., and J. Birn, Magnetosphere-ionosphere coupling during plasmoid evolution: First results, *J. Geophys. Res.*, in press, 1990b.

Hones, E. W., Jr., Substorm processes in the magnetotail: Comments on 'On hot tenuous plasma fireballs and boundary layers in the Earth's magnetotail' by L. A. Frank, L. L. Ackerson, and R. P. Lepping, *J. Geophys. Res.*, 82, 5633, 1977.

Hones, E. W., Jr., Association of plasma sheet variations with auroral changes during substorms, *Adv. Space Res.*, 8, (9)129, 1988.

Hones, E. W., Jr., Transient phenomena in the magnetotail and their relation to substorms, *Space Sci. Rev.*, 23, 393, 1979.

Hones, E. W., Jr., and K. Schindler, Magnetotail plasma flow during substorms: a survey with IMP 6 and IMP 8, *J. Geophys. Res.*, 84, 7155, 1979.

Hones, E. W., Jr., S. J. Bame, and J. R. Asbridge, Proton flow measurements in the magnetotail plasma sheet made with IMP 6, *J. Geophys. Res.*, 81, 227, 1976.

Huang, C. Y., and L. A. Frank, A statistical study of the central plasma sheet: Implications for substorm models, *Geophys. Res. Lett.*, 13, 652, 1986.

Iijima, T., and T. A. Potemra, The amplitude distribution of field-aligned currents at northern high latitudes observed by Triad, *J. Geophys. Res.*, 81, 2165, 1976.

Iijima, T., and T. A. Potemra, Large-scale characteristics of field-aligned currents associated with substorms, *J. Geophys. Res.*, 83, 599, 1978.

Kettmann, G., T. A. Fritz, and E. W. Hones, Jr., CDAW 7 revisited: further evidence for the creation of a near-Earth substorm neutral line, *J. Geophys. Res.*, 95, 12,045, 1990.

Lee, L. C., Z. F. Fu, and S.-I. Akasofu, A simulation study of forced reconnection processes and magnetowpheric storms and substorms, *J. Geophys. Res.*, 90, 10,869, 1985.

Lopez, R. E., and A. T. Y. Lui, A multisatellite case study of the expansion of a substorm current wedge in the near-Earth magnetotail, *J. Geophys. Res.*, 95, 8009, 1990.

Lopez, R. E., A. T. Y. Lui, D. G. Sibeck, R. W. McEntire, L. J. Zanetti, T. A. Potemra, and S. M. Krimigis, The longitudinal and radial distribution of magnetic reconfigurations in the near-Earth magnetotail as observed by AMPTE/CCE, *J. Geophys. Res.*, 93, 997, 1988a.

Lopez, R. E., D. N. Baker, A. T. Y. Lui, D. G. Sibeck, R. D. Belian, R. W. McEntire, T. A. Potemra, and S. M. Krimigis, The radial and longitudinal propagation characteristics of substorm injections, *Adv. Space Res.*, 8, (9)91, 1988b.

McPherron, R. L., C. T. Russell, and M. A. Aubry, Satellite studies of magnetospheric substorms on August 15, 1968, 9. Phenomenological model for substorms, *J. Geophys. Res.*, 78, 3131, 1973.

McPherron, R. L., A. Nishida, and C. T. Russell, Is near-Earth current sheet thinning the cause of auroral substorm onset?, in *Quantitative Modeling of Magnetosphere-Ionosphere Coupling Processes*, edited by Y. Kamide and R. A. Wolf, p. 252, Kyoto Sangyo University, Kyoto, Japan, 1987.

Min, K., H. Okuda, and T. Sato, Numerical studies on magnetotail formation and driven reconnection, *J. Geophys. Res.*, 90, 4035, 1985.

Mitchell, D. G., D. J. Williams, C. Y. Huang, L. A. Frank, and C. T. Russell, Current carriers in the near-Earth cross-tail current sheet during substorm growth phase, *Geophys. Res. Lett.*, 17, 583, 1990.

Otto, A., K. Schindler, and J. Birn, Quantitative study of the nonlinear formation and acceleration of plasmoids in the Earth's magnetotail, *J. Geophys. Res.*, 95, 15,023, 1990.

Petschek, H. E., Magnetic field annihilation, *Nasa SP-50*, 425, 1964.

Rostoker, G., S.-I. Akasofu, W. Baumjohann, Y. Kamide, and R. L. McPherron, The roles of direct input of energy from the solar wind and unloading of stored magnetotail energy in driving magnetospheric substorms, *Space Sci. Rev.*, 46, 93, 1987.

Schindler, K., D. N. Baker, J. Birn, E. W. Hones, Jr., J. A. Slavin, and A. B. Galvin, Analysis of an extended period of Earthward plasma sheet flow at ~220 R_E: CDAW 8, *J. Geophys. Res.*, 94, 15,177, 1989.

Siscoe, G. L., D. G. Sibeck, J. A. Slavin, E. J. Smith, B. T. Tsurutani, and D. E. Jones, ISEE 3 magnetic field observations in the magnetotail: implications for reconnection, in *Magnetic Reconnection in Space and Laboratory Plasmas, Geophys. Monogr. Ser.*, vol. 30, edited by E. W. Hones, Jr., p. 240, AGU, Washington, D.C., 1984.

Slavin, J. A., E. J. Smith, D. G. Sibeck, D. N. Baker, R. D. Zwickl, and S.-I. Akasofu, An ISEE 3 study of average and substorm conditions in the distant magnetotail, *J. Geophys. Res.*, 90, 10875, 1985.

Tsurutani, B. T., D. E. Jones, J. A. Slavin, D. G. Sibeck, and E. J. Smith, Plasma sheet magnetic fields in the distant tail, *Geophys. Res. Lett.*, 11, 1062, 1984.

Vasyliunas, V. M., Theoretical models of magnetic field line merging, *Rev. Geophys. Space Phys.*, *13*, 303, 1975.

Walker, R. J., and T. Ogino, Global magnetohydrodynamic simulations of the magnetosphere, *IEEE Trans. Plas. Sci.*, *17*, 135, 1989.

Walker, R, J., T. Ogino, and M. Ashour-Abdalla, A magnetohydrodynamic simulation of reconnection in the magnetotail during intervals with southward interplanetary magnetic field, in *Magnetotail Physics*, edited by A. T. Y. Lui, p. 183, Johns Hopkins Univ. Press, Baltimore, MD, 1987.

Wiechen, H., and K. Schindler, Quasi-static theory of the Earth's magnetotail, including the far tail, *J. Geophys. Res.*, *93*, 5579, 1988.

A MAGNETOSPHERE WAGS THE TAIL MODEL OF SUBSTORMS

G. Atkinson

Canadian Space Agency, P.O. Box 7275, Vanier Postal Station,
Ottawa, Ontario K1L 8E3, Canada

Abstract. Substorm behaviour in the tail is controlled by processes in the dipole-like region. Consider these processes. There is a steady-state location for the near-earth edge of the plasma sheet which is determined by the physics of Alfvén layers. It is closer to the earth for faster dayside merging rates. In the growth phase, in response to an enhancement of dayside merging, the near-earth plasma sheet moves earthward towards the steady-state location corresponding to the enhanced merging rate. The motion is slow because energy must be extracted from the solar wind. Frequently, the steady-state location is not reached before an expansion is triggered. A subsequent decrease in dayside merging triggers the expansion if the decrease is large enough that the new steady-state location for the near-earth plasma sheet is tailward of its actual location at the time of decrease. Thus a large decrease creates an unstable situation with energetic plasma too close to the earth and too much flux in the tail. The resulting instability can be described as an interchange/ballooning/merging instability with westward drift of energetic particles removing energy from the midnight sector and allowing rapid merging to occur in a localized time slot and causing ballooning to the west. An implication of the model is that merging in the tail is coupled to or controlled by plasma processes on dipole-like flux tubes. There is too much energetic plasma on tail-like flux tubes for them to collapse into a dipole-like configuration. Thus for convection to occur into the dipole-like region, it is necessary that there be an X line near the boundary between tail-like and dipole-like flux tubes. Thus the X line always remains near the boundary and the merging rate must match the earthward convective flow, which is determined by convection processes in the dipolar region.

Introduction

In this paper we shall consider the plasma sheet behaviour in some detail. Closed flux tubes within the plasma sheet and nightside magnetosphere vary from dipolar to greatly distorted into a tail-like form. We need a clear definition for the discussions here and hence classify them according to the value of β at the equatorial crossing. They are classified as:
1. tail-like, when $\beta > 1$ at the neutral sheet (that is at their neutral-sheet crossing), and 2. dipole-like, when $\beta < 1$ at the equatorial plane. In using these definitions, we consider only the macro-scale properties and ignore any localized departures from the conditions given on β.

Magnetospheric Substorms
Geophysical Monograph 64
Published in 1991 by the American Geophysical Union.

Further, it is clear that "dipole-like" includes flux tubes that are quite distorted as beta approaches unity. The author could have introduced a third set of flux tubes: "transition" tubes on which beta is near unity at the equatorial crossing. This was not done in order to keep the model as simple as possible. It may be necessary in future development. The regions and the boundary between them are shown in figure 1. It follows from the above definitions that ß = 1 at the equatorial crossing of the boundary.

For readers not familiar with English sayings, the title of the paper is chosen to emphasize that the proposed model is based on the belief that substorm events in the tail are controlled by physical processes closer to earth. In particular, earthward flow in the tail (and also merging) is controlled by the region consisting of dipole-like flux tubes. The behaviour of this region can be compared to a shut-off valve (or a water tap) which controls both the upstream and downstream flow in a water system. Accordingly, we shall spend much of the earlier part of this paper discussing processes in the dipolelike region. Only near the end will we discuss how the dipole-like region controls merging in the tail.

To understand the dipolarization that occurs in substorms we must understand why the tail existed in the first place. That is, we must explain either the existence of tail-like flux tubes or of the cross-tail current in terms of plasma stresses. (It cannot be over-emphasized that the relationship curl B = $\mu_0 j$ indicates that the variables B and j are simply equivalent descriptions of the same physical phenomenon and hence explaining one in terms of the other is meaningless.) Clearly, the radial component of the plasma stresses (plasma pressure gradient) is important since it is this component that prevents the collapse of tail-like flux tubes to the dipole-like state. It follows that in order to explain

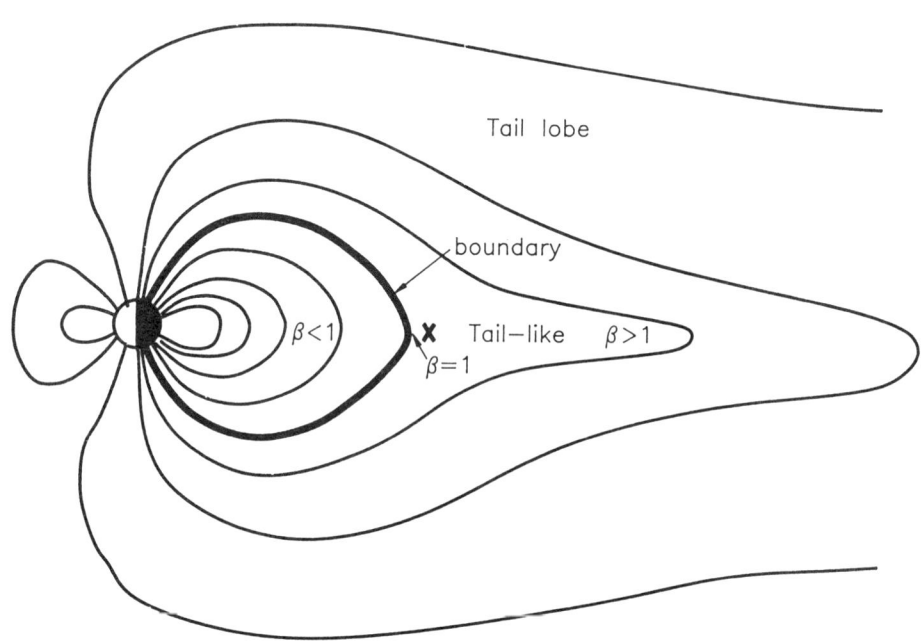

Fig. 1. Noon-Midnight section showing regions of tail-like and dipole-like flux tubes and the boundary between. The value of β applies at the equatorial plane.

dipolarization/current diversion (again these are equivalent descriptions of the same physical phenomenon) we must explain how the plasma stresses, and hence energy, are removed in the near-midnight sector.

We shall, in the next three sections, discuss the physics of the dipole-like flux tubes, considering in turn: energy, momentum (stress balance) and particle behaviour. We shall look at existing morphological (observation based) models of processes and arrive at some conclusions as to requirements on a substorm theoretical model, as well as reviewing some of the basic physics of this region. These discussions provide necessary background for the substorm model developed in the following sections.

Energy Considerations

During the growth phase of substorms the plasma sheet approaches the earth, the magnetic field becomes more tail-like at distances of 6 - 20 R_E and the auroral oval moves equatorwards. These all represent an increase in the stored energy of the system. Processes can be explained in terms of the Poynting vector energy flow and its divergence.

As indicated in the table, in the growth phase the Poynting vector energy flow is from the high-latitude and tail magnetopause (which is a source), with negative divergence due to magnetic energy storage in the nightside magnetosphere and tail, ionospheric dissipation, and plasma energization as it is compressed and convected to smaller L values on the nightside. The last three are sinks for the energy flow. The growth phase is slow because energy must be extracted from the solar-wind interaction at the magnetopause.

In the expansion phase or intensification (see the table) the stored magnetic energy decreases and hence there is a source of energy available on a short time scale (Alfvén-wave travel time). In addition, the plasma may also be a source of energy in some locations since it can expand and release energy stored by compression in the growth phase. Energy release of this nature is typical of interchange and ballooning instabilities.

Current Closure/Stress Balance

Figure 2 shows commonly accepted morphological models of current systems schematically, as heavy lines: the region 1 currents, their closure in the ionosphere to the region 2 currents and the closure of the region 2 currents in the outer magnetosphere by the partial ring current. This closure pattern has recently been verified by Iijima et al, 1990. In reality the currents should be sheets, but for simplicity we show them as lines.

TABLE

Equation	$\nabla \cdot \dfrac{E \times B}{\mu_o}$	$= \dfrac{\partial}{\partial t}\left(\dfrac{B^2}{2\mu_o}\right)$	$+ J \cdot E$
Growth	Energy Flow	Energy Sink	Magnetopause - Source Ionosphere - Sink Plasma - Sink
Expansion	Energy Flow	Energy Source	Magnetopause - Source Ionosphere - Sink Plasma - Both

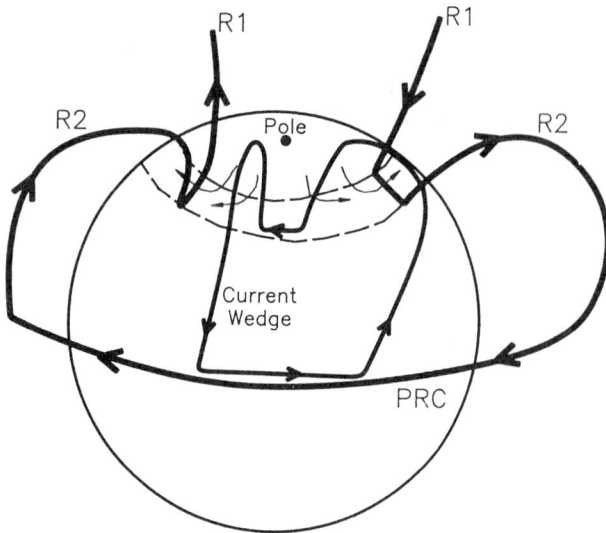

Fig. 2. Nightside view showing: (1) heaviest lines, region 1 and 2 currents and their closure by ionospheric and partial ring current; (2) medium lines, the current wedge system which is added in expansions and reduces the partial ring current in a local time sector; (3) light arrows, convection streamlines into and along the oval.

This current system can be related to the stresses acting on flux tubes. (If flux tubes are considered to be fluid elements, the force acting on them due to plasma is $-J \times B$, since $J \times B$ is the force of the fields on the plasma.) Convective flow of flux tubes equatorward from the polar cap into the oval and then eastward and westward toward the dayside is shown by the thin arrows. The major stresses controlling the flow are the line tying effect (ion-neutral collisions producing $-\underset{\sim}{J} \times \underset{\sim}{B}$ in the ionosphere opposite to the convection) limiting the convection rate towards the dayside, and the partial ring current (radial $-\underset{\sim}{J} \times \underset{\sim}{B}$) which limits the depth of penetration of the flow into the nightside magnetosphere (that is the L value reached).

The currents just discussed are the dominant ones in the growth phase. They are related to the stresses controlling flux tube convection as discussed. In expansions (intensifications), according to a commonly accepted morphological model, some of the partial ring current is diverted to the ionosphere, corresponding to the so-called current wedge indicated by the mediumthickness lines in figure 2. This creation of a near-midnight reduction in the partial ring current (Iijima et al, 1990) and appearance of an intense ionospheric westward current implies that plasma stresses and hence energy is reduced in a local time sector. Stress balance limiting convection to lower L values is maintained by the increased westward ionospheric current.

Particle Behaviour

Now we look at particle behaviour and must consider the physics of Alfvén layers and associated current systems. Alfvén layers are layers of electrical charge which change the convection electric field and thereby limit the convection of particles of a given energy and pitch angle to lower L values in the magnetosphere. For a distribution of energies and pitch angles they become thick layers. In the presence of an ionosphere there are discharging currents along field lines to the ionosphere (region 2 currents) and currents in the magnetosphere exist (the partial ring current) to maintain the electric fields and the shielding of convection from the inner magnetosphere.

The Alfvén layer occurs at L values where other types of particle drift become as large as the $\underset{\sim}{E} \times \underset{\sim}{B}/B^2$ drift, and hence become important in moving particles and particle energy within the magnetosphere (Fejer, 1964; Vasyliunas, 1972; Southwood, 1977). We write the above condition

$$v_{E \times B} \approx v_{other}$$

The other drifts include curvature and gradient drift, and could also include serpentine orbits if neutral sheet geometries were sufficiently close to the earth. The last possibility is considered less likely for the following reasons. From the observed strength of the region 2 currents and cross-polar cap potentials, Atkinson (1985) concluded that $\sim 10^{21}$ particles were involved on each unit (1 weber) of flux. This number is typical of flux tubes within the dipole-like region, and hence it appears likely that the Alfvén layers (and region 2 currents) are connected to these regions and not to the tail-like region. This viewpoint is supported by the observed latitudes of region 2 currents (e.g. Bythrow et al, 1984).

For a given merging rate, or cross polar cap potential, the above equation defines a steady-state location of the Alfvén layer, or depth of penetration for hot plasma. For larger merging rates, E is greater and hence the steady-state Alfvén layer must be at lower L values in order to have larger V_{other}. Thus an increase in the dayside merging rate should start the Alfvén layer moving to lower L values toward the new steady-state location, and a decrease (assuming it started at the steady-state location) would initiate an outward movement. There is ample evidence for these latitudinal motions as seen in particles, fields, auroras and electrojets (e.g. Akasofu, 1964; Vallance Jones et al, 1982; Pellinen et al, 1982; McPherron and Manka, 1985; Rostoker and Phan, 1986).

Growth, Triggering and Expansion

With the background discussions provided by the previous three sections, we are now in a position to present a fairly complete qualitative model of the growth phase, triggering and expansion of substorms in the dipole-like region. (We shall consider the response of the tail in later sections of the paper.)

From energy considerations, based primarily on observations, we have concluded that energy is stored in the growth phase both as magnetic and particle energy. The growth phase is slow because the energy must be extracted from the solar wind. In expansions or intensifications, this energy is released on a much shorter time scale. From stress and current closure considerations, based primarily on observations, we concluded that, in expansions, a reduction of plasma energy occurs in a localized time sector. From our knowledge of Alfvén layers and particle behaviour, we concluded that for each dayside merging rate there is a steady-state location for the Alfvén layer, or depth of penetration of the plasma. An increase in dayside merging initiates an earthward motion of the Alfvén layer toward the new steady-state location at lower L values, and a decrease initiates an outward motion, assuming the system was initially close to the steady-state.

The growth phase of substorms is simply explained in terms of the above discussion. An increase in the dayside merging rate from V_1 to V_3 at a time, t_1, causes the Alfvén layer, oval and plasma sheet inner edge to start moving to lower L values as illustrated by the line AB in figure 3 (from $L(V_1)$ towards $L(V_3)$ where $L(V)$ is the steady-state value of L for a given V). The movement toward $L(V_3)$ is slow because energy must be extracted from the solar wind.

The expansion is triggered when a decrease in dayside merging rate at t_2 (to V_2) occurs which is sufficiently large to place $L(V_2)$ above the line AB; that is the Alfvén layer must move outward. It can be seen from figure 3 that a larger value for the change $V_3 - V_2$ is required for the expansion to occur earlier (consistent with the observations reported by Pudovkin, 1990, that a stronger northward turning is required to trigger onset when the field is less tail-like

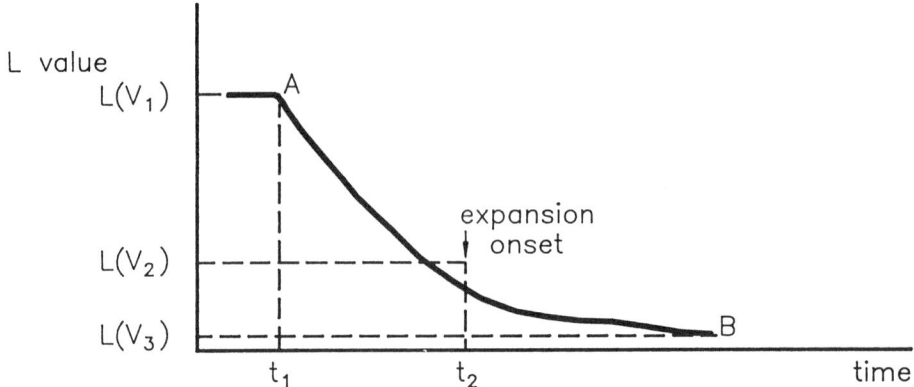

Fig. 3. L value of plasma sheet earthward edge versus time. $L(V)$ is the steady-state location corresponding to a merging rate, V. Line AB shows the actual location of the edge if merging increases from V_1 to V_3 at t_1.

at geostationary orbit). For sufficiently long $t_2 - t_1$, only a small perturbation is needed, producing non-triggered substorms. Other solar wind parameter changes may also trigger expansion if they produce conditions that require the Alfvén layer to move outward.

In the growth phase, a continued energy input is required from the solar wind and hence there is no energy available to drive instabilities. In the expansion, free energy is available from the compressed plasma and the magnetic field as discussed earlier, and hence instabilities are likely. Hence the expansion may exhibit across-tail variations and, in addition, may overshoot the steady-state location, $L(V_2)$.

Consider the nature of the instability. The change in the location of the steady-state Alfvén layer means that: hot plasma is trapped too close to the earth for the new merging rate and there is too much magnetic flux in the tail lobes. It is well known that the first condition leads to the interchange instability, and since tail-like flux tubes are involved, it is likely to be an interchange/ballooning instability. Korth (1990) reports that observed conditions are indeed consistent with this instability. However, the second condition (too much flux in the tail lobes) requires that magnetic flux must also be merged, and hence merging must also occur. The term interchange/ballooning/merging is somewhat cumbersome, but describes events. Figure 4, in the author's view, is a schematic of the major processes. Plasma near midnight (the point of deepest penetration in the growth phase) drifts westward transporting thermal energy and causing ballooning to the west of the westward surge. The removal of energy from the midnight sector allows inward convection, and rapid merging, thereby providing additional energy input to the instability. Thus there is growth phase behaviour west of the surge and expansion east of it, consistent with observation (Gelpi et al, 1987).

The above provides a basic model for substorms which is quite consistent with generally accepted views on processes in the magnetosphere. The author proposes it as a theory of substorm growth, triggering and expansion.

The remaining sections of the paper discuss how the tail is controlled by these nearer-earth processes.

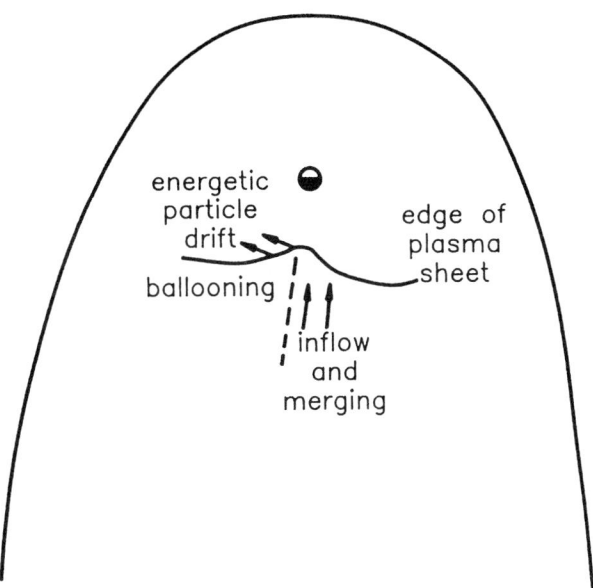

Fig. 4. Schematic of Interchange/-ballooning/merging instability. Energetic plasma near midnight (the point of deepest penetration in the growth phase) drifts westward. The resulting energy transport causes ballooning west of the westward surge and allows inflow and merging near midnight.

Magnetospheric Control of Tail Processes

There is an energy problem in convecting flux tubes earthward in the tail-like region, and into the dipole-like region. Erickson and Wolf (1980) and Schindler and Birn (1982) showed that the convection cannot occur without removal of energy (presumably energetic plasma) from the flux tubes. Atkinson (1988) arrived at a similar conclusion for dipolar flux tubes if they start with $\beta = 1$ at the tail-dipolar boundary and convect earthwards. Hence, he deduced that the boundary between tail-like and dipole-like flux tubes must coincide with the outer boundary of the Alfvén layer. (In the Alfvén layer, the drift of energetic plasma off inwardly convecting flux tubes transports energy away and enables the condition $\beta \lesssim 1$, to be maintained). In the remainder of this paper we shall refer to this problem as the beta problem. It is central to the arguments for control of the tail by the dipole-like region. Consider the role of X lines on closed flux-tubes in the tail. If an X line appeared at a significant distance tailward of the boundary between dipole-like and tail-like flux tubes, the merged flux tubes cannot convect into the nightside of the dipole-like region because of the beta/energy problem. (Collapse to a dipole-like form would produce $\beta > 1$.) The only place an X line could occur with significant earthward flow is near the boundary between tail-like and dipole-like flux tubes, as indicated by the X in figure 1. An X line at this location would produce dipole-like flux tubes on its earthward side with $\beta \approx 1$, and these could convect earthward because of the Alfvén layer physics of this region as discussed earlier. In fact, the existence of an X line near the boundary appears to be a necessary condition for convection into the dipole-like region to occur. It would not move down the tail until the boundary did. In general, merging would not be the "fast merging" discussed in the literature with outflow limited by inertial effects. Downstream boundary conditions would limit outflow rates. Nishida et al (1986) discuss stagnant plasmoids in this context. In summary, the boundary between tail-like and dipole-like flux tubes occurs at the outer edge of the thick Alfvén layer and an X line exists just outside the boundary.

The author wishes to propose two possibilities for the behaviour of the tail. The first is that the X line appears at the onset of expansion near the dipole-like/tail-like boundary in response to the onset of the interchange/ballooning/merging instability discussed in the previous section of the paper. In the case of a simple single-intensification substorm, the dipole-like region would expand downtail and would probably overshoot the steady-state prediction for its location because 1. tail-lobe field lines become involved in the merging (they contain very little plasma and hence the beta problem would be reduced) and 2. the steady-state criterion is probably invalid when the expansion occurs rapidly in a localized time slot. One can speculate that multiple intensifications are due to continued variable dayside merging, or possibly due to the localized merging stopping before sufficient lobe flux has been merged to restore the tail-wide steady-state. The effects of this near-earth neutral line would be expected to be quite similar to the Hones' morphological model (e.g. Hones et al, 1986).

The second possibility that the author wishes to propose for the behaviour of the tail is that the X line behaves as described above in the expansive phase but exists also in the growth phase. The suggestion is that a very slow merging rate in the growth phase is greatly increased in the local-time slot where expansion occurs. This possibility is attractive because it explains the behaviour of aurora as discussed later.

One can speculate on the reason that substantial southward fields are not observed in the tail in the growth phase. The author suggests that O lines, magnetic islands and flux ropes are efficient dissipators of energy and that transport of energy and plasma across the tail becomes important. Thus only very weak islands are produced when merging is slow. For example, if the stresses driving cross-tail convection are balanced by the ionospheric line-tying forces, one can calculate typical sizes of these stresses. If, after disconnection from the ionosphere by merging, the stresses remain the same, then the magnetic islands accelerate across the tail at a rate that removes them in tens to hundreds of seconds. This will be further investigated in a separate publication.

Summary

The basic thesis of this substorm model is that merging at near-earth neutral lines is controlled by processes in the dipolar region since this region controls the convective flow rate into the nightside and eastward and westward toward the dayside. Since it is radial plasma pressure gradients that prevent collapse of tail-like plasma-sheet flux tubes in the growth phase, plasma energy must be reduced in a local time slot allowing dipolarization in the expansion phase.

In the growth phase, in response to enhanced dayside merging, the thick Alfvén layer (and associated Partial Ring Current and region 2 currents in the near-earth plasma sheet) moves earthward toward the steady-state location that is consistent with the enhanced dayside merging rate. The earthward motion is slow because energy must be extracted from the solar wind to provide for both dissipation in the ionosphere and for the increase in energy stored in the magnetic field and the compressed plasma. Frequently, the steady-state is not attained because insufficient time elapses before expansion is triggered.

A subsequent decrease in dayside merging, if large enough, will require the Alfvén layer to move outward and hence trigger the expansion. When the growth phase is long, the system has approached the steady-state and a small perturbation can trigger expansion. A short growth phase results if a sufficiently large-amplitude perturbation occurs earlier in the growth phase.

The decrease in dayside merging (or other change in the solar wind that requires the Alfvén layer to move outward) creates an unstable state with hot plasma trapped too close to the earth and too much flux in the tail lobes. Thus both magnetic and plasma energy are available on a short timescale and can cause instabilities. The suggested instability is an interchange/ballooning/merging instability in which the hot plasma near midnight (the point of deepest penetration) drifts westward, causing ballooning of previously dipolar tubes into a more tail-like form to the west, and simultaneously allowing a more rapid inflow and merging near midnight.

Finally, the response of the tail to the above processes on dipole-like flux tubes was considered. Unless there is an X line near the boundary between tail-like and dipole-like flux tubes, too much energetic plasma is convected into the nightside dipole-like region for it to remain dipole-like. Thus it is concluded that it is a necessary condition for convection into the dipolar region that an X line remains close to the boundary. It should not move downtail until the dipole-like region expands downtail. The author proposes two possibilities for tail behaviour: (1) the X line appears at the onset of expansion in the local-time slot of expansion and the expansion of dipole-like flux tubes downtail overshoots the steady-state location because of very low beta lobe flux tubes being involved in the merging and (2) same as the above except that the X line exists in the growth phase also with a slow merging rate which is greatly enhanced in a localized time slot during expansion.

Epilogue

In Atkinson et al (1989) it was found that quite complicated behaviour and structure of discrete auroral arcs could be explained if it was assumed that there was a one-to-one relationship between arcs and X lines. Atkinson (1991) has proposed a mechanism causing such a relationship. The objective of this paragraph is to point out that the foregoing substorm model is consistent with the X line-arc relationship. Consistency includes the brightening of the most-equatorward arc, rapid poleward motion of the arc in a localized time sector consistent with enhanced merging rate, and slow drift equatorward in the growth phase (including west of the westward surge). There are three ways in which the most active X line (the one with the highest merging rate) can move away from the earth: 1. a smooth motion; 2. the formation of a new X line at a greater distance, and 3. the transfer of activity to a pre-existing X line further out. The last two are capable of producing magnetic islands trapped in the dipole-like flux which, in Atkinson et al (1989), were used to account for the equatorward running arcs inside the substorm bulge. Some further examples of these arcs are shown in Pudovkin (1990). They are a common phenomenon. The trapped islands may also account for the short-duration earthward travelling plasma events reported by Baumjohan et al (1990).

References

Akasofu, S.-I., The development of the auroral substorm, Planet. Space Sci., 12, 273, 1964.

Atkinson, G., Time-dependent flows in the renovated model of the magnetosphere, J. Geophys. Res., 90, 10843-10850, 1985.

Atkinson, G., The location of the high-latitude boundary for magnetospheric convection is not arbitrary, J. Geophys. Res., 93, 11533-11535, 1988.

Atkinson, G., F. Creutzberg, R.L. Gattinger, J.S. Murphree, Interpretation of complicated discrete arc structure and behaviour in terms of X lines, J. Geophys. Res., 94, 5292-5302, 1989.

Atkinson, G., Mechanism by which merging at X lines causes discrete arcs, Submitted to J. Geophys. Res., 1991.

Baumjohann, W., G. Paschmann, H. Luhr, Characteristics of high-speed ion flows in the plasma sheet, J. Geophys. Res., 95, 3801-3809, 1990.

Bythrow, P.F., T.A. Potemra and L.J. Zanetti, Variation of the Auroral Birkeland current pattern associated with the north-south component of the IMF, in Magnetospheric Currents, Editor T.A. Potemra, Geophysical Monograph 28, American Geophysical Union, Washington, D.C., 1984.

Erickson, G.M. and R.A. Wolf, Is steady convection possible in the earth's magnetotail?, Geophys. Res. Lett., 1, 897, 1980.

Fejer, J.A., Theory of the geomagnetic daily disturbance variations, J. Geophys. Res., 69, 123-135, 1964.

Gelpi, C., H.J. Singer, W.J. Hughes, A comparison of magnetic signatures and DMSP auroral images at substorm onset: Three case studies, J. Geophys. Res., 92, 2447-2460, (1987).

Hones, E.W. Jr., T.J. Rosenburg and H.J. Singer, Observed associations of substorm signatures at South Pole, at the auroral zone and in the magnetotail, J. Geophys. Res., 91, 3311-3320, 1986.

Iijima, T., T.A. Potemra and L.J. Zanetti, Large-scale characteristics of magnetospheric equatorial currents, J. Geophys. Res., 95, 991-999, 1990.

Korth, A., Statistical study of substorm onset conditions at geostationary orbit, Paper presented at Chapman Conference on Magnetospheric Substorms, Hakone, Japan (1990).

McPherron, R.L., R.H. Manka, Dynamics of the 1054 UT March 22, 1979, substorm event: CDAW 6, J. Geophys. Res. 90, 1175-1190, 1985.

Nishida, A., M. Scholer, T. Terasawa, S.J. Bame, G. Gloeckler, E.J. Smith and R.D. Zwickl, Quasi-stagnant plasmoid in the middle tail: a new pre-expansion phase phenomenon, J. Geophys. Res., 91, 4245-4255, 1986.

Pellinin, R.J., W. Baumjohann, W.J. Heikkila, V.A. Sergeev, A.G. Yahnin, G.K. Marklund, A.O. Melnikov, Event study on pre-substorm phases and their relation to the energy coupling between solar wind and magnetosphere, Planet. Space Sci. 30, 271-388, 1982.

Pudovkin, M.I., Physics of Magnetospheric substorms: A review paper presented at Chapman Conference on Magnetospheric Substorms, Hakone, Japan (1990).

Rostoker, G. and T.D. Phan, Variation of auroral electrojet spatial location as a function of the level of magnetospheric activity, J. Geophys. Res. 91, 1716-1722, 1986.

Schindler, K. and J. Birn, Self-consistent theory of time-dependent convection in the earth's magnetotail, J. Geophys. Res., 87, 2263-2275, 1982.

Southwood, D.J., The role of hot plasma in magnetospheric convection, J. Geophys. Res., 82, 5512-5520, 1977.

Vallance Jones, A., F. Creutzberg, R.L Gattinger, and F.R. Harris, Auroral Studies With a Chain of Meridian-Scanning Photometers, 1. Observation of Proton and Electron Aurora in Magnetospheric Substorms, J. Geophys Res. 87, 4489-4503, 1982.

Vasyliunas, V.M., The interrelationship of magnetospheric processes, in Earth's Magnetospheric Processes, edited by B.M. McCormac, D. Reidel, Hingham, Massachusetts, 1972.

Role of the Near Earth Plasmasheet at Substorms

A. Roux,[1] S. Perraut,[1] A. Morane,[1] P. Robert,[1] A. Korth,[2] G. Kremser,[2,3] A. Pederson,[4] R. Pellinen,[5] and Z.Y. Pu[6]

Recent observations performed onboard GEOS-2 and AMPTE CCE and IRM have renewed the interest for the role played by the inner plasmasheet at substorms. GEOS data are used to show that (i) electron injection at breakup is dispersionless and is therefore due to a local process, (ii) strong earthward gradients in the flux of energetic ions are observed prior to breakup; after breakup, the direction of this gradient oscillates, and (iii) oscillations are also identified in electric and magnetic field data; they correspond to an azymuthally-propagating wave. These results are shown to be consistent with a ballooning instability developing in the highly-stressed magnetic geometry that builds up in the Central Plasma Sheet (CPS) prior to substorms. As it grows, this instability drives a system of transient field-aligned currents, hence leading to the partial cancellation of the tail current. This results in an increase of the H component of the magnetic field (the dipolarization) and to the corresponding induced electric field resulting in particle injection. Comparison with the ground-based ASC (All-Sky Camera) suggests that the surges observed simultaneously on the ground are the image of this instability drawn onto the upper atmosphere by precipitating electrons. According to this interpretation, the northward expansion of the auroral arcs reflects the radial expansion of the region where the ballooning instability develops. The validity of this interpretation is checked against data from AMPTE CCE and IRM and from ISEE-1,2.

1. INTRODUCTION

On All-Sky Camera (ASC) pictures, the substorm breakup is usually characterized by a rapid brightening and development of a surge, which later propagates or expands to the West. The Westward Travelling Surge (WTS) is formed at the poleward edge of the diffuse auroral region, at magnetic latitudes in the range 65 to 70°. While the southern edge of the region where bright auroras develop is essentially unchanged, the arcs and bulges expand to the West, to the East and, at a slower velocity, to the North. This asymmetric expansion is very clear on images from DE1 and Viking. About 30 mn after breakup, various arcs cover from ~ 65-70° to 75-80°.

A strong upward field-aligned current, associated with precipitating electrons, flows near the leading edge of the WTS [*Opgenoorth et al.*, 1983] and the signature of reverse currents, flowing towards the Earth, has also been identified east and west of the WTS [*Kozelova and Lyatskiy*, 1984]. Then, a system of field-aligned currents develops, as the WTS is formed. The relationship between the formation of the surge and the field-aligned current system is discussed in the present work.

Following *Dungey* [1961], several authors have discussed the possible connection between substorm development and the formation of an X point or line in the central part of the geomagnetic tail. According to a widely spread scenario, the thinning of the CPS produces a very small B_z component, which favours the development of tearing modes. In the collisionless CPS plasma, however, the development of the tearing mode instability is not granted: *Lembège and*

---

[1]CNET/CRPE/CNRS, Issy-les-Moulineaux-Cedex, France
[2]Max-Planck Institut für Aeronomie, Katlenburg-Lindau, Germany
[3]Department of Physics, University of Oulu, Oulu, Finland
[4]ESTEC/SSD, Noordwijk, The Netherlands
[5]Finnish Meteorological Institut, Helsinki, Finland
[6]Department of Geophysics, Beijing University, Beijing, P.R.C.

Magnetospheric Substorms
Geophysical Monograph 64
Copyright 1991 American Geophysical Union

Pellat [1982] have shown, indeed, that even a very small B_z component can stabilize the tearing modes. Should B_z be locally small enough to allow the tearing instability to develop, then B_z would grow, anyhow, as a consequence of the instability. The saturation level of the tearing mode instability is therefore expected to be quite small, hence raising questions as to its importance in the dynamics of the geomagnetic tail.

Substorm onset in the near Earth plasmasheet at the geosynchronous orbit is characterized by a reconfiguration of the magnetic field, initially tail-like, towards a more dipole-like configuration [*Sauvaud and Winckler*, 1980] and by an injection of energetic particles [*Mauk and McIlwain*, 1974]. This magnetic reconfiguration is associated to a reduction, at least locally, of the cross tail current. *McPherron et al.* [1973] interpreted this reduction in terms of a current wedge that diverts a fraction of the tail current into the ionosphere, via field-aligned currents. Injection of energetic particles is often dispersionless, hence it must be achieved by a local acceleration process. *Mauk and Meng* (1987) suggest that an induced electric field, associated with the reconfiguration of the magnetic field, does produce the required acceleration. In the present paper, we show that the dipolarization, the current wedge and the injection are due to a single cause: the development of the ballooning instability in the near Earth plasmasheet.

This work is based on the interpretation of data from the geostationary satellite GEOS-2. These data are presented and discussed in Section 2. Experimental evidence for the development of a ballooning instability is given in Section 3. The theoretical framework of the ballooning instability is discussed in Section 4. In Section 5, the consequences of the development of the ballooning instability are discussed in the context of ISEE and AMPTE measurements performed beyond the geostationary orbit.

2. SUBSTORMS AT THE GEOSTATIONARY ORBIT

The present section is mainly based on data registered in the near Earth plasmasheet, close to the geographic equator, by the ESA/GEOS-2 spacecraft. This section is a summary of a work by *Roux et al.* (1991), who studied in detail an isolated dispersionless substorm that took place on January 25, 1979. About 20 similar isolated dispersionless events, observed on GEOS-2, are analysed in a companion paper by *Korth et al.* [1991, this issue].

Full advantage is taken of the stationary position of GEOS-2, to make a comparison with ground-based data. On January 25, 1979, ground-based magnetometers at Kiruna and riometers at Kilpisjärvi show the typical signature of an isolated substorm [see *Roux et al.*, 1991, for more details].

For future discussions, the signature of the breakup on ASC's is important to describe. Figure 1 shows defolded images built from pictures gathered simultaneously by two ASC's, one at Kevo (69.8° N, 29.0°E) and one at Kilpisjärvi (69.0°N, 20.8°E) in northern Scandinavia, close to the magnetic footprint of GEOS-2. Time intervals between two successive frames is 20 s. An intense surge (Surge I) first appears in the central part of the figure, between 20.17.43 and 20.18 UT, and later propagates to the west. Other surges develop further to the north (Surges II and III). At 20.21, a well-defined surge develops in the same region as Surge I. It is not clear, however, that it propagates afterwards; hence, it is labelled DAF (Discrete Auroral Form) in what follows. In view of its very clear westward motion, Surge I is labelled WTS (Westward Travelling Surge) hereafter. Such a label is also consistent with the current idea that the WTS is the first surge to develop at breakup.

Figure 2 gives an overview of the various parameters measured at GEOS-2 over 1 hour. It shows the magnetic field in the VDH frame (V is radial outward, D is azimuthal and H is essentially parallel to the Earth's rotation axis, at GEOS orbit). Panel 2 shows the radial (E_R) and azymuthal (E_A) components of the electric field. The following panels show low energy and high energy particles. Let us describe the successive phases of the substorm.

The pre-substorm (before 20.17 UT) is characterized by a slow decrease of the H component of the magnetic field B accompanied by a slow increase of the V component until they reach comparable amplitudes, at 20.10 UT; accordingly, a "tail-like" configuration progressively builds up. Given the low magnetic latitude (~ 3°) of GEOS-2, the equality between the V and H components corresponds to a highly stressed magnetic field configuration, which is often observed for substorms developing in the pre-midnight sector at the geostationary orbit. The build-up of such a configuration implies an increase of the tail current and/or a motion of this current earthwards. Simultaneously, the radial component of the quasi-static electric field (E_R) progressively decreases, reaching -2mV.m^{-1} at 20.15 UT. Hence, prior to the breakup, the electric field is radially earthward; the plasma motion ExB is directed azymuthally to the East.

The number density of low energy electrons (100-500 eV) is typical for the plasmasheet, thus indicating that, prior to the substorm breakup, GEOS was located within the plasmasheet.

The integral fluxes of energetic electrons (E > 22 keV) and ions (E > 27 keV) are plotted on the lowest panel. The ion flux progressively decreases until substorm breakup; the behavior of electrons is essentially the same but there is a

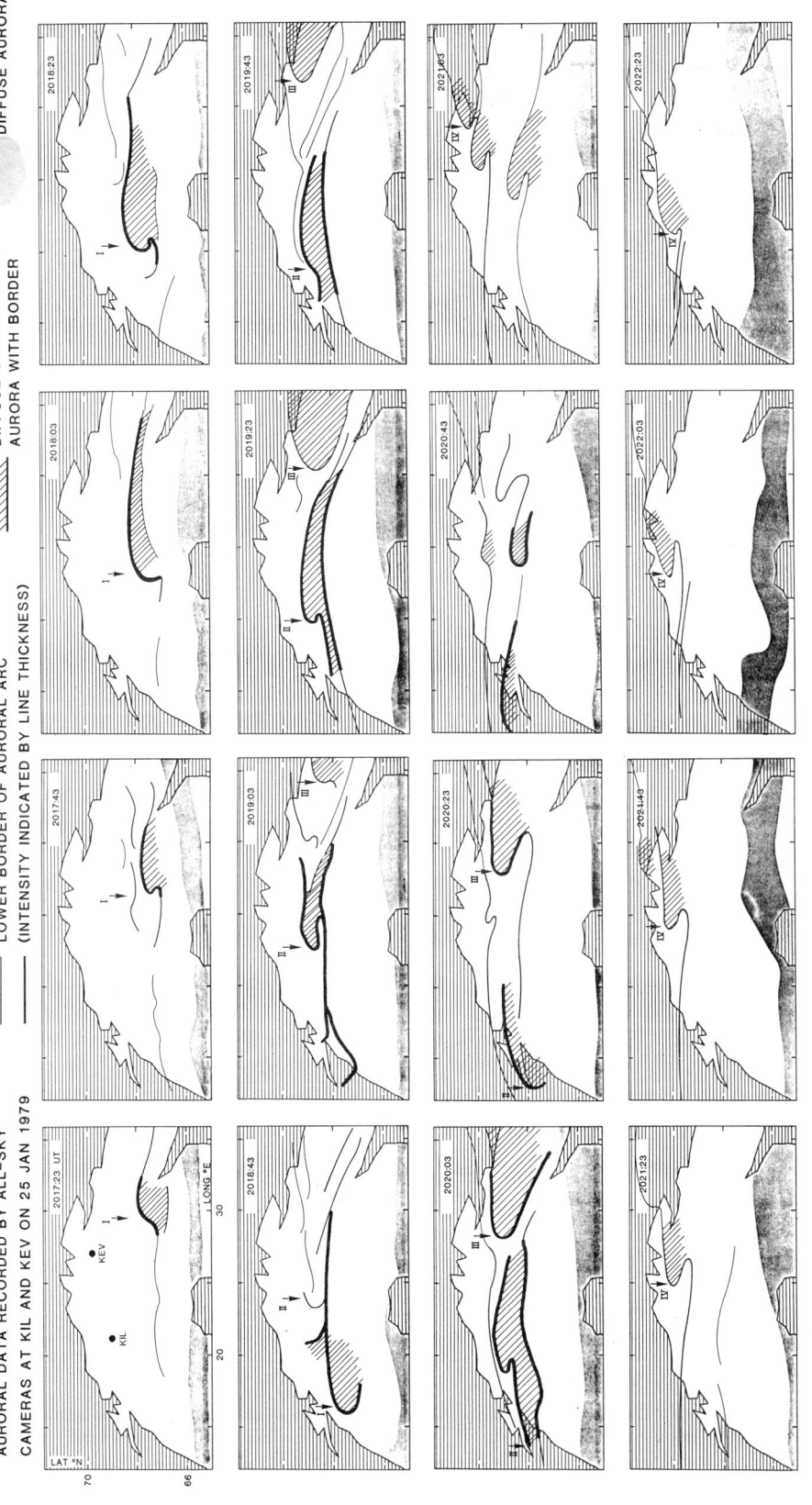

Fig. 1 (from Roux et al., 1991). This defolded picture has been built from the photographs taken by two All-Sky Cameras, one at Kilpisjärvi and one at Kevo. One frame is taken every 20 s, from 20.17.23 UT to 20.22.23 UT. Notice the strong intensification between 20.17.43 and 20.18.03, as the first WTS (WTS I). Most of the following surges (WTS II and III) develop further to the north. A Discrete Auroral Form (DAF) appears at ~ 20.21 at the same locations as WTS I.

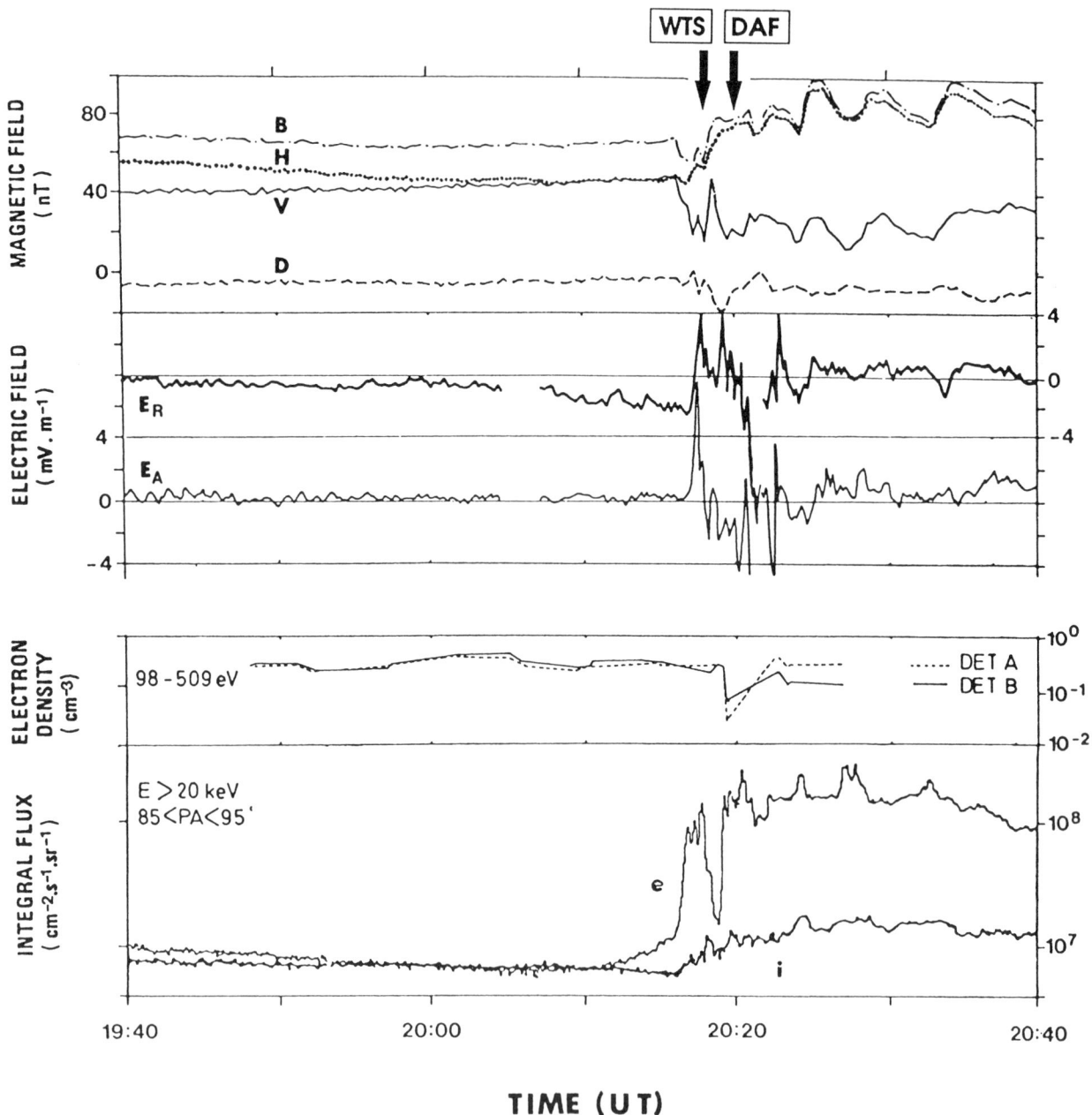

Fig. 2 (from Roux et al., 1991). Composite view showing one hour of data. From top to bottom: (i) the 3 components (VDH) of the magnetic field plus the total magnetic field (B), (ii) the radial and azymuthal component of the electric field, E_R and E_A, respectively, (iii) the electron density from detector A measuring perpendicular to the spin axis and B along the spin axis, and (iv) the integral flux of energetic electrons and ions above 20 keV for electrons and 27 keV for ions.

weak increase prior to the substorms. This increase is only observed, however, in the 2 lower energy channel (20-30 keV), as will be discussed later.

The <u>active phase</u> starts at 20.17 UT and lasts until 20.25; it is characterized by a sharp transition from a tail-like to a dipole-like configuration of the magnetic field. This transition is not monotonic; it is characterized by large fluctuations of the three components of the magnetic field. The total magnetic field (B) starts decreasing from 20.16 to 20.18 UT. Then, after 20.18 UT, it increases above its initial value. The first stage, i.e. the decrease of B, is associated with a decrease of the V component, H staying approximately constant. Later, after 20.18, the increase of B is due to an increase of the H component, V staying approximately constant, on average. Similarly, transient electric field spikes with very large amplitudes (E ≥ 10 mV.m^{-1}) are observed. During the active phase of the substorm, the number density of low energy electrons (> 500 eV) shows a deep minimum, as already pointed out by *Shepherd et al.* [1980]. Finally, the fluxes of both energetic electrons and ions suddenly rise up to values comparable to (for ions), or larger (for electrons) than the value they had well before the substorm. As noted by *Sauvaud and Winckler* [1980], the flux of energetic electrons is strongly correlated with the magnetic configuration; low fluxes correspond to tail-like and high fluxes to dipole-like configurations.

At the <u>end of dipolarization</u>, after 20.25, the dipolar configuration is established, since the H component is close to the total magnetic field B. Fluctuations of the magnetic field are still present but they now have longer periods and are more regular. These pulsations are predominantly compressional. B and the flux of energetic electrons are in antiphase. Low (E < 500 eV) and medium (0.5 < E < 20 keV) energy fluxes remain typical of the plasmasheet population; GEOS is still in the plasmasheet, but the magnetic configuration is more dipole-like, thus indicating that the tail current has been dissipated.

Figure 2 already gave evidence for a slow increase in the integral flux of electrons; this increase starts at ~ 20.00, about 20 mn before the breakup. Differential fluxes displayed in Figure 3 not only confirm this trend, but also indicate that it is limited to the lowest energy channels (essentially from 20 to 27 keV). At higher energies, the differential fluxes keep on decreasing slowly until the breakup. We believe this increase in the intensity of the lower energy channel is due to the inward motion of the plasmasheet and/or to its heating prior to the breakup. Such an interpretation is consistent with the model of *Goertz and Smith* [1989] who suggest that, prior to the breakup (that they attribute to a thermal catastrophe), the CPS is heated via resonant absorption of Alfven waves. An enhanced convection can also drive the plasmasheet earthward: there is indeed an increase in the modulus of the electric field, from 20.00 to 20.17, but this electric field is earthward-directed and not westward, as one would expect for an enhanced convection. This deviation from the westward direction is probably due to the location of the spacecraft, close to the Harang discontinuity, as evidenced by the radial electric field that builds up prior to breakup.

At ~ 20.17, differential fluxes start increasing in all energy channels, at the same time as the magnetic field configuration abruptly changes; this change is better seen in Figure 2, last panel, where the time resolution for the integral intensity is 5.5 s instead of 1 mn for Figure 3. At ~ 20.19, a second dispersionless increase of the electron intensity occurs in all energy channels.

Differential intensities of ions are displayed in Figure 3b. Here again, no significant energy dispersion can be measured; most of the energy channels show a peak at ~ 20.18. On the lowest energy channel (~ 27 keV), however, there is a decrease at 20.18 followed by two increases. This delay is not related to an energy dispersion; the observed transient oscillations in the flux of the lowest energy ions in Figure 3b are likely to be due to the development of the instability discussed in the rest of the paper. Time resolution, unfortunately, is not good enough on differential plots to ascertain this interpretation.

In summary, the study of the differential intensities of energetic electrons and ions leads to the identification of four time periods: (i) before 20.00 UT, in all energy channels, the differential intensities of energetic electrons and ions is constant or slowly decreasing, as the tail-like magnetic configuration builds up, (ii) from ~ 20.00 to ~ 20.17, the highly stressed tail-like configuration remains essentially the same, the intensities of the most energetic electrons and ions keep on decreasing slowly but the electron intensity at low energies increases, suggesting an heating of the plasmasheet, (iii) from 20.17 to ~ 20.20, dramatic increases in the intensities of electrons and ions occur. Differential intensities give evidence for transient oscillations, with a phase shift between electrons and ions, and some time lag between peaks at the lower and the higher energies, but no significant energy dispersion, and (iv) after 20.20, more regular oscillations of electrons and ions are observed, with little, if any, energy dispersion.

The main conclusion of this Section is that the dramatic increase observed at breakup in the differential intensities of energetic electrons and ions is not consistent with an energy dispersion. *Roux et al.* [1991] have shown that there is no pitch angle dispersion either (with a very high time resolution of 5.5 s). Then, energetic electrons and ions

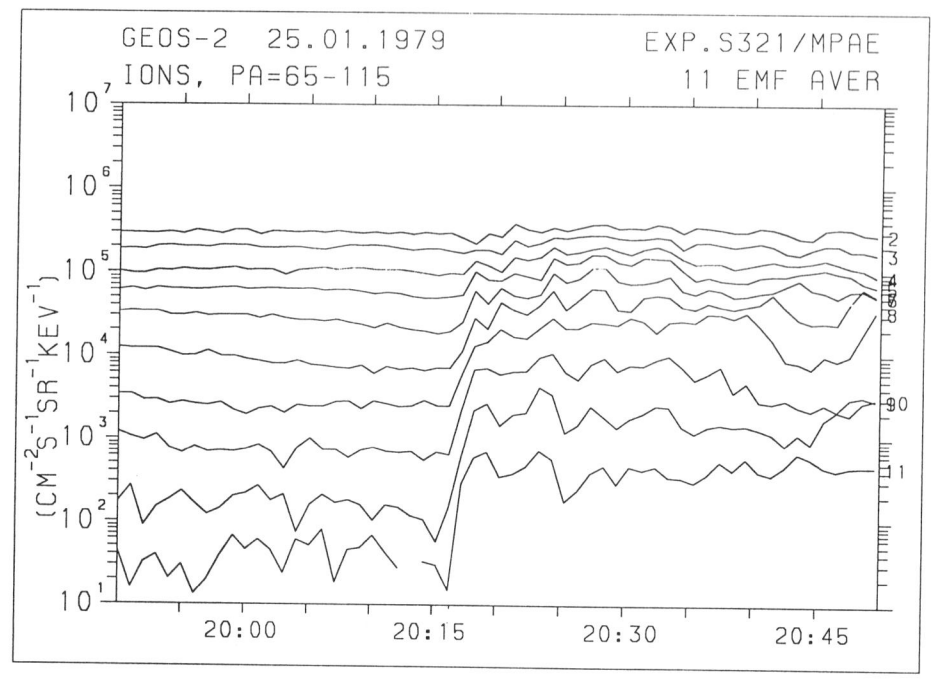

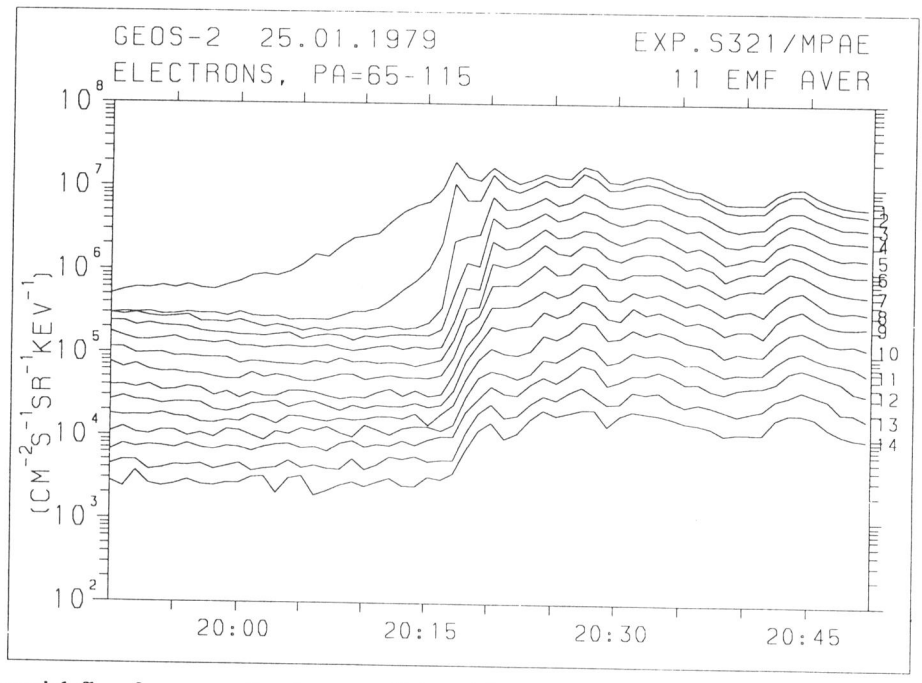

Fig. 3. Differential flux for energetic electrons (3a) and ions (3b), plotted over one hour, with ~ 1 mn time resolution. Notice that no clear energy dispersion shows up.

must have been accelerated locally or/and the spacecraft must have crossed a region with a sharp gradient in the intensities of energetic ions and electrons. Let us now use high time resolution measurements (5,5 s) to give evidence for the development of a ballooning instability.

3. BALLOONING INSTABILITY: EXPERIMENTAL EVIDENCE

Most of the parameters displayed in Figure 4 are the same as in Figure 2, but a higher time resolution is used : only

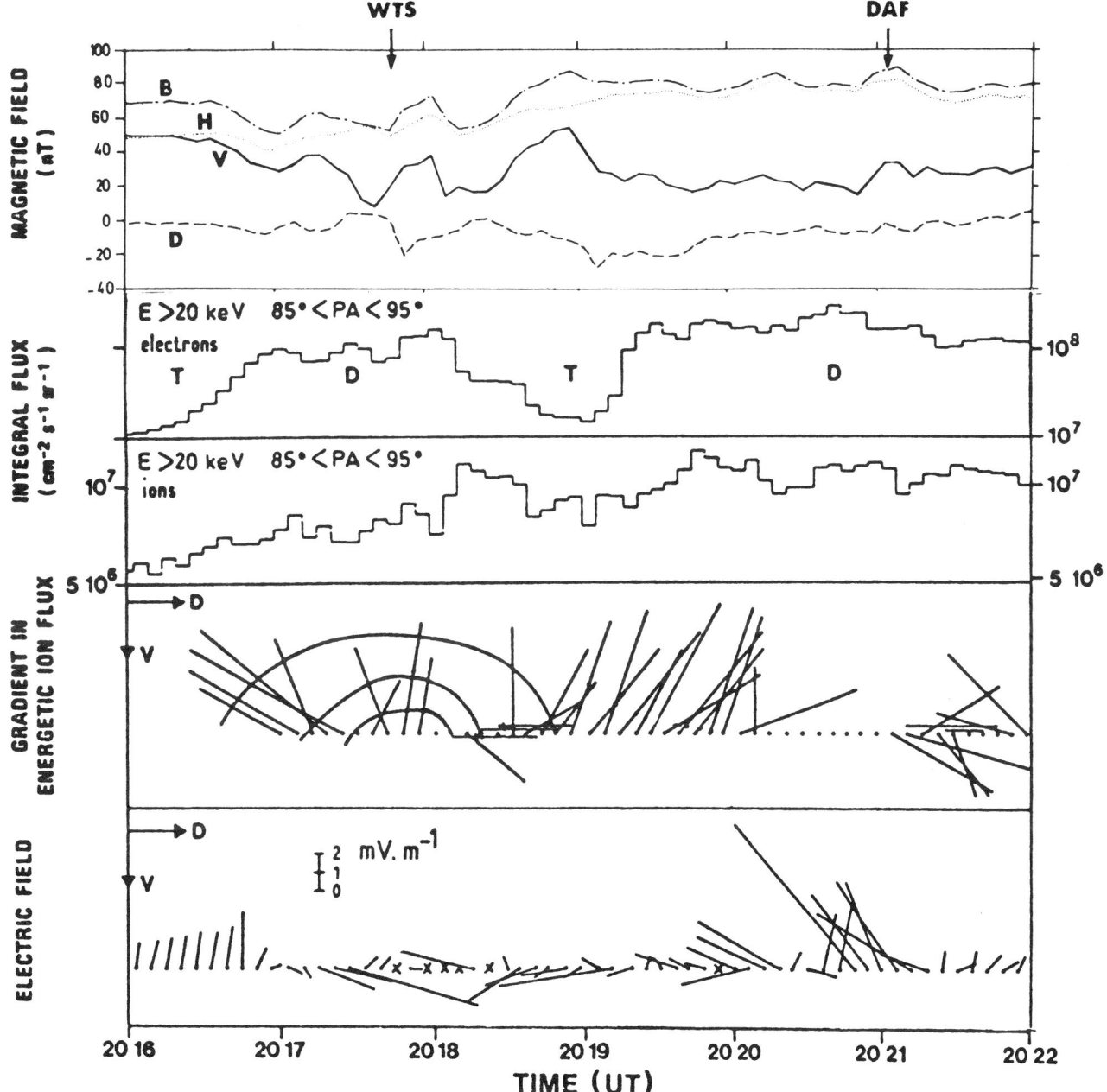

Fig. 4. Detailed composite figure showing (i) the magnetic field data in the VDH frame with 5.5 s resolution, (ii) flux of energetic electrons, (iii) flux of energetic ions, (iv) gradient in the flux of energetic ions, projected onto the V,D plane, and (v) direction of the spin-averaged electric field, also projected onto the V,D plane. Full lines on panel IV suggest the tentative isopressure contours; the size of the arrows is proportional to intensity of the gradient in ions.

six minutes of data are displayed. Magnetic field data show large fluctuations, together with a change from a tail-like to a dipole-like configuration. The flux of energetic electrons (panel 2) is anti-correlated with the magnetic configuration; as BV decreases, the energetic electron flux increases and vice versa. T and D in the figure mean that the magnetic field is, in average, Tail- or Dipole-like. In panel 3, the flux of energetic ions is plotted. Notice that the scale is not the same as for electrons. While the general trend is the same, the flux of energetic ions is not really well correlated with that of electrons; there is even a maximum in the ion flux from 2018 to 2019 UT, whereas the electron flux is minimum.

The flux of energetic ions is not azimuthally symmetric. Because the Larmor radii of energetic ions are large (≥ 200 km), this asymmetry reflects the presence of a spatial gradient with a typical scale of a few Larmor radii. We have estimated the size and the direction of the gradients in the ion flux, as sketched in Figure 4, panel 4. The arrows indicate the direction and amplitude of the energetic ion flux gradient in the spin plane (roughly the V,D plane). Note that, on the average, the gradient is earthward, which suggests that the source of energetic ions is located earthward and not tailward, as is often assumed. This seems to be the more frequently encountered situation, as demonstrated in a companion paper by *Korth et al.* [this issue]. There are however significant departures from the earthward direction of the gradients; in particular the gradients just after 2018 and 2021 UT are eastwards. Tentative iso-contours of the flux have been plotted, whenever possible; they are perpendicular to the direction of the gradient in the ion flux. These changes in the direction of the gradient also suggest the existence of a wave-like structure passing by the spacecraft. This idea is supported by the good correlation existing between the variations of the magnetic configuration, the flux of energetic electrons and the changes in the direction of the gradient in the ion flux.

In the fifth panel of Figure 4, electric field vectors projected onto the spin plane (V,D) are plotted. These vectors have been determined from raw data, assuming that they are quasi-stationary over a spin period of 6 s. This assumption is certainly fulfilled before 2017 UT, that is before break up, where the E field was steady and directed earthwards. At break up the E field changes more quickly and the data shown here represent estimates. Clearly the size and direction of E change quite fast, especially around 2018 and 2021 UT when a WTS (WTS I) and a Discrete Auroral Form (DAF) are observed at GEOS 2 magnetic footprint. Crosses have been drawn instead of arrows, whenever the preamplifier of the E field experiment was saturated by too large voltages. Intense E fields spikes are observed around 2018 and 2021 UT; at these times, there is a reversal in the direction of these spikes, from eastward to westward. Between these two times and later on, the electric field is essentially westwards.

Figure 4, panel 1, gives evidence for large, transient fluctuations in the magnetic field components. These data have been analysed with the help of the minimum variance analysis, applied to eight minutes of the data. This time interval comprises the most active part of the substorm. The result is shown at the top of Figure 5 where the 3 components B_1, B_2 and B_3 corresponding respectively to maximum, intermediate and minimum variance are plotted. The variances along B_1, B_2, B_3 directions are $\sigma_1 = 15$ nT, $\sigma_2 = 5.5$ nT and $\sigma_3 = 3.8$ nT, respectively, for the sliding averaged data over 1 mn. For the 5.5 s resolution data, $\sigma_1 = 15.5$ nT, $\sigma_2 = 8.2$ nT and $\sigma_3 = 5.2$ nT. Minimum variance is essentially in the azimuthal direction, $\theta \sim 80°$, $\phi \sim 92°$, where θ and ϕ are the polar and azimuthal angles, measured with respect to the VDH frame of reference. Maximum variance is therefore essentially in a meridian plane, in a direction more or less perpendicular to the initial direction of the magnetic field. In the B_2 component in particular, large amplitude, quasi periodic fluctuations are observed, with a pseudo period T of approximately 1 min.

In the lower left panel of Figure 5, an hodogram of the (sliding averaged) magnetic field is displayed, in the B_1, B_2 plane. For reference, the projection of B_1 B_2 plane on the V H plane is also sketched on the lower right panel (the angle between these planes is small but not zero). This hodogram gives evidence for a wave that is essentially polarized in the B_1, B_2 plane. The B_3 is not negligible, however. As discussed in Roux et al., 1991, the B_3 component which is along the D direction is indicative of field-aligned currents associated with the development of the instability (see Fig. 6). In addition to the signature of the wave which propagates azymuthally along the direction of minimum variance, there is a steady increase of the component along B_1, which is essentially perpendicular to the direction that the magnetic field had prior to the breakup.

Then minimum variance analysis applied to magnetic field data suggests azimuthally-propagating transient oscillations. Oscillations with the same pseudo-periods are found in (i) the direction of the gradient in the flux of energetic ions, (ii) the flux of energetic electrons, and (iii) the direction of the electric field. These observations suggest that an instability develops at breakup, in the near Earth plasmasheet. The observed gradient in the pressure of energetic ions is a clue to the identification of this instability. This gradient is, in average, directed earthwards, that is to say in the same direction as the gradient in the

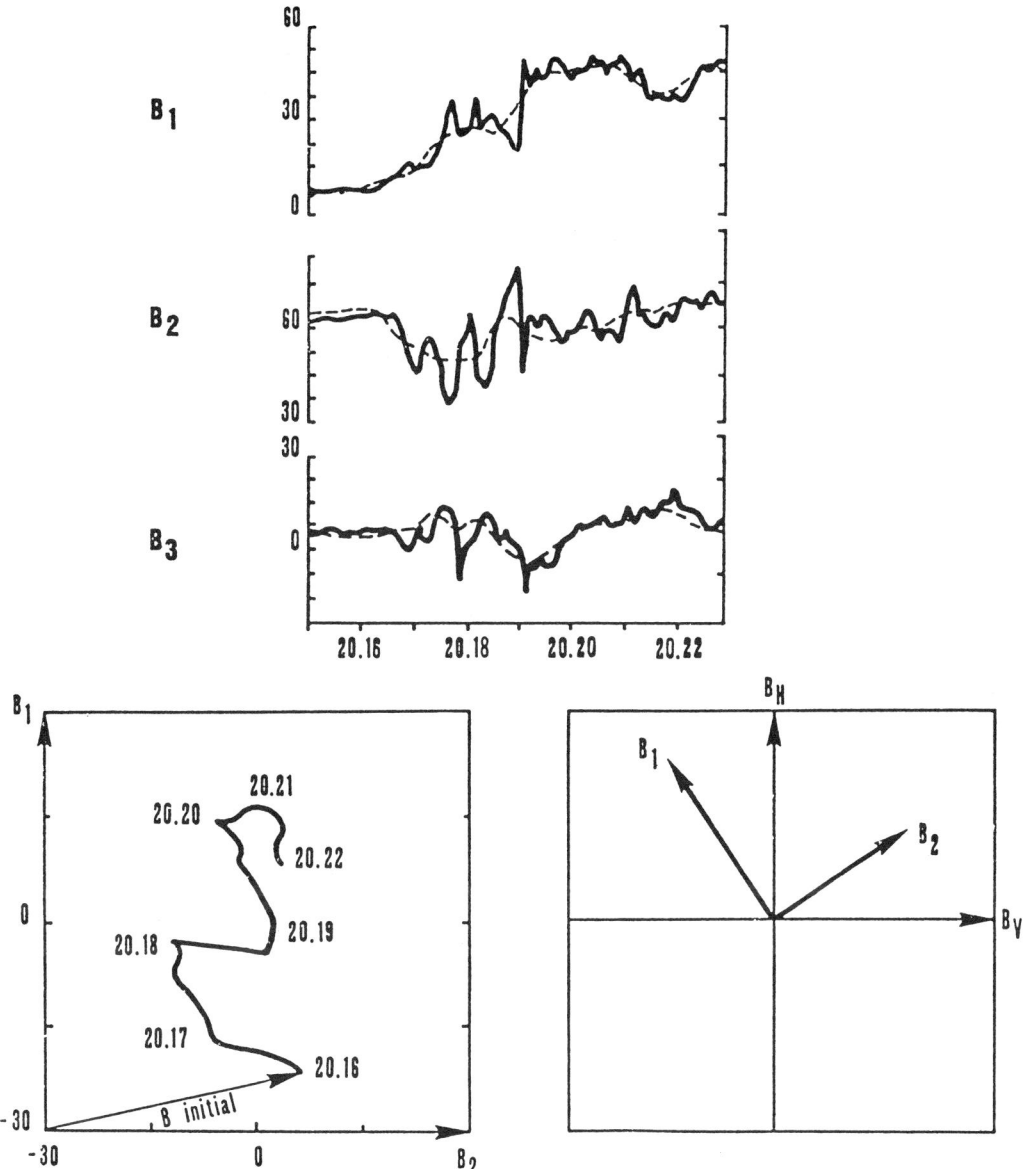

Fig. 5. Top: minimum variance analysis applied to 5.5 s resolution magnetic field data (full line). Maximum variance is for $\theta \sim 39°$, $\phi = \sim 165°$, minimum variance $\theta \sim 80°$, $\phi \sim 92°$. Dashed line corresponds to minimum variance applied to sliding averaged data; then maximum variance (B_1) corresponds to $\theta \sim 33°$, $\phi \sim 164°$ and minimum (B_3) variance to $\theta \sim 90°$, $\phi \sim 106°$. Bottom left: hodogram of sliding averaged data (B_1, B_2). Bottom right: projection of $B_1 B_2$ in the VH frame. Amplitudes are $\sigma_1 \cong 15$ nT (maximum variance), $\sigma_2 \cong 5.5$ nT (mean), $\sigma_3 \cong 3.8$ nT (minimum).

magnetic field. This situation is known to be unstable to the ballooning modes. *Korth et al.* [this issue], show that earthward-directed pressure gradients, similar to these reported here, are indeed regularly observed at GEOS orbit, prior to breakup, at least for isolated dispersionless substorms.

4. BALLOONING INSTABILITY: THEORETICAL FRAMEWORK

The idea that an earthward-directed pressure gradient is unstable to interchange or ballooning modes is not new. Long ago, *Swift* [1967] suggested that the outer edge of the ring current is unstable and that this instability could play a role in substorm dynamics. *Liu* [1970] studied the role of electrostatic low frequency drift instability in the behaviour of the ring current outer edge. He concluded that the drift flute ($k_{//} = 0$) mode is unstable. In his study, however, Liu assumed that electrostatic perturbations develop in a low β ($\beta < m_e/m_i$) plasma and that the ionosphere is a perfect conductor. These hypotheses are not applicable to present data; the perturbations are indeed electromagnetic and they take place in a large β plasma ($\beta \sim 1$). There is no obvious reason why the ionosphere could be considered as a perfectly conducting medium. More recently, *Miura et al.* [1989] have included finite β effects in their analysis and considered electromagnetic perturbations. They concluded that ballooning mode can be destabilized, provided that the scale of the pressure gradient (L_p) is small enough and that the perpendicular wave length is larger than the ion Larmor radius (ρ_i). These two conditions can be expressed as [see *Roux et al.*, 1991]

$$\beta > \frac{L_p}{R_c} > \frac{\pi^2}{16} k_\perp^2 \rho_i^2 \qquad (1)$$

where R_c is the curvature radius and β the ratio of the kinetic to the magnetic pressure. The above inegalities suggest that an arbitrary large value of β would meet the instability condition. In fact, β is constrained by the stress balance which can be expressed as

$$\frac{2L_p}{R_c} > \beta \qquad (2)$$

Then

$$\frac{2L_p}{R_c} > \beta > \frac{L_p}{R_c} \qquad (3)$$

A detailed comparison between the actual values of β, L_p, R_c, deduced from measurements made at substorm onsets, has been carried out by *Korth et al.* [this issue]. The conclusion that emerges from their study is that the above conditions are fulfilled for the 22 cases studied. In the case of the January 25th event studied here, Korth et al. obtain $\beta \sim 2.1$, $L_p = 2950$ km, $R_c = 2480$ km. Then, $2L_p/L_c = 2,4$, $\beta = 2,1$ and $L_p/R_c = 1,2$; therefore condition (3) is satisfied.

Notice that the above values are not the measured values, but the ones projected at the magnetic equator [see *Korth et al.* for discussions].

Ohtani et al. [1989] have used the bi-fluid approximation to study the ballooning instability, which allows to include the coupling between the shear Alfven and the slow magnetosonic mode. The criteria that can be deduced from their study are essentially similar, but not identical to those given above.

Even the use of the bifluid approximation is questionable, under the extreme conditions that prevail prior to substorm breakup. The large angles between the measured magnetic field and the dipole field indicate that thin current sheets localized close to the geomagnetic equator, do develop prior to substorm breakup. Then, with such a thin layer, the ion bounce frequency can easily exceed the wave frequency. A kinetic description is then needed to account for the changes in the instability conditions along the field line. In particular, it is well known, from fusion devices, that reversals in the curvature that necessarily develop if the current is pinched in the equatorial region, do play a stabilizing role. All these effects can only be described by a kinetic analysis.

5. DISCUSSION

5.1. Consequences of the ballooning instability

While the diamagnetic current is divergence-free, the currents associated with gradient and curvature drift are not divergence-free, which leads to the generation of positive and negative charges at the leading and trailing edges of the azimuthally (here westward) propagating wave. Of course, the plasma does not support charge excesses, then field-aligned currents have to flow along field lines to maintain charge neutrality. This results in a bimodal field-aligned current system, with the current flowing successively towards the Earth, carried by upflowing electrons, and from the Earth, carried by precipitating electrons.

These precipitating electrons lead to the formation of the auroral structures (WTS and DAF) observed simultaneously on the ground. Then, the field-aligned current system, associated with the development of the ballooning instability, produces an upward current above the WTS, and downward currents both to the west and to the east of the surge. This is indeed what *Opgenoorth et al.*, [1983] and *Kozelova and Lyatskiy* [1984] have deduced from their measurements.

This field-aligned current system has to close, in the equatorial region and in the ionosphere. Figure 6 shows the first loop of this current system; the closure in the plasmasheet of the loop implies an eastward current, thereby leading to a reduction or even a cancellation of the tail current, at least locally. As suggested in Figure 6, this will result in an increase of the H component that tends to recover the dipole value. Such an increase produces an induced electric field oriented to the west. Figure 7 shows that it is indeed the case; the increase of the H component does correspond to an enhanced westward electric field. Then, the dipolarization and the injection result from the current system imposed by the development of the ballooning instability.

5.2. Comparison with other observations

The idea that dispersionless injection can occur quite close to the Earth is not new; *Mauk and McIlwain* [1974] have already shown several examples of dispersionless injections of electrons and ions at substorm breakup, that they also interpret as a local injection process. *Mauk and Meng* [1987] have suggested that this injection mechanism is different from the mechanism of X line formation further out in the tail. *Roux* [1985] suggested that the injection at substorm could be due to an instability developing in the near Earth plasmasheet. Analyzing AMPTE CCE data,

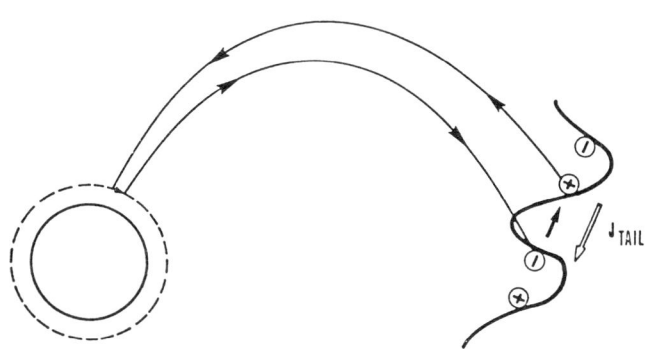

Fig. 6 (from Roux et al., 1991). Sketch showing the closure of the field-aligned current circuit and its effects on the reduction/cancellation of the tail current. Reduction in tail current implies $\partial B_H/\partial t > 0$ and hence an increase of the westward electric field.

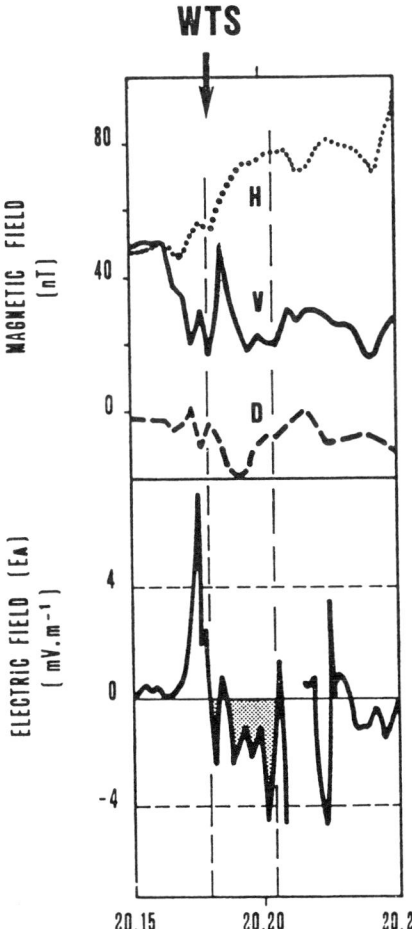

Fig. 7 (from Roux et al., 1991). Top: the 3 components of the magnetic field in a VDH frame. Notice the increase of the H component. Bottom: azymuthal component of the electric field. The shaded area corresponds to the period where the H component drastically increases.

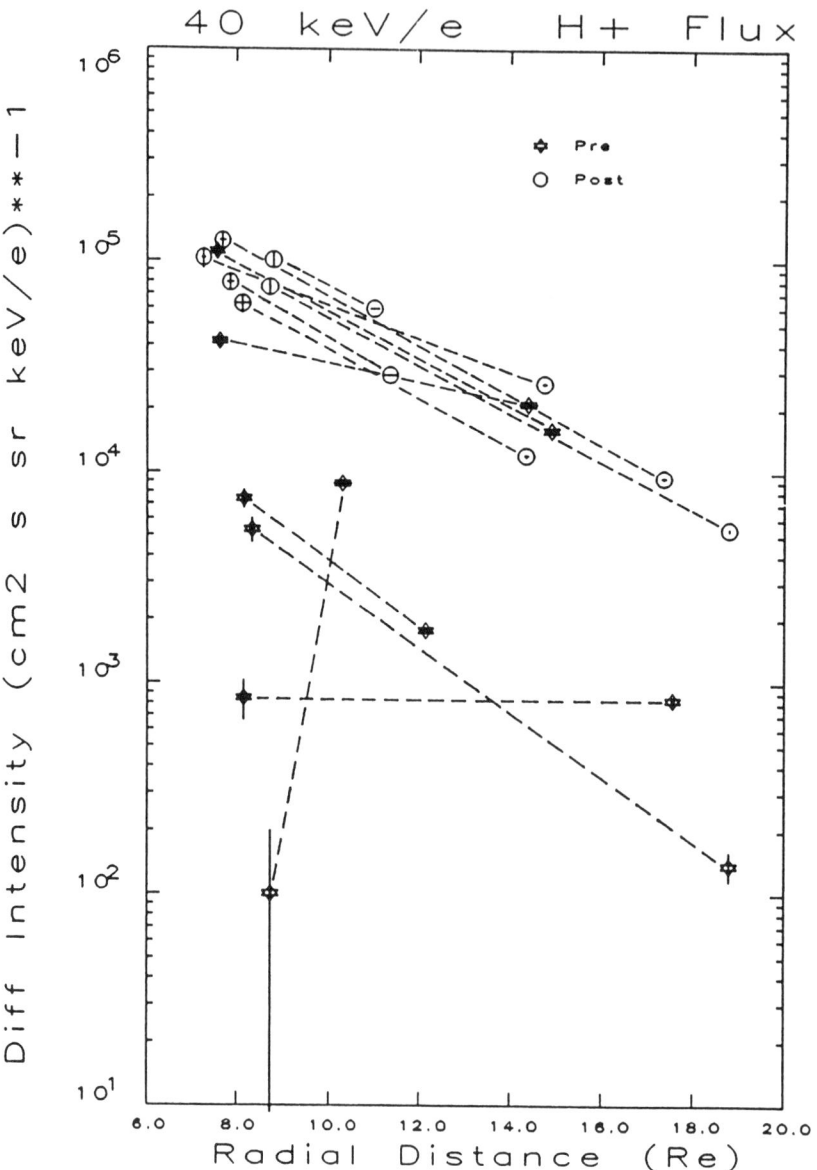

Fig. 8 (from Kistler et al., 1990). Differential intensity of protons with energies E > 40 keV/e versus radial distance. Data from both AMPTE IRM and CCE have been used to build this figure.

Lopez et al. [1988] found that dipolarization events, similar to the one reported here, can occur at all radial distances between 6.4 and ~ 9 R_E. In a more recent study, *Lopez et al.,* [1990], have analyzed the azymuthal asymmetry of energetic ions, at breakup. Results displayed in plate 3 of their paper, for instance, can be interpreted as evidence for for a strong gradient in the ion pressure pointing successively earthward and tailward. Then these authors also found that the dipolarization of the magnetic configuration is directly linked to oscillations in the direction of the ion pressure gradient. For a more detailed account of these results, see *Lui* [1991, this issue]. *Kistler et al.* [1990] have combined data from AMPTE IRM and CCE when these spacecraft were located in the night sector at different radial distances, but at similar longitudes. *Kistler et al.* have shown that the ion bursts observed at substorm breakup had very similar spectra at both locations. Post-injection fluxes at energies 40 keV were found to depend only on radial distances, and the same composition ratios O^+/H^+ and He^{++}/H^+ were found at both locations.

Figure 8, taken from *Kistler et al.* [1990], shows the pre-breakup fluxes (stars) and the post breakup fluxes (circles) for energetic ions. These fluxes are measured between ~ 7 and ~ 19 R_E by AMPTE IRM and CCE. While the pre-breakup fluxes are at random, the post breakup fluxes are well organized by the radial distance and show coherent values at the two spacecraft locations. These data are not consistent with ions being accelerated in a given region: the far plasmasheet or the inner plasmasheet, and convected afterwards, to the observation point. Instead, these data suggest that the same acceleration process can be operative over a broad region in the tail. If the ballooning instability discussed here is the process that leads to this energization of ions, then it should develop simultaneously over a broad range of radial distances, or propagate quite fast across the plasmasheet.

In a recent paper, *Jacquey et al.* [1991] have used magnetic field data from the ISEE spacecraft located in the tail lobes, to study the possible motion of dipolarization events. They found that the dipolarization usually starts quite close to the Earth, at 6-8 R_E, and later propagates or expands at ~ 300 km/s towards the tail. This analysis suggests then that the dipolarization that follows the current disruption propagates/expands tailwards at a very large velocity.

CONCLUSION

An alternative approach to the problem of the dissipation of magnetic energy stored in the geomagnetic tail has been proposed. It is based on the ballooning instability fed by strong earthward gradients in the flux of energetic ions. This instability, which does not require that the vertical component of the magnetic field goes to zero, can explain the dipolarization/injection occurring at breakup. If, as suggested by *Jacquey et al.* [1991], the dipolarization events first develop at R ~ 6-8 R_E and later propagate or expand tailwards at a large velocity, then the northward motion or expansion of the arcs are the natural consequence of the radial motion/expansion of the instability that lead to the dipolarization.

Acknowledgements. We acknowledge helpful discussions with D. Le Quéau, S. Ohtani, R. Pellat, and R. Tamao.

REFERENCES

Dungey, J.W.., Interplanetary magnetic field and the auroral zones, *Phys. Rev. Lett., 6,* 47, 1961.

Goertz, C.K., and R.A. Smith, The thermal catastrophe model of substorms, *J. Geophys. Res., 94,* 6581, 1989.

Jacquey, C., J.A. Sauvaud, and J. Dandouras, Location and propagation of the magnetotail current disruption during substorm expansion: analysis and simulation of an ISEE multi-onset event, *Geophys. Res. Lett., 18,* 389, 1991.

Kistler, L.M., E. Möbius, B. Klecker, D. Hovestadt, G. Glöcker, F.M. Ipavitch, and D.C. Hamilton, Spatial variations in the suprathermal ion distributions during substorms in the plasmasheet, *J. Geophys. Res.,* submitted to, 1990.

Korth, A., Z.Y Pu, G. Kremser, and A. Roux, A statistical study of substorm onset conditions at geostationary orbit, *this issue,* 1991.

Kozelova, T.V., and V.B. Liatskiy, A longitudinal current in front of a westward travelling surge, *Geomagn. and Aeron., 24,* 191, 1984.

Lembège, B., and R. Pellat, Stability of a two-dimensional quasineutral sheet, *Phys. Fluids, 25,* 1995, 1982.

Liu, C.S., Low frequency drift instabilities of the ring current belt, *J. Geophys. Res., 75,* 3789, 1970.

Lopez, R.E., D.G. Sibeck, A.Y.T. Lui, K. Takahashi, R.W. McEntire, T.A. Potemra, and D. Klumpar, Substorm variations in the magnitude of the magnetic field: AMPTE/CCE observations, *J. Geophys. Res., 53,* 14,444, 1988.

Lopez, R.E., H. Lühr, B.J. Anderson, P.T. Newell, and R.W. McEntire, Multipoint observations of a small substorm, *J. Geophys. Res., 95,* 18,897, 1990.

McPherron, R.L., C.T. Russell, and M.P. Aubry, Satellite studies of magnetospheric substorms on August 15, 1968. Phenomenological model for substorms, *J. Geophys. Res., 78,* 3131, 1973.

Mauk, B.H., and C.E. McIlwain, Correlation of k_p with the substorm-injected plasma boundary, *J. Geophys. Res., 79,* 3193, 1974.

Mauk, B.H., and C.I. Meng, Plasma injection during substorms, *Physica Scripta, T-18,* 128, 1987.

Miura, A. S. Ohtani, and T. Tamao, Ballooning instability and structure of diamagnetic hydromagnetic waves in a model magnetosphere, *J. Geophys. Res., 94,* 231, 1989.

Ohtani, S., A. Miura, and T. Tamao, Coupling between Alfven and slow magnetosonic waves in an inhomogeneous finite-b plasma. Coupled equations and physical mechanism, *Planet. Space Sci., 37*, 567, 1989.

Opgenoorth, H.J., R.J. Pellinen, W. Baumjohann, E. Nielsen, G. Marklund, and L. Eliasson, Three-dimensional current flow and particle precipitation in a westward travelling surge (observed during the BA-GEOS experiment), *J. Geophys. Res., 88,* 3138, 1983.

Pellat, R. Une nouvelle approche de la reconnexion magnétique : sous-orages magnétiques et vent stellaires, *C.R. Acad. Sci.*, to be published, 1991.

Roux, A., Generation of field-aligned current structures at substorm onset, *ESA SP-235*, 151, 1985.

Roux, A., S. Perraut, P. Robert, A. Morane, A. Pedersen, A. Korth, G. Kremser, B. Aparicio, D. Rodgers, and R. Pellinen, Plasma instability related to the westward travelling surge, *J. Geophys. Res.*, to be published, 1991.

Sauvaud, J.A., and J.R. Winckler, Dynamics of plasma energetic particles and fields near geosynchronous orbit in the night-time sector during magnetospheric substorms, *J. Geophys. Res., 85*, 2043, 1980.

Shepherd, G.G., R. Boström, H. Derblom, C.G. Fälthammar, R. Gendrin, K. Kaila, A. Korth, A. Pedersen, R. Pellinen, and G.L. Wrenn, Plasma and field signatures of poleward propagating auroral precipitation observed at the foot of the GEOS 2 field line, *J. Geophys. Res., 85*, 4587, 1980.

Swift, D.W., The possible relationship between the auroral breakup and the interchange instability of the ring current, *Planet. Space Sci., 15*, 1225, 1967.

PROPERTIES OF THE GEOTAIL PLASMA SHEET – THEORY AND OBSERVATION

C.J. Owen* and S.W.H. Cowley

Blackett Laboratory, Imperial College, Prince Consort Road, London, SW7 2BZ, U.K.

I.G. Richardson

NASA Goddard Space Flight Center, Code 661, Greenbelt, MD 20771, U.S.A.

Abstract. The properties of the plasma sheet formed by reconnection in the deep geomagnetic tail may be determined from consideration of the field and plasma stresses on the reconnected field lines threading a one-dimensional current sheet. In particular, we assume that the field tension is balanced by the anisotropic plasma pressure of the combined plasma populations flowing into, and out of, the tail current sheet, and calculate the speed of the outflowing plasma sheet needed to maintain this balance. Unlike previous models, we also include the possibility that the plasma is heated during its interaction with the current sheet, so that account can be taken of the diamagnetic depression of the plasma sheet magnetic field. The analysis also provides estimates of the thickness of the plasma sheet at a given downtail distance from the neutral line, and of the magnitude of the magnetic field component threading the current sheet. We briefly review differences in the plasma sheet properties expected if the flow were driven by a viscous transfer of momentum from the magnetosheath, and present an example of data which includes spacecraft encounters with both reconnection associated plasma sheet and viscously driven plasma sheet, indicating the need for both processes in explaining the configuration in the deep tail.

1. Introduction

The convection of plasma within the Earth's magnetosphere and the processes that drive it have been a source of debate for some thirty years. Dungey [1961] first outlined a model of the magnetosphere based on reconnection processes, whilst in the same year Axford and Hines [1961], suggested that plasma convection resulted from a viscous-like interaction operating at the magnetopause. The reconnection model of the magnetotail has been further developed by Coroniti and Kennel [1972, 1973], Russell and McPherron [1973] and Hones [1979 and references therein] in relation to substorm behaviour, and by Cowley [1980, 1984] in relation to plasma structures. These models have been highly successful in explaining both near-Earth and distant tail observations. The neutral line model of substorms [Hones, 1979] suggests that during geomagnetically quiet times a single neutral line exists in the deep tail at which reconnection occurs at a slow and steady rate. Plasma from the tail lobes ExB drifts into the tail current sheets lying downstream from the neutral line, and is accelerated along the reconnected field lines, forming jets of plasma which flow away from the neutral line on either side [Cowley, 1980, 1984]. A substorm sequence begins with a coherent development of the magnetosphere caused by an increased reconnection rate at the dayside magnetopause, which results in increased amounts of magnetic flux and plasma being added to the tail lobes. After a period of ~1 hour, this drives an instability of the near-Earth tail, and a new neutral line forms within the near-Earth plasma sheet. The reconnection rate at this new neutral line is sufficiently large to disconnect the pre-substorm plasma sheet from the Earth after a period of a few minutes, thus forming a closed loop plasmoid structure, which is subsequently accelerated down the tail under the influence of the magnetic tension of field lines reconnected at the substorm neutral line, leaving a new "post-plasmoid plasma sheet," a region of tailward accelerated plasma lying downstream of the new neutral line. The properties of both the quiet time, and "post-plasmoid" plasma sheets may be determined from considerations of stress balance on the reconnected field lines [e.g., Cowley, 1980]. In this paper, we will review the problem of stress balance in a one-dimensional current sheet, and extend previous models of this type by including the effect of plasma heating at the current sheet. We determine the plasma sheet outflow velocity, the thickness of the current sheet, and the value of the magnetic field component threading the current sheet, assuming reasonable values of tail parameters under different geomagnetic conditions.

The "viscous" models of the tail structure are less well developed, but in general, involve the transportation of closed magnetic field

---

* Now at: NASA Goddard Space Flight Center, Code 696, Greenbelt, MD 20771, U.S.A.

lines into the tail region by processes that transfer mass and momentum from the magnetosheath. Several processes have been suggested (e.g., Heikkila, [1982, 1987]; Pu and Kivelson, [1983]; Lemaire, [1987]), and in this paper we briefly review the general expectations of plasma sheet structure based on these processes.

Finally, we present an example of data (and review more extensive work) in which we observe two plasma sheet regions, one conforming to the expectation of reconnection theory, and one which can only be explained in terms of a "viscous" mechanism. Although we cannot identify the exact nature of the process driving the plasma convection in this latter region, we show that the process transfers momentum, but not mass, into the magnetosphere, and that the plasma populations involved most likely originate in the hot, near-Earth plasma sheet.

2. Theory

a) Reconnection-Associated Plasma Sheet

The occurrence of reconnection at a neutral line within the tail current sheet results in the formation of sharply bent magnetic field lines threading through the current sheet. In the distant tail, where the plasma pressure gradient along the current sheet is small, the magnetic tension of these structures is balanced by the stress imposed by anisotropic plasma pressure of the combined plasma populations flowing into and out of the tail current sheet. Under these circumstances one-dimensional equilibrium models may be set up (e.g., Cowley, [1980]; Cowley and Southwood, [1980]), in which the variation of the system parameters in the direction perpendicular to the current sheet is rapid compared to the variations along the current sheet. In the rest frame of the reconnected field lines, in which the electric field transverse to the plane of the current sheet has been transformed away [deHoffman and Teller, 1950], the general stress balance conditions for a one-dimensional current sheet are

$$M_{\parallel} - P_{\perp} = \frac{B^2}{\mu_0} \quad (1)$$

$$P_{\perp} + \frac{B^2}{2\mu_0} = K \quad (2)$$

(e.g., Owen and Cowley, [1987b]), where $M_{\parallel}$ and $P_{\perp}$ are the components of the plasma stress tensor parallel and perpendicular to the field direction respectively, B is the field strength outside the current sheet, and K is a constant. Previous models based on one-dimensional stress balance have made the simplifying assumption that the plasma flowing into the current sheet from each lobe may be represented by a cold field-aligned beam of particles, which is then turned around within the current sheet, and emerges as a cold beam of equal speed. Under these assumptions, it can be shown that the plasma must enter and exit the current sheet at the Alfven speed just outside of the current sheet, V_A, and that the plasma sheet flow speed in the Earth's frame is of order $V_0 = 2V_A \pm v_L$, where v_L is the field-aligned downtail flow speed of the lobe plasma, taken to be positive when directed away from the Earth. The positive sign in this expression applies to regions tailward of the neutral line, where

positive V_0 represents tailward flow, whilst the negative sign applies to the region Earthward of the neutral line where the flow is Earthward for V_0 positive.

However, these models do not take into account any heating of the plasma during its interaction with the current sheet, which in general results in substantial diamagnetic depressions in the plasma sheet field strengths. Under these circumstances, the plasma outflow speed V_0 given above is an overestimate. In this paper, we shall reconsider the one-dimensional stress balance problem to include, in a simplified manner, the effect of heating of the outflow plasma. The field structure assumed in this work is shown in Figure 1(i), in which the magnetic field (solid arrowed lines) first becomes depressed in strength on entering the plasma sheet (shown dotted) from the lobe, and then reverses in direction across a thin current sheet (shown hatched). In common with other models, we shall assume that in the field line rest frame (Figure 1(ii)) the plasma flowing into the current sheet from the lobes on each side may be considered a cold field-aligned beam of density n_L, and speed $v_{\parallel L}$, such that the perpendicular component of the plasma pressure $P_{\perp} = 0$ in the lobe regions. Also

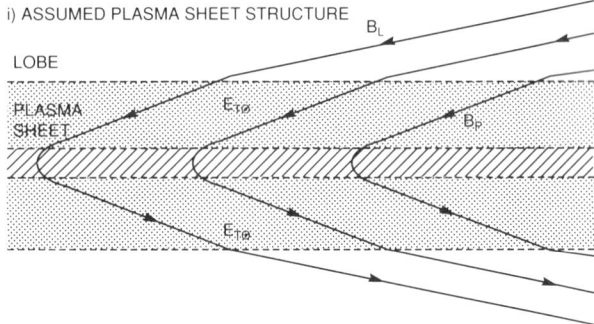

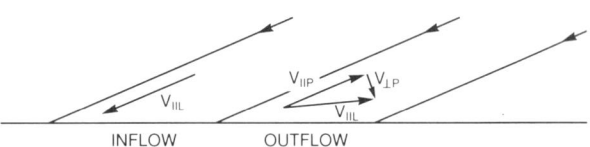

Fig. 1. (i) Sketch of the fields of the 1-D current sheet system employed in this paper. Within the plasma sheet (shown dotted) the component of the magnetic field transverse to the thin current sheet (shown hatched) is reduced in strength compared to that in the lobe due to the diamagnetic effect of the heated plasma in this region, and reverses in direction across the current sheet. Throughout the system the field component normal to the current sheet is small compared to the transverse components. In the context of the deep geomagnetic tail, the Earth is to the left of the figure. (ii) The plasma flows in the frame in which the cross-tail electric field has been transformed away. The particles move into the central current sheet at speed $v_{\parallel L}$ along the field direction, but are scattered in pitch angle and emerge with velocity components $v_{\parallel P}$ and $v_{\perp P}$. In the absence of the electric field, the particles experience no change in speed during their interaction with the current sheet.

in common with the earlier models, we assume that there is no change of the plasma speed during its interaction with the current sheet due to the absence of the tangential electric field in this frame, but we do allow for scattering to larger pitch angles, resulting in an outflow beam with velocity components both parallel and perpendicular to the field direction. For simplicity, we assume the outflow plasma has velocity components $v_{\|P}$ and $v_{\perp P}$ respectively, as shown in Figure 1(ii), where $v_{\perp P}$ can be considered a mean perpendicular velocity providing a perpendicular pressure component in the plasma sheet

$$P_{\perp} = \frac{1}{2} n_P m_i v_{\perp P}^2 \quad , \tag{3}$$

where n_P is the density of the outflowing plasma population, and m_i is the ion mass (the electron contribution is neglected throughout this analysis). Since we have assumed that the kinetic energy of the ions is conserved during the interaction with the current sheet, we also have

$$v_{\|L}^2 = v_{\|P}^2 + v_{\perp P}^2 \quad , \tag{4}$$

We further require that the combined inflow and outflow populations provide no net field-aligned flux of ions into the current sheet. Using this condition, the relationship between the density of the inflow beam and that of the outflow beam is simply given by

$$n_L v_{\|L} = n_P v_{\|P} \quad , \tag{5}$$

Finally, in setting up this problem, we shall assume that the system is in perpendicular pressure balance through the entire tail structure, and that in particular, equation (2) also holds across the boundary between the lobe and the plasma sheet. If the field strength in the lobe is given by B_L, and that in the plasma sheet by B_P (where $B_L > B_P$ due to the diamagnetic depression of the latter), we obtain a second expression for the perpendicular pressure exerted by the plasma sheet outflow population

$$P_{\perp} = \frac{(B_L^2 - B_P^2)}{2\mu_0} \quad , \tag{6}$$

Substituting from equations (3) and (4) we obtain the expression

$$v_{\|L}^2 - v_{\|P}^2 = \frac{(B_L^2 - B_P^2)}{\mu_0 m_i n_P} \quad , \tag{7}$$

whilst equation (1), expressing the balance of the magnetic field tension with the stresses exerted by the plasma, yields

$$M_{\|} = \frac{(B_L^2 + B_P^2)}{2\mu_0} \tag{8a}$$

$$= m_i(n_L v_{\|L}^2 + n_P v_{\|P}^2) \quad , \tag{8b}$$

Eliminating n_L by use of equation (5), and after some algebra, we obtain solutions for the field aligned velocities of the inflow and outflow populations in the field line rest frame

$$v_{\|L} = \frac{(3B_L^2 - B_P^2)}{B_L(8\mu_0 m_i n_P)^{1/2}} \quad , \tag{9}$$

$$v_{\|P} = \frac{(B_L^2 + B_P^2)}{B_L(8\mu_0 m_i n_P)^{1/2}} \quad , \tag{10}$$

and $$n_P = \frac{n(3B_L^2 - B_P^2)}{4B_L^2} \quad , \tag{11}$$

where $n = n_P + n_L$ is the total density in the plasma sheet, due to the simultaneous presence of both inflow and outflow populations. Note that in the limit $B_L = B_P$, such that there is no diamagnetic depression of the plasma sheet magnetic field, the solutions reduce to $v_{\|L} = v_{\|P} = V_A$, in which the plasma flows both into and out of the current sheet at the Alfven speed just outside the current sheet, which is the solution found in previous models.

Having determined the nature of the field-aligned flow of the plasma moving into and out of the current sheet in the field line rest frame which is consistent with the required stress balance conditions (equations (1) and (2)), and a diamagnetic depression of the plasma sheet field, we now need to transform back into the rest frame of the Earth. We assume here that the field-aligned flow is closely parallel to the plane of the current sheet (i.e. the component of the magnetic field, B_n, normal to the current sheet is negligible compared to the transverse components). Performing the transformation, we obtain a plasma sheet outflow speed

$$V_0 = v_{\|L} + v_{\|P} \pm v_L \tag{12}$$

$$= \frac{2B_L}{(2\mu_0 m_i n_P)^{1/2}} \pm v_L \quad ,$$

where again v_L is the lobe bulk plasma flow speed, taken to be positive in the direction away from the Earth. As before, the positive sign in this expression applies to regions tailward of the neutral line where positive V_0 indicates tailward flow, and the negative sign applies to regions lying Earthward of the neutral line, with positive V_0 indicating flow towards the Earth. Taking typical values of lobe field strength $B_L = 9.2$ nT, plasma sheet field strength $B_P = 3.7$ nT, total plasma sheet density $n = 0.2$ cm^{-3} and lobe bulk flow velocity $v_L = 150$ km s^{-1} in the deep tail (e.g., Slavin et al., [1985]; Zwickl et al., [1984]), equations (9), (10) and (12) combine to give a plasma sheet flow speed (assuming a location tailward of the neutral line) of $V_0 \simeq 900$ km s^{-1} giving excellent agreement with observed average plasma sheet flow speeds of 890 km s^{-1} [Richardson and Cowley, 1987]. Note that previous models which neglect the perpendicular pressure of the outflow plasma would estimate this value to be 1085 km s^{-1} in this case, an increase of about 20%. Note also that we can simply define the strength of the normal component of the magnetic field B_n threading the current sheet since the contraction speed of the sharply bent magnetic field lines in the Earth rest frame is given by

$$V_F = v_{\|L} \pm v_L = \frac{E_T}{B_n} \tag{13}$$

where E_T is the electric field tangential to the current sheet [deHoffman and Teller, 1950]. Since $v_{||L}$ is fixed by the requirements of stress balance, the value of B_n is directly proportional to the value of the cross-tail electric field E_T, which, applying Faraday's law, is also a measure of the rate of reconnection at the neutral line. In the geomagnetic tail we expect B_n to be positive in the region Earthward of the neutral line and negative tailward thereof.

We may also determine the thickness of the plasma sheet at a given distance downstream from the neutral line. In the time that the plasma flows a distance L from the neutral line at speed V_0 it moves a distance

$$D = \frac{L}{v_0} \cdot \frac{v_{||P} B_n}{B_P} \quad (14)$$

away from the current sheet on each side. The full thickness of the plasma sheet may thus be written

$$W_{PS} = 2D = \frac{2E_T L(V_0 - V_F)}{V_0 V_F B_P}, \quad (15)$$

Note that the plasma sheet thickness is also proportional to the reconnection rate E_T. It is not generally possible to determine a value for this parameter from the deep tail data, so reasonable values must be assumed based on the prevailing geomagnetic conditions. If we assume the above typical values, and take a value of E_T corresponding to a quiet-time voltage of 35 kV across a tail of width ~ 50 R_E (~ 0.1 mV m^{-1}), we obtain a value $|B_n| = 0.14$ nT, which is much less than the field strengths in either the lobe or the plasma sheet, as previously assumed. At a distance of 100 R_E tailward of the neutral line position (corresponding, for example, to the relative position of the ISEE-3 spacecraft at -200 R_E to a quiet time neutral line at -100 R_E), equation (15) implies a plasma sheet thickness $W_{PS} = 1.9 R_E$. On this basis, we expect the quiet-time plasma sheet in the deep tail to be a relative thin structure, such that encounters with it should be relatively rare and confined to brief traversals across the tail centre plane.

However, during disturbed times, we may expect the voltage across the tail to be significantly enhanced, to values of order 100 kV (corresponding to $E_T \sim 0.3$ mV m^{-1}), and for the neutral line to lie much closer to the Earth. Under these circumstances, the theory outlined above gives $|B_n| = 0.4$ nT, in good agreement for example with the results of Slavin et al. [1987], and in the deep tail (e.g., ISEE-3 located at 200 R_E tailward of the neutral line position) the plasma sheet has a thickness 11.4 R_E. Thus we expect the plasma sheet to be much thicker during periods of enhanced geomagnetic activity compared to that expected for quieter times.

Furthermore, due to the finite outflow speed of the plasma sheet ions, there must exist a region of reconnected field lines which the outflowing plasma sheet ions have yet to reach, such that the local field and thermal plasma characteristics remain those of the tail lobe. These field lines thread the plasma sheet in their equatorial regions, so that energetic particles from the plasma sheet population may, by virtue of their higher speed, stream out into this region forming an energetic particle "plasma sheet boundary layer" located between the outer edge of the plasma sheet population and the field separatrix mapping back to the neutral line. There is an inherent velocity filter effect within this system, such that successively faster particles may be expected to fill successively larger volumes from the edge of the plasma sheet out to the separatrix (Hill, [1975]; Cowley, [1980]). Applying the principle of magnetic flux conservation, we find that the thickness of this layer on either side of the plasma sheet at distance L from the neutral line is

$$W_{BL} = \frac{E_T L}{V_0 B_L} \quad (16)$$

Under the conditions outlined above, we find that the thickness of this layer in the deep tail is of order 1.2 R_E during quiet periods, and of order 7.2 R_E during more disturbed times.

The theory outlined above thus gives theoretical estimates for the outflow speed of the plasma sheet, the strength of the field component threading the tail current sheet, and the thickness of the plasma sheet and its boundary layer. We have assumed that the current sheet is one dimensional, and we have included in a simple fashion the effect of plasma heating and the consequent diamagnetic depression of the plasma sheet field strength. The model can be generalized further to include the effects of slow variations of parameters along the current sheet (e.g., Cowley and Southwood, [1980]), the effect of time-dependent reconnection rates and neutral line motion [Owen and Cowley, 1987a], and the effects of asymmetries in the tail lobe plasma conditions [Owen and Cowley, 1987b]. The effects of including time-dependent reconnection, and neutral line motion consistent with the neutral line model of substorms are shown in Figure (2), where we have assumed the typical lobe and plasma sheet magnetic fields and plasma densities discussed above. The top two panels of this figure show the assumed time-dependent reconnection rates E_T and position of the neutral line x_N consistent with the neutral line model of substorms. The dashed lines represent the reconnection rate and position of the pre-existing quiet-time neutral line, whilst the solid line represents the parameters associated with the new substorm-related neutral line. The quiet-time neutral line is located at 100 R_E with $E_T = 0.1$ mV m^{-1}, whilst the substorm-associated neutral line is formed at 20 R_E at time t = 0, with $E_T = 0.3$ mV m^{-1}. For simplicity, we have further assumed that reconnection at the quiet-time neutral line ceases at t = 5.5 min. This is the time required for the new neutral line to reconnect the flux within the quiet time plasma sheet, ending in the disconnection of a plasmoid in the neutral line model. The new neutral line is taken to reside at 20 R_E for 30 min. after its formation, after which time it begins to move tailwards and the reconnection rate declines (the recovery phase of the substorm) until it reaches 100 R_E at t = 90 min., when the quiet time conditions are re-established.

The lower two panels of Figure (2) indicate the thickness of the plasma sheet and its boundary layer (third panel) and the magnitude of B_n threading the current sheet at a position 200 R_E from the Earth. During the quiet time (t < 0) the plasma sheet at 200 R_E is relatively thin, and the component of the field threading the current sheet is small, as described above. On the basis of the neutral line

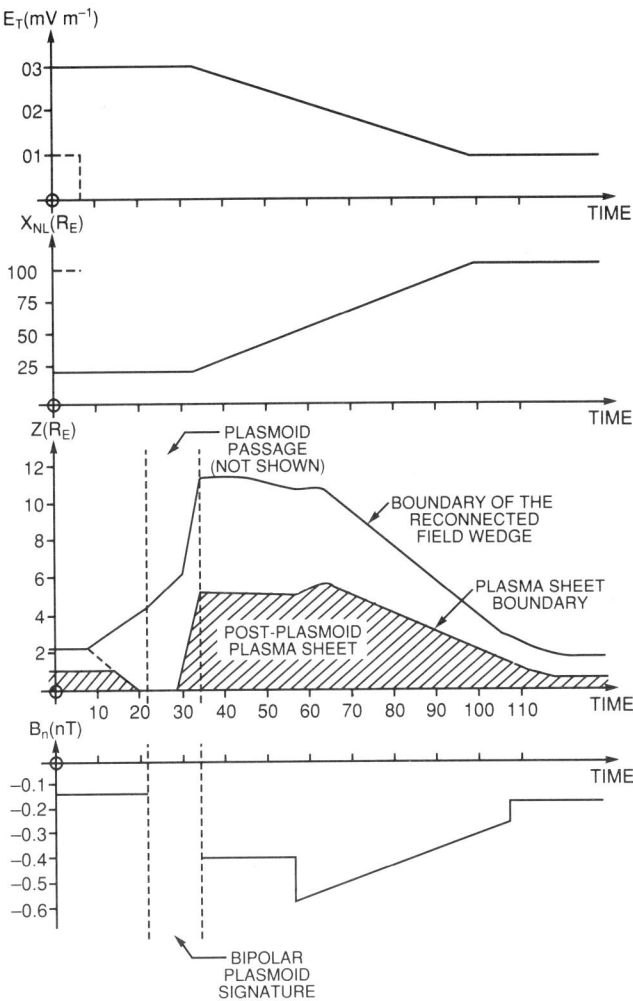

Fig. 2. Time development of the structure of the deep-tail plasma sheet during substorm activity, predicted from models based on one-dimensional stress balance of reconnected field lines. The upper two panels show the assumed reconnection rates and locations of both the quiet-time (dashed traces) and substorm (solid traces) neutral lines, which are consistent with the neutral line model of substorms. The third panel shows the half thickness of the plasma sheet and the thickness of the plasma sheet boundary layer at 200 R_E downtail from the Earth during the substorm sequence, whilst the lower panel indicates the expected component of the field threading the tail current sheet at this location. Between t = 21 mins and t = 33.5 mins we expect the signatures of the passage of a plasmoid past 200 R_E which are not described by our model. Adapted from Owen and Cowley [1987a].

model of substorms, a closed loop plasmoid structure is located between the last reconnected field line from the quiet-time neutral line, and the first open field line reconnected at the substorm neutral line. The passage of this structure past 200 R_E occurs between t = 21 and t = 33.5 minutes in our model, consistent with

observations made by Baker et al., [1987]. During this period, B_n is expected to show a bipolar signature, and the plasma sheet layers will be displaced outwards in a manner dictated by the size and shape of the plasmoid. These effects are not described by our model, and so are not included in Figure (2).

After the passage of the plasmoid past 200 R_E, there remains a thick "post-plasmoid plasma sheet" and boundary layer in which the component of the magnetic field threading the current sheet is somewhat larger than that during the quiet period. This structure persists until the effects of the recovery phase of the substorm appear at 200 R_E, when the plasma sheet and its boundary layer begin to return to their quiet-time thicknesses. After an initial jump in B_n associated with the onset of neutral line motion, this parameter also shows a steady decline to quiet time values. We note that this plasma sheet B_n variation is not inconsistent with the B_z signature observed in the lobes following the passage of "travelling compression regions", the lobe signature of plasmoids (e.g., Slavin et al., [1984]). Full details of this time-dependent plasma sheet structure may be found in Owen and Cowley [1987a].

b) Viscously Driven Plasma Sheet

As mentioned in the introduction, it has been suggested that the plasma sheet flow may also be driven by viscous effects operating along the tail magnetopause. The exact mechanism which results in the transfer of magnetosheath momentum (and possibly mass) into the geomagnetic tail is still the subject of current debate. Suggested mechanisms include, for example, impulsive penetration of magnetosheath plasma through the magnetopause [Heikkila, 1982, 1987; Lemaire, 1987], wave-driven spatial diffusion (e.g., Tsurutani and Thorne, 1982], and the Kelvin-Helmholtz instability [Pu and Kivelson, 1983 and references therein]. It is not our purpose here to review each of these mechanisms in detail, but we point out below some of the general properties expected if one or more of these mechanisms is operational in the tail plasma sheet. Firstly, in the absence of reconnection, we expect that the tail plasma sheet will be populated entirely by closed flux tubes which are carried tailward by the transfer of momentum from the magnetosheath. Since the mechanism involved must work against the tension of the magnetic field (i.e. energy is being stored in the field), we expect the tailward plasma flow speeds to be less than those observed in the adjacent magnetosheath, and in general much less than those expected from the reconnection-associated plasma sheet described above. In addition, energetic ions located in the boundary layer of such a viscously driven structure should always flow Earthward, whilst energetic electrons should be isotropic, consistent with the closed field structure. Finally, we expect that the component of the magnetic field threading through the central current sheet should be northward at all locations in the tail.

These properties of the plasma sheet are sufficiently different from those expected if the plasma sheet is driven by reconnection processes, so that examination of data from the deep tail should reveal whether each of these processes occurs in the tail, and if so, which is dominant. In the next section, we present an example of data in which both types of plasma sheet may be seen, and summarise other work on this topic.

3. Observations

Figure (3) shows 4 hours of data recorded on 28 January, 1983 by the ISEE-3 spacecraft located some 215 R_E downtail from the Earth. The first four panels show data from the Energetic Particle Anisotropy Spectrometer [e.g., Balogh et al., 1978]. The first of these panels shows the direction-averaged flux of 35 keV protons, whilst the remaining three panels indicate their bulk flow speed and direction (latitude and longitude). In the plasma sheet these particles are convected with the thermal plasma [e.g., Daly et al., 1984; Richardson et al., 1987], such that these plots provide an indication of the speed and direction of the plasma sheet flow. In these plots $\phi_p = 180°$ represents tailward flow, and $\phi_p = 90°$ is duskward, while $\theta_p = 90°$ is northward. The next four panels show data from the Los Alamos electron plasma analyser, indicating (from top to bottom) the ecliptic plane bulk speed and azimuth, temperature, and number density. Finally, the lower three panels show the magnetic field strength and its GSM latitude and longitude at 64 s resolution, as determined by the JPL magnetometer. The regions encountered by the spacecraft are indicated in the top panel (PS = plasma sheet, L = lobe, MS = magnetosheath).

Before examining this period in detail, we point out that this interval is characterised by low levels of geomagnetic activity, except for the occurrence of substorm activity whose expansion phase onset occurred towards the end of this period, at 0740 UT (see also Baker et al., [1988]). On the basis of the discussion above, we expect brief encounters with relatively thin plasma sheet structures, in which the plasma flow speeds are of order 900 km s^{-1}. The properties of the plasma sheet observed during this period do not agree with these expectations. In particular, the long encounters with the plasma sheet suggest a thickness much greater than the 1.9 R_E anticipated, and the flow speeds are well below 900 km s^{-1} for the entire period. However, we note that these values were determined on the basis of average values of lobe and plasma sheet magnetic field strength and plasma density. During the first encounter with the plasma sheet (0400 - 0450 UT) and with the lobe (0450 - 0520 UT), we note lower values of the magnetic field strengths, and an unusually high plasma sheet density. Using observed values of $B_L = 7$ nT, $B_p = 3$ nT, and $n = 0.5$ cm^{-3}, together with a lobe bulk flow speed $v_L = 100$ km s^{-1}, we obtain a value of $V_0 = 465$ km s^{-1}, a value of $B_n = -0.28$ nT, and a plasma sheet thickness of 4.4 R_E at a distance of 100 R_E tailwards of the neutral line (assuming $E_T = 0.1$ mV m^{-1}, as above). With the exception of the value of B_n, these theoretically derived values are not inconsistent with the observations made during the first encounter with the plasma sheet. The increased thickness may explain the extended nature of the encounter, and the flow velocity is close to that determined from the ion data, although the electron velocities are usually slightly lower. One difficulty with the identification of this region as a reconnection-associated plasma sheet tailward of the neutral line is the predominantly positive value of the observed B_z ($\theta > 0$). However, Richardson et al. [1989] have shown that the plasma sheet is tilted through approximately 90° during this period, such that B_z does not correspond to the component of the magnetic field B_n normal to the current sheet in this case.

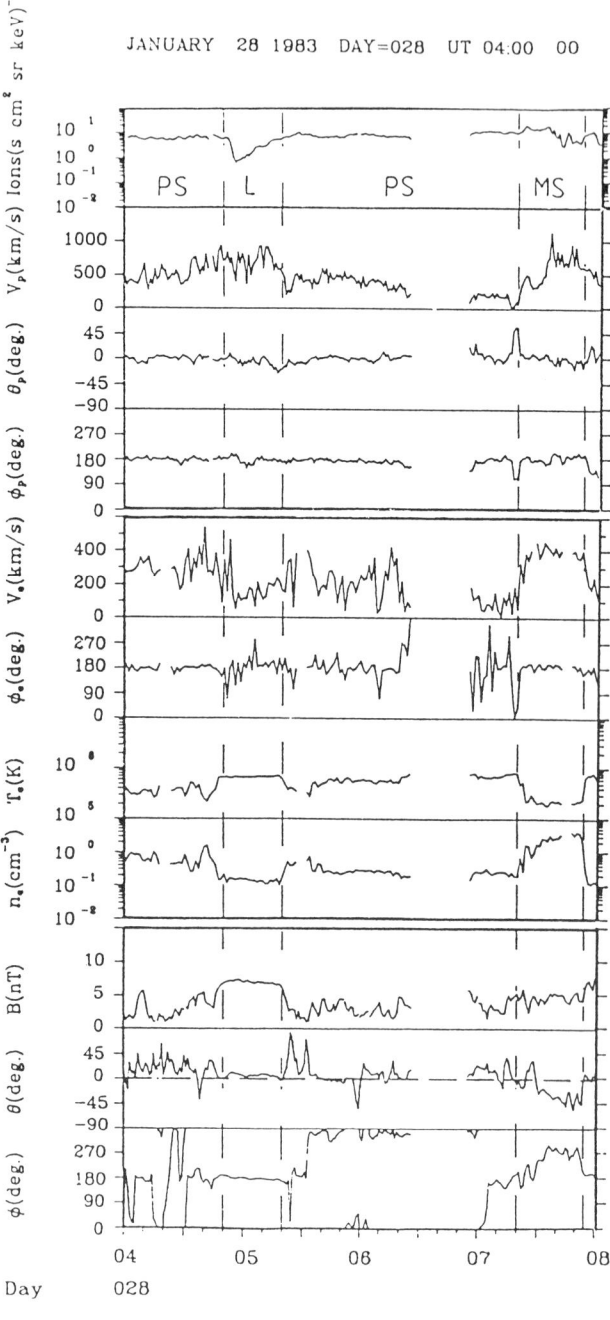

Fig. 3. ISEE-3 data for the period 0400 - 0700 UT on 28 January 1983, when the spacecraft was located at GSM coordinates (-215, 3, -7) R_E. The first four panels show EPAS energetic ion data, from the top, the direction-averaged intensity of 35-56 keV ions, and the bulk ion flow velocity and polar and azimuthal directions. The next four panels show the Los Alamos electron plasma ecliptic plane flow speed and azimuth, temperature and density, whilst the JPL magnetic field magnitude and GSM direction are shown in the lower three panels. The regions encountered by the spacecraft are indicated in the top panel, as described in the text.

After the spacecraft returns to the plasma sheet at 0520 UT, we note that the tailward plasma sheet flow speeds of both the ions and electrons begin to decline significantly from those observed during the previous plasma sheet encounter, reaching values of order 100 km s^{-1} just prior to crossing the magnetopause at 0720 UT. At the same time, there is a decline in the number density to ~ 0.3 cm^{-3}, resulting in an increase in the flow speed expected for a reconnection-driven plasma sheet. Hence it seems unlikely that the plasma sheet observed during this period, particularly that after the data gap, can be reconciled with the hypothesis of reconnected field lines lying tailward of the neutral line. Since the flow remains in a predominantly tailward direction, these data suggest the existence of a separate population located on the tail flanks in which some form of viscous interaction is driving the antisunward flow. The low flow speeds may then be accounted for on the basis that this layer is located on closed field lines being dragged tailwards against the opposing field tension (see also Slavin et al., [1987]). The plasma temperature and densities observed in this region are essentially the same as those observed during the earlier reconnection-associated plasma sheet, and show no evidence of significant admixture of the cooler, denser plasma from the magnetosheath. The process driving this layer therefore transfers momentum into the tail without at the same time transferring mass. The properties of this layer are therefore not those of the low latitude boundary layer observed inside the near-Earth magnetopause (e.g., Eastman et al., [1976]).

The above example of data therefore confirms that there are at least two processes which may drive the plasma sheet convection in the deep tail. One of the processes results in plasma sheet properties which agree closely with the expectations of stress balance of reconnected field lines, whilst the other requires an alternative explanation, probably based on viscous-like interactions with the magnetosheath flow. In a more extended study of this type, Richardson et al. (1989) considered the properties of the plasma sheet under a wide range of geomagnetic conditions. In general, these authors find good agreement between the observed properties of the plasma sheet and the expectations based on current sheet stress balance. In many of these cases the plasma sheet flow speed significantly exceeds that in the magnetosheath, such that an explanation in terms of a viscous interaction may be completely ruled out. However, they also find several examples, during periods of low or intermittent geomagnetic activity, of slow tailward plasma sheet flow just inside the tail magnetopause, which cannot be accounted for on the basis of stress balance of reconnected field lines. In some of these cases, energetic electron data indicates that the field lines in these regions are indeed closed. The plasma density and temperature in these regions do not differ significantly from those typical of the plasma sheet in general, nor from those of adjacent reconnection-associated plasma, and do not indicate the entry of plasma from the magnetosheath. Richardson et al. [1989] conclude on the basis of these results that the slow plasma sheet flux tubes originate in the hot plasma sheet region earthward of the tail neutral line (and not in the low latitude boundary layer). After reconnection and an interval of Earthward flow, these flux tubes retreat tailwards once more under the influence of a viscous transfer of momentum from the magnetosheath, as shown schematically in Figure (4). Although the nature of the viscous interaction is not categorically determined, several of the intervals of slow plasma sheet indicate the possible occurrence of the Kelvin-Helmholtz instability, in common with evidence presented by Sibeck et al. [1987]. Note also that the tail configuration presented in Figure (4) is also compatible with the magnetic field observations made by Slavin et al. [1985], which indicate that in the deep tail the B_n component in the plasma sheet is on average negative in the centre of the tail,

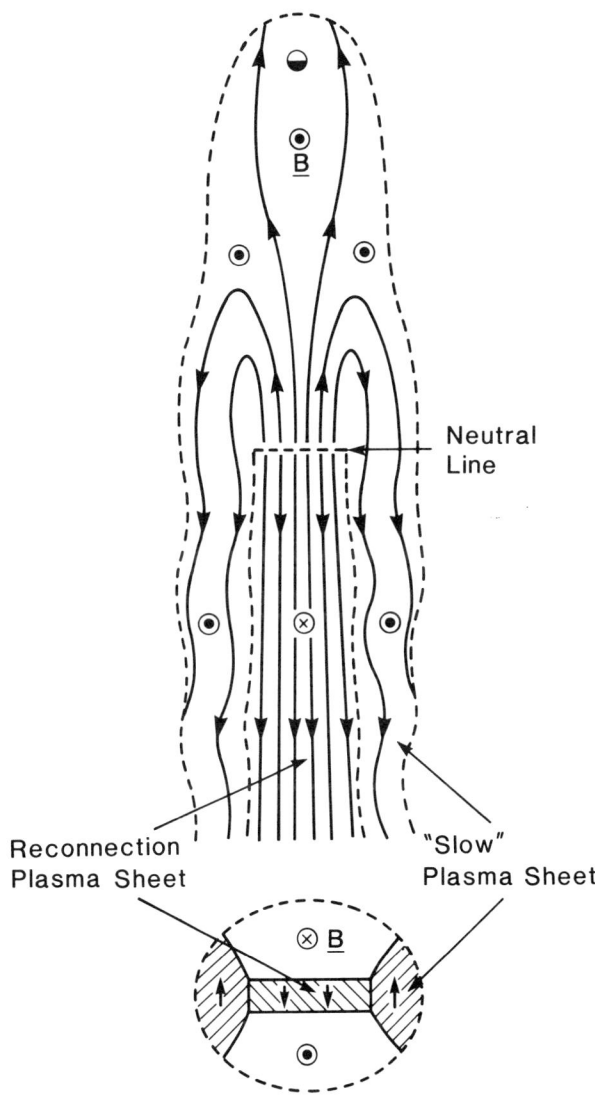

Fig. 4. Sketches showing the structure of the extended tail during magnetically quiet periods. The upper figure shows the plasma flow in the equatorial plane, whilst the lower figure shows a cross-section through the tail at a location tailward of the neutral line. The arrows in the latter diagram indicate the direction of the equatorial magnetic field. From Richardson et al. [1989].

consistent with location tailwards of the neutral line, and positive on the flanks, indicating the presence of closed magnetic flux tubes. The average flow is tailwards in both regions, but slower on the flanks.

4. Summary

In this paper, we have extended existing models of the plasma sheet structure based on the balance of stresses on reconnected field lines threading a one-dimensional current sheet. We have included in a simple way the effects of heating of the plasma during its interaction with the current sheet and the resulting diamagnetic depression generally observed in the deep-tail plasma sheet. Previous models which do not include these effects result in an overestimate of the plasma sheet outflow speed. The thickness of the plasma sheet at a given distance downstream from the neutral line, as well as the component of the magnetic field threading the current sheet can also be determined from this analysis, providing that a reasonable estimate of the cross-tail electric field is made. For a given set of lobe and plasma sheet conditions, the thickness of the plasma sheet should be greatest during periods of high levels of geomagnetic activity.

The properties of the plasma sheet determined from one-dimensional stress balance considerations cannot be duplicated if one assumes that the plasma sheet is driven by a viscous transfer of momentum. In particular, the retreat of magnetic field lines from the neutral line and the associated release of magnetic energy results in high speed plasma flows along the field lines away from the neutral line on either side. The tailward flow of such plasma typically exceeds the magnetosheath flow by a factor of two. In contrast, tailward flow speeds of viscously driven plasma must remain below those in the magnetosheath since work must be done against the tension of the closed field lines. We have shown, however, that there is a requirement for both reconnection and viscous processes to fully explain the structure of the deep geomagnetic tail, especially during periods of low geomagnetic activity. Although the nature of the viscous interaction cannot be explicitly determined from the data we have discussed, we have shown that the process transfers momentum, but not mass, across the tail magnetopause, resulting in the tailward convection of plasma on closed field lines and plasma from the previously reconnected plasma sheet Earthward of the neutral line.

References

Axford, W.I., and C.O. Hines, A unifying theory of high latitude geophysical phenomena and geomagnetic storms, *Can. J. Phys., 3,* 1433, 1961.

Baker D.N., R.C. Anderson, R.D. Zwickl, and J.A.Slavin, Average plasma and magnetic field variations in the distant magnetotail associated with near Earth substorm effects, *J. Geophys. Res., 92,* 71, 1987.

Baker D.N., J.D. Craven, R.C. Elphic, D.H. Fairfield, L.A. Frank, H.J. Singer, J.A. Slavin, I.G. Richardson, C.J. Owen, and R.D. Zwickl, The CDAW-8 substorm event on 28th January 1983: A detailed Global study, *Adv. Space Res., 8(9),* 113, 1988.

Balogh A., G. van Dijen, J. van Genechten, J. Henrion, R. Hynds, G. Korfmann, T. Iverson, J. van Rooijen, T. Sanderson, G. Stevens, and K.-P. Wenzel, The low energy proton experiment on ISEE-C, *IEEE Trans. Geosci. Electron., GE-16(3),*176, 1978.

Coroniti, F.V. and C.F. Kennel, Changes in the magnetospheric configuration during the substorm growth phase, *J. Geophys. Res., 77,* 3361, 1972.

Coroniti, F.V. and C.F. Kennel, Can the ionosphere regulate magnetospheric convection?, *J. Geophys. Res., 78,* 2837, 1973.

Cowley, S.W.H., Plasma populations in a simple open model magnetosphere, *Space Sci. Rev., 25,* 217, 1980.

Cowley, S.W.H., The distant geomagnetic tail in theory and observation, in *Magnetic Reconnection in Space and Laboratory Plasmas, Geophys. Monogr. Ser.,* Vol. 30 edited by E.W. Hones Jr., p228, AGU, Washington, D.C.,1984.

Cowley, S.W.H. and D.J. Southwood, Some properties of a steady state geomagnetic tail, *Geophys. Res. Lett., 7,* 833, 1980.

Daly, P.W., T.R. Sanderson, and K-P. Wenzel, Survey of energetic (E > 35 keV) ion anisotropies in the deep geomagnetic tail, *J. Geophys. Res., 89,* 10733, 1984.

deHoffman, F., and E. Teller, Magneto-hydrodynamic shocks, *Phys. Rev., 80,* 692, 1950.

Dungey, J.W., Interplanetary magnetic field and the auroral zones, *Phys. Rev. Lett., 6,* 47, 1961.

Eastman, T.E., E.W. Hones Jr., S.J. Bame, and J.R. Asbridge, The magnetospheric boundary layer: Site of plasma, momentum and energy transfer from the magnetosheath into the magnetosphere, *Geophys. Res. Lett., 3,* 685, 1976.

Heikkila, W.J., Inductive electric field at the magnetopause, *Geophys. Res. Lett., 9,* 877, 1982.

Heikkila, W.J., Neutral sheet crossings in the distant magnetotail, in *Magnetotail Physics,* edited by A.T.Y. Lui, p65, the Johns Hopkins University Press, Baltimore, Md., 1987.

Hill, T.W., Magnetic merging in a collisionless plasma, *J. Geophys. Res., 80,* 4689, 1975.

Hones, E.W., Jr., Transient Phenomena in the magnetotail and their relationship to substorms, *Space Sci. Rev., 23,* 393, 1979.

Lemaire, J., Interpretation of the northward Bz (NBZ) Birkland current system and polar cap convection pattern in terms of the impulsive penetration model, in *Magnetotail Physics,* edited by A.T.Y. Lui, p65, the Johns Hopkins University Press, Baltimore, Md., 1987.

Owen, C.J. and S.W.H. Cowley, Simple models of time dependent reconnection in a collision free plasma with an application to substorms in the deep geomagnetic tail, *Planet. Space Sci.,35,* 451, 1987a.

Owen, C.J. and S.W.H. Cowley, A note on current sheet stress balance in the geomagnetic tail for asymmetric tail lobe plasma conditions, *Planet. Space Sci., 35,* 467, 1987b.

Pu., Z-U., and M.G. Kivelson, Kelvin Helmholtz instability at the magnetopause: solution for compressible plasmas, *J. Geophys. Res., 88,* 841, 1983.

Richardson I.G. and S.W.H. Cowley, Plasmoid associated energetic ion bursts in the distant geomagnetic tail, in *Magnetotail Physics,* edited by A.T.Y. Lui, p65, the Johns Hopkins University Press, Baltimore, Md., 1987.

Richardson I.G., S.W.H. Cowley, E.W. Hones Jr., and S.J. Bame, Plasmoid-associated energetic ion bursts in the deep geomagnetic tail: Properties of plasmoids and the post plasmoid plasma sheet, *J. Geophys Res., 92,* 9997, 1987.

Richardson I.G., C.J. Owen, S.W.H. Cowley, A.B. Galvin, T.R. Sanderson, M. Scholer, J.A. Slavin and R.D. Zwickl, ISEE-3 observations during the CDAW-8 Intervals: Case studies of the distant geomagnetic tail covering a wide range of geomagnetic activity, *J. Geophys Res., 94,* 15189, 1989.

Russell, C.T. and R.L. McPherron, The magnetotail and substorms, *Space Sci. Rev., 15,* 205, 1973.

Sibeck, D.G., J.A. Slavin, and E.J. Smith, ISEE-3 magnetopause crossings: Evidence for the Kelvin-Helmholtz instability, in *Magnetotail Physics,* edited by A.T.Y. Lui, p65, the Johns Hopkins University Press, Baltimore, Md., 1987.

Slavin, J.A., E.J. Smith, B.T. Tsurutani, D.G. Sibeck, H.J. Singer, D.N. Baker, J.T. Gosling, E.W. Hones and F.L. Scarf, Substorm related travelling compression regions in the distant tail: ISEE-3 geotail observations, *Geophys Res. Lett., 11,* 657, 1984.

Slavin, J.A., E.J. Smith, D.G. Sibeck, D.N. Baker, R.D. Zwickl, and S-I. Akasofu, An ISEE-3 study of average and substorm conditions in the distant magnetotail, *J. Geophys Res., 90,* 10875, 1985.

Slavin, J.A., P.W. Daly, E.J. Smith, T.R. Sanderson, K-P. Wenzel, R.P. Lepping, and H.W. Kroehl, Magnetic configuration of the distant plasma sheet: ISEE-3 observations, in *Magnetotail Physics,* edited by A.T.Y. Lui, p59, the Johns Hopkins University Press, Baltimore, Md., 1987.

Tsurutani B.T. and R.M. Thorne, Diffusion processes in the magnetopause boundary layer, *Geophys Res. Lett., 9,* 1247, 1982.

Zwickl, R.D., D.N. Baker, S.J. Bame, W.C. Feldman, J.T. Gosling, E.W. Hones Jr., D.J. McComas, B.T. Tsurutani, and J.A. Slavin, Evolution of the Earth's distant magnetotail: ISEE-3 electron plasma results, *J. Geophys Res., 89,* 11007, 1984.

MID-TAIL ION DYNAMICS AT SUBSTORM ONSET

D. C. Delcourt

Space Science Laboratory, NASA/MSFC, Hunstville, Alabama 35812

Abstract. Features of ion trajectories during transient "tail-like" to "dipole-like" reconfigurations of the geomagnetic field are investigated by means of three-dimensional single particle codes. It is shown that the large electric fields induced by the collapse of the geomagnetic tail result in rapid injections of the mid-tail (~10-15 R_E) populations into low L shells. During this transport, the computations demonstrate a possible transient violation of the particle adiabatic invariants due to local field variations on the time scale of the gyro-period. This effect, which depends upon the particle charge-to-mass ratio, can yield intense (up to the hundreds of kilo-electron volt range) ion accelerations and pitch angle variations. The circulation of terrestrial outflows is further examined within this process, and it is shown that the cleft ion fountain is a potential plasma source for the storm-time ring current. The high-latitude upflowing polar wind also appears of particular interest in this context, since it yields dense and energetic proton shells in inner magnetospheric regions. A detailed analysis of the trajectory results furthermore reveals a clear earthward boundary for the newly-created energetic populations, which corresponds to the post-dipolarization location of the ion adiabatic-nonadiabatic separatrix in the geomagnetic tail.

Introduction

Numerous studies [e.g., Cummings et al., 1968; McPherron, 1979; Sauvaud and Winckler, 1980; Moore et al., 1981] have pointed out the dynamic character of the magnetosphere at substorm onset and presented evidences for a rapid reconfiguration of the magnetic field lines from a distorted "tail-like" geometry to a more dipolar one. Associated with such events, large electric fields have been observed [e.g., Shepherd et al., 1980; Aggson et al., 1983] very likely of an inductive nature, since, according to Faraday's law, they correlate well with the temporal variations of the magnetic field. The existence of such intense transients is of major importance for the understanding of magnetospheric plasma transport since they yield particle accelerations well above those resulting from large-scale convection

Magnetospheric Substorms
Geophysical Monograph 64
This paper is not subject to U.S. copyright. Published in 1991 by the American Geophysical Union.

[Pellinen and Heikkila, 1978]. Indeed, in disturbed times, a variety of measurements displays bursts of energetic particles [e.g., Hones et al., 1976; Krimigis and Sarris, 1979] and injections into the ring current region [e.g., Geiss et al., 1978; Lennartsson et al., 1981].

By means of systematic trajectory calculations based on the guiding center approximation, Mauk [1986] quantitatively analyzed the "convection surge" resulting from the collapse of the geomagnetic tail and demonstrated clear correlations between field line dipolarization and flux enhancements in the near-Earth region. The trajectory-based study of Cladis and Francis [1985] also brought out a feasible connection between the high-latitude ionosphere and the ring current during active times. More recently, Delcourt et al. [1990a] further examined the storm-time particle orbits and pointed out a two-step behavior for the ions with transient violation of their adiabatic invariants in the mid-tail region (~10-15 R_E). This effect is due to local field variations on the time scale of the particle gyro-period. The implications of this outcome on the circulation of specific terrestrial outflows were examined in two subsequent reports [Delcourt et al., 1990b; Delcourt and Sauvaud, 1991], and the present paper is a review of these latter studies. A brief description of the storm-time trajectory tracing technique will be first given, followed by an illustration of transient non-adiabatic behaviors. It will then be shown that, under such conditions, high-latitude ionospheric outflows can substantially contribute to the near-Earth ring current, and it will be demonstrated that the newly-created energetic populations can be delineated by means of simple non-adiabatic arguments.

Three-Dimensional Particle Code

The storm-time trajectory calculations were carried out using the magnetic field model of Mead and Fairfield [1975]. The geomagnetic field **B** was thus considered to be the sum of the Earth dipole, $\mathbf{B}^{dip}$, and an external contribution at four different levels of geomagnetic activity. Accordingly, the storm-time collapse of the geomagnetic tail was modeled by a gradual variation of this latter term from a given Kp level to a less disturbed one. Such a procedure may be questionable in some respects since the model of Mead and Fairfield [1975] is statistical in nature, representing average rather than typical instantaneous states of the geomagnetic field. The four states of this model were somewhat arbitrarily set to cover the full range of mean field variability, and so transitions between them are likewise somewhat arbitrary rather than physical. Nevertheless, it does seem clear that partial dipolarization is a fundamental feature of substorms, and the field variations derived hereinafter are in qualitative agreement with those measured in situ (see, for instance, Figure 1 of Delcourt et al. [1990a]).

Throughout expansion phase, the magnetic and induced electric fields in a given position $\mathbf{r}$ were respectively obtained from

$$\mathbf{B}(\mathbf{r},t) = \mathbf{B}^{dip}(\mathbf{r}) + curl[\mathbf{A}(\mathbf{r},t)] \tag{1}$$

and

$$\mathbf{E}^{ind}(\mathbf{r},t) = -\frac{\partial \mathbf{A}(\mathbf{r},t)}{\partial t} \tag{2}$$

Here, $\mathbf{A}$ represents the instantaneous vector potential associated with the Mead and Fairfield external magnetic contribution, the temporal dependence of which was modeled by means of a polynomial of degree 5 in order to account for the $\mathbf{E} = 0$ boundary conditions (the reader can consult Appendix A of Delcourt et al. [1990a] for a detailed description of this calculation). Also, it must be pointed out that calculation of $\mathbf{E}^{ind}$ according to (2) results in a substantial component in the direction parallel to $\mathbf{B}$. In the present computations, this component was cancelled by assuming, in first approximation, highly mobile electrons along the field lines. To conform with this assumption (see Appendix B of Delcourt et al. [1990a]), an additional electric field $\mathbf{E}^{pol}$ was superimposed, which redistributes the field in the perpendicular direction and accounts for plasma polarization [Heikkila and Pellinen, 1977]. In practice, this electrostatic field was calculated by integration of neighboring magnetic field lines down to the planet's surface (assumed to be equipotential [e.g., Birmingham and Jones, 1968]), and by making use of the analytical expression: $\mathbf{E}^{pol}_{\parallel} = \mathbf{b} \cdot \partial \mathbf{A}/\partial t$, where $\mathbf{b}$ denotes a unit vector in the direction of $\mathbf{B}$. As is, the particle code features only two adjustable parameters, namely: the collapse duration and amplitude. In the following, a 2-minute magnetic transition from highly disturbed to ground state configurations was chosen. As mentioned above, this yields peak-induced electric fields in qualitative agreement with those observed (specifically, a maximum magnitude of ~5 mV/m at the location of Aggson et al. [1983]) and insures realistic influences on the particles. Note at last that, owing to its relatively weak magnitude, the large-scale convection electric field was neglected throughout the dipolarization process.

As for the equation of motion, it must be stressed that a general condition attached to the validity of the guiding center approximation is that the field variations are negligible during the particle cyclotron turn [Northrop, 1963]. For a time-dependent magnetic field as in the present case, this condition is equivalent to

$$\chi = \tau_g \frac{\dot{B}}{B} = \frac{2\pi m \dot{B}}{qB^2} \ll 1 \tag{3}$$

where τ_g is the particle gyro-period, m its mass, and q its charge. It follows from equation (3) that the guiding center approximation must be questioned in presence of large temporal variations of the magnetic field. As will be seen in the following sections, precise trajectory calculations indeed reveal non-adiabatic transport features in regions as close as $L \sim 10$, and this has made it necessary to trace all particles using the full equation of motion

$$\ddot{\mathbf{r}} = \frac{q}{m}[\mathbf{E} + \dot{\mathbf{r}} \times \mathbf{B}] + \mathbf{g} \tag{4}$$

where $\mathbf{r}$ is the instantaneous position of the particle ($\mathbf{g}$ being the gravitational acceleration). Equation (4) was integrated by means of a fourth-order Runge-Kutta technique with a variable time step Δt corresponding to a $10°$ gyration for H+ (i.e., $\Delta t \ll \tau_E$, noting τ_E the electric field variation time scale). Convergence of (4) using such temporal interval for particle pushing was checked by means of backward integration in time.

Violation of Adiabatic Invariants

In order to illustrate the perpendicular behavior of the particles, Figure 1 presents the results of the calculations for two equatorially mirroring O+ initiated with 100 eV energy and opposite gyration phases at 8.5 R_E geodistance on the tail axis. As for their drift path, it is apparent from Figures 1a and 1b that both ions are not tied to the initial field line, the simultaneous displacement of which is spotted by triangles. Indeed, prior to travelling earthward, the particles undergo a short-lived westward transport under the effect of an intense polarization drift [e.g., Chen, 1974]. On the other hand, their energy variations in Figures 1c and 1d clearly demonstrate a substantial breaking of their first adiabatic invariant. Indeed, while, in the present case, magnetic moment conservation should yield betatron energization by about a factor 3, one of the O+ (Figure 1c) experiences a net energy gain of the order of 6, the other one being in contrast decelerated by about a factor 2 (Figure 1d). These features result from particle motion in directions alternatively parallel and antiparallel to the rapidly growing $\mathbf{E}$, which alternatively yields ion acceleration and deceleration; hence, the strong sensitivity to initial gyro-phase.

As for the ion magnetic moment, Figures 1e and 1f present the temporal evolution of its first order magnitude ($\mu = m(\dot{\mathbf{r}} - \mathbf{U}_E)^2/2B$, $\mathbf{U}_E$ being the $\mathbf{E} \times \mathbf{B}$ drift velocity) normalized to the final value. It is apparent from both figures that fast μ changes occur at the onset of expansion phase, which are followed by damped oscillations about the final value. While this latter effect relates to local $\mathbf{B}$ variations within a cyclotron turn, the former suggests a determining role of the instantaneous polarization drift in the violation of the first adiabatic invariant. This was demonstrated in Figure 8 of Delcourt et al. [1990a] and led these authors to conclude to a two-step behavior for the particles, namely: (1) at large geodistances where the field varies on the time scale of the ion gyro-period, a non-adiabatic transport is obtained, characterized by significant energization (or, possibly, de-energization) in the perpendicular direction. As will be more apparent in Figure 3 hereinafter, this conversely results into a significant scattering of the particle pitch angle. (2) Subsequently, as the particles travel into closer L shells, this transient behavior progressively evolves into a more adiabatic one, owing to weaker $\mathbf{B}$ changes within a cyclotron turn (shorter gyro-periods). As implicit in equation (3), these trajectory features directly depend upon particle mass and charge state (via gyro-period) and, given a

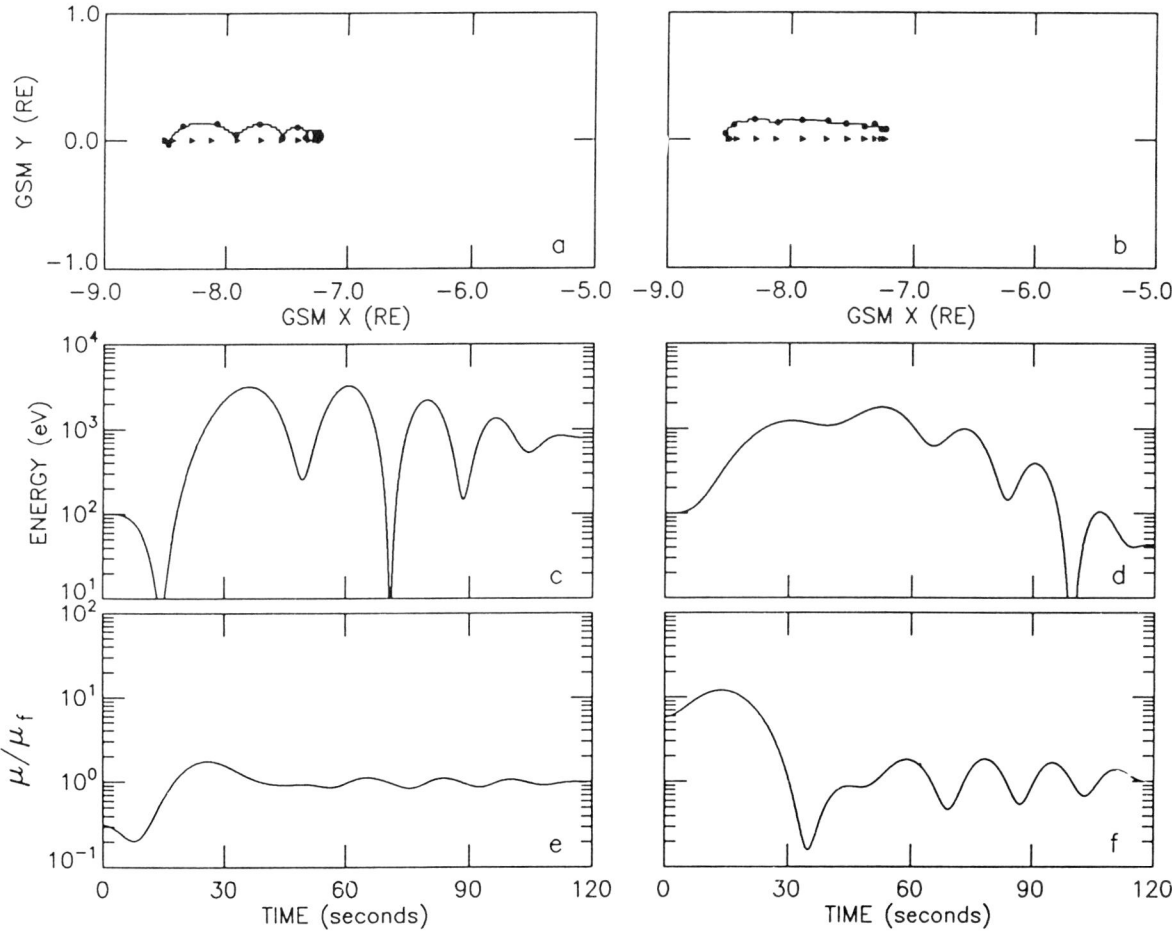

Fig. 1. Model trajectories of two equatorially mirroring O+ (a) and (b): trajectory projection in the GSM XY plane; (c) and (d): kinetic energy versus time; (e) and (f): normalized magnetic moment versus time. The ions were started with 100 eV energy and opposite gyro-phases at 8.5 R_E geodistance on the tail axis. In (a) and (b), circles and triangles denote the time in 10-second steps, respectively for the particles and the initial field line.

magnetic transition, will occur further out into the geotail for higher gyro-frequencies.

Storm-Time Injection of Energetic Particles

In view of these results, it appeared of interest to examine the storm-time transport of high-latitude ionospheric outflows, as they are located upstream of the mid-tail region [e.g., Cladis, 1986]. In the following, emphasis will be placed on the behavior of O+ ions from the cleft fountain and on that of polar wind protons. In both cases, test particles were initiated with parameters representing mean characteristics of the ion flows, as listed in Table 1 of the data-based numerical study of Delcourt et al. [1989].

Consistently with in situ measurements [e.g., Moore et al., 1986], this region was assumed to extend from 1000 to 1400 magnetic local time (MLT), and from 74° to 82° invariant latitude (ILAT). This modeled cusp was then divided in bins of 1-hour width in MLT and 2° width in ILAT, each of them being attributed a test particle (see Figure 3 of Delcourt et al. [1989]). Prior to substorm onset, the ion transport in the steady state magnetosphere was tracked by means of equi-distant steps of 1 R_E, from ejection into the magnetosphere until precipitation into the conjugate hemisphere. Subsequently, a test particle was assigned to each of these trajectory steps, initialized with the corresponding local energy and pitch angle. This collection of test particles was then traced within the dipolarizing field lines, adopting collapse characteristics similar to those in the previous section. The results of these calculations are summarized in Figure 2.

Using a hemispherical MLT-latitude format, this figure presents the initial and final coded altitudes (top panels) and energies (bottom panels) of the particles. Prior to the geotail collapse, the left panels in Figure 2 portray several ion streams anchored in the dayside cusp and connecting the conjugate auroral zone (note

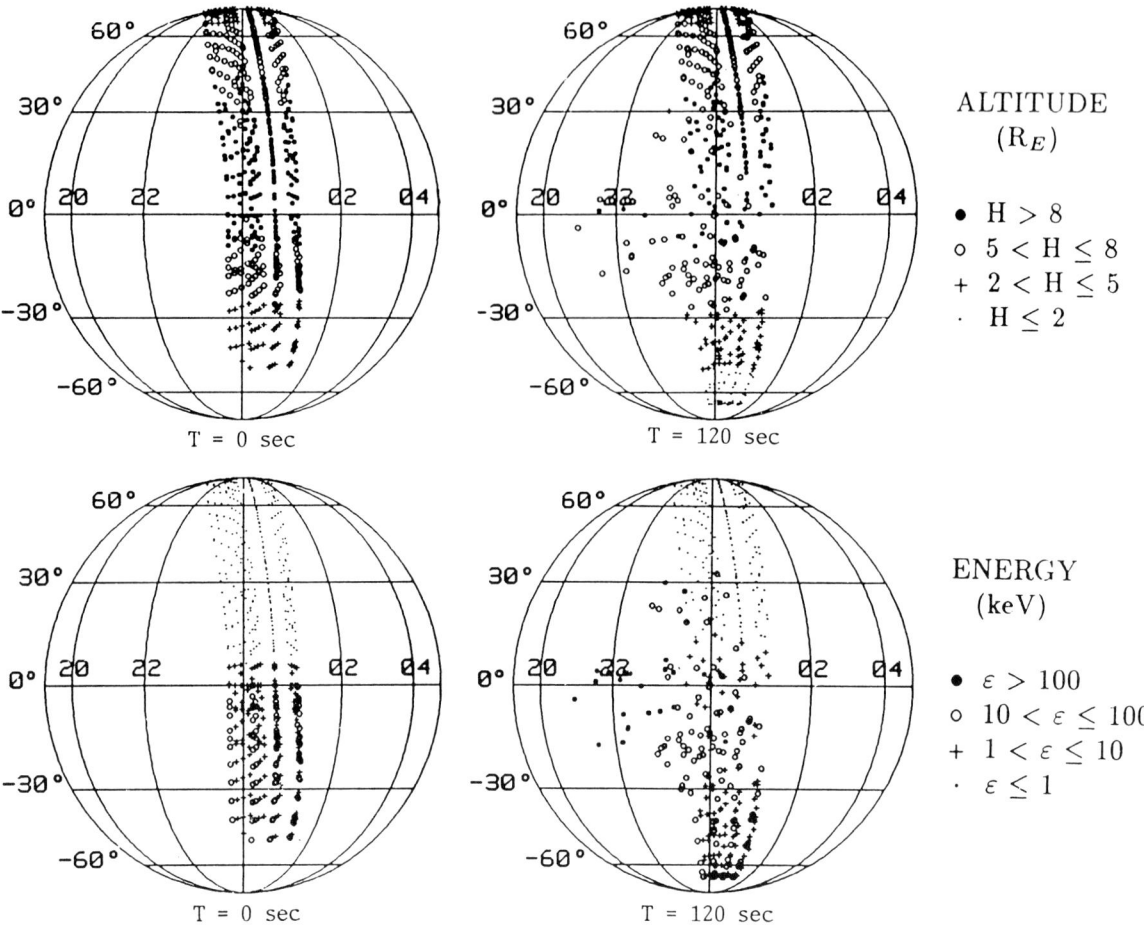

Fig. 2. MLT-latitude diagrams showing the coded altitudes (upper panels) and energies (lower panels) of cleft originating O+ before (left panels) and after (right panels) field line dipolarization. The particles were expelled from the cleft region with 10 eV energy and 125° pitch angle (northern hemisphere) at 5000 km altitude.

that only O+ originating from the northern hemisphere and initially located above 2 R_E altitude are considered in this figure). As demonstrated in previous numerical studies [e.g., Cladis, 1986], these drift paths are characterized by large (several kilo-electron volts) parallel accelerations in the equatorial region, which relate to enhanced curvature of the local magnetic field. A progressive increase of the O+ velocity is indeed apparent in the lower left panel of Figure 2 which depicts energy variations from tens of electron volts up to the kilo-electron volt range as the particles fall onto the center plane.

On the other hand, several features of interest can be pointed out after the dipolarization process (right panels). First, as stated by Mauk [1986], a strong latitudinal dependence is noticeable, since the populations residing above ~20° or below ~-20° latitude remain fairly unaffected. In contrast, a scattering of the well-defined ion streams is noticeable at low latitudes. Here, Figure 2 reveals prominent (a few R_E) earthward drifts, in conjunction with dramatic accelerations which possibly reach the hundreds of kilo-electron volt energy range. An angular analysis of these trajectories indicates that such accelerations occur primarily in the perpendicular direction, relating as discussed earlier to transient violation of the first adiabatic invariant. In addition, it can be seen in this figure that, as the particles are jetted into low L shells, they substantially travel westward under the influence of gradient and curvature drifts; hence, the formation of a delocalized plasma cloud in the vicinity of the center plane, which extends throughout the evening sector at the end of the collapse. In view of the source intensity in the ionosphere (over 10^{25} ions s^{-1}), these ions which have been strongly isotropized should significantly contribute to the trapped populations of the ring current. As a matter of fact, precise density calculations, as performed in Figure 5 of Delcourt et al. [1990b], reveal possible enhancements up to a few tenths of ions cm^{-3} in the 6-8 R_E altitude range.

Energetic Ion Boundary

Another interesting aspect of the storm-time ion transport is the unique spatial distribution produced for the newly created energetic populations. This feature can be visualized in Figure 3 which presents the computation results for polar wind protons. The morphology of this low-energy H+ flow in the "tail-like" configuration preceeding onset was evaluated by means of steady state trajectory calculations identical to those of Delcourt et al. [1989]. In order to focus on the ion behavior in the equatorial region where field line dipolarization is the most effective (see Figure 2), test protons were initiated at $20°$ latitude on the midnight meridian and traced down to a symmetrical position in the conjugate hemisphere (note that, assuming negligible energy change since their ejection from the ionosphere, only northern hemisphere originating H+ were considered here, starting along the magnetic field with 1 eV energy and from 5 R_E to 15 R_E geodistance by steps of 0.1 R_E). The pre-dipolarization distribution obtained is presented in Figures 3a and 3b which show, respectively, the color-coded H+ energy and pitch angle. In these figures, an overall folding of the H+ drift paths toward close L shells is noticeable, as expected from large-scale plasma convection. On the other hand, though the particles experience a progressive acceleration as they fall onto the center plane, it can be seen from Figure 3a that they remain in the hundreds of electron volt range. As this curvature-related acceleration occurs in the parallel direction [e.g., Cladis, 1986], these ions subsequently remain aligned with the magnetic field (Figure 3b).

The test protons were then traced throughout substorm expansion phase, their post-dipolarization distribution being presented in Figures 3c and 3d (respectively, in terms of energy and pitch angle). The earthward plasma compression induced by the collapse of the geomagnetic tail is clearly apparent in these figures which display the following distinct regimes: (1) a relatively low-energy (a few hundreds of electron volts at most) core in the innermost region (Figure 3c), where the Earth's dipole largely exceeds the external magnetic contribution. For this population, Figure 3d indicates negligible angular deviations from the magnetic field. (2) Further out, Figure 3c displays high-energy protons (up to a few tens of kiloelectron volts) distributed at the edge of the inner shells. In this case, Figure 3d reveals significant isotropization during transport, the final pitch angles possibly reaching the $45°$ range. As mentionned earlier, this relates to transient violation of the particle adiabatic invariants.

To support this latter result, Delcourt and Sauvaud [1991] further examined the adiabaticity criterion in equation (3) and

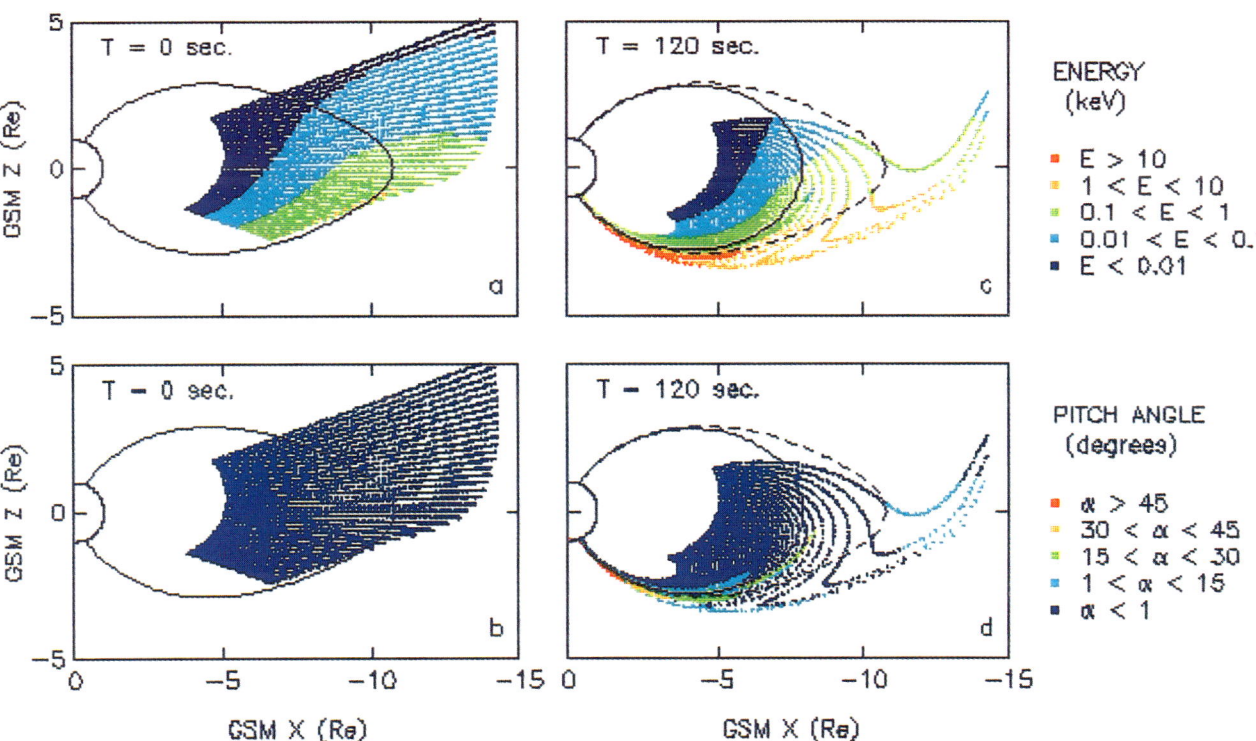

Fig. 3. Distributions of polar wind H+ in the GSM XZ plane before (left panels) and after (right panels) field line dipolarization. The upper panels present the color-coded particle energy, the lower panels, their color-coded pitch angle. The ions were initiated at $20°$ latitude along the magnetic field with 1 eV energy, from 5 R_E until 15 R_E geodistance by steps of 0.1 R_E. The solid line in (a) and (b) depicts the pre-dipolarization field line intercepting the H+ non-adiabatic "threshold" at $L \sim 10.8$. In (c) and (d), this field line is presented together with its earthward-shifted "image" after dipolarization (dotted and solid lines, respectively).

estimated an empirical χ value of ~5 %, above which the guiding center approximation fails. In the "tail-like" configuration preceeding onset, and for the 2-minute collapse considered, this "threshold" value corresponds to an H+ injection depth at ~10.8 R_E along the tail axis. Hence only protons initially located beyond this L shell will execute transient non-adiabatic motions, and thus experience dramatic energization as well as magnetic moment variation leading to trapping. In other words, if storm-time particle trapping follows from the above non-adiabatic heating in the perpendicular direction, this should evidently occur tailward of the adiabatic-nonadiabatic separatrix. In Figures 3c and 3d, the separatrix at $L \sim 10.8$ is presented together with its earthward-shifted "image" after collapse (dotted and solid lines, respectively). Keeping in mind that the particles were initially aligned with the magnetic field, it is indeed apparent from these figures that this separatrix precisely delimits regions of cold and hot plasma, with the presence of energetic and isotropized H+ exclusively on its tailward edge.

A further consequence of this can be appreciated in Figure 4 which displays the temporal evolution of the polar wind H+ during the minute following expansion phase. Note that this post-dipolarization flow was calculated only for the kilo-electron volt population, assuming that the geomagnetic field persists in its final ground state configuration. As apparent from the distinct "snapshots" performed, Figure 4 illustrates the gradual formation of a northward travelling "tongue" which reaches the conjugate mirror points within a few tens of seconds. This feature results from initial accumulation of protons with various energies at low altitudes (see Figure 3c), which subsequently spreads out with prominent time-of-flight effects. In other words, while the particles are transported earthward without much energy dispersion during the collapse of the geotail, prominent separation effects can be expected during recovery phase.

Most notably, it is evident in Figure 4 that, as recovery phase is initiated, the newly-created kilo-electron volt H+ travel tailward of the post-dipolarization location of their adiabatic-nonadiabatic

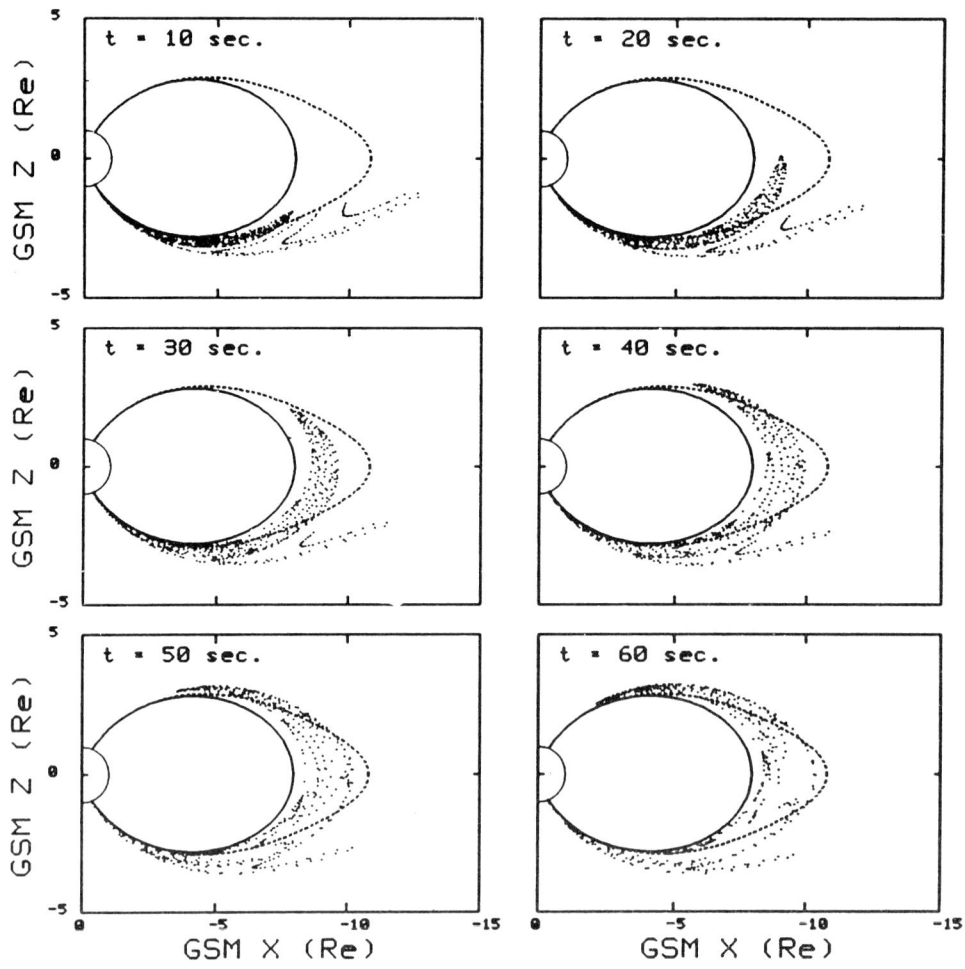

Fig. 4. Post-dipolarization evolution of the newly-trapped polar wind protons (only kilo-electron volt particles are considered). The successive H+ positions are displayed in the GSM XZ plane during 1-minute (by steps of 10 seconds from top to bottom) after substorm expansion phase. In each panel, dotted and solid lines depict the location of the H+ adiabatic-nonadiabatic separatrix in a format similar to Figure 3. From Delcourt and Sauvand [1991].

separatrix (solid lines). Conversely, this separatrix forms a clear inward boundary for the trapped and energetic populations, which is reminiscent of the "substorm injection boundary" model. Indeed, this concept introduced by McIlwain [1974] postulates a common boundary for the particles during storm times, regardless of their energy. These initially dispersionless particles subsequently drift at various speeds from the injection region, causing particularly specific signatures at geosynchronous altitudes [e.g., Mauk and Meng, 1983]. The features displayed in Figure 4 obviously support such a transport scenario and could provide a plausible explanation for the "injection boundary" itself.

Conclusion

The three-dimensional trajectory calculations performed have illustrated that the expansion phase of substorms is associated with rapid earthward injections of low-latitude magnetospheric populations. They have demonstrated that, for ions initiated in the mid-tail (~10-15 R_E geodistance), this "convection surge" may go together with dramatic acceleration and pitch angle variation, owing to transient violation of the particle adiabatic invariants. This effect depends upon the rate of change of the magnetic field within a cyclotron turn, and thus varies according to particle mass per charge (via gyroperiod). On such basis, the simulations suggest a substantial contribution from high-latitude ionospheric outflows to the energetic populations which are trapped at low L shells. For these populations, a clear earthward boundary can be identified, which corresponds to the post-dipolarization "image" of their non-adiabatic threshold in the geomagnetic tail.

Acknowledgments. Support for this work came from the ESA Research Fellowship Program and from NASA grant NAS8-38405. I am thankful to B. Debray for providing me with color graphics.

References

Aggson, T.L., J.P. Heppner, and N.C. Maynard, Observations of large magnetospheric electric fields during the onset phase of a substorm, *J. Geophys. Res., 88,* 3981, 1983.

Birmingham, T.J., and F.C. Jones, Identification of moving magnetic field lines, *J. Geophys. Res., ,* 5505, 1968.

Chen, F.F., *Introduction to Plasma Physics,* p. 35, Plenum, New York, 1974.

Cladis, J.B., Parallel acceleration and transport of ions from polar ionosphere to plasma sheet, *Geophys. Res. Lett., 13,* 893, 1986.

Cladis, J.B., and W.E. Francis, The polar ionosphere as a source of the storm time ring current, *J. Geophys. Res., 90,* 3465, 1985.

Cummings, W.D., J.N. Barfield, and P.J. Coleman, Jr., Magnetospheric substorms observed at synchronous orbit, *J. Geophys. Res., 73,* 6687, 1968.

Delcourt, D.C., C.R. Chappell, T.E. Moore, and J.H. Waite, Jr., A three-dimensional numerical model of ionospheric plasma in the magnetosphere, *J. Geophys. Res., 94,* 11,893, 1989.

Delcourt, D.C., J.A. Sauvaud, and A. Pedersen, Dynamics of single-particle orbits during substorm expansion phase, *J. Geophys. Res., 95,* 20,853, 1990a.

Delcourt, D.C., J.A. Sauvaud, and T.E. Moore, Cleft contribution to ring current formation, *J. Geophys. Res., 95,* 20,937, 1990b.

Delcourt, D.C., and J.A. Sauvaud, Generation of energetic proton shells during substorms, *J. Geophys. Res., 96,* 1585, 1991.

Geiss, J., H. Balsiger, P. Eberhardt, H.P. Walker, L. Weber, D.T. Young, and H. Rosenbauer, Dynamics of magnetospheric ion composition as observed by the GEOS mass spectrometer, *Space Sci. Rev., 22,* 537, 1978.

Heikkila, W.J., and R.J. Pellinen, Localized induced electric field within the magnetotail, *J. Geophys. Res., 82,* 1610, 1977.

Hones, E.W., I.D. Palmer, and P.R. Higbie, Energetic protons of magnetospheric origin in the plasma sheet associated with substorms, *J. Geophys. Res., 81,* 3866, 1976.

Krimigis, S.M., and E.T. Sarris, Energetic particle bursts in the Earth's magnetotail, in *Dynamics of Magnetosphere,* edited by S.-I. Akasofu, D. Reidel, Hingham, Massachusetts, 1979.

Lennartsson, W., R.D. Sharp, E.G. Shelley, R.G. Johnson, and H. Balsiger, Ion composition and energy distribution during 10 magnetic storms, *J. Geophys. Res., 86,* 4628, 1981.

Mauk, B.H., Quantitative modeling of the "convection surge" mechanism of ion acceleration, *J. Geophys. Res., 91,* 13,423, 1986.

Mauk, B.H., and C.-I. Meng, Characterization of the geostationary particle signatures based on the "injection boundary" model, *J. Geophys. Res., 88,* 3055, 1983.

McIlwain, C.E., Substorm injection boundaries, in *Magnetospheric Physics,* edited by B.M. McCormac, D. Reidel, Hingham, Massachusetts, 1974.

McPherron, R.L., Magnetospheric substorms, *Rev. Geophys. Space Phys., 17,* 657, 1979.

Mead, G.D., and D.H. Fairfield, A quantitative magnetospheric model derived from spacecraft magnetometer data, *J. Geophys. Res., 80,* 523, 1975.

Moore, T.E., R.L. Arnoldy, J. Feynmann, and D.A. Hardy, Propagating substorm injection fronts, *J. Geophys. Res., 86,* 6713, 1981.

Moore, T.E., M. Lockwood, M.O. Chandler, J.H. Waite, Jr., C.R. Chappell, A. Persoon, and M. Sugiura, Upwelling O+ ion source characteristics, *J. Geophys. Res., 91,* 7019, 1986.

Northrop, T.G., *The Adiabatic Motion of Charged Particles,* Wiley Interscience, New York, 1963.

Pellinen, R.J., and W.J. Heikkila, Energization of charged particles to high energies by an induced substorm electric field within the magnetotail, *J. Geophys. Res., 83,* 1544, 1978.

Sauvaud, J.A., and J.R. Winckler, Dynamics of plasma, energetic particles, and fields near synchronous orbit in the nighttime sector during magnetospheric substorms, *J. Geophys. Res., 85,* 2043, 1980.

Shepherd, G.G., R. Böstrom, H. Derblom, C.-G. Fälthammar, R. Gendrin, K. Kaila, A. Korth, A. Pedersen, R. Pellinen, and G. Wrenn, Plasma and field signatures of poleward propagating auroral precipitation observed at the foot of the GEOS-2 field line, *J. Geophys. Res., 85,* 4587, 1980.

5. AURORAL SUBSTORM MORPHOLOGY

AURORAL SUBSTORMS OBSERVED BY UV-IMAGER ON AKEBONO

Eisuke KANEDA and Tatsundo YAMAMOTO

Department of Earth and Planetary Physics, University of Tokyo
Bunkyo-ku, Tokyo 113, Japan

Abstract. We have studied global aurora dynamics by the analyses of high time resolution UV-images acquired on the AKEBONO satellite. The substorm expansion onset is one of challenging targets in studies of global aurora dynamics. It is confirmed that the expansion onset is constituted by succeeding two stages. The first stage is the initial brightening within a localized region of 1 hour longitudinal extent and continues for several minutes. The succeeding one is the flaring-up, which follows the former and accompanies breakup aurora displays including bulge - surge formation.

Introduction

The study of global aurora dynamics had seen a new epoch by virtue of snapshot imaging by VUV-TV camera [Kaneda et al., 1977] on 'KYOKKO' satellite. This new method enabled space imaging of aurora to take its global displays instantaneously. Thus, after 1980, UV-snapshot imagers have been intensively used for observations of global aurora displays by DE-1 [Frank et al., 1981] and Viking [Anger et al., 1987a] satellites with specific modulations respectively. In the project of 'AKEBONO' satellite [Oya and Tsuruda, 1990], the aurora imager is charged with the mission coordinating on-board observations with macroscopic activities of the magnetosphere, manifested as global aurora displays. UV aurora imager (ATV-UV) has been developed, on the base of that on 'KYOKKO' satellite, with intensive improvements in shot rate and coverage of aurora oval [Oguti et al., 1990].

The auroral substorm [Akasofu, 1964] is described in the form of model substorm illustrated by sequence of schematic diagrams. This model has been proposed by analyses of a large amount of ground observation data. However, those observations do not cover simultaneously the entire auroral region. Many discussions have been made with signatures of the auroral substorm. Meanwhile space imaging has been succeeded in seizing some of its real features. But, further studies are necessary with several problems concerning to dynamical aspects of phenomena relating to the substorm, e.g. controversy about the developing mode of westward travelling surge [cf. Rostoker et al., 1987, and Craven et al., 1989]. The substorm expansion onset is the very important one among these problems. The observation of onset is a hard task for imaging on low altitude satellites owing to rare probabilities of meeting it. High altitude satellites are favored with observing chance, but detailed features of expansion onset are not recorded by low time resolution in image data. By these reasons, substorm expansion onset still remains as one of challenging targets in studies of global aurora dynamics.

In the AKEBONO project, we have obtained supports of four ground stations in telemetry reception, KSC Uchinoura, Price-Albert Saskatchewan, ESRANGE Kiruna and Syowa Station Antarctica. Their geographical distributions enable ATV-UV to take global aurora images on every pass over polar regions of the both hemi-sphere in apogee mode. This data coverage gives ATV-UV better probability of observing expansion onset during this solar maximum phase. It is noteworthy that the shot rate of ATV-UV is rapid enough to seize detailed processes of substorm expansion onset. In this study, we demonstrate two events of substorm expansion onset, each of which is selected as an example under the storm-time condition on November 17, 1989, and the non storm-time on February 23, 1990.

Observations

In order to pick up signatures of aurora displays corresponding to precipitations of various energy ranges of electrons, and protons, ATV-UV is sensitized to emissions in wave-length range 1150 - 1390 A with efficiencies more than 1% of the peak value at about 1230 A. Thus, observed aurora images can contain emissions of atomic oxygen lines at 1304 and 1356 A, nitrogen line at 1200 A, molecular nitrogen LBH bands and hydrogen Lyman-alpha. All observational results discussed here are obtained over the arctic region at altitudes more than 5,000 km. The operational mode of imaging is a standard one, consecutive 4 shots with 8 second intervals in every 64 seconds. Every aurora image presented here is restored from distortion due to image formation by ATV-UV optics and mapped into an ordinary format of isometric projection. The overlaid grid on some of aurora images refers to the corrected geomagnetic coordinates, and is drawn with latitude circles from 80° to 40° in 10° steps and meridional lines in MLT (magnetic local time) 2 hour pitch with enhancement of the noon-midnight line. In Plates 1 and 2, images are aligned time-sequentially from the left to the right with 4 images per line. On the leftmost of each line, shot-times (in UT) of the first and forth images are denoted with two figures of 6 digits respectively. Those of intermediate two images are obtained from linear trisectary interpolation of these figures.

Expansion Onset of November 17, 1989

Substorm expansion onset is observed during 185656 - 192800 UT operation interval and its features are shown in Plates 1a and 1b with 36 consecutive aurora images between 191814 and 192704 (3 images among them are missed by lack of header label in telemetry data). In these plates, intensity of aurora is coded by pseudo-colors, from dark blue, pale magenta, red, yellow and to white as its level increases. The gridded version of the second image of each line are shown in the rightmost position of the same line. These gridded images delineate that auroral activation take place in the unusually expanded oval with low latitude border nearly on 50°, and no auroral forms exist beyond this border.

In this event, slight signatures of brightness intensification are found in the midnight aurora within images of 191918 and 191926 UT shot. The midnight meridian can be identified within gridded images as one traversing Novaya Zemlya Is. in the Arctic Ocean. We can see undoubted brightness increases with the midnight aurora in 191942 UT image and their continuation in the succeeding images. There can not be detected tendencies of steady auroral intensification in the same location before brightening start at 191918 UT. These facts show that the very beginning of expansion occurs at 191926 UT with ambiguities less than 10 seconds. Note that the longitudinal extent of brightened area is initially less than one hour MLT width.

This brightening continues after 191942 UT without appreciable spatial expansion. Though the eastern edge of the bright region fluctuates occasionally, the location of its center remains almost still around the midnight meridian. Above mentioned signatures are seen in images from 192015 to 192432 UT. They make clear contrast to those of spatial developments witch start at 192440 UT. We consider from these evidences as follows that at around 192440 UT auroral displays proceed from the initial stage of brightening into the next stage of spatial developments accompanying further intensifications in brightness. We hereafter call the former stage as 'initial brightening' and the latter 'flaring-up' respectively.

In the flaring-up stage, the activated region spreads rapidly more than 3 hour MLT width within a few minutes. This region shows latitudinal expansion in both directions polewards and

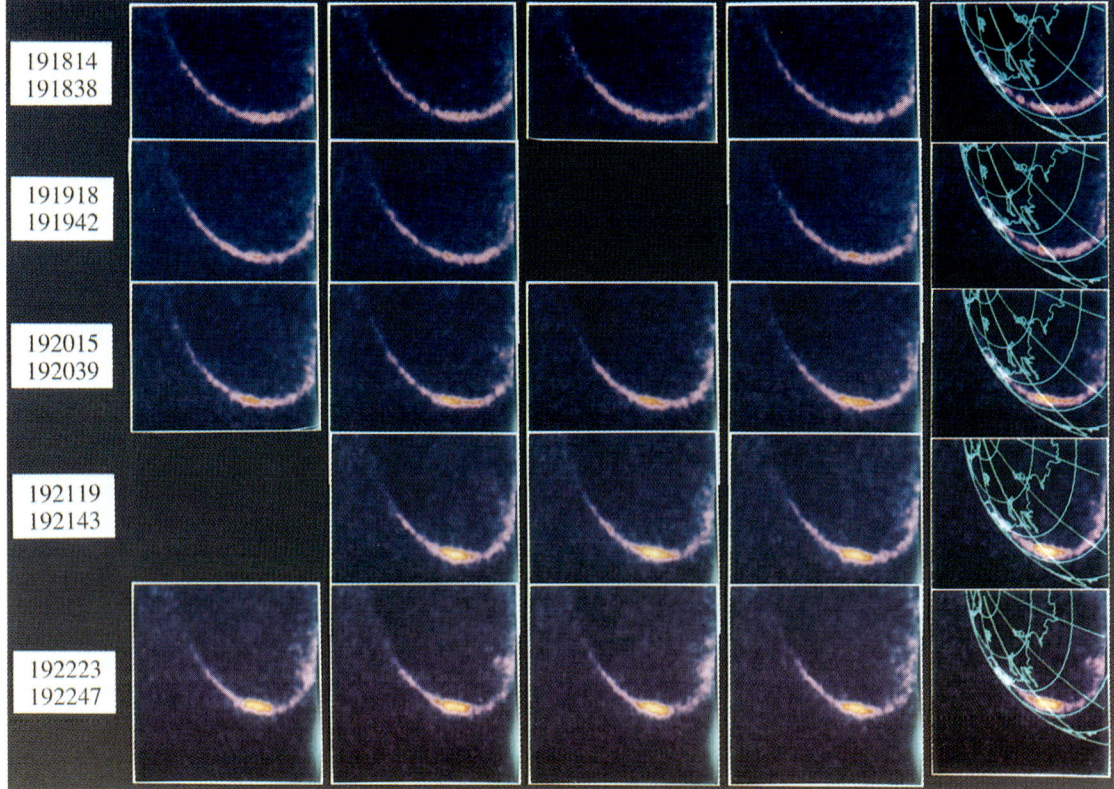

Plate 1a. Sequence of 20 (2 missed) consecutive false color images during 191814 - 192247 UT on November 17, 1989. Image on rightmost of each line is gridded one of the second from left on the line.

equatorwards. Its low latitude border crosses over 50° circle by 192704 UT. Detailed spreading features in its western end are not clear by its location near the limb, but its eastern end is evidently seen with extending into morning hour regions. Auroral displays resided in dawn branch of the oval have been activated preceding to intrusion of the eastern end of midnight auroras.

Expansion Onset of February 23, 1990

Aurora displays including substorm expansion onset is observed during 064011 - 071818 UT operational interval on February 23, 1990. Results are summarized in Plate 2 with sequential 24 images. In illustration of this event, required images exceed 90 in number for full coverage of the event observed during more than 20 minutes. For cutting off lengthy exposition, each of 24 images in Plate 2 is prepared by superposition of consecutive 4 images in each 64 second interval, and its shot time determined by averaging those of 4 shots. In this plate, the monotone is employed for brightness expression. Sunlit regions on the earth appear as the white portion in top of several initial images. First five images suffer telemetry noise contaminations whose traces are seen as nearly vertical arches with irregular brightness changes. Coordinate grids are unable to calculate with first 8 images due to the lack of attitude data from difficulties in telemetry receptions indicated by these noise traces.

The expansion onset to have already started in limited portion around midnight of the oval within 064526 UT image. There are no traces of systematic aurora brightening indicating the start of expansion in the first 3 images in Panel 2. Its very beginning is found within the previous on 064422 UT. For confirmation of start time, a check is made with original 4 images of this averaged 064422 UT one. It turns out that increases of brightness from background oval are detected for the first time in the forth shot at 064434 UT. The east-west extent of auroral enhancements in the midnight is nearly identical with 064526 UT and 064630 UT images, but proceeding of enhancements is clearly seen in these images. In contrary to this trend, after 064733 UT developments of auroral enhancements become prominent in spatial extents and are accompanied by appearances of active forms within this enhanced region. We can summarize two stage constitution of expansion onset from these facts in the following way. The initial brightening stage starts at about 064422 UT, and continues until 064630 UT, without essential growth in spatial extents. The expansion phase proceeds into flaring-up stage at about 064733 UT. In flaring-up stage, azimuthal extents of the enhanced auroral

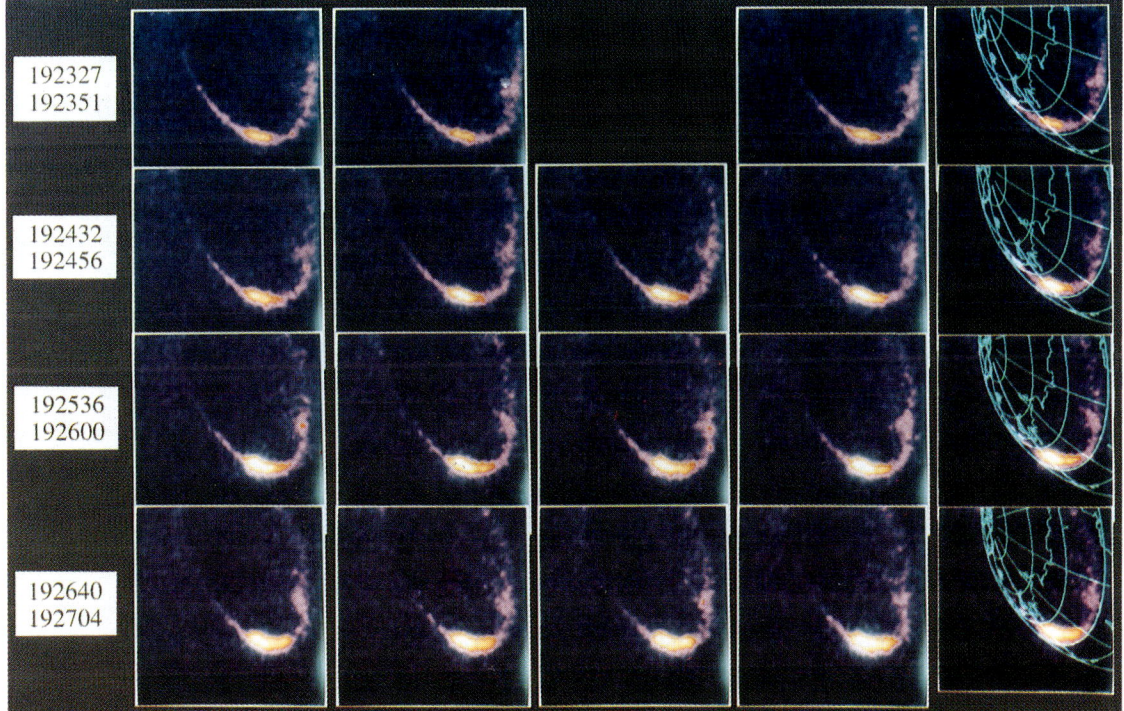

Plate 1b. Sequence of 16 (1 missed) consecutive false color images during 192327 - 192704 UT on November 17, 1989, succeeding to images in Plate 1a. The format is the same as for Plate 1a.

238 AURORAL SUBSTORMS OBSERVED BY AKEBONO

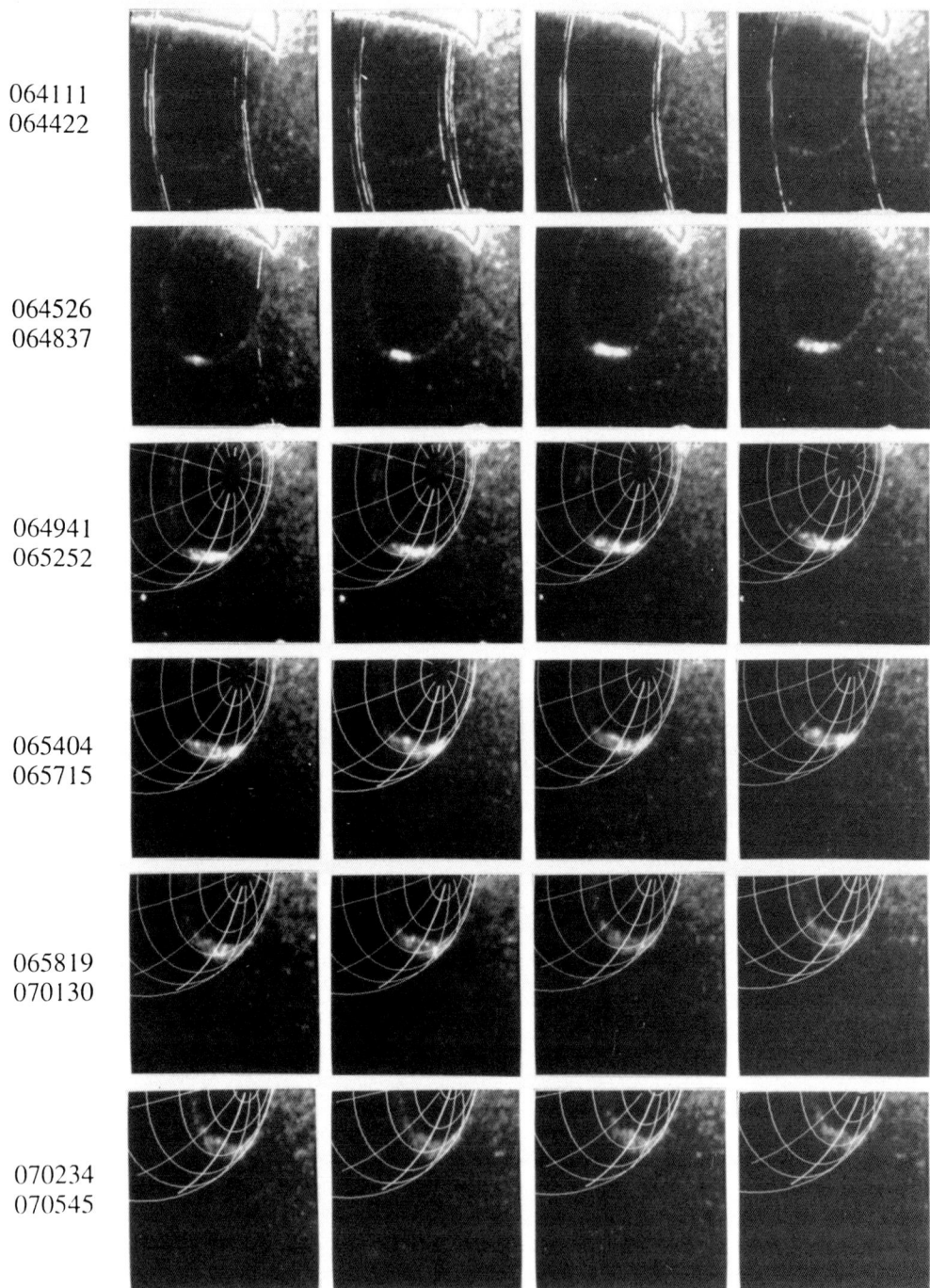

Plate 2. Sequence of 24 consecutive superposed images during 064111 - 070545 UT on February 23, 1990.

region spread towards east and west directions from about 2 hours to 4 hours in MLT width within initial several minutes.

Meanwhile active auroral forms emerge in this region, and grow up into westward travelling surge. In the duskside oval, some brightness increases are observed within dusk branch of the oval prominently in images from 064941 to 065252 UT. The sufficient time resolution between succeeding images assures these increases free from effects of westward travelling surge propagation. They are different from those extending from the head of westward travelling surge seen in 070130 - 070545 UT images. Further, they can not be taken account by the long distance propagation mode of westward travelling surge [Craven et al., 1989]. It is necessary for full seizings of this increases to investigate relations with aurora activities in dayside regions where is masked by strong albedos with this event.

Another interesting fact is found by sequential inspections of images after 064941 UT. In spite of auroral developments as a whole, there exists in the region of auroral breakup a demarcation line around 1 hour MLT which is the boundary of auroral activities proceeding toward opposite directions. Another example of this demarcation under stormtime condition can be seen with images in Figure 2 of Oguti et al. [1990].

Summary

The expansion onset of substorm consists of two stages, initial brightening and the flaring-up. The start of the initial brightening is the beginning of auroral expansion, and this stage lasts several minutes with increases in brightness but no appreciable expansion of activated region. In the flaring-up stage, this region expands rapidly within a few minutes and grows into the auroral bulge filled up with active forms. Demarcation line emerges sometimes evidently in the bulge, on the both sides of which directions of activity propagation are opposite.

On expansion onset, there take place sometimes, on the dusk branch of the oval, auroral activities unable to understand in relation with westward travelling surge propagation.

Discussion

The present result clarifies that at the expansion onset of auroral substorm, explosive displays (flaring-up) do not directly emerge from the quiet background oval, but take place on a previously enhanced region (initial brightening) with further activations. We identify the start of initial brightening as the onset of auroral substorm expansion. It is described in the model substorm [Akasofu, 1964; 1968] that the expansive phase starts with brightening of quiet arc several minutes before developments of auroral bulge and other active displays. Though illustrations by schematic diagram leave ambiguities with longitudinal extents of brightening in this model, features in his diagram are considerably different from those in the initial brightening stage. Another model substorm is proposed by Starkov and Fel'dstein [1971], that is somewhat different especially in expansion onset from Akasofu model. Their schematic diagrams show direct emerging of active aurora on expansion onset. This is totally different from the present result.

We find a favorable example to two stage constitution of expansion onset in UV observations by the Viking satellite [Rostoker et al., 1987]. Onset-time auroras in midnight region are delineated in their study of the event 184755 - 190741 UT on April 1, 1986. The first three images in this event endorse the present view of expansion onset, namely the first showing features of pre-expansion, the second, of 180 seconds later shot, corresponding to the initial brightening stage, and the third, of further 40 seconds later shot, belonging to the flaring-up stage. We can see auroral enhancements in the second image spanning only over about 1 hour LT extents, but in the third spreading into twice of that in the second with considerable increases in brightness. This is an evidence of the transition from the initial brightening to the flaring-up stage. It is confirmed by referring to AE(12)-index (World Data Center C_2 for Geomagnetism) that this event on April 1, 1986 is manifestation of substorm expansion. However, about other events observed by Viking UV imager [Anger et al., 1987; Shepherd et al., 1987, 1990; Lyons et al., 1990], comments are impossible with features of expansion onset owing to large time interval in illustrations of succeeding images and/or shortage in time coverages. In examples of imaging on the DE-1 satellite [Craven and Frank, 1985, 1987; Frank and Craven, 1988], features of brightening in early stage of expansion phase is clearly shown in confinements within small portion of the oval. But it is impossible by large time intervals with delineated images to trace detailed processes of expansion onset with these examples.

The two-stage development of the breakup aurora is a remarkable feature presented in this paper. Similar signatures are found in the near-tail region of the magnetosphere. In recent analyses of AMPTE/CCE data, Ohtani et al. (unpublished manuscript, 1991) find the increase in the energetic particle fluxes in association with the signatures of tail-current intensifications, which is preceding to the magnetic dipolarization by a few minutes. These flux increases are observed statistically in the midnight sector with an extension to premidnight hours. Simultaneous observations at GOES 5 and 6 show that abrupt growth of the tail-current takes place in a localized MLT area of 1 Re or less in extent [Ohtani et al., this issue]. Nagai [this issue] also notes that the magnetic dipolarization occurs in a localized region at its onset, then is followed by spreading of the region. These observations, together with our results, state that the energy release in the plasma sheet starts in the localized region of 1 or 2 hours local time extent at its beginning. The initial brightening stage of the auroral expansion should be the ionospheric manifestation of the physical processes in the magnetotail as the onset of the auroral substorm.

As discussed above, the substorm expansion does not progress simultaneously and explosively in a wide area, but is much more likely to start in a relatively limited region and goes on in successive steps, both in the energetic particle precipitation as aurora display and in the magnetospheric signature of field configuration change. This implies that simple view of parallel developments among various manifestations of substorm phenomena is not kept on the critical point of expansion onset. This breakdown of the general parallelism has a possibility contributing to reconsideration of current crude recognitions with substorm signatures [Rostoker et al., 1980], because they are endorsed by perfect parallelism between auroral display and ground magnetic disturbances [Wiens and Rostoker, 1975; Kawasaki and Rostoker, 1979]. Extensive data analyses in near future are indispensable for identifications of concurrence with phenomena in the magnetotail.

Concerning to activity increases on dusk branch of the oval on expansion onset, it has been exemplified that their occurrence precedes to arrival of westward travelling surge [Kaneda, 1973;

Kaneda et al., 1981]. These previous results suggest also needs of simultaneous observation with dayside region, especially with the cusp and its neighborhoods. Further analyses should be made for clarifications of this phenomena as well as of demarcation line in the region filled up with breakup auroras.

Acknowledgements. The authors express their hearty thanks to Prof. T. Oguti of the STE Laboratory, Nagoya University, the principal investigator of ATV-UV and -VIS project, for his valuable guidances. We are also grateful to Prof. K. Tsuruda of the Institute of Space and Astronautical Science, the manager of the Akebono Project, for his kind encouragements. We wish to thank all other members of ATV-team for their endeavors on the project. Thanks are also due to every person who backs up the ATV/Akebono project.

References

Akasofu, S.-I., The development of the auroral substorm, *Planet. Space Sci., 12,* 273-282, 1964.

Akasofu, S.-I., *Polar and Magnetospheric Substorms,* D. Reidel, Hingham, Mass., 1968.

Anger, C. D., S. K. Babey, A. Lyle Broadfoot, R. G. Brown, L. L. Cogger, R. Gattinger, J. W. Haslett, R. A. King, D. J. McEwen, J. S. Murphree, E. H. Richardson, B. R. Sandel, K. Smith, and A. Vallance Jones, An ultraviolet auroral imager for the Viking spacecraft, *Geophys. Res. Lett., 14,* 387-390, 1987a.

Anger, C. D., J. S. Murphree, A. Vallance Jones, R. A. King, A. L. Broadfoot, L. L. Cogger, F. Creutzberg, R. L. Gattinger, G. Gustafsson, F. R. Harris, J. W. Haslett, E. J. Llewellyn, J. C. McConnell, D. J. McEwen, E. H. Richardson, G. Rostoker, B.R. Sandel, G. G. Shepherd, D. Venkatesan, D. D. Wallis and G. Witt, Scientific results from the Viking ultraviolet imager: An introduction, *Geophys. Res. Lett., 14,* 383-386, 1987b.

Craven, J. D., and L. A. Frank, The temporal evolution of a small auroral substorm as viewed from high altitude with Dynamics Explorer 1, *Geophys. Res. Lett., 12,* 465-468, 1985.

Craven, J. D., and L. A. Frank, Latitudinal motions of the aurora during substorms, *J. Geophys. Res., 92,* 4565-4573, 1987.

Craven, J. D., L. A. Frank, and S.-I. Akasofu, Propagation of a westward travelling surge and the development of persistent auroral features, *J. Geophys. Res., 94,* 6961-6967, 1989.

Frank, L. A., J. D. Craven, K. L. Ackerson, M. R. English, R. H. Eather and R. L. Carovillano, Global auroral imaging instrumentation for the Dynamic Explore mission, *Space Sci. Instrum., 5,* 369-393, 1981.

Frank, L. A., and J. D. Craven, Imaging results from Dynamic Explorer 1, *Rev. Geophys., 26,* 249-283, 1988.

Kaneda, E., Dayside auroral activity and its relation to substorm, *Rep. Ionosph. Space Res. Japan, 27,* 209-212, 1973.

Kaneda, E., M. Takagi, and N. Niwa, Vacuum ultraviolet television camera, *Proc. 12th Intnl. Symp. Space Tech. Sci.,* 233 -238, Agne, Tokyo, 1977.

Kaneda, E., T. Mukai, and K. Hirao, Synoptic features of auroral system and corresponding electron precipitation observed by KYOKKO in *Physics of Auroral Arc Formation,* edited by S.-I. Akasofu and J. R. Kan, pp. 24-30, AGU Geophys. Monogr. Ser., Vol.25, 1981.

Kawasaki, K., and G. Rostoker, Auroral motions and magnetic variations associated with the onset of auroral substorms, *J. Geophys. Res., 84,* 7117-7122, 1979.

Lyons, L. R., O. de la Beaujardiere, G. Rostoker, J. S. Murphree, and E. Friis-Christensen, Analysis of substorm expansion and surge development, *J. Geophys. Res., 95,* 10575-10589, 1990.

Nagai, T., An empirical model of substorm-related magnetic field variations at synchronous orbit, *this issue.*

Oguti, T., E. Kaneda, M. Ejiri, S. Sasaki, A. Kadokura, T. Yamamoto, K. Hayashi, R. Fujii, and K. Makita, Studies of aurora dynamics by aurora-TV on the Akebono (EXOS-D) satellite, *J. Geomag. Geoelectr., 42,* 555-564, 1990.

Ohtani, S., K. Takahashi, L. J. Zanetti, T. A. Potemra, R. W. McEntire, and T. Iijima, Tail current disruption in the geo-synchronous region, *this issue.*

Oya, H., and K. Tsuruda, Introduction to the Akebono (EXOS-D) satellite observations, *J. Geomag. Geoelectr., 42,* 367-370, 1990.

Rostoker, G., S.-I. Akasofu, J. Foster, R. A. Greenwald, Y. Kamide, K. Kawasaki, A. T. Y. Lui, R. L. McPherron, and C. T. Russell, Magnetospheric substorms - definition and signatures, *J. Geophys. Res., 85,* 1663-1668, 1980.

Rostoker, G., A. Vallance Jones, R. L. Gattinger, C. D. Anger, and J. S. Murphree, The development of the substorm expansive phase: The "EYE" of the substorm, *Geophys. Res. Lett., 14,* 399 -402, 1987.

Shepherd, G. G., C. D. Anger, J. S. Murphree and A. Vallance Jones, Auroral intensifications in the evening sector observed by the Viking ultra violet imager, *Geophys. Res. Lett., 14,* 395 - 398, 1987.

Shepherd, G. G., A. Steen, and J. S. Murphree, Auroral boundary dynamics observed simultaneously from the Viking spacecraft and from the ground, *J. Geophys. Res., 95,* 5845-5865, 1990.

Starkov, G. V., and Ya. I. Fel'dstein, Substorm in auroras, *Geomag. Aeronomy, 11,* 478-479, 1971 (English Ed.).

Wiens, R. G., and G. Rostoker, Characteristics of the development of the westward electrojet during the expansive phase of magnetospheric substorms, *J. Geophys. Res., 80,* 2109-2128, 1975.

Viking Optical Substorm Signatures

J. S. MURPHREE, R. D. ELPHINSTONE, L. L. COGGER, AND D. HEARN

Department of Physics and Astronomy, University of Calgary, Calgary, AB T2N 1N4

The classical description of a substorm focuses on the brightening and poleward expansion of an arc system in a localized time sector followed by either continuous or episodic westward motion of large scale spiral forms. When viewed in the context of the entire auroral distribution however, other signatures are sufficiently repeatable as to ultimately require their inclusion in substorm modeling schemes. Prior to the major optical onset a common feature is the rapid appearance and disappearance of localized intensifications in the general area of the subsequent expansion. These indicate that a large region on the magnetotail ($> 10\ R_E$) is involved in the growth phase process although the final onset region may occupy an ionospheric extent of less than 500 km. The growth phase, however, does not affect the character of the diffuse aurora prior to onset in a consistent manner. The latitudinal width can increase, decrease or stay the same prior to onset although in general it seems that the diffuse aurora moves equatorward. The positions of onset in the magnetotail, as determined from magnetic field mapping, places them at locations very near the Earth ($< 15\ R_E$ in all cases) which is consistent with previously published measurements of injection boundaries. As well there can exist emissions well poleward of the onset latitude implying that onset occurs deep within the closed field line region.

INTRODUCTION

The traditional view of the evolution of a substorm based on optical measurements [*Akasofu*, 1964] has been successful in putting very disparate measurements into a simple, consistent context. Classically the description focuses on the onset wherein the most equatorward discrete arc brightens and moves poleward [*Rostoker et al.*, 1980] initiating the classical signature of a substorm, the westward travelling surge. Numerous ground based and satellite measurements have added to this picture of the auroral substorm [e.g., *Pellinen and Heikkila*, 1978; *Craven and Frank*, 1987; *Rostoker et al.*, 1987; *Craven et al.*, 1989], but not surprisingly each new technological advance in measuring capability has brought with it important modifications. In the case of the Viking UV Imager [*Anger et al.*, 1987] the instrumental improvements focus on the ability to simultaneously expose the entire field of view with an exposure time of approximately one second while retaining reasonable spatial and temporal resolution between images.

Before discussing the actual optical signatures as routinely measured by the Viking UV Imager, it should be stressed that we take here the global concept of the substorm as our framework, rather than any local context. This viewpoint is facilitated by the large scale perspective afforded by satellite imagers and presumes that the observed auroral distribution represents in some essential way the actual temporal and spatial evolution of a substorm. A substorm in these terms has a clear temporal history independent of the location and/or biases of a particular satellite or ground based observer. For example, if a ground based observer is located eastward of the initial ionospheric activation region (i.e., the start of the westward travelling surge), they might predominantly observe auroral morphology consistent with the recovery phase, but the actual recovery phase for this particular substorm will not have begun until the expansion phase ends.

Figure 1 illustrates the context in which optical substorm features as observed by Viking will be discussed in this paper. The twelve images (which are actually only portions of the original data) are from the LBH camera (sensitive to emissions between 1400 and 1800 Å) and are presented in pseudo-color with black representing no (or low) intensity and red the highest intensity. In this case the images have only been corrected for non-uniformity in the optical system and some background subtraction has been performed. Overlaid on all the images are one or more Magnetic Local Time (MLT) meridians. These (and all other magnetic coordinates unless specified to the contrary in this paper) utilize the eccentric dipole coordinate system as determined from IGRF 1985.0 and the instantaneous anti-solar vector to compute their position. This set of data was chosen to illustrate a working optical definition of a substorm.

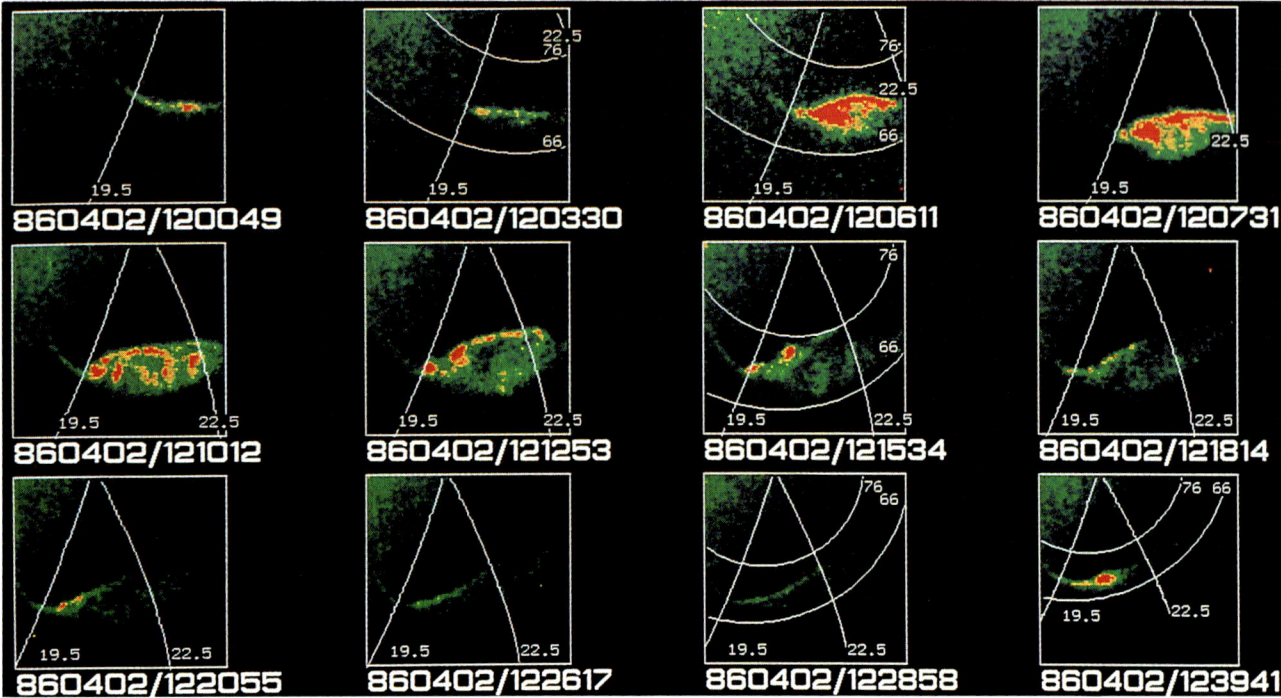

Figure 1. Portions of 12 images acquired on April 2, 1986 by the Viking UV Imager. The data are corrected for non-uniformity and background and color coded with red indicating highest LBH intensities. The data are limited to the premidnight sector (meridian lines are 19.5 and 22.5 MLT and magnetic latitudes are 66 and 76 Mlat) and show the evolution of a temporally confined substorm.

In this paper we shall consider the growth phase of the substorm to be a period before onset of relatively quiescent auroral forms in the evening sector (see Table 1). Although there may be some activity at this time, there is a definitive time at which this auroral distribution brightens and subsequently moves poleward. For example, the image at 1203:30 UT in Figure 1 is taken before the substorm onset which has occurred by 1206:11 UT. The expansion phase is characterized by the formation of the auroral bulge (both east and west of the onset location), a surge at the westward head of the expansion (often called a westward travelling surge, but as can be seen in this example no motion occurs) and the chaotic North-South auroral features which fill this bulge. This phase is illustrated in Figure 1 between the times 1206:11 UT and 1210:12 UT. At 1212:53 UT the auroral forms within the auroral bulge fade and the poleward progression of the bulge stops, leaving a discrete arc system at its poleward edge. This fading of auroral forms presumably results in the so-called poleward leap of the electrojet and signals the start of the substorm recovery phase. This phase is characterized by the end of the bulge's expansion east, west and north and a gradual retreat equatorward of the poleward auroral boundary (1212:53 UT to 1228:58 UT). This poleward edge can sometimes reactivate with a new westward travelling surge. This latter form of substorm will not be discussed in this paper. The following points should be made concerning this phenomenological description of the substorm:

1) The evolution of a substorm is considered to take place within a relatively confined period of time and is not, for the purposes of this paper, the several hour episode of activity commonly presented by indices such as AE. Thus in this particular example the expansion and recovery

TABLE 1

UT	Substorm Phase
1200:49	Growth
1203:30	Growth
1206:11	Expansion
1207:31	Expansion
1210:12	Expansion
1212:53	Expansion
1215:34	Recovery
1218:14	Recovery
1220:55	Recovery
1226:17	Recovery
1228:58	Recovery
1239:41	Growth?

phases cover only ≈ 30 minutes which is admittedly unusually short. A growth phase may have been in existence considerably before 1200:49 UT and a complete substorm will be assumed to consist of all three phases. This substorm was reflected in the AE index by a relatively small peak lasting about 30 minutes.

2) Only substorms whose onset (i.e., the time when the expansion phase begins) occurs when the auroral distribution is essentially quiet (i.e., few discrete emissions in the midnight sector) will be stressed. This is done to avoid exceedingly complex auroral distributions which can occur during multiple onsets.

3) The data presented in Figure 1 is classified as a substorm solely on the basis of the optical data. It may be argued that this data agrees with the standard definition of optical characteristics during a substorm, but that by itself does not mean that other signatures (e.g., negative H bays, enhanced westward electrojet, Pi 2's etc.) necessarily occurred. This is not felt to be a limitation because there would seem to be no reason for prioritizing signatures.

SOME COMMENTS ON SATELLITE AURORAL IMAGES

Images such as provided by the Viking UV Imager yield significant information about the instantaneous auroral distribution. With reasonable temporal and spatial resolution they can pinpoint the ionospheric response (and control?) of magnetospheric processes which can only fortuitously be measured in situ. However, there are important limitations to this set of data which must be kept in mind when attempting to interpret them. First an almost trivial observation: the image data provide information only on regions where significant fluxes of particles precipitate into the ionosphere. This is significant for at least three reasons:

1) Movements of optical features reflect motion of regions of precipitation which may not necessarily be related to source population motion [e.g., *Murphree et al.*, 1989]. Extensive particle populations may exist on field lines with no significant loss cone fluxes until some mechanism causes the loss cone to be filled.

2) Precipitation may in fact be occurring at significant levels, but at energies such that the emissions are in a wavelength region not being measured. This of course is the rationale behind attempting to utilize combinations of various auroral emissions [e.g., *Rees et al.*, 1988], but it also means that caution must be exercised when using results from a single wavelength regime.

3) Instrument sensitivity must be sufficient to distinguish when a precipitating source population is absent and when some relative precipitation minimum occurs. It is unlikely that any single instrument will ever be able to absolutely eliminate this problem because of such uncontrollable factors as background contamination.

Another well known limitation of optical data (specifically as acquired from satellite imagers) is the spatial resolution since it affects the determination of discrete versus diffuse features. Individual discrete arcs can have widths well below 1 km [*Maggs and Davis*, 1968], but current satellite imagers have much worse resolution. Ultimately the distinction between discrete and diffuse aurora rests on particle characteristics. However, their independent identification in satellite images is important because of the significance attached to where different physical processes (e.g., acceleration) are operative. In the same vein it is necessary to recognize that temporal resolution can affect the interpretation of satellite images. There are actually two aspects to this, the first being whether or not the image is acquired in an instantaneous manner. In most cases (e.g., ISIS 2, DE-1) they are not, with the result that the image represents a particular time sequence of measurements of optical emissions with an unknown temporal history. Secondly, image repetition rate is significant because of the rapidity with which the auroral signature of the various magnetospheric and ionospheric processes occur. For example, in the case of Viking which operated typically with an image every minute, there is still a large degree of uncertainty particularly in the timing of events at various spatial positions (Viking measurements are generally made only one second out of sixty). Finally and in common with any instrument that samples large regions of space or time (e.g., particle detectors which effectively measure both local and non-local conditions [*Mitchell*, 1987]), the images provide information about features which may or may not be related in terms of their occurrence. Thus it is conceivable a single image could measure the effects of a substorm onset, a flux transfer event and a detached arc at the same time. The image then effectively represents the integration of the results of what could be totally unrelated processes. It is up to multiple event studies to show that a relationship does indeed exist.

GROWTH PHASE

There is general consensus that there is a consistent period of activity in the Earth's magnetosphere prior to the explosive onset of a substorm. By ascribing the time between the southward turning of the IMF and substorm onset as the growth phase, its length has been statistically shown to be on the order of 1 to 2 hours [*Bargatze et al.*, 1985]. In situ magnetotail measurements have shown conclusively that the magnetic field within the tail becomes more tail-like [*McPherron*, 1970; *Baker et al.*, 1981] which is indicative of an increase in magnetotail currents [*Kaufmann*, 1987] and is thought to reflect a buildup of magnetic energy in the lobes due to enhanced merging on the dayside of the Earth [*Nishida*, 1983]. Such a modification of the field affects particle distribution functions so as to create butterfly type distributions near geo-synchronous orbit [*Baker*, 1984] and to result in electron and ion precipitation near their respective trapping boundaries [*Kirkwood and Eliasson*, 1990]. Optically however, it is not usually considered to be a particularly active period of the substorm process. Aside from general statements such as the auroral oval expanding equatorward (e.g., the results of *Zverev et* al. [1979] show variations in boundaries with B_z up to onset time) and ground based, high spatial resolution observations which indicate the growth phase is characterized by slowly equatorward drifting discrete auroral arcs [*Vorobjev et al.*, 1976 and *Pellinen and Heikkila*, 1984] there seems to be little emphasis on growth phase optical activity.

As noted above the polar cap should expand in size during the substorm growth phase, or as is commonly assumed to be an equivalent statement, the poleward boundary of auroral emissions should move equatorward. Measurements of high energy electron precipitation by riometers [*Ranta et al.*, 1981] and balloon-borne X ray detectors [*Pytte et al.*, 1976] have shown that in the afternoon to midnight sector equatorward drift speeds of 40 to 600 m/s occur while *Kirkwood and Eliasson* [1990] report speeds of 30 to 60 m/s using EISCAT electron density measurements. (These low speeds are comparable to the apparent speed of features fixed in the ionosphere which rotate into the field of view of a corotating Earth fixed observer). More recently *Tanskanen et al.* [1987] have shown with a variety of measurement techniques that not only are drift speeds on the order of 100 m/s observed, but also that the energy spectrum of the precipitating particles hardens during the equatorward motion. Quite reasonably, such observations have been identified as the ionospheric signature of an enhanced magnetotail lobe field (e.g., *Maezawa* [1975] reported a 30% increase in magnetotail flux prior to onset) with concomitant plasma sheet thinning and enhancement of cross tail current. *Sergeev et al.* [1990] have modeled such effects and have found that 80% of the equatorward expansion is due to the temporal variation of the magnetic field in the tail (i.e., the evolution to a more tail-like field). It is commonly supposed that this enhanced magnetic field pressure ultimately leads to an unstable situation wherein the field energy is through some process converted into particle energy. Establishing a consistent equatorward motion of the auroral distribution is therefore an important part of understanding the magnetospheric processes which lead to substorm onset.

Data from the Viking UV Imager have been selected to investigate specifically the character of the aurora in the MLT region where substorm onset ultimately occurs. When substorm activity has been absent for a sufficient period of time, the large scale auroral distribution consists primarily of featureless diffuse-like aurora. The results of *Ashour-Abdalla and Thorne* [1978] concerning resonant scattering by electrostatic waves have indicated that such diffuse aurora may be the result of strong particle diffusion into the loss cone. Alternatively *Ejiri* [1978] have shown that the diffuse aurora could arise strictly from particle trajectory considerations. *Fontaine et al.* [1986] have reported the simultaneous measurement of diffuse auroral particle signatures at geostationary orbit and in the ionosphere. These signatures measured by GEOS-2 and ARCAD-3 and also computed from the EISCAT E-region electron density showed the expected Maxwellian shape at $\approx$ 2KeV. *Lui et al.* [1977] identified the equatorward boundary of the diffuse aurora as being consistent with a mapping to the inner edge of the plasma sheet with the link between the ionosphere and magnetosphere accomplished by (in the evening sector) the Region 2 downward field-aligned current [*Klumpar*, 1979]. However, *Feldstein and Galperin* [1985] have argued that the diffuse aurora is mapped to the region of the outer radiation belt between the plasmapause and the plasma sheet. As noted above examples have been chosen here wherein the auroral distribution is quiet and hence dominated by what appears on a spatial scale of a hundred kilometers or so to be diffuse aurora. Figure 2 is an example image acquired on April 1, 1986. The entire evening sector is characterized by a thin region of emission, with apparently randomly varying intensities at different local times. It is possible that at a finer spatial resolution these variations may be related to discrete arc systems embedded in this diffuse distribution. The subsequent onset (which occurs by 1850 UT) meridian (23.6 MLT) is indicated. Figure 3a illustrates the change in boundary locations of the auroral emissions at this MLT with time. From each available Viking image the boundaries were quantitatively determined by carefully taking into account background emissions and are accurate to within .2 degrees magnetic latitude. The locations are presented in terms of time (in minutes) from the first image and in this case was some 12 minutes prior to onset. Overlaid on each of the poleward and equatorward boundaries shown in Figure 3 are linear fits of the motion with time.

The first apparent result is that in the short time period before onset, the changes to the auroral boundaries were relatively small compared to the changes after onset (times after onset are depicted by the thick line in the figure). The motion of the equatorward boundary before onset is relatively slow at 100 m/s equatorward although shortly after onset it moves equatorward at speeds on the order of 900 m/s. This equatorward motion of the equatorward boundary prior to onset was a consistent trend for all cases investigated (see Figures 3b and 3c). Using a small sample of 6 good onset cases the average equatorward motion was 235 ± 95 m/s. Note that as mentioned previously this refers to motion relative to the assumed fixed magnetic coordinate system. Such motion is consistent with an enhancement of the cross tail current during a growth phases.

The motion of the poleward boundary was found to be much more variable. In the example shown in Figure 3a the poleward boundary moves equatorward at a speed of 170 m/s before onset while after onset this boundary moves rapidly poleward at speeds between 1 to 2 km/s. Thus before onset the latitudinal width of the auroral distribution does not change by more than 10% of its original value (about 6 ° latitude). More pronounced narrowing is sometimes observed at least for short periods of time. In Figure 3b while both boundaries move equatorward once again, the poleward boundary moves much more slowly (<70 m/s) than the equatorward boundary (180 m/s) resulting in an increase of the width of the aurora if a linear variation is

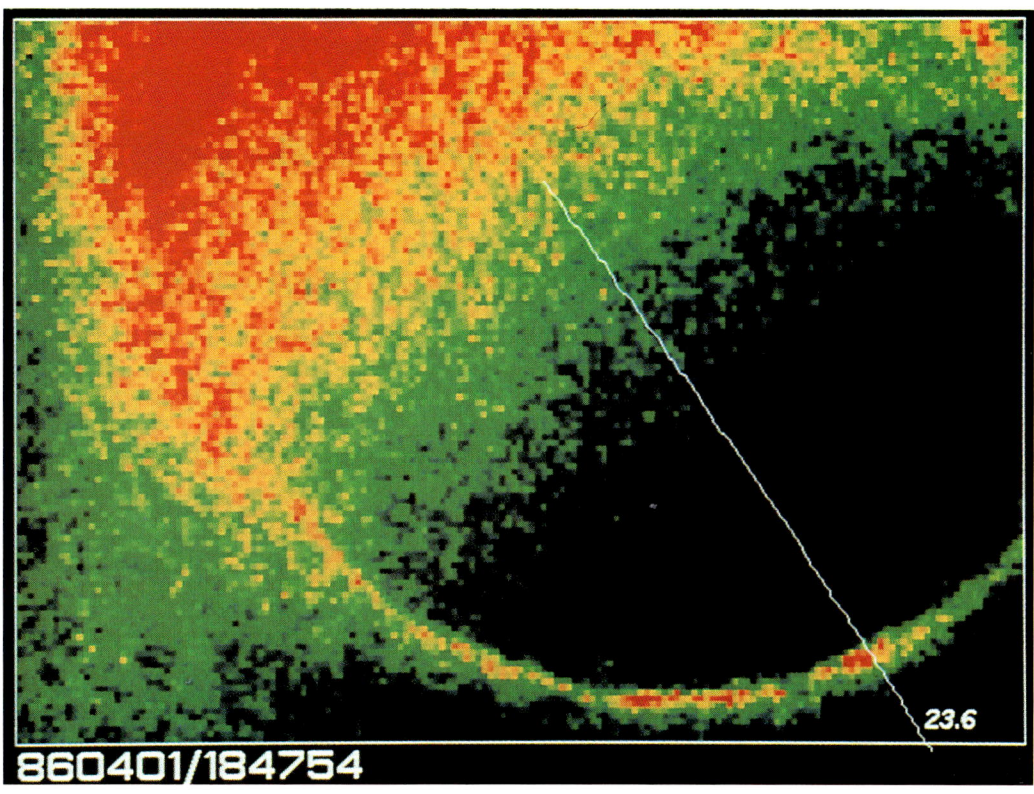

Figure 2. Single image acquired on April 1, 1986 illustrating conditions just prior to onset. The meridian locates the MLT of subsequent onset.

assumed. However, the correlation coefficient for the fit of the poleward boundary is very low (<.3) indicating such a fit is not reasonable. Part of the reason for this is the more rapid equatorward motion of the poleward boundary at approximately 44 minutes which occurs just at the same time as the equatorward boundary deviates poleward from its more general trend of equatorward motion. A similar effect can be seen in Figure 3c at around 12 minutes. With the limited data set available however, it is not possible to determine whether this is a common growth phase effect. In fact 2 cases showed a consistent poleward motion of the poleward boundary prior to onset. Thus while the equatorward boundary moves consistently equatorward the poleward boundary does not. The width therefore, of the diffuse aurora is variable from growth phase to growth phase.

The motions illustrated in Figure 3 might be used as 'evidence' that the amount of open flux in the magnetotail increases during the growth phase. However, such a process if consistently operating would affect the poleward boundary which is noted above has a highly variable character during the growth phase. In particular it is difficult to reconcile poleward motion of the boundary with enhanced lobe flux if the poleward boundary of emissions is identified with the open/closed field line boundary. Another piece of information which bears on this question is the observation that the polar regions can become devoid of activity [e.g., Davis, 1963] and this has also sometimes been regarded as an increase in the polar cap area. Such variations need not occur at the location of subsequent onset and in fact such variations are more pronounced near dawn and dusk since this is where the effects of northward B_z are largest [Murphree et al., 1982]. This is illustrated in Figure 4 which shows a sequence of images of the northern hemisphere auroral distribution on April 1, 1986. During the 40 minutes covered by these images the auroral distribution varied from a typical northward B_z topology with an expanded evening sector to a classic substorm onset wherein the polar arc had disappeared. It is apparent that the polar cap area as defined by optical emissions increased considerably. However, this is not necessarily the same as saying that open magnetic flux increased in the tail lobes. That the latter need not occur while the former does, is evidenced from the observation that the changing size of the auroral distribution can be accounted for by changes to the magnetospheric current system and that the boundary of open/closed field lines not be necessarily associated with the poleward boundary of optical emissions [Elphinstone et al.,

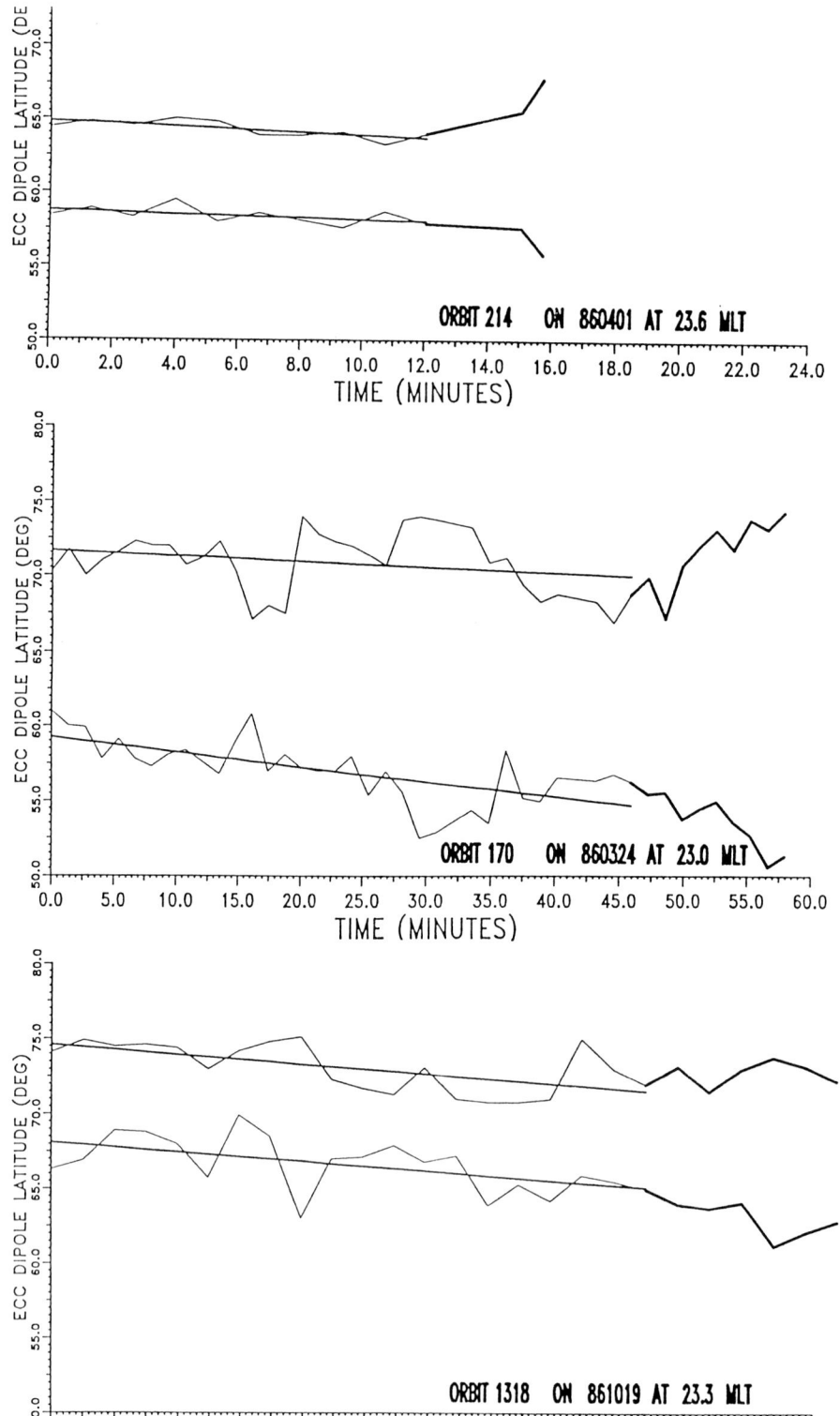

Figure 3. Plots of the poleward and equatorward boundaries of diffuse auroral emission prior to onset for three substorms. a) April 1, 1986 around 1850 UT. b) March 24, 1986 around 0123 UT. c) October 19, 1986 around 1132 UT.

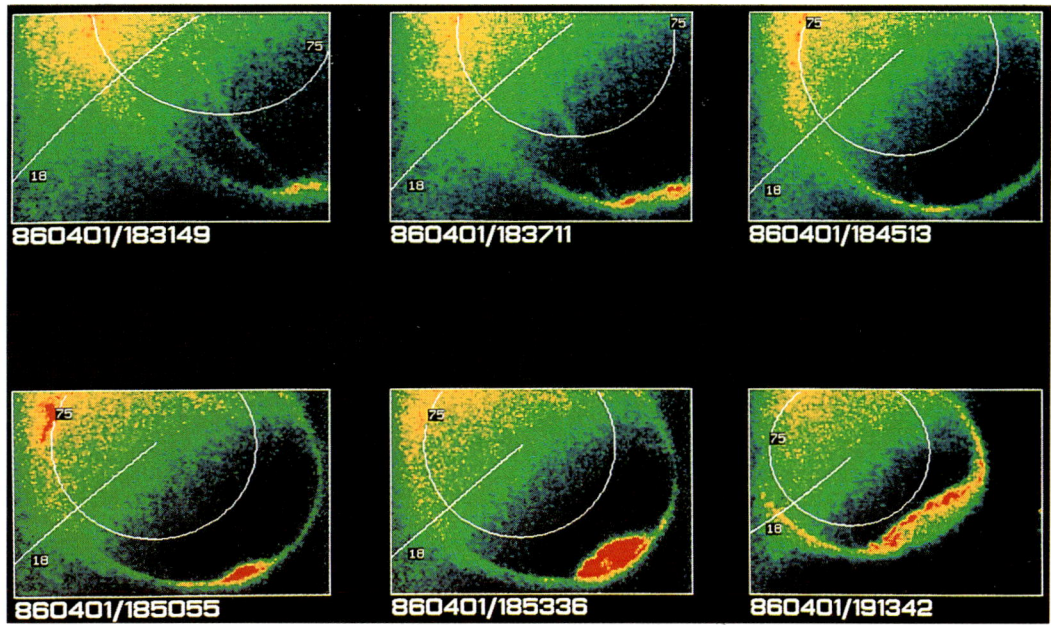

Figure 4. Sequence of 6 images acquired on April 1, 1986 illustrating the clearing of the polar cap of duskside high latitude emissions prior to substorm onset.

1991a]. In fact the upper panels in Figure 4 indicate that the polar arc simply fades away rather than moving rapidly equatorward and merging with the pre-existing auroral "oval". The significance of this observation will be dealt with further below.

Another feature of the optical emissions during the growth phase is the evidence for precursor activity prior to onset [Shepherd and Murphree, 1988]. This precursor activity in the evening sector is clearly seen at 1837 UT in Figure 4 and is further illustrated in Figure 5 with data taken from October 11, 1986 around 12 UT. The six images are color coded so as to emphasize the variation of intensity of the evening sector emissions prior to the onset near 1204:29 UT. (For a discussion of the relationship of pre-onset activity at other local times such as those in the afternoon sector see Elphinstone et al. [1991b]). The MLT meridians are shown to provide a reference for the intensifications which occur for some 15-20 minutes prior to onset. At 1148:18 UT there was localized regions of emission near 19 and 21 MLT. The enhancement at 19 MLT essentially fades away never to reappear (at least prior to onset) and is replaced by highly variable activity at 20 MLT. At this MLT in the 16 minutes prior to onset the intensity changes by a factor of 2. The image at 1154:15 UT for example shows a relative peak in intensity while at 1203:30 a minimum is reached. Throughout this period however, there is evidence of auroral activity. Just to the east, at 21 MLT, the variations in intensity are not so pronounced, but just as erratic. The peak at 1148:18 UT is replaced by a relative minimum in the auroral distribution at 1157:13 UT. After this the auroral distribution throughout this entire region decreases in intensity and an essentially quiet distribution exists at 1203:30 UT, but within 1 minute approximately three hours of MLT are involved in what continues into a poleward expanding bulge (i.e., substorm onset). This intensification at onset represents (at 21 MLT) more than a factor of 3 enhancement over the pre-onset intensity variation. It is not possible from this example to specify the precise local time of onset or the local time extent of the onset region. What can be said, however, is that precursor activity does occur outside the onset region delineated in the image at 1204:29 UT and that the onset region (see for example the image at 1157:13 UT) has gone through a fading [Pellinen and Heikkila, 1978] prior to the actual expansive phase. However, the fading time is clearly dependent on the MLT under consideration. It might be noted that the events shown in Figures 1 and 4 both display similar activity.

This behaviour of intensifications over an extended local time sector has important consequences for theories regarding substorm onset. First the signatures in the auroral distribution imply a considerable local time extent for pre-onset activity. For the example shown in Figure 5, approximately four hours of MLT are involved which corresponds to approximately a $10R_E$ extent (primarily in the Y GSM direction) in the minimum B surface of the magnetotail. This is to be compared to the approximately 3.5 R_E extent in the radial direction reported by Lui [1978]. Other multiple intensifications at several local time positions suggest that perhaps a steady progression to substorm onset does

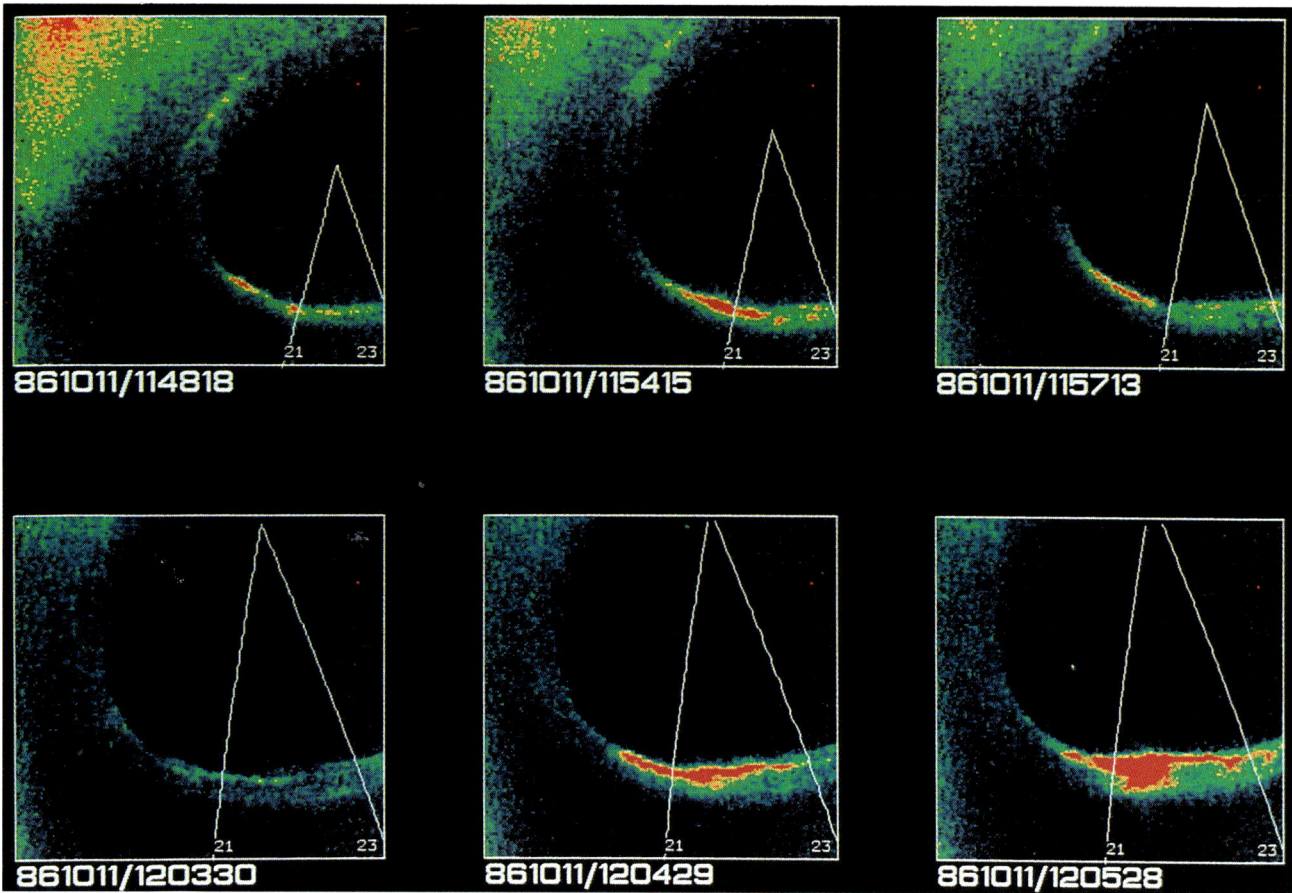

Figure 5. Sequence of images acquired on October 11, 1986. The data are corrected for non-uniformity and background, and red indicates highest intensities. The meridians indicate the MLT and provide a reference for the temporal variation of the premidnight auroral emissions prior to onset at 1204:29 UT.

not occur. It is generally recognized that a number of conditions must be satisfied before onset will actually happen. For example, *Lui et al.* [1990] suggest a combination of instability activity at the neutral sheet and the plasma sheet boundary layer to instigate substorm onset. This current disruption process depends strongly on the region in the tail where the relative drift between ions and electrons is maximized. However, it is likely that the substorm process is controlled at least in part by the inability of the ionosphere to divert current [*Kan et al.*, 1988]. Under this scenario not only must magnetospheric convection be enhanced, but the convection reversal region must map to the region of the poleward gradient of conductance near the diffuse auroral boundary. It thus may be the case that over an extended region of the magnetotail convection is enhanced, but sufficient conductance to divert significant current through the ionosphere is not available. The example of Figure 5 shows that since the subsequent onset does not necessarily occur at the same position as the previous intensifications (specifically other local time regions had significant particle precipitation prior to onset), particle produced conductance enhancements may not be the only requirement. It appears that the coupling between the tail and ionosphere is not a simple progression toward instability breakups (see also comments by *Lui et al.*, [1990]), but neither is it a process which tries to bootstrap itself by episodically initiating the onset from a single tail location.

ONSET

Historically, optical substorm onset has been discussed in terms of the brightening, and subsequent poleward expansion of the most equatorward discrete arc [*Rostoker et al.*, 1980]. With regard to models of substorm evolution it is important to identify the location of this optical onset within the topology of the magnetosphere. Recently *Kirkwood and Eliasson* [1990] have suggested that the onset arc brightens near the poleward boundary of the trapping region which would place the mapping of such a process quite near the earth. Indeed *Elphinstone et al.* [1991a] have determined that the main auroral distribution can be associated with a single source region in the nightside near-Earth magnetotail. They found the aurora to be coincident with the ionospheric projection of the nightside magnetotail current density maximum. This is the region of the magnetosphere where the magnetic field lines undergo a transition from a dipole-like to tail-like configuration. Measurements reported by *Sergeev et al.* [1990] show that in this region (at 9 R_E in the tail) the current sheet thickness can become very thin (0.1 R_E) during the growth phase suggesting further that onset occurs in this same region. This relationship is illustrated in Figure 6 where the current density maximum is projected as the most equatorward thin white line onto the UV auroral distribution observed on October 19, 1986 at 1131:01 UT. This line overlays the auroral distribution quite well given the fact that no Interplanetary Magnetic Field effects have been included. A point to be emphasized is that although there is a large degree of uncertainty in mapping of features between the ionosphere and the equatorial plane in the midnight sector, the relationship at other local times supports the above identification [see also *Elphinstone et al.*, 1991b]. Thus the source region at most local times (excluding a region centered about 12 MLT) appears to be associated with what has been termed as the "nightside cusp" by *Lui and Burrows* [1978]. This region encompasses the major distortion of field lines from dipolar to tail-like and represents the transition region from central plasma sheet to boundary plasma sheet signatures as measured at low altitudes. This is consistent with the results of *Feldstein and Galperin* [1985].

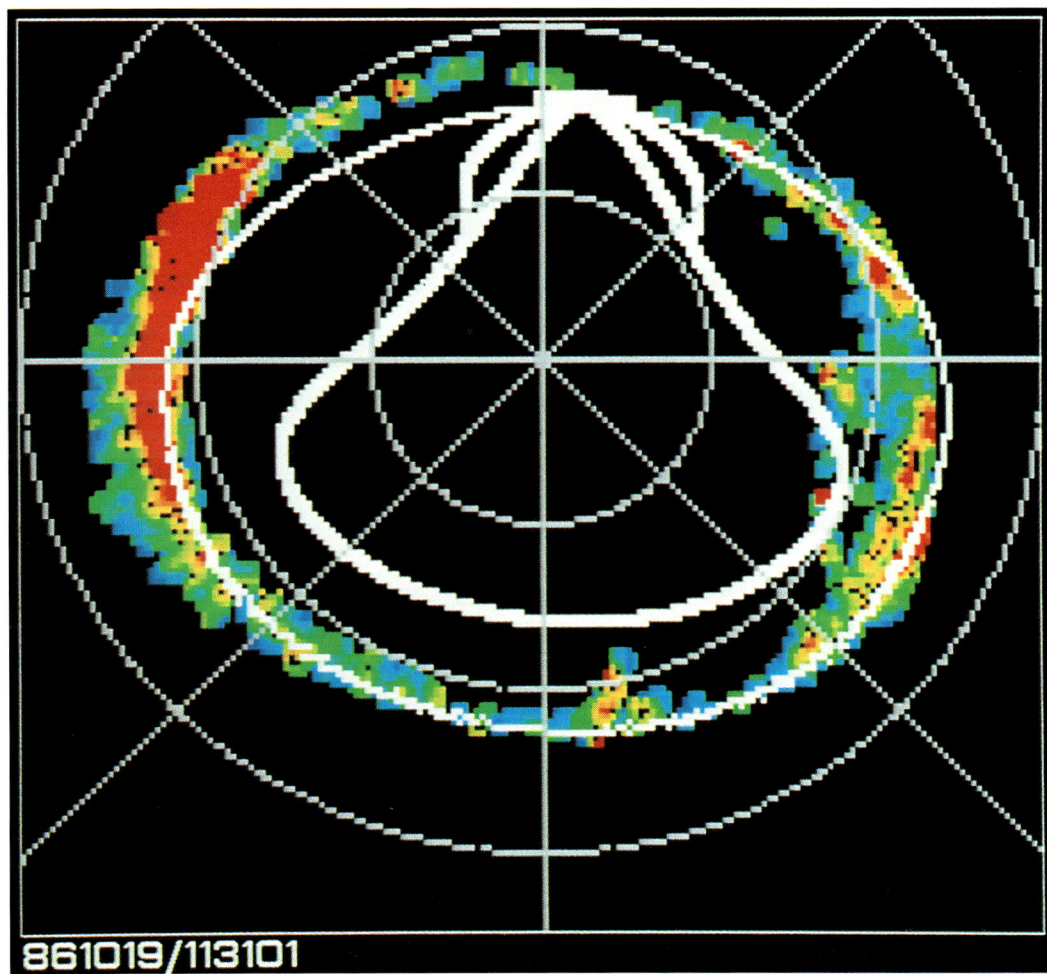

Figure 6. Ionospheric projection for October 19, 1986 at 1131:01 UT of the auroral distribution. The data have been transformed into eccentric dipole coordinates on a plane tangent to the earth's surface. Overlaid on this image are two boundaries obtained by a mapping of the Tsyganenko (1987) model under these geophysical conditions. The lower latitude thin white line which overlays the auroral distribution represents the mapping of the cross tail current density maximum and the thick white line at higher latitudes the open/closed model field line boundary.

TABLE 2 - Onset Times & Locations

Orbit	Date	U.T.	mlt	mlat ecc 85	Kp	X GSM	Y GSM	Z GSM
214	860401	185055	23.6	62.0	1 or 3	-6.2	1.3	1.3
236	860405	181821	22.0	64.9	2	-6.8	4.4	1.1
382	860502	073423	23.8	63.8	3	-7.5	0.9	0.6
383	860502	121538	23.1	64.0	3	-4.4	0.7	0.9
386	860503	011210	21.3	57.1	5	-3.7	2.8	0.5
398	860505	052529	23.3	60.4	4	-6.2	1.2	0.3
1177	860923	205345	0.4	62.2	5-	-7.0	0.1	0.7
1182	860924	182657	24.0	65.2	3-	-13.4	0.4	0.4
1188	860925	205537	0.6	65.6	3+	-10.1	-0.3	-0.5
1204	860928	184336	23.4	64.6	1 or 3	-8.3	1.9	1.2
1226	861002	183309	23.8	60.6	5	-6.7	0.8	0.1
1315	861018	223502	1.6	65.4	4	-12.9	-3.0	-0.7
1318	861019	113201	23.2	69.1	3	-7.9	1.2	-0.6
1554	861201	090240	23.0	67.0	3	-7.8	1.9	-2.

Associating the presubstorm auroral distribution with the nightside cusp at most ionospheric local times leads one naturally to interpret substorm expansion phase onset as a near-Earth phenomenon. We have investigated this possibility by cataloguing a number of substorm expansion phase onsets as seen in the Viking images during 1986. The initial arc brightening was chosen to define the ionospheric location of expansive phase onset. The temporal accuracy of the imager was 1 minute so that substorm poleward expansion speeds of 2 km/s gives spatial inaccuracies on the order of $\approx$100 km. The onset locations were determined in the ionosphere and mapped along the field line (using a combination of the *Tsyganenko* [1987] external model and the IGRF [1985] internal field model) to the nightside minimum B surface. The results are summarized in Table 2 and Figure 7 shows the distribution of onset location in the XY GSM plane. The ionospheric eccentric dipole (1985) magnetic local time of onset was between 21.3 to 1.6 MLT while the latitude range varied between 57 ° and 69 °. The projection of these points into the tail gives onset locations confined in Y GSM position between -3 R_E and +4.4 R_E and in X GSM between -3.7 R_E and -13.4 R_E. The main

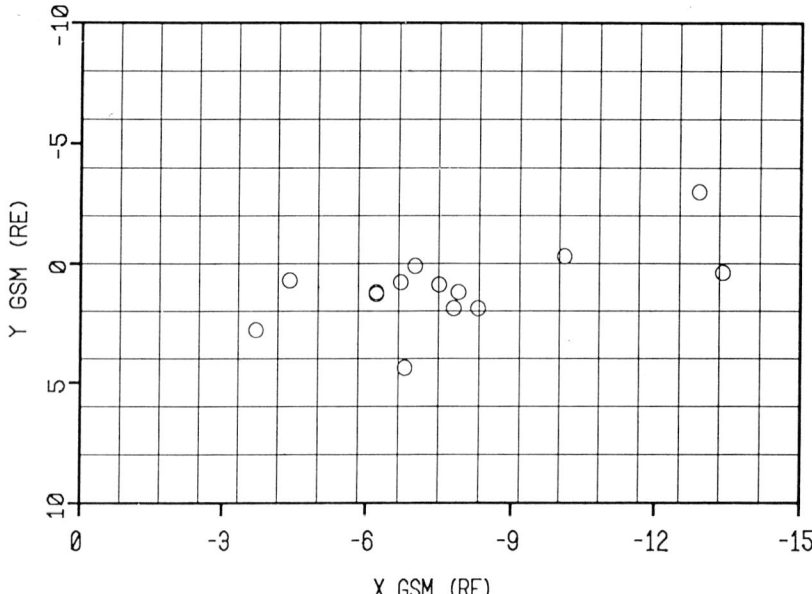

Figure 7. Plot of the XY GSM locations of substorm onset as determined from the Viking optical onset signature mapped out to the equatorial plane using the Tsyganenko (1987) model.

TABLE 3

Kp	CGM latitude of the maximum	Radial Distance (R_E) TSY 87 Current density maximum at appropriate MLT (0 tilt; 6 UT)			Radial Distance (R_E) Lopez et al. (1990) (140-17 Kp)/MLT-10) Energetic Ion Injection Boundary	Radial Distance (R_E) Mauk and McIlwain (1974) (122-10 Kp)/(LT-7.3) Low Energy Substorm Injection Boundary
		R_E	X GSM	Y GSM		
0 MLT						
0	68.0	8.9	-8.9	0.0	10.0	7.3
2	64.4	6.1	-6.1	0.0	7.6	6.1
3	64.4	6.24	-6.2	0.0	6.4	5.5
4	61.8	5.0	-5.0	0.0	5.1	4.9
21 MLT						
0	68.5	9.24	-6.9	6.1	12.7	8.9
2	65.8	7.1	-5.2	4.8	9.6	7.5
3	65.8	7.5	-5.5	5.0	8.1	6.7
4	64.1	6.46	-4.8	4.3	6.5	6.0
1 MLT						
0	67.8	9.1	-8.8	-2.2	10.7	6.9
2	64.9	6.3	-6.1	-1.6	8.2	5.8
3	64.6	6.3	-6.1	-1.6	6.8	5.2
4	62.3	5.2	-5.0	-1.3	5.5	4.6
22 MLT						
0	68.2	9.1	-8.1	4.1	11.7	8.3
2	65.2	6.6	-5.8	3.1	8.8	6.9
3	64.9	6.67	-5.9	3.1	7.4	6.3
4	62.7	5.5	-4.8	2.7	6.0	5.6
23 MLT						
0	67.8	8.7	-8.4	2	10.7	7.8
2	64.4	6.0	-5.8	1.5	8.2	6.5
3	64.5	6.3	-6.1	1.5	6.8	5.9
4	62.0	5.1	-5.0	1.2	5.5	5.2

grouping is concentrated between X GSM -6 R_E and -8 R_E with Y GSM limited between 0 and +2 R_E.

It is possible the mapping of pre-substorm auroral features is inaccurate due to differences between the static model and the topology associated with the growth phase of a substorm. To test for this, a comparison was made between the model and empirical observations in the magnetotail. *Mauk and McIlwain* [1974] determined there was a correlation between Kp and the low energy substorm-injection plasma boundary. *Lopez et al.* [1990] arrived at a similar result and interpreted the energetic ion injection boundary as the inner edge of the cross-tail current disruption associated with the expansion phase of a substorm. Figure 11 in *Lopez et al.* [1990] schematically outlines their interpretation of this boundary and its relation to the substorm expansion phase. They suppose that the injection boundary corresponds to a minimum contour of a parameter δ representing a criteria for current disruption. δ was defined to be the ratio of the magnetic perturbation of the current sheet to the background field. This parameter will be larger in the vicinity of the nightside cusp. In this region the current density maximizes and the perturbation magnetic field starts to play a significant role in relation to the background internal field. It is reasonable to expect then that the maximum in the nightside current density as defined in detail by *Elphinstone et al.* [1991a] should correspond to the injection boundary. Using a variety of Kp levels and determining the current density maximum, a comparison could be made between the model and the injection boundary found by *Lopez et al.* [1990]. Table 3 shows their injection boundaries (and as well those from *Mauk and McIlwain* [1974]) for various Kp levels and MLT as well as the locations of the maximum current density found using the model. To facilitate ground based comparisons the Corrected Geomagnetic (CGM) latitudes for epoch 1980.0 are also given. Figure 8 shows a comparison of the two estimates of the injection boundary to the model current density maxima which are associated with the pre-substorm auroral distribution. *Mauk and McIlwain* [1974] use the local time variability of the plasmapause to find their relation, so it is reasonable to expect their result to be somewhat earthward of the model calculation. Both results however put the injection boundary within about 3 R_E of the model current density maximum. The agreement is best earthward of 7 R_E (i.e., the midnight sector).

The current density maximum in the model appears to be linked to the injection boundary as well as to the overall auroral distribution in the ionosphere. The equatorial shift of the auroras during more active conditions can be seen as an intensification and earthward shift of tail current sheet which sometimes becomes unstable and results in an auroral substorm expansive phase. These results suggest that the motion of the auroras can be explained without requiring changes to the amount of open flux in the magnetotail or by invoking any reconnection theory. In Figure 6 the open/closed field line boundary as estimated from the Tsyganenko model is shown as the tear-drop shaped white line. Any modifications to open flux would therefore alter a boundary well poleward of the observed auroral distribution. Hence results such as illustrated in Figure 3 may be due to more direct physical processes which affect the cross tail current density.

This mapping of the onset region to near Earth provides supportive information only on the position within the magnetosphere where onset is initiated because of the high degree of uncertainty in mapping individual field lines in the midnight sector. It is perhaps more fundamental to look for topological information in the optical data themselves. Thus a simple question that might be asked is "what is the spatial relationship between the onset location

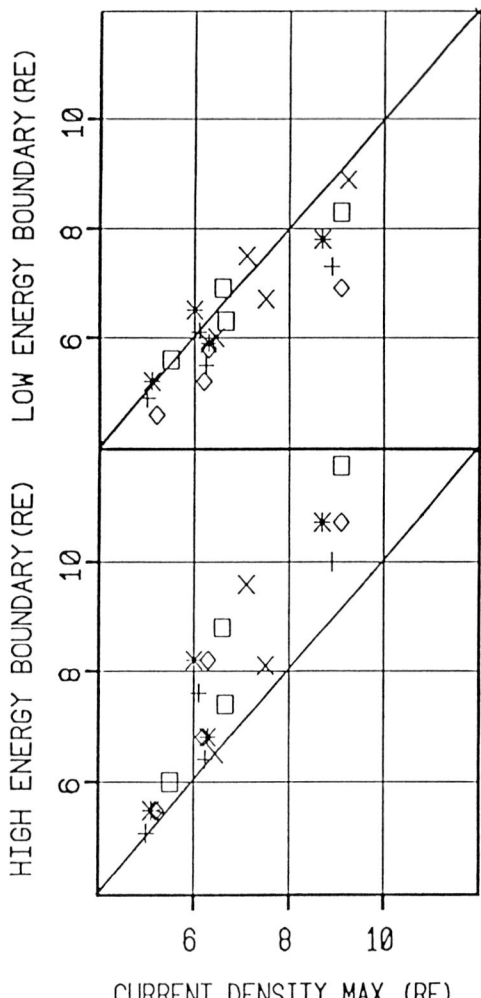

Figure 8. Plots of the relationship between the ion injection boundary of Lopez et al. (1990) (bottom) and plasma boundary of Mauk and McIlwain (1974) (top) with the position of the current density maximum derived from the Tsyganenko (1987) model. The different symbols represent different MLT throughout the midnight sector: 0 MLT (plus sign), 21 MLT (cross), 1 MLT (diamond), 22 MLT (square), 23 MLT (star).

and the most poleward extent of the optical emissions at that same local time?" The current understanding of substorm onset (e.g., *Rostoker et al.*, 1980] places it at the most equatorward discrete arc which implicitly recognizes that other discrete features may exist poleward. The significance of this latitudinal spread poleward of the onset region is that it defines the <u>minimum</u> separation between the onset location and the open/closed field line boundary. There is no evidence to suggest that any optical emissions poleward of the onset location (including diffuse aurora) must occur on open field lines and in fact the global context provided by satellite images suggests exactly the opposite. Some confusion arises because in the ionospheric signature of substorm onsets it is sometimes the case that the latitudinal scale size of the onset region is so small that it is possible to presume that the onset occurs near the open/closed field line boundary. This presupposes that the poleward extent of optical emissions is identified with that boundary. While it has been suggested above that such an equivalence is not necessarily correct, for many substorm models that presupposition is in fact irrelevant. However, this is not the case for models such as a near Earth neutral line model wherein the reconnection of the last closed field line is related to substorm expansion phase. Irrespective of mapping considerations it is possible to show that this signature does not always (if ever) agree with the optical observations because substorm onsets can occur significantly equatorward of the most poleward emissions. Figure 9 shows a sequence of images acquired on October 2, 1986 where a substorm onset occurs around 1836 UT. This particular period of time is not preceded by as extended a period of quiet as the previous examples and thus gives an indication of the poleward extent of optical emissions beyond the diffuse aurora particularly near 21 MLT. That such a region of precipitation can exist is well documented [*Eather and Mende*, 1971]. The substorm onset occurs right at midnight and as discussed above some precursor activity takes place prior to the onset at 1833:08 UT. A few hours local time are involved in this process although it seems most evident at the same local time. This particular example indicates that the actual substorm onset region is confined in local time to a region no bigger than .5 MLT, and it may well be smaller, but the one minute interval between images precludes a more accurate description. The latitudinal position of the substorm onset is seen to occur in the main auroral distribution but there are emissions poleward of this region by as much as 5-8° latitude. This is a common observation when conditions are such that the region poleward has significant fluxes. Previous reports of auroral bulge expansion [e.g., *Sergeev and Yahnin*, 1979] have noted the existence of emissions poleward and that the subsequent expansion will proceed to, but not exceed that position. This is perhaps a reflection of an expansion up to, but not exceeding the boundary of the plasma sheet.

SUMMARY

The optical signatures of the auroral substorm process have been investigated and the following points summarize the essential elements in this paper:

1) The latitudinal width of the auroral distribution in the vicinity of substorm onset does not undergo a systematic variation during substorm growth phase.

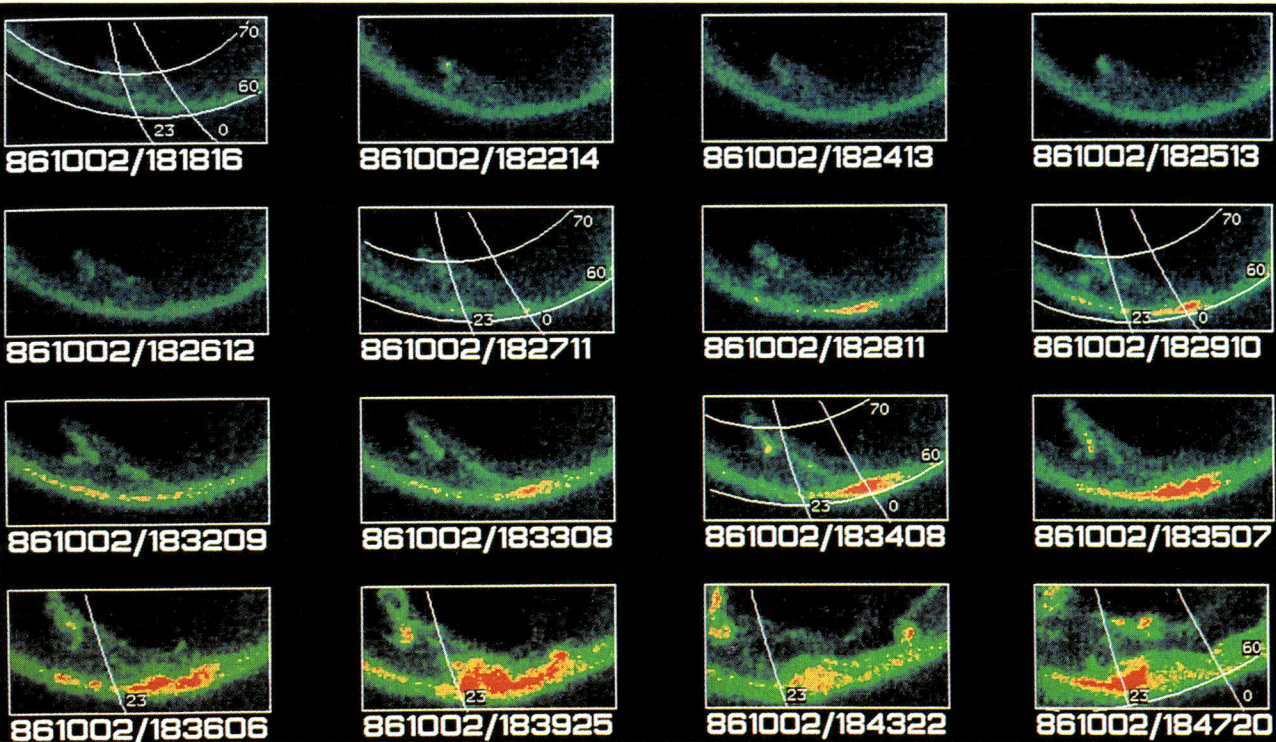

Figure 9. Sequence of images acquired on October 2, 1986 at about 1830 UT illustrating a localized substorm onset occurring a few degrees equatorward of the poleward edge of the auroral distribution.

2) Motion of the equatorward boundary of the diffuse aurora is generally equatorward at speeds less than a few hundred m/s.

3) The substorm expansion phase is preceded by auroral intensifications covering up to several hours of local time which fade shortly before onset. The onset region itself is very localized being less than 500 km across in the ionosphere.

4) Auroral observations under moderately active conditions indicate that auroral emissions can extend several degrees of latitude poleward of the onset location. These observations suggest that the onset region can be well separated from the boundary between open and closed field lines.

5) Mapping of onset positions into the equatorial plane is consistent with the location of onset positions previously deduced from injection boundary studies.

Acknowledgements. The authors would like to thank Greg Enno for assistance with the figures and Ann Marie Morris for typing the manuscript. The Viking project was managed by the Swedish Space Corporation under contract to the Swedish Board for Space Activities. The UV imager was built as a project of the National Research Council of Canada and this work was supported under grants from the Natural Sciences & Engineering Research Council of Canada.

REFERENCES

Akasofu, S.-I., The Development of the Auroral Substorm, *Planet. Space Sci., 12*, 273, 1964.

Anger, C. D., S. K. Babey, A. L. Broadfoot, R. G. Brown, L. L. Cogger, R. Gattinger, J. W. Haslett, R. A. King, D. J. McEwen, J. S. Murphree, E. H. Richardson, B. R. Sandel, K. Smith, and A. Vallance Jones, An Ultraviolet Auroral Imager for the Viking Spacecraft, *Geophys. Res. Lett., 14*, 387-390, 1987.

Ashour-Abdalla, M. and R. M. Thorne, Toward a Unified View of Diffuse Auroral Precipitation, *J. Geophys. Res., 83*, 4755-4766, 1978.

Baker, D. N., E. W. Hones, Jr., P. R. Higbie, R. D. Belian and P. Stauning, Global Properties of the Magnetosphere During a Substorm Growth Phase : A Case Study, *J. Geophys. Res., 86*, 8941-8956, 1981.

Baker, D. N., Particle and Field Signatures of Substorms in the Near Magnetotail, in *Magnetic Reconnection in Space and Laboratory Plasmas*, edited by E. W. Hones, Jr., American Geophysical Union Monograph 30, Washington, D. C., p. 193-202, 1984.

Bargatze, L. F., D. N. Baker, R. L. McPherron, and E. W. Hones, Jr., Magnetospheric Impulse Response for Many Levels of Geomagnetic Activity, *J. Geophys. Res., 90*, 6387-6394, 1985.

Craven, J. D., and L. A. Frank, Latitudinal Motions of the Aurora During Substorms, *J. Geophys. Res., 92*, 4565-4573, 1987.

Craven, J. D., L. A. Frank, and S.-I. Akasofu, Propagation of a Westward Traveling Surge and the Development of Persistent Auroral Forms, *J. Geophys. Res., 94*, 6961-6967, 1989.

Davis, T. N., Negative Correlation Between Polar-Cap Visual Aurora and Magnetic Activity, *J. Geophys. Res., 68*, 4447-4453, 1963.

Eather, R. H. and S. B. Mende, Airborne Observations of Auroral Precipitation Patterns, *J. Geophys. Res., 76*, 1746-1755, 1971.

Ejiri, M., Trajectory Traces of Charged Particles in the Magnetosphere, *J. Geophys. Res., 83*, 4798-4810, 1978.

Elphinstone, R. D., D. Hearn, J. S. Murphree and L. L. Cogger, Mapping Using the Tsyganenko Long Magnetospheric Model and its Relationship to Viking Auroral Images, *J. Geophys. Res., 96*, 1467-1480, 1991a.

Elphinstone, R. D., J. S. Murphree, L. L. Cogger, D. Hearn, M. G. Henderson and R. Lundin, Observations of Changes to the Auroral Distribution Prior to Substorm Onset, (*this volume*), 1991b.

Feldstein, Y. I. and Yu. I. Galperin, The Auroral Luminosity Structure in the High-Latitude Upper Atmosphere: Its Dynamics and Relationship to the Large-Scale Structure of the Earth's Magnetosphere, *Rev. Geophys., 23*, 217-275, 1985.

Fontaine, D., S. Perraut, N. Cornilleau-Wehrlin, B. Aparicio, J. M. Bosqued, and D. Rodgers, Coordinated Observations of Electron Energy Spectra and Electrostatic Cyclotron Waves During Diffuse Auroras, *Annales Geophysicae, 4*, 405-412, 1986.

Kan, J. R., L. Zhu and S.-I. Akasofu, A Theory of Substorms : Onset and Subsidence, *J. Geophys. Res., 93*, 5624-5640, 1988.

Kaufmann, R. L., Substorm Currents: Growth Phase and Onset, *J. Geophys. Res., 92*, 7471-7486, 1987.

Kirkwood, S., and L. Eliasson, Energetic Particle Precipitation in the Substorm Growth Phase Measured by EISCAT and Viking, *J. Geophys. Res., 95*, 6025-6037, 1990.

Klumpar, D. M., Relationships Between Auroral Particle Distributions and Magnetic Field Perturbations Associated with Field-Aligned Currents, *J. Geophys. Res., 84*, 6524-6532, 1979.

Lopez, R. E., D. G. Sibeck, R. W. McEntire and S. M. Krimigis, The Energetic Ion Substorm Injection Boundary, *J. Geophys. Res., 95*, 109-118, 1990.

Lui, A. T. Y., D. Venkatesan, C. D. Anger, S.-I. Akasofu, W. J. Heikkila, J. D. Winningham and J. R. Burrows, Simultaneous Observations of Particle Precipitations and Auroral Emissions by the ISIS 2 Satellite in the 19-24 MLT Sector, *J. Geophys. Res., 82*, 2210-2226, 1977.

Lui, A. T. Y., Estimates of Current Changes in the Geomagnetotail Associated with a Substorm, *Geophys. Res. Lett., 5*, 853-856, 1978.

Lui, A. T. Y., and J. R. Burrows, On the Location of Auroral Arcs Near Substorm Onsets, *J. Geophys. Res., 83*, 3342-3348, 1978.

Lui, A. T. Y., A. Mankofsky, C.-L. Chang, K. Papadopoulos and C. S. Wu, A Current Disruption Mechanism in the Neutral Sheet : A Possible Trigger for Substorm Expansions, *Geophys. Res. Lett.*, *17*, 745-748, 1990.

McPherron, R. L., Growth Phase of Magnetospheric Substorms, *J. Geophys. Res.*, *28*, 5592-5599, 1970.

Maezawa, K., Magnetotail Boundary Motion Associated with Geomagnetic Substorms, *J. Geophys. Res.*, *80*, 3543-3548, 1975.

Maggs, J. E. and T. N. Davis, Measurements of the Thicknesses of Auroral Structures, *Planet. Space Sci.*, *16*, 205-209, 1968.

Mauk, B. H. and C. E. McIlwain, Correlation of Kp with the Substorm-Injected Plasma Boundary, *J. Geophys. Res.*, *79*, 3193-3196, 1974.

Mitchell, D. G., Kinetic Aspects of Magnetotail Physics - Observations, in *Magnetotail Physics*, edited by A. T. Y. Lui, The Johns Hopkins University Press, Baltimore, p. 207-224, 1987.

Murphree, J. S., C. D. Anger and L. L. Cogger, The Instantaneous Relationship Between Polar Cap and Oval Auroras at Times of Northward Interplanetary Magnetic Field, *Can. J. Physics*, *60*, 349-356, 1982.

Murphree, J. S., R. D. Elphinstone, L. L. Cogger and D. D. Wallis, Short-Term Dynamics of the High-Latitude Auroral Distribution, *J. Geophys. Res.*, *94*, 6969-6974, 1989.

Nishida, A., IMF Control of the Earth's Magnetosphere, *Space Sci. Rev.*, *34*, 185, 1983.

Pellinen, R. J., and W. J. Heikkila, Observations of Auroral Fading Before Breakup, *J. Geophys. Res.*, *83*, 4207-4217, 1978.

Pellinen, R. J. and W. J. Heikkila, Inductive Electric Fields and Their Relation to Auroral and Substorm Phenomena, *Space Sci. Rev.*, *37*, 1-61, 1984.

Pytte, T., H. Trefall, G. Kremser, L. Jalonen and W. Riedler, On the Morphology of Energetic (> 30 KeV) Electron Precipitation During the Growth Phase of Magnetospheric Substorms, *J. Atmos. Terr. Phys.*, *38*, 739-755, 1976.

Ranta, H., A. Ranta, P. N. Collis and J. K. Hargreaves, Development of the Auroral Absorption Substorm : Studies of Pre-Onset Phase and Sharp Onset Using an Extensive Riometer Network, *Planet. Space Sci.*, *29*, 1287-1313, 1981.

Rees, M. H., D. Lummerzheim, R. G. Roble, J. D. Winningham, J. D. Craven and L. A. Frank, Auroral Energy Deposition Rate, Characteristic Electron Energy, and Ionospheric Parameters Derived from Dynamics Explorer 1 Images, *J. Geophys. Res.*, *93*, 12841-12860, 1988.

Rostoker, G., A. Vallance Jones, R. L. Gattinger, C. D. Anger and J. S. Murphree, The Development of the Substorm Expansive Phase: The 'Eye' of the Substorm, *Geophys. Res. Lett.*, *14*, 399-402, 1987.

Rostoker, G., S.-I. Akasofu, J. Foster, R. A. Greenwald, Y. Kamide, K. Kawasaki, A. T. Y. Lui, R. L. McPherron and C. T. Russell, Magnetospheric Substorms - Definition and Signatures, *J. Geophys. Res.*, *85*, 1663-1668, 1980.

Sergeev, V. A., and A. G. Yahnin, The Features of Auroral Bulge Expansion, *Planet. Space Sci.*, *27*, 1429-1440, 1979.

Sergeev, V. A., P. Tanskanen, K. Mursula, A. Korth and R. C. Elphic, Current Sheet Thickness in the Near-Earth Plasma Sheet During Substorm Growth Phase, *J. Geophys. Res.*, *95*, 3819-3828, 1990.

Shepherd, G. G., and J. S. Murphree, Diagnosis of Auroral Dynamics Using Global Auroral Imaging with Emphasis on Localized and Transient Features, *Proc. of the International Conf. on Auroral Physics*, Cambridge, England, p. 265-273, 1988.

Tanskanen, P., J. Kangas, L. Block, G. Kremser, A. Korth, J. Woch, I. B. Iversen, K. M. Torkar, W. Riedler, S. Ullaland, J. Stadnes and K.-H. Glassmeier, Different Phases of a Magnetospheric Substorm on June 23, 1979, *J. Geophys. Res.*, *92*, 7443, 1987.

Tsyganenko, N. A., Global Quantitative Models of the Geomagnetic Field in the Cislunar Magnetosphere for Different Disturbance Levels, *Planet. Space Sci.*, *35*, 1347-1358, 1987.

Vorobjev, V. G., G. V. Starkov and Y. I. Feldstein, The Auroral Oval During the Substorm Development, *Planet. Space Sci.*, *24*, 955, 1976.

Zverev, V. L., G. V. Starkov and Y. I. Feldstein, Influences of the Interplanetary Magnetic Field on the Auroral Dynamics, *Planet. Space Sci.*, *27*, 665-667, 1979.

Observations of Changes to the Auroral Distribution Prior to Substorm Onset

R. D. ELPHINSTONE, J. S. MURPHREE, L. L. COGGER,
D. HEARN, AND M. G. HENDERSON

Department of Physics and Astronomy, University of Calgary, Calgary, Alberta

R. LUNDIN

Swedish Institute of Space Physics, Kiruna, Sweden

Abstract. The temporal development of the auroral distribution is described for a time period ($\approx$40 minutes) before substorm expansion phase onset. Initial auroral activity is observed to begin in the dayside ionosphere and then shift to the nightside. The dayside precursor events are associated with the growth of eastward (afternoon) and westward (morning) electrojets shortly after the interplanetary magnetic field turns southward. During this time, at both noon and midnight, equatorward motions of the peak intensity aurora are observed with speeds of 300 to 500 m/s. Longitudinal motions within the auroral "oval" during this "growth" phase tend to be eastward in the afternoon sector and westward in the morning. These motions are very rapid, sometimes exceeding 15 km/s in the ionosphere. It appears that the stable afternoon sector aurora near 16 MLT is likely to be associated with nightside magnetospheric activity. On some occasions, however, the dayside auroral dynamo may be associated with transient auroral forms superimposed on the more stable aurora originating from the nightside magnetotail. A remarkable set of combined observations by the Viking satellite and ground magnetometer data illustrates this just a few minutes before the start of an auroral substorm. A brief outline of a possible magnetospheric interpretation to these observations is presented whereby a solar wind disturbance separately affects the dayside and nightside magnetospheric dynamos. This first creates the observed dayside auroral phenomena, and eventually perturbs the cross-tail current circuit triggering the substorm expansion phase in the near Earth region.

INTRODUCTION

Increased knowledge of the auroral distribution has consistently led to the development of theories concerning the magnetosphere. *Feldstein* [1963] showed there existed an asymmetry between the dayside and nightside latitudes of auroral occurrences. This later became good evidence for the asymmetric shape of the magnetosphere. *Akasofu* [1964], through the study of all-sky camera data, described the development of the auroral substorm. Since then, a large amount of literature regarding substorms, and the associated magnetospheric mechanism for them, has been written (see for example the review by *McPherron*, [1979]).

The major portion of this literature has concentrated on times after the onset of the expansion phase in the midnight sector. The period previous to this onset, the growth phase, has been less studied in the past, partly due to the difficulty in knowing exactly when the expansion phase begins [*Akasofu and Snyder*, 1972]. Nevertheless, there does seem to be evidence for some period of "growth" before the main breakup of arcs in the late evening sector. The noon sector precipitation region is thought to move equatorward prior to the expansion phase [*Hoffman and Burch*, 1973] and the dayside magnetopause may move earthward [*Aubry et al.*, 1970]. The tail field becomes more taillike [*Pytte et al.*, 1976] and there is sometimes a slow change in the magnetic field measured on the ground during the time preceding an auroral substorm [*Untiedt et al.*, 1978; *Nishida and Kamide*, 1983]. *Frank et al.* [1988] have investigated the polar region bounded by the auroral oval and have associated changes within this area directly to changes in the tail lobe flux. The results of *Elphinstone et al.* [1991] and this paper show that alternative interpretations to these observations are possible.

The growth phase has been thought to be a period of limited optical auroral activity although some effects have been reported in the literature. *Pellinen and Heikkila* [1978] have reported the fading of auroral intensity in the night sector just prior to the expansion phase. More recently *Shepherd and Murphree* [1991] have noted the common occurrence of brightenings throughout the night sector prior to onset which might reflect magnetospheric precursor activity attempting to initiate a closure current through the ionosphere. Previous studies have indicated that the midday auroras show morphological changes during an auroral substorm [*Feldstein and Starkov*, 1967; *Akasofu*, 1972]. *Vorobjev et al.* [1976] and the references therein investigated the intensity of day and night sector auroras using all-sky camera data. They concluded that auroral activity intensified in the dayside ionosphere about one hour before the expansion phase onset. These forms of correlative global studies are difficult to perform from the ground and also from satellite imagery which involves scanning technology to acquire the data. Unambiguous auroral observations are, however, possible from the global imaging provided by the Viking UV camera (see *Anger et al.*, 1987b for details of the instrumentation). These images were acquired by simultaneously exposing the two dimensional field of view of the cameras. The exposure time was 1 s with a separation between images of about 1 minute. This eliminated the uncertainties normally inherent in relating effects at different local times. Although the investigation of dayside morphology during substorm periods is still of interest [e.g., *Lui et al.*, 1987] it is more fundamental to attempt to relate the dynamics of the aurora on both the dayside and nightside in order to address the processes leading to substorm occurrence. A reference time to relate these dynamic changes is needed, and for this we shall adopt the universal time at which the midnight sector aurora (including local times from 21 MLT to about 2 MLT) brightens, and subsequently moves poleward. This shall be used as a working definition of the expansion phase onset of a substorm. This excludes from the expansion phase both pseudo-breakups [*Shepherd and Murphree*, 1991], where evening sector arcs brighten and then fade again (without the poleward motion), and also auroral activations at other local times such as the "midday auroral breakups" [*Sandholt et al.*, 1990]. This paper reports the results of a study to identify systematic changes to the auroral distribution prior to this expansion phase onset.

OBSERVATIONS

Plate 1 shows a sequence of six images of the northern hemisphere auroral distribution taken in the Lyman-Birge-Hopfield bands ($\approx$1400-1800 Å) on September 23, 1986. This sequence was selected from a broader set of observations to illustrate certain features of the aurora which appear to be involved in the substorm process. In the Plates displaying auroral data, the color coding is such that blue represents low intensity LBH emissions and increasing intensities are shown by progressing through the colors of the rainbow eventually to red and finally to black for the most intense emissions. The color scale is shown to the left of the upper left panel. Horizontal white lines from bottom to top of the color bar delineate raw data number values 10, 50, 100, 150 and 200. The magnetic meridians are local time meridians measured in hours from midnight. Both the magnetic local time (MLT), and magnetic latitudes (Mlat) shown are in eccentric dipole (1985) coordinates. The bright regions on the dayside are due to dayglow associated with the sunlit portion of the Earth. Also shown are 60° zenith angle field of views (at 120 km altitude) of Barrow, Murmansk and Sodankyla ground stations.

The upper left panel of Plate 1 at 2022:22 UT shows an auroral distribution (the 1-6 MLT auroral region is missing due to unfavorable viewing conditions at this time) which, with the exception of the afternoon sector poleward of 70° Mlat, had no particular activation regions. For the next nine minutes, the aurora at all other local times remained essentially unchanged until 2031:18 UT (top center panel) when there were two bright intensifications in the prenoon sector near 10.5 MLT just eastward of the Barrow ground magnetometer station. These forms disappeared after one minute (not shown in the Plate). A few minutes later (see image at 2038:14 UT) a larger scale auroral form appeared near the equatorward boundary of emissions between 9 and 10.5 MLT just poleward of the Barrow ground station. This form was the first indication of a series of arc systems which developed over the next five minutes. From about 2030 UT to 2110 UT the Barrow magnetometer showed pulsations with a period of about 3 min coincident with a gradual increase in the negative X component of the magnetic field (the deflection being about 200 nT by 2110 UT). This slow development of a westward electrojet appears to be associated with a morning sector auroral enhancement. The lower left panel in Plate 1 at 2043:51 UT shows the extension of this enhancement to times between about 6 and 10 MLT, in a formation that we morphologically refer to as "fan" arcs. These "fan" arcs are reminiscent of dayside auroral arcs first schematically represented by *Akasofu* [1981]. The first four panels of Plate 1 show that these features are the result of westward development of discrete auroral activity from 10 MLT to 6 MLT. For this event, a specific intensification was not observed to propagate westward, but rather the entire region became filled with discrete aurora. In other events studied (see for example Figure 4 of *Anger et al.*, 1987a), however, the motion can be associated with a single discrete form. The activation is generally observed to advance westward at speeds exceeding 5 km/s.

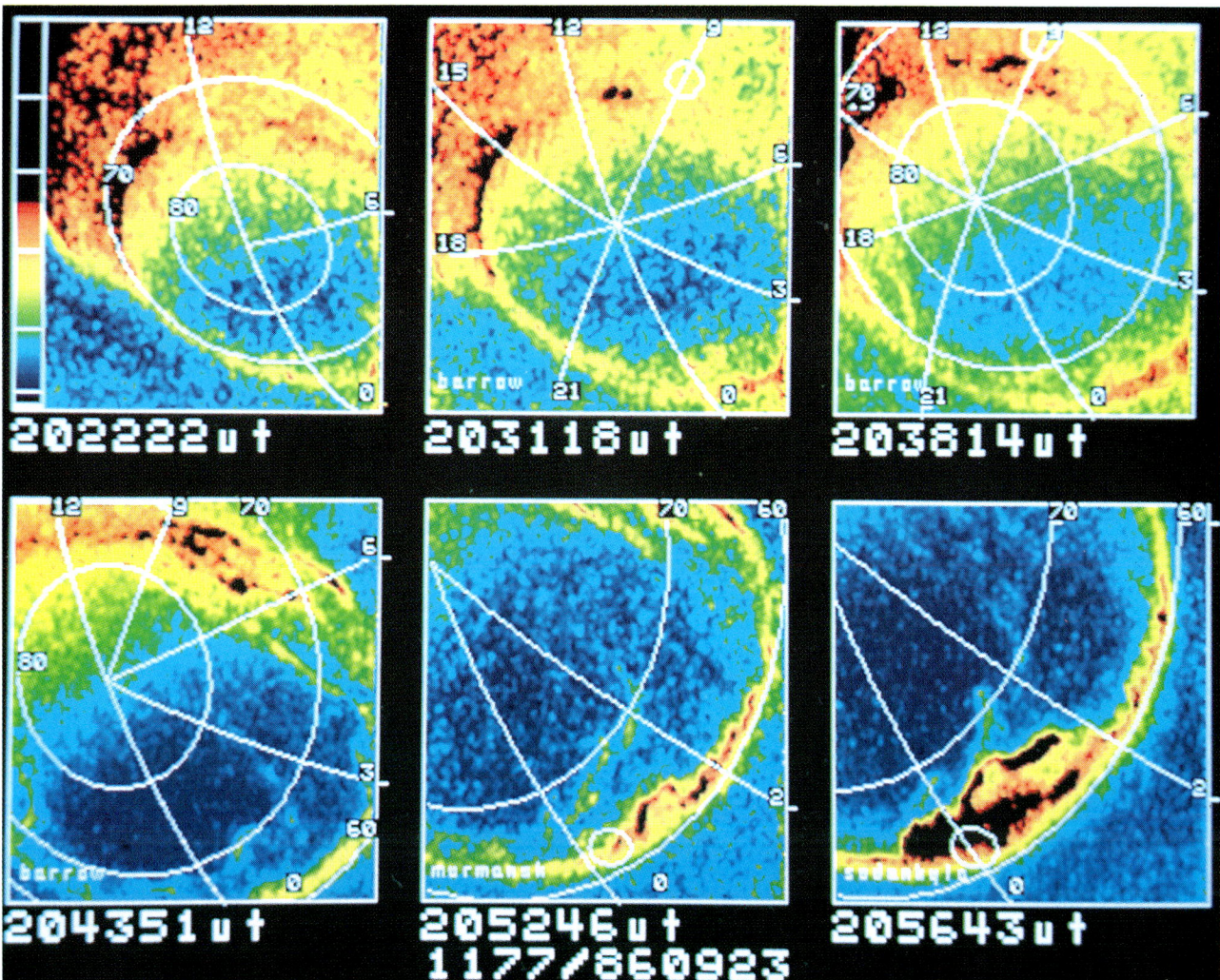

Plate 1. The development of "fan" arcs in the morning sector is shown in the time period 20-30 minutes before the evening sector substorm expansion phase onset. The high latitude arc system at 1 MLT indicates that the substorm onset occurred deep within the closed field line region probably near the nightside "cusp". The local time meridians and latitudes shown are in eccentric dipole magnetic coordinates (1985). The spatial dimension of each pixel in the images corresponds to 37 km at a 120 km altitude. A color scale is shown in the upper left panel. The horizontal white lines represent increasing raw data number levels 0, 10, 50, 100, 150, and 200 starting from the bottom. 60° zenith angle field of views at 120 km altitude for the ground stations Barrow, Murmansk and Sodankyla are shown for reference.

The lower left and center panels of Plate 1 illustrate an auroral morphology which has important implications regarding magnetospheric topology and the location of substorm onset. At 2043:51 UT an extended, but non-uniform in intensity, morning sector arc poleward of the fan arcs is apparent which defines the poleward extent of emissions between about 4 and 10 MLT. At 2052:46 UT (but also apparent in the previous panel) a less intense, distorted arc system also existed well poleward of the main auroral distribution in the midnight sector (just equatorward of 70° Mlat). The arc system occurred some 5 to 7° poleward of the newly activated region near midnight. With regard to the morning sector arc system, *Eliasson et al.* [1987] have shown that such features are likely to be on closed field lines. If one assumes that the midnight sector arc system was also on closed field lines, then it becomes difficult to imagine a topology of the magnetosphere which could allow the newly activated region near 62° Mlat (see panel at 2052:46 UT) to be near the open-closed field line boundary. This example serves to illustrate very clearly that the substorm onset is occurring well within the closed field line region of the nightside magnetotail.

The circles shown in the last two panels of Plate 1 show 60° zenith angle field of views from ground stations, Murmansk and Sodankyla. Magnetometer data from this Soviet sector shows that the activation near 1 MLT (which has progressed substantially poleward by 2056:43 UT as illustrated in the bottom right panel) corresponded to an expansion phase onset of a new substorm ($\approx$ 2050 UT) during a magnetically active time (Kp = 5-). The magnetometer station, Arkangelsk, just south of Murmansk recorded a development of a westward electrojet which was overhead at that station by the peak of the disturbance at 2215 UT (G. Rostoker, private communication, 1988). These observations confirm that the event was indeed a substorm and not simply a localized region of enhanced precipitation. Implications of night sector observations such as this are discussed by *Murphree et al.* [1991].

This event is just one example from a set of observations which illustrates the activation of the dayside morning sector. In a number of examples found in the Viking data set (see Table 1), this seems to occur some 10 to 30 minutes before the expansion phase onset of an auroral substorm. In some cases at least, there is a rapid (5-10 km/s) westward progression of the prenoon activity either in the form of "fan" arcs or westward propagating discrete activity embedded in the pre-existing diffuse aurora. This activity occurs not at the highest latitude of observed emissions but well within what must be considered the closed field line region. The activity is not a localized dayside transient, but rather a large scale phenomenon lasting more than 15 minutes and is probably linked to enhancements of electrojet activity in the appropriate sectors.

In the second example (shown in Plates 2 and 3) we have tried to illustrate the development of the entire northern auroral configuration on October 19, 1986 in the 40 minutes previous to an auroral expansion phase onset. In order to better show the true changes to the aurora for this event, the images have been corrected for the satellite viewing angle and have been put into a magnetic eccentric dipole (1985) polar coordinate system. The color scale is shown at the bottom of the image taken at 1055:20 UT. Noon is at the top of all these images and dusk to the center left. The small white circles in Plate 2 represent ground station field of views corresponding to a 45° zenith angle at a 120 km altitude. The upper right image (1112:11 UT) has displayed on it the distribution of ground stations whose X-component magnetometer traces are given in Figure 1. They are (clockwise from 9 MLT); Thule, Amderma, Dixie Island, Tixie Bay, Cape Schmidt and Yellowknife.

The upper panels of Plate 2 at 1055:20 UT and 1103:16 UT show the first brightening at the location of the pre-existing auroral distribution between 13 and 15 MLT near 75° Mlat. It then progressed both east and west over the next eight minutes resulting in the configuration shown in the upper right panel of Plate 2 at 1112:11 UT. At this time, the dayside aurora showed enhanced emissions between about 11 and 16 MLT implying a longitudinal expansion speed of about 1-2 km/s since the time of the initial brightening (1103:16 UT). Some discrete activity also commenced in the more equatorward regions near 9 MLT.

When this longitudinal expansion in the afternoon sector ended, the discrete auroral activity in the morning sector started in a manner similar to that described for the September 23, 1986 event. Here however, rather than the distinctive fan arcs shown in Plate 1, the prenoon intensification and subsequent westward development took the form of isolated intensifications generally near the equatorward boundary of emissions. These emissions were not as intense in this case as the afternoon sector enhancements, but nevertheless represented significant changes to the aurora over several hours of local time. Beginning near 9 to 10 MLT, equatorward of the poleward edge of the aurora, these intensifications developed more and more westward, reaching 6 MLT by 1114:10 UT (the bottom center panel of Plate 2). The region between 5 and 6 MLT also contained enhanced emissions by 1115:10 UT (lower right panel). The speed of the activation (more than 5 km/s) was much greater than for the afternoon sector, but was similar to the speeds noted for the September 23 morningside event.

Plate 3 continues the sequence of observations on October 19, 1986 through the time period up to and including the auroral expansion phase onset (at 1132 UT). The afternoon sector appears to have faded somewhat by 1117:09 UT (following the initial enhancement which lasted for about 12 min), but was reactivated at 1119:08 UT simultaneously with the morning sector near 6 MLT. By 1123:06 UT the aurora between 9 and 15 UT had acquired a broken up appearance with short (<500 km) discrete arcs scattered chaotically throughout this local time sector. (This is difficult to see in the large scale format of Plate 3). In the early morning sector, the discrete forms which had developed were confined to the main auroral distribution although near 1 and 3 MLT there is evidence of the development of an extended morning arc system farther poleward. The early evening sector had some minor intensifications (see 1123:06 UT panel), but the premidnight region was actually the least active area in the entire distribution. The main auroral brightenings associated with this growth phase began between 1125 to 1126 UT (middle row). The post midnight sector brightened between about 2.5 and 4 MLT (1125:05 UT) and expanded longitudinally eastward and westward reaching 0 and 6 MLT by 1130 UT. The westward propagation of this brightening was again quite rapid, covering 2.5 hours of local time in just 5 minutes ($\approx$5 km/s).

The afternoon sector during this time also displayed significant intensity enhancements which expanded eastward from 16 MLT to 20 MLT in just a few minutes (middle row). The implied eastward velocity at which the progression took place was between 5 to 20 km/s. It can also

TABLE 1 - Preonset Dayside Auroral Activity

Orbit	Date (yymmdd)	Expan. Onset (t2)	1st Activation morning (t1) UT(hhmm) *	1st Activation afternoon (t1) UT(hhmm) *	Time of IMF or Pressure Change t0 (hhmm) #	t2-t1 (min)	t1-t0 (min)
175	860325	1743	Y-1730	Y-1727	ND	16	-
842	860724	2311	Y-2242	W-2244	IMFN-2311	29	-29
					P-2329		-47
866	860729	0730	N-N	Y-O-0658	ND	32	
868	860729	1556	W-1549	Y-O-1529	ND	27	
871	860730	0521	Y-0457	Y-O-0457	IMFS-0437	24	20
					IMFN-0456		1
					P-0450		7
873	860730	1352	W-1322	Y-O-1341	IMFS-1255	30	27
					IMFN-1311		11
877	860731	after 0750 before 0807	Y-0721	Y-0727	IMFS-0649	29-46	32
					IMFN-0657		24
885	860801	1829	Y-1810	Y-1809	ND	20	-
1160	860920	1827	Y-1821	Y-O-1815	ND	12	-
1177	860923	2052	Y-2031	U	IMFS-2011,2027	21	20,4
					IMFN-2016		15
1182	860924	1828	U	Y-1809	IMFS-1743	19	26
1205	860928	2252,2309	Y-2246	Y-O-2226	ND	26	-
1226	861002	1834	W-1821	Y-O1808	ND	26	-
1296	861015	1146	Y-1111	Y-O-U	IMFS-1052,1113	35	19,-2
					IMFN-1054,1139		17,-28
1318	861019	1132	Y-1112	Y-1103	IMFS-1048,1113	29	15,-10
					IMFN-1051,1132		12,-29

*
- U - data not available
- Y - enhancement in the sector
- N - no enhancement
- W - weak enhancement
- O - nightside precursor activity and/or simultaneous dayside enhancement

\#
- P - pressure enhancement
- IMF - Bz change in IMF (N-North, S-South)
- ND - no solar wind data from IMP-8 available

be seen that in addition to a net expansion of the activation region, there was also a longitudinal displacement of the brightest intensities. These shifted eastward from between 12 and 16 MLT, at 1126:04 UT, to between about 14 and 20 MLT, by 1130:02 UT. This dayside intensification coincided approximately with the post-midnight enhancement and preceded the substorm expansion phase onset in the evening sector by just a few minutes. The start of the expansion phase as shown in the last two panels of Plate 3 occurred between 1131:01 UT and 1132:01 UT. It began near 23 MLT and (depending on the threshold intensity chosen) was confined to between 1/2 and 3 hours local time within one minute of the onset. This expansion in the midnight sector did not result in the disappearance of the morning sector high latitude arc system. This system can be seen in Plate 3 extended from 1 to at least 6 MLT reaching latitudes as high as 85 degrees.

The pre-onset activity manifested in the auroral distribution can be linked directly to the substorm growth phase by investigating ground based magnetometer data. The X

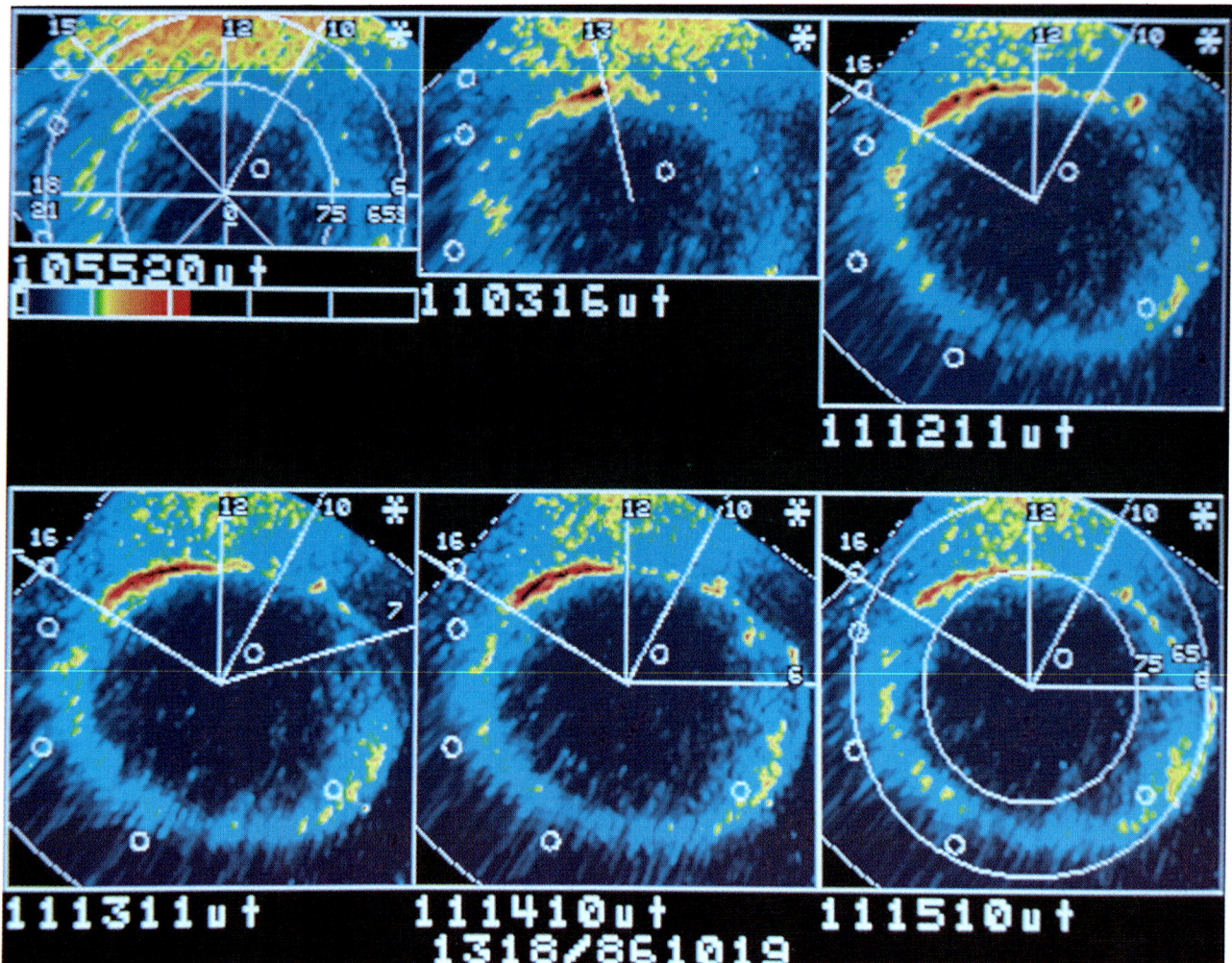

Plate 2. The development of the afternoon and morning sector auroras before an expansion phase onset which occurred at about 1132 UT on October 19, 1986. The meridians delineate regions of interest in the panels. The upper left and center panels were taken by the imager when the midnight sector was near the limb of the Earth resulting in considerable distortion of the auroral data on the nightside. This data has been eliminated from the two panels. The color scale and the spatial resolution are the same as for Plate 1. 45° zenith angle field of views of magnetometer stations are displayed. Clockwise from 9 MLT the stations shown are Thule, Amderma, Dixon Island, Tixie Bay, Cape Schmidt and Yellowknife.

component magnetometer traces displayed in Figure 1 are from the ground stations shown in Plates 2 and 3. The afternoon sector magnetometers (AMD, DIK, and TIK) all show a growth of an eastward electrojet beginning approximately at the time of the first dayside enhancement at 1103 UT (see Plate 2). The high latitude morning sector station, Thule, observed a negative X deflection (i.e., an enhancement of a westward electrojet) at about this time whereas the morning sector station, Yellowknife, observed a negative deflection much later (about 1120 UT), but well before substorm onset. The term growth phase has been used to describe just such an enhancement of the eastward (afternoon) and westward (morning) current systems prior to substorm onset [Nishida and Kamide, 1983]. The IMF for this event turned southward first at 1048 UT for a few minutes and then again at 1112 UT (see Table 1). The observations given here are consistent with ionospheric changes to both the auroral distribution and currents occurring due to a directly driven system resulting from a southward turning of the interplanetary magnetic field.

The panels in Plate 3 show only the large scale distribution of the aurora (one side of each of the approximately

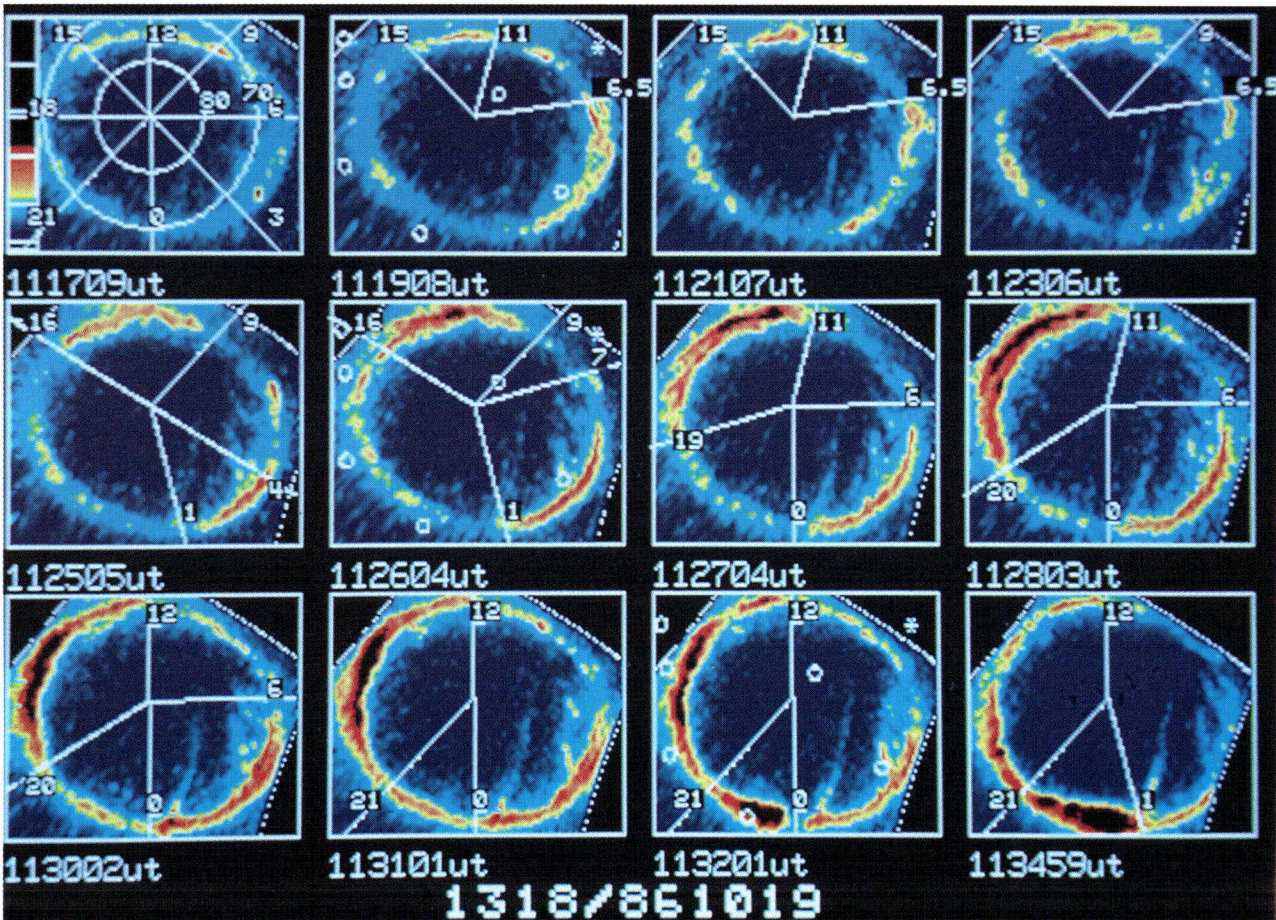

Plate 3. A continuation of Plate 2 illustrating the subsequent development of the auroras up to and including the substorm expansion phase onset. Of particular interest are the activation regions in both the afternoon and postmidnight sectors at about 1128:00 UT. The spatial dimension of each pixel in the images corresponds to 52 km at a 120 km altitude. The color scale has the same meaning as in Plate 1. The ground stations shown are the same as in Plate 2.

120^2 pixels in the image represents a spatial extent of 52 km at 120 km altitude) so it is difficult to pick out the fine structure which is available from the Viking imager. To illustrate in more detail what occurred between 1125:05 UT and 1130:02 UT, the data were re-examined at finer spatial and temporal scales. The results are illustrated in Plate 4 where each of the 160^2 pixels represents a 21 km × 21 km square at 120 km altitude. At this spatial scale, different aspects of the afternoon enhancement appear. It can be seen that the intensification of the aurora at 1126:04 UT was localized to the 13 MLT sector in the form of a spiral-like feature, with two arc systems on the east side. The arc segment attached to the dayside aurora was located equatorward of the segment connected to the nightside portion of the auroral configuration. A careful measurement at 1125:05 UT of the pre-existing auroral distribution near 14 MLT put the equatorward and poleward edges of the arc at 72.8° Mlat and 74.4° Mlat respectively. After the intensification (1126:04 UT) the arc connected to the nightside aurora (i.e., the poleward arc system) was found to be at the same location as the original arc (i.e., the one at 1125:05 UT). The equatorward edge of the system connecting to the dayside spiral was located at 71.8° Mlat. The implied motion of the boundary was due to an intrusion of an arc from earlier local times into the 14 MLT sector. The image at 1127:04 UT and the subsequent images in Plate 4 show the development of this intrusion around to the dusk sector.

A study of Plate 4 indicates that as the dayside arc structure penetrated towards dusk, the entire auroral region between 13 and 19 MLT contained enhanced emissions. The arc connected to the dayside reached to about 18 MLT

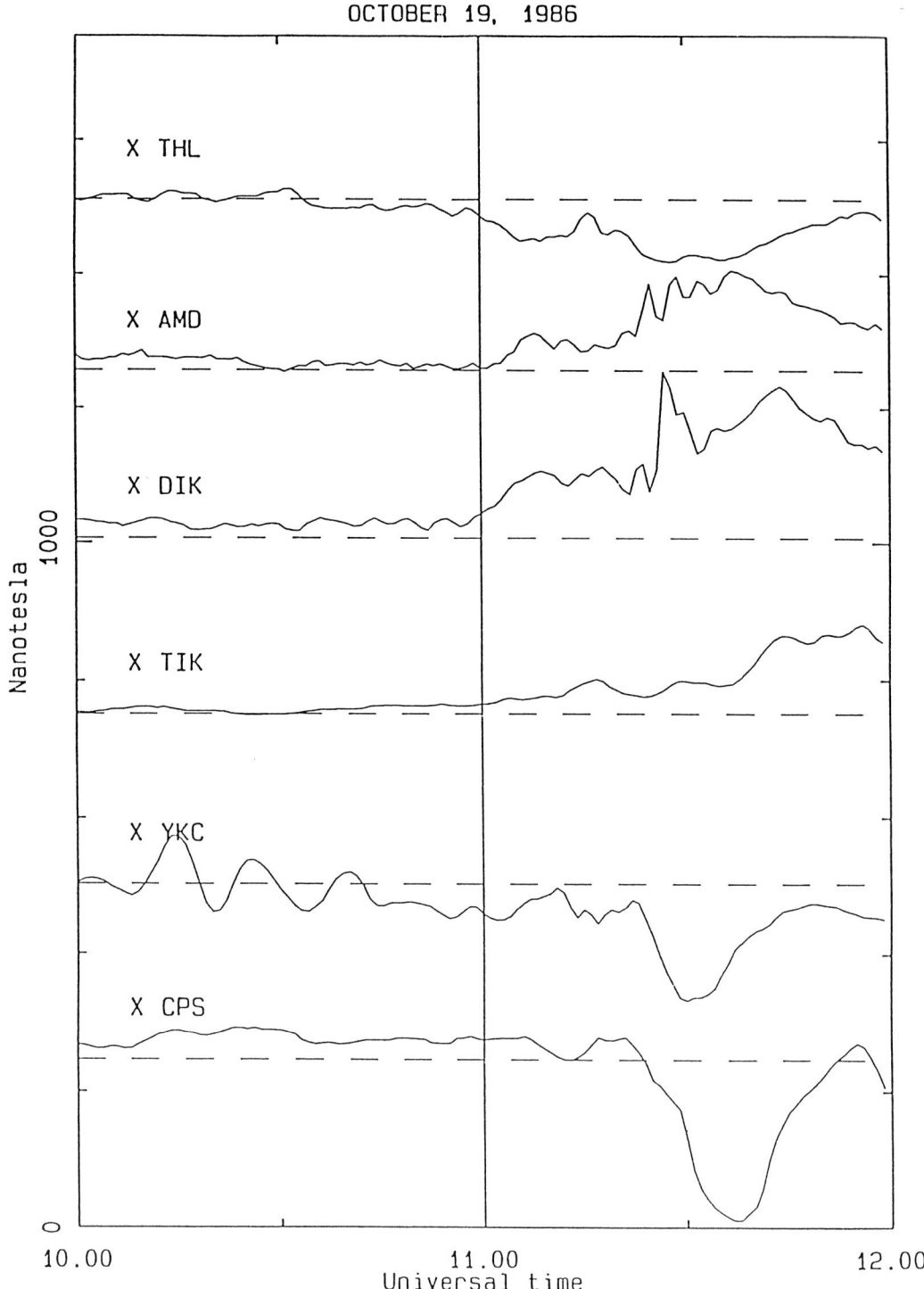

Figure 1. The x component of the magnetic field as a function of time derived from magnetometer stations at various magnetic local times for the October 19, 1986 event shown in Plates 2, 3 and 4. The approximate local times of the stations from the top to the bottom are: Thule, 9 MLT; Amderma, 16 MLT; Dixon Island, 17 MLT; Tixie Bay, 19.5 MLT; Yellowknife, 3.5 MLT; and Cape Schmidt, 23 MLT. The exact locations of the stations have been plotted in Plates 2 and 3. The vertical scale is correct for the Cape Schmidt data. The zero lines for the other stations are represented by the dashed lines.

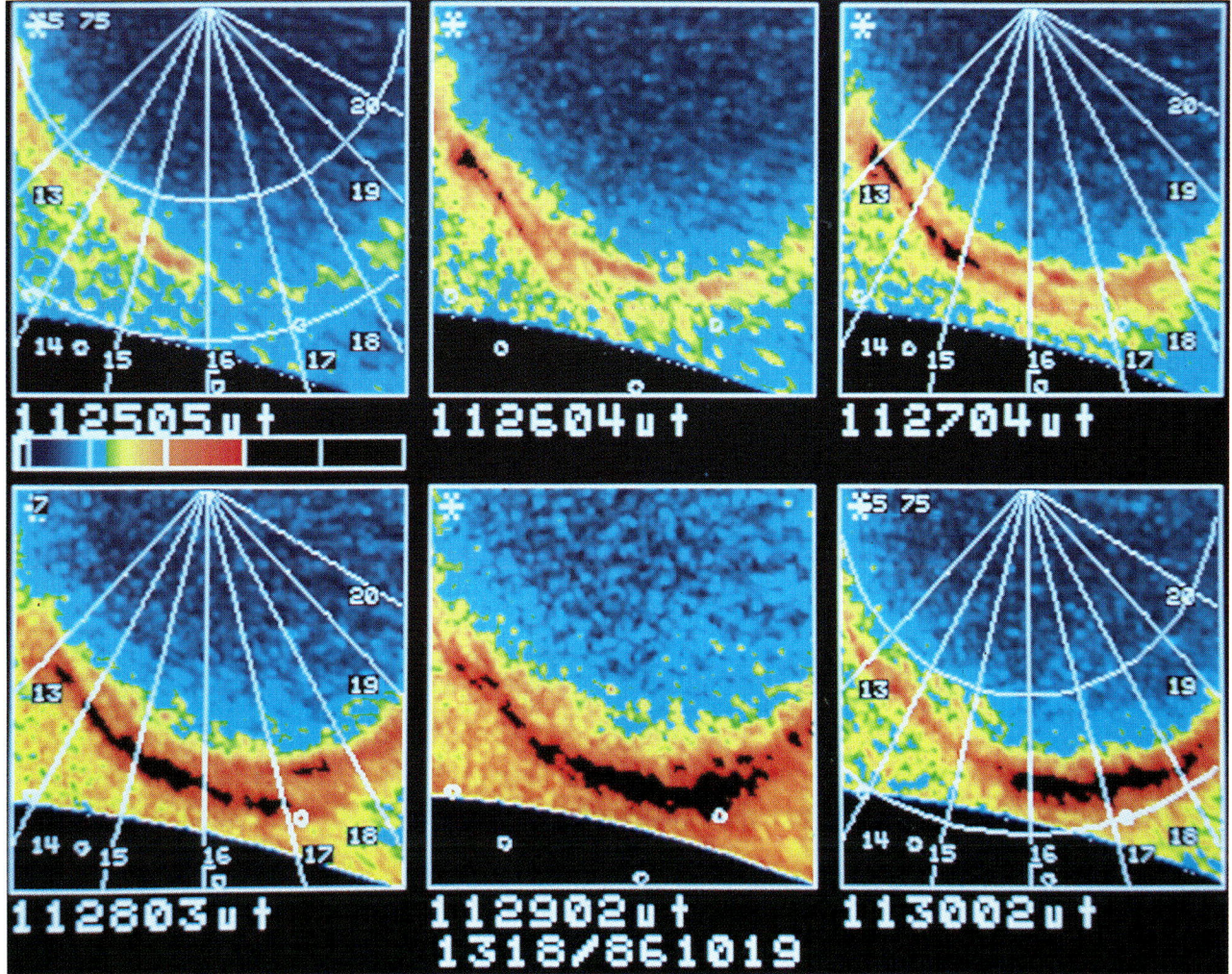

Plate 4. Expanded views of the auroral morphology in the afternoon sector for the October 19 event. The development and intrusion eastward of a dayside arc segment separated from the nightside distribution can be seen. The spatial dimension of each pixel in the images corresponds to 21 km at a 120 km altitude. The four ground stations 20° zenith angle field of views are shown. In order of increasing local time they are Abisko, Murmansk, Amderma and Dixon Island. The color scale located below the image at the upper left has the same meaning as the one in Plate 1.

by 1128:03 UT. Also by this time, the spiral at 13 MLT had faded, and a new intensification had appeared, centered on the region where the night and day sector arcs overlapped. By 1129:02 (bottom center panel) the intensification moved farther east, and the distinction between the day and night sectors can no longer be made. Comparing the top center and bottom right panels it is evident that the main active region of the aurora shifted from between 13 and 14 MLT over to 16 and 18 MLT, coinciding with a net intensification of the entire distribution. The shift in the peak emission implies an eastward progression of the most intense auroral region of between 3 to 10 km/s.

Speeds for the eastward intrusion of the dayside arc system varied from 9 to 24 km/s dependent on the exact time and intensity threshold used.

Also shown in Plate 4 are 20° zenith angle field of views at 120 km altitude for four ground magnetometer stations, Abisko, Murmansk, Amderma and Dixon Island (from left to right in each image). The X component magnetometer trace from each station is presented in Figure 2. Beginning about 1104 UT the stations all observe a 100 nT positive X deflection in the field. This corresponds very well to the initial dayside enhancement shown in Plate 2. This exists until about 1121 to 1126 UT when an additional large posi-

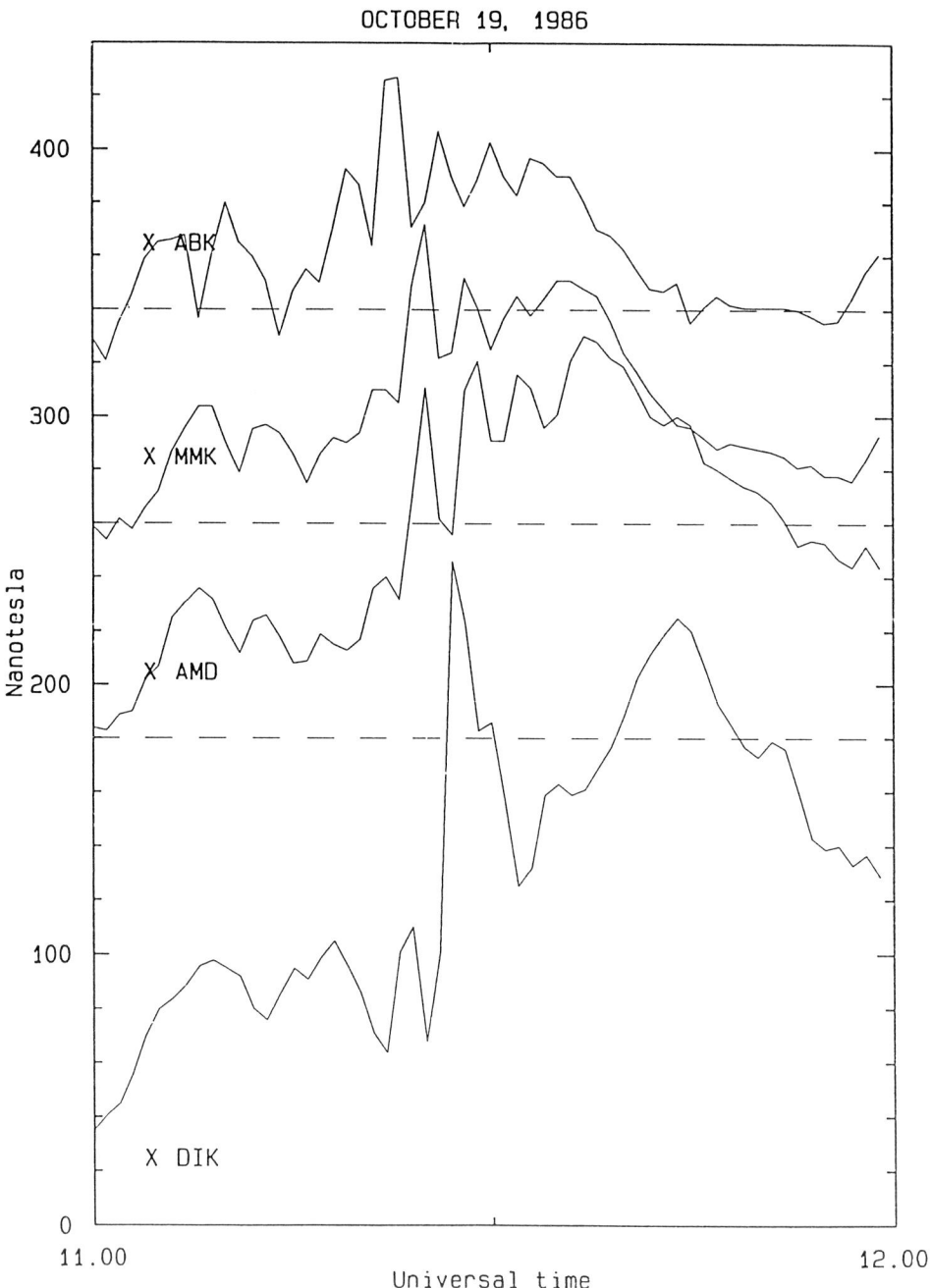

Figure 2. Same as Figure 1 except the time scale is one hour and the magnetometer stations correspond to the stations plotted for the afternoon sector event in Plate 4. The approximate magnetic local times of the stations from top to bottom are: Abisko, 14 MLT; Murmansk, 15 MLT; Amderma, 16 MLT; and Dixon Island, 17 MLT. The positive spike in the x component has a time shift corresponding to a eastward progression of 9 km/s ± 1.5 km/s.

tive X spike (about 150 nT at Dixon Island) is observed at all these stations. The delay times for the arrival of the spike at different stations can be used in a linear regression fit to calculate an eastward speed. The implied eastward progression of the impulse was 9 ± 1.5 km/s (the value of f associated with the regression was 34). This speed is similar to the eastward motion of the auroral transient. It appears that the apparent enhancement (implied by the positive X bay in the magnetometer data) of the eastward electrojet in the afternoon sector was reflected directly in the dynamics of the auroral distribution. Similar pre-substorm enhancements of the eastward afternoon sector auroral electrojet have been observed by *Troshichev et al.* [1974], *Kamide and Richmond* [1984] and *Untiedt et al.* [1978]. *Nishida and Kamide* [1983] attribute such pre-onset ionospheric current changes to activity directly driven by the solar wind. Our results are consistent with that interpretation and provide a link between auroral and magnetometer observations which has not been made previously. Plate 4 illustrates an afternoon auroral feature which is distinctly different from the morning sector "fan" arcs. It is likely that this is an example of the aurora first observed by *Vorobjev et al.* [1976] and later described by *Meng and Lundin* [1986] as showing a separation of the day and night sector aurora. The Viking imagery has provided the opportunity to observe the formation and dissolution of one of these systems only 2 to 7 minutes before the expansion phase onset of an auroral substorm. Coincident with this, a positive X spike in the magnetometer traces in this sector was associated with an eastward propagation speed of the transient at about 9 km/s.

The above description sought to outline in detail observations for 2 specific events. While these two events are sufficient to demonstrate the dramatic dayside changes which can occur before a substorm, they leave open the question of how general a phenomenon this activity represents. A partial answer to this has been obtained by directly associating these auroral signatures to growth phase magnetometer signatures. A survey of the Viking data set was also performed to independently determine the frequency of occurrence of such events. The results of this study are summarized in Table 1. Due to satellite viewing angles and orbital characteristics it was difficult to obtain views of both the day and night sectors for an extended time period. Interpretation of the available events is further complicated by the necessity to define what activity would constitute an event similar to those described above. The list given in Table 1 includes only those events which have an unambiguous change in auroral intensity in the dayside portion of the ionosphere prior to substorm onset. The criteria were further constrained by imposing the condition that when the dayside region activated, it remained active, so that auroral transients would be eliminated from the list. This latter restriction ensured that only large scale phenomena were included and that more random activity was not spuriously identified with the substorm process. The total available sample was somewhat fewer than 25 to 30. Of those, the 15 cases listed in Table 1 showed well defined pre-substorm activations on the dayside. Five to ten others showed ambiguous signatures which merit further study and fewer than 5 could be said to have no dayside signatures involved. Eight of the events in Table 1 had IMP-8 interplanetary magnetic field and solar wind data available. Of these, two showed pressure pulses although only one occurred prior to both the onset and dayside activation. Seven of the eight showed B_z go northward in the 2 hour interval prior to onset. Seven of the eight also showed B_z turn southward. For the cases when B_z turned northward, the turning took place at a variety of times from 0 to 53 minutes before onset and from -29 to +24 minutes "before" the first dayside activation. Of the events when B_z turned southward, every one could be attributed to a turning taking place 15 to 32 minutes prior to the first dayside activation and 41 to 57 minutes before substorm onset. (The times quoted do not take into account the satellite location relative to the magnetosphere). Of the three possibilities (pressure pulse and north/south magnetic field changes), the IMF turning southward in the 40 to 60 minutes prior to substorm onset seems to be the most likely solar wind parameter associated with both the dayside and nightside activity.

KEOGRAMS AND LATITUDINAL MOTIONS

In order to quantify both the latitudinal and intensity variations of the auroral distribution at all local times, a computer algorithm was designed to objectively locate auroral boundaries and to remove background emissions (such as dayglow) from the image data. The first step involved producing a one hour magnetic local time average of equivalent 3914 Å intensity (kR) for each universal time (the Universal Time resolution was typically 1 minute). At this point, dark current backgrounds and vignetting due to the imager optics were corrected for. An iterative procedure was then used on the averaged profile to remove the UV dayglow contamination. This was done by fitting the dayglow contribution to a cubic function and iteratively removing the aurora. On the basis of the fit the results were accepted or rejected by the algorithm. Accepted fits were used to generate residual auroral intensity versus latitude profiles in 3914 Å (kR) equivalent intensity. The peak intensity and its latitude were stored along with the equatorward and poleward boundaries of the aurora at that local time for that Universal Time (3 standard deviations above the background dayglow noise was used to define the auroral boundaries). For all the acceptable Universal Times, the residual auroral profiles were put together to form a keogram. Plate 5 illustrates the results for 3 magnetic local times (15, 19 and 21) on September 24, 1986. The horizontal axis shows Universal Time and the vertical axis represents eccentric dipole (1985) magnetic latitude. The intensities have been color coded to equivalent 3914 Å intensities and the color scale is given on the right of the Plate. The dayside auroral distribution for this event is shown and described in *Elphinstone and Murphree* [1989] (their Figure 7) and the nightside distribution, in *Nielsen et al.* [1988]. At the time (about 1809-1812 UT) of the first dayside intensification (at 15 MLT) the auroral distribution at 15, 19 and 21 MLT began to shift equatorward. The equatorward motions found by a linear regression fit to the data (between 1806 UT and 1826 UT) of the peak intensity

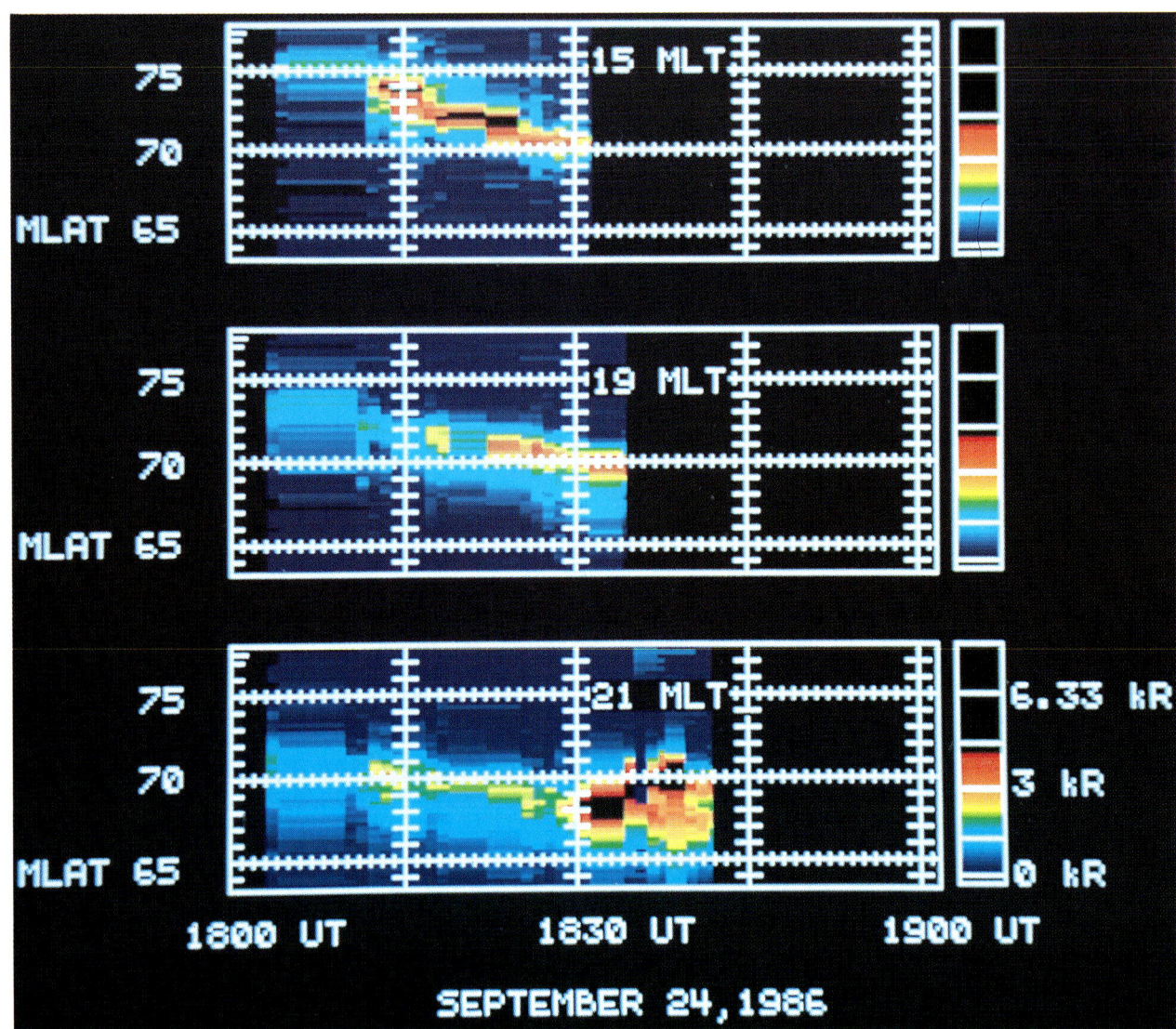

Plate 5. An event on September 24, 1986 is shown after dayglow emissions have been removed and the data reformatted into a keogram analysis. The color scales on the right represent equivalent 3914 Å kR intensities. The vertical scale is eccentric dipole magnetic latitude and the horizontal scale is universal time. Each keogram is generated from an average of one hour magnetic local time centered on 15 (top), 19 (middle) and 21 (bottom) MLT. The activation occurs later at progressively greater local times and is a sustained process involving more than just single transient dayside events. The first activation occurs coincident with the beginning of equatorward motion of the auroral boundaries typically associated with the growth phase.

aurora at 15, 19 and 21 MLT were 450 ± 51 (an f value of 77), 270 ± 46 (f = 35), and 240 ± 23 (f = 111) m/s respectively. The dayside event began with the equatorward and poleward boundaries of the aurora at 15 MLT undergoing impulsive motions, broadening the auroral zone by 2° to 6° in 7 or 8 minutes. Over the same time the peak 15 MLT intensity doubled to 5.2 kR (3914 Å equivalent) and a small auroral transient rotated clockwise (when viewed parallel to the magnetic field) into the region poleward of the auroral distribution. The enhancement spread to later local times (middle panel of Plate 5) and at 1828 UT the 21 MLT sector (bottom panel) shows the first signs of the expansion phase of the substorm. The peak intensity there moved poleward and changed by a factor of 3 from about 2 kR to 6.3 kR by 1838 UT.

The above analysis was also performed at all magnetic local times for the event presented in Plates 2, 3 and 4 and the latitudinal motions derived are summarized in Table 2.

TABLE 2 - Equatorward Motions During Growth Phase

Magnetic Local Time	Velocity (m/s)	Correlation Coefficient	F Value of Fit	Time Range (hhmm)
0	430 ± 56	.90	59	1115-1130
1	260 ± 18	.93	223	1103-1145
2	190 ± 36	.82	30	1115-1130
3	230 ± 57	.75	17	1116-1130
4	60 ± 60	.28	.9	1112-1128
5	-90 ± 70	.4	1.7	1112-1130
6	70 ± 150	.14	.2	1112-1127
7	40 ± 140	.11	.9	1116-1132
8	10 ± 40	.05	.04	1107-1133
9	-660 ± 60	.95	110	1112-1124
10	260 ± 85	.71	9	1121-1131
11	360 ± 122	.76	9	1121-1131
12	360 ± 88	.81	16	1121-1131
13	490 ± 44	.96	121	1121-1132
14	660 ± 111	.91	36	1121-1130
15	190 ± 29	.81	45	1103-1130
16	130 ± 33	.78	14	1103-1136
17	210 ± 90	.68	6.2	1112-1127
18	170 ± 33	.74	25	1103-1136
19	270 ± 18	.94	220	1103-1144
20	310 ± 14	.96	452	1103-1149
21	310 ± 14	.96	449	1107-1149
22	390 ± 54	.89	51	1115-1130
23	370 ± 40	.93	86	1115-1130

The Universal Time range used for calculating latitudinal motions is given in the last column. The linear fits in the morning sector between 4 and 8 MLT were very poor. At 9 MLT the 660 m/s poleward motion is due to the transient event which started at 1112 UT (Plate 2). At all other local times the motion was equatorward and was highest in the 11 to 14 MLT sector and the 19 to 1 MLT sector. The highest speed at 14 MLT reflects the formation of the dayside arc system (Plate 4). The evening sector had equatorward motions of about 300 to 400 m/s while the noon/afternoon sector had motions between 360 and 660 m/s. These observations are relatively consistent with values found by *Vorobjev et al.* [1976].

MAPPING RESULTS

In order to interpret the above observations it is useful to attempt to map features from the ionosphere to the magnetosphere. To do this, the *Tsyganenko* [1987] external magnetic field model was used. The *Tsyganenko* [1987] external magnetic field model consists of analytical functions to describe the magnetopause/field aligned currents, the ring current and the cross-tail current. The parameters involved in the analytic functions were then found from fits to a large set of observed magnetic field vectors. In *Elphinstone et al.* [1991], this model was used to show that most of the auroral distribution is due to near-Earth nightside phenomena. It is important to note that the model used in that paper and in this work is the 1987 version and not the later version developed by *Tsyganenko* [1989]. The models have quite different formulations and the results from one are not transferrable to the other. The *Tsyganenko* [1987] magnetospheric field model (Kp = 5) in combination with the IGRF (1985) internal field were used to model the September 23, 1986 event. Various regions of the magnetosphere have been colored and projected into the ionosphere in Plate 6 in order to facilitate comparisons with the auroral data in Plate 1. The ionospheric projection of the peak current density from the nightside magnetotail is

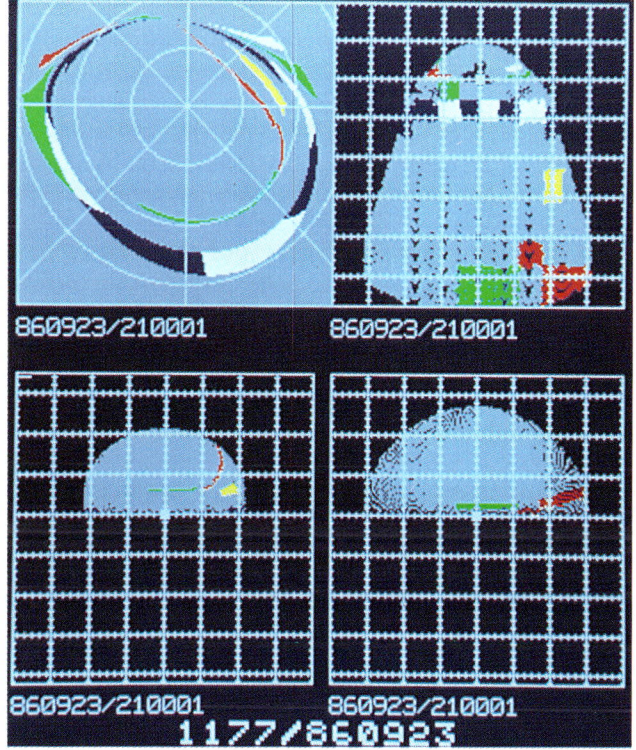

Plate 6. Model calculations for the event of September 23, 1986 shown in Plate 1. The upper left panel represents the northern ionosphere in eccentric dipole coordinates (1985). Noon is at the top center, dusk to the center left, and 60°, 70°, and 80° Mlat are also shown. The upper right panel shows the minimum B surface near the equatorial plane of the magnetosphere. Each large square represents 10 Earth radii. Positive xGSM is to the top of the panel, positive yGSM to the left. The Earth is represented by the solid white square. The lower left panel shows a view from the deep tail of the y-z GSM plane at xGSM = -15 R_E. Only the portion of the magnetotail for positive zGSM has been colored. The lower right panel is similar to the lower left except the cross-section is at xGSM = -45 R_E. The coloring has been used to illustrate how various auroral features in the ionosphere project into the magnetotail.

known to match relatively well to the auroral "oval" [Elphinstone et al., 1991]. This projection as a function of local time has been approximated in what follows as the projection of the xGSM = -5R_E line from the minimum B surface of the tail. In the upper left panel of Plate 6 it appears as the equatorward boundary of the blue and white striped oval shaped region. This panel represents the model northern ionosphere in eccentric dipole coordinates (noon is at the top and dusk to center left). The poleward boundary of the blue white oval region has been arbitrarily set to the projection of the line at xGSM = -10R_E from the minimum B surface of the tail (see upper right panel of Plate 5 for the view of this surface in the x-yGSM plane).

The equatorward boundary of the oval shaped region in Plate 6 matches relatively well the main observed auroral "oval" in Plate 1 (see for example the image at 2043:51 UT). In particular, it should be noted that the region in Plate 1 near the onset location ($\approx$1 MLT, 62 °Mlat) maps quite close to the Earth. This is in agreement with Lui and Burrows [1978] who put the onset near the "nightside cusp". One might be concerned whether this mapping is a strong function of the model used and/or the current strength implied by the model. For this case, however, the ambiguity is partially removed due to the observed second arc system poleward of the activation region. This is further substantiated when one projects a region at a distance of about 50-60 R_E in the minimum B surface into the ionosphere (the thin green region between 21 and 3 MLT in the upper left panel of Plate 6). This boundary matches relatively well with the poleward arc system and is very close to the model open closed field line boundary. Another feature supporting the mapping is the relative thicknesses of the regions when projected to the ionosphere. Zelenyi et al. [1990] have pointed out that a likely signature of the plasma sheet boundary layer (PSBL) could be the appearance of a narrow region ($\approx$1°) of ion precipitation at the poleward edge of the main boundary plasmasheet (BPS) (see Feldstein and Galperin, 1985 for a review of the controversy centering around relating the entire BPS to the PSBL). The narrowness of the PSBL signature in the ionosphere is supported by the combination of the Viking images shown in Plate 1 and the model presented in Plate 6. The narrow line in the upper left panel of Plate 6 (in green) representing the high latitude arc in the ionosphere, implies a 10 to 15 R_E slab in the minimum B surface, a 2-3 R_E thick region in the cross-section at xGSM = -45 R_E (lower right panel), and about .5 R_E at xGSM = -15R_E (lower left panel). Plate 6 also illustrates another aspect of the mapping from the magnetotail. A conservation of flux argument can be used to explain why the 10 to 15 R_E slab in the deep tail maps to a thin line in the ionosphere. The same argument applied to a similar slab near the Earth yields a much broader ionospheric signature. This is confirmed both by the width of the model regions in the ionosphere and by the thickness of the auroral oval relative to the high latitude arc.

Since the aurora is a dynamic phenomenon during the growth phase, it is useful to investigate how sensitive the above results are to changes in the external magnetic fields. Sergeev et al. [1990] have presented how these changes associated with the growth phase are manifested in ionospheric projections. To do this they modified the Tsyganenko [1987] external field model to conform with near-Earth satellite observations. Using their results, a comparison was made between the static model and their disturbed model just prior to the expansion phase onset. The general results described above are not affected by the current changes found by Sergeev et al. [1990]. Mappings involving the other event on October 19, 1986 also support this conclusion. Whether one uses the auroral location at the start of the October event or just before onset, the mapping of the midnight sector peak intensity aurora is consistently between xGSM = -6R_E and = -11R_E. Altering the peak current density by a factor of 1.5 changes the mapping of the onset region by only a small amount (modelled by changing the Kp from 3 to 4). This lack of any strong dependence is likely to be due to the auroral region mapping to near the earthward edge of the nightside magnetotail. Varying the strength of the cross-tail current system may drastically alter the mapping of features tailward of it, without changing much the mappings of the earthward edge.

Moving away from the midnight sector, the high latitude morning sector arc (between 4 MLT, 65 Mlat and 10 MLT, 78 Mlat) can be shown as mapping to the deep flanks of the magnetotail (the red coloring in Plate 6) much as was schematically shown by Lundin et al. [1991a]. We associate the even higher latitude arc system in Plate 1 (at 2052:46 UT) with the region where the plasma becomes decoupled from the ionosphere, somewhere near 150 R_E in the deep tail (W. Heikkila, private communication, 1990).

The morning sector fan arcs have in the past been attributed to both flux transfer events [Crooker, 1990] and to plasma injection events [Lundin and Evans, 1985]. Rather than addressing a particular magnetospheric mechanism, we shall simply indicate where the features map to, as illustrated in Plate 6 (compare the colored segments to the real arc systems shown in the lower left panel of Plate 1). It was noted previously that these features started near 10 MLT (as modelled by the white colored segment equatorward of the main "oval" at 9-10 MLT in the upper left panel of Plate 6) and then moved around to 6 MLT, eventually also including the more poleward region (colored green, blue, and yellow in the panel.) The mapping of these colored regions to the minimum B surface implies a disturbance propagating at Alfven speeds (>5 km/s in the high latitude ionosphere generally will translate to greater than 100 km/s in the equatorial plane and more likely on the order of 1000 km/s) from the regions on the dayside magnetopause (at relatively high y values) around to the region where the field becomes tail-like, eventually reaching the flanks of the deep tail. The longitudinal extension of the arcs is likely to be related to the depth of penetration of the disturbance into the magnetosphere, as well as the magnitude of any normal component of the magnetic field at the magnetopause [*Crooker*, 1990]. The latter effect may partially explain the varying character of these features for different events (i.e., a single discrete rounded form or multiple fan arc structures).

Shifting now to the afternoon sector, we note that the character of the aurora there is much different from the morning sector, generally being more confined in latitude. This is probably due to the different sense of region 1 current at dawn and dusk [*Reiff et al.*, 1978]. The afternoon sector separation of the day and night sector arcs is illustrated in Plate 6 by the intrusion of the red segment (near 16 MLT) just equatorward of the segment which is part of the auroral "oval". The observations that the arc connected to the nightside oval (at 1126:04 UT in Plate 4) is at the same location as the main auroral oval (at 1125:05 UT) then implies that the stable aurora at 16 MLT maps to the nightside. Transient processes equatorward from this nightside arc (i.e., the red segment in Plate 6 attached to the spiral represented by the green segment at 13 MLT) are linked to dayside processes penetrating deep within the magnetosphere (perhaps via wave propagation rather than direct plasma motion). The actual disconnection of the dayside and nightside auroral systems has previously been attributed to separate dayside and nightside magnetospheric processes [*Rezhenov et al.*, 1979]. They found that the division corresponded with a peak in the DP-2 current system. *Meng and Lundin* [1986] later also associated such systems with separate dayside and nightside processes. The observations and modelling presented in this paper support the view that typically the aurora at 16 MLT originates from a nightside magnetospheric dynamo, but that these transient auroral morphologies shown in Plate 4 are due to the dayside dynamo becoming an active process. Their short-lived nature is likely to be related to a damping of the dayside process resulting from the linkage through the ionosphere of the two widely separated magnetospheric domains. It also appears that the 14 MLT enhancement and the associated spirals are associated with the meeting point of the two dynamos in their ionospheric projections. This hypothesis is supported by the observations in Plate 4 which show the appearance of the bright enhancement at the time of the arc separation, and another bright enhancement again when the arcs coalesce at a much later local time. Other Viking observations show this separation of afternoon sector arcs is associated with spiral formation. If this were true, it would give very strong support for dayside auroral spirals being a near-Earth indirect result of magnetospheric processes, rather than being a direct signature of them.

DISCUSSION

The above set of observations indicate that the auroral distribution at local times other than midnight can undergo significant changes prior to the expansion phase. These effects may not occur for every substorm and the exact details for each event may differ (i.e., a slightly different ordering or only one of the dayside sectors being affected). Nevertheless, dayside sector enhancements like those described above are a frequent occurrence before substorms. The exact characteristics of the afternoon sector for the October event (see Plate 4) are observed less frequently, but the apparent relationship to the morning sector enhancement and the timing of the expansion phase imply that this day-night separation may be related to magnetospheric processes which occur just before onset. Figure 3 illustrates the ionospheric signatures which one might expect to see before, and up to, the expansion phase. At T_0, the afternoon sector intensifies, associated with spirals or beads of emissions often observed by Viking [*Potemra et al.*, 1990, and reference therein to earlier work]. This enhanced level of emission expands at 1-2 km/s from a region of about one hour of local time to encompass the longitudinal sector between about 11 and 16 MLT. Sometime during or after this expansion ($T_1 = T_0 + 5$-10 min), the morning aurora shows the development of either fan arcs (upper right panel of Figure 3) or a single discrete feature which propagates rapidly westward. The speed of activation in the morning sector seems to point to an associated wave activity in the magnetosphere whereas the afternoon expansion speeds of 1-2 km/s could be accounted for by direct plasma motion.

The postmidnight sector may then brighten, possibly in association with the separation of day/night arcs in the dusk sector (lower left panel of Figure 3). Another possibility is that the region near or at the onset location can brighten and then fade [*Untiedt et al.*, 1978; *Shepherd and Murphree*, 1991]. These auroral features occur 1-15 minutes before the evening expansion phase onset. All of the longitudinal speeds associated with this time period are rapid enough to assume waves to be the dominant carriers of this information. The wave activity can be further supported by observing "motion pictures" of the Viking auroral imagery. In these, we can see typical slow, westward, and eastward motions of the large scale auroral forms. Superimposed on these is a small scale "flickering" of aurora which appears to propagate from early afternoon to the evening sector.

During the time period of the October 19, 1986 event shown in Plate 3 there were equatorward motions of the peak intensities of the auroras both at noon and midnight (see Table 2). This increase in the polar area bounded by the aurora is commonly attributed to an increase in the amount of open flux in the lobes of the magnetosphere [*Frank et al.*, 1988]. *Maltsev and Lyatsky* [1975] have shown that an alternative view, for the noon sector at least, is to attribute magnetopause and ionospheric boundary changes to variations in the 3-dimensional current system. This may be true of the auroral oval at other local times as well. *Elphinstone et al.* [1991] showed that the auroral oval can be correlated quite well to the region near the Earth where

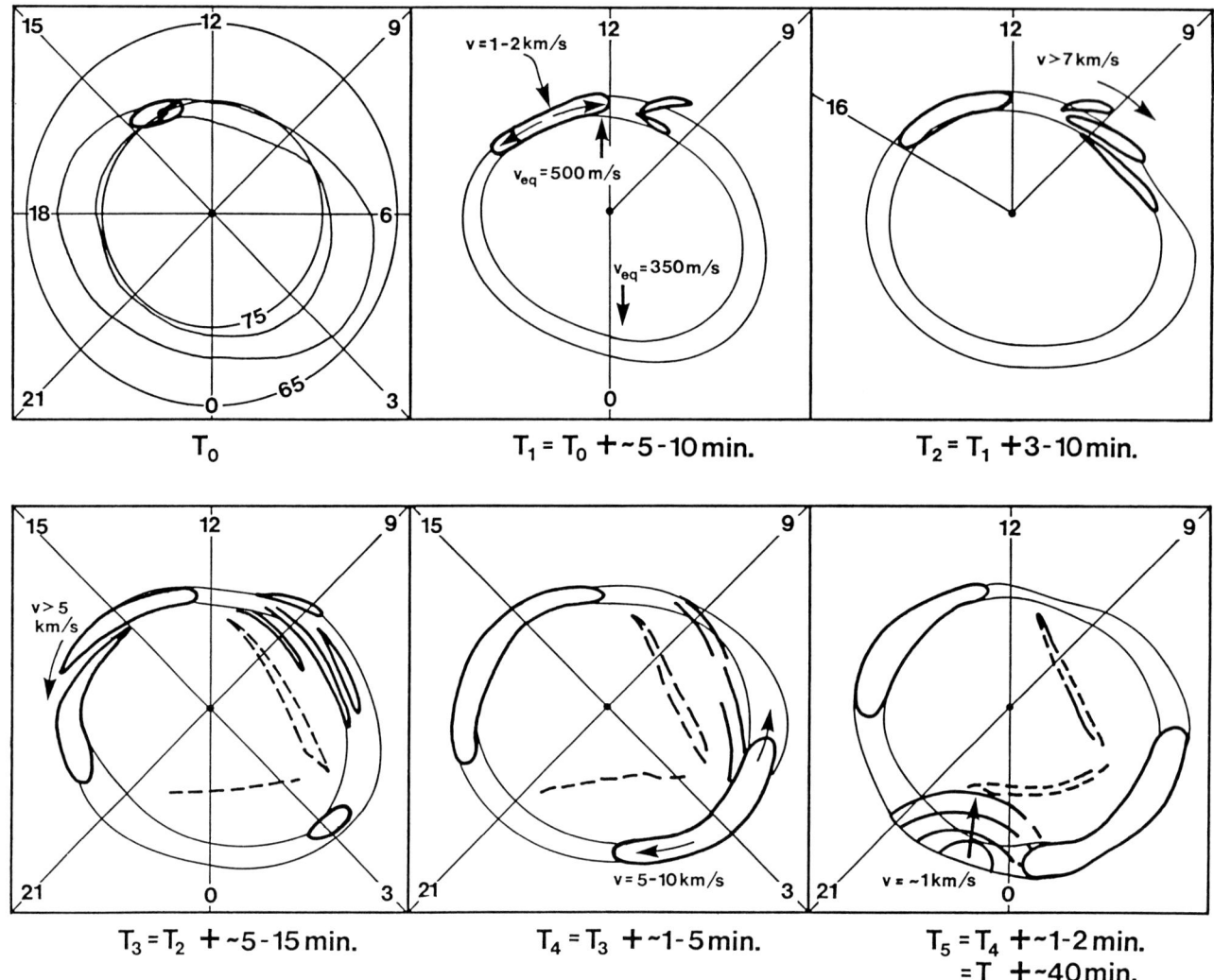

Figure 3. A schematic illustrating the development of auroral morphology prior to the evening sector expansion phase onset.

the cross-tail current maximizes. They demonstrated that as the magnetic activity increased (Kp in the *Tsyganenko* [1987] model), the cross-tail current density maximum moved earthward from xGSM $\approx$ -8 - 12 R_E to about xGSM = -5R_E. The projection of this into the ionosphere resulted in an auroral oval which increased in diameter with increasing magnetic activity without requiring explicit reconnection.

In order to summarize the mapping results, and to illustrate their context in the pre-substorm temporal evolution, we have schematically represented in Figure 4 the sequence of events as they might occur in the minimum B surface of the magnetosphere. A solar wind disturbance perturbs the dayside magnetopause propagating rarefaction waves into the magnetosphere [*Coroniti and Kennel*, 1973]. Accompanying the wave activity, presumably, are field aligned currents resulting in the rapidly westward (eastward) moving morning (afternoon) sector arcs. The propagating waves eventually are associated with changes to the cross-tail current circuit and, having altered the balance between the dayside and nightside dynamos may be related to the unusual transient auroral phenomena shown in Plate 4. Eventually, the perturbations to the current circuit reach the near-Earth nightside "cusp" region and "trigger" the expansion phase onset there. Details of the possible magnetospheric processes can be found in *Lundin et al.* [1991b].

While all the observations above may not occur for every substorm, enough events have been found in the Viking data base to conclude that auroral features at local times other than midnight are sometimes part of a general scenario of pre-onset magnetospheric activity. To determine how widespread or common these features are would

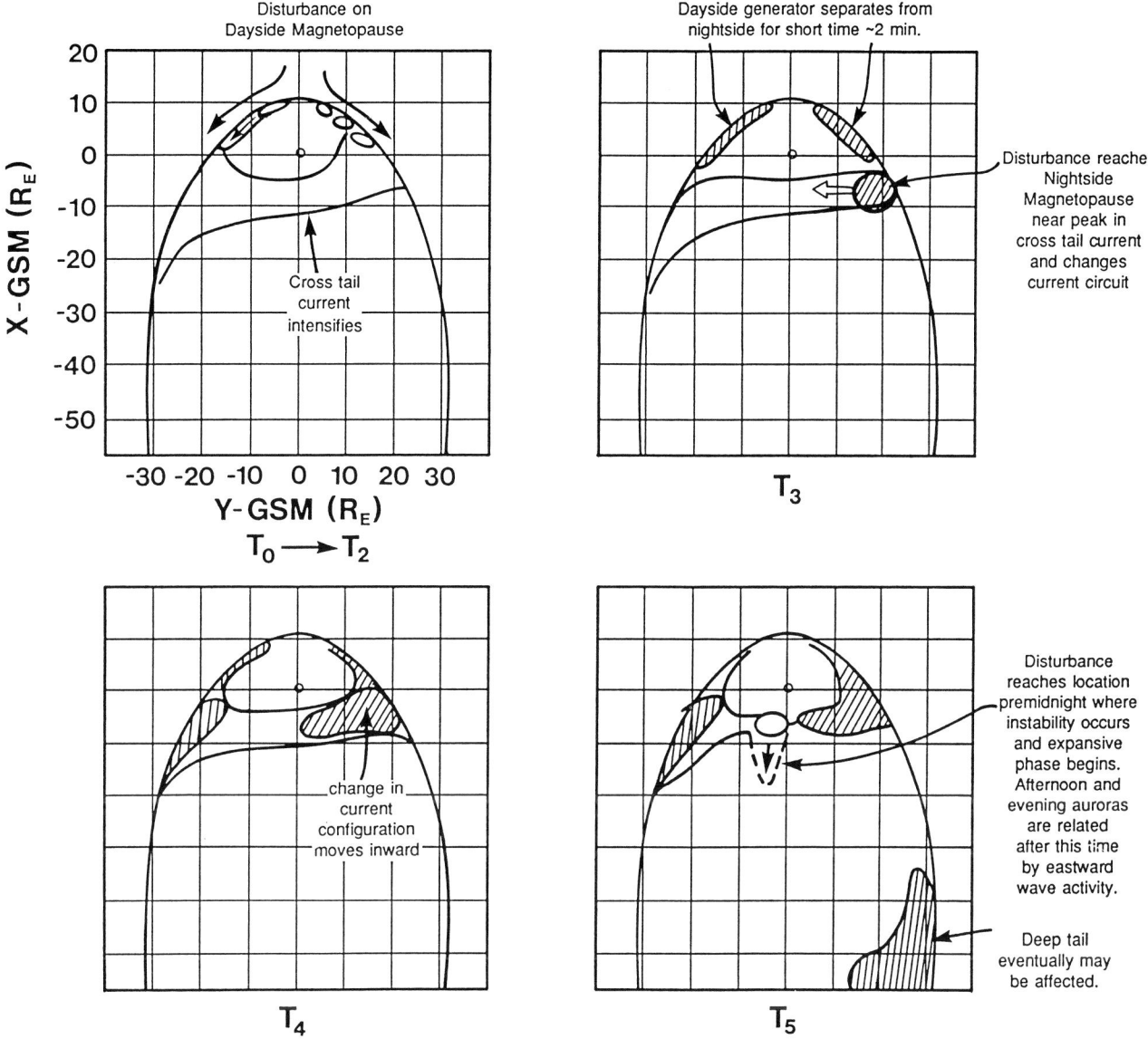

Figure 4. A schematic representation of a possible pre-onset development of a disturbance in the minimum B surfaces of the magnetosphere. The times T_0 to T_5 represent approximately the times corresponding to the schematic in Figure 1.

require a careful re-examination of pre-onset ionospheric data at high time resolution.

Acknowledgements. The authors would like to thank Ann Marie Morris for helping in the production of this paper and V. Petrov, from IZMIRAN, U.S.S.R., who supplied a valuable magnetometer analysis system to the University of Calgary. Discussions with Y.-I. Feldstein made us aware of important contributions made by Soviet researchers in this field. The Viking UV imager was built as a project of the National Research Council of Canada and this work was supported under a grant by the Natural Science and Engineering Research Council of Canada.

REFERENCES

Akasofu, S.-I., The Development of the Auroral Substorm, *Planet. Space Sci., 12,* 273, 1964.

Akasofu, S.-I., Midday Auroras and Magnetospheric Substorms, *J. Geophys. Res., 77,* 244, 1972.

Akasofu, S.-I. and A. L. Snyder, Comments on the Growth Phase of Magnetospheric Substorms, *J. Geophys. Res., 77,* 6275, 1972.

Akasofu, S.-I., Auroral Arcs and Auroral Potential Structure, in *Physics of Auroral Arc Formation,* Geophysical

Monograph 26, edited by S.-I. Akasofu and J. R. Kan, p. 1, 1981.

Anger, C. D., J. S. Murphree, A. Vallance Jones, R. A. King, A. L. Broadfoot, L. L. Cogger, F. Creutzberg, R. L. Gattinger, G. Gustafsson, F. R. Harris, J. W. Haslett, E. J. Llewellyn, J. C. McConnell, D. J. McEwen, E. H. Richardson, G. Rostoker, B. R. Sandel, G. G. Shepherd, D. Venkatesan, D. D. Wallis, and G. Witt, Scientific Results from the Viking Ultraviolet Imager: An Introduction, *Geophys. Res. Lett.*, *14*, 383, 1987a.

Anger, C. D., S. K. Babey, A. L. Broadfoot, R. G. Brown, L. L. Cogger, R. Gattinger, J. W. Haslett, R. A. King, D. J. McEwen, J. S. Murphree, E. H. Richardson, B. R. Sandel, K. Smith, and A. Vallance Jones, An Ultraviolet Auroral Imager for the Viking Spacecraft, *Geophys. Res. Lett.*, *14*, 387, 1987b.

Aubry, M. P., C. T. Russell, and M. G. Kivelson, On Inward Motion of the Magnetopause Before a Substorm, *J. Geophys. Res.*, *75*, 7018, 1970.

Coroniti, F. V. and C. F. Kennell, Can the Ionosphere Regulate Magnetospheric Convection, *J. Geophys. Res.*, *78*, 2837, 1973.

Crooker, N. U., Flux Transfer Event Footprint Patterns and Implications for Convection, *J. Geophys. Res.*, *95*, 10,567, 1990.

Eliasson, L., R. Lundin and J. S. Murphree, Polar Cap Arcs Observed by the Viking Spacecraft, *Geophys. Res. Lett.*, *14*, 451, 1987.

Elphinstone, R. D. and J. S. Murphree, Dayside Auroral Forms as Observed from Above by the Viking UV Imager, GEM, *Report of the Workshop on Ionospheric Signatures of Cusp, Magnetopause and Boundary Layer Processes*, edited by T. J. Rosenberg, p. 21, University of Maryland at College Park, U.S.A., 1989.

Elphinstone, R. D., D. Hearn, J. S. Murphree and L. L. Cogger, Mapping Using the Tsyganenko Long Magnetospheric Model and Its Relationship to Viking Auroral Images, *J. Geophys Res.*, *96*, 1467, 1991.

Feldstein, Y. I., Some Problems Concerning the Morphology of Auroras and Magnetic Disturbances at High Latitudes, *Geomag. and Aeron.*, *3*, 227, 1963.

Feldstein, Y. I., and G. V. Starkov, Dynamics of Auroral Belt and Polar Geomagnetic Disturbances, *Planet. Space Sci.*, *15*, 209, 1967.

Feldstein, Y. I. and Yu. I Galperin, The Auroral Luminosity Structure in the High-Latitude Upper Atmosphere: Its Dynamics and Relationship to the Large-Scale Structure of the Earth's Magnetosphere, *Rev. Geophys.*, *23*, 217, 1985.

Frank, L. A. and J. D. Craven, Imaging Results from Dynamics Explorer 1, *Rev. Geophys.*, *26*, 249, 1988.

Hoffman, R. A. and J. L. Burch, Electron Precipitation Patterns and Substorm Morphology, *J. Geophys. Res.*, *78*, 2867, 1973.

Kamide, Y., and A. D. Richmond, Estimation of Electric Fields and Currents from Ground-Based Magnetometer Data, *in Magnetospheric Currents*, edited by T. A. Potemra, Geophys. Mono. 28, American Geophysical Union, Washington, D.C., U.S.A., p. 67, 1984.

Lui, A. T. Y. and J. R. Burrows, On the Location of Auroral Arcs Near Substorm Onset, *J. Geophys. Res.*, *83*, 5342, 1978.

Lui, A. T. Y., D. Venkatesan, G. Rostoker, J. S. Murphree, C. D. Anger, L. L. Cogger and T. A. Potemra, Dayside Auroral Intensifications During an Auroral Substorm, *Geophys. Res. Lett.*, *14*, 415, 1987.

Lundin, R., D. S. Evans, Boundary Layer Plasmas as a Source for High-Latitude Early Afternoon Auroral Arcs, *Planet. Space Sci.*, *32*, 1389, 1985.

Lundin, R., L. Eliasson and J. S. Murphree, The Quiet Time Aurora and the Magnetospheric Configuration, *in Auroral Physics*, edited by C.-I. Meng, M. J. Rycroft and L. A. Frank, Cambridge University Press, Cambridge, U.K., p. 177, 1991a.

Lundin, R., I. Sandahl, J. Woch and R. D. Elphinstone, The Contribution of the Boundary Layer EMF to Magnetospheric Substorms, *this volume*, 1991b.

Maltsev, Yu. P. and W. B. Lyatsky, Field-Aligned Currents and Erosion of the Dayside Magnetosphere, *Planet. Space Sci.*, *23*, 1257, 1975.

McPherron, R. L., Magnetospheric Substorms, *Rev. Geophys. and Space Phys.*, *17*, 657, 1979.

Meng, C.-I., and R. Lundin, Auroral Morphology of the Midday Oval, *J. Geophys. Res.*, *91*, 1572, 1986.

Murphree, J. S., R. D. Elphinstone, L. L. Cogger, M. G. Henderson, and D. Hearn, Viking Optical Substorm Signatures, *this volume*, 1991.

Nielsen, E., J. Bamber, Z.-S. Chen, A. Brekke, A. Egeland, J. S. Murphree, D. Venkatesan and W. I. Axford, Substorm Expansion into the Polar Cap, *Ann. Geophys.*, *6*, 559, 1988.

Nishida, A., and Y. Kamide, Magnetospheric Processes Preceding the Onset of an Isolated Substorm: A Case Study of the March 31, 1978 Substorm, *J. Geophys. Res.*, *88*, 7005, 1983.

Pellinen, R. J., and W. J. Heikkila, Observations of Auroral Fading Before Breakup, *J. Geophys. Res.*, *83*, 4207, 1978.

Potemra, T. A., H. Vo, D. Venkatesan, L. L. Cogger, R. E. Erlandson, L. J. Zanetti, P. F. Bythrow, and B. J. Anderson, Periodic Auroral Forms and Geomagnetic Field Oscillations in the 1400 MLT Region, *J. Geophys. Res.*, *95*, 5835, 1990.

Pytte, T., R. L. McPherron, M. G. Kivelson, H. I. West and E. W. Hones, Multiple-Satellite Studies of Magnetospheric Substorms: Radial Dynamics of the Plasma Sheet, *J. Geophys. Res.*, *81*, 5921, 1976.

Reiff, P.H., J. S. Burch, and R. A. Heelis, Dayside Auroral Arcs and Convection, *Geophys. Res. Lett.*, *5*, 331, 1978.

Rezhenov, B. V., V. G. Vorobjev and Y. I. Feldstein, The Interplanetary Magnetic Field B_z-Component Influence on the Geomagnetic Field Variations and on the Auroral Dynamics, *Planet. Space Sci.*, *27*, 699, 1979.

Sandholt, P. E., M. Lockwood, B. Lybekk and A. D. Farmer, Auroral Bright Spot Sequence Near 1400 MLT: Coordinated Optical and Ion Drift Observations, *J. Geophys. Res., 95*, 21,095, 1990.

Sergeev, V. A., P. Tanskanen, K. Mursala, A. Korth, and R. C. Elphic, Current Sheet Thickness in the Near-Earth Plasma Sheet During Substorm Growth Phase, *J. Geophys. Res., 95*, 3819, 1990.

Shepherd, G. G. and J. S. Murphree, Diagnosis of Auroral Dynamics Using Global Auroral Imaging with Emphasis on Localized and Transient Features, in *Auroral Physics*, edited by C.-I. Meng, M. J. Rycroft and L. A. Frank, Cambridge University Press, Cambridge, U.K., p. 289, 1991.

Troshichev, O. A., B. M. Kuznetsov and M. I. Pudovkin, The Current Systems of the Magnetic Substorm Growth and Explosive Phases, *Planet. Space Sci., 22*, 1403, 1974.

Tsyganenko, N. A., Global Quantitative Models of the Geomagnetic Field in the Cislunar Magnetosphere for Different Disturbance Levels, *Planet. Space Sci., 35*, 1347, 1987.

Tsyganenko, N. A., A Magnetospheric Magnetic Field Model with a Warped Tail Current Sheet, *Planet. Space Sci., 37*, 5, 1989.

Untiedt, J., R. Pellinen, F. Kuppers, H. J. Opgenoorth, W. D. Pelster, W. Baumjohann, H. Ranta, J. Kangas, P. Czechowsky, and W. J. Heikkila, Observations of the Initial Development of an Auroral and Magnetic Substorm at Magnetic Midnight, *J. Geophys., 45*, 41, 1978.

Vorobjev, V. G., G. V. Starkov and Y. I. Feldstein, The Auroral Oval During the Substorm Development, *Planet. Space Sci., 24*, 955, 1976.

Zelenyi, L. M., R. A. Kovrahzkin and J. M. Bosqued, Velocity-Dispersed Ionbeams in the Nightside Auroral Zone: Aureol 3 Observations, *J. Geophys. Res., 95*, 12119, 1990.

EXOS-D OBSERVATIONS OF CHARGED PARTICLE PRECIPITATION AND ACCELERATION PROCESSES

T. Mukai

The Institute of Space and Astronautical Science, Sagamihara 229, JAPAN

N. Kaya

Faculty of Engineering, Kobe University, Kobe 657, JAPAN

W. Miyake

Communications Research Laboratory, Koganei 184, JAPAN

Abstract. The LEP (low energy particle) instrument onboard EXOS-D makes comprehensive measurements of energy and pitch-angle distributions of electrons and ions along with mass per charge analysis of positive ions. In this paper we discuss three topics on particle distributions in the polar region. (1) The clear energy dispersion of ions observed in the polar cusp can be interpreted in terms of the modified velocity filter effect in spite of complexities in the electric field. (2) With regard to the upflowing ion conics, altitude variations of the cone angle and the temperature demonstrate that the perpendicular ion energization often takes place over a wide altitude range. (3) The existence of the parallel electric field simultaneously below and above the spacecraft is often suggested by observations of ions and electrons. In order to explain the difference in the energy dependencies of upflowing ion beams by the species, we propose a new acceleration model, in which the region of the parallel electric field is confined in a narrow region, which moves upward and/or downward and whose potential drop is smaller at lower altitudes.

Introduction

The Japanese EXOS-D (Akebono) satellite was launched on February 22, 1989, into an elliptical orbit with an inclination of 75° and with initial apogee and perigee heights of 10,482 km (in the southern hemisphere) and 272 km, respectively. It carries a comprehensive set of instrumentation for particle and field measurements; see Mukai et al.[1990] and Kaya et al.[1990] for instrumentation of the low-energy particle measurement, Hayakawa et al.[1990] for the electric field measurement, and Fukunishi et al.[1990] for the magnetic field measurement. Extensive studies are currently in progress on various particle precipitation and acceleration features at altitudes below ~1.6 R_e over the auroral oval and the polar cap region. Of particular interest in this paper are (a) injection and transport of magnetosheath plasma in the cusp, (b) upflowing ion conics and beams on the dayside, and (c) particle acceleration due to the field-aligned electric field. Initial reports on (a) and (b) have appeared in GRL special issue for EXOS-D [Mukai et al., 1991; Matsuoka et al., 1991; Miyake et al., 1991a]. We report on some new results; especially, a characteristic energy dispersion signature of ions in the polar cusp, altitude variations of energy and pitch angle of ion conics, and electron and ion distribution functions in the acceleration region.

Injection and Transport of Magnetosheath Plasma in the Polar Cusp

The signatures of magnetosheath plasma injection and transport in the polar cusp/cleft have hitherto been studied extensively [e.g., Heikkila and Winningham, 1971; Frank, 1971; Reiff et al., 1977, 1980; Burch et al., 1980, 1982, 1986; Kaya et al., 1985; Lundin, 1988], and it seems to be established that the energy dispersion of the precipitating ions can be interpreted by the "velocity filter" effect due to poleward ($\mathbf{E} \times \mathbf{B}$) convection of magnetosheath plasma injected in a narrow region. However, comparison of the simultaneously observed electric field with the electron and ion precipitations in the polar cusp reveals that the convection flow vectors are neither simply poleward nor uniform but contain shears and intense fluctuations [Mukai et al., 1991]. Nevertheless, the ion precipitation shows a clear energy dispersion. Mukai et al.[1991] have also estimated the effect of the electric field fluctuations on the ion energy dispersion in terms of a simplified model assuming a narrow region of the plasma entry. If frequency of fluctuations is lower than the ion gyrofrequency and the magnetic field lines are equipotentials, the convected path length ℓ across magnetic field lines is given by

$$\ell = \int_0^t V dt, \tag{1}$$

where V is the convection speed, and t is the transit time from the injection point to the observation point.

$$t = s/v_\parallel, \tag{2}$$

where s is the distance to the injection point along the magnetic field line and $v_\parallel = \sqrt{(2E/m)}$ for ions with energy of E and pitch angle of $0°$. If the fluctuations have frequency ω and are expressed as

$$V = a + b\cos(\omega t), \tag{3}$$

then

$$\ell = as/v_\parallel + (b/\omega)\sin(\omega s/v_\parallel). \tag{4}$$

The energy dispersion, $dE/d\ell$, is always negative for $a > b$. The factor of b/ω of the second term represents the effect of the fluctuating component on the observed energy dispersion, and it becomes smaller with increasing frequencies. The estimated fluctuation in the energy dispersion signature is less than the observed energy spread, for example, for $a = 800m/s$, $b = 600m/s$, $s = 16R_e$ and $\omega = 0.1 rad/s$, as shown in Figure 1. These values of a and b correspond to the intensities of DC- and fluctuating electric fields of 15 mV/m and 11 mV/m, respectively, observed at altitude of 1 R_e. Actually the electric field fluctuations have a wideband frequency spectrum [e.g., Matsuoka et al., 1991]. This means that the oscillations with various frequencies are superposed and the energy spread is widened, but the dispersion pattern is not significantly affected. Hence it is reasonable that fluctuations in the electric field do not have a significant influence on the dispersion signature.

If this is correct, the source distance along the field line can be estimated using the ion dispersion curve and the convection velocity obtained by averaging the electric field data. It can also be estimated independently from the characteristic relation (so-called 'V' pattern) between ion energy and pitch angle [Burch et al., 1982; Menietti and Burch, 1988]. It should be noted that the latter method can be used for the observation in short periods of time (e.g., in one spin period), whereas the overall shape of the dispersion has to be used for the former method. Comparison between the

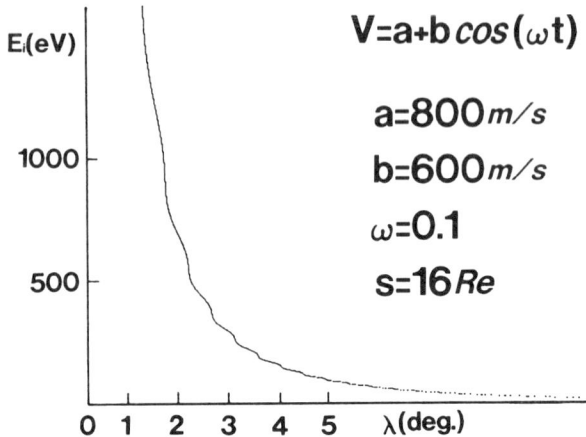

Fig. 1. Numerical result of the ion energy dispersion with existence of electric field oscillations in the case of $a = 800ms^{-1}$, $b = 600ms^{-1}$, $s = 16R_e$, and $\omega = 0.1s^{-1}$ in Eq.(3). The vertical scale is the proton energy in eV. The horizontal axis is the magnetic latitude of the field line assuming the dipole field, in which the $0°$ stands for the latitude corresponding to the injection point at the magnetopause.

results obtained by these two methods can be used to test the validity of the velocity filter concept. Figure 2 shows an example of such a comparison. The upper six panels shows energy (per unit charge) spectra of electrons and positive ions in pitch-angle sorted E-t diagrams, of which the upper and the lower three panels correspond to the electron and ion data, respectively. The $0°$ ($180°$) pitch angle corresponds to downward (upward) component along magnetic field lines. The lowest panel shows temporal/spatial variation of energy versus pitch-angle distributions of ions in an expanded time scale, in which ticks are marked on the horizontal axis every 8 seconds, and the distribution between tick marks shows the energy versus pitch angle relation. Absence of data at the largest pitch angle is due to the ion loss cone. It can be seen that the ion energies are higher at larger pitch angles in each time interval between tick marks.

The source distance obtained by the first method is 11 R_e for the case of Figure 2. At several points of this Figure we have also indicated the source distances which are obtained by the latter method using the energy dependence on pitch angles. Both results are in a rough agreement, though the latter estimates are somewhat scattered. Similar tendencies are seen in other examples observed by EXOS-D, where the source distance are mostly 8-10 R_e away from the spacecraft. This suggests that the injection region for the cusp plasma is located at high latitudes in the entry layer [Burch et al., 1986; Menietti and Burch, 1988]. If the plasma were precipitated from the low-latitude boundary layer, the source distance might be larger. There are also small number of instances where the estimated distances are as large as 16

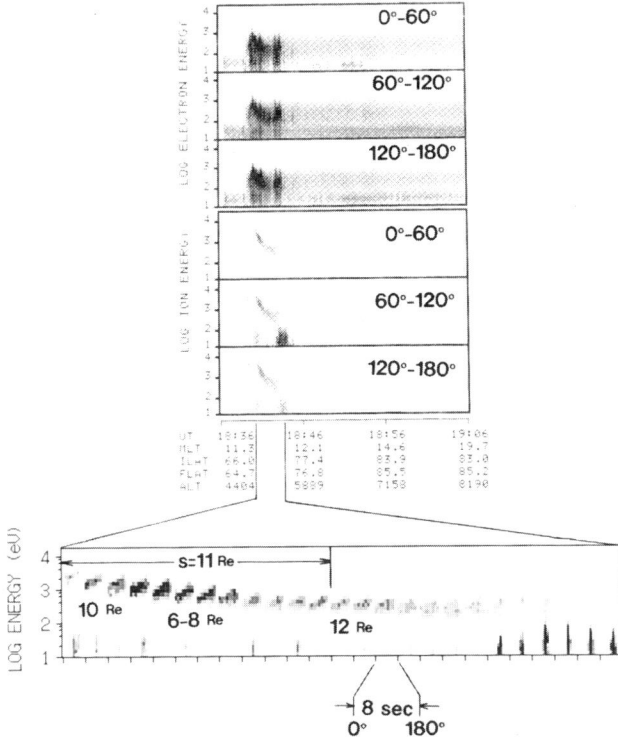

Fig. 2. Upper six panels show pitch-angle sorted energy-time spectrograms for electrons and positive ions observed at a cusp crossing in a nearly noon-midnight orbit on February 7, 1990. The lowest panel shows temporal/spatial variation of energy versus pitch-angle distributions of ions in an expanded time scale, in which the source distances of plasma are also indicated; see text.

R_e. We have looked in details of the electron and ion distributions for these exceptional cases, but they also show signatures of the cusp proper. This suggests that sometimes particles originate from the low-latitude magnetosphere or the low-latitude boundary layer.

Thus, in spite of the high variability of the electric field, particle and electric field observations in the cusp region can still be considered to be compatible with the view that the energy dispersion of precipitating ions is produced by the velocity filter effect. However, Possibilities of other interpretations are not denied and have to be pursued.

Upflowing Ions on the Dayside

The upflowing ions are basically divided into two categories: the conics including TAI [Klumper, 1979], and the beams, based on their characteristic pitch angle distributions. Upward ion beams are usually believed to be accelerated by the parallel electric field, whereas wave heating is generally thought to produce ion conics. The ion bowl distributions (or, elevated conics) is interpreted in terms of the combined effect of perpendicular and parallel accelerations of ions [Klumper et al., 1984; Lundin et al., 1987; Hultqvist et al., 1988], though the combination may not be necessary [Horwits, 1986; Temrin, 1986]. The frequent occurrence of low-energy upflowing ions on the dayside is well known as a source for the cleft ion fountain [e.g., Lockwood et al., 1985]. On EXOS-D, upflowing ion conics and beams have also been observed at all local times over the auroral oval, but most frequently with low energies around the dayside cusp/cleft region [Mukai et al., 1990]. One of the interesting

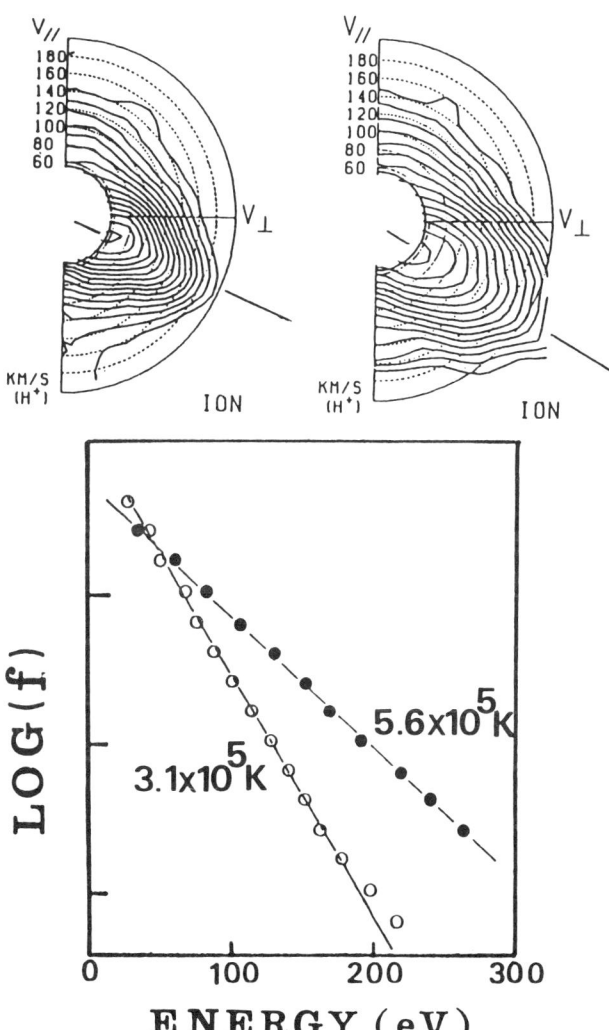

Fig. 3. Two examples of phase space density distributions of ion conics. In the upper panel the velocity scale is made by assuming that all ions are H^+. In the lower panel, the open and the solid circles represent the distribution functions along the cone axis as shown in the upper-lefthand and upper-righthand panels, respectively [after Miyake et al., 1991b].

results reported by Miyake et al.[1991a] is that the occurrence of the field-aligned velocity shift (elevated conics and upward ion beams) of the upflowing ions is well correlated to electron precipitation (and the upward field-aligned current region) even at small scales in time/space, indicating that the field-aligned velocity shift is caused by the acceleration by parallel electric field near the spacecraft. On the other hand, the anisotropy (perpendicular heating) of upflowing ions does not respond to small-scale variations in the electron precipitation. These two energization processes are completely independent.

In order to study the evolution of the ion perpendicular energy with altitude, Miyake et al.[1991b] have defined two parameters; the cone angle and the cone temperature. In Figure 3, two examples of the phase space density distribution of the ion conics are shown in the upper panel, where the cone angle is defined as the angle between the cone axis and the magnetic field. The cone temperature is derived from the slope of the distribution function (exactly, log(f) versus ion energy) in the lower panel, which shows the section of the distribution function along the cone axis. Altitude dependence of the observed cone angle is shown in Figure 4, in which the vertical and horizontal bars indicate the spread or change during each continuous conics event. For reference, four curves show the cone angle variations with the altitude, which are expected if ions were perpendicularly heated at altitudes of 2,000, 4,000, 6,000 and 8,000 km, respectively, and move upward conserving the adiabatic invariant. The observed cone angles tend to decrease with increasing altitude, but not so much as expected from the adiabatic model.

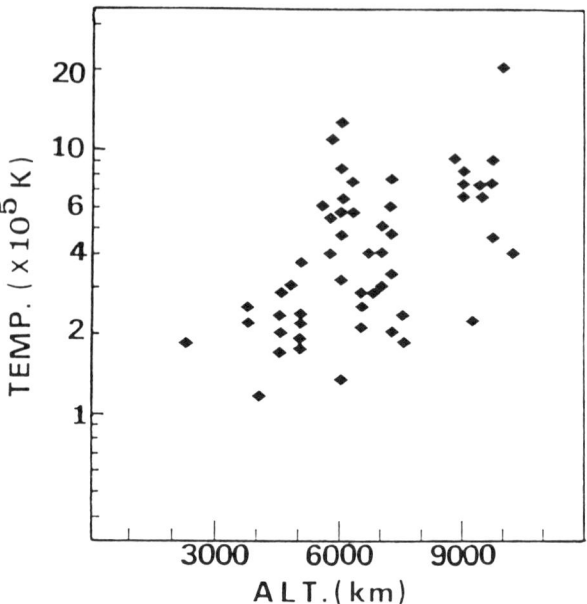

Fig. 5. Altitude variation of temperatures of ion conics [after Miyake et al., 1991b].

Figure 5 shows altitude variation of the cone temperature, in which the maximum temperature measured during each continuous event is used. (Since the data are taken from the LEP energy range above 13 eV/Q, the derived temperature may be erroneous when it is very low.) It can be seen that the temperature of the conics tends to be higher at higher altitudes. These two figures suggest that the ion conics do not simply flow adiabatically upward, but continue to be heated perpendicularly as they flow to higher altitudes [e.g., Temrin, 1986].

Particle Accleration Due to the Parallel
Electric Field

Figure 6 shows E-t diagrams for electrons and ions observed at altitudes of ~8,300 km over the eveningside auroral oval in the southern hemisphere. Only the data from the detectors with the same viewing direction as that of the energetic ion mass spectrometer are used here in order to compare the energy per charge spectra of all the ions with the spectra sorted by the ion species. In the lower panel, accelerated ions are seen intermittently in time/space, varying their energies slowly. These accelerated ions represent the upward ion beams (UFI), since they are observed only when sampled in the upward field-aligned direction. Figure 7 shows that the main constituents of these ions are O^+ and He^+, both of which have nearly equal energies [Kaya et al., 1990]. It is interesting to note that the similar abundance of He^+ and O^+ has rarely been observed previously [Collin et al., 1984]; that is, the He^+ ions are unusually rich in the present case. When the upflowing H^+ ions are also seen,

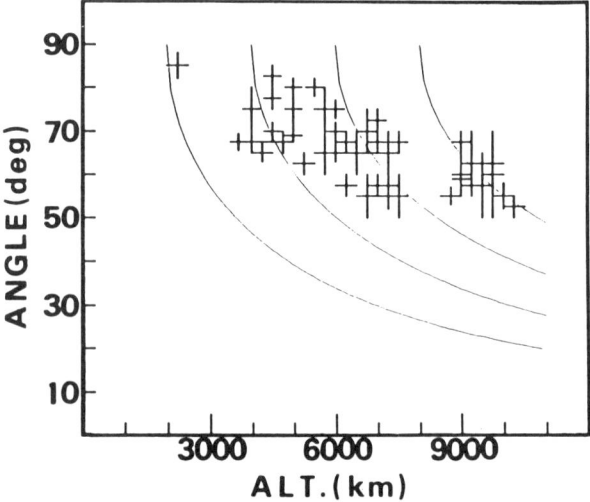

Fig. 4. Altitude variation of cone angles of ion conics. Four curves represent the variations expected from the adiabatic motion in the magnetic mirror force field. The vertical and horizontal bars stand for the spread or change during each continuous conic event [after Miyake et al., 1991b].

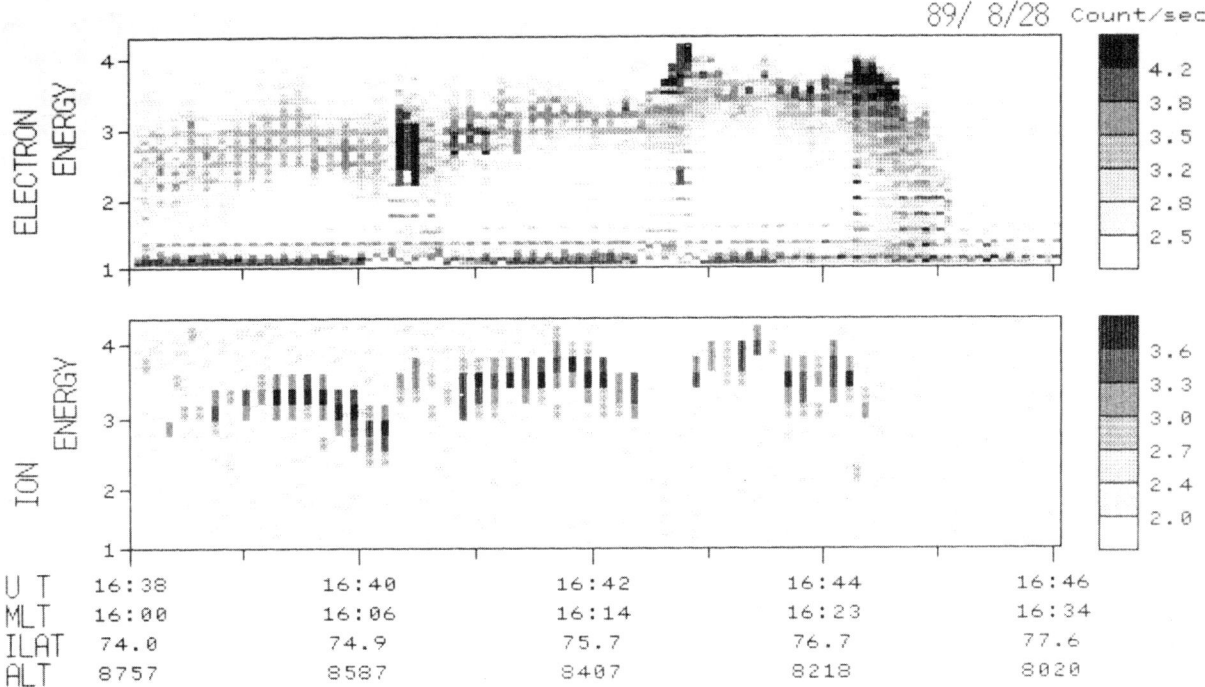

Fig. 6. Energy-time spectrograms of electrons and ions for a pass through the eveningside auroral oval in the southern hemisphere. Here are used only the data from the detectors in the same viewing direction as that of the energetic ion mass spectrometer [after Kaya et al., 1990].

their energies correspond to the higher-energy part of the energies of upflowing O^+ and He^+. Similarity of upflowing ion energies of different species suggests that they are accelerated by a potential drop below the spacecraft altitude. During most of the time from 1641 to 1644 UT, the electron distribution also shows a monoenergetic peak at several keV, suggesting existence of a potential drop above the spacecraft. Thus, the acceleration regions exist both above and below the EXOS-D satellite, and the total potential drop between the ionosphere and the distant magnetospheric region is the sum of the peak energies of precipitating electrons and upgoing ions.

An energy versus pitch-angle distribution of electrons during the period of 1639:15-1639:45 UT (at altitude of 8,630 km) is shown in Figure 8. The abscissa is pitch angle, and the ordinate is electron energy with a logarithmic scale. All the sampled data are displayed including data of very low count rate, and no interpolation of data has been made. It can be seen that the electron loss cone angle is wider than expected by a simple mirror motion (18° for mirror altitude of 100 km) and is dependent on the electron energy. If a parallel potential drop exists below the spacecraft, the critical loss-cone angle, α, of an electron with the kinetic energy, E, is given by

$$\sin^2\alpha = (B/B_b)\{1 + (e\Delta\Phi/E)\}, \qquad (5)$$

where B and B_b are the magnetic field intensities at the observed altitude and the base of a field line, respectively, and $\Delta\Phi$ is the field-aligned potential difference between the two altitudes. We have carried out test particle simulations in order to ascertain whether the energy-dependent widening of the electron loss cone is consistent with the parallel potential drop below the spacecraft which is estimated independently from the UFI energy. Figure 9 shows an example of the result, in which the observed downward flux is used as the input distribution, and the parallel potential drop is set at 1.5 kV in accordance with the observed peak energy of UFI beams. The calculated distribution of the upward flux, especially the shape of the energy-dependent loss cone, is in fairly good agreement with the observed one as shown in Figure 8. This seems to be a reasonable result, but it is contrary to the DE-1 result obtained by Reiff et al.[1988]. They reported a somewhat surprising result that the UFI's peak energy was frequently 30% (and in one case 70%) less than the potential drop below the DE-1 observation as inferred from the electron distributions. The difference between these results may originate from the difference in altitudes between these spacecraft. The widening of the electron loss cone can be more clearly identified at lower altitudes as long as the spacecraft is still above the acceleration region. Since the altitude of EXOS-D is lower than that of DE-1, the present data could be better suited

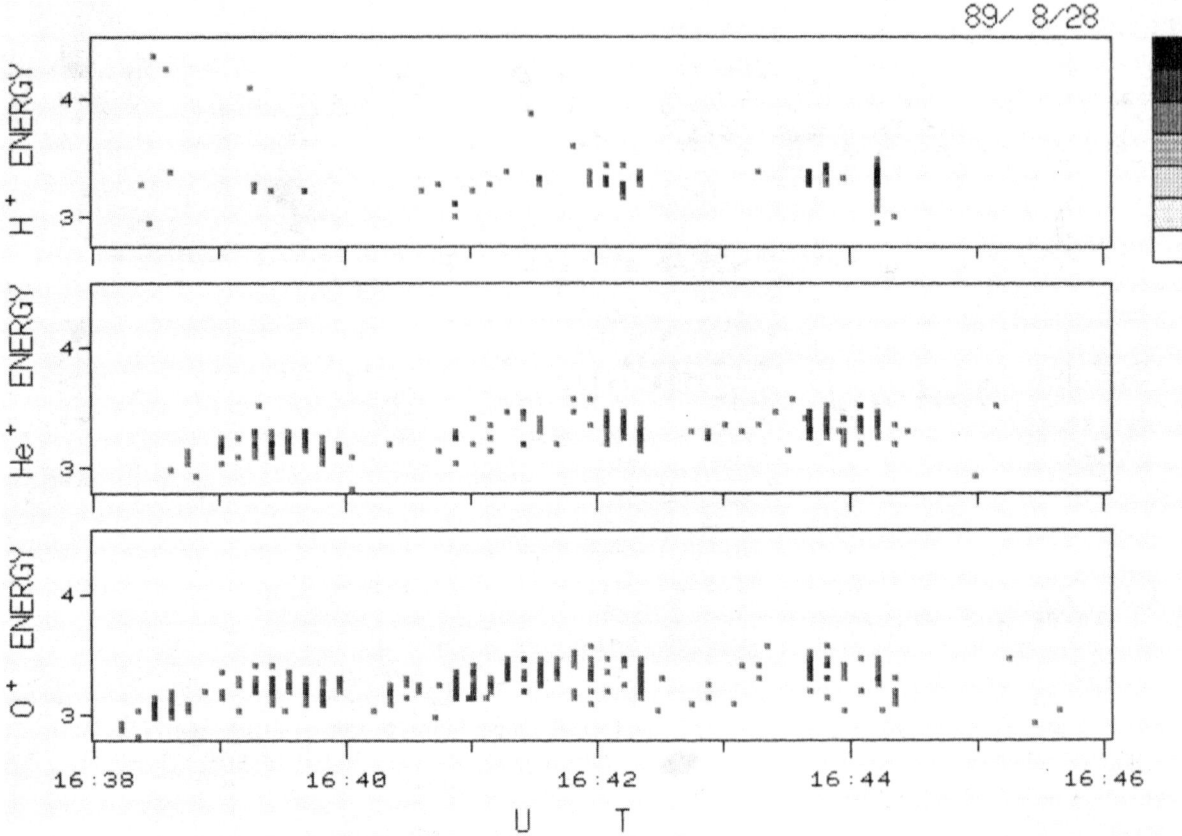

Fig. 7. Energy-time spectrograms of H$^+$ (top panel), He$^+$ (middle panel), and O$^+$ (bottom panel) observed by the energetic ion mass spectrometer in the same time period as in Figure 6.

to study the particle accleration due to the parallel electric field. Some thermalization processes, such as suggested by Reiff et al.[1988], may alter (decrease) the distribution and the peak energy of UFI when transported over a long distance.

As has been noted earlier, all species of upflowing ions seem to have been accelerated by the same energy per charge. However, Figure 7 also shows that the ions have energy spread which is dependent on ion species. The max-

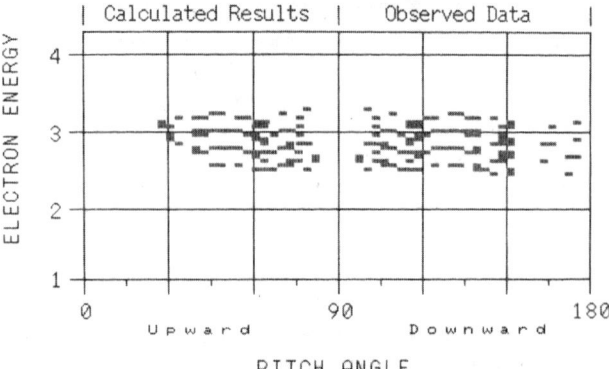

Fig. 8. Energy versus pitch-angle distribution of electrons during 1639:15-1639:45, when the UFI beams are observed. The 0° (180°) pitch angle corresponds to upward (downward) component along the magnetic field lines, since this observation was in the southern hemisphere. It can be seen that the electron loss cone is widened, depending on the electron energy.

Fig. 9. Result of the test particle simulation during the same period as in Figure 8 under the condition that the parallel electric field with potential drop of 1.5 kV exists below the spacecraft altitude. The upward flux is calculated from the observed downward data with count rate of $> 10^{3.3}$ as the input distribution.

imum energies of different species are nearly equal, but the minimum energies of heavier ions (e.g., O^+) are lower than that of lighter ions. This observation is inconsistent with any purely static model of the parallel acceleration, since the minimum energies, not the maximum energies, of different ion species must be equal, if the parallel electric field is assumed to be distributed over a wide altitude region. In addition, the average energy of O^+ must be higher than that of H^+, because the density of cold O^+ ions decrease more rapidly with increasing altitude than that of H^+. However, this expectation is contrary to the observed result. In order to explain the observed result, we propose as a possibility that the region of the parallel electric field is confined within a narrow region, which moves upward and/or downward, decreasing the potential drop with decreasing altitude. The higher potential drop is sustained at higher altitudes, where H^+ as well as He^+ and O^+ are the major ion species, and thus these three ion species are accelerated simultaneously to equal energy per charge. We propose that the field intensity becomes weaker with increasing density of ambient plasma, as the parallel field region moves downward to lower altitudes where O^+ is the major ion component. Hence, the parallel potential drop becomes lower with decreasing altitudes, where mostly the O^+ ions are accelerated to lower energies. The threshold effect of the instrument is taken into consideration in the above discussion. With higher sensitivity of the instrument, slight amount of the H^+ ions might also be observed to be accelerated to similar energies as O^+. The moving speed of the parallel field region should be discussed further quantitatively in terms of the nature of the parallel field and the efficiency of the ion acceleration. However, it should be noted here that the electric field must have a quasistatic nature in the sense that heavy ions experience a unidirectional (upward) DC field by which they are accelerated. Such a moving parallel field may be ensemble of solitons, or multiple flickering double layers [Sharp et al., 1980].

Concluding Remarks

We have described some new results that have been obtained by LEP onboard EXOS-D on the injection and transport of plasma in the polar cusp, on the upflowing ion conics and beams, and on the particle acceleration due to the parallel electric field. Common to these results is the dynamic nature of plasma at altitudes of 1-1.5 R_e, where the electric field and the field-aligned structure are highly complex and fluctuating. The observed particle distribution is the result of the particles' interaction with these fields. Their exact natures must be clarified in relation to their mechanism(s) and their source(s) of free energy. This is important for better understanding of magnetosphere-ionosphere coupling and substorm processes. For this purpose, we feel the orbit of EXOS-D is well suited. Extensive studies are currently in progress and will be published in the near future.

Acknowledgments. We thank all members of the EXOS-D project team for their extensive efforts to the success of EXOS-D, especially T. Tsuruda (Project Manager) and H. Oya (Project Scientist), and the LEP team member, especially S. Machida, T. Obara, M. Hirahara, and Y. Saito. We would also appreciate A. Nishida for critical reading of the manuscript and useful comments.

References

Burch, J. L., P. H. Reiff, R. W. Spiro, R. A. Heelis, and S. A. Fields, Cusp region particle precipitation and ion convection for northward interplanetary magnetic field, *Geophys. Res. Letters*, 7, 393, 1980.

Burch, J. L., P. H. Reiff, R. A. Heelis, J. D. Winningham, W. B. Hanson, C. Gurgiolo, J. D. Menietti, R. A. Hoffman, and J. N. Barfield, Plasma injection and transport in the mid-altitude polar cusp, *Geophys. Res. Letters*, 9, 921, 1982.

Burch J.L., J.D. Menietti, and J.N. Barfield, DE-1 observations of solar wind-magnetosphere coupling processes in the polar cusp, in *Solar Wind-Magnetosphere Coupling*, edited by Y. kamide and J.A. Slaven, Terra, Tokyo, 1986.

Collin, H.L., R.D. Sharp, E.G. Shelley, The magnitude and composition of the outflow of energetic ions from the ionosphere, *J. Geophys. Re.*, 89, 2185, 1984.

Frank, L.A., Plasma in the earth's polar magnetosphere, *J. Geophys. Res.*, 76, 5202, 1971.

Fukunishi, H., R. Fujii, S. Kokubun, K. Hayashi, T. Tohyama, Y. Tonegawa, S. Okano, M. Sugiura, K. Yumoto, I. Aoyama, T. Sakurai, T. Saito, T. Iijima, A. Nishida, and M. Natori, Magnetic field observations on the Akebono (EXOS-D) satellite, *J. Geomag. Geoelectr.*, 42, 385, 1990.

Hayakawa, H., T. Okada, M. Ejiri, A. Kadokura, Y. -I. Kohno, K. Maezawa, S. Machida, A. Matsuoka, T. Mukai, M. Nakamura, A. Nishida, T. Obara, Y. Tanaka, F. S. Mozer, G. Haerendel, and K. Tsuruda, Electric field measurement on the Akebono (EXOS-D) satellite, *J. Geomag. Geoelectr.*, 42, 371, 1990.

Heikkila, W. J., and J. D. Winningham, Penetration of magnetosheath plasma to low altitudes through the dayside magnetospheric cusps, *J. Geophys. Res.*, 76, 883, 1971.

Horwitz, J. L., Velocity filter mechanism for ion bowl distributions, *J. Geophys. Res.*, 91, 4513, 1986.

Hultqvist, B., R. Lundin, K. Stasiewicz, L. Block, P. -A. Lindqvist, G. Gustafsson, H. Koskinen, A. Bahnsen, T. A. Potemra, and L. J. Zanetti, Simultaneous observation of upward moving field-aligned energetic electrons and ions on auroral zone field lines, *J. Geophys. Res.*, 93, 9765, 1988.

Kaya, N., T. Mukai, H. Matsumoto,, and T. Itoh, Characteristics of auroral particles observed by EXOS-C, *J. Geomag. Geoelectr.*, 37, 347, 1985.

Kaya, N., T. Mukai, and E. Sagawa, Preliminary results

from new type ion mass spectrometer onboard the Akebono (EXOS-D) satellite, *J. Geomag. Geoelectr., 42*, 497, 1990.

Klumper, D.M., Transversely accelerated ions: an ionospheric source of magnetospheric ions, *J. Geophys. Res., 84*, 4229, 1979.

Klumper, D.M., W.K. Peterson, and E.G. Shelley, Direct evidence for two-stage (bimodal) acceleration of ionospheric ions, *J. Geophys. Res., 89*, 10779, 1984.

Lockwood, M., M.O. Chandler, J.L. Horwitz, J.H. Waite,Jr, T.E. Moore, and C.R. Chappell, The cleft ion fountain, *J. Geophys. Res., 90*, 9736, 1985.

Lundin, R., L. Eliason, B. Hultqvist, and K. Stasiewics, Plasma energization on auroral field lines as observed by the Viking spacecraft, *Geophys. Res. letters, 14*, 443, 1987.

Lundin, R., Acceleration/heating of plasma on auroral field lines: Preliminary results from the Viking satellite, *Ann. Geophys., 6*, 143, 1988.

Matsuoka, A., T. Mukai, H. Hayakawa, Y. -I. Kohno, K. Tsuruda, A. Nishida, and T. Okada, EXOS-D observations of electric field fluctuations and charged particle precipitation in the polar cusp, *Geophys. Res. Letters, 18*, 305, 1991.

Menietti, J. D., and J. L. Burch, Spatial extent of the plasma injection region in the cusp-magnetosheath interface, *J. Geophys. res., 93*, 105, 1988.

Miyake, W., T. Mukai, N. Kaya, and H. Fukunishi, EXOS-D observations of upflowing ion conics with high time resolution, *Geophys. Res. Letters, 18*, 341, 1991a.

Miyake, W., T. Mukai, N. Kaya, and H. Fukunishi, Ion conics observed by Akebono satellite, *Proc. NIPR Symp. Upper Atmos. Phys., in press*, 1991b.

Mukai, T., N. Kaya, E. Sagawa, M. Hirahara, W. Miyake, T. Obara, H. Miyaoka, S. Machida, H. Yamagishi, M. Ejiri, H. Matsumoto, and T. Itoh, Low energy charged particle observations in the "auroral" magnetosphere: first results from the Akebono (EXOS-D) satellite, *J. Geomag. Geoelectr., 42*, 479, 1990.

Mukai, T., A. Matsuoka, H. Hayakawa, S. Machida, K. Tsuruda, A. Nishida, and N. Kaya, Signatures of solar wind injection and transport in the dayside cusp: EXOS-D observations, *Geophys. Res. Letters, 18*, 333, 1991.

Reiff, P. H., T. W. Hill, and J. L. Burch, Solar wind plasma injection at the dayside magnetospheric cusp, *J. Geophys. Res., 82*, 479, 1977.

Reiff, P. H., J. L. Burch, and R. W. Spiro, Cusp proton signatures and the interplanetary magnetic field, *J. Geophys. Res., 85*, 5997, 1980.

Reiff, P.H., H.L. Collin, J.D. Craven, J.L. Burch, J.D. Winningham, E.G. Shelley, L.A. Frank, and M.A. Friedman, Determination of auroral electrostatic potentials using high- and low-altitude particle distribution, *J. Geophys. Res., 93*, 7441, 1988.

Sharp, R. D., E. G. Shelley, R. G. Johnson, and A. G. Ghielmetti, Counterstreaming electron beams at altitudes of $\sim 1~R_E$ over the auroral zone, *J. Geophys. Res., 85*, 92, 1980.

Temrin, M, Evidence for a large bulk ion conic heating region, *Geophys. Res. Letters, 13*, 1059, 1986.

AURORA AND ENERGETIC PARTICLE SIGNATURES DURING A SUBSTORM WITH MULTIPLE EXPANSIONS

Rumi Nakamura[1], Takasi Oguti[2], Tatsundo Yamamoto[3], Susumu Kokubun[3], Daniel N. Baker[4], and Richard D. Belian[5]

Abstract. We compare geosynchronous particle signatures with aurora and geomagnetic disturbances during a substorm with multiple expansions, which began at 0852 UT on January 18, 1986. There were three localized expansions whose onset region moved successively westward. Energetic particle signatures observed from the geosynchronous satellite 1984-129 at the premidnight sector differed significantly for these three expansions in accordance with the transition of the onset region from the east to the west of the satellite foot-point. The results suggest that the source region of the energetic particle injection is confined within the local time sector of the expansion aurora. It was estimated that during this event an injection associated with an expansion started from a localized region with less than 28° in longitudinal width and was still confined within a region of about 40° even at the maximum of the expansion. This study also confirms that an injection boundary, if it ever exists, would be produced due to multiple localized injections, rather than due to a continuous expansion of a source region.

1. Introduction

Substorm-associated injection of energetic particles and plasma has been reported by many authors [e.g. DeForest and McIlwain, 1971; Erickson et al., 1979]. It is important to identify the original source of the injected particles temporarily and spatially in order to discuss the dynamics and energetics of substorms.

Particle signatures near the geosynchronous orbit are often dispersed in energy. Energy dispersion associated with substorms has been considered to be caused by energy-dependent drifts of the particles from a source region, where particles of all energy were originally accelerated or transported. McIlwain [1974] calculated the drift path of the particles by using a model of the electric field and the magnetic field and concluded that the particles originated at a common azimuthally extended boundary, which was called the injection boundary. This model was applied to explain the energy dispersion events observed near the geosynchronous region [Konradi et al., 1975; Greenspan et al., 1985; Mauk and Meng, 1983].

Spatial structures of the injection region have been also discussed using particle data during dispersionless events. Arnoldy and Moore [1983] concluded that the plasma injection has an onset region which expands continuously toward dusk and dawn from a relatively narrow sector near midnight and forms an wide extended boundary ($\geq$ 8h). Both radially inward and tailward motion of the dispersionless injection front were reported [Moore et al., 1981; Lopez et al., 1989]. Kistler et al. [1990] compared ion observations at the near-Earth plasma sheet (7-9 Re) and farther down the tail (10-19 Re) and reported spectral similarity at both locations during the post-injection period. Lopez et al. [1990] have suggested the Kp dependence of the occurrence of the dispersionless injection which was consistent with an injection boundary. He identified the injection boundary with the earthward edge of the region of the magnetotail where current sheet disruption takes place during the substorm expansive phase.

As for the mechanism of the injection, Moore et al. [1981] proposed that hot plasma, which penetrates in association with an earthward propagating compressional wave, builds an injection boundary at the inner front. This is, however, not consistent with tailward propagation of the disruption of the tail current [Lopez et al., 1989]. Local acceleration of the particles due to the induced electric field at the near-Earth neutral line [Baker, 1984] or at the current disruption region [Lui et al., 1988], or combined acceleration at both regions [Kistler et al., 1990] have been considered as an alternative explanation for the particle injection.

Energization of particles is manifested also as auroral particle precipitation into the ionosphere. Global morphology of the aurora during the expansive phase was first discussed by Akasofu [1964]. The first indication of an auroral substorm is a sudden brightening of a part of a quiet arc. This is followed by a rapid poleward motion of the arc resulting in a "bulge" around the midnight sector and westward propagating folds, called the "westward travelling surge", at the evening sector. Several studies compared ATS 5 particle flux with auroral signatures obtained by an all-sky camera near the ATS foot point [Akasofu et al., 1974; Mende and Shelley, 1976].

---

[1] National Institute of Polar Research, Kaga 1-9-10, Itabashiku, Tokyo, 173 Japan

[2] Solar Terrestrial Environment Laboratory, Nagoya University, Honohara 3-13, Toyokawa, Aichi, 442 Japan

[3] Department of Earth and Planetary Physics, University of Tokyo, Bunkyoku, Tokyo 113 Japan

[4] NASA / Goddard Space Flight Center, Laboratory for Extraterrestrial Physics, Greenbelt, Maryland 20771

[5] Los Alamos National Laboratory, Los Alamos, New Mexico 87545

The results showed that an injection at ATS 5 correlated very well to some particular auroral activities, such as the initial brightening of an auroral arc at the midnight sector. It is, therefore, very likely that energy conversion of the particle injection as well as the auroral expansion originally takes place at the same onset region.

In this study we compare auroral expansions and flux enhancements of energetic electrons and protons during ground-satellite coordinated observations in order to determine the spatial extent and temporal evolution of the energy conversion region in the magnetosphere. Aurora obtained from all-sky TV observations and energetic electrons (30 keV - 300 keV) and protons (145 keV - 560 keV) obtained from the geosynchronous satellites 1984-129 and 1982-019 are used to trace the substorm evolution. Multi-point auroral observations enable us to determine not only the temporal but also spatial structure of the substorm-associated disturbances, which is crucial in substorm studies.

2. Data

Ground based data used in this study were obtained during the Global Aurora Dynamics Campaign (GADC) period when multi-station network observations of aurora and magnetic fields were carried out in the northern Canada, Alaska and Norway [Oguti et al., 1988]. The distribution of the stations are given in Figure 1 in the corrected geomagnetic coordinates. During the event reported in this study, all-sky TV cameras were operated at Poker Flat (PFT), Shamattawa (SHM), and La Ronge (LRG), with aurora detected at PFT and SHM. At each station, total light images of aurora were recorded at the normal sampling rate of 30 frames per second on videotapes. The threshold of the system was about 100 R at the green line (5557 Å) equivalent. The video records of aurora were processed by means of a television image processor system (TVIP) which digitizes the video signals and registers the digital data in the frame memories of 512 * 512 * 8 bits. Further descriptions of the TVIP system are given by Yamamoto [1984]. Geomagnetic field data were available from 32 stations with time resolution between 1 sec and 5 sec at most of the stations. Further specifications of the instruments are described by Oguti et al. [1988].

We examined energetic particle data obtained by the geosynchronous satellites 1982-019 (38°W) and 1984-129 (156°W). Electron flux data in the energy range from 30 keV to 300 keV and proton flux data in the range from about 100 keV to 600 keV were obtained from the Charged Particle Analyzer (CPA) experiment. The CPA instruments are described in detail by Higbie et al. [1978]. Estimated conjugate points of the geosynchronous satellites at the ionospheric level (100 km) are also marked in Figure 1. They are calculated by using IGRF85 [IAGA I-1, 1985] for the internal geomagnetic field and by the Tsyganenko magnetic field model [Tsyganenko, 1987] for the external origin part. The field line configuration used here is the model field for Kp 2-, 2 and 2+ in accordance with the Kp value on January 18; Kp=2, 3 during 6-12 UT. According to the model, the foot point moves 0.5° poleward for Kp 0, and 0.3° equatorward for Kp greater than 5+, whereas the longitudinal difference among all Kp value is less than 0.1° for the foot point of 1984-129 and 0.3° for that of 1982-019. Note that the foot point of the satellite 1984-129 (156°W) was located about 11° east of PFT; just east of the field of view of the all-sky TV camera at PFT.

3. Observations

3.1. Ground Observations

Figure 2a shows magnetic records of XM (geomagnetic northward), YM (geomagnetic eastward), and Z (downward) components obtained from six auroral zone stations, Great

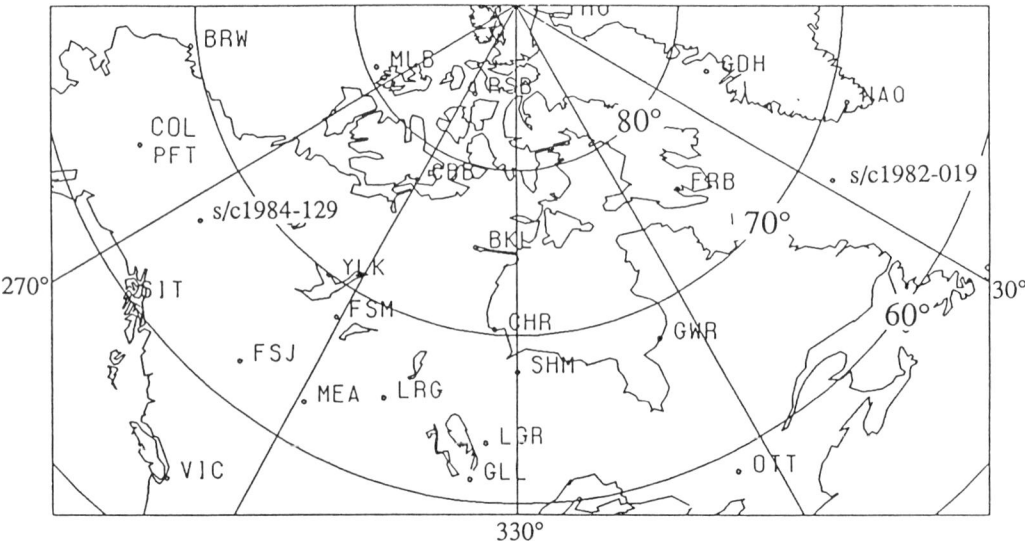

Fig. 1. Distribution of the stations given in the corrected geomagnetic coordinates. Calculated conjugate points of the geosynchronous satellites at the ionospheric level (100 km) are also marked.

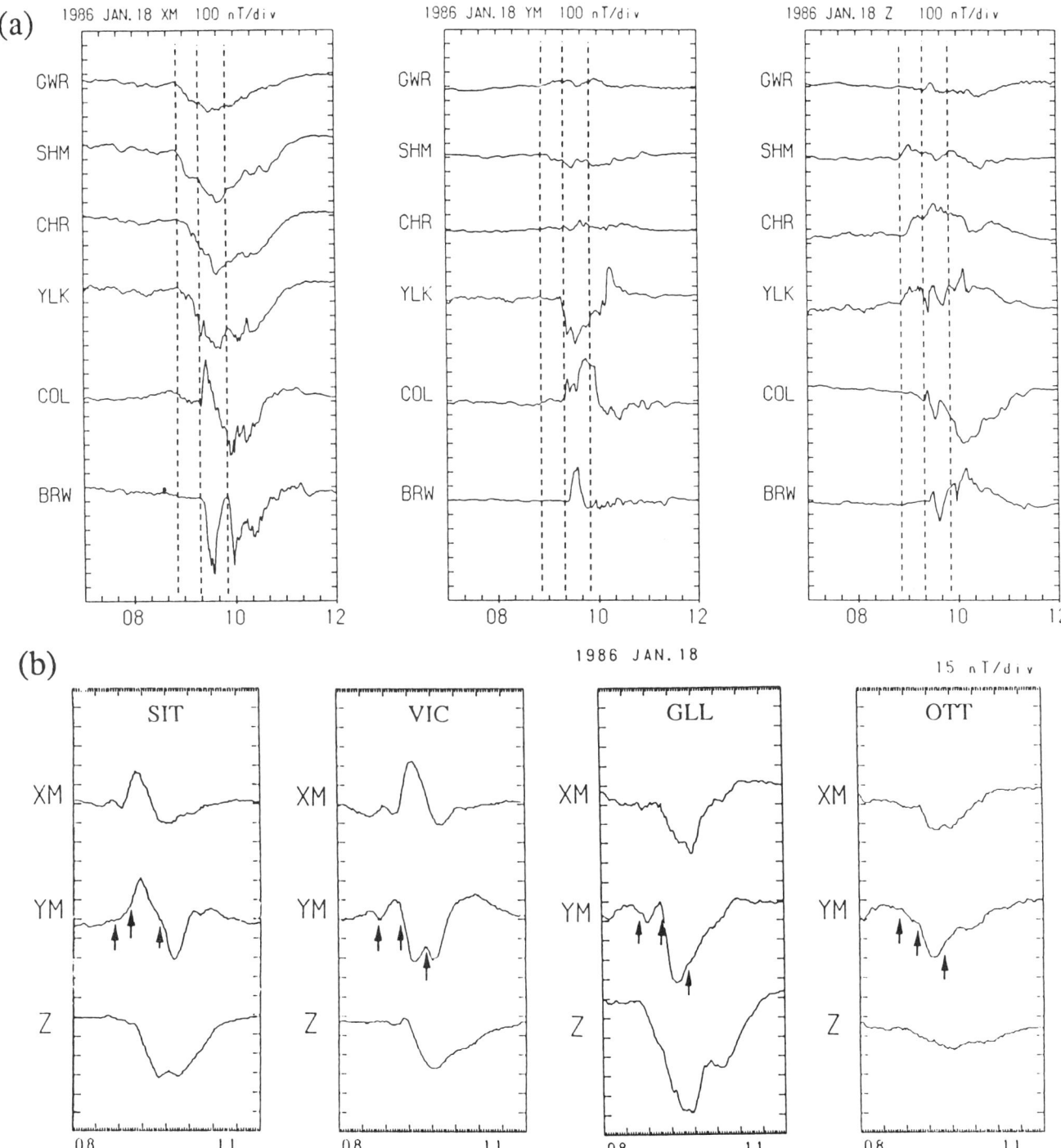

Fig. 2. (a) Magnetic records obtained from six auroral zone stations during the event reported in this study. The three panels show temporal variation of the XM (geomagnetic northward), YM (geomagnetic eastward), and Z (downward) components of the geomagnetic field, from the left to the right. The six stations are ordered from the east to the west, with the most easterly one at the top of each panel. Note that COL is located close to PFT (0.3° north and 0.3° west of PFT) where auroral data were obtained. There were three expansion onsets during the events at the times indicated by the dashed lines. (b) Magnetic records obtained from midlatitude and subauroral zone. The stations are ordered from the west to the east, with the most westerly one at the left. The three onsets of the auroral zone electrojets are indicated by the arrows in the XM trace.

Whale River (GWR), Churchill (CHR), Shamattawa (SHM), Yellowknife (YLK), College (COL), and Barrow (BRW) during the event. In the figure, the stations are ordered from east to west, with the most easterly one at the top of the figure. Note that COL is located close to PFT (0.3° north and 0.3° west of PFT), the latter being one of the auroral observatories. Three electrojet enhancements occurred at 0852 UT, 0919 UT and 0950 UT after a prolonged magnetically quiet period. The onset times are indicated by the dashed lines. Figure 2b shows magnetic records from Sitka (SIT), Victoria (VIC), Glenlea (GLL), and Ottawa (OTT), which are located at the midlatitudes and in the subauroral zone. The stations are ordered from west to east, with the most westerly one at the left. The three onsets of the auroral-zone electrojets are indicated by the arrows in the YM component traces.

The auroral-zone XM plot (Figure 2a) shows that the first westward electrojet activation started at ~0852 UT at SHM, GWR, CHR and YLK but not at the two other stations to the west. A positive deviation of the Z components at YLK, CHR and SHM indicates that the westward electrojet was located south of these stations. The YM components in the midlatitude stations deflected positively at SIT and VIC but negatively at GLL and OTT. The positive YM deviations are typical for the west of the onset meridian, while the negative ones are typical for the east of the onset meridian [Rostoker et al., 1980]. The onset region of the first event, therefore, was located between the VIC and GLL meridian. At 0919 UT, the next negative bay intensification was observed at YLK, SHM and BRW, while a positive bay started at COL. Clear negative YM deviations at YLK and VIC, and positive YM deviations at COL, BRW, and SIT were observed. This indicates that the center of the activity moved westward between the VIC meridian and the SIT meridian. A negative XM deviation was also observed at the higher latitude station such as BKL (not shown), indicating poleward expansion of the electrojet activity. The third intensification of the westward electrojet was detected from 0950 UT at BRW, COL and YLK. The negative YM deviation at SIT and VIC clearly shows that the active region of the westward electrojet moved further westward to beyond the SIT meridian. As for the eastern stations, both in the auroral zone and in the midlatitude zone, this interval corresponds to the recovery phase, and no significant change occurred associated with the electrojet activity at the western stations. To summarize, the longitude where an intense westward electrojet was observed moved successively westward in the course of the three expansions.

Auroral activity during the three expansions is shown in Figure 3. The upper two panels are the meridian plots of the aurora observed at SHM and PFT. These plots are obtained by sampling the all-sky TV data along the magnetic meridian line that passes the zenith of the stations. The auroral luminosity is presented in a negative image. The shaded area in the PFT plots shows where we have no data due to the moonlight protector. The bottom three rows of the sequential auroral images present auroral activity during the three expansions observed at PFT, which is located close to the subsatellite point of the 1984-129 satellite. The displays are transformed images from all-sky TV data, and are plotted in the ionospheric coordinates at an altitude of 110 km. They show images as viewed from the top of the ionosphere. The shaded area in the three leftmost plots represents again where we have no data due to the moonlight protector.

From 0859 UT, the discrete auroral region at SHM expanded poleward, whereas the diffuse auroral region expanded both poleward and equatorward (indicated by an arrow in the SHM meridian plot). This auroral activation was associated with the first electrojet enhancement. (The white area south of 67° between 0905 UT and the data gap in the SHM meridian plot is where the luminosity was saturated and indicates a bright, diffuse, auroral region.) We identified from the TV data that discrete auroral structures at SHM propagated eastward and turned into a diffuse or pulsating aurora (not shown). This is a typical auroral feature to the east of the onset [Nakamura, 1990]. The PFT meridian plot shows that a discrete auroral arc near the zenith intensified after 0852 UT. Poleward expanding signatures in the aurora can be seen only in the sequential plots (A) but not in the meridian plot, because the auroral activity did not develop sufficiently westward to the PFT meridian. The sequential plots show the western edge of the expansion, i.e. a westward traveling surge from 0858 UT. This surge was detected first at 0854 UT at the very eastern end of the field of view of the all-sky TV data (not shown), which cover a wider area than the sequential plots. The surge extended westward to about 150 km east of the PFT meridian at 0902 UT. Note that the aurora then shrank and had disappeared to the east by 0908 UT.

The next poleward expansion of auroras started at 0920 UT near the PFT meridian (indicated by an arrow in the PFT meridian plot). The sequential images (B) show that a westward traveling surge developed from the east of PFT, just north of the discrete auroral arc at the zenith. This aurora extended westward as well as poleward and reached the northern horizon around 0924 UT. From 0928 UT, after the active aurora had reached the northwestern horizon, an intense aurora expanded southward from the north and formed an N-S shape. Such N-S aligned auroras were also identifiable in the meridian plot until 0949 UT. These auroral signatures are typically observed at the center of the bulge [Oguti, 1981]. On the other hand, the SHM meridian plot shows discrete auroral activation until 0924 UT and diffuse aurora thereafter. Active pulsating auroral patches covered the SHM field of view from 0935 UT. These auroral features clearly show that the center of the auroral expansion had moved farther westward compared to the former auroral activation. This is consistent with the ground magnetic observation where the onset region moved westward to the west of the VIC meridian (Figure 2).

The PFT meridian plot again shows poleward expansion of a discrete aurora from ~0950 UT (indicated by an arrow in the plot). This corresponds to the third electrojet activation centered west of the PFT. In fact, the sequential images of PFT (C) show that a northwest aligned aurora was activated from the west and extended eastward and poleward from 0950 UT. This is a characteristic auroral feature to the east of the onset [Nakamura, 1990]. The SHM meridian plot shows that there was no discrete auroral activation at SHM; but bright pulsating auroral patches were observed during this period.

3.2. Satellite Observations

Figure 4 shows the locations of the foot points of the geosynchronous satellites 1982-019 (38°W), 1984-129 (156°W), together with the active region of the electrojet and aurora for the three events starting from 0852 UT, 0919 UT

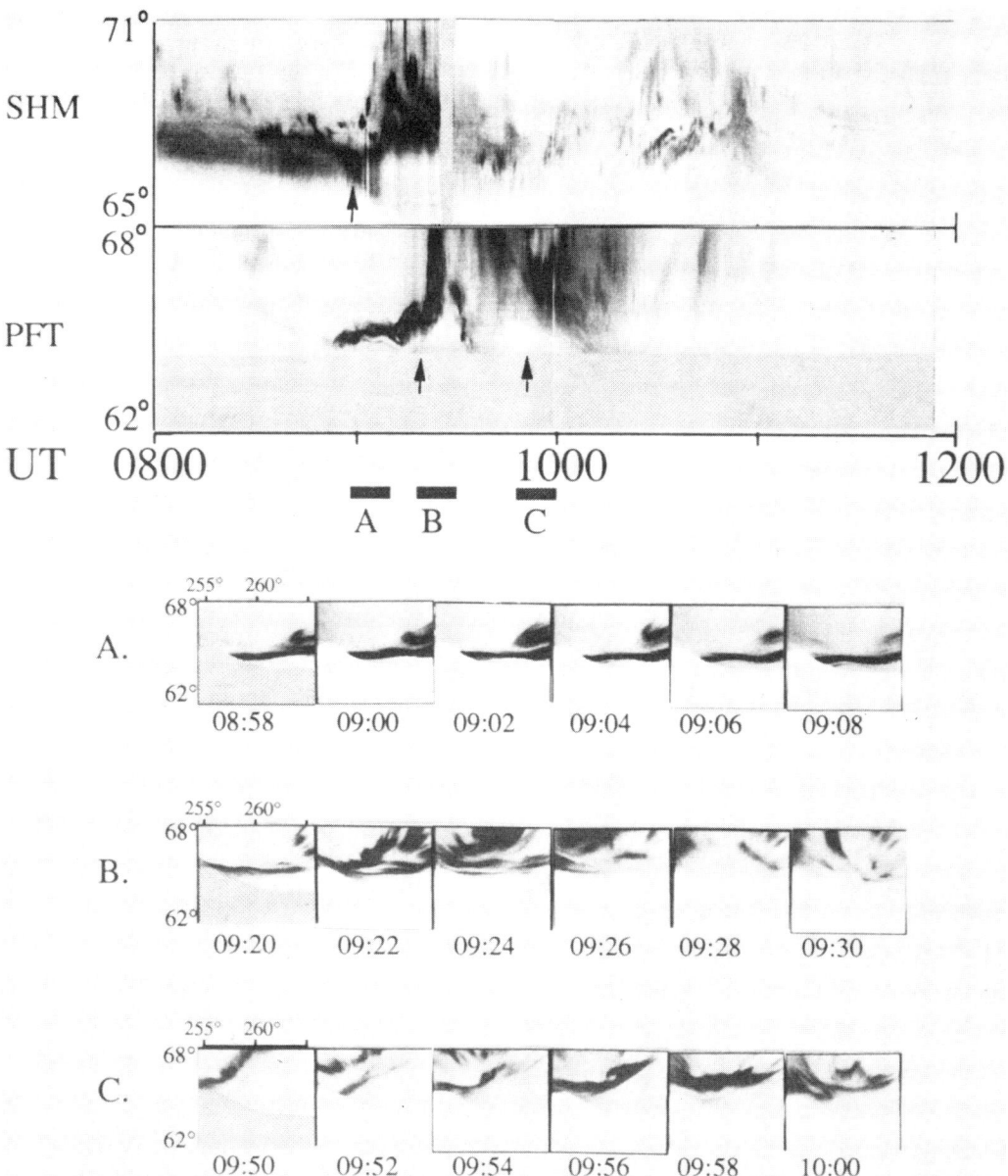

Fig. 3. Auroral activity during the three expansions. The upper two panels are the meridian plots of the aurora observed at SHM and PFT. The sequential auroral images show the auroral activity at PFT associated with the three expansions. The presented period A, B and C corresponds to the time interval given at the bottom of the PFT meridian plot. The displays are transformed images from all-sky TV data and are plotted in the ionospheric coordinates at an altitude of 110 km. The shaded area in the left three panels represents where we have no data due to the moonlight protector.

and 0950 UT. Auroral distribution at the maximum of each expansion is schematically illustrated in each panel. The hatched area represents the center of the electrojet inferred from the geomagnetic data, which were obtained at the stations indicated by the small circles. The western (eastern) end of the hatched area corresponds to the meridian of the easternmost (westernmost) station, where the positive (negative) YM was observed at the midlatitude or at the subauroral zone stations. The northern and the southern ends of the active area were determined from the Z component of the magnetogram of the auroral zone stations, where the westward electrojet was observed. The northern (southern) end of the

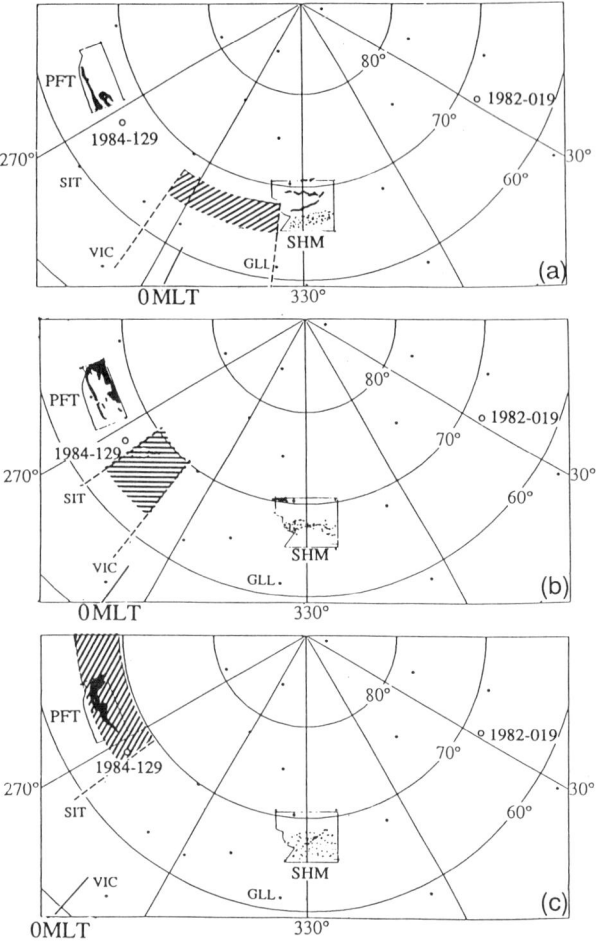

Fig. 4. Distribution of the foot points of the geosynchronous satellites 1982-019 (38°W), 1984-129 (156°W) and the active region of the electrojet and aurora for the three expansions starting from 0852 UT, 0919 UT and 0950 UT. Auroral distribution at the maximum epoch of each expansion is schematically illustrated. The hatched area represents possible center of the electrojet inferred from geomagnetic disturbances.

hatched area corresponds to the latitude of the southernmost (northernmost) station, where a positive (negative) Z deviation associated with negative XM was observed; this indicates that the station was located north (south) of the westward electrojet.

The relative location of the foot point of 1984-129 to the center of the ground disturbances changed significantly during the three events. For the first event, the center of the expansion inferred from the auroral distribution and from the YM deviation at the mid-latitude was located east of VIC and west of GLL. The foot point of the satellite was, therefore, located at least 25° west of the center, close to the western end of the expansion. During the next event, auroral expansion started in the vicinity of the western horizon of PFT. The satellite foot point was, therefore, close to the onset region.

The third expansion started from the west of PFT and the center of the electrojet was located also west of PFT. Due to the sparse distribution of observatories at the midlatitude, the foot point of 1984-129 is plotted within the hatched area. We believe that the foot point of the satellite was actually located farther east of the onset region inferred from the eastward expansion of aurora. On the other hand, the foot point of 1982-019 was located far east of the disturbed region during the whole period.

The satellite at the premidnight sector. Figure 5 shows energetic particle data from the 1984-129 satellite during the auroral events. The two upper panels are meridian plots of aurora at SHM and PFT, while the bottom two panels show the

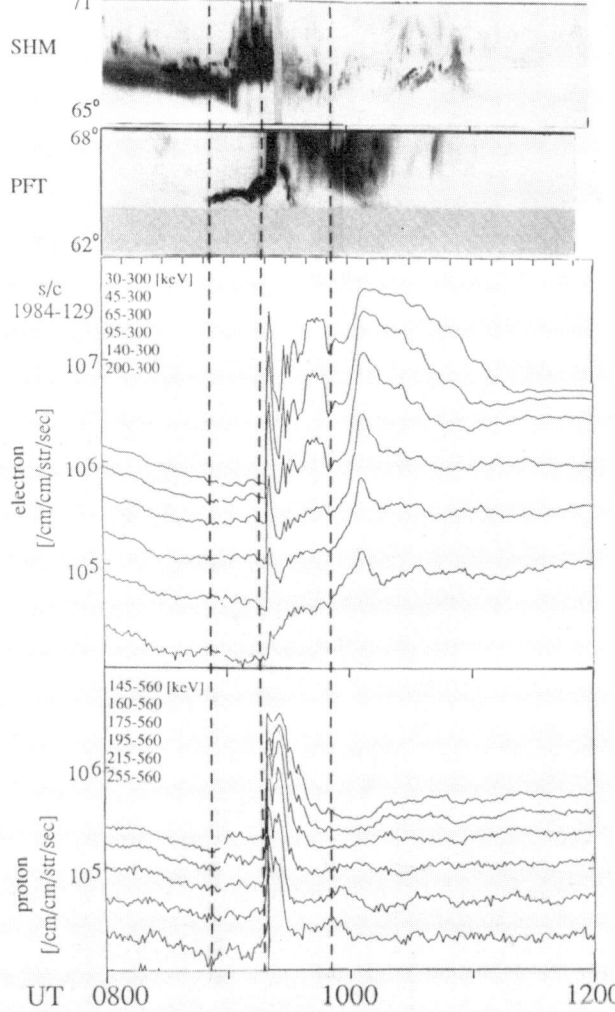

Fig. 5. Energetic particle data observed by the 1984-129 satellite during the auroral expansions. The two upper panels are meridian plots of aurora at SHM and PFT, while the bottom two panel show the electron and proton flux variation observed by 1984-129. The onsets of the three expansions are presented by the dashed lines.

electron and proton flux data observed by the satellite. The onsets of the three expansions are presented by the dashed lines. Both the electron flux of all energy channels and the proton flux of the lower three energy channels decreased until the onset of the first expansion, showing typical growth phase features [e.g., Baker, 1984]. . A small enhancement in the proton flux started at ~0854 UT in association with the first expansion. No enhancement was observed in the electron flux. In association with the second expansion, transient enhancements were observed in both electrons and protons of all energy channels, starting first at ~0919 UT in the proton flux. The most prominent enhancement was observed at 0920 UT in all channels, followed by several dispersionless fluctuations within the time scale of a few minutes in both particle species. The flux level of protons had decayed with slight energy dispersion to a quiet level by 0948 UT. In association with the last expansion, the electron flux of all energy channels increased from ~0950 UT rather gradually compared to the sharp onset of the second expansion. This enhancement manifested energy dispersion when the flux reached its maximum and also when the flux level reached its recovery level.

The satellite at the dawn sector. Electron differential flux data from the 1982-019 satellite during the auroral events are shown in Figure 6. Clear flux enhancement started at the time between 0907 UT and 0916 UT with energy dispersion, beginning with the 140-200 keV channels and progressing to the lower energy channels. This enhancement started between the onset of the first and the second expansion, and hence, was in association with the first expansion. It should be noted that electron enhancement was observed only by 1982-019 but not by 1984-129, suggesting a localized region of the injection. From ~10 min after the second expansion onset, the flux increased again with energy dispersion. The time when the flux reached its maximum level was ~0932 UT in the 200-300 keV channel, and ~1000 UT in the 45-65 keV channel. Thus, the energy dispersion was larger in this case than for the former event. After the third expansion onset, the flux increased again with even larger energy dispersion and larger time delay relative to the third expansion onset. These three individual enhancements in the flux profiles would not be caused by particle drift echoes [e.g. Chanteur et al., 1977; Belian et al., 1978], because the complete drift periods of the electrons around their drift shells at 6.6 Re are much longer than that expected to produce the observed dispersion: the drift period is ~90 min for 45 keV particles and ~14 min for 300 keV particles. Thus, the observed enhancements represent three separate injection events.

4. Discussion

The substorm on January 18, 1986 developed in the form of three successive intensifications of auroral activity as well as similar electrojet enhancements. The onset region moved westward in each successive event. This is consistent with the stepwise expansions of the westward electrojet reported by Wiens and Rostoker [1975]. While the substorm activity propagated significantly westward, the onset region from which the aurora expanded hardly moved poleward (less than 1°) as shown in the sequential plots in Figure 3. It is, therefore, expected that the substorm onset region in the magnetosphere also propagated mainly azimuthally. The longitudinal difference among the three events would have caused significantly different signatures in the energetic particles observed by 1984-129 in the premidnight sector.

The satellite was located nearest to the onset local time during the second expansion. Transient enhancements in both the electron and the proton flux were observed. Recent CCE observations reported such transient proton flux variations in association with rapid changes in the local magnetic field that were consistent with an in-situ observation of the disruption of the cross-tail current [Lui et al., 1988; Lopez et al., 1989]. Energetic particles would be accelerated at these current disruption regions due to the induced turbulent electric field [Lui et al., 1988]. Lopez et al. [1989] showed from east-west gradient anisotropy data that during such events, an enhanced proton flux was observed both tailward and earthward of the satellite and suggested that the acceleration was due to both dawn-to-dusk and dusk-to-dawn electric fields. We have inspected the east-west gradient anisotropy of ions at 1984-129 during the period of the expansion. It indicated that the ions came from both earthward and tailward between 0900 UT and 0940 UT; this period coincides with the second expansion period. The particle signatures during the second expansion are quite similar to that observed by Lopez et al. [1989]. The 1984-129 could, therefore, be located at the current disruption region.

The onset of the dipersionless electron and proton enhancement associated with the second event was about 1 minute later than the onset of the westward electrojet. If we assume that this delay was caused by the finite expansion speed of the disturbance from the initial onset region, we can estimate the onset location. Because the satellite foot point was located at about the same latitude of the auroral arc from which the expansion started, the discrepancy could be caused mainly by an azimuthal expansion of the injection front [Arnoldy and Moore, 1983]. The western end of the injection region at the onset is then estimated to be about 3° east of the

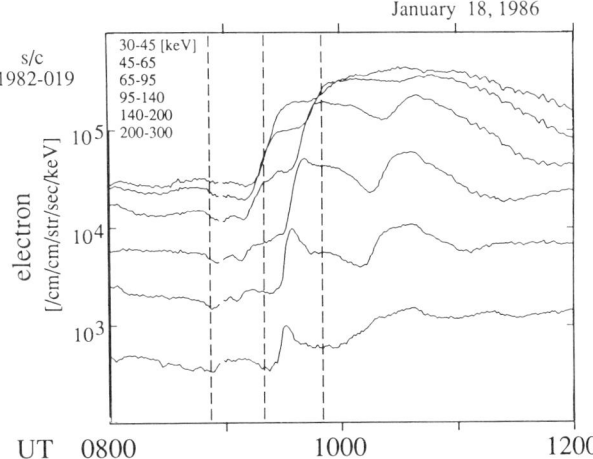

Fig. 6. Electron differential flux data observed by the 1982-019 satellite during the auroral events. The onsets of the three expansions are presented by the dashed lines.

foot point; the expansion speed was assumed to be the propagation speed of the surge at PFT, which was about 3°/min. This estimated injection front is within the onset region inferred from the midlatitude magnetogram (the hatched area in Figure 4a). If we assume that an injection front expands also eastward with about the same speed of an westward expansion, the initial longitudinal width of the injection region can be estimated to be less than 28°.

The electron flux observed at 1984-129 did not increase during the first expansion when the foot-point of the satellite was located west of the estimated onset region. It is, however, confirmed from the electron enhancement observed at 1982-019 that the electrons were injected during this event. The injected electrons would drift eastward away from the onset region [Baker, 1984], and would therefore not be observed at 1984-129 when the onset region was east of its foot point. Note that the first expansion started from nearly the same latitude of the second expansion, when clear dipersionless injection was observed. This indicates that particle injection for both events occurred close to the same latitude; at the same radial distance in the magnetosphere. Moreover, the tailward and earthward propagation of the injection front [Moore et al.,1981; Lopez et al.,1989; Kistler et al., 1990] would make the small latitudinal difference between the two events negligible. We, therefore, would expect to observe injection signatures also during the first event if the injection region had expanded azimuthally to the satellite meridian. Hence, the lack of the electron injection during the first event strongly suggests that the injection region did not expand westward until the longitude of the satellite but was confined to the eastern region of the satellite. It should be noted that although the western end of the surge reached the field of view of PFT, the electron injection was not observed. This suggests that the injection region was more longitudinally confined than the expansion aurora. The azimuthal width of the expansion for the event was estimated to be about 60° from the auroral data. So the injection region was less than about 40°. Proton, on the other hand, drifted westward and could, therefore, encounter the satellite meridian. The encounter of the proton of the lowest energy channel (145-560 keV) was 2 minutes later than the electrojet onset. If we consider that this delay was due to the drift motion of a 145 keV proton, which has a drift speed of ~11.6°/min, the western end of the onset region is estimated to be about 6° east of VIC meridian. This is consistent with the onset location inferred from the midlatitude magnetogram (the hatched are in Figure 4a)).

During the third expansion, enhancements only in the electrons should be expected because the 1984-129 satellite was just east of the injection region. A characteristic signature for the event was that the electrons enhanced rather gradually compared to the injection during the second event. We consider that this could be caused by the integrated effect of the eastward drifted electrons from the structured source region. Proton enhancement was not observed suggesting that the satellite was located close to the east of the injection region. Since we have no information from any region further west, we cannot determine the onset region or discuss the longitudinal extent of the onset region for this event.

As for the electron enhancements observed by the 1982-019 satellite at the dawn sector, their temporal and spatial relationships of the three auroral expansions have been examined by calculating the drift trajectory of the electrons. The results show that the energy dispersion and the delay time between the electron events and the expansions could be interpreted as due to the magnetic drift of the injected electrons from the local time sector of the expansion with about 6 minutes ambiguity. This is consistent with previous multi-spacecraft observations [e.g. Belian et al., 1978], which have shown energy dispersion observed at different local time sectors to be associated with injection observed at local midnight.

As noted before, it has been suggested that particles suddenly appear behind an injection boundary extending from the dusk to the dawn sector [McIlwain, 1974; Mauk and Meng, 1983; Konradi 1975]. This model was invoked in order to explain energy dispersion observed from a single satellite by assuming a global convection electric field. Moreover, Arnoldy and Moore [1983] proposed that a localized substorm onset, i.e. a current disruption region, expands continuously in the azimuthal direction from a relatively narrow sector and develops an injection boundary. Such a spatially distributed boundary, however, has not been confirmed from observations within the source region, i.e., right in the dispersionless injection region. Our present observations showed that the source region of the injected particles, and therefore the acceleration region of the particles, is confined in longitude not only at the onset but also at the maximum time within the localized region of the expansion aurora. The longitudinal range of an injection region was about less than 28° at the onset (during the second event) and was about 40° even at the maximum (during the first event). Arnoldy and Moore [1983] concluded from geosynchronous magnetic field observations that the onset region was about 3h (45°) wide, which is about the maximum value obtained in this study.

Lopez et al. [1990] have shown the radial and local time distribution of the injection occurrence, which was extended in a wide local time range. Their results, however, would not necessarily mean that an injection region during an individual expansion is azimuthally extended. Kistler [1990] showed similar ion spectra observed at 7-9 Re and 10-17 Re during the post-injection period (after 40 min from the onset) and suggested a global mechanism to explain the injection. We have given an example when a substorm activity expanded in the course of three discrete intensifications with a finite longitudinal expansion. In accordance with the multiple intensifications, multiple localized injections of the energetic particles took place. Hence, on a large temporal scale, the onset regions may become widely distributed, and an extended injection boundary may be identified in accordance with the previous studies. The important point here is that each dispersionless injection region is confined within the conjugate area of the relevant expansion aurora. The observation, therefore, supports an in situ acceleration mechanism of the particles at the onset region.

5. Conclusion

A moderate substorm with three successive expansions commenced at 0852 UT on January 18, 1986 after a magnetically quiet period. The temporal as well as spatial evolution of the substorm was inferred from a multi-point network of observations of aurora and the geomagnetic field. Geosynchronous satellites 1984-129 located in the premidnight sector and 1982-019 at the dawn sector observed dynamic change in the energetic particle flux. The three successive onsets of the electrojet activity were accompanied by distinct auroral expansions, which started from a localized region and expanded only within a finite longitudinal range. In the course of the three expansions, the onset region moved westward. Accordingly, the relative location of the estimated foot point of 1984-129 changed from the west to the east of the relevant onset region. Drastic changes in the energetic particle signatures were observed by 1984-129. The observation could be interpreted that the source region of the energetic particles, i.e. where dispersionless and transient injections are observed, are confined within the local time sector of the expansion aurora. It is estimated that during this event, an injection associated with an expansion started from a localized region with less than 28° in longitudinal width and was still confined within a region of about 40° even at the maximum point of the expansion. The energetic electrons injected at the conjugate region within the expansion aurora drifted eastward and were observed by the 1982-019 satellite at the dawn sector. A fundamental feature in the injection process is the longitudinal confinement to the auroral expansion region. An azimuthally extended source region, such as the injection boundary, would be produced by multiple localized injections rather than by a continuous expansion of a single injection front.

Acknowledgements. We are greatly indebted to those members of the Global Dynamics Auroral Campaign Group and the Geophysics Research Laboratory, University of Tokyo who were involved in constructing the instruments and collecting the aurora and magnetic data. We would like to thank K. Hayashi, and T. Iijima for their valuable comments. The work of R. Nakamura was supported financially by the fellowships from the Japan Society for the Promotion of Science for Japanese Junior Scientists. This work was also supported by NASA and by the U.S. Department of Energy.

References

Akasofu, S.-I., The development of auroral substorms, *Planet. Space Sci., 12,* 273, 1964.

Akasofu, S.-I., S. DeForest and C. McIlwain, Auroral displays near the 'foot' of the field line of the ATS-5 satellite, *Planet. Space Sci., 22,* 25, 1974.

Arnoldy, R.L., and T.E. Moore, Longitudinal structure of substorm injections at synchronous orbit, *J. Geophys. Res., 88,* 6213, 1983.

Baker, D.N., Particle and field signature of substorms in the near magnetotail, in *Magnetic reconnection in space and laboratory plasmas,* E.N. Hones, ed.,1984.

Belian, R.D., D.N. Baker, P.R. Higbie, and E.W. Hones, Jr., High-resolution energetic particle measurements at 6.6 Re 2. High-energy proton drift echoes, *J. Geophys. Res., 83,* 4857, 1978.

Chanteur, G., R. Gendrin, and S. Perraut, Experimental study of high-energy electron drift echoes observed on board ATS 5, *J. Geophys. Res., 82,* 5231, 1977.

DeForest, S.E. and C.E. McIlwain, Plasma clouds in the magnetosphere, *J. Geophys. Res., 76,* 3587, 1971.

Erickson, K.N., R.L. Swanson, R.J. Walker, and J.R. Winckler, A study of magnetosphere dynamics during auroral electrojet events by observations of energetic electron intensity changes at synchronous orbit *J. Geophys. Res., 84,* 931, 1979.

Greenspan, M.E., D.J. Williams, B.H. Mauk, and C.-I. Meng, Ion and electron energy dispersion features detected by ISEE1, *J. Geophys. Res., 90,* 4079, 1985.

Higbie, P.R., R.D. Belian, and D.N. Baker, High-resolution particle measurements at 6.6 Re, 1. electron micropulsations, *J. Geophys. Res., 83,* 2177, 1978.

IAGA Division I Working Group 1, International geomagnetic reference field revision 1985, *J. Geomag. Geoelectr., 37,* 1157, 1985.

Kistler, M., E. Moebius, B. Klecker, G. Gloeckler, F.M. Ivavich, and D.C. Hamilton, Spatial variations in the suprathermal ion distributions during substorms in the plasma sheet, *J. Geophys. Res., 95,* 18871, 1990.

Konradi, A.C., L. Semar, and T.A. Fritz, Substorm-injected protons and electrons and the injection boundary model, *J. Geophys. Res., 80,* 543, 1975.

Lopez, R.E., A.T.Y. Lui, D.G. Sibeck, K. Takahashi, R.W. McEntire, L.J. Zanetti, and S.M. Krimigis, On the relationship between the energetic particle flux morphology and the change in the magnetic field magnitude during substorms, *J. Geophys. Res., 94,* 17105, 1989.

Lopez, R.E., D.G. Sibeck, R.W. McEntire, and S.M. Krimigis, The energetic ion substorm injection boundary, *J. Geophys. Res., 95,* 109, 1990.

Lui, A.T.Y, R.E. Lopez, S.M. Krimigis, R.W. McEntire, L.J. Zanetti, and T.A. Potemra A case study of magnetotail current sheet disruption and diversion, *J. Geomag. Geoelectr., 15,* 721, 1988.

Mauk, B.H., and C.-I. Meng, Characterization of geostationary particle signatures based on the "injection boundary" model, *J. Geophys. Res., 86,* 3055, 1983.

McIlwain, C.E., Substorm injection boundaries, in *Magnetospheric Physics,* B.M. McCormac, ed., D. Reidel, Hingham, Mass., 143, 1974.

Mende, S.B., and E.G. Shelley, Coordinated ATS 5 electron flux and simultaneous auroral observations, *J. Geophys. Res., 81,* 97, 1976.

Moore, T.E., R.L., Arnoldy, J. Feynman, and D.A., Hardy, Propagating substorm injection fronts *J. Geophys. Res., 86,* 6713, 1981.

Nakamura, R., Aurora dynamics and particle injection associated with magnetospheric substorms, *Ph. D. Thesis, Univ. of Tokyo,* 1990.

Oguti, T., TV observations of auroral arcs, in A. G. U., Geophys. Monograph Ser. 25, 31, 1981.

Oguti., T., T. Kitamura, and T. Watanabe, Global aurora dynamics campaign, 1985-1986, *J. Geomag. Geoelectr., 40,* 485, 1988

Rostoker, G., S.-I. Akasofu, J. Foster, R.A. Greenwald, Y. Kamide, K. Kawasaki, A.T.Y. Lui, R.L. McPherron, and C.T. Russell, Magnetospheric substorms: Definition and signatures, *J. Geophys. Res., 85,* 1663, 1980.

Tsyganenko, N.A., Global quantitative models of the geomagnetic field in the cislunar magnetosphere for different disturbance levels, *Planet. Space Sci., 35,* 1347, 1987.

Wiens, R.G., and G. Rostoker, Characteristics of the development of the westward electrojet during the expansive phase of magnetospheric substorms, *J. Geophys. Res., 80,* 2109, 1975.

Yamamoto, T. Temporal and spatial characteristics of pulsating auroras and possible mechanisms, *Ph. D. Thesis, Univ. of Tokyo,* 1984.

Auroral Substorms

WALTER J. HEIKKILA

University of Texas at Dallas FO22, Richardson, TX 75083-0688

Many models of the substorm process assume a uniform current sheet before the auroral breakup and the onset of the expansion phase; in fact, a lower energy state for the cross-tail current is a set of filamentary currents. We hypothesize that such filaments are connected to auroral arcs during the growth phase. We must have an arc for it to break up, an essential part of the substorm. This means that we should look at instabilities of current filaments in the magnetotail. We have proposed that the appropriate instability is a simple meander of the current filament in the equatorial plane. An outward meander will be caused by the current carriers, undergoing curvature drift, becoming demagnetized. We take the inductive electric field as $\mathbf{E}^{IND} = -\partial A/\partial t$, using the Coulomb or transverse gauge. This inductive electric field will in general have a component parallel to the magnetic field. We take the response of the plasma to be reflected in a scalar potential, $\mathbf{E}^{ES} = -\nabla\phi$; that response must be such as to diminish the actual (or net) $E_\parallel$. Part of the response is the formation of field-aligned currents producing the well-known substorm current diversion. At the same time the plasma will enhance the transverse component of the induction electric field. Other work has indicated that a substorm reconnection X-line will form. The enhanced induction electric field near the emerging X-line will cause a discharge, again to decrease $E_\parallel$. After subsequent betatron acceleration even zero energy particles can be energized to MeV energies in a matter of seconds in a two-step process. A plasmoid will be created which will move in the direction of least magnetic pressure, namely tailward.

1. BACKGROUND

Magnetospheric substorms are periods of enhanced geomagnetic and auroral activity lasting one to a few hours during which there is a major departure from equilibrium conditions in the magnetotail. Satellite data in the magnetotail have suggested that during a substorm part of the plasma sheet is isolated from the earth by magnetic merging to form a *"plasmoid"*, a body of plasma with closed magnetic loops, a magnetic island. This plasmoid then travels antisunward to leave the magnetotail. Recently this picture has been dramatically confirmed by observations with the ISEE-3 spacecraft in the magnetotail recording plasmoids passing that location in a clearly delayed response to substorms [*Baker et al.*, 1987]. It now appears that plasmoid release is the basic process whereby the magnetosphere gives up its excess stored energy, much as with solar flares and comets [*Harrison et al.*, 1990]. The substorm phenomenon seen at earth is an example of a fundamental cosmic plasma process.

It is appropriate to point out right away that the invalidity of steady-state reconnection in the magnetotail implied by the results of *Heikkila* [1988] is not relevant to this discussion; steady-state theory assumes that curl $\mathbf{E}=0$, while any time-dependent process must have a finite value of the curl appropriate to local changes in the magnetic field. An inductive electric field given by Faraday's law is required so that energy stored in the magnetic field is accessible (see the next section for more details). We need to define *merging* as time-dependent reconnection in three dimensions, with a finite value of the curl [*Heikkila*, 1990]. Some theories using MHD do exist [e.g *Birn and Hesse*, 1991], but nevertheless the usual way of addressing the physics still seems to be based on Dungey (Petschek) style reconnection. There are significant qualitative changes between the two processes (reconnection and merging) such as the sense of plasma convection [see *Pellinen and Heikkila*, 1984]. Another important point is that substorm expansion phase begins deep inside the plasma sheet, perhaps at a distance of 10 – 20 R_E, while any steady-state X-line must be much farther away. We take the view that a substorm neutral line of limited extent within the plasma sheet is a reality; that feature is highly localized in space (at least at first) and the associated electric field topology must be consistent with an appropriate value of the curl.

The goal of the present paper is to review our earlier work [see especially the reviews by *Heikkila et al.* [1979] and *Pellinen and Heikkila* [1984], and to offer some modifications and updates based on new observations and theoretical work.

2. THE CREATION OF AURORAL ARCS

We can divide auroral phenomena into two main categories: quiet-time auroral arcs and substorm type active auroral forms (there is a third important type as well, the diffuse aurora, but not of direct interest here). There is considerable energy dissipation involved, and in both cases the energies of auroral particles have grown by large factors from their values in the solar wind. In the case of substorms some particles achieve very high energies (Mev), and in the case of electrons even relativistic energies.

The energy of charged particles can be increased only by means of electric fields, either electrostatic or inductive, d.c. or a.c., since the Lorentz force due to the magnetic field is normal to the velocity vector. This energy can be redistributed among the particles by collisions or wave-particle interactions; however, in the magnetosphere the former mechanism is not important. The specific mechanisms that dominate the energization of particles may be different in different phases of auroral development. In any phase, Poynting's theorem (with the usual definitions of quantities) can be used to elucidate the exchange of energy between the electromagnetic field and the kinetic energy of particles [*Heikkila and Pellinen*, 1977]:

$$\int_{vol} \mathbf{E} \cdot \mathbf{J} d\tau = - \oint_{surf} \mathbf{E} \times \frac{\mathbf{B}}{\mu_0} \cdot d\mathbf{S}$$

$$- \int_{vol} (\varepsilon_0 \mathbf{E} \cdot \frac{\partial \mathbf{E}}{\partial t} + \frac{\mathbf{B}}{\mu_0} \cdot \frac{\partial \mathbf{B}}{\partial t}) d\tau$$

All the energy gained (or lost) by particles is included in the left hand volume integral, given by the energy gained by the current carriers in the electric field. The possible sources (or sinks) for this energy are described by the terms on the right hand side (the surface element vector $d\mathbf{S}$ is positive outwards). Steady state reconnection theories consider only the first term, which (as will be shown below) corresponds to a flow of energy from an external source. Any internal magnetospheric source must be described by a volume integral, and both those integrands vanish unless the field is time-dependent. These two terms specify the conversion of stored electric and magnetic energies respectively.

Neglecting the last two righthand terms for steady-state effects, we have a situation as shown in Figure 1 (ignore the induction electric field for the moment). Since the cross-tail current (shown by the solid lines) is in the same direction as the electrostatic field (the dot-dash lines) the integrand on the left is positive, and we do have a net gain in energy by the plasma particles. In this case we can readily derive the following relation:

$$\int_{vol} \mathbf{E} \cdot \mathbf{J} d\tau \approx - \oint_{surf} \phi \mathbf{J} \cdot d\mathbf{S} \quad (growth\ phase)$$

where ϕ is the scalar potential. This equation shows that the energy supply to uphold the $\mathbf{E} \cdot \mathbf{J}$ dissipation requires that the volume considered must be connected to an external source by currents $\mathbf{J}$ that enter and leave its surface at different potentials. We should note that this dissipation need not be Joule (heat) dissipation but can be direct energization of charged particles, for example through curvature and gradient drifts in the presence of the cross-tail electrostatic field.

The current fed into the plasma sheet must be connected to a generator somewhere outside, i.e. a region in which $\mathbf{E} \cdot \mathbf{J} < 0$, so that the particles lose energy to create the electric field by a polarization current. Such a generator, driven by the solar wind, may be located on the sunward side of the magnetosphere, probably in the low latitude boundary layer, but in the present context its exact location is irrelevant.

From the observed structure of the magnetic field it is well established that there is a dawn-to-dusk cross-tail current of the order of 30 mA/m. It has also been experimentally verified that there is a dawn-to-dusk cross-tail potential difference across the plasma sheet supporting earthward convection. Although the strength of the field varies from very low values to 2 mV/m, it has an average value of the order of 0.2 mV/m corresponding to a cross-tail potential of 50 kV (the extremes correspond to 5 and 500 kV). The average values just mentioned should correspond to conditions during a growth phase of a substorm. The currents and potentials (30mA/m and 50kV) imply that a power of 1.5 kW/m; a total of 10^{12} W is supplied to the plasma sheet, taking the length of the plasma sheet as $100 R_E$. As an energy supply for the growth phase aurora this is more than sufficient: hence the possibility of some instability.

Let us now turn to look at the volume integrals on the right hand side of the first equation. As is well known, the energy stored in the electric field is small. It is apparent that the only significant internal source is that described by the last term, which specifies the rate of conversion of stored magnetic energy:

$$\int_{vol} \mathbf{E} \cdot \mathbf{J} d\tau \approx - \int_{vol} (\frac{\mathbf{B}}{\mu_0} \cdot \frac{\partial \mathbf{B}}{\partial t}) d\tau = -\frac{d}{dt} \int_{vol} \frac{B^2}{2\mu_0} d\tau$$

$(expansion\ phase)$

It is this equation we must use if we wish to describe the conversion of stored magnetic energy as appropriate during the expansion phase. Since the integrand on the right-hand side involves the factor $\partial \mathbf{B}/\partial t$ it is clear that there is an inductive electric field given by Faraday's law. Thus the induction electric field essentially acts as an intermediary in the transfer of energy. Again, it should be noted that it is not included in steady-state reconnection models.

During the growth phase of a substorm the magnetic energy gradually increases, so any changes in $\mathbf{B}$ also require an input of energy. This additional energy must also be supplied by the external generator. Actually, a uniform current sheet is unstable to the pinch effect; *Zwingman* [1983] has shown that filamentation of the cross-tail current lowers the total energy. *Heikkila* [1981] has proposed that an auroral arc maps to such a filament. This means that the cross-tail current must increase during the growth phase, resulting in an inductive electric field directed from dusk-to-dawn in the equatorial plane in agreement with Lenz's law, opposing the increase in the current density. If at the same time there is a tendency towards increased filamentation we would have a situation as shown in Figure 1. The return part of the induction electric field loop must be from dawn to dusk; such an orientation would convect plasma into the arc structure, as has indeed been found by *Doolittle et al.* [1990]. *Opgenoorth* (private communication) has also shown a film showing the same type of motion.

It is now acknowledged that some additional energy is supplied by an electric field parallel to the magnetic field lines $E_\parallel$ above auroral arcs. In our view, this latter effect is secondary in nature as far as the magnetotail is concerned, and is associated with the ionospheric-magnetospheric interaction.

We should also note that the last two terms on the right hand side include the energy residing in waves, but their contribution to the energy density is negligible.

3. AURORAS DURING SUBSTORMS

Thus during quiet times, and during the main part of the growth phase of a substorm, the electric field is essentially electrostatic; however, an inductive component has to exist also for the buildup of magnetic energy. The strengthening cross-tail current is trying to go towards a lower energy state, becoming filamentary in nature. Close to the onset time of the expansion phase this inductive field plays a significant role in the triggering and expansion mechanism, and must not be neglected.

We note that although the term "induced" has been often used in the literature, nevertheless there has been little acknowledgement of the fact that the induction field has a curl (locally), and an electromotance $\oint \mathbf{E} \cdot \mathbf{dl}$ around any closed contour (usually called an electromotive force). Also, the term $\mathbf{V} \times \mathbf{B}$ is often called an induced electric field; it is just a frame transformation. The curl must play a vital part in the physics: we need a finite curl $\mathbf{E}$ to draw upon stored magnetic energy.

A magnetospheric substorm represents a rapid release of large amounts of energy, which shows up in various forms. The cross-tail current is diverted to the polar ionosphere, accompanied by large increases in the auroral electrojets. Auroral particle fluxes in the 1-50 keV range are enhanced, especially in the ring current region; and bursts of energetic particles (of both signs) occur up to at least 1 MeV. All these phenomena are highly localized at the time of substorm onset, although they do grow in extent with time so as to eventually encompass most of the magnetosphere.

The electric field in the magnetotail can be quite large, up to many mV/m (i.e. more than an order of magnitude larger than the average strength of the cross-tail electrostatic field estimated above). But even this in a simplistic explanation is not enough to explain the very high energies that a few particle can achieve, with the scale size of the inital disturbance; 10 mV/m over $5R_E$ yields only 300,000 eV. Yet particle energies reach at least several MeV. As we shall see, the answer lies with the curl of the electric field.

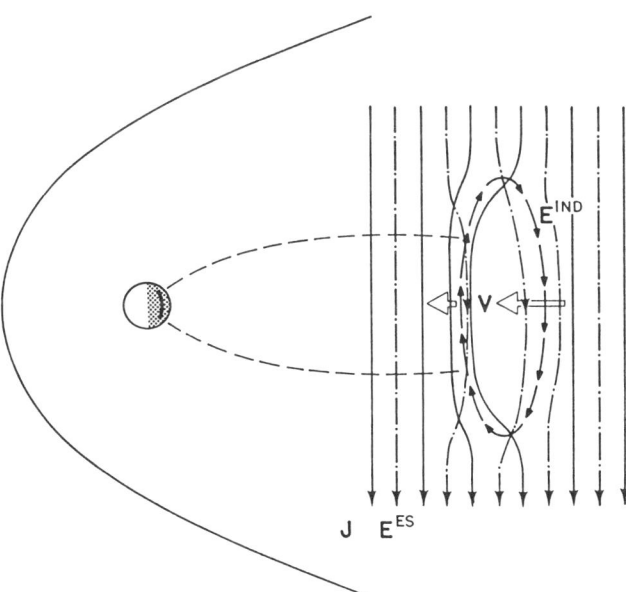

Figure 1. The cross-tail current (shown here by solid lines) is most likely in the form of current filaments, since filamentation lowers the total energy of the system. The cross-tail electrostatic field (dot-dash lines) energizes the current carriers since $\mathbf{E} \cdot \mathbf{J} > 0$. We propose that auroral arcs are a footprint of these filaments. Changes in the local current density will be associated with an induction electric field (bold-face arrows) whose direction we can get from Lenz's law.

Another feature of the substorm is the deflation of the magnetotail in $10^3 - 10^4$s [*Russell and McPherron*, 1973] providing an obvious source of energy, namely the magnetic energy represented by the last righthand term in Poynting's theorem.

Sometime ago we have proposed a simple model that can explain most of these features, a model which is based directly on fundamental physical principles. We suggested that the long-sought-for instability that sets off a substorm is simply a meander of a cross-tail current filament (Figure 2). We can contemplate either an outward or an inward meander in the equatorial plane; a meander toward the lobes of the magnetotail faces increased magnetic pressure and is not likely. A tailward meander, characterized by a southward orientation of its magnetic moment indicated by μ_- in Figure 3, will be produced when the current carriers become sufficiently energized to become demagnetized; they would tend to flow in the direction of least magnetic pressure, which is tailward *Schindler* [1974]. We can deduce the characteristics of the inductive electric field by Lenz's law, opposing the change in current everywhere, as indicated in Figure 3.

It is worth repeating that a tailward meander of a current filament is the key to explain substorms; it will reproduce complex patterns in a variety of observables demonstrated by data acquired over the years, as will now be shown. However, as a prelude we must consider the response of a plasma to an inductive electric field. Some of this is primitive, but apparently not well appreciated.

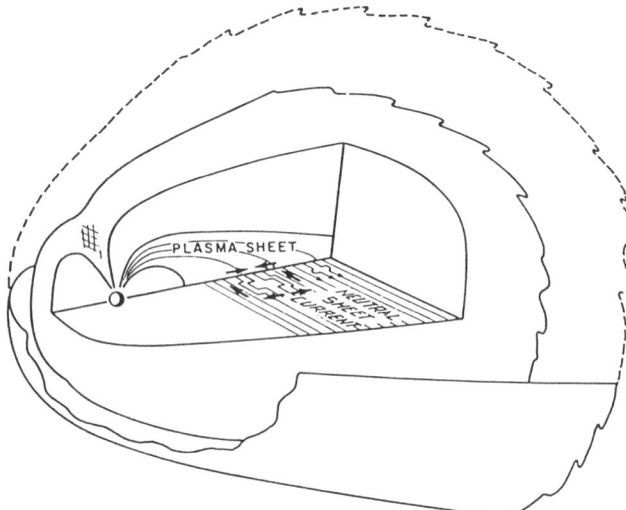

Figure 2. We propose that the instability that sets off a substorm is a localized displacement of a current filament, a current meander. An outward meander is possible if the current carriers become non-magnetized due to energization by the cross-tail electrostatic field.

4. Theoretical Digression

Since our analysis given below makes extensive use of potential functions it may prove helpful to make some preliminary remarks. It has been persistently asserted that the vector potential **A** has no physical meaning. "The outcome ...is that potential descriptions, by ϕ and **A**, of electromagnetic fields are at least as meaningful as descriptions by **E** and **B**. Whereas **E**, **B** describes a field in terms of forces the field can exert on charged matter, ϕ and **A** describe the same field in terms of energies and momenta the entire field makes available to the matter" [*Konopinski*, 1981, p159].

Another comment concerns the use of the Coulomb (or transverse) gauge. This gauge is fully causal [*Jackson*, 1975, p223; *Cragin and Heikkila*, 1981] in spite of some claims to the contrary. Here the inductive electric field $\mathbf{E}^{IND}$ is given by $-\partial \mathbf{A}/\partial t$ with the choice of div **A**=0. It describes the effect of changing currents, quite apart from plasma effects due to charge separation. The latter are entirely due to the electrostatic potential $\mathbf{E}^{ES}$ given by $-\nabla\phi$. The advantage of using this gauge is that it recognizes that the electric field, unlike the magnetic field, has two types of sources. Use of the Coulomb gauge achieves "a more complete separation between 'longitudinal' and 'transverse' fields (for that Lorentz system) than any other choice of gauge would allow" [*Morse and Feshbach*, 1953, p.211]. This gauge, however, is not invariant under Lorentz transformation.

A well-known example of the profitable use of the vector potential is for an analysis of a long solenoid. A current flowing through the solenoid produces only a weak magnetic field **B** outside: **A** is large, but curl **A** is small. However, if we suddenly try to cut-off the current, a large inductive electric field appears given by $\mathbf{E}^{IND} = -\partial \mathbf{A}/\partial t$ that extends for a considerable distance outside the solenoid. The vector direction at any point in space is completely due to the vectorial change in the current, integrated over the source region. The reality of this effect can readily be demonstrated.

In the usual MHD treatment **E** and **J** are eliminated from the equations, being recovered by

$$\mathbf{J} = \nabla \times \mathbf{B}/\mu_0; \quad \mathbf{E} = -\mathbf{v} \times \mathbf{B} + \mathbf{J}/\sigma$$

If σ is high, as appropriate to most plasmas, then

$$\mathbf{E} = -\mathbf{v} \times \mathbf{B}$$

Thus, the distinction between the induction electric field $\mathbf{E}^{IND} = -\partial \mathbf{A}/\partial t$ and the plasma response $\mathbf{E}^{ES} = -\nabla\phi$, given above, is completely lost. In particular, in MHD the electric field is transverse to **B**, and $E_\parallel \sim 0$, in complete contrast to $\mathbf{E}^{IND} = -\partial\mathbf{A}/\partial t$. In fact $\mathbf{E} = -\mathbf{v} \times \mathbf{B}$ is sometimes referred to as an induced effect; it is just a frame transformation. The actual electric field $\mathbf{E} = \mathbf{E}^{IND} + \mathbf{E}^{ES}$ includes two

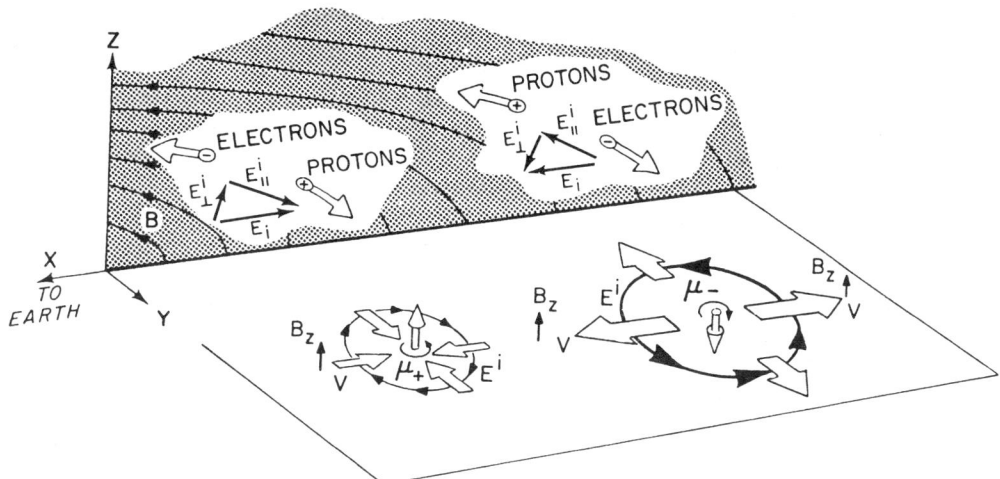

Figure 3. The two possible choices for the current meanders in the equatorial plane are outward (shown by the magnetic moment μ_-), or inward (shown by μ_+). The outward meander is explosive in that the **ExB** direction is everywhere out of the disturbance. The induction electric field has only x and y components everywhere, due to its source which is assumed to be the current meander in the equatorial plane. Because of the tail-like geometry this means that the electric field has a large component parallel to the magnetic field above and below the current plane. This leads to field-aligned forces on charged particles.

completely different types of effects, and one can be the cause of the other.

Another comment may be helpful with regard to the search for anomalous resistivity thought to be important for the initiation of reconnection. *Jackson* [1975, p.14] has noted that $\mathbf{J}=\mathbf{J}[\mathbf{E},\mathbf{B}]$. The square brackets are intended to signify that the relation may not be necessarily simple, and may depend on gradients, past history, may be non-linear, etc. With a hot plasma, currents may be driven by magnetic field structure, either with or against the electric field (an electrical load versus dynamo). A simple relation such as $\mathbf{J} = \sigma\, \mathbf{E}$ can be highly misleading, as in the case of the search for an "anomalous" resistivity.

5. Response of a Plasma to an Inductive Electric Field

Now we are ready to begin our analysis. If we could entirely disregard the charged particles as contributors to the electric field, the inductive electric field $\mathbf{E}^{IND}$ could be calculated from the change in the current-density distribution $\delta \mathbf{J}$ during a given time interval δt in a chosen Lorentz frame by using the Coulomb gauge to define a unique vector potential. As the induction field will cause differential motion of ions and electrons, there will, due to charge polarization, arise a secondary (irrotational) electric field $\mathbf{E}^{ES} = -\nabla \phi$ that adds to the "pure" induction field $\mathbf{E}^{IND}$, so that the actual electric field is

$$\mathbf{E} = -\nabla\phi - \partial \mathbf{A}/\partial t$$

The potential ϕ and hence the actual electric field $\mathbf{E}$, are extremely difficult to calculate, as (1) the plasma responds on many different time scales (electron plasma period, electron gyro period, electron bounce period, the corresponding ion periods, Alfvén wave resonance times, etc., and (2) the potential depends in an exceedingly sensitive way on the actual differential displacements of positive and negative particles. In the volume considered above, a relative charge imbalance of 10^{-10} over any substantial part of the volume would be enough to generate electric fields of a strength comparable to the "pure" induction field. However, regardless of the complexities involved in calculating the actual electric field we can identify several important features.

(a) The secondary electric field due to polarization of charges can never extinguish the induction field. Polarization and induction electric fields are topologically different; an electrostatic field has a vanishing line integral around every closed contour, so it cannot affect the electromotive force of the induced electric field. We can see this by forming the line integral around any closed contour enclosing a finite flux change:

$$\oint \mathbf{E} \cdot \mathbf{dl} = \oint (\mathbf{E}^{IND} + \mathbf{E}^{ES}) \cdot dl$$

$$= \oint \mathbf{E}^{IND} \cdot \mathbf{dl} + \oint \mathbf{E}^{ES} \cdot \mathbf{dl} = -\frac{d\Phi^M}{dt} + 0$$

where Φ^M is the total magnetic flux enclosed by the contour. Whatever the distribution of the secondary field $\mathbf{E}^{ES}$ is, the resultant field must remain finite and large enough to make the line integral finite and equal to $-(d\Phi^M/dt)$. A plasma is powerless to influence the electromotive force due to changing magnetic fields by any redistribution of charge (but not so with currents).

(b) If the polarization of charges leads to suppression of the magnetic-field-aligned electric field component $E_\parallel$, the above mentioned requirement on the line integral around an arbitrary contour implies that the value of the transverse component $E_\perp$ is *enhanced* instead [*Heikkila and Pellinen*, 1977]. The truth of this assertion can be confirmed by looking at Figure 3b and Figure 4 (to be discussed in the next section; when the polarization electric field is such that it can lead to partial or complete cancellation of the parallel component of the induced electric field, it will be such as to cause enhancement of the transverse component. This statement rests completely on the different topologies of rotational and irrotational fields, and does not depend on such things as non-linear plasma properties. The tendency toward suppression of $E_\parallel$ is due to the relative freedom of motion of charged particles along, as compared with transverse to, the magnetic field lines. The opposite acceleration of positive and negative particles under the influence of a transient magnetic-field-aligned component of the inductive field may explain the observations of large opposite magnetic-field-aligned flow velocities of ions and electrons in the plasmasheet [*Lui et al.*, 1978].

The enhancement of the transverse electric field implies a corresponding increase in the $\mathbf{E} \times \mathbf{B}/B^2$ convection speeds, which will be especially large near the neutral sheet where B is small. As in some regions the transverse electric field must be directed dusk-to-dawn (finite curl) it may cause expulsion of plasma away from the Earth, something intimately related to the plasma expulsion known to take place in plasmoids.

(c) When the net $E_\parallel$ becomes small the transverse electric field $E_\perp$ in the ionosphere and in the magnetotail may be in opposite senses. For example, we might have a westward electric field on the nightside at low altitudes (in agreement with a dawn-dusk magnetotail electric field), while at greater distances in the magnetotail along the same magnetic field lines there would be a dusk-dawn electric field (see Figures 3b and 4). Even if some unknown electrostatic field of external origin is added, the transverse electric field can vary in both magnitude and direction along magnetic field lines. Thus, with an induced electric field we cannot map electric fields from one region of the magnetotail to another, *even if $E_\parallel = 0$ everywhere.*

6. STATIC AND INDUCED ELECTRIC FIELDS – THE AGENTS FOR PARTICLE ENERGIZATION

As already pointed out, only the electric field (and not the magnetic field) is able to impart energy to charged particles. As its own energy is negligible, it acts as an intermediary, tapping energy either from an external source (static case) or from the magnetic field (inductive case). There are characteristic differences between the energization process in the two cases.

As electric potential fields are conservative, the acceleration that any individual particle can experience is limited by the potential difference transversed by its orbit, perhaps by gradient and curvature drifts. Thus an upper limit to the energy gain is of the order of 50 keV corresponding to the total potential across the magnetosphere. However, the magnetic field greatly reduces the range of energies available to most particles, because it prohibits the particle to move across the magnetic field parallel or antiparallel to $\mathbf{E}$ (as would be needed for a change in its energy) and instead causes it to drift transverse to both the electric and magnetic fields as an $\mathbf{E} \times \mathbf{B}/B^2$ motion. An exception to this is provided by a particle in the vicinity of a magnetic neutral line.

Induction electric fields are in several ways more efficient for particle energization than conservative fields. For one thing the induction fields are larger and hence allow stronger acceleration. Another and even more important feature of induction fields is the ability to energize charged particles locally by betatron acceler-

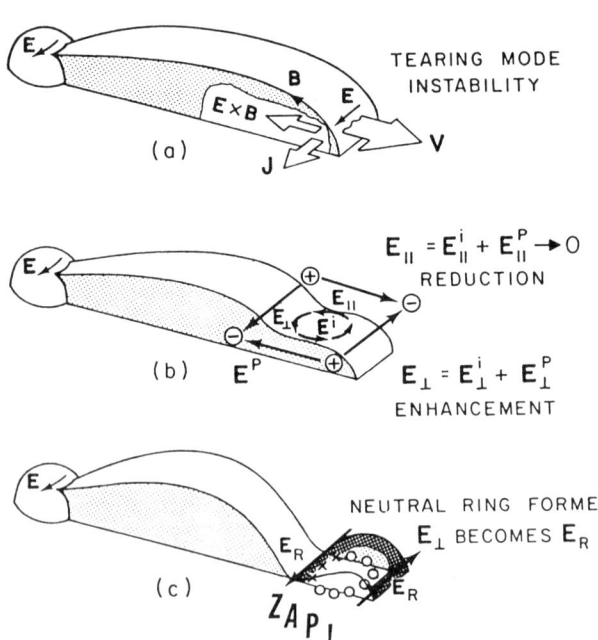

Figure 4. The outward current meander is really a tearing mode instability. The parallel component of the induction electric field will cause polarization of the plasma, trying to suppress the total $E_\parallel$. This must enhance the transverse component as shown in (b). When a substorm neutral line is formed this causes a discharge along it. Further energization is caused by betatron acceleration, with the result that a few particles can reach Mev energies in a few seconds.

ation (i.e. magnetic compression). A third factor is that the inductive electric field can appear wherever **A** has a finite value, far removed from its sources because of the weak decay with distance of the vector potential $\sim 1/r$; such a property may explain the local in-situ energization of both electrons and ions during a substorm [*McIlwain*, 1975; *Roux et al.*, 1990].

There are some limitations. One is that the ratio of final to initial energy is given by the ratio of final to initial magnetic field strength if the magnetic moment is conserved. Hence the largest relative energy increases are achieveable with particles starting out near a magnetic neutral line, where the initial field strength can be very small, and going to regions of greater magnetic field strength. Another limitation is due to the dependence on the particle's magnetic moment (for relativistic particles the relevant quantity is instead the flux enclosed by the gyro orbit). Only particles with a substantial magnetic moment, i.e. particles that already have a rather high energy, are efficiently energized. In particular, betatron acceleration is ineffective on "cold" plasma particles.

7. A Mechanism for Production of Very High Energies

The induction electric field, enhanced by the plasma (as pointed out above), will be present across a portion of the magnetotail near the developing disturbance, as shown in Figure 4b. Our previous work, and that of others [*Birn*, 1984; *Walker and Sato*, 1984], have shown that a neutral line may be formed; in this case it will be an X-type neutral line joined to an O-type neutral line, as shown by Figure 4c by the sequences of x and o. With reduction, even vanishing, of the qv × **B** force, this enhanced electric field will drive particles unimpeded by the Lorentz force. It becomes literally a discharge along the neutral line ("the equalization of a difference of electric potential between two points"). Our calculations have indicated that particles are energized by 10-100 keV at this stage.

A second stage of acceleration follows immediately. The acceleration along the neutral line (the discharge) provides adequate particle energies so that, when the particles leave the neutral line (with its small value of B) and enters into regions of higher B in the surrounding magnetic field, it has sufficient magnetic moment to benefit from the betatron acceleration. As shown by *Pellinen and Heikkila* [1978; 1984] and *Galeev* [1982], for reasonable values of the magnetic field changes during the substorm, initially low energy particles can be energized to MeV energies in a few seconds. This process was found independently by *Bulanov and Sasorov*, 1975. The dependence of the mechanism on the neutral line region also illustrates how to achieve selective acceleration, rather than an equitable distribution of available energy among the whole particle population, which is a firm requirement.

8. Tailward Retreat of a Plasmoid and its Footprint in the Polar Cap

Many other observed features of auroral substorms seem to follow from the negative meander (at least on qualitative grounds) as will now be discussed. However, these ideas are tentative, and more work needs to be done.

The first feature that should be noted closely is the following. Since **A** has only x and y components (due to the assumed meander being entirely in the equatorial plane) so does the induction electric field $\mathbf{E}^{IND} = -\partial \mathbf{A}/\partial t$, even above and below the current sheet (note the weak decay with distance in the definition of the vector potential, $\sim 1/r$). Because of the tail-like feature, the magnetic field lines have strong components in the x-direction well away from the field-reversal region, as indicated by the backdrop in Figures 3 and 5. The net result is that the induction electric field has large components parallel to the magnetic field oppositely directed on the dawn and dusk sides. The result of this geometry is that the plasma particles along the dawn and dusk flanks (the dawn flank is shown by the backdrop) are propelled by $E_{\parallel}^{IND}$ along the field lines. They now become current carriers; their motion constitutes a field-aligned current. The sense of the induction electric field is exactly the one to drive Birkeland currents in the manner first proposed by *Boström* [1964] shown in the inset of Figure 5.

Because of the mirror force due to the converging magnetic topology the field lines are not perfect, or ideal, conductors; a force on the current carriers is required to cause a field-aligned current. Any diversion of the cross-tail current must be *driven*; the onset of a high resistivity along the path of the cross-tail current, assumed in many reconnection models, is simply unphysical, and not even observed [*Coroniti*, 1985]. This is especially true on the dusk side with its upward current; any plasma electrons (of any energy) will be accelerated towards the ionosphere by $E_{\parallel}^{IND}$ to produce the well-known westward traveling surge (see second paragraph below). On the morning side the downward current can be carried by upward going ionospheric electrons, but here too we find that energized positive ions do carry a substantial amount of the current [*Shepherd et al.*, 1980]. Again it should noted that the net, or actual, electric field is $\mathbf{E} = \mathbf{E}^{IND} + \mathbf{E}^{ES}$; both the terms on the right hand side can be large, but as they are oppositely oriented their result can eventually be small. Because of the finite inertia of the current carriers there may be time delays and/or overshoots; this is a case of dynamic equilibrium. Just because $E_{\parallel} \sim 0$ does not mean that nothing is going on!

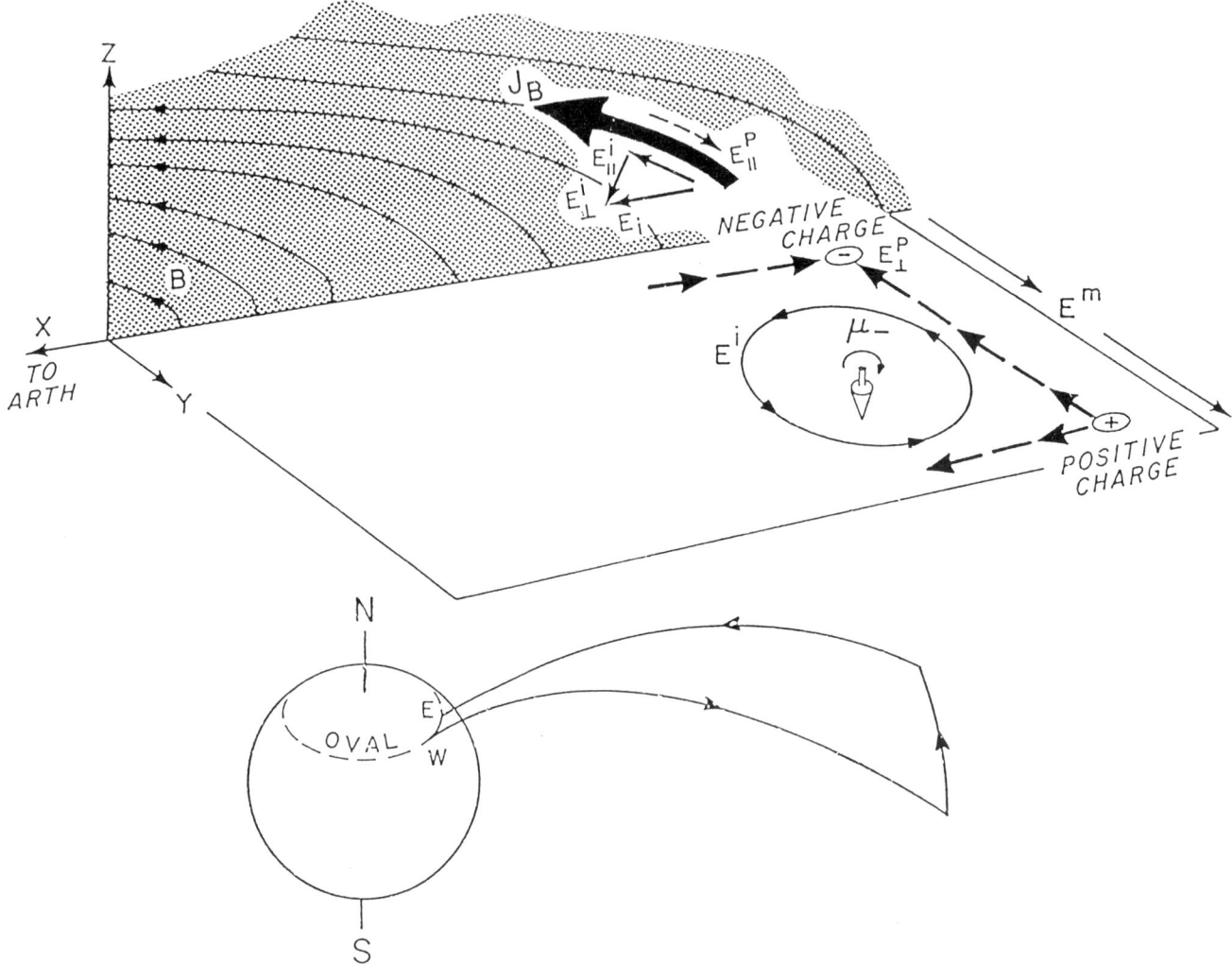

Figure 5. The parallel component of the induction electric field will cause field aligned currents to flow, in the manner postulated by Boström shown in (b). Due to the explosive nature of the outward meander the footpoints of the field-aligned currents will move as a westward traveling surge shown as W. A similar form (but with different characteristics) will travel eastward.

The second aspect to note about the inductive electric field for an outward meander of a cross-tail current filament is that $\mathbf{E} \times \mathbf{B}/B^2$ is explosive, i.e. the flow is outward everywhere from the region of the disturbance (see Figure 3). That is the reason for the westward motion of the WTS referred to above. A similar feature is an eastward traveling auroral form coincident with the downward current.

The third property of the negative meander is the tailward motion at the tailward extremity of the disturbance. The instability that caused the meander provides only the initial outward motion. This motion is enhanced by the plasma, as was discussed at length in Section 3.2. Once it begins it is hard to stop, a case of positive feedback. A plasmoid is expected to develop.

Confirmation has come from plasma simulations, for example by *Birn* [1984] and *Walker and Sato* [1984]. The plasmoid will go in the direction of least magnetic pressure, which is tailward (Figure 6). It makes little difference whether the field lines are closed or open, because the magnetic pressure in the tailward direction at this point is inconsequential.

Our final point is that the ionospheric signature of the escaping plasmoid in the polar cap will vanish when the plasmoid reaches the distant X-line, and goes onto open field lines [*Heikkila and Treilhou*, 1985]. The X-line in this model is well beyond ISEE-3 apogee of 220 R_E, perhaps even beyond 1000 R_E judging by the findings of *Villante* [1976].

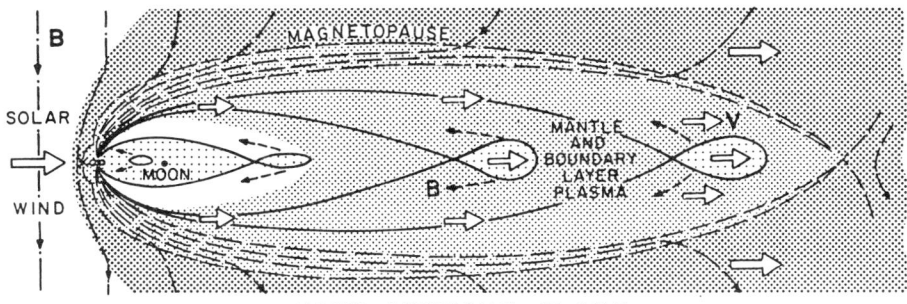

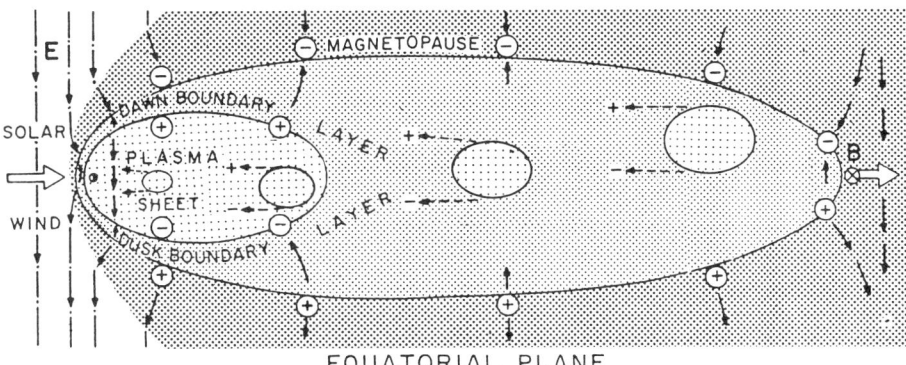

Figure 6. The proposed geometry of the tail during substorms. Plasmoids are shown traveling tailward in a region of closed field lines. The magnetic pressure is inconsequential compared to the pressure in the plasmoid. Field-aligned currents continue to flow on closed field lines, as has been confirmed observationally as shown by the inset. The auroral forms disappear when the plasmoid reaches interplanetary field lines.

9. CONCLUSIONS

The starting point of our work is the recognition that the cross-tail current is filamentary, and the auroral arcs are connected to individual filaments. Hence, we must look at instabilities of current filaments in the magnetotail to understand the triggering mechanism that initiates a substorm. An outward meander will be caused by the current carriers (undergoing curvature drift) becoming demagnetized. The inductive electric field $\mathbf{E}^{IND} = -\partial A/\partial t$ (using the Coulomb gauge) will in general have a component parallel to the magnetic field. We take the response of the plasma to be reflected in a scalar potential, $\mathbf{E}^{ES} = -\nabla\phi$; that response must be such as try to make the actual (or net) $E_\parallel$ approach zero. Charged particles are driven along the geomagnetic field lines (the plasma response) by the induction electric field (the cause); the result is the formation of field-aligned currents producing the well-known substorm current diversion. Due to the finite inertia of the current carriers (even electrons), overshoots can arise that undoubtedly complicate the physics.

A little know fact of plasma physics is that the response by the plasma to the sudden appearance of an induction electric field, in trying to remove the parallel component, enhances the transverse component. This is due entirely to the different topologies of the two types of fields. The enhanced induction field near the emerging X-line will cause a discharge; after subsequent betatron acceleration even zero energy particles can be energized to MeV energies in a matter of seconds in a two-step process.

The substorm process derives its energy from stored magnetic energy which is released in the deflation of the tail field, and fed to the charged particles by the associated electric induction field. The enhancement of the transverse component will also cause the plasma near the disturbance to form a plasmoid; it will move in the direction of least magnetic pressure, namely tailward.

Acknowledgments. I thank a very conscientious and thoughtful referee, and my previous co-workers especially Risto Pellinen. This work was supported by the National Science Foundation.

The editor thanks a referee for his assistance in evaluating this paper.

REFERENCES

Baker, D. N., S. J. Bame, D. J. McComas, R. D. Zwickl, J. A. Slavin, *Plasma and magnetic field variations in the distant magnetotail associated with near-earth substorm effects*, in Magnetotail Physics, ed. A. T. Y. Lui, Johns Hopkins University Press, 1987.

Birn, J., *Three-Dimensional Computer Modeling of Dynamic Reconnection in the Magnetotail: Plasmoid Signatures in the near and Distant Tail*, in Magnetic Reconnection in Space and Laboratory Plasmas, ed E. Hones, AGU Monograph 30, p. 264, 1984.

Birn, J. and M. Hesse, *MHD Simulations of Magnetic Reconnection in a Skewed Three-Dimensional Tail Configuration*, J. Geophys. Res., 96, 23, 1991.

Boström, R., *A Model of the Auroral Electrojets*, J. Geophys. Res., 69, 4983, 1964.

Bulanov, S. V. and P. V. Sasorov, Astron. Zh. 52, 763, 1975.

Coroniti, F. V., *Space Plasma Turbulent Dissipation: Reality or Myth?*, Space Sci. Rev. 42, 399, 1985.

Cragin, B. and W. J. Heikkila, *Alternative Formulations of Magnetospheric Plasma Electrodynamics*, Rev. Geophys. Space Phys., 19, 223, 1981.

Doolittle, J. H., S. B. Mende, R. M. Robinson, G. R. Swenson, and C. E. Valladares, *An Observation of Ionospheric Convection and Auroral Arc Motion*, J. Geophys. Res. 95, 19123, 1990.

Galeev, A., *Magnetospheric Tail Dynamics*, p. 143 in Magnetospheric Plasma Physics, ed. A. Nishida, D. Reidel, 1982.

Harrison, R. A., E. Hildner, A. J. Hundhausen, D. G. Sime and G. M. Simnett, *The Launch of Solar Coronal Mass Ejections: Results From the Coronal Mass Ejection Onset Program*, J. Geophys. Res. 95, 917, 1990.

Heikkila, W. J. and R. J. Pellinen, *Localized Induced Electric Field Within the Magnetotail*, J. Geophys. Res. 82, 1610, 1977.

Heikkila, W. J., R. J. Pellinen, C.-G. Fälthammar and L. P. Block, *Potential and Inductive Electric Fields in the Magnetosphere During Auroras*, Planet. Space Sci., 27, 1383, 1979.

Heikkila, W. J., *Formation of Auroral Arcs by Plasma Sheet Processes*, in Physics of Auroral Arc Formation, AGU Mono. 25, ed. Akasofu and Kan, p. 266, 1981.

Heikkila, W. J. and J.-P. Treilhou, *The Distant Magnetotail*, in Results of the Arcad 3 Project, Cepadues-Editions, p.281, 1985.

Heikkila, W. J., *Current Sheet Crossings in the Distant Magnetotail*, Geophys. Res. Lett., 15, 299, 1988.

Heikkila, W. J., *Magnetic Reconnection, Merging, and Viscous Interaction in The Magnetosphere*, Space Science Rev. 53, 1, 1990.

Jackson, J. D., *Classical Electrodynamics*, Second Edition. John Wiley, 1975.

Konopinski, E. J., *Electromagnetic Fields and Relativistic Particles*, McGraw Hill, 1981.

Lui, A. T. Y., L. A. Frank, K. L. Ackerson, C.-I. Meng, and S.-I. Akasofu, *Plasma Flows and Magnetic Field Vectors in the Plasma Sheet During Substorms*, J. Geophys. Res. 83, 3849, 1978.

McIlwain, C., *Auroral Electron Beams near the Magnetic Equator*, in Physics of the Hot Plasma in the Magnetosphere, B. Hultqvist and L. Stenflo (eds.), Plenum Press, p. 91, 1975.

Morse, Philip M. and Herman Feshbach, *Methods of Theoretical Physics*, McGraw Hill, 1953.

Pellinen, R. J. and W. J. Heikkila, *Energization of Charged Particles to High Energies by an Induced Substorm Electric Field Within the Magnetotail*, J. Geophys. Res. 83, 1544, 1978.

Pellinen, Risto J. and Walter J. Heikkila, *Inductive Electric Fields in the Magnetotail and Their Relation to Auroral and Substorm Phenomena*, Space Science Rev. 37, 1, 1984.

Roux, A., S. Perraut, P. Robert, A. Morane, A. Pedersen, A. Korth, G. Kremser, and S. Y. Pu, *Role of the Near Earth Plasmasheet at Substorms: a Synthesis*, presented at the Chapman Conference on Magnetospheric Substorms, Hakone, Japan, September 3-7, 1990.

Russell, C. T. and R. L. McPherron, *The Magnetotail and Substorms*, Space Science Rev. 15, 205, 1973.

Schindler, K., *A Theory of the Substorm Mechanism*, J. Geophys. Res. 79, 2803, 1974.

Shepherd, G., R. Boström, H. Derblom, C.-G. Fälthammar, R. Gendrin, A. Korth, A. Pedersen, R. Pellinen, and G. Wrenn, *Plasma and Field SIgnatures of Poleward Propogating Auroral Precipitation Observed at the Foot of the Geos 2 Field Line*, J. Geophys. Res. 85, 4587, 1980.

Villante, U., *Neutral Sheet Observations at 1000 R_E*, J. Geophys. Res., 81, 212, 1976.

Walker, R. and T. Sato, *Externally Driven Magnetic Reconnection*, in Magnetic Reconnection in Space and Laboratory Plasmas, ed E. Hones, AGU Monograph 30, p. 272, 1984.

Zwingmann, W., *Self-Consistent Magnetotail Theory: Equilibrium Structures Including Arbitrary Variation Along the Tail Axis*, J. Geophys. Res., 88, 9101. 1983.

Toward a Better Understanding of the Global Auroral Electrodynamics Through Numerical Modeling Studies

GÖRAN MARKLUND AND LARS BLOMBERG

Department of Plasma Physics, Alfvén Laboratory, Royal Institute of Technology, Stockholm

Results from numerical model studies are presented focusing on various aspects of the auroral electrodynamics. Studies of specific events, by means of extensive sets of observations which are fed into a numerical model to provide snapshots of the global electrodynamics, have been conducted in parallel with theoretical model studies. This has proven to be an efficient way to obtain a better physical understanding of the events in general and of the interrelationships between the physical parameters in particular. Of special interest are the electrodynamical features of northward IMF phenomena, to which we pay particular attention in this review. Examples are presented of transpolar arc events with particular emphasis on the convection signatures and how these relate to the auroral distribution. The influence on the global potential distribution of localized Birkeland currents associated with transpolar arcs is shown to be quite different from that of the more extended large-scale NBZ currents. For the former case the potential pattern is typically of a two cell type with modifications showing up as a poleward expansion of one of the cells, a local electric field reversal or possibly an electric field reduction at the center of the arc depending on how the arc-associated currents close in the ionosphere. For the case with large-scale NBZ currents, presumably related to extended regions of dim auroral features, the potential pattern is shown to have two, three, or four large-scale cells depending on the ratio between the NBZ and the oval currents, on the sign of the IMF B_y, and on the ionospheric conductivity.

Of importance to these studies are also the statistical properties of the global electric field, current, and conductivity distributions and the relation between these for various conditions, a subject that will be briefly discussed. The effect of parallel potential drops in regions of upward field-aligned current has been taken into account in the model by assuming a linear relationship between these. Distortions of the equipotential contours above the potential drop occur as a result of this assumption as illustrated by projecting the ionospheric potential to the magnetosphere.

1. INTRODUCTION

The understanding of the ionosphere-magnetosphere system in general, and of the auroral phenomena in particular has improved considerably in recent years. The rich data material and high-quality scientific outcome from recent polar-orbiting spacecraft such as Dynamics Explorer and Viking, as well as from coordinated ground-based investigations have contributed greatly to this end. The ability to image the global auroral distribution while measuring in situ auroral particles, electric fields, currents, and wave emissions has played a key role for the success of these missions.

The combination of UV imaging and in situ observations has proven to be particularly useful for studies of the auroral electrodynamics on a global scale. Of fundamental importance are the relationships between particle precipitation, the induced auroral emissions, and ionospheric conductivities, and how these relate to electric fields and currents for various activity levels, phases of auroral substorms, and conditions of the interplanetary medium.

The average, large-scale signatures of the various electrodynamical parameters are today relatively well known. Field-aligned current distributions for different K_p have been determined by Iijima and Potemra [1976a, b; 1978] and for northward IMF by Iijima et al. [1984] and Iijima and Shibaji [1987]. Average high-latitude potential (or convection) patterns for various IMF B_y and B_z components are presented by Heppner and Maynard [1987] derived from satellite electric field data and by Foster [1987] derived from incoherent scatter radar data. The seasonal dependence and IMF B_y dependence of the high-latitude convection patterns have recently been studied by de la Beaujardière et al. [1990] using Søndre Strømfjord radar data. The distribution and intensity of ion and electron precipitation and the corresponding induced ionospheric conductivity distribution as a function of K_p have been estimated by Hardy et al. [1985; 1989] and earlier by Wallis and Budzinski [1981].

The electrodynamic state of the auroral ionosphere is highly variable and influenced by a large number of factors such as, the magnetic activity level, the orientation of the IMF, the season, and the universal time, just to mention a few examples. Due to the different assumptions, data sets, and experimental techniques used to derive them, empirical statistical patterns are not consistent with each other even in a global context. For the reasons given above, the field-aligned current distribution of Iijima and Potemra [1976a] for $K_p < 3$, for example, is not fully consistent with the conductivity distribution of Wallis and Budzinski [1981] and together these distributions correspond to a potential pattern which is different in many respects from the Heppner and Maynard [1987] pattern. Statistical studies of the various electrodynamical parameters using the same data sets and same type of binning are therefore required to obtain more accurate patterns as well as to improve our understanding of the different relationships. This will help to reduce the artificial features which arise due to the inconsistences between the involved quantities.

Whereas the average behavior of the global auroral electrodynamics is relatively well understood, this is not the case for the instantaneous behavior. Because of the highly variable state of the auroral ionosphere, it is not likely that a statistical pattern, such as the distorted two cell pattern of Heppner and Maynard [1987] for northward IMF, could be used to describe a particular event. To understand the temporally and spatially varying electrodynamical parameters and their interrelationships it is necessary to find some way to "image" these various patterns, as is done optically by auroral UV imagers. Different alternative techniques to derive such snapshots have been developed recently as briefly discussed below.

Mishin et al. [1979; 1986] and Kamide et al. [1981] used a technique to derive instantaneous electric fields, currents, and conductivities in the high-latitude ionosphere, using as input instantaneous equivalent current patterns derived from ground-based magnetometer data. The instantaneous patterns derived in this way are, however, relatively uncertain mainly due to uncertainties in the conductivity model and to spatial data gaps in the magnetic data. A further development of this technique is the AMIE (Assimilative Mapping of Ionospheric Electrodynamics) procedure by Richmond and Kamide, taking into account many different types of measurements as well as statistical information in the modeling procedure [Richmond and Kamide, 1988; Richmond et al., 1988].

An alternative technique to derive snapshots of auroral electrodynamical parameters has been described in detail by Marklund et al. [1987; 1988]. An important ingredient in this technique is the qualitative information on the global auroral electrodynamics that can be extracted from a UV image of the auroral distribution. The UV images are used to give the appropriate geographical frame for the other observations and to infer the instantaneous distributions of the particle-induced ionospheric conductivity and the upward field-aligned currents in a qualitative sense. In situ satellite measurements of particles, currents, and electric fields are then used to calibrate and refine these patterns so that they also agree quantitatively with the localized measurements wherever these are available.

Another important feature of this technique is the use of certain relationships between the electrodynamical parameters, namely: (1) the upward field-aligned currents and the particle-induced conductivity, and (2) upward field-aligned currents and the parallel potential drops through which the auroral particles are accelerated. The influence of the first of these relationships on the global potential distribution has been studied in detail by Blomberg and Marklund [1988] and is briefly discussed here. The effect of the linear relationship between upward field-aligned current and parallel potential drop is studied here by applying it to a single event. As a result of using this relationship distortions of the equipotential contours occur above the potential drops, as illustrated by projecting the ionospheric potential to the magnetospheric equatorial plane.

The technique has so far been applied to three different events, the first of which is treated in detail by Marklund et al. [1987; 1988] and not further discussed here. The other two studies focus on electrodynamics associated with transpolar arc events during northward IMF conditions. Of particular interest are the localized electric field and field-aligned currents associated with the polar arcs and how these are related to and may influence the global potential distribution.

These studies have been conducted in parallel with model studies of northward IMF phenomena treating idealized transpolar arc events as well as situations associated with large-scale NBZ current systems. This has proven to be an efficient way to obtain a better physical understanding of auroral electrodynamics in general and of the interrelationships between the electrodynamical parameters in particular. The results from these combined event/model studies which focus on northward IMF electrodynamics will be given particular attention in this paper.

2. Model Description

The model used in these studies is basically the same as that described by Marklund et al. [1988, 1991] and Blomberg and Marklund [1988, 1991b]. The major difference lies in the way the input field-aligned currents and conductivities have been chosen and represented in the model. For the event studies these have been inferred from the actual observations, by a combination of UV imager information on large-scale qualitative features and in situ measurements of magnetic fields and particles. For the pure model studies idealized distributions have been used to represent the region 1/2 system, NBZ currents, cusp currents, and transpolar arc currents, and the associated conductivity distributions. The basic equation given below is used to calculate the ionospheric potential from given field-aligned current and conductivity distributions.

$$\nabla \cdot (\Sigma \nabla \Phi) = -j_\parallel \sin I \qquad (1)$$

Here Σ is the height-integrated conductivity tensor, Φ the electrostatic potential, $j_\parallel$ the field-aligned current, and I the magnetic field inclination. For a more complete description of the general characteristics of the model, see *Blomberg and Marklund* [1991b].

3. THE $j_\parallel - \Sigma$ RELATIONSHIP

One of the four terms contributing to the height-integrated Pedersen and Hall conductivities, given as equations (6a) and (6b) in *Blomberg and Marklund* [1988], relates the conductivity enhancement produced by precipitating particles to the upward field-aligned current.

For $j_\parallel$ upward

$$\Sigma_{j_\parallel} = k(MLT) \cdot |j_\parallel| \qquad (2a)$$

For $j_\parallel$ downward

$$\Sigma_{j_\parallel} = 0 \qquad (2b)$$

The proportionality given by equations (2a) and (2b) is motivated by observations of discrete auroral arcs. The discrete aurora has been found to appear mostly within regions of upward field-aligned currents and the diffuse aurora has, in particular on the evening side, been associated with downward field-aligned currents [e.g., *Kamide and Akasofu*, 1976]. *Lyons et al.* [1979] found that in the vicinity of an auroral arc the net downward electron energy flux typically varied as V^2 and the field-aligned current as V, for the case a field-aligned potential drop, V, could be identified from the electron spectra. Since Σ_P has been found to be roughly proportional to the square root of the electron energy flux [e.g., *Harel et al.*, 1981], the relations above imply a direct proportionality between Σ_P and $j_\parallel$.

The influence of such a relationship on the global potential distribution is illustrated by Figure 1, taken from *Blomberg and Marklund* [1988]. To the left are shown two idealized and symmetric field-aligned current distributions having a region 1 to region 2 current ratio of 3:1 (a) and 3:2 (d), respectively. In the middle are shown the corresponding potential patterns calculated using equation 1 and no current-conductivity coupling term (b) and (e). To the right are shown the potential patterns that result from including

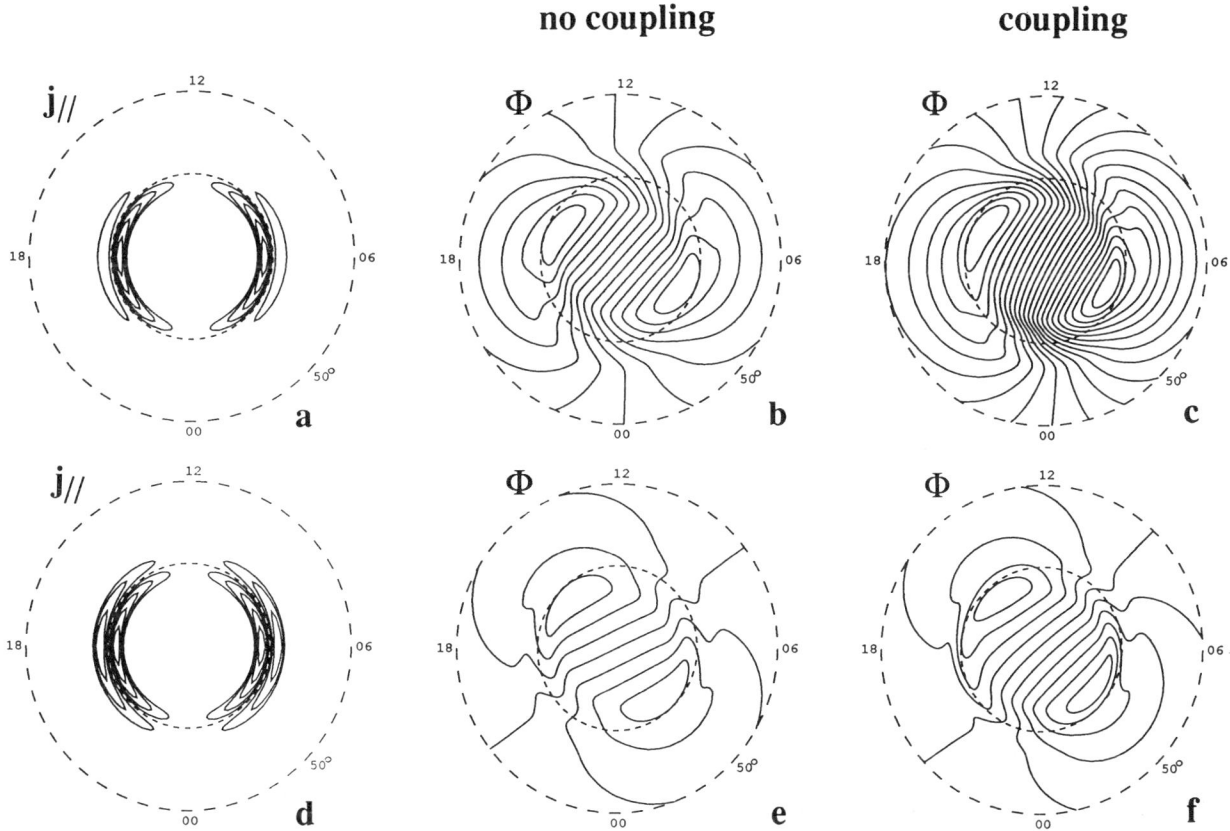

Fig. 1. Estimate of the influence of the $j_\parallel - \Sigma_P$ coupling on high-latitude potential distributions.

the coupling term (c) and (f). The major effects of including the current-conductivity coupling term may be summarized as follows: (1) the clockwise rotation of the polar cap electric field becomes less pronounced, (2) the convection cell related to a net upward current (dusk cell) is associated with a weaker electric field than the cell related to a net downward current (dawn cell).

The clockwise rotation of the convection pattern is a characteristic feature that has been found in many previous studies [e.g., *Yasuhara et al.*, 1983; *Nopper and Carovillano*, 1979; *Marklund et al.*, 1985]. Using only an azimuthally symmetric conductivity distribution, representing the diffuse aurora, but no coupling term representing the discrete aurora, will typically lead to an overestimate of this rotation. The latter effect, the reduction of the electric field in regions of upward field-aligned currents, is well confirmed by observations in and above auroral arcs in the evening and morning sectors [e.g., *Evans et al.*, 1977; *Marklund et al.*, 1982; 1983].

The results of this study suggest that the assumption concerning current-conductivity coupling is very important and should therefore be taken into account when modeling auroral electrodynamics on global and auroral scale sizes.

4. $j_\| - V_\|$ Relationship

In the technique to derive snapshots of the auroral electrodynamics, in situ measurements of the different electrodynamical parameters made at various altitudes are projected along the magnetic field lines to a common altitude (e.g., the E layer maximum at 120 km) under the assumption of a perfect mapping, i.e., no field-aligned potential drop below the altitude of the measurements. This assumption is, however, known to fail on auroral flux tubes associated with discrete aurora, i.e., in regions of intense upward field-aligned currents. For nightside arcs *Lyons et al.* [1979] and others have found that a linear relationship generally exists between $j_\|$ and $V_\|$. In other situations the linear relationship does not hold and instead there will be a saturation current which is independent of the potential drop as shown by e.g., *Brüning et al.* [1990]. For the limiting case $eV_\|/T \ll B_i/B_v$ the current-potential relationship derived assuming strong pitch angle scattering [e.g., *Knight*, 1973; *Lyons*, 1980], reduces to the linear relationship:

$$j_\| = en \left(\frac{T}{2\pi m_e} \right)^{1/2} \left(1 + \frac{eV_\|}{T} \right) \quad (3a)$$

Here T represents the thermal energy of the Maxwellian particle distribution, and B_i/B_v the ratio between the magnetic field in the ionosphere and that at the top of the acceleration region. Thus, given a field-aligned current distribution and a corresponding conductivity distribution, equation (1) combined with a relationship such as expressed by equation (3a) will allow the calculation of both the ionospheric potential and the high-altitude potential above the acceleration region. To illustrate this we have applied the simplified linear relationship given below to a field-aligned current and potential distribution reconstructed for a specific event treated by *Marklund et al.* [1988].

$$V_\| = k_\| j_\| \quad (3b)$$

Here $k_\|$ is the proportionality factor (expressed in units of kV/(μA/m^2)).

The top panel of Figure 2 shows the model (dotted line) and measured (solid line) field-aligned current density profiles along the ionospheric projection of the Viking satellite orbit between 7:00 UT and 9:40 UT on April 28, 1986. The bottom panel shows the calculated model potential profiles (dotted lines) corresponding to four different values of the proportionality factor $k_\|$ (0, 2, 4 and 6 kV/(μA/m^2)) together with the measured potential profile (solid line). As can be seen the local potential minima seen in the measured potential profile in the upward current region are roughly reproduced by a proportionality factor of 6 kV/(μA/m^2). Figure 3 shows for the same set of $k_\|$-values the corresponding potential patterns projected to the magnetospheric equatorial plane using a dipolar magnetic field. Distortions or vortices in the equipotential contours are seen to occur on the auroral flux tubes which are associated with intense upward currents and thus significant parallel potential drops. The local enhancements of westward convection associated with these hot spots are believed to play a key role in a recent theory for substorm onset (Rothwell, private communication, 1990.)

5. A Transpolar Arc Event

The prime goal of this study is to reconstruct the ionospheric electrodynamical state of a transpolar arc event. The transpolar arc was viewed by the auroral UV imager experiment on the Viking satellite at 2046 UT on September

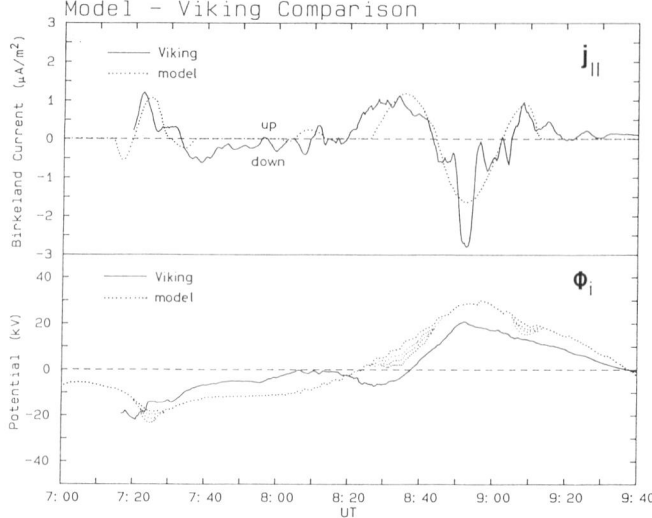

Fig. 2. Reproduction of the high-altitude potential profile along the Viking orbit for various assumed field-aligned potential drops using the $j_\| - V_\|$ relationship. For more information see text.

MAGNETOSPHERIC CONVECTION

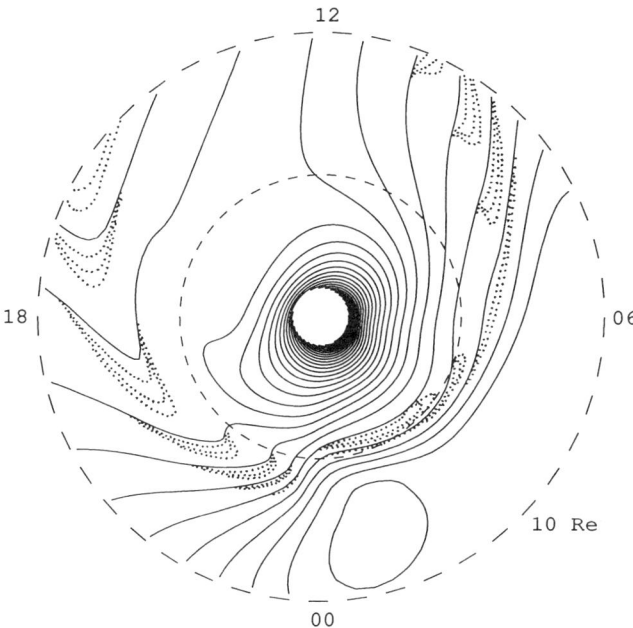

Fig. 3. Distortions of the equipotential contours in the magnetospheric equatorial plane for various assumed field-aligned potential drops between the ionosphere and the magnetosphere using the $j_\parallel - V_\parallel$ relationship.

25, 1986. A detailed description of the event reconstruction is given by *Marklund et al.* [1991]. Only a brief summary of the modeling results is given here with particular focus on the convection signatures with respect to the auroral distribution.

The left part of Figure 4 shows the particle-induced conductivity distribution (top) and the field-aligned current distribution (bottom) that are qualitatively consistent with the UV image. They are calibrated so as to be quantitatively consistent with magnetic field and particle measurements from the Viking, the DMSP F7, and the Hilat satellites. To the right is shown the corresponding potential distribution calculated using equation (1). It is superposed on the auroral distribution to visualize the close connection between the two patterns.

The convection pattern can be characterized as a distorted two cell pattern or possibly as a three cell pattern. It has a large evening cell with two local minima and a minor crescent shaped dawn cell. A large variation is found in the convection signatures along the arc. Close to the dayside where it is most intense, the arc forms a local transition between sunward and antisunward convection on the dusk and dawn sides, respectively. Toward the nightside, where the transpolar arc is very faint the arc has only a minor influence on the convection which is mainly controled by the nightside oval convection.

A variation of the convection signatures along the arc, which depends on the balance between the ambient and the arc-associated electric fields is believed to be a characteristic rather than uncommon feature for this kind of events. We therefore suggest that experiments on polar orbiting satellites passing across transpolar arcs at different times and different locations will observe a multitude of different electric field, current, and particle signatures. These can only be understood having the information on the global auroral situation as provided by, e.g., auroral UV imagers. These problems are further discussed in the next section.

6. Transpolar Arc Electric Fields and Field-aligned Currents

This study [*Marklund and Blomberg*, 1991] focuses on how various field-aligned current distributions representative of transpolar arcs may influence the global potential distribution. The localized field-aligned currents associated with the transpolar arcs are for these cases assumed to exist without the simultaneous presence of large-scale NBZ currents poleward of the main auroral oval. The other extreme situation, with extended NBZ currents but no distinct bright polar arcs is treated in the next section.

Figure 5 shows three alternative model representations of transpolar arc field-aligned current systems (case 1 (b), case 2 (c) and case 3 (d)) and a reference system (a) together with the corresponding potential distributions calculated using equation (1) and a conductivity model described below. A description is first given of the specific choice of the model input (field-aligned current and conductivity) distributions followed by a discussion on the corresponding potential patterns.

The integrated sheet current magnitudes are indicated by the numbers given in units of hundred kA. A region 2 to region 1 current ratio of 0.5 has been used throughout the calculations. At the top right (b) the transpolar arc is represented by one single upward current sheet of the same magnitude as the net current from the region 1/2 system. As can be noted, the total upward and downward currents do not balance for this case. It would have been easy to modify the current magnitudes slightly so as to balance the total upward and downward currents. Since the general characteristics of the solution was found to be roughly unaltered after such a modification it was, however, found better to retain the region 1/2 current fixed throughout the calculations. At the bottom left (c) the transpolar arc is represented by two equal but opposite current sheets directed in the opposite sense as compared to the adjacent region 1 currents. At the bottom right (d) the directions of the transpolar arc current sheets are reversed as compared to the previous case (c).

For a general description of the ionospheric conductivity model the reader is referred to *Marklund et al.* [1988]. The specific model that has been used throughout (a)-(d) include the contributions from: the solar EUV radiation, represented by $\Sigma_{UV} = 5.0(10.0)\sqrt{\cos\chi}$, where the numbers

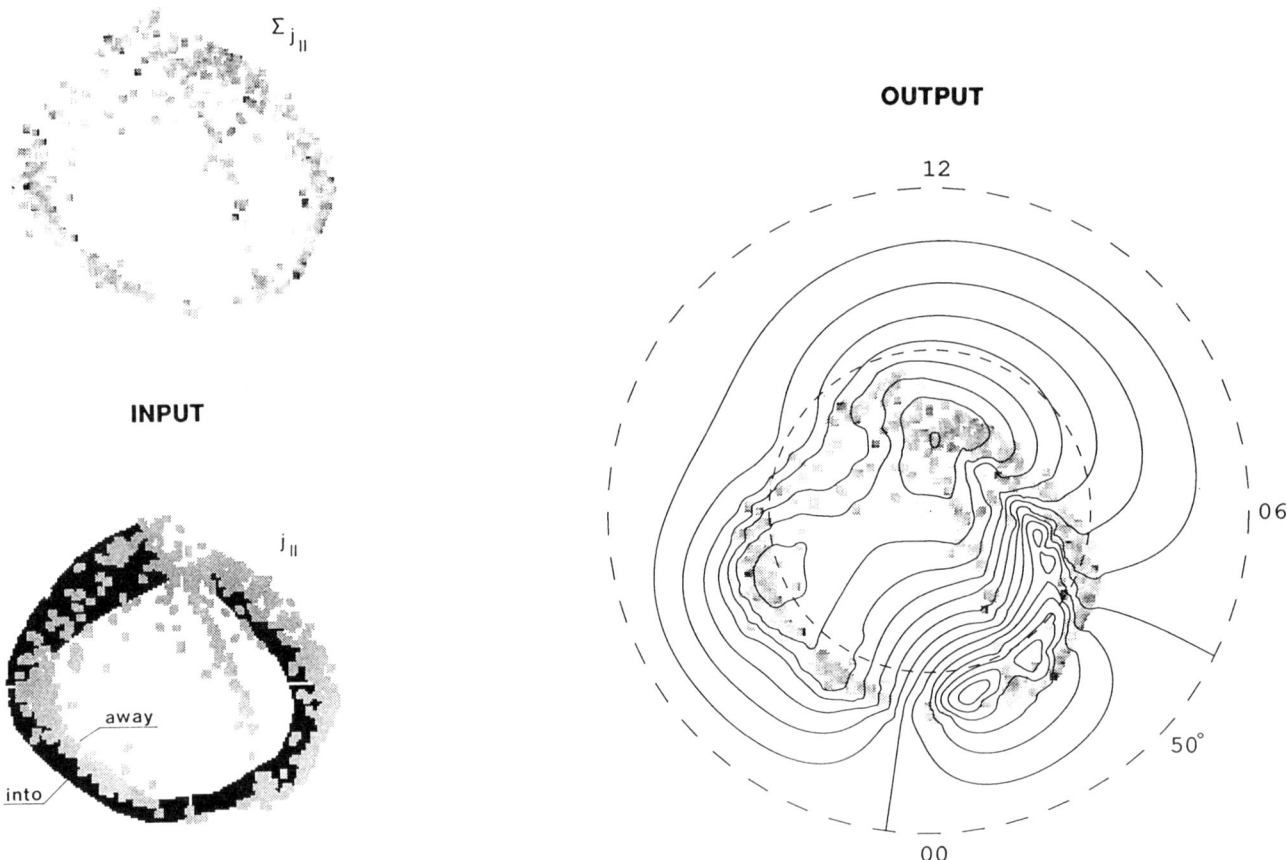

Fig. 4. Input ionospheric conductivity and field-aligned current distributions (left) and output potential distribution (right) reconstructed for a transpolar arc event.

refer to the Pedersen and Hall conductivities in Siemens and χ is the solar zenith angle; the cosmic and galactic radiation, represented by a constant background term, $\Sigma_0 = 0.5$ S; and the diffuse auroral particle precipitation represented by a local time independent but latitude dependent Gaussian distribution, Σ_G, peaking at a value of 5 S in the center of the auroral oval.

Case 1 (Fig. 5b)

The convection pattern calculated for Case 1 with a single net upward current sheet is characterized by a large evening cell with a minimum almost at the pole and a smaller crescent shaped morning cell somewhat displaced toward noon. The pattern can be seen to be quite similar to what is expected for positive IMF B_y conditions. The region of antisunward convection is concentrated on the dawnside of the noon-midnight meridian and the flow is more intense than in the reference case shown in Figure 5a. The displacement of the dusk potential minimum toward the pole implies that the region of sunward convection reaches to very high latitudes.

To exemplify this we show in Figure 6 a drift pattern with an expanded region of sunward drift associated with a polar arc event. The data shown were obtained by the UV imager experiment and the double-probe electric field experiment on the Viking satellite around 18:30 UT on April 1, 1986. The polar arc is seen to be connected at its two ends to the dusk-side auroral oval. It is quite obvious for this case that the arc forms a part of an expanded dusk-side auroral oval. This is also confirmed by the particle observations. The large-scale convection of the morning cell is qualitatively consistent with the normal pattern depicted in Figure 5b, although a lot of fine structure can be seen with small-scale reversals probably related to faint arcs.

Case 2 (Fig. 5c)

The convection pattern for this case is essentially the same as the reference two cell pattern shown in Figure 5a except for a narrow region in the immediate vicinity of the transpolar arc where a reversal from the general antisunward flow to sunward flow can be seen. The reversal results from the fact that the duskside and dawnside balanced current sheets

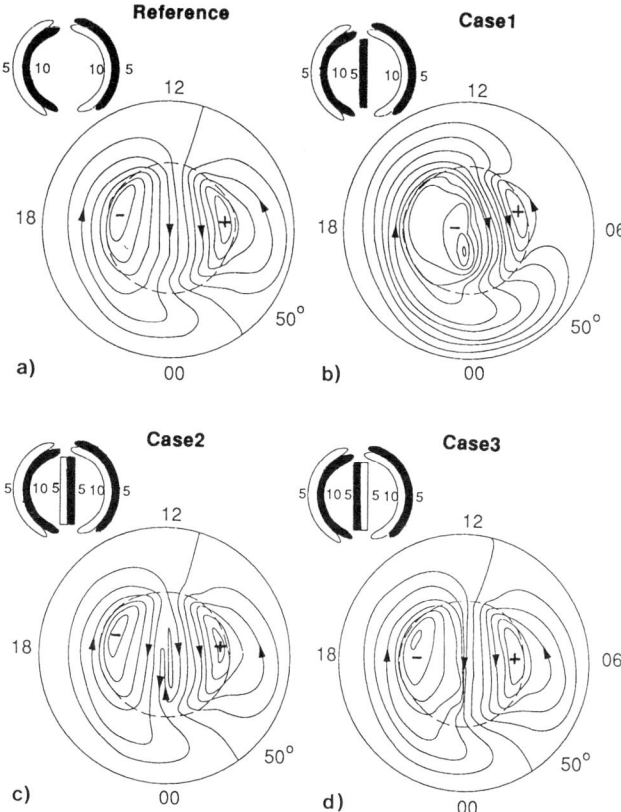

Fig. 5. Model representations of field-aligned current distributions associated with idealized transpolar arcs and the corresponding potential distributions.

associated with the transpolar arc have opposite signs to the adjacent region 1 currents, similar to the current orientations typical for the much more extended NBZ current sheets. This kind of pattern and orientation of the transpolar arc current sheets appears to be relatively common as is illustrated below using Viking electric field data.

Case 3 (Fig. 5d)

By reversing the orientations of the transpolar arc current sheets the dusk and dawn cells remain essentially unchanged except for the immediate vicinity of the arc where the equipotential contours are compressed and the electric field intensified in contrast to the previous case where the electric field was weakened and/or reversed. The arc-associated convection is locally intensified in the antisunward direction in the center of the arc which added to the ambient antisunward convection results in a very strong antisunward flow along the arc. This kind of pattern appears to be relatively uncommon and no clear case has been found in the analyzed Viking electric field data.

Figure 7 shows Viking electric field observations to illustrate the various cases discussed above. For each frame the upper panel shows the dawn-dusk potential profile as a function of time, the middle panel the dawn-dusk electric field used to calculate the potential and the bottom panel the standard deviation of the electric field which has proven to be an excellent indicator of auroral activity. The dotted lines show for comparison, the dawn-dusk potential profiles derived from the model patterns in Figure 5. The measured potential profile consistent with the reference pattern is recognized at the top left. The profile consistent with the expanded dusk-side oval (case 1) is shown at the top right and two profiles corresponding to case 2 with localized electric field reversals at the position of the arc are shown at the bottom. As mentioned above no clear example of case 3 is found in the database studied so far.

7. INFLUENCE OF LARGE-SCALE NBZ CURRENTS

Characteristic for persistent northward IMF conditions is significant auroral activity poleward of the classical auroral oval. In some situations the activity appears in the form of relatively distinct and localized polar and transpolar arcs, a situation treated in section 6, above. In other situations, which tend to be more common, in particular during more prolonged periods of northward IMF, extended regions of diffuse auroral activity cover a large portion of the "polar cap" region. The latter situations are likely to be associated with an extended field-aligned current system, called the NBZ current system. This study [*Blomberg and Marklund*, 1991a] models the high-latitude convection as a function of the large-scale field aligned currents, consisting of three main parts, the region 1, the region 2, and the NBZ current systems. Of particular interest is the evolution of the convection pattern as the current system changes from a typical IMF southward to a typical IMF northward configuration.

Figure 8 presents such an evolution of the convection patterns from a reference pattern (top left) as the region 2 to region 1 current ratio is increased from top to bottom and as the NBZ to region 1 current ratio is increased from left to right. The attached numbers denote the region 2, region 1, and NBZ current magnitudes, respectively, given in units of MA. The results shown here are representative of IMF $B_y < 0$ conditions, for which the dividing line between the upward and the downward NBZ current regions is displaced toward dawn. An individual balancing of the currents within the NBZ system and within the region 1/2 system has been assumed here. Since the main emphasis in this study is on the shape rather than the strength of the convection, a region 1 current magnitude of 1 MA has been used throughout the cases shown. The conductivity model is the same as the one used for the transpolar arc model study.

It can be seen that as the NBZ current increases from 0 MA (left column) to 0.8 MA (right column) the convection pattern changes from a two cell, to a three cell, and finally to a four cell pattern. This evolution is seen to be more pronounced for the case with a region 2 current magnitude of 0.5 MA (bottom row) than for the case with no region 2 current (top row). The reason for this is that the global con-

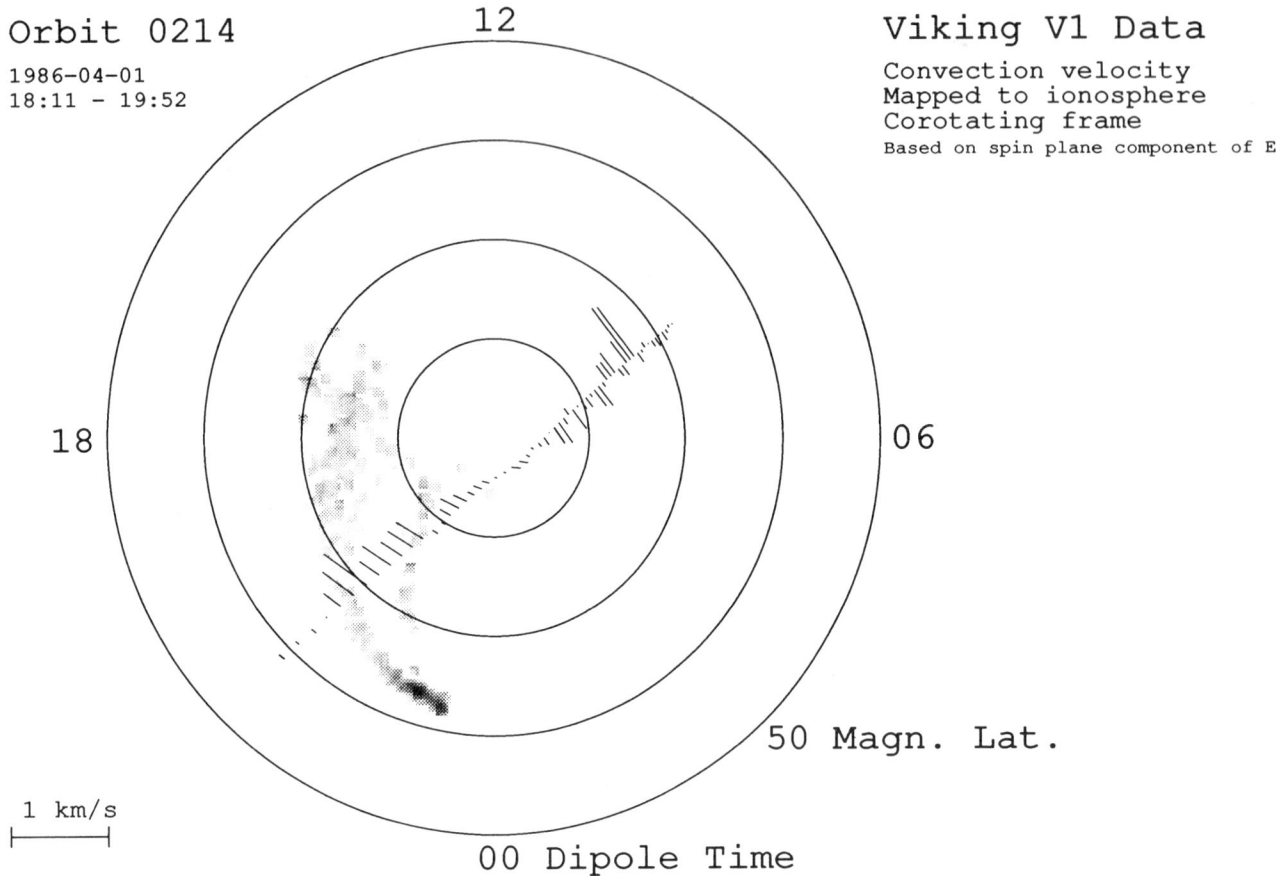

Fig. 6. Duskside auroral distribution as viewed by the UV imager experiment on the Viking satellite during Viking orbit 214 at 1830 UT on April 1, 1986. Superposed on this is the $\mathbf{E} \times \mathbf{B}$ drift pattern calculated from the Viking electric field data and plotted along the ionospheric projection of the Viking orbit.

vection pattern is largely controled by the interplay between the NBZ currents and the net current from the region 1/2 system which decreases with increasing region 2 current.

The characteristics of this evolution critically depends on the conductivity model and the way the currents are balanced. If the NBZ currents do not perfectly balance which is likely to be the normal situation, but close partly through the adjacent region 1 current, the dawn-dusk asymmetry of the convection pattern is further pronounced. By introducing the current-conductivity coupling term the convection cells associated with upward field-aligned currents weakens compared to the cells associated with the downward currents as discussed in Section 3. For a given current distribution, an increase in the UV conductivity contribution (seasonal effect) results in a weakening of the "polar convection" relative to the ambient convection associated with the region 1/2 system.

8. Discussion

The modeling studies presented here serve several purposes. They provide the scientific community with realistic and self consistent global snapshots of the auroral electrodynamical parameters for various events. Such information, which is necessary for a quantitative understanding of the ionospheric effects of magnetospheric processes and for tests of theoretical models, has not been available to date. The studies also provide valuable explicit examples of the quantitative relationships between the ionospheric conductivities, field-aligned currents, and electric fields with respect to the auroral distribution.

Consider, for example, the basic hypothesis that more intense auroral emissions are related to upward field-aligned currents. As regard to discrete auroral structures there should be no problem with this assumption. For the diffuse aurora, the relationship to field-aligned currents is more complicated. The diffuse aurora has, in particular on the evening side, been associated with downward field-aligned currents [Kamide and Akasofu, 1976], but above a certain intensity threshold the emissions are more likely associated with upward currents. For the event studies referred to above a good qualitative agreement was found between the locations of upward field-aligned currents and locations of

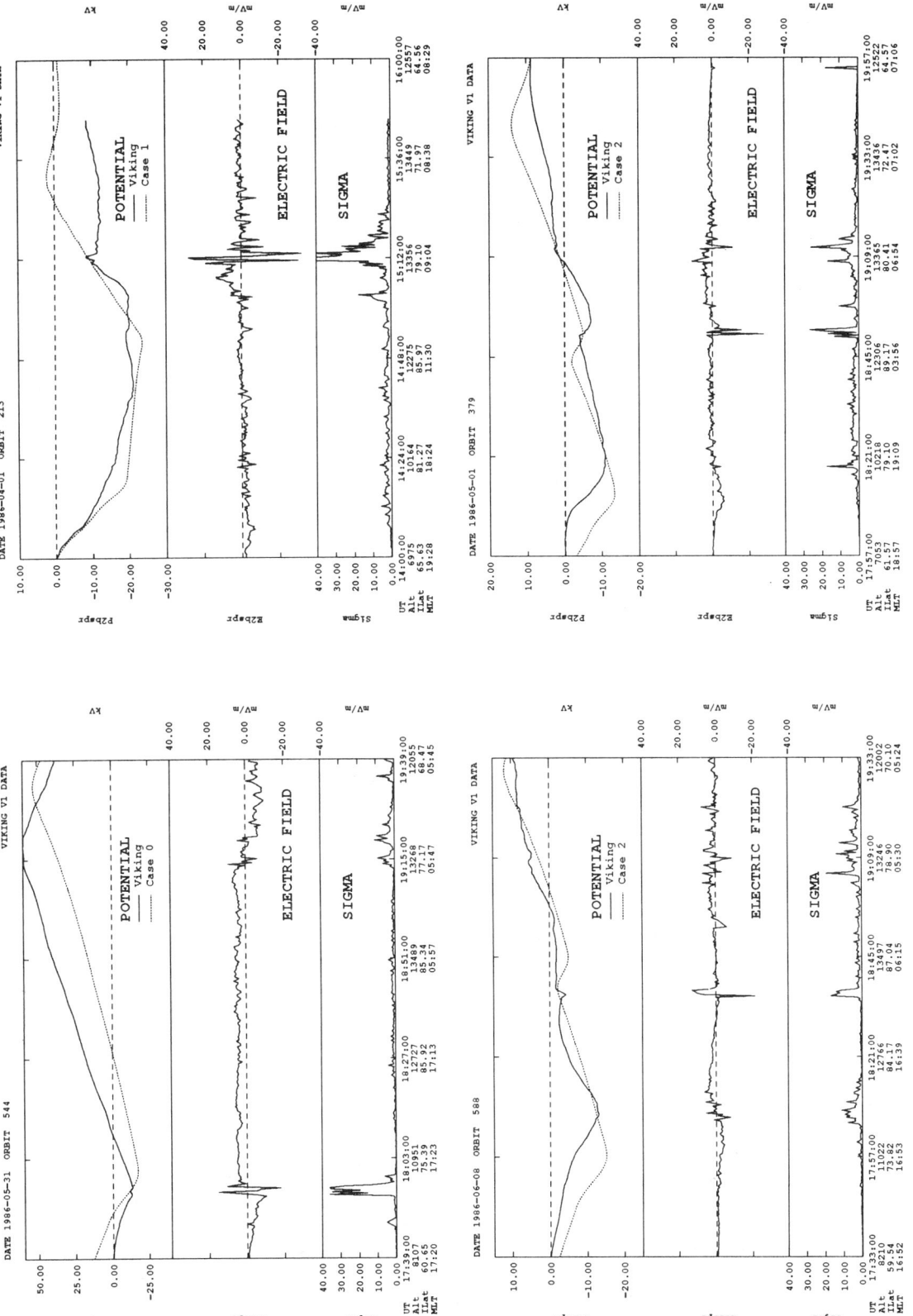

Fig. 7. Viking electric field data from four different dusk-dawn orbits illustrating cases 0, 1 and 2 in Figure 5. The panels show as a function of time: the dawn-dusk potential profile (top), the dawn-dusk electric field (middle) and the standard deviation of the electric field (bottom). The dotted curves in the top panels show for comparison the corresponding model potential profiles along the ionospheric footprint of Viking.

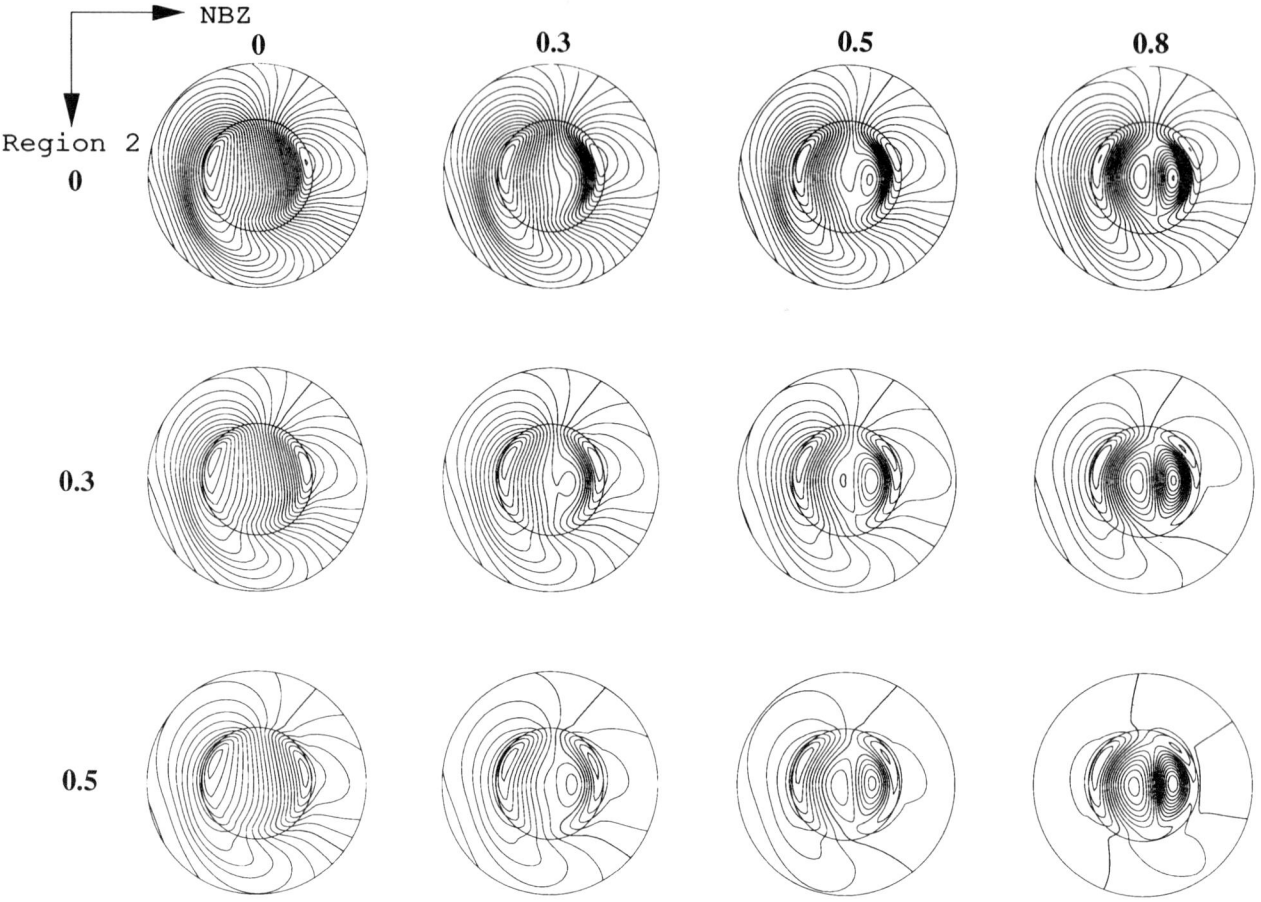

Fig. 8. High-latitude potential distributions for various region 2/region 1 (top to bottom) and NBZ/region 1 (left to right) current ratios. The figure also illustrates the possible evolution as the IMF changes from a southward to a northward orientation and IMF $B_y < 0$.

more intense auroral emissions concerning the large-scale features. This was found to be the case also for the relatively structured and wide morningside auroral oval prevailing during the transpolar arc event presented above (cf. Figure 4).

A more direct connection exists, of course, between the auroral emissions and the ionospheric conductivities, which has been used by e.g., *Kamide et al.* [1986] to derive instantaneous conductivity patterns from DE 1 UV images. The relationships between the auroral emissions, the conductivities, and the upward currents which in reality are very complicated are represented in the model in very simplified forms. Nevertheless, the use of these simplified relationships when studying auroral UV images makes it possible both to perform a quantitative modeling and to obtain a qualitative understanding of the auroral electrodynamics on a global scale.

How does the electric field, current, and conductivity patterns that are associated with the large variety of auroral distributions viewed by the UV imager compare with statistical patterns for similar geomagnetic conditions, or for similar orientations of the IMF? During periods of southward IMF the auroral activity and associated field-aligned currents and conductivities are essentially concentrated to the classical auroral oval. The global convection pattern which is typically of the two cell type varies in a rather predictable way with the auroral distribution and the IMF B_y component. The rotation of the convection pattern with increasing particle-induced conductivity [e.g., *Yasuhara et al.*, 1983], local electric field distortions due to polarisation fields in the vicinity of bright auroral features associated with intense conductivity gradients [e.g., *Kan et al.*, 1984; *Marklund et al.*, 1984] as well as characteristic IMF B_y-asymmetries [cf. *Mozer and Gonzalez*, 1973] are features which today are relatively well understood.

Figure 9 shows three convection patterns, all derived for IMF $B_z < 0$ and IMF $B_y > 0$ conditions and similar geomagnetic activity. The instantaneous convection pattern

INSTANTANEOUS CONVECTION PATTERN

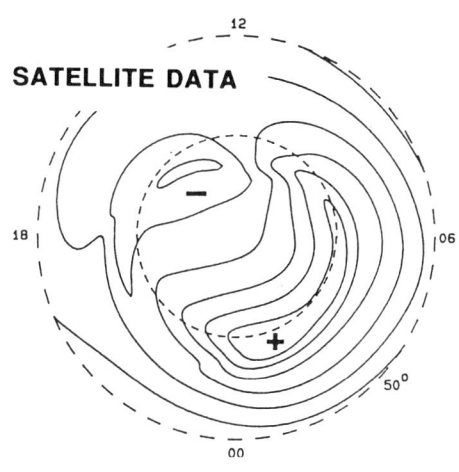

Marklund et al, 1988

STATISTICAL CONVECTION PATTERNS

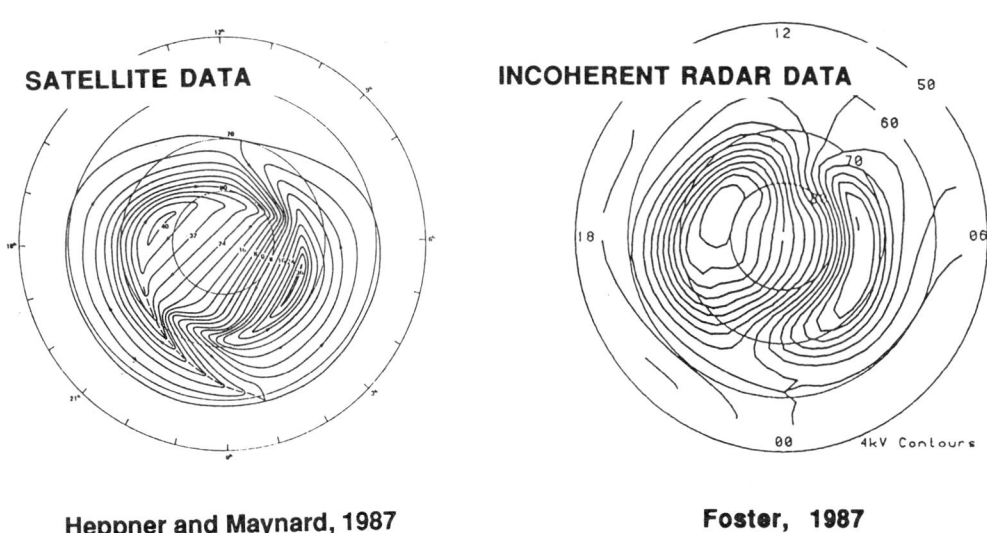

Heppner and Maynard, 1987   Foster, 1987

Fig. 9. Comparison between instantaneous and statistical convection patterns for southward IMF conditions and IMF $B_y > 0$.

shown at the top has been taken from *Marklund et al.* [1988] and the two statistical patterns shown at the bottom have been taken from *Heppner and Maynard* [1987] (left) and from *Foster* [1987] (right) based on DE 2 electric field data and Millstone Hill radar data, respectively. The relatively large differences between the two statistical patterns are believed to be mainly a result of the different instrumental techniques and methods used to derive the patterns. Note that the instantaneous pattern and the average (BC) pattern of Heppner and Maynard are quite similar, as regards to the general form of the convection cells and to the orientation of the polar cap electric field (rotation of the

INSTANTANEOUS CONVECTION PATTERN

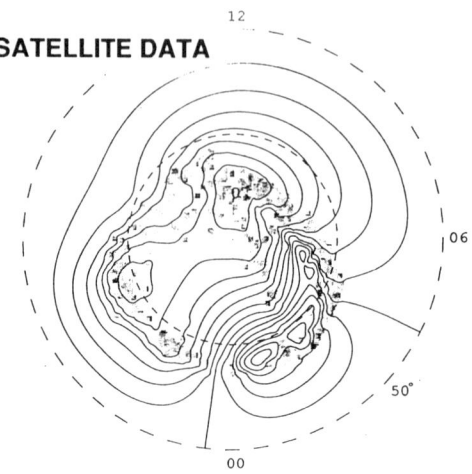

STATISTICAL CONVECTION PATTERNS

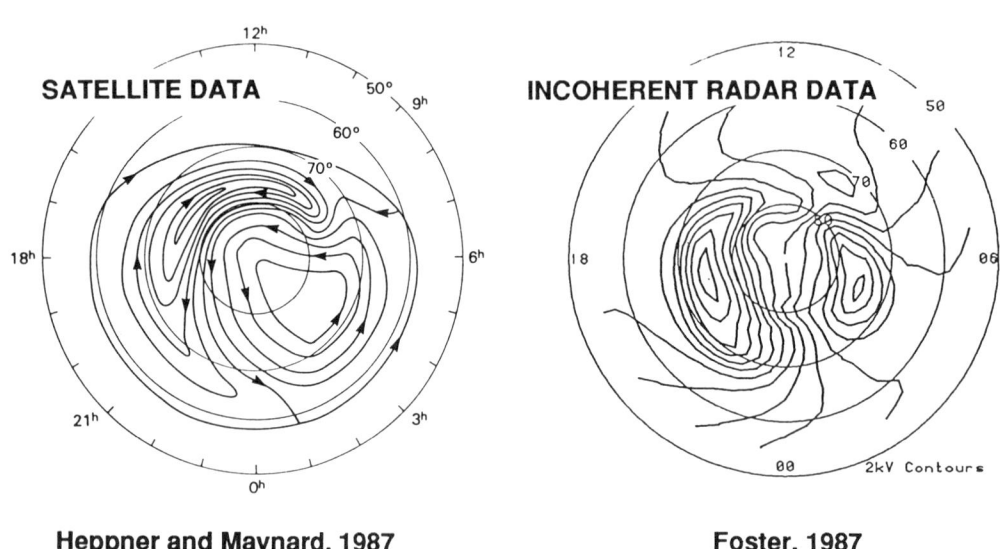

Fig. 10. Same as Figure 9 but for northward IMF conditions and IMF $B_y < 0$.

convection pattern). The main difference can be seen in the polar cap where the convection streamlines are severely curved for the instantaneous pattern but straight for the statistical pattern.

The ionospheric electrodynamics during northward IMF events are much more difficult to predict or reconstruct. Figure 10 shows a comparison similar to that of Figure 9 but for IMF $B_y < 0$ and IMF $B_z > 0$ conditions. Note that the differences between the various patterns are here much more pronounced than for the patterns representative of IMF $B_z < 0$ conditions. Another type of pattern that has been suggested for northward IMF conditions is the multicell pattern and in particular the four cell pattern [e.g., Burke et al., 1979; Heelis et al., 1986] which further serves to strengthen this point. This reflects both the larger variety of auroral electrodynamical conditions that exist for IMF

$B_z > 0$ and the larger difficulties involved in inferring the electrodynamical patterns when the auroral activity extends over the entire polar region. It is therefore unlikely that a statistical pattern should be representative of a specific northward IMF event. The large differences seen between the different patterns (and even between the two statistical patterns) also suggest that improved methods, better data sets and data coverage are needed to reduce artificial features in the derived patterns.

The approach chosen here to investigate the complicated electrodynamics associated with northward IMF phenomena has been to conduct event and idealized model studies in parallel. The distinction made between situations characterized by extended NBZ currents and situations characterized by one major transpolar arc (or theta aurora) is to account for both large-scale and smaller-scale phenomena and their influence on the global electrodynamics. Events with large-scale NBZ currents are difficult to reconstruct on the basis of UV images and magnetometer data. The coupling between the relatively diffuse auroral structures at high latitudes and the current is unclear and the coverage by in situ magnetometer measurements or ground-based magnetometer observations in the polar cap region is not sufficient to reconstruct the extended current sheets. This is in contrast to the situation with a well defined transpolar arc, where the location and orientation of the current sheets are likely to be given by the visible arc. This assumption was used to derive the field-aligned current distribution and potential pattern associated with the transpolar arc event treated in Section 5 above (cf. Figure 4). The transpolar arc was here modeled by an upward field-aligned current closed by the downward region 1 and cusp currents. Another possibility would have been to model the arc by two oppositely directed current sheets, partly or completely balancing each other.

Comparing our basic model situations to Viking electric field data we conclude that the case with a net upward transpolar arc-associated current and the case with balanced transpolar arc current sheets, respectively, are the situations most commonly occuring in reality. As illustrated in Figure 5 the influence on the global convection pattern is much more pronounced for the case with a single polar current sheet. This is natural since in this case the current can not close within the "polar cap" as in the case with balanced current sheets. The basic effect of a single upward current sheet will be to introduce an additional convection cell. However, a more interesting result is that if the current is sufficiently strong this additional cell will "merge" with the evening cell, giving a large cell having its minimum (or reversal) at an extremely high latitude. This is in clear contrast to the case with two (or more) locally balanced current sheets where the modifications to the convection pattern are localized. If the transpolar arc is associated with (or embedded in) a region of net downward current the associated convection cell might "merge" with the morning cell, resulting in the "mirror image" of the convection pattern discussed above. Although such situations are not a priori ruled out we have so far not found observational evidence in support of them. It should be mentioned that the modifications of the reference pattern due to transpolar arc currents should be more pronounced for a higher value of the region 2/region 1 current ratio than has been used here (i.e., 0.5).

9. Summary and Conslusions

This paper reviews a number of interrelated modeling studies of global auroral electrodynamics that have been completed or initiated over the last few years. The common aims of these studies are: to improve our understanding of the typical distributions of the various parameters of interest to large-scale auroral electrodynamics (i.e., electric field, current, conductivity, ohmic power dissipation etc); to gain insight into the relationships between these parameters and their respective boundaries; to study the influence of ionospheric current closure on ionospheric convection; to provide snapshots of the electrodynamical situation prevailing at a given instant; and to improve our feeling for how the ionosphere interacts with the magnetosphere.

In particular the influences of assumed relationships such as between auroral emissions and particle-induced conductivity, between particle-induced conductivity and upward field-aligned current and between field-aligned current and parallel potential drops have been carefully examined in idealized model studies and then utilized in an interactive way for the reconstruction of selected events. The ultimate goal has been to construct realistic and self consistent global images of the temporally and spatially varying auroral electrodynamical parameters from a given set of multi point observations. Northward IMF phenomena such as large-scale NBZ currents and transpolar arcs and their associated electrodynamics, for which the common understanding is still in a relatively immature state, have been given particular attention in this overview. A brief summary of some important results from these studies is given below.

1. The inclusion of the experimentally justified linear relationship between $j_\parallel$ and Σ_P in the modeling has profound influence on the resulting potential patterns, which become more realistic and consistent with observations. A less pronounced clockwise rotation of the polar cap electric field and a reduction of the electric field in regions of upward currents are the major features resulting from the coupling assumption.

2. By taking into account the $j_\parallel - V_\parallel$ relationship it is possible to reconstruct also the high-altitude (magnetospheric) potential in addition to the ionospheric potential. In this way it has been shown that even the localized potential minima measured on auroral flux tubes above the acceleration region could be accurately reproduced by using a physically reasonable value for the coupling parameter. A modification or distortion in the magnetospheric flow pattern occurs above these potential drops, being consistent with the requirement leading to substorm onset according to a recent theory.

3. As illustrated by the results from the snapshot studies the large variety of auroral configurations that are observed, in particular during northward IMF conditions, correspond to an equally large variety of convection patterns which in many respects differ significantly from predictions based on statistical results or theoretical considerations. The information gained from such individual event studies is essential for a quantitative understanding of the dynamic ionosphere-magnetosphere system, as a critical test of various theories, and as a reference frame necessary to understand smaller-scale phenomena.

4. The global potential patterns associated with "ideal" polar arc (or theta aurora) events, without the simultaneous presence of NBZ currents, are found to be basically of the two cell type. The modifications typically show up as a poleward expansion of one of the cells (causing sunward flow at very high latitudes) or as a local convection reversal in the center of the arc, depending on how the polar arc-associated currents close in the ionosphere. The balance between the ambient and the arc-associated electric field and thus the convection signatures are typically found to vary significantly along the arc.

5. For the case with large-scale NBZ currents, associated with more extended regions of dim auroral features the potential pattern is more drastically modified and can be shown to have two, three, or four cells depending on the ratio between the NBZ currents and the region 1/2 currents, on the closure of these currents, on the direction of the IMF, and on the conductivity model. A qualitative illustration is given of the evolution from a two cell pattern, representative of southward IMF, through a transitional state that may be of the three cell type, to a four cell pattern, that may be representative of persistent northward IMF.

6. More generally the interrelated model and event studies have provided valuable experience on the physical relationships between the various electrodynamical parameters both from the interactive way the model data are calibrated or checked against measurements in event studies and from investigations of the basic assumptions in idealized model studies.

Although a lot of effort has been put into the work discussed above, we are only at the beginning of gaining a deeper understanding of the global auroral dynamics, and much remains to be investigated more thoroughly. Among the studies to be conducted in the near future are: inclusion of time evolution in our snapshot scheme; use of UV images and in situ measurements from both hemispheres simultaneously for snapshot modeling; a joint event study using our snapshot technique and the AMIE technique; and a joint statistical study of electric fields, field-aligned currents, and, if possible, conductivity where all data are obtained from the same spacecraft or set of spacecraft.

Acknowledgments. The authors are grateful to the large number of people who have contributed to these studies either in terms of delivery of data or in terms of valuable comments or suggestions to the work. We are especially grateful to the scientists at: Johns Hopkins University (T. A. Potemra, L. J. Zanetti, R. E. Erlandson and others); University of Calgary (R. D. Elphinstone, J. S. Murphree); Swedish Institute of Space Physics (K. Stasiewicz, I. Sandahl, R. Lundin, L. Eliasson and others); Air Force Geophysics Laboratory, (F. J. Rich, D. A. Hardy, P. L. Rothwell) and to O. de la Beaujardière, SRI International; M. B. Silevitch, Northeastern University, Boston; and R. A. Heelis, University of Texas at Dallas. This research was supported by the Swedish National Space Board. The Viking Project was managed and operated by the Swedish Space Corporation under contract from the Swedish National Space Board.

References

Blomberg, L. G., and G. T. Marklund, The influence of conductivities consistent with field-aligned currents on high-latitude convection patterns, *J. Geophys. Res.*, *93*, 14493, 1988.

Blomberg, L. G., and G. T. Marklund, High-latitude convection patterns for various large-scale field-aligned current configurations, *Geophys. Res. Lett.*, *18*, 717, 1991a.

Blomberg, L. G., and G. T. Marklund, A numerical model of ionospheric convection derived from field-aligned currents and the corresponding conductivity, *Rep. TRITA-EPP-91-03*, Royal Inst. of Technol., Stockholm, 1991b.

Brüning, K., L. P. Block, G. T. Marklund, L. Eliasson, R. Pottelette, J. S. Murphree, T. A. Potemra, and S. Perraut, Viking observations above a postnoon aurora, *J. Geophys. Res.*, *95*, 6039-6049, 1990.

Burke, W. J, M. C. Kelley, R. C. Sagalyn, M. Smiddy, and S. T. Lai, Polar cap electric field structures with a northward interplanetary magnetic field, *Geophys. Res. Lett.*, *6*, 21, 1979.

de la Beaujardière, O., D. Alcaydé, J. Fontanari, and C. Leger, Seasonal dependence of high-latitude electric fields, *J. Geophys. Res.*, *96*, 5723, 1991.

Evans, D. S., N. C. Maynard, J. Troim, T. Jacobsen, and A. Egeland, Auroral vector electric field and particle comparisons. 2. Electrodynamics of an arc, *J. Geophys. Res.*, *82*, 2235, 1977.

Foster, J. C., Radar-deduced models of the convection electric field, in Quantitative modeling of magnetosphere-ionosphere coupling processes, edited by Y. Kamide and R.A. Wolf, *Kyoto, March 9 - 13, 1987*.

Hardy, D. A., M. S. Gussenhoven, and E. Holeman, A statistical model of auroral electron precipitation, *J. Geophys. Res.*, *90*, 4229, 1985.

Hardy, D. A., M. S. Gussenhoven, and D. Brautigam, A statistical model of auroral ion precipitation, *J. Geophys. Res.*, *94*, 370, 1989.

Harel, M., R. A. Wolf, P. H. Reiff, R. W. Spiro, W. J. Burke, F. J. Rich, and M. J. Smiddy, Quantitative simulation of a magnetospheric substorm, 1, model logic and overview, *J. Geophys. Res.*, *86*, 2217, 1981.

Heelis, R. A., P. H. Reiff, J. D. Winningham, and W. B. Hanson, Ionospheric convection signatures observed by the DE2 during northward interplanetary magnetic field, *J. Geophys. Res.*, *91*, 5817, 1986.

Heppner, J. P., and N. C. Maynard, Empirical models of high-latitude electric fields, *J. Geophys. Res.*, *92*, 4467, 1987.

Iijima, T., and T. A. Potemra, The amplitude distribution of field-aligned currents at northern high latitudes observed by Triad, *J. Geophys. Res.*, *81*, 2165, 1976a.

Iijima, T., and T. A. Potemra, Field-aligned currents in the dayside cusp observed by Triad, *J. Geophys. Res.*, *81*, 2165, 1976b.

Iijima, T., and T. A. Potemra, Large-scale characteristics of field-aligned currents associated with substorms, *J. Geophys. Res.*, *83*, 599, 1978.

Iijima, T., T. A. Potemra, L. J. Zanetti, and P. F. Bythrow, Large-scale Birkeland currents in the dayside polar region during strongly northward IMF: A new Birkeland current system, *J. Geophys. Res.*, *89*, 7441, 1984.

Iijima, T. and T. Shibaji, Global characteristics of northward IMF-associated (NBZ) field-aligned currents, *J. Geophys. Res.*, *92*, 2408, 1987.

Kamide, Y., J. D. Craven, L. A. Frank, B.-H. Ahn, and S.-I. Akasofu, Modeling substorm current systems using conductivity distributions inferred from DE auroral images, *J. Geophys. Res.*, *91*, 11235, 1986.

Kamide, Y., A. D. Richmond, and S. Matsushita, Estimation of ionospheric electric fields, ionospheric currents and field-aligned currents from ground magnetic records, *J. Geophys. Res.*, *86*, 801, 1981.

Kamide, Y., and S.-I. Akasofu, The location of the field-aligned currents with respect to discrete auroral arcs, *J. Geophys. Res.*, *81*, 3999, 1976.

Kan, J. R., R. L. Williams, and S.-I. Akasofu, A mechanism for the westward travelling surge during substorms, *J. Geophys. Res.*, *89*, 2211, 1984.

Knight, S. Parallel electric fields, *Planet. Space Sci.*, *21*, 741, 1973.

Lyons, L. R., Generation of large-scale regions of auroral currents, electric potentials, and precipitation by the divergence of the convection electric field, *J. Geophys. Res.*, *85*, 17, 1980.

Lyons, L. R., D. S. Evans, and R. Lundin, An observed relation between magnetic field aligned electric fields and downward electron energy fluxes in the vicinity of auroral forms, *J. Geophys. Res.*, *84*, 457, 1979

Marklund, G. T., I. Sandahl, and H. Opgenoorth, A study of the dynamics of a discrete auroral arc, *Planet. Space Sci.*, *30*, 179, 1982.

Marklund, G. T., W. Baumjohann, and I. Sandahl, Rocket and ground-based study of an auroral breakup event, *Planet. Space Sci.*, *31*, 207, 1983.

Marklund, G. T., L. G. Blomberg, T. A. Potemra, J. S. Murphree, F. J. Rich, and K. Stasiewicz, A new method to derive "instantaneous" high-latitude potential distributions from satellite measurements including auroral imager data, *Geophys. Res. Lett.*, *14*, 439, 1987.

Marklund, G. T., R. A. Heelis, and J. D. Winningham, Rocket and satellite observations of electric fields and ion convection in the dayside auroral ionosphere, *Can. J. Phys.*, *64*, 1417, 1986.

Marklund, G. T., L. G. Blomberg, K. Stasiewicz, J. S. Murphree, R. Pottelette, L. J. Zanetti, T. A. Potemra, D. A. Hardy, and F. J. Rich, Snapshots of high-latitude electrodynamics using Viking and DMSP F7 observations, *J. Geophys. Res.*, *93*, 14479, 1988.

Marklund, G. T., L. G. Blomberg, J. S. Murphree, R. D. Elphinstone, L. J. Zanetti, R. E. Erlandson, I. Sandahl, O. de la Beaujardière, H. Opgenoorth, and F. J. Rich, On the electrodynamical state of the auroral ionosphere during northward IMF: A transpolar arc case study, *J. Geophys. Res.*, *96*, 9567, 1991.

Marklund, G. T. and L. G. Blomberg, On the influence of localized electric fields and field-aligned currents associated with polar arcs on the global potential distribution, *J. Geophys. Res.*, in press, 1991.

Marklund, G. T., M. A. Raadu, and P.-A. Lindqvist, Effects of Birkeland current limitation on high-latitude convection patterns, *J. Geophys. Res.*, *90*, 10864, 1985.

Mishin, V. M., S. B. Lunyushkin, D. Sh. Shirapov, and W. Baumjohann, A new method for generating instantaneous ionospheric conductivity models using ground-based magnetic data, *Planet. Space Sci.*, *34*, 713, 1986.

Mishin, V. M., A. D. Bazarzhapov, and G. B. Shpynev, Electric fields and currents in the earth's magnetosphere, in Dynamics of the magnetosphere, edited by S.-I. Akasofu, *D. Reidel, Hingham, Mass.*, pp. 249, 1979.

Mozer, F. S. and W. D. Gonzalez, Response of polar cap convection to the interplanetary magnetic field, *J. Geophys. Res.*, *78*, 6784, 1973.

Nopper, R. W. Jr., and R. L. Carovillano, On the orientation of the polar cap electric field, *J. Geophys. Res.*, *84*, 6489, 1979.

Richmond, A. D., and Y. Kamide, Mapping electrodynamic features of the high-latitude ionosphere from localized observations: Technique, *J. Geophys. Res.*, *93*, 5741, 1988.

Richmond, A. D., Y. Kamide, B.-H. Ahn, S.-I. Akasofu, D. Alcayde, M. Blanc, O. de la Beaujardière. D. S. Evans, J. C. Foster, E. Friis-Christensen, T. J. Fuller-Rowell, J. M. Holt, D. Knipp, H. W. Kroehl, R. P. Lepping, R. J. Pellinen, C. Senior, and A. N. Zaitsev, Mapping electrodynamic features of the high-latitude ionosphere from localized observations: Combined incoherent scatter radar and magnetometer measurements for January 18-19, 1984, *J. Geophys. Res.*, *93*, 5760, 1988.

Wallis, D. D., and E. E. Budzinski, Empirical models of height-integrated conductivities, *J. Geophys. Res.*, *86*, 125, 1981.

Yasuhara, F., R. Greenwald, and S.-I. Akasofu, On the rotation of the polar cap potential pattern and associated polar phenomena, *J. Geophys. Res.*, *88*, 5773, 1983.

6. CHARACTERISTICS OF SUBSTORM PHASES

Satistical Features of The Substorm Expansion-Phase as Observed by The AMPTE/CCE Spacecraft

I. A. Daglis[1,3], N. P. Paschalidis[2,3], E. T. Sarris[3], W. I. Axford[1], G. Kremser[4,1], B. Wilken[1], and G. Gloeckler[5]

Abstract. We present a statistical study of the substorm expansion phase in terms of variations of the energy density of the major ions (H^+, O^+, He^{++}, He^+). The data were collected by the Charge-Energy-Mass (CHEM) spectrometer on board the AMPTE/CCE satellite in the equatorial (within ± 16° magnetic latitude) nightside magnetosphere, at geocentric distances ≤ 9 R_E, and cover the period from March 1985 to December 1987 (which makes a total of 1255 CCE orbits with the apogee located on the nightside). Previous statistical studies of magnetospheric populations were limited to energies less than 32 keV/e at geosynchronous altitudes, less than 17 keV/e at other altitudes, and energies greater than about 200 keV/nucleon; the CHEM data (energies ~1 to 300 keV/e) fill the energy gap, in which the bulk of the region's energy density is contained. Furthermore, there are two major differences between this study and previous ones: first we explicitly used time intervals just after substorm onset and not active times in general; second we examined the energy density instead of the number density, since we were mainly interested in the energy budget evolution and not in the composition of the ion population. Our observations show a remarkable difference between the energy-densities of O^+ and He^{++} ions in the early expansion phase of the substorm: the ionospheric-origin O^+ energy density grows non-linearly with AE, while the solar wind-origin He^{++} energy density seems almost uncorrelated with AE. The mixed-origin H^+ and He^+ energy-densities exhibit an intermediate behavior by increasing with AE, but in a lesser degree than O^+. We suggest that the trend of O^+ energy density is associated with increased direct feeding of the near-Earth magnetotail with energetic ionospheric ions, during highly active periods. The outflowing energetic ionospheric plasma can participate in the cross-tail current enhancement during late growth-phase (*Daglis et al.*,1990,1991) and subsequently favour the growth of instabilities in the plasma sheet (*e.g. Baker et al.*, 1982). Another interesting feature is the tendency of the He^{++} energy density to higher values towards dawn; this may be interpreted as a result of increased penetration of solar wind ions from the magnetosheath through the dawnside magnetopause (*Lennartson and Sharp*, 1982).

---

[1]Max - Planck - Institut für Aeronomie, Katlenburg - Lindau, Federal Republic of Germany

[2]The Johns Hopkins University, Applied Physics Laboratory, Laurel, MD, 20723

[3]Section of Telecommunications and Space Science, Demokritos University of Thrace, Xanthi, Greece

[4]Department of Physics, University of Oulu, SF-90570 Oulu, Finland

[5]Department of Physics and Astronomy, University of Maryland, College Park, MD 20742, U.S.A.

Magnetospheric Substorms
Geophysical Monograph 64
Copyright 1991 American Geophysical Union

INTRODUCTION

The energization processes involved in substorms, the composition and source of the magnetospheric plasmas are current problems in magnetospheric research. A number of studies concerned with variations of the ion number density in correlation with the level of magnetospheric activity and solar flux, have been published during the last years. Most of these studies used data from the ISEE-1 and GEOS spacecrafts, limiting the investigation to energies below 17 keV/e. The central point of discussion was usually the composition of the magnetospheric plasma, and the relative contribution of the solar wind and the ionosphere.

Peterson et al. [1981] examined a group of six time intervals and showed that during highly active times O+

ions can represent more than half of the plasma sheet number density.

Lennartsson and Sharp [1982] presented a study on the ion composition in the near equatorial magnetosphere with data from the plasma composition experiment on ISEE 1. They found that at L>5 all four ion species may be present simultaneously in highly variable ratios, but the H+ is the most abundant, particularly during quiet conditions. The O+/H+ and He+/H+ ratios decrease with increasing L, while the He++/H+ ratio slowly increases. They reported typical values of 1 to 3% for the He++/H+ and He+/H+ ratios during both quiet and disturbed times, and 1 to 200% for the O+/H+ ratio, depending on the magnetic activity level.

Lennartsson and Shelley [1986] found that there is a linear relation between the ratio O+/H+ and the AE index on a time scale of several hours. Furthermore they found that O+ number density is rising with rising AE, while H+ and He++ number-densities are falling.

Gloeckler and Hamilton [1987] reported the following quiet-time number densities, observed by the CHEM instrument in the plasma sheet (L~ 9 R_E): for H+ ~ 0.7 cm-3, for O+ ~ 0.034 cm-3, for He++ ~ 0.006 cm-3 and for He+ ~ 0.0098 cm-3. Thus, protons are a dominant constituent of the quiet-time plasma sheet (92.2% of the total number density). The contribution of the O+ ions increases with decreasing geocentric distance and with increasing geomagnetic activity.

Lennartsson [1989] reported that the O+ number density increased by one order of magnitude during the rising phase of solar cycle, 21 (mid 1977 to late 1979).

All these results have confirmed the existence of an ionospheric source of O+ and a solar wind source of He++. What remains unclear is the contribution of the two sources to the most abundant species of the magnetosphere, that is H+. Also unclear are the access paths of the different ion species to the acceleration regions.

This study is concerned with statistical features of the energetic particle (1 to 300 keV/e) energy density in the early expansion phase of substorms.

DATA

The AMPTE/CCE spacecraft has been described by *Dassoulas et al.* [1985], and the CHEM spectrometer has been described in detail by *Gloeckler et al.* [1985]. However, we will summarize some of the features of the spacecraft orbit and the spectrometer here.

The CCE spacecraft is in an elliptic orbit with apogee at ~ 8.8 R_E, perigee at 1.2 R_E, and an orbital inclination of ~ 4.8°. The orbital period is ~ 16 hrs; the spacecraft is spin stabilized with the spin axis roughly pointing along the Earth-Sun line and has a spin period of ~ 6 sec. The orbit apogee moves through all local times within ~ 16 months. The CHEM spectrometer determines the energy (E), mass per charge (M/Q) ratio, mass (M) and pitch angle (α) of ions from ~ 1 to 300 keV/e, using a combination of electrostatic deflection, post acceleration of up to 30 kV, and time-of-flight and energy measurements. The energy range of ~ 1 to 300 keV/e is sampled in 32 logarithmic steps with a 18.9% spacing.

An onboard classification system assigns to each pair of energy and time-of-flight signals a M and M/Q value and stores the counts in the corresponding register of the M versus M/Q matrix; the sections of this matrix correspond to specified ion species (H+, O+ etc.) and the accumulated counts of each section are read out at each energy step as matrix rates.

To convert the rates into the physically meaningful differential flux intensity j (E,α) we divide the count rates by conversion factors which include the geometric factor of CHEM and the energy bandwidth. Although the expected background is very low, we apply a background-subtraction algorithm (provided by *L. Kistler*) in order to have the best quality data possible. For this study we used the energy density, which is calculated via the formula:

$$ED = \int \frac{1}{2} m v^2 f(\bar{v}) \, d\bar{v} =$$

$$= \pi \sqrt{2m} \int \int j(E,\alpha) \sqrt{E} \sin\alpha \, dE \, d\alpha \qquad (1)$$

where f(v) is the velocity space distribution function and is related to the differential flux intensity as follows (*e.g. Lyons and Williams*, 1984):

$$f(\bar{v}) = \frac{m^2 j(E,\alpha)}{2E} \qquad (2)$$

Our data come from orbits with the apogee located in the nightside magnetosphere, anywhere between late afternoon and early morning (i.e., between 19:00 LT and 07:00 LT). The data cover the periods January 11 to October 15, 1985, April 16, 1986 to January 17, 1987 and July 30, 1987 to April 29, 1988; that makes a total of 1255 orbits.

To select substorm-onset events we used temporal evolution plots of the onmidirectional differential flux intensity of H+ the energy density of the four ion species H+ O+, He++, He+, and the magnetic field (data from the magnetic field experiment onboard AMPTE/CCE, details by *Potemra et al.*, 1985). The onset-selection criteria were:

a. dispersionless (on a time-scale of 186 sec, which is the duration of the instrument's full energy cycle) injection of high-energy ($\geq$ 15 keV/e) ions

b. reconfiguration of the magnetic field from a tail-like to a dipole-like geometry

c. substantial increase (by a minimum of 50% of the pre-substorm level) of the particle energy density

The fulfillment of these criteria implies that the spacecraft is located directly within the region of maximum substorm activity just after the start of explosive dissipation of energy.

OBSERVATIONS

The resulting group consisted of 131 events. Then the energy density behavior of each ion species with respect to AE index was considered. For AE index we took the mean value of the index during the first 15 minutes after substorm onset as defined above. Unfortunately the AE index has not been available for the whole time period of interest. Thus the study was performed only for the January 1985 to August 1986 period and included 79 events. The use of the 1-minute AE index instead of the 3-hours Kp index distinguishes this work from past ones. The obvious difference between these two indices is that the AE is much more precise than the Kp; two substorm onsets occurring within 3 or 4 hours will have the same Kp index, while they will have two distinct AE values.

The energy density calculations obtained for periods after the expansion phase. As we mentioned in the data section the CHEM view cone is almost perpendicular to sun-earth line. This precludes the measurements of field aligned ions during the growth phase. However the pitch angle coverage after the expansion phase and the dipolarization of the field, is almost complete (periods were the study is performed). Then most part of the ion distribution is taken into account and only narrow field aligned beams are unestimated. Nevertheless we do not expect significant contributions to the bulk energy density from beams and low energy ions during/after substorm expansion phase. Any instrumental biasing of the data due to incomplete pitch-angle coverage should be common to all ion species and thus becomes of minor scientific importance when we consider the relative behavior of the ion energy densities and not only the absolute energy density of each species.

Next we took the mean value of each ion-species' energy density during the first 15 minutes after local substorm-onset (identified as stated above by the high-energy particle injection and the magnetic field dipolarization). The 15-minutes interval was necessary in order to maintain good statistics for all ion species. The energy density was calculated by considering 9 pitch-angle bins (0-20, 20-40,...,160-180) and summing over the 32 energy channels and over the 9 pitch-angle bins. In order to examine the mean energy density characteristics with respect to location and substorm size, we used three sorting parameters: magnetic local time (MLT), radial distance (r/R_E), and AE index.

No significant dependence of the energy density upon the radial distance was found. The relation between energy density and local time was examined by plotting the expansion-phase mean energy density of each event versus the magnetic local time position of the spacecraft at substorm onset. No special trend was found, except for He++, where the linear fitting showed a tendency to higher energy density values towards dawn (Figure 1). In order to check the confidence level of the correlation for all fits of this study, we used the correlation coefficient table given by *Bevington* [1969] and the F-test as described by *Stoodley et al.* [1980]. In the case of He++ energy density versus magnetic local time, the correlation coefficient for the 131 events is r = 0.21 and there is a probability between 0.05 and 0.01 that the parameters are uncorrelated. From the F-test we obtained (for 131 observations and an F-ratio value of 6.04) a confidence level of 97.5% (which means that there is a probability less than 0.025 that the variables are uncorrelated).

The highest values of O+ energy density occur in the pre-midnight region, but there is no strong bias in mean energy density versus local time. We obtain a different picture if we plot the same variables for strong substorms only, that is with an AE value greater than 500 nT (Figure 2). In this case, an exponential fitting can be applied, with a confidence level between 95% and 97.5%, and a tendency for higher energy density towards dusk. This tendency is absent in the weak-substorm group (*Daglis*, 1991], not shown here).

Figures 3 to 6 show the dependence of the mean expansion-phase energy density of each event on the corresponding AE index; the best fit to the data is presented in every plot. The best correlation between energy density and AE level was found to exist for the O+ ions; the exponential fitting gave a correlation coefficient r = 0.63, with an F-ratio value of 49.99, which means a confidence level > 99.9%. The regression coefficient is b = 2.46 x 10^{-3}, so that we have ED(O+) = 0.14 x $e^{2.46 \times 10^3 E}$,

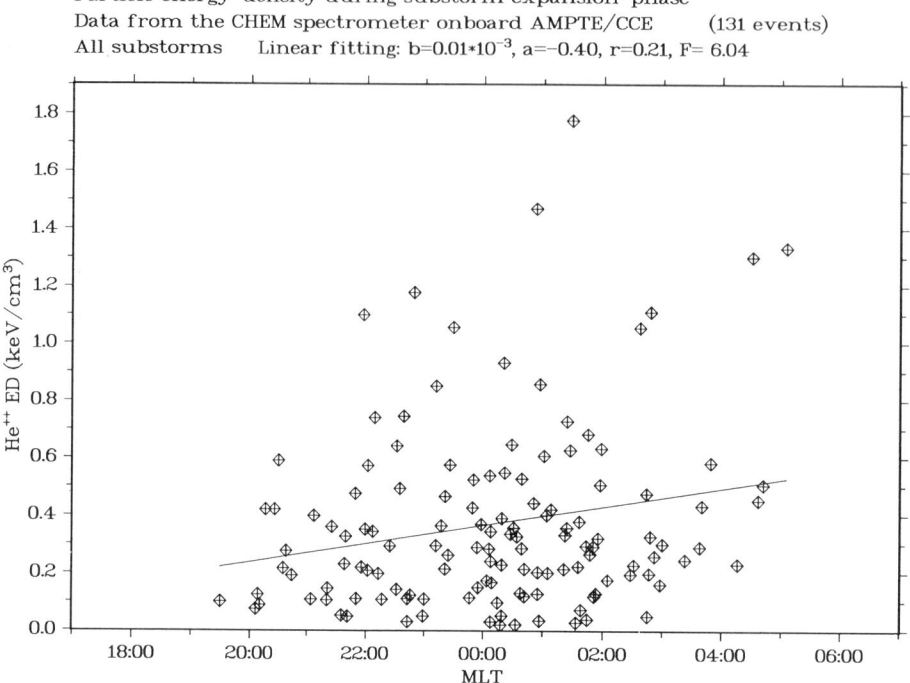

Fig. 1. Energy density of He^{++} ions versus magnetic local time.

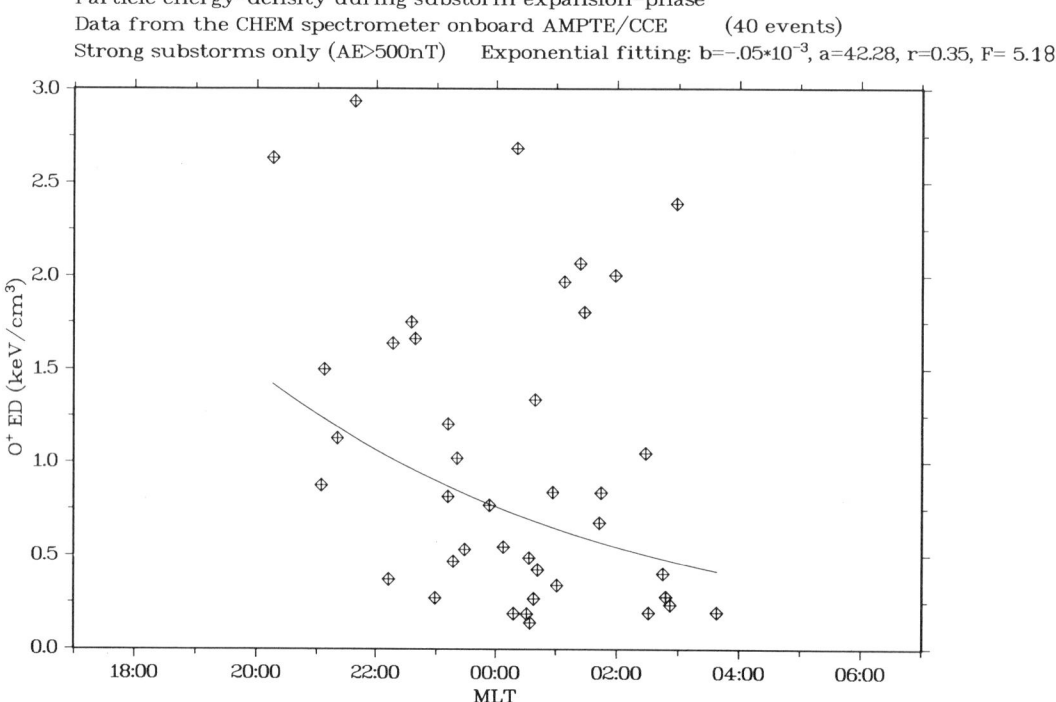

Fig. 2. Energy density of O^+ ions versus magnetic local time for strong substorms (AE > 500 nT).

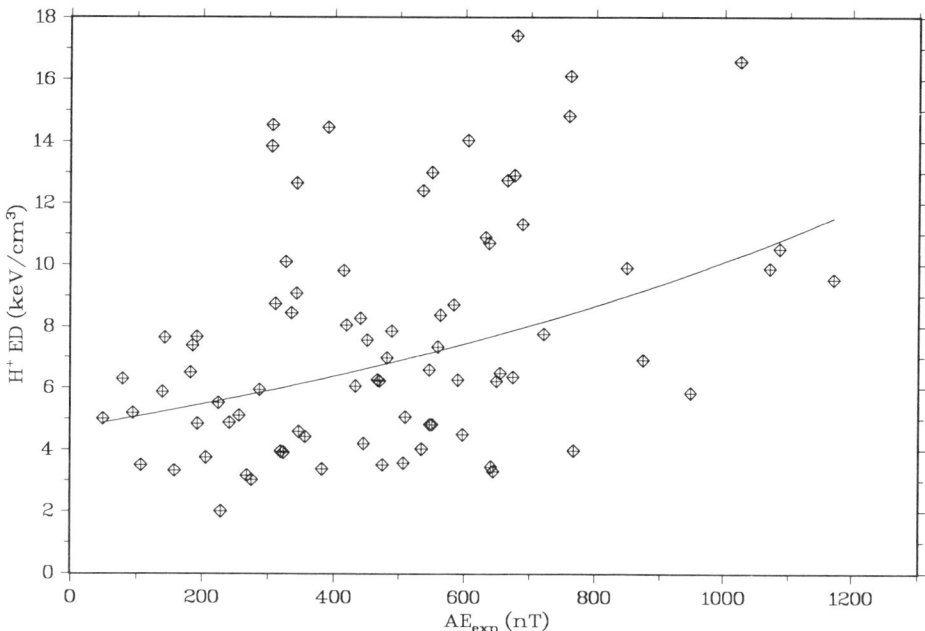

Fig. 3. Dependence of the H+ energy density on the value of the AE index.

where AE is given in nT and ED in keV/cm^3. The weakest correlation was found for the He^{++} ions: the best fitting was a linear fitting with a rather small regression coefficient of 0.28×10^{-3} and a correlation coefficient of only 0.24 with the questionable confidence level of 95% (F-ratio = 4.53). The corresponding values for the H$^+$ and He$^+$ ions lie between those for the ionospheric origin O$^+$ and the solar wind origin He^{++}. For H$^+$ the exponential fitting gave a rather small regression coefficient of 0.77×10^{-3} (He$^+$: 1.35×10^{-3}) and a correlation coefficient of 0.39 (He$^+$: 0.42) with a confidence level of 99.9% (same for He$^+$).

DISCUSSION

Our study of the early expansion-phase in the near-Earth equatorial magnetotail with AMPTE/CCE data revealed some interesting statistical features. The longitudinal energy density pattern of the major ion species exhibits a trend only for the ionospheric-origin O$^+$ and the solar wind-origin He^{++} ions. The He^{++} ions tend to have higher energy density levels towards dawn, while the O$^+$ ions tend to have higher energy density levels towards dusk.

In Figure 7 the average energy spectra of He^{++} show that up to 50 keV/e there are considerably higher flux-intensities in the postmidnight group of events (> 02:30 MLT) than in the pre-midnight group (< 22:00 MLT). The same feature (although weaker) is observed in the average energy spectra of H$^+$ and O$^+$, with the difference that the spectra of H$^+$ and O$^+$ are harder in the pre-midnight group, i.e. the flux-intensities in the high-energy range are higher in pre-midnight than in post-midnight, while for He^{++} this feature is very weak. This constitutes a major difference between He^{++} and the other ions. Numerical simulations of ion drifts in the tail (*Ejiri et al., 1980; Takahashi and Iyemori, 1989*) predict no local time asymmetry at low energies if the source ions are equally distributed throughout the tail. Thus the source of the He^{++} must be concentrated on the dawnside by, for example, increased penetration of solar wind ions from the magnetosheath through the dawnside magnetopause (*e.g. Lennartson and Sharp, 1982*) in order to explain the observations.

The dawn-dusk asymmetry observed for the O$^+$ energy density during strong substorms, with maximum at duskside, is most probably due to the westward polarization drift (originating from impulsive electric fields associated with the depolarizing magnetic field at substorm

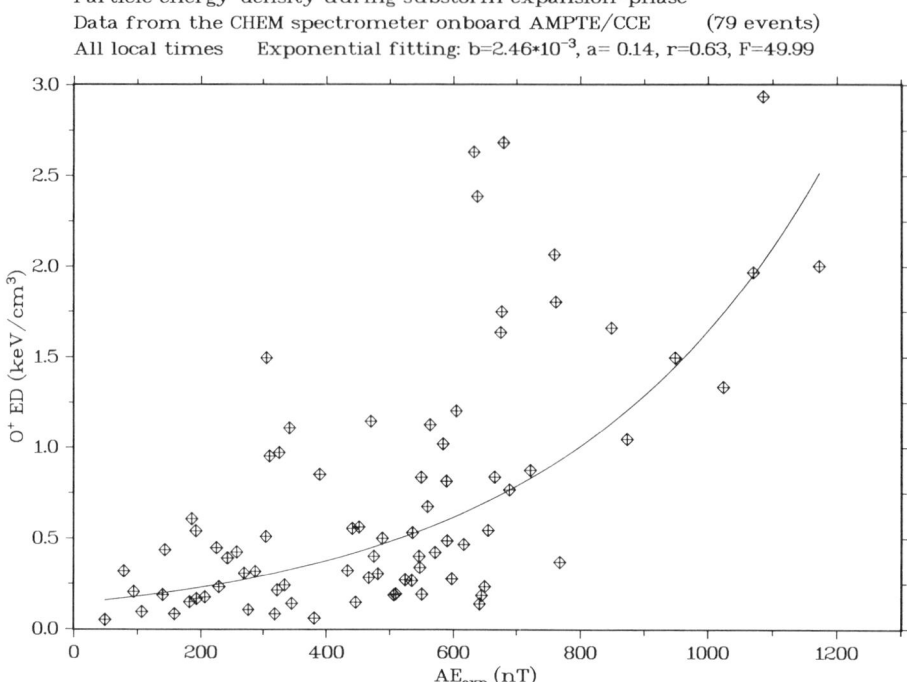

Fig. 4. Dependence of the O$^+$ energy density on the value of the AE index.

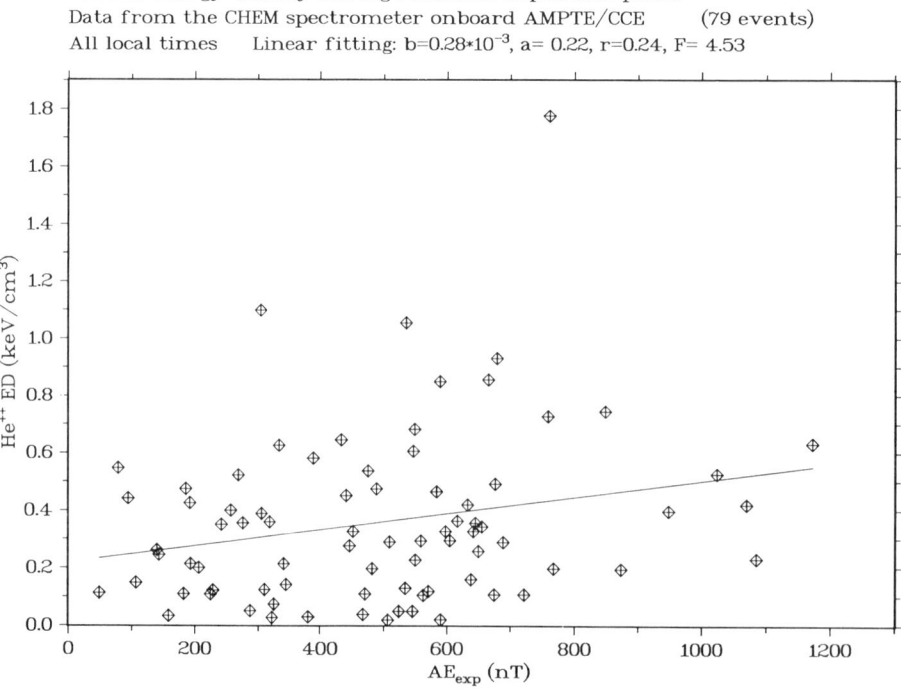

Fig. 5. Dependence of the He^{++} energy density on the value of the AE index.

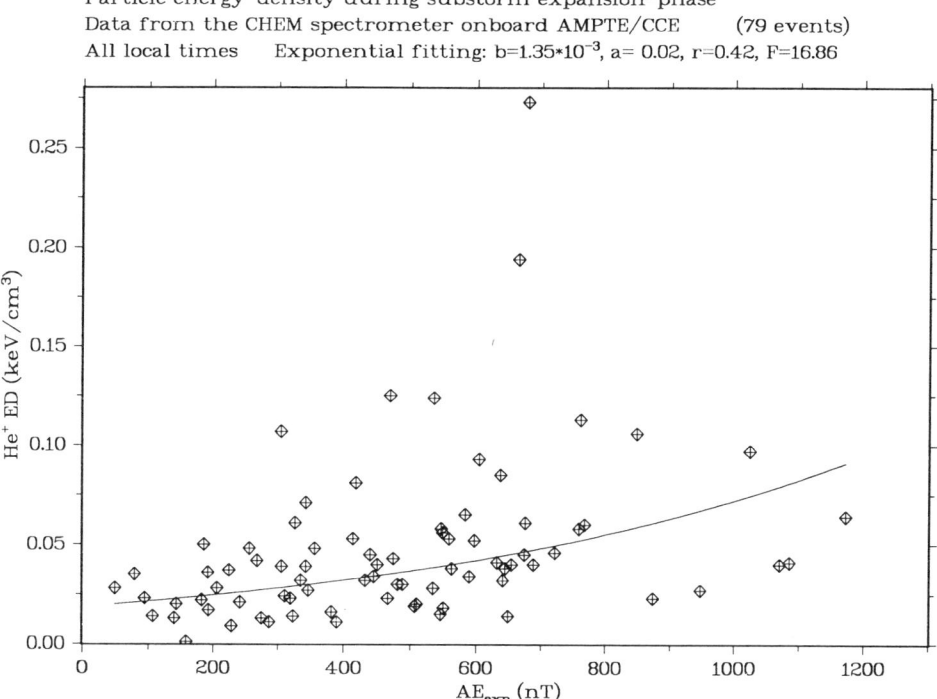

Fig. 6. Dependence of the He+ energy density on the value of the AE index.

onset), which is large for the heavy O+ ions, and the expected concentration of energetic upward flowing ions in the premidnight region (*e.g. Ghielmetti et al.*, 1978).

The part of our study concerned with the AE dependence of the energy density showed that the correlation of O+ energy density with the AE index value is much stronger than the correlation between the other ion species and AE. Our result recalls previous studies concerning the correlation of plasma composition in the low-energy range with the geomagnetic activity averaged over longer time-periods. In those studies (*e.g. Peterson et al.*, 1981; *Young et al.*, 1982; *Lennartsson and Shelley* 1986) an increase of the absolute O+ number density (within the energy range 0.1 - 17 keV/e) and also of the percentage of total number density due to O+ with geomagnetic activity was reported. Furthermore, it was reported that during strongly disturbed conditions O+ often becomes the most abundant ion species. In accordance with *Gloeckler and Hamilton* [1987] we found that although the above cited features hold, the number density (not shown here) and energy density of O+ are always less than those of H+ within the energy range of CHEM (1 to 300 keV/e) and within geocentric distances of 7 to 9 R_E. Presumably the difference is mainly due to the wider energy range of our study and, to a minor extent, due to the geocentric distances ($\geq 7 R_E$).

The observed correlation between energy density and AE index suggests that there are acceleration mechanisms operating in the high-altitude ionosphere which are much more effective than measurements of upflowing ions have indicated in the past. The nonlinear enhancement of the ionospheric-origin O+ energy density with the increase of the AE index implies that the near-Earth magnetotail is fed with energetic ionospheric ions during early expansion-phase. We suggest that the observed strong correlation is at least partly due to enhanced extraction and acceleration of ions from the ionosphere. We note that a high AE level results from increased magnetospheric earthward convection (during both the growth and the expansion phase) and thus an increased activity of the internal magnetospheric dynamo. The magnetospheric dynamo is connected with the auroral potential structures above the ionosphere (*e.g. Akasofu*, 1983), where ionospheric ions can be accelerated most effectively by a combination of quasi-electrostatic parallel and fluctuating perpendicular electric fields (*e.g. Lundin and Hultqvist*, 1989). Thus, the enhanced and temporally variable earthward convection during late growth phase and early expansion phase of the substorms can effectively contribute to the feeding of the near-Earth magnetotail with energetic ionospheric ions (see also *Daglis et al.*, 1991).

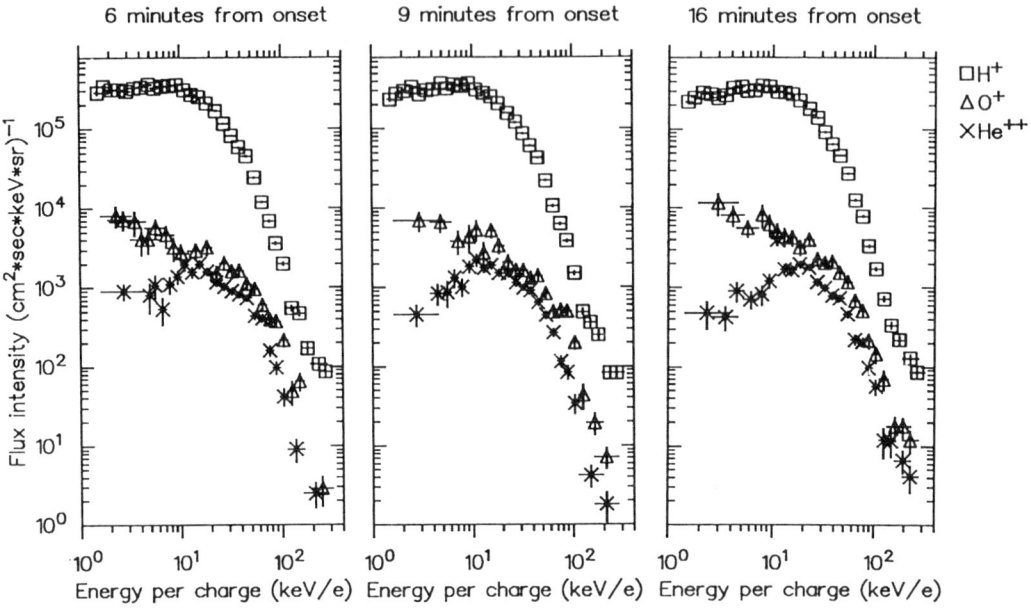

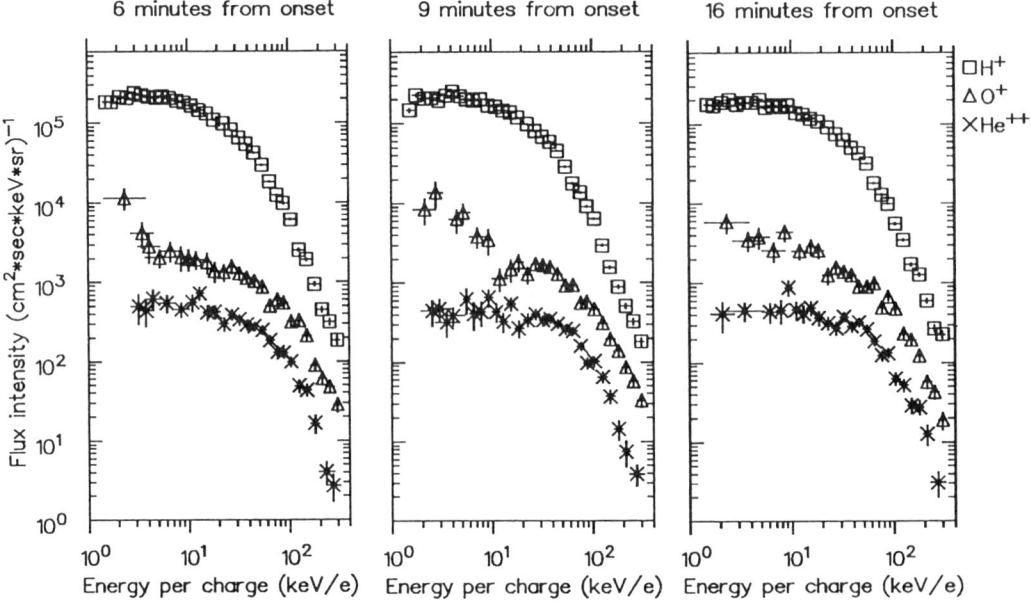

Fig. 7. Average dawnside and duskside spectra of the three ion species H^+, O^+, He^{++}. The differences are obvious, especially for He^{++}.

The energy density of the solar wind origin He^{++} ions is practically unaffected by the processes which contribute to the increase of the AE value. The two mixed-origin ion species H^+ and He^+ exhibit a statistically confident dependence on the AE level, but the regression coefficient is much smaller than for O^+. The fact that both the correlation and the regression coefficient are higher for He^+ than for H^+ indicates that either the acceleration mechanisms are more effective for ionospheric He^+ than for ionospheric H^+ or the ionospheric-source-percentage is higher for He^+ than for H^+.

The validity of this scenario can only be checked by multipoint observations, because the morphology pattern of substorms (tailward stretching of the magnetic field and gradual flux dropout followed by explosive flux increase) used for one-point measurements in the near-Earth magnetotail does not permit the undoubted recognition of increased outflow of energetic ionospheric ions during the late growth phase/early expansion phase. Simultaneous fast particle and electric fields measurements by a group of spacecrafts, together with ground data on ionospheric parameters can provide insight into this major interest aspect of substorm evolution. We hope that the four-spacecraft Cluster mission (*Schmidt and Goldstein*, 1988) will be a significant step towards the full understanding of these processes. Furthermore, in order to clarify the relative significance of the growth phase feeding and the expansion phase feeding of the near-Earth magnetotail with energetic ionospheric ions, the dependence of particle energy density upon the AU and the AL index, which represent the growth phase current system and the expansion phase current system respectively, (e.g. *Troshichev et al.*, [1974; *Gizler et al.*, 1976; *Troshichev*, 1982) will be examined in the near future.

SUMMARY

We used data from the CHEM experiment on AMPTE/CCE to investigate statistical features of the substorm expansion phase in the near-Earth equatorial magnetotail. We examined a group of 79 events within the time period extending from January 1985 to August 1986. The ions included in our study had energies in the range 1 to 300 keV/e. The main points of interest may be summarized as follows:

o The energy density of O^+ increases non-linearly with increasing AE index. We suggest that a large fraction of the energized O^+ ions comes directly from the ionosphere, thus giving rise to the observed strong correlation of its energy-density with the AE index: during the enhanced earthward plasma convection (i.e. during late growth phase and early expansion phase), and the development of the substorm current wedge at substorm onset, (which are both reflected in the AE index), the near-Earth magnetotail is increasingly fed with energetic ionospheric ions. This may be regarded as an indirect coupling of the solar wind with the ionosphere (through the internal magnetospheric dynamo), which influences substorm dynamics both during growth phase (*Daglis et al.*, 1991) and during expansion phase.

o The correlation of the He^{++} energy density with AE is very weak. Most probably the He^{++} energy density increases (if at all) only slightly with increasing AE index, having no relation to the mechanisms that enhance the energy density of ionospheric ions and depending almost totally on the acceleration by inductive fields at the site of substorm initiation (either instability growth region or neutral line region).

o The H^+ and He^+ ions exhibit intermediate features confirming their mixed origin. According to our scenario we would expect He^+ to have a stronger dependence on AE because in addition to its ionospheric source it has not a direct solar wind source (like He^{++} and H^+) but rather an indirect solar wind source, resulting from He^{++} through charge exchange. In other words, we would expect a larger ionospheric component for the He^+ ions rather than the H^+ ions. Indeed, the He^+ correlation with AE is stronger than that of H^+.

o A bias is seen in He^{++} energy density versus magnetic local time, with higher values towards dawn. The bias may be explained by preferential entry of solar wind ions through the dawnside magnetopause and enhanced dawnward drift under the influence of the convection electric field. In the group of strong substorms, a trend for higher values towards dusk is observed for O^+ energy density. Presumably this is due to the enhanced polarization drift because of strong inductive electric fields at substorm onset. The dawnside and duskside spectra of O^+ and He^{++} exhibit significant differences compatible with our interpretation of the observed asymmetry in local time: for He^{++} the dawnside spectra have considerably higher low-energy flux than the duskside spectra, while for O^+ the duskside spectra are considerably harder than the dawnside spectra.

Acknowledgements. The authors wish to thank Drs. D. C. Hamilton and F. M. Ipavich (University of Maryland) for their contribution to the development and construction of the CHEM spectrometer, Dr. T. A. Potemra (Johns Hopkins University Applied Physics Laboratory) for the supply of magnetic field data and Dr. L. M. Kistler (University of New Hampshire) for the background-subtraction algorithm. We are grateful to Prof. T. Araki and T. Kamei (World Data Center C2 for Geomagnetism at the Kyoto University) for providing tapes with the one-minute AE index. We thank Dr. R. E. Lopez (Johns Hopkins University Applied Physics Laboratory) for helpful discussions. The work on the CHEM instrument has been supported by the Bundesministerium für Forschung und Technologie (F. R. Germany) under grant No. 010M158 0 and by NASA (USA) under grants NAG5-716 and NAGW-101. One of us (IAD) benefited from a Max-Planck-Gesellschaft (F. R. Germany) scholarship.

REFERENCES

Akasofu, S.-I., Solar wind-magnetosphere coupling, in High-latitude space plasma physics, edited by B. Hultqvist and T. Hagfors, pp. 205-223, *Plenum Press, New York*, 1983.

Baker, D. N., E. W. Hones, Jr., D. T. Young, and J. Birn, The possible role of ionospheric oxygen in the initiation and development of plasma sheet instabilities, *Geophys. Res. Lett., 9*, 1337-1340, 1982.

Bevington, P. R., Data reduction and error analysis for the physical sciences, pp. 310-312, *McGraw-Hill Book Company, New York*, 1969.

Daglis, I. A., A study of substorm dynamics in the Earth's magnetosphere with AMPTE/CCE observations, *PhD Thesis, Demokritos University of Thrace, Xanthi*, 1991.

Daglis, I. A., G. Kremser, W. Studemann, B. Wilken, G. Gloeckler, D. C. Hamilton, and F. M. Ipavich, Observations of the ion distribution in the nightside magnetosphere during substorm-associated dropout events, paper presented at the *XXVII Plenary Meeting of COSPAR*, Espoo, Finland, July 18-29, 1988.

Daglis, I. A., E. T. Sarris, and G. Kremser, Indications for ionospheric participation in the substorm process from AMPTE/CCE observations, *Geophys. Res. Lett., 17*, 57-60, 1990.

Daglis, I. A., E. T. Sarris, and G. Kremser, Ionospheric contribution to the cross-tail current during the substorm growth phase, J. *Atmos. Terr. Phys., 53*, on print, 1991.

Dassoulas, J., D. L. Margolies, and M. R. Peterson, The AMPTE CCE spacecraft, *IEEE Trans. Geosci. Remote Sens., GE-23*, 182- 191, 1985.

Ejiri, M., R. A. Hoffman, and P. H. Smith, Energetic particle penetration into the inner magnetosphere, *J. Geophys. Res., 85*, 653-663, 1980.

Ghielmetti, A. G., R. G. Johnson, R. D. Sharp, and E. G. Shelley, The latitudinal, diurnal, and altitudinal distributions of upward flowing energetic ions of ionospheric origin, *Geophys. Res. Lett., 5*, 59-62, 1978.

Gizler, V. A., B. M. Kuznetsov, V. A. Sergeev, and 0. A. Troshichev, The sources of the polar cap and low latitude bay-like disturbances during substorms, *Planet. Space Sci., 24*, 1133-1139, 1976. Gloeckler, G., and D. C. Hamilton, AMPTE ion composition results, Physica Scripta, T18, 73-84, 1987.

Gloeckler, G., F. M. Ipavich, W. Studemann, B. Wilken, D. C. Hamilton, G. Kremser, D. Hovestadt, F. Gliem, R. A. Lundgren, W. Rieck, E. 0. Tums, J. C. Cain, L. S. MaSung, W, Weiss, and H. P. Winterhoff, The charge-energy-mass (CHEM) spectrometer for 0.3 to 300 keV/e ions on the AMPTE/CCE, *IEEE Trans. Geosci. Remote Sens., GE-23*, 234-240, 1985.

Lennartsson, W., Energetic (0.1- to 16-keV/e) magnetospheric ion composition at different levels of solar FlO.7, *J. Geophys. Res., 94*, 3600-3610, 1989.

Lennartsson, W. and R. D. Sharp, A comparison of the 0.1-17 keV/e ion composition in the near equatorial magnetosphere between quiet and disturbed conditions, *J. Geophys. Res., 87*, 6109-6120, 1982.

Lennartsson, W. and E. G. Shelley, Survey of 0.1- to 16-keV/e plasma sheet ion composition, *J. Geophys. Res,, 91*, 3061-3076, 1986.

Lundin, R. and B. Hultqvist, Ionospheric plasma escape by high altitude electric fields: magnetic moment pumping, *J. Geophys. Res., 94*, 6665-6680, 1989.

Lyons, L. R., and D. J. Williams, Quantitative aspects of magneto-spheric physics, *D. Reidel Publishing Company*, Dordrecht, 19114.

Peterson, W. K., R. D. Sharp, E. G. Shelley, R. G. Johnson, and H. Balsiger, Energetic ion composition of the plasma sheet, *J. Geophys. Res., 86*, 761-767, 1981.

Potemra, T. A., L. J. Zanetti, and M. H. Acuña, The AMPTE/CCE magnetic field experiment, *IEEE Trans. Geosci. Remote Sensing, GE-23*, 246-249, 1985.

Schmidt, R. and M. L. Goldstein, Cluster - a fleet of four spacecraft to study plasma structures in three dimensions, in *The Cluster mission, ESA SP-1103*, pp. 7-13, 1988.

Stoodley, K. D. C., T. Lewis, and C. L. S, Stainton, *Applied Statistical Techniques, Ellis Horwood Ltd. Publishers*, Chichester, 1980.

Takahashi, S. and T. Iyemori, Three-dimensional tracing of charged particle trajectories in a realistic magnetospheric model, *J. Geophys. Res., 94*, 5505-5509, 1989.

Troshichev, 0. A., Polar magnetic disturbances and field-aligned currents, *Space Sci. Rev., 32*, 275-360, 1982.

Troshichev, 0. A., B. M. Kuznetsov, and M. I. Pudovkin, The current systems of the magnetic substorm growth and explosive phase, *Planet. Space Sci., 22*, 1403-1412, 1974.

Auroral Signatures of Substorm Recovery Phase: A Case Study

T.I. Pulkkinen,[1] R.J. Pellinen,[1] H.E.J. Koskinen,[1] H.J. Opgenoorth,[2] J.S. Murphree,[3] V. Petrov,[4] A. Zaitzev,[4] and E. Friis-Christensen[5]

Danish Meteorological Institute, Department of Geophysics, Copenhagen, Denmark

We present an event where an entire substorm recovery phase has been recorded by the ultraviolet imager onboard the Viking spacecraft, all-sky cameras in Finland, and a variety of global ground-based instrumentation. By these data we have witnessed the following development: Towards the end of the substorm expansive phase a large scale auroral surge was observed west of the onset location. The surge was stationary and faded gradually. During the recovery phase, the auroral activity retreated slowly eastward, and a faint arc formed promptly about ten degrees poleward of the previous activity in the premidnight sector. The morningside auroral oval brightened, and, finally, eastward moving omega bands – or auroral hooks as seen by the UV imager – appeared during the last half hour of the recovery phase. With the global imaging it is possible to study the global dynamics of the substorm recovery phase that has rarely been addressed thus far. The imager recorded throughout one hour of the recovery phase with less than one minute temporal resolution. It seems evident that the magnetospheric mechanisms producing the recovery phase auroras are different from those responsible for the auroral bulge development. Magnetic field-aligned mapping using Tsyganenko's 1989 model relates these auroral structures with the current sheet region between 6 and 13 R_E in the morning sector, where large gradients in both electric and magnetic fields provide effective means for particle acceleration during the refilling of the plasma sheet.

Introduction

Since Akasofu's pioneering work [*Akasofu*, 1964], mainly based on ground-based all-sky camera data, the substorm sequence has been divided into expansion and recovery phases that are separated both in space and time. Later, it was found that the onset of the expansive phase is often preceded by a growth phase [*McPherron*, 1970], during which the energy and momentum transfer from the solar wind (SW) into the magnetosphere is enhanced due to the southward turning of the interplanetary magnetic field (IMF) Z-component. It is now generally accepted that substorms may include both the loading-unloading component manifested by the growth-phase development and a component directly driven by changes in the SW. During the last decades, satellite-borne auroral imaging has provided further details on auroral behavior during substorms, such as the global scale surge development. Magnetospheric survey programs have enhanced our knowledge about the causal relationship between dynamic magnetospheric processes and substorm auroras. The substorm onset separating the growth and expansive phases is characterized by an abrupt auroral breakup following the slow equatorward motion of the growth-phase arcs. The breakup takes place in a confined region usually close to magnetic midnight, and and spreads both latitudinally and longitudinally during the expansion phase. However, the transition from the expansive phase to the recovery phase does not have such a clear signature. The recovery phase auroras are not localized but appear in a wide range of longitudes along the morningside auroral oval. Furthermore, recovery phase features may be observed when fragments of the expansive phase auroras still are present in the evening sector auroral oval. Vice versa, during a recovery phase a new expansive phase may be initiated in the pre-midnight sector.

Typically, the decay of the electrojet system characteristic of the recovery phase starts 60–120 min after the onset of the expansion phase. From its maximum value, it usually takes about one hour for the AE-index to recover its quiet-time value. The auroral activity during this period moves

[1] Finnish Meteorological Institute, Helsinki, Finland
[2] Swedish Institute of Space Physics, Uppsala, Sweden
[3] Department of Physics, University of Calgary, Calgary, Alberta, Canada
[4] USSR Academy of Sciences, Troitsk, USSR
[5] Danish Meteorological Institute, Department of Geophysics, Copenhagen, Denmark

Magnetospheric Substorms
Geophysical Monograph 64
Copyright 1991 American Geophysical Union

eastward to the morning sector, and diffuse pulsating patches, pulsating aurora [*Johnstone*, 1983; *Sandahl*, 1984; *Davidson*, 1990], wavy boundaries, and torch-like structures or auroral omega bands [*André and Baumjohann*, 1982; *Opgenoorth et al.*, 1983] appear along the oval. Ps 6 magnetic pulsations that are observed during large substorms [*Nielsen and Sofko*, 1982; *Buchert et al.*, 1988, 1990], can be associated with the appearance of large size omega bands [*Gustafsson et al.*, 1981; *André and Baumjohann*, 1982; *Opgenoorth et al.*, 1983]. A discrete auroral arc is often observed poleward of the poleward boundary of the morningside auroral oval. This arc is usually connected to the main oval in its western edge, but is elsewhere separated from it by a wide region void of precipitation.

The ionospheric conductivity enhances rapidly during the expansion phase. In the morning side the enhanced conductivity is retained by continuing precipitation during the recovery phase. Rostoker et al. [1985] studied the relationship between hard diffuse precipitation and the westward electrojet in the morning sector. They found a very clear correlation between the position of the maximum electrojet and the hardest particle precipitation. It appears that in the morning sector the major portion of the Hall current electrojet flows in the region of diffuse aurora, which is produced by harder precipitation than the diffuse evening aurora. This continued high conductivity combined with the unloading of energy and momentum stored in the ionosphere (e.g. in the neutral wind system) during the expansion phase can therefore be expected to contribute to the long temporal scale of the recovery phase.

At the end of the expansive phase, the near-Earth tail has a dipolarized magnetic field configuration. The refilling of the plasma sheet usually takes place rather rapidly, within about 12 minutes from the maximal magnetic disturbance recorded on ground, and is associated with Earthward plasma flow parallel to the equatorial plane [*Lui et al.*, 1977; *Hones et al.*, 1986]. As the morningside precipitation is harder than the typical evening sector precipitation spectrum [*Rostoker et al.*, 1985; *Lyons and Fennell*, 1986; *Pudovkin et al.*, 1989], there must be an efficient acceleration mechanism acting in the magnetosphere during the recovery phase. At present, both the source region and the acceleration mechanism of the precipitating particles are open questions.

Although the recovery phase as a part of the substorm process has been identified from the very beginning of the substorm studies, the related physical processes have not been extensively studied. As a relaxation process from an excited state toward a ground state of the magnetosphere, the recovery phase can reveal fundamental new information on the properties of the underlying physical system. These issues are preferably attacked using simultaneous *in situ* measurements and global auroral imaging with temporal and spatial resolution appropriate for the details of auroral substorm development. The Swedish Viking satellite mission in 1986 has provided the most recent opportunity to study some of the gaps in our understanding in auroral substorm physics.

In this work we describe the development of an entire substorm, with special emphasis on the recovery phase, which was monitored in global scale with 20 sec temporal resolution by the two UV cameras onboard the Viking satellite. An extensive set of ground-based observations is used to study the time intervals not covered by the spacecraft, and to record the development of some electrodynamical parameters. The global images are used to study how the surge deforms into the recovery phase patchy auroras, and the appearance and duration of the wavy boundaries. The Tsyganenko 1989 model [*Tsyganenko*, 1989, hereafter referred to as T89] is utilized in mapping of the morningside auroral structures into the tail current sheet. The possible source region of the particles and precipitation mechanisms are discussed.

OBSERVATIONS

On March 25, 1986 three consecutive substorm activations took place between 1900 – 0200 UT. The first two onsets occurred around 1900 and 2000 UT. They were initiated to the east of Scandinavia near magnetic midnight, and their westward expansion features could be traced by several ground-based instruments located in Scandinavia. During the recovery of the second activation the Viking UV imager was observing the auroral development along major part of the auroral oval with high temporal resolution. The third activation started around 2200 UT over Greenland, and expanded mainly eastward.

The upper panel of Fig. 1 shows the development of the AU and AL indices after 1200 UT. Between 1230–1630 UT there was substorm activity with 500 nT average AE disturbance. Presumably after 1800 UT the IMF B_Z component turned southward (IMF data not available) leading to enhanced convection currents (growth phase) which later led to the first substorm activation at around 1900 UT. The lower panel shows the development of the magnetic X component at the northernmost of the EISCAT magnetometer cross stations, Sorøya (SOR; 67.30°N, 107.90°E in corrected geomagnetic coordinates). The three subsequent activations having about 500 nT magnetic bays are indicated by 1, 2, and 3. All-sky camera (ASC) data from Kilpisjärvi (KIL; 65.97°N, 105.57°E) are available during the first two activations, 1857–2120 UT, after which cloud coverage prevented further observations. After some overlapping with the ASCs the Viking imager took over the observations, covering the entire recovery of the second activation (2047–2153 UT).

During the first activation, the auroral break-up took place to the east of KIL field of view. Before the onset, until 1903:23 UT, a southward moving quiet arc was observed. After this a precursor of the main auroral bulge, the auroral horn [*Yahnin et al.*, 1983], propagated into the field of view. It was followed by a westward traveling surge (WTS), which was overhead at 1910:23 UT. By five minutes later the activation disappeared toward the western horizon. The second activation took place further to the east at about

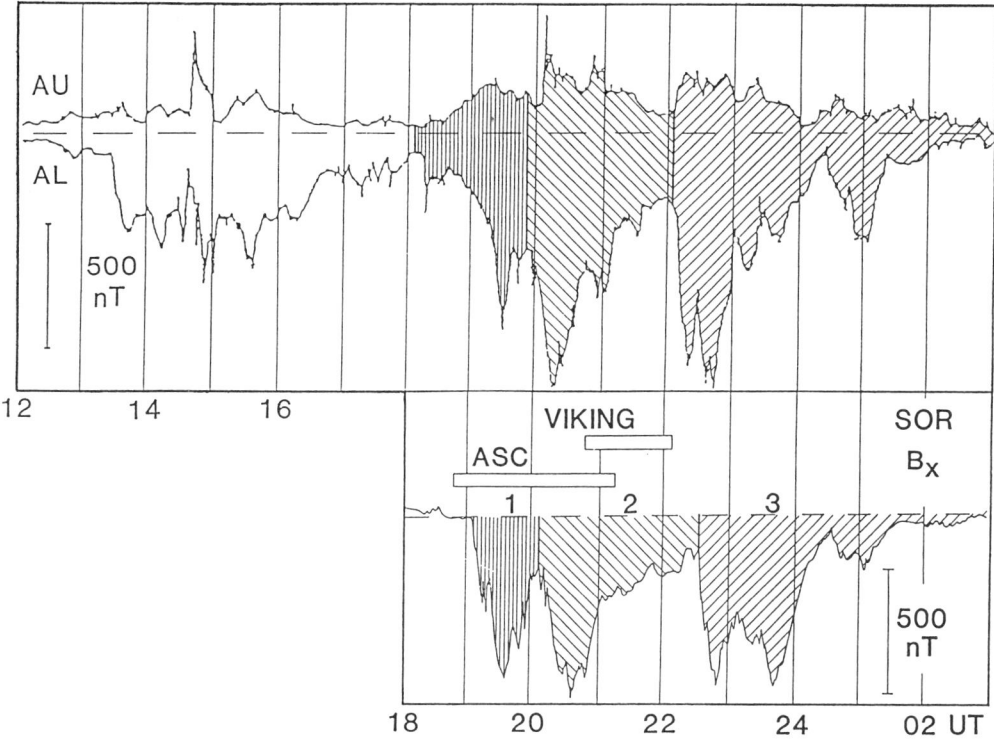

Fig. 1. Upper panel: AU and AL indices from 1200 UT to 0300 UT on March 25 - 26, 1986. Lower panel: The magnetic X component recorded at the northernmost station of the EISCAT magnetometer cross, Sorøya. The observation periods of the ASC in Kilpisjärvi and the Viking imager are indicated in the picture, and the three different activations studied are indicated with different hatchings.

2000 UT. The development was quite similar to the first one, but the structures were larger and not fully covered by the ASC field of view.

From magnetic data recorded at seven stations in Greenland (data not shown here) it can be concluded that the first activation hardly reached the east coast of Greenland, whereas the second activation reached the west coast of Greenland before it stopped and started to fade. First magnetic signatures of the second activation were recorded on the east coast of Greenland between 2010–2011 UT, and on the west coast three minutes later, between 2013–2014 UT. The equivalent current pattern around the surge was typical for an evening sector surge: Eastward electrojet to the south of the surge, strong westward current within the surge (SCW), and sunward equivalent current across the polar cap in the far north.

While the AE index at 2100 UT was still disturbed by the decaying surge over Greenland, influencing the local AE stations, in the Scandinavian sector the recovery of the electrojet system had already started, and continued until the eastward expansion of the third activation reached Scandinavia.

Fig. 2 shows the global development of auroral activity during the second substorm activation. Some of the auroral zone magnetic stations, the field of view of the KIL all-sky camera (ASC), and the ionospheric footprint of the Viking satellite for this orbit (176) are shown on the map. Times in UT (left) and satellite altitudes in km (right) are given at selected points along the orbit. The onset location ($T = 0$, corresponding to 1959 UT) has been determined from magnetic observations, and after two minutes ($T = +2$) the KIL ASC recorded the appearance of dynamic auroral structures. At $T = +5$ the activity reached its maximum, and seven minutes later the activation propagated out of the field of view of the camera ($T = +12$). After this the sky was void of auroras until 2117 UT when an auroral arc appeared in the zenith of KIL. Later cloud coverage prevented further observations.

The first Viking image at $T = +47$, shows a stationary surge above Greenland that had started to decay. The disintegration went from its western boundaries eastward, and from the poleward edge equatorward during about 15 min. Simultaneously, diffuse auroral structures limited by a wavy poleward boundary over Spitzbergen (associated with enhanced magnetic H bay activity at Ny Ålesund, commencing at $T = +65$) appeared further to the east, as shown in the frame taken at $T = +60$. At $T = +80$ the surge had disappeared and the activity had transferred to the east, with a bright poleward structure appearing in the premidnight sector. The KIL ASC field of view was again filled with

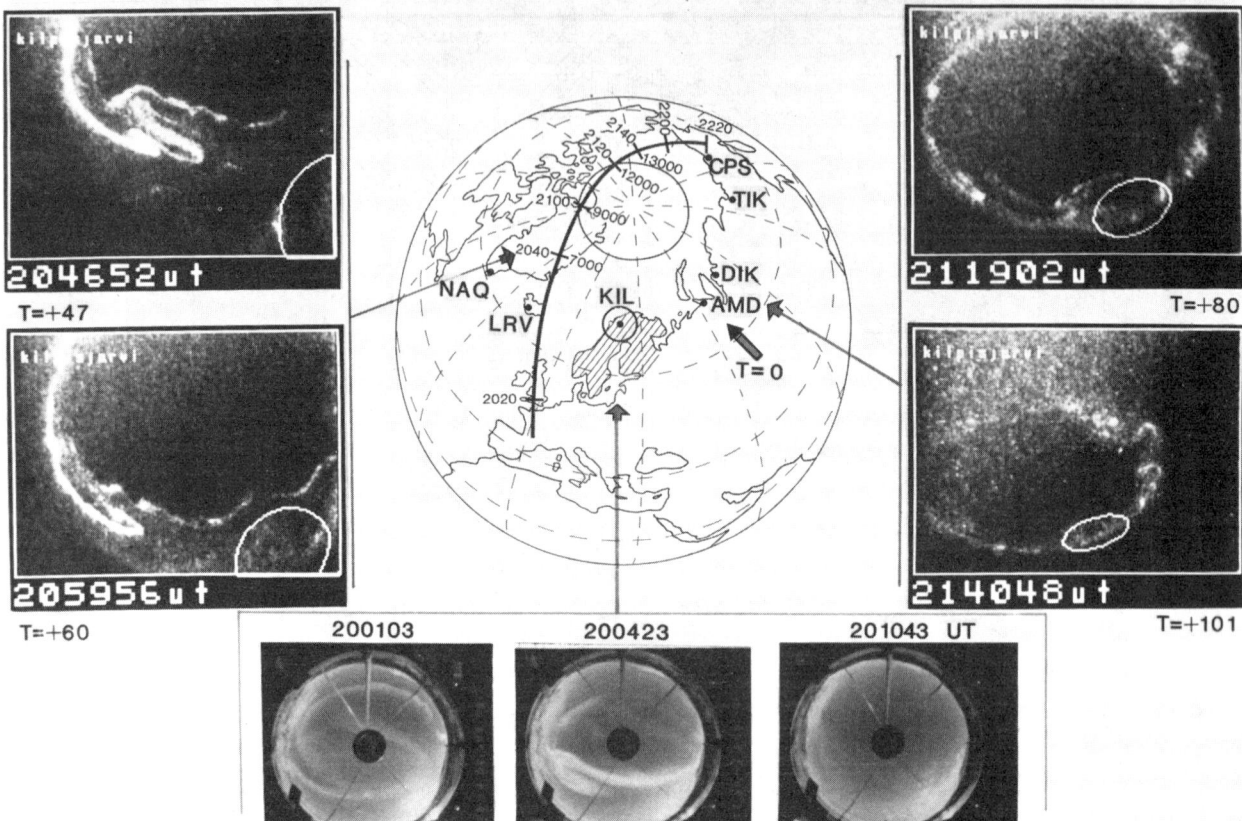

Fig. 2. Auroral substorm development from $T = 0$ to $T = +101$ min. The footprint of the Viking orbit 176 is shown on the map, with times in UT on the left and satellite altitudes on the right given at selected points along the orbit. Some of the auroral zone magnetic stations and the Kilpisjärvi ASC field of view are indicated on the map. The successive times are counted from the onset moment at 1959 UT.

irregular auroras at about 2110 UT, as seen by the Viking imagers that are more sensitive than ASC observations. Twenty minutes later, at about 2130 UT, the auroral activity to the west had subsided while along the Siberian coast over Dikson (DIK) and 500 km east of it "hook-type" active forms appeared. We interpret these hooks as omega bands, moving eastward along the active morningside auroral oval. The last frame ($T = +101$) shows that almost all auroral activity was in the eastern sector near the end of the recovery phase at 2200 UT. Characteristic of this activity was a narrow unusually bright morningside oval where the hooks appeared and remained for 20–30 min, becoming gradually more unstructured and finally merging with the bright auroras along the oval.

Magnetic data from some key observatories at auroral latitudes are presented in Fig. 3. Overlaid in the figure is a schematic drawing of the auroral hooks as recorded by Viking at 2141 UT. The magnetic activity over Greenland ceased quite rapidly when the surge decayed after 2100 UT.

The cessation of the activity was quite definite at Leirvogur (LRV) somewhat later. However, in the eastern sector at Alta (ALT), and more definitely at Amderma (AMD), Dikson (DIK), and Tiksie Bay (TIK), a more gradual decay was observed. Thus the recovery of the AE-index is controlled by the prolonged depression of the magnetic H-component in the morning sector.

The third activation at 2200 UT took place close to Narssarssuaq (NAQ), and caused the strongest disturbance in the AE-index. The auroral activity of the previous recovery phase had at that time shifted to late morning hours, whereas the new activation took place in the evening sector at almost opposite side of the oval. The nearest station at the same latitude in Canada (data not shown here) recorded only enhanced eastward electrojet effects after the start of activation at southern Greenland, which indicates that if a surge was formed it was fairly stationary with no dramatic westward motion. The activation expanded clearly eastward to DIK with decreasing intensity.

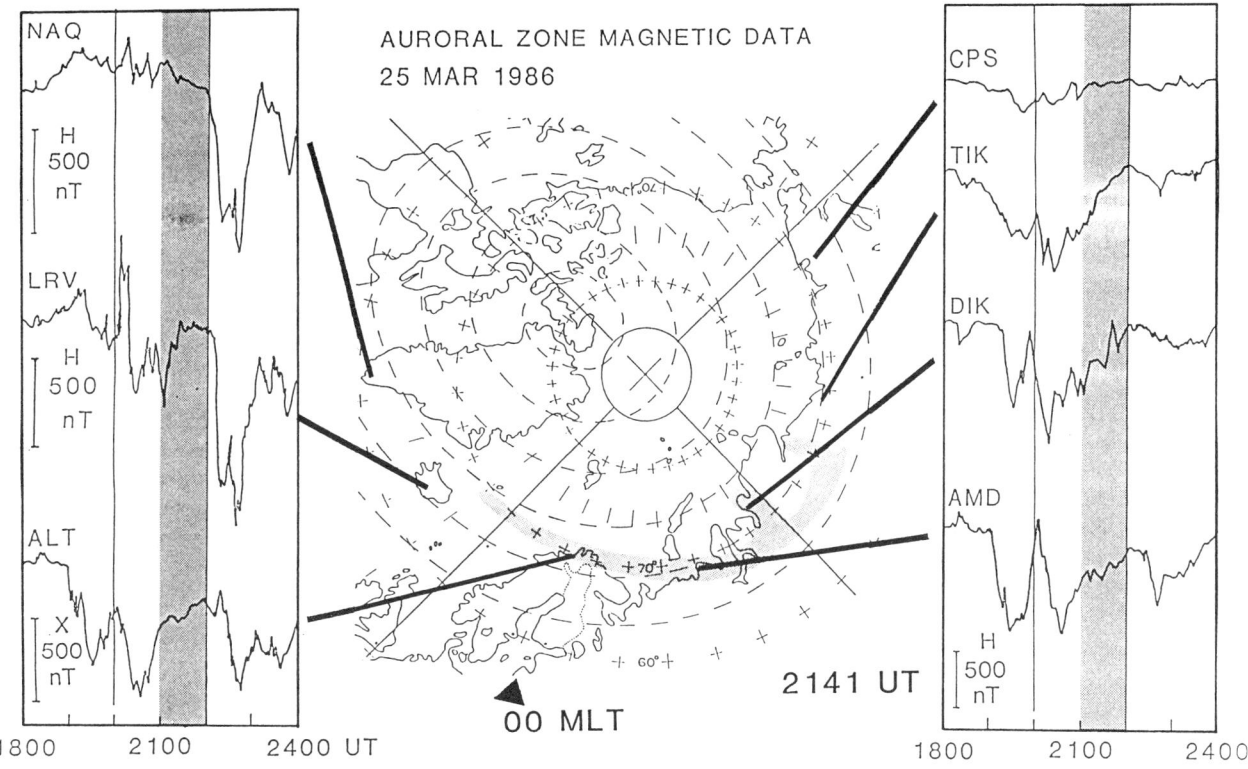

Fig. 3. The magnetic X components from selected stations on the auroral zone. The recovery phase period is indicated with hatching. The location of the stations and a schematic picture of the auroral hooks as seen by the Viking imager at 2141 UT are drawn on the map.

Mapping

We estimate the magnetospheric source region of the precipitating particles by magnetic field-aligned mapping of the auroral forms recorded in the ionosphere into the current sheet in the equatorial plane. The UV images, all-sky camera observations, and magnetic recordings are used to locate the precipitation region in the ionosphere. The accuracy of the mapping results is critically dependent on the model for the geomagnetic field: We use the statistical field model developed by Tsyganenko [1989] for the external field, together with the IGRF coefficients for a given epoch for the internal field. The T89 model gives the field as a function of geomagnetic activity parametrized by the Kp index and the tilt angle between the GSM Z-axis and the dipole axis. We have also made some further adjustment of the model to better represent the local conditions by varying the model ring current intensity.

Fig. 4 shows the mapping of two omega band events recorded by the Viking imager (one of them being the event described in this study), and six either omega-band or Ps 6 pulsation events reported in earlier literature. The mapping in each case has been conducted by using the appropriate tilt angle and the highest level of magnetic activity allowed by the model ($Kp \geq 5-$). Because omega bands are associated with high magnetic activity, and occur during periods when the equatorial Dst index describing the enhancement of the ring current is at maximum, this seems to be the best choice of the basic model configuration. In Fig. 4, the mapping of the omega bands shows only the center of the hook-like structure in the UV image. Omega bands have a scale size of about 500–1000 km in the ionosphere, which mapped to the magnetosphere gives a scale size of several Earth radii. Thus the mapping of omega bands actually covers much larger areas in the current sheet than indicated in the picture. For the Ps 6 events, the location of the recording station closest to the estimated maximum of the upward current has been mapped at a time in the middle of the period when the pulsations were recorded. As the pulsations are often observed over a range of latitudes covering a few degrees in the ionosphere, the mapping of the whole region of disturbance also covers larger areas in the current sheet.

The mapping of these eight events locates the source region in the current sheet between 6 and 13 R_E in the morning sector and between about 0200 and 0700 in magnetic local time. This region outside of the ring current maximum is associated with high magnetic and electric field gradients. The arrows in the picture (arbitrary units) show the current density in the center of the current sheet computed from the Tsyganenko model. Also overlaid is the electric

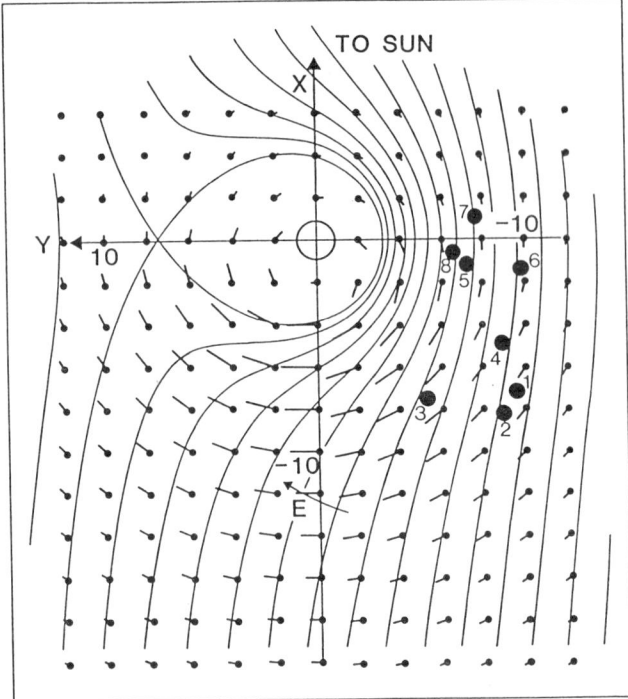

Fig. 4. Field-aligned mapping from the ionosphere to the current sheet of two omega band events recorded by the Viking imager, and additional Ps 6 pulsation events taken from earlier literature. The arrows indicate the cross-tail current intensity and orientation in the center of the current sheet in arbitrary units. Because the omega bands have a scale size of 500 – 1000 km in the ionosphere, the mapping corresponds to much larger area (several R_E) than indicated in the figure (see text). The dashed lines show the equipotential contours in an electric field model with corotation and convection electric fields [*Lyons and Williams*, 1984]. The events are numbered as follows: 1) 25. 3. 1986 (Viking V5). 2) 26. 9. 1986 (Viking V5). 3) 22. 10. 1979 (ASC) [*Opgenoorth et al.*, 1983]. 4) 16. 2. 1977 (ASC) [*André and Baumjohann*, 1982]. 5) 10. 01. 1983 (Ps6) [*Steen et al.*, 1988]. 6) 21. 4. 1985 (Ps6) [*Buchert et al.*, 1988]. 7) 28. 8. 1978 (Ps6) [*Nielsen and Sofko*, 1982]. 8) 8. 7. 1975 (Ps6) [*Gustafsson et al.*, 1981.*]

field equipotential pattern resulting from a simple model for the corotation and convection electric fields [*Lyons and Williams*, 1984]. This model reveals only the large-scale static features of the magnetospheric electric fields and is based on much less observational information than T89. In fact, very little is known about the actual electric field distribution during the substorm recovery phase.

During substorms, the magnetospheric field undergoes severe deformation, causing uncertainties in the field-aligned mapping. After the expansive phase onset, the tail field dipolarizes and the ring current reaches its maximum during the refilling of the plasma sheet. We modified the T89 model to take into account the increase in the ring current intensity, and tested the mappings by increasing the ring current term in the T89 model. The enhanced current shifts the mappings further away from the Earth but, even for fairly large enhancement of the ring current (factor of 3 from the maximal activity level allowed by the basic model), the mapped field-lines in all cases crossed the current sheet closer than 20 R_E from the Earth. In only one of the Ps 6 pulsation events we have mapped, the field-line did not reach the current sheet before crossing the model boundary at $Z = 25\ R_E$.

DISCUSSION

We have presented an extensive data set describing the global development of the recovery phase of an auroral substorm. The global images recorded by the UV imager onboard the Viking spacecraft together with the ground-based observations reveal the following development of the auroral activity: The recovery phase started after a fading stationary surge was observed, after which the auroral activity shifted eastward to the morning sector. The morningside auroral oval brightened, and towards the end of the recovery phase omega bands, or bright spots or hooks as seen by the UV imager, appeared during a limited time period. Near midnight, a discrete boundary poleward of the auroral oval appeared after the fading of the westward traveling surge. A new stationary surge-type structure appeared for a short time interval in the premidnight sector, poleward of the discrete boundary. The recovery phase ended with a new activation initiated in the premidnight sector.

It is difficult to identify the beginning of the recovery phase from the AE index alone. Individual auroral features, such as surges or discrete arcs, all contribute to the index if appearing sufficiently close to any of the AE stations. Hence, as a lot of auroral activity may still be in progress along the oval when the AE index is at maximum, we suggest that the recovery phase on global scale starts only when the AE index starts to return monotonically towards quiet (or considerably less active) level. This occurs typically 60–120 min after breakup, and, in practice, is an indication that the westward traveling surge has disappeared, the current wedge switched off, and even the large sheet-type field-aligned currents begin to relax. Data from individual magnetic stations reveals that in the evening sector stations, which observe the H-component depression due to the expansion phase activity, recover rapidly after the end of the expansion phase. However, the morningside auroral activity during the recovery phase causes depression of the H-component of the morningside magnetic stations. Thus, the maintenance of the high AE-values during substorm recovery phase are due to the ongoing particle precipitation into the morning sector auroral oval.

Similar observations of wavy boundaries on the poleward side of the diffuse aurora and fully developed omega bands during substorm recovery have been recorded by DE 1 [*J. Craven, personal communication*] and the DMSP spacecraft [*Buchert*, 1988]. During substorm recovery in the morning sector, diffuse and patchy eastward drifting structures appear with a distinct poleward boundary. Omega bands are observed to develop along this boundary at some later stage of the recovery phase, particularly after large substorms. Poleward of the omega bands is often a gap of weak luminosity, 5°–10° wide in latitude, separating a region of irregular discrete auroras, which seem to expand towards the morning sector in time. Here we have only discussed the auroral development in the equatorward part of the oval, even though we recognize the importance of the more poleward discrete structures for a complete understanding of the recovery phase precipitation mechanisms.

The relation between ground-based optical all-sky camera pictures and Viking UV images in the Lyman-Birge-Hopfield (LBH) band may be different for the morningside recovery phase auroras as compared to the midnight substorm expansive phase auroras. While the correspondence of ground-based optical recording and Viking LBH images is fairly good in the evening-midnight sector [Shepherd et al., 1990], it is expected that differences may be larger in the morning sector. This is because the LBH emission rate as seen by a satellite imager depends on the characteristic energy of the precipitating particles [Rees et al., 1988]. Auroras in the morning sector tend to be produced by harder electron precipitation than in the evening-midnight sector [Lyons and Fennell, 1986; Pudovkin et al., 1989], and the increasing precipitation energy lowers the altitude from which the main luminosity originates. As a result the upward propagating UV light experiences more atmospheric absorption than emissions produced at a higher altitude, and this may create differences in images recorded from the ground and space. When the diffuse part is lost in the UV images, structures that can be identified as omega bands in the ASC pictures, show as bright spots or hooks in the UV images. This shows particularly clearly in a case study of a substorm on September 26, 1986, where omega bands were recorded between 0116–0138 UT both by the Viking imager and KIL ASC. The ground-based recordings showed a large omega band with a diffuse "tongue" bordered by a discrete boundary passing eastward over the field of view. The corresponding structure recorded by the Viking imager was a hook-type feature with no diffuse environment.

During the recovery phase the plasma sheet expands rapidly at all distances and strong Earthward plasma flows have been observed along the expanding boundaries [Lui et al., 1977]. In the inner magnetosphere, particles are energized and injected into the ring current by the enhanced electric field due to both enhanced convection and the temporal changes in the magnetic field giving rise to an inductive electric field. Hones [1985] relates the thickening of the plasma sheet with the sudden jump of the auroras and the auroral electrojets to high latitudes in the end of the expansion phase. In a case study, Hones et al. [1986] interpreted ISEE 1 and 2 plasma and magnetic field observations in terms of the near-Earth neutral line model, where the start of the recovery phase is signalled by a rapid tailward retreat of the near-Earth neutral line. The refilling of the plasma sheet is thereafter maintained by enhanced reconnection at the retreating neutral line. In this case, it took only five minutes for the plasma sheet to return to the location of the ISEE spacecraft which was estimated to be ~0.5 R_E above the neutral sheet at a distance about 20 R_E in the tail. The difference in time scales of the recovery in the ionosphere and in the magnetotail may reflect the fact that whereas the AE-index monitors the global dynamics of the oval, the spacecraft measures only very local properties of the tail plasma sheet. On the other hand, the relaxation time scales of the magnetosphere and ionosphere may be very different. Assuming that the plasma sheet recovery takes place within 10–20 min, the slower recovery of the ionosphere may lead to a more active role of the ionosphere in the ionosphere - magnetosphere coupling during late recovery phase than during the earlier phases of the substorm.

The ionospheric electric field and conductance patterns giving rise to omega bands have been modeled by several authors [Gustafsson et al., 1981; Opgenoorth et al., 1983; Buchert et al., 1990]. The conductivity and electric field patterns that correspond to the ionospheric observations result in a field-aligned current pattern with upward current density of a few $\mu A/m^2$ in the bright areas, and downward current in the dark areas. Lyons and Walterscheid [1985] proposed a mechanism, where the omega bands are created by neutral wind acceleration in the ionosphere that is enhanced due to the enhanced electron precipitation. Shear flows that appear due to the acceleration may be unstable to the Kelvin-Helmholtz instability, producing omega bands drifting eastward with the local drift speed. One of the advantages of this model is that the neutral wind activity may be sustained independent of the recovery of the plasma sheet. However, according to Buchert et al. [1990], the model is not fully supported by either observations of the phase velocity of the pulsations or the velocity of the neutral motion deduced from electric field observations.

It is not clear where the particles precipitating into the omega bands originate in the magnetosphere. Gustafsson et al. [1981] and Opgenoorth et al. [1983] did not address the question of the source region in the magnetosphere. Saito [1978] suggested that the precipitation arises due to a modulation of the cross-tail current, whereas Rostoker and Samson [1984] proposed the Kelvin-Helmholtz instability at the low-latitude boundary layer in the magnetotail as a mechanism to produce precipitation into omega bands. Buchert et al. [1990] referred to the Kelvin-Helmholtz instability as a good candidate to explain their observations, but did not commit themselves strongly to this model. Our field-aligned mapping of omega bands into the current sheet shows that the current source is located in the near-Earth region, not very far from the current maximum where the brightest part of the oval maps to [Elphinstone et al., 1990]. Of the cases we have studied, the mapping of omega bands or Ps 6 pulsation events located eight out of nine events within 13 R_E from the Earth. Thus it seems fairly likely that the formation of these structures is not directly connected with the low-latitude boundary layer processes.

There are several other possibilities for the mechanism generating particle precipitation into the morningside auroral ionosphere. As the newly injected particle population drifts around the Earth it may become unstable to wave generation, or interact with waves generated elsewhere, resulting in particle precipitation into the ionosphere. On the other hand, electrons injected close to the ring current maximum in the morning sector experience large local magnetic field gradients that effectively transport them further in towards morning sector [Roederer and Hones, 1974, Takahashi and Iyemori, 1990]. Here the gradients of superimposed corotation and cross-tail electric field are large, resulting in efficient perpendicular energization of the electrons while they are drifting due to the magnetic field gradients. Because the acceleration process is adiabatic (conserves the first adiabatic invariant), the energy increase is directly transferred to the longitudinal energy component, hence lowering particle mirror points closer to the ionosphere [Pellinen and Heikkila, 1984]. This energization mechanism is most powerful for particles with large pitch angles, mostly resulting in diffuse auroras.

Conclusions

The recovery phase has a large number of observable signatures, whose complexity makes it a difficult object for a rigorous study. In this work we have addressed the ionospheric signatures of the recovery phase. However, some key issues, like the relationship between plasma sheet refilling and recovery phase auroras, have not been discussed due to lack of appropriate data. We have documented the whole sequence of events leading from the WTS to finally appearing omega bands in the global scale with good resolution. The mutual relationship between the plasma sheet refilling and the recovery phase associated auroras, such as omega bands and pulsating auroras, still remains poorly understood. The mapping of omega bands presented in this study relates the origin of the most intense particle precipitation to omega bands and Ps6 pulsations to the region inside about 13 R_E, pointing out the importance of processes in the near-Earth tail.

The timing of the recovery phase poses a difficult problem. Localized processes in the ionosphere-magnetosphere system need a finite time to communicate with other parts of the system. The characteristic time scales are also vastly different for processes in various regions of the geospace. Thus, the division of the substorm to the growth, expansion, and recovery phases is difficult during the transition between the different phases. The recovery of the ionosphere appears to take much longer than the refilling of the plasma sheet. Also, quite often the recovery phase is followed by a new expansive phase onset in the evening-midnight sector before the recovery in the morning sector ionosphere has been completed. Thus the timing of the recovery phase from the auroral electrojet indices only is often ambiguous and obviously needs more consideration to be better understood. Consideration of individual magnetic observations from the auroral region rather than the AE-index, which allows localization of the activity in the auroral oval, may help to reveal some of this ambiguity.

Substorms have been studied now for more than a quarter of a century. A lot has been learned, but the Chapman Conference on Substorms in Japan, where this paper was presented, called for new ways of attacking the substorm question. We propose that more thorough consideration of the relaxation from the substorm expansion, i.e., the recovery phase, can provide a fresh viewpoint and lead to a better understanding of the still intriguing substorm phenomenon.

Acknowledgements. This study has greatly benefitted from the wide cooperation organized around the operations of the Viking satellite, leading to stimulating discussions and good working relationships between various research groups. The satellite project was funded by the Swedish Board of Space Activities and managed by the Swedish Space Corporation. The work by T.P., R.P., H.K., V.P., and A.Z. was partially supported by the Soviet-Finnish working group on geophysics.

References

Akasofu, S.-I., The development of the auroral substorm, *Planet. Space Sci.*, *12*, 273, 1964.

André, D. and Baumjohann, W., Joint two-dimensional observations of ground magnetic and ionospheric electric fields associated with auroral currents 5. Current system associated with eastward drifting omega bands, *J. Geophys.*, *50*, 194, 1982.

Buchert, S., Untersuchung über die Dynamik der Ionosphäre im Bereich des Nordlichts mit Hilfe des EISCAT Ionosphärenradars, Ph.D. Thesis, TU München, 1988.

Buchert, S., Baumjohann, W., Haerendel, G., La Hoz, C., and Lühr, H., Magnetometer and incoherent scatter observations of an intense Ps 6 pulsation event, *J. Atmosp. Terr. Phys.*, *50*, 357, 1988.

Buchert, S., Haerendel, G., and Baumjohann, W., A model for the electric fields and currents during a strong Ps 6 pulsation event, *J. Geophys. Res.*, *95*, 3733, 1990.

Davidson, G. T., Pitch-angle diffusion and the origin of temporal and spatial structures in morningside aurorae, *Space Sci. Rev.*, *53*, 45, 1990.

Elphinstone, R. D., Hearn, D., Murphree, J. S., and Cogger, L. L., Mapping using the Tsyganenko long magnetospheric model and its relationship to Viking auroral images. *J. Geophys. Res. 96*, 1467, 1991.

Gustafsson, G., Baumjohann, W., and Iversen, I., Multi-method observations and modeling of the three-dimensional currents associated with a very strong Ps6 event, *J. Geophys.*, *49*, 138, 1981.

Hones, E. W., Jr., The poleward leap of the auroral electrojet as seen in auroral images, *J. Geophys. Res, 90*, 5333, 1985.

Hones, E. W., Fritz, T. A., Birn, J., Cooney, J., and Bame, S. J., Detailed observations of the plasma sheet during a substorm on April 24, 1979, *J. Geophys. Res.*, *91*, 6845, 1986.

Johnstone, A. D., The mechanism of pulsating aurora, *Ann. Geophysicae*, *1*, 397, 1983.

Lui, A. T. Y., Hones, E. W., Yasuhara, F., Akasofu, S.-I., and Bame, S. J., Magnetotail plasma flow during plasma sheet expansions: Vela 5 and 6 and IMP observations, *J. Geophys. Res.*, *82*, 1235, 1977.

Lyons, L. R. and Fennell, J. F., Characteristics of auroral electron precipitation on the morningside, *J. Geophys. Res.*, *91*, 11 225, 1986.

Lyons, L. R. and Walterscheid, R. L., Generation of auroral omega bands by shear instability of the neutral winds, *J. Geophys. Res., 90,* 12 312, 1985.

Lyons, L. R. and Williams, D. J., Quantitative Aspects of Magnetospheric Physics, D. Reidel, Dordrecht, 1984.

Nielsen, E. and Sofko, G., Ps 6 spatial and temporal structure from STARE and riometer observations, *J. Geophys. Res., 87,* 8157, 1982.

McPherron, R. L., Growth phase of magnetospheric substorms, *J. Geophys. Res., 75,* 5592, 1970.

Opgenoorth, H. J., Oksman, J., Kaila, K. U., Nielsen, E., and Baumjohann, W., Characteristics of eastward drifting omega bands in the morning sector of the auroral oval, *J. Geophys. Res., 88,* 9171, 1983.

Pellinen, R. J. and Heikkila, W. J., Inductive electric fields in the magnetotail and their relation to auroral and substorm phenomena, *Space Sci. Rev., 37,* 1, 1984.

Pudovkin, M. I., Semenov, V. S., Kornilova, T. A., and Koselova, T. V., The development of a magnetospheric substorm: Theory and experiment, *Researches on geomagnetism, aeronomy and solar physics, issue 89, Physics of substorms,* Moscow, Nauka, 1989.

Rees, M. H., Lummerzheim, D., Roble, R. G., Winningham, J. D., Craven, J. D., and Frank, L. A, Auroral energy deposition rate, characteristic electron energy, and ionospheric parameters derived from Dynamics Explorer 1 images, *J. Geophys. Res., 93,* 12841, 1988.

Roederer, J. G. and Hones, E. W. Jr., Motion of magnetospheric particle clouds in a time-dependent electric field model, *J. Geophys. Res., 79,* 1432, 1974.

Rostoker, G. and Samson, J. C., Can substorm expansive phase effects and low frequency Pc magnetic pulsations be attributed to the same source mechanism?, *Geophys. Res. Lett., 11,* 271, 1984.

Rostoker, G., Kamide, Y., and Winningham, J. D., Energetic particle precipitation into the high latitude ionosphere and the auroral electrojets, 3. Characteristics of electron precipitation into the morning sector auroral oval, *J. Geophys. Res., 90,* 7495, 1985.

Saito, T., Long-period irregular micropulsations, Pi 3, *Space Sci. Rev., 21,* 427, 1978.

Sandahl, I., Pitch angle scattering and particle precipitation in a pulsating aurora - An experimental study, *KGI Report 185,* Kiruna Geophysical Institute, Kiruna, 1984.

Shepherd, G. G., Steen, A, and Murphree, J. S., Auroral boundary dynamcis observed simultaneously from the Viking spacecraft and from the ground. *J. Geophys. Res., 95,* 5845, 1990.

Steen, Å, Collis, P. N., Evans, D., Kremser, G., Capelle, S., Rees, D., and Tsurutani, B. T., Observations of a gradual transition between Ps 6 activity with auroral torches and surgelike pulsations during strong geomagnetic disturbances, *J. Geophys. Res., 93,* 8713, 1988.

Takahashi, S. and Iyemori, T., Simulation of charged particle motions in realistic model magnetospheres and the effect of corotation electric field, *Annales Geophysicae,* 8, 503, 1990.

Tsyganenko, N.A., Magnetospheric magnetic field model with a warped tail current sheet, *Planet. Space Sci., 37,* 5, 1989.

Yahnin, A. G., Sergeev, V. A., Pellinen, R. J., Baumjohann, W., Kaila, K. U., Ranta, H., Kangas, J., and Raspopov, O. M., Substorm time sequence and microstructure on 11 November 1976. *J. Geophys., 53,* 182, 1983.

A Statistical Study of Substorm Onset Conditions at Geostationary Orbit

A. Korth,[1] Z.Y. Pu,[1,2] G. Kremser,[1,3] and A. Roux,[4]

GEOS-2 observations by the energetic particle spectrometer and the magnetometer are used to investigate the magnetospheric plasma and field conditions at substorm onsets close to the geostationary orbit. 22 events were used for a statistical study. It was found that earthward pressure gradients of energetic ions with scale lengths $< 1\ R_E$ and a tail-like magnetic field configuration exist at the inner edge of the plasmasheet prior to the onsets. The plasma β increases to a critical value for the generation of the ballooning mode instability. The appearance of Pi2 pulsations on the ground and variations of the ion fluxes and the magnetic field are consistent with the rapid growth of the ballooning mode. At substorm onset (appearance of Pi2 pulsations) the radial ion pressure gradient decreases to zero ($\nabla\ P_{ET} \rightarrow 0$). After onset highly turbulent fluctuations of the radial gradient are observed with scale lengths of a few ion gyroradii. We conclude that important substorm-associated processes occur as close to the Earth as the geostationary orbit, and are consistent with the development of the ballooning instability.

1. INTRODUCTION

Magnetospheric substorms are transient processes initiated on the nightside of the Earth in which a significant amount of energy is deposited in the auroral ionosphere and magnetosphere. It is now generally accepted that the nightside magnetic field is stretched out tailward during the growth phase of a substorm due to the enhancement of the plasmasheet current. During the expansion phase, the auroral electrojet increases. This can be explained by the formation of a current wedge, in which the cross-tail currents are partly diverted into the ionosphere [McPherron et al., 1973]. The first auroral break-up and each intensification of the electrojet are associated with Pi2 pulsation bursts, and sometimes with westward travelling surges [Rostoker et al., 1987]. Hot plasma clouds are created in the nightside magnetosphere known as plasma injection [DeForest and McIlwain, 1971]; some injection events do not show any energy dispersion [Mauk and Meng, 1983]. At the time of injection the tailward-stretched geomagnetic field returns to a more dipole-like configuration [Baker et al., 1981]. All these phenomena are related to each other, presenting a complicated scenario for the substorm analysis.

Substorm dynamics has been a subject of interest for a long time [McPherron, 1979; and references therein]. In the past two decades considerable effort has been devoted to the investigation of characteristic features of substorm related phenomena. A large amount of details has been found out [Baker and McPherron, 1990; Baker et al., 1984; Roux, 1985; Lopez et al., 1988; Nagai, 1987; Nagai et al., 1987; Smits et al., 1986; Kremser et al., 1982, 1988]. However, substantial uncertainties exist on how these phenomena are created. For instance, why do Pi2 pulsations appear at substorm onsets and how are they generated? What causes the diversion of cross-tail currents, the association between dipolarization and highly fluctuating magnetic fields, and the formation of the so-called injection boundary [McIlwain, 1974; Lopez et al., 1990] ?

Recently it has been recognized that the near-earth plasmasheet plays an important role in the substorm de-

[1]Max-Planck-Institut fur Aeronomie, Katlenburg-Lindau, Germany
[2]Department of Geophysics, Peking University, Beijing, China
[3]Department of Physics, University of Oulu, Oulu, Finland
[4]CNET/CRPE, Issy-les-Moulineaux, France

Magnetospheric Substorms
Geophysical Monograph 64
Copyright 1991 American Geophysical Union

velopment [Baker, 1984; Roux et al., 1991; Kaufmann, 1987; Lui et al., 1990, 1988; Nielsen, 1988; Goertz and Smith, 1989; Mitchell et al., 1990]. There exists evidence that the current wedge occurs close to the geostationary orbit, near the region of plasma injection [Kelly et al., 1984; Barfield et al., 1986; Mauk and Meng, 1987]. AMPTE/CCE observed a localized and turbulent current disruption region at 8.8 Re, associated with a dispersionless energetic particle injecton [Takahashi et al., 1987; Lui et al., 1987]. More recently Roux et al. [1991] found in a case study that an earthward intensity gradient of energetic ions existed at the GEOS-2 orbit at substorm break-up with a typical scale length of several ion gyroradii. Particle, electric, and magnetic field data seem to be consistent with the development of an instability at the onset of the expansion phase, to which the westward travelling surge and field-aligned currents appear to be related.

In order to obtain further more detailed information on the substorm development we have carried out a statistical study of substorm onset conditions with energetic particle and magnetic field data observed by GEOS-2.

2. INSTRUMENTATION AND DATA

This study is based on data from the magnetometer and the particle spectrometer onboard the ESA geosynchronous satellite GEOS-2. GEOS-2 was launched in August 1978 and positioned at $37.5°$ E, which corresponds to a nominal dipole latitude of about $3.5°$ S. Latitudinal shifts were carried out before 25 August 1978 and after 10 July 1979. The spin period is 6 s.

The particle spectrometer onboard GEOS-2 [Korth and Wilken, 1978; Korth et al., 1978] measures electrons in an integral channel with energies above 22 keV and in 16 differential channels between 14 and 213 keV. Ions are detected at energies above 27 keV and in 10 energy channels between 28 and 402 keV. A full energy spectrum is obtained every 5.5 s over a pitch angle range of about $120°$ for electrons and ions. No identification of different ion species is possible. For the data interpretation we assume that most of the ions are protons.

The integral particle intensities ($E_e > 22$ keV, $E_i > 27$ keV) are measured simultaneously in all directions with a time resolution of 0.172 s. The ion fluxes show strong azimuthal asymmetries in the fixed pitch angle range from $60°$ to $120°$. This fairly broad pitch angle range was admitted in order to minimize statistical fluctuations. Therefore we distinguish the ions according to the location of their gyrocenters in the plane perpendicular to the ambient magnetic field. We determined the intensities of ions with gyrocenters to the east or to the west of the satellite. Their difference provides the east-west component of the ion intensity gradient. In an analogous way the radial intensities are defined as those of ions with gyrocenters earthward or radially outward of the satellite. Their difference provides the radial component of the ion intensity gradient [see Kremser et al., 1982; Kremser et al., 1988].

The magnetic field data are measured by a triaxial flux gate magnetometer and are presented in a despun spacecraft coordinate system. In this system B_H is the component parallel to the axis of rotation, B_V points radially outward, B_D is the component in the azimuthal direction, positive eastward, and B_T is the total magnetic field. The time increment of the magnetic field measurements used for the investigations is 5.5 s.

3. OBSERVATIONS

The data base for this study consists of one year of data obtained between August 1978 and July 1979. Quick-look plots were scanned visually for substorm onsets. The selection criteria were as follows:

- The injection of electrons should show no dispersion in energy on a time scale of 1 min.

- Only isolated substorms were taken into account. In the case of multiple onset substorms only the first onset was analyzed.

A typical example is shown in Fig. 1. The electron and ion intensities are displayed in the pitch angle range $85°-95°$ together with the geomagnetic field recorded on GEOS-2. The data are averaged over 1 min (10 spin periods). The intensity increases (injections) are seen for electrons and ions in all energy channels. The injection for electrons appears simultaneously (within 1 min) in all energy channels. This dispersionless increase suggests that the acceleration or injection occurred locally close to the satellite. If these electrons had been accelerated far away from the spacecraft, they should show energy dispersion signatures in the energy range of our observations. The intensity increase is associated with an increase of the H-component and the decrease of the V-component of the magnetic field, indicating the return from an inflated (tail-like) field to deflated (dipolar) field. This dipolarization is typical for the onset of substorms.

We found 22 similar events during the period between August 1978 and July 1979 which are listed in Table 1. The substorm onsets were determined from the onsets of Pi2 bursts in groundbased pulsation recordings, as proposed by Rostoker et al. [1980]. We used

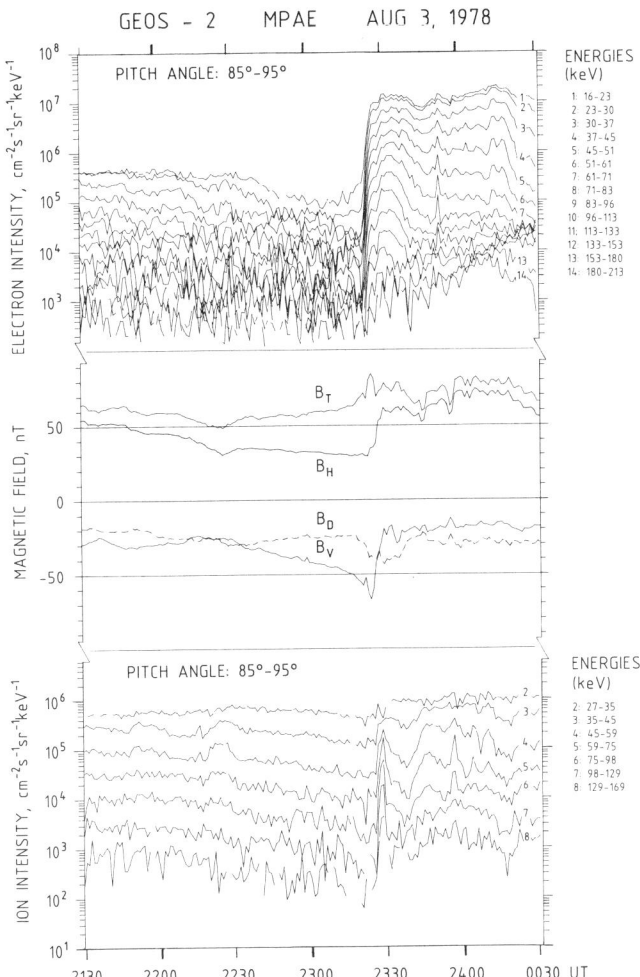

Fig. 1. The top panel presents differential electron intensities in 14 energy channels in the pitch angle range 85°–95°. The middle panel contains variations of the geomagnetic field in the VDH coordinate system. The bottom panel shows differential ion intensities in 7 energy channels in the pitch angle range 85°–95°.

are injected before the electrons. The time difference amounts up to a few minutes.

The middle column of Table 1 provides the times at which the radial component of the ion intensity gradient decreases to zero. These times were derived from plots similar to Fig. 2, which displays the earthward ion intensity (faint line, the gyrocenters are located earthward of the satellite) and the tailward ion intensity (solid line, the gyrocenters are located radially outward of satellite) for ions with energies > 27 keV. The 16-min time interval starts 8 min before the onset of the substorm (onset of Pi2 pulsations). The timing accuracy obtained from the Pi2 pulsation recordings is about ± 0.5 min and is indicated by broad arrows at the times at which the radial (earthward–tailward) ion intensity gradient (∇P_{ET}) decreases to zero and the electron and ion injections occur. The timing accuracy of the injections is about ± 0.5 min.

Figure 2 shows for three examples that earthward pressure gradients existed before substorm onset (during the growth phase). At substorm onset the radial pressure gradients (∇P_{ET}) decrease to zero and thereafter exhibit strong temporal variations. About 0.5 to 1 min before $\nabla P_{ET} = 0$ the tailward ion intensity starts to increase. The other 19 events behave in a very similar way (not displayed). A possible explanation of this behaviour is provided in the discussion section.

Some of the substorms show an east-west or azimuthal gradient before substorm onset in addition to the radial gradient that is always observed. Fig. 3 shows an ex-

TABLE 1. Selected substorms between August 1978 and July 1979

Event No.	Date	Pi2 Onset Time (Göttingen) [UT]	Ion Gradient ⟶ 0 [UT]	Plasma Injections		
				Electrons [UT]	Ions [UT]	Electrons [LT]
1	Aug. 03, 1978	23:22	23:22	23:22	23:22	23:22
2	Oct. 17, 1978	19:48	19:48	19:48	19:48	22:18
3	Nov. 11, 1978	20:09	20:09	20:10	20:09	22:40
4	Nov. 21, 1978	20:59*	20:59	21:00	20:59	23:30
5	Nov. 26, 1978	19:03	19:03	19:04	19:04	21:34
6	Dec. 04, 1978	19:39	19:39	19:39	19:39	22:09
7	Jan. 22, 1979	19:44	19:45	19:46	19:45	22:16
8	Jan. 25, 1979	20:16	20:17	20:17	20:17	22:47
9	Jan. 26, 1979	19:18	19:18***	19:18	19:18	21:48
10	Jan. 27, 1979	20:59	20:59	20:59	20:59	23:29
11	Feb. 06, 1979	21:24	21:24	21:24	21:24	23:54
12	Feb. 18, 1979	20:28	20:28	20:28	20:28	22:58
13	Feb. 26, 1979	20:25	20:25	20:26	20:26	22:56
14	Mar. 04, 1979	22:35**	22:36	22:37	22:36	01:07
15	Mar. 24, 1979	20:12	20:13	20:13	20:13	22:43
16	Mar. 25, 1979	19:56	19:56	19:58	19:56	22:28
17	May 23, 1979	20:28	20:30	20:35	20:30	23:05
18	May 27, 1979	18:25	18:25	18:32	18:25	21:02
19	June 20, 1979	20:21	20:21	20:21	20:21	22:51
20	July 03, 1979	21:49	21:50	21:51	21:51	00:21
21	July 16, 1979	20:59	20:59	21:03	21:00	23:09
22	July 27, 1979	00:39	00:43	00:46	00:46	02:06

* Borok
** Sodankylä
*** Oscillation starts

recordings from the mid-latitude station Göttingen with the exception of two events, for which the Pi2 response was very small in the Göttingen recordings. In these cases recordings from the auroral zone stations Borok and Sodankylä were used instead. The last three columns in Table 1 show the injection time for electrons and ions in universal time [UT] and local time [LT]. All substorm events occurred in the midnight sector at local times between 2100 and 0200 LT.

The plasma injections started simultaneously with or later than the Pi2 pulsations; they never started earlier than the Pi2 burst. The injections of electrons and ions coincide in 14 cases. In the other 8 cases the ions

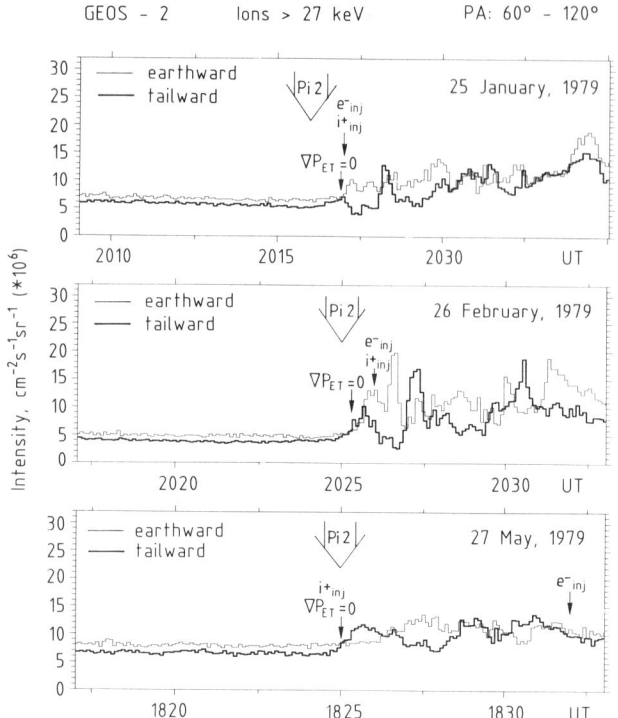

Fig. 2. Variations of the spatial gradient of the ion intensity displayed by its earthward and tailward (radial) components for three selected events.

ample for the event of 25 January 1979. The east-west gradient shows large oscillations after substorm onset ($\nabla P_{ET} = 0$) indicating the occurrence of field-aligned currents, and these fluctuations are in close connection with the variation in the D-component of the magnetic field. A similar relationship was found by *Hruska* [1986]. 15 out of the selected 22 events display no visible east-west gradients before substorm onset; 5 events have no complete data coverage; only 2 events have a clear westward component of the ion intensity gradient before onset. However, after substorm onset all events show in the east-west ion gradient and this gradient then exhibits temporal variations similar to those at the radial gradient. Both of them are regarded as the result of the same mechanism as will be explained further in the discussion.

It should be mentioned here that during magnetically quiet periods no ion gradient was found at the geostationary orbit in the midnight sector. This is not astonishing since the maximum of the running current particle flux is located for quiet times close to the geostationary orbit. The maximum moves inward during disturbed periods. It is also important to note that asymmetries due to magnetospheric convection are about one order of magnitude lower than the gradients reported here [*Pu et al.*, 1991].

A comparison between the radial component of the ion intensity gradient and the three magnetic field components is shown in Fig. 4 for the event of 26 February 1979. The Pi2 burst started at 2025:20 UT. In this case the onset time could be determined very accurately. Precisely at that time the radial intensity gradient decreased to zero. The magnetic field in the upper panel is characterized before substorm onset by a constant H-component and a slowly decreasing V-component. The magnitude of the V-component is larger than the H-

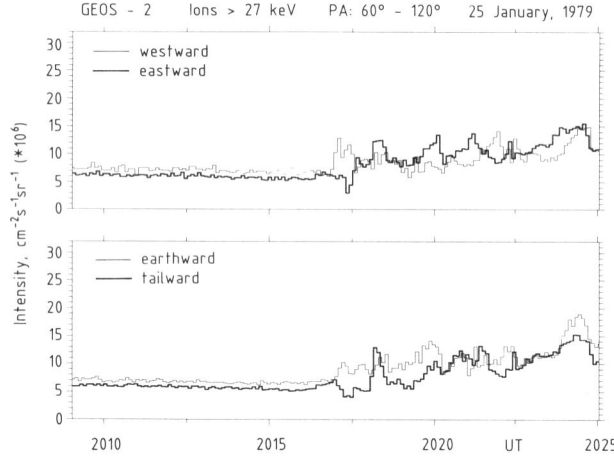

Fig. 3. Variations of the spatial gradient of the ion intensity presented by its azimuthal as well as radial components for the 25 of January, 1979, event.

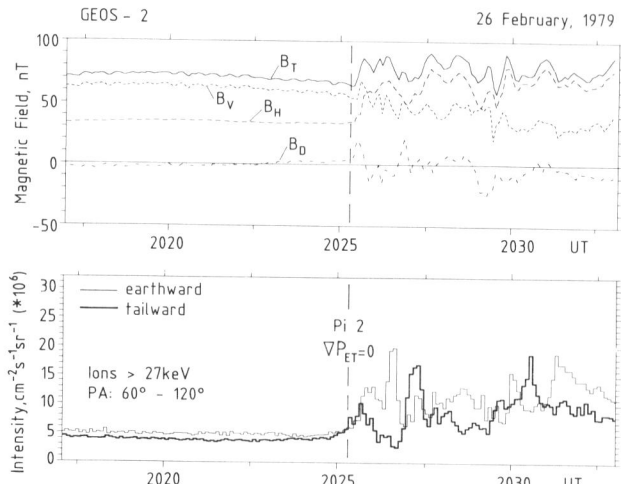

Fig. 4. Magnetic field components and the radial component of the ion intensity gradient for the event of 26 February 1979. The substorm onset is indicated by a dashed vertial line.

component, which corresponds to a very tail-like geomagnetic field. Simultaneously with the appearance of the Pi2 pulsations and $\nabla P_{ET} = 0$ the H-component increases sharply and the V-component fluctuates and decreases. This means that a rapid transition from a tail-like to a dipole-like geomagnetic field configuration occurred. The D-component shows large variations after onset which indicates the generation of field-aligned currents. The other events show similar changes of the geomagnetic field configuration after $\nabla P_{ET} = 0$.

The timing of the substorm onset (vertical dashed line) and the particle injection is shown in Fig. 5 for the same event as in Fig. 4. Electrons with energies above 22 keV are displayed in the top panel in the pitch angle range $85° - 95°$. In the middle panel ions with energies above 27 keV are presented in the same pitch angle range. The time resolution for both panels is 5.5 s. We see that the electron and ion injections occur almost simultaneously at the onset ($\nabla P_{ET} = 0$). The ion injection is preceded by a weak gradual intensity increase that is also observed in the tailward flux (bottom panel). The other events listed in Table 1 show a similar behaviour. The particle injections either occur immediately after the substorm. On in a few cases (3 out of 22) the electron injection is delayed by up to a few minutes.

It was mentioned already that the radial pressure gradient turns to zero at substorm onset and becomes highly turbulent. Compressional and transverse Alfvén waves have been observed, whose structure provides evidence for the drift-bounce resonance of hot ions. Details about these waves will be discussed in a follow-up paper.

4. PARAMETERS DERIVED FROM THE OBSERVATIONS

a. Curvature radius of the geomagnetic field lines at the equator

Usually a continuous decrease of the H-component of the geomagnetic field and a slow increase of the V-component are observed during the substorm growth phase, indicating that a tail-like configuration is progressively built up. During 19 of our events GEOS-2 was located at $3.5°$ below the equator and during the remaining three events even closer to the equator. We estimated the field line curvature, R_C, in the equatorial plane by using the observed V, D, H components of the magnetic field and a model that was first presented by *Quinn and McIlwain* [1979] and later on used by *Mauk* [1986]. In the model of Quinn and McIlwain R_C is calculated by the formular $R_C = R_o (3+4k)^{-1}$ whereby R_o is the equatorial distance of the magnetic field line and the parameter k describes the deviation from the dipole field line. The parameter k can be determined from the observed magnetic field components and the geomagnetic latitude of the satellite and R_o can be determined from the geocentric distance R of the satellite, its magnetic latitude and k. The calculated values for R_C are listed in Table 2. For most of the events, R_C is considerably smaller than $2.2\ R_E$, the curvature radius in the dipole field at geostationary orbit. The average of R_C is about $1\ R_E$. There are a few events with R_C-values as low as $0.25\ R_E$. These values must be regarded with caution, since the above model may not be applicable for such extremely tail-like cases.

b. Scale length of the ion pressure gradient

The scale length for the ion pressure gradient is defined by

$$L_{Pi} = |\frac{Pi}{\nabla Pi}|$$

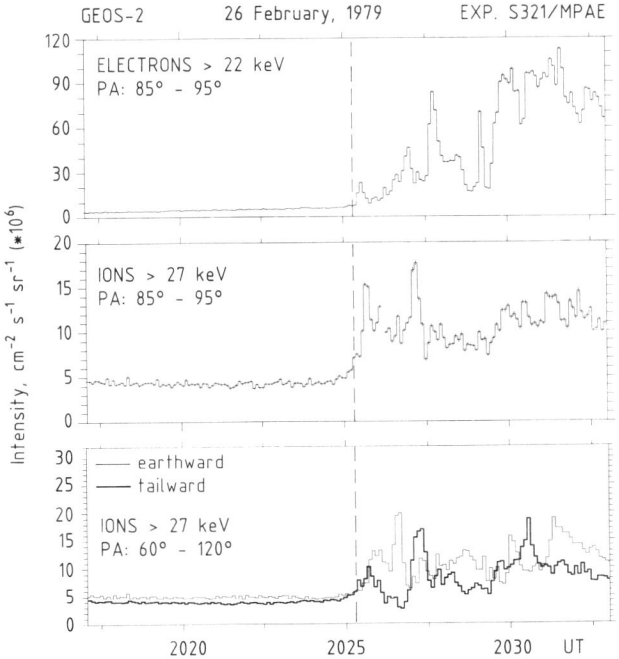

Fig. 5. Integral electron (> 22 keV) and ion (> 27 keV) intensities in the pitch angle range $85°$-$95°$ and radial component of the ion intensity gradient for the event on 26 February 1979. The substorm onset is indicated by a dashed vertical line.

and can be expressed approximately at the equator by the following formula

$$L_{Pio} = \frac{I_E + I_T}{I_E - I_T} \cdot \frac{B}{B_o} \cdot r_i$$

TABLE 2. 3-min averages of plasma parameters prior to substorm onsets

Event No.	Date	R_c [km]	L_{Pio} [km]	β_e	β_i	β_{io}
1	Aug. 03, 1978	2228	4433	0.01	0.51	3.5
2	Oct. 17, 1978	5350	6286	0.01	1.50	2.2
3	Nov. 11, 1978	2977	2965	0.01	0.77	1.6
4	Nov. 21, 1978	5076	2469	0.04	0.67	0.9
5	Nov. 26, 1978	3312	5895	0.04	0.69	3.0
6	Dec. 04, 1978	2426	4236	0.03	1.27	3.1
7	Jan. 22, 1979	2976	3462	0.01	0.71	2.1
8	Jan. 25, 1979	2479	2950	0.08	0.87	2.1
9	Jan. 26, 1979	1537	3908	0.03	1.04	4.2
10	Jan. 27, 1979	3581	4082	0.15	0.66	1.7
11	Feb. 06, 1979	1618	3259	0.01	0.58	3.5
12	Feb. 18, 1979	2025	1754	0.01	0.42	1.6
13	Feb. 26, 1979	2479	2720	0.06	0.55	1.9
14	Mar. 04, 1979	2306	1290	0.02	0.22	1.0
15	Mar. 24, 1979	3542	4871	0.03	1.80	2.6
16	Mar. 25, 1979	2119	3675	0.03	1.09	2.8
17	May 23, 1979	7034	5385	0.04	1.19	1.0
18	May 27, 1979	7410	4112	0.08	0.96	1.0
19	June 20, 1979	4026	4472	0.02	2.04	1.8
20	July 03, 1979	10532	5094	0.01	0.67	0.9
21	July 16, 1979	7363	6189	0.01	0.56	1.3
22	July 27, 1979	1777	4268	0.01	0.27	4.2

I is the integral ion intensity; the subscripts E and T refer to the earthward and tailward component. r_i denotes the average ion gyroradius and B_o is the total magnetic field strength at the equator, calculated with the magnetic field model from *Quinn and McIlwain* [1979].

The computed values of L_{Pio} are listed in Table 2. The calculations were performed for an average particle energy of 40 keV. This is a relatively high energy. We know, however, for the event of Jan. 25, 1979, the number density at energies below 1 keV is comparable to the number density at energies above 30 keV [*Pu et al.*, 1991]. This means that the pressure exerted by the high energy ions must be considerably higher than the pressure of the low energy particles. We assume a similar situation for the other events, too. In summary, for most of the events the scale length of the ion pressure gradient amounts to 10-15 ion gyroradii before substorm onset. After onset the scale length decreases to a few (3-5) ion gryoradii (not listed in Table 2).

c. Plasma β

The plasma β value is defined as the ratio of the plasma pressure over the magnetic field pressure. Table 2 provides β-values for electrons (β_e) and ions (β_i) for the position of GEOS-2, which means at a distance of 6.6 R_E and a latitude of $\leq -3.5°$. The corresponding value in the equatorial plane, β_{io}, is calculated from the formula $\beta_{io} = Pi/(B_o^2/2\mu_o)$, where B_o is the field strength at the equator. It can be seen clearly that the electron energy density is about 10% of the ion density before substorm onset at the GEOS-2 po-

sition. Most of the β-values ($\beta_e + \beta_i$) range between 0.6 and 1.5, whereas the β_{io}-values vary between 1 and 3, since $B/B_o > 1$. In reality the β-values are slightly larger than listed in Table 2, since the integration was performed over energies above 27 keV, corresponding to the energy range of the instrument.

In Fig. 6 the β-values ($\beta_e + \beta_i$) for the events No. 3, 9, and 18 are plotted with 1-min time resolution. Before substorm onset the β-values are constant and close to one and increase strongly at the onset.

5. DISCUSSION

We have investigated in detail temporal variations of the distribution of electrons and ions in the tens of keV energy range at substorm onsets close to the geostationary orbit. The 22 selected events are listed in Table 1. They occurred at local times between 2100 and 0200 LT. Several plasma parameters have been determined. The numbers listed in Table 2 should be regarded as characteristic values. The substorm onsets were determined from Pi2 recordings.

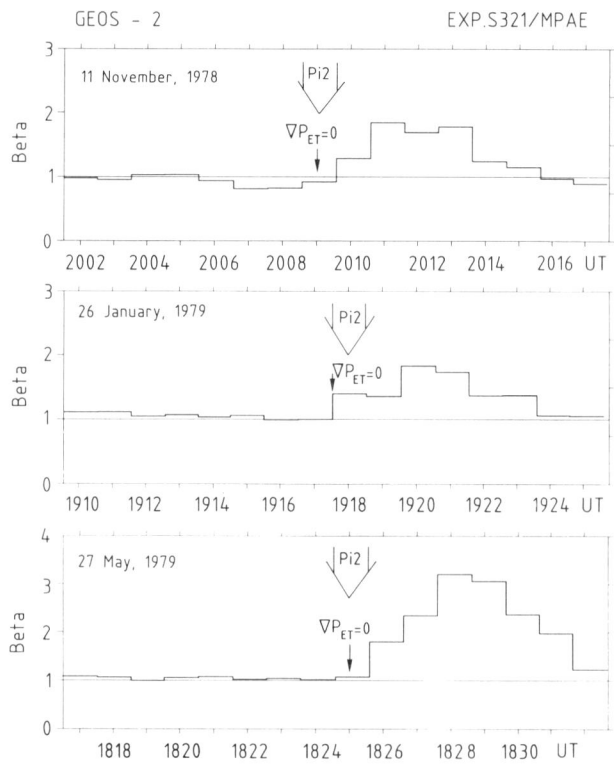

Fig. 6. β-values for three selected events before and after the substorm onset.

The investigations yielded the following results:

- Prior to substorm onsets we observe an earthward ion pressure gradient with a scale length of 10 - 15 ion gyroradii. The earthward gradient decreases to zero at substorm onset and its direction starts to oscillate thereafter. The scale length then amounts to 3–5 ion gyroradii.

- The geomagnetic field becomes tail-like during the growth phase of the substorm and returns to a dipolar configuration [Spence et al., 1989] when the ion pressure gradient decreases to zero. The parameters characterizing the tail-like field configuration are the curvature radius of the field lines in the equatorial plane (Table 2) and the distance to which a field line is stretched out at the magnetic equator. The formation of the tail-like geomagnetic field can be explained by an increase and an inward motion of the plasma sheet current. The reconfiguration at substorm onset can be related to the current disruption and diversion into the ionosphere near the inner edge of the plasma sheet.

- The plasma injections which appear even up to a few minutes after onset could result from acceleration processes associated with inductive electric fields during the dipolarization of the geomagnetic field [Mauk, 1986; Roux et al., 1991a].

- The β-values before substorm onset are close to 1.

In the following we will show that the above obtained results are consistent with the development of the ballooning mode instability, as suggested in a case study by Roux et al. [1991b].
The pressure gradients observed before substorm onset are necessary to balance the curvature force in a tail-like magnetic field configuration. In this configuration the ion pressure gradient, ∇P_i, and the magnetic field gradient, ∇B, as well as the curvature vector of the magnetic field lines all point towards the Earth. Such a condition is potentially unstable and therefore favourable for triggering the ballooning mode instabilities. Another important parameter is the plasma β. If it exceeds a certain value and is not too large, the ballooning mode instability can occur in a tail-like magnetic field. A detailed computation of the instability criteria will be given in Roux et al. [1991b] and Pu et al. [1991]. In a first approximation β_o (at the equator) must satisfy the condition

$$\frac{L_{poi}}{R_c} < \beta_o < \frac{2L_{poi}}{R_c} \tag{1}$$

where R_c is the curvature radius of the field lines and L_{poi} is the scale length of the ion pressure gradient at the equator. Both values are given in Table 2. This necessary condition for the excitation of the ballooning mode instability is fulfilled for all 22 selected events in Table 1. Hence, we expect that at substorm onset the ballooning mode instability is generated at the equator in the midnight sector near the geostationary orbit. Once the ballooning mode is excited the development of the instability tends to smooth the ion gradient; this is observed in the tailward ion intensity which starts to increase about 0.5 to 1 min before $\nabla P_{ET} = 0$ (see Fig. 2, all events). The characteristic time for the growth is about one minute. Thereafter the development of the instability leads to large oscillations in the pressure gradient and the magnetic field. These large oscillations may result from the non-linear development of the ballooning instability. A feedback from the ionosphere following electron precipitation may also take place leading to enhanced oscillations [Kan, this issue].

After substorm onset the variations of the particle fluxes and the magnetic field near the equator, and the Pi2 pulsations on the ground may be produced by the ballooning instability. The Pi2 response from the development of the ballooning instability propagates with the Alfvén velocity and reaches the ground within less than 1 minute. Therefore, it can be understood that in most of the selected events Pi2 bursts occur right at the time when $\nabla P_{ET} = 0$ (Fig. 2). In most cases strong oscillations are clearly observed in the east-west component of the ion gradient just after the time when $\nabla P_{ET} = 0$ (Fig. 3) and finite amplitude variations are seen in the D-component (Fig. 4), which are both known to be closely related with field-aligned currents in the magnetosphere. These field-aligned currents may result as a consequence of the non-linear development of the ballooning mode instability.

We can conclude that important substorm-associated processes occur as close to the Earth as the geostationary orbit, and are consistent with the development of the ballooning instabiliy.

Acknowledgements. The magnetic field measurements were provided by F. Mariani. The particle spectrometer was designed and constructed with financial support from the Max-Planck-Gesellschaft zur Förderung der Wissenschaften and the Bundesministerium für Forschung und Technologie through the DFVLR-PT under contract RV 14-B12/73 (WRK 243)-SF-21. The GEOS data reduction was performed by F. Both, H. Michels, and J. Wallbrecht.

REFERENCES

Baker, D.N., Particle and field signatures of substorms in the near magnetotail, in *Magnetic Reconnection in Space and Laboratory Plasmas, Geophys. Monogr. Ser.*, vol. *30*, edited by E. W. Hones, Jr., p. 193, AGU, Washington, D.C., 1984.

Baker, D.N., E.W. Hones, Jr., P.R. Higbie, R.D. Belian, and P. Stauning, Global properties of the magnetosphere during a substorm growth phase: A case study, *J. Geophys. Res., 86,* 8941, 1981.

Baker, D.N., S.J. Bame, R.D. Belian, W.C. Feldman, J.T. Gosling, P.R. Higbie, E.W. Hones, Jr., D.J. McComas, and R.D. Zwickl, Correlated dynamical changes in the near-Earth and distant magnetotail regions: ISEE 3, *J. Geophys. Res., 89,* 3855, 1984.

Baker, D.N. and R.L. McPherron, Extreme energetic particle decreases near geostationary orbit: A manifestation of current diversion within the inner plasma sheet, *J. Geophys. Res., 95,* 6591, 1990.

Barfield, J.N., N.A. Saflekos, R.E. Sheehan, R.L. Carovillano, T.A. Potemra, and D. Knecht, Three-dimensional observations of Birkeland current, *J. Geophys. Res., 91,* 4393, 1986.

DeForest, S.E. and C.E. McIlwain, Plasma clouds in the magnetosphere, *J. Geophys. Res., 76,* 3587, 1971.

Goertz, C.K. and R.A. Smith, The thermal catastrophe model of substorms, *J. Geophys. Res., 94,* 6581, 1989.

Hruska A., Field-aligned currents in the Earth plasma sheet, *J. Geopys. Res., 91,* 371, 1986.

Kelly, T.J., C.T. Russell, and R.J. Walker, ISEE 1 and 2 observations of an oscillating outward moving current sheet near midnight, *J. Geophys. Res., 89,* 2745, 1984.

Kaufmann, R.L., Substorm currents: Growth phase and onset, *J. Geophys. Res., 92,* 7471, 1987.

Korth, A. and B. Wilken, New magnetic electron spectrometer with directional sensitivity for a satellite application, *Rev. Sci. Instr., 49,* 1435, 1978.

Korth, A., G. Kremser, and B. Wilken, Observations of substorm-associated particle flux variations at $6 \leq L \leq 8$ with GEOS-1, *Space Sci. Rev., 22,* 501, 1978.

Kremser, G., J. Bjordal, L.P. Block, K. Brønstad, M. Håvåg, I.B. Iversen, J. Kangas, A. Korth, M.M. Madsen, J. Niskanen, W. Riedler, J. Stadsnes, P. Tanskanen, K.M. Torkar, S.L. Ullaland, Coordinated balloon-satellite observations of energetic particles at the onset of a magnetospheric substorm, *J. Geophys. Res., 87,* 4445, 1982.

Kremser, G., A. Korth, S.L. Ullaland, S. Perraut, A. Roux, A. Pedersen, R. Schmidt, and P. Tanskanen, Field-aligned beams of energetic electrons (16 keV $\leq$ E $\leq$ 80 keV) observed at geosynchronous orbit at substorm onsets, *J. Geophys. Res., 93,* 14,453, 1988.

Lui, A.T.Y., R.W. McEntire, and S.M. Krimigis, Evolution of the ring current during two geomagnetic storms, *J. Geophys. Res., 92,* 7459, 1987.

Lui, A.T.Y., R.E. Lopez, S.M. Krimigis, R.W. McEntire, L.J. Zanetti, and T.A. Potemra, A case study of magnetotail current sheet disruption and diversion, *Geophys. Res. Lett., 15,* 721, 1988.

Lui, A.T.Y., A. Mankofsky, C.-L. Chang, K. Papadopoulos, and C.S. Wu, A current disruption mechanism in the neutral sheet: A possible trigger for substorm expansions, *Geophys. Res. Lett., 17,* 745, 1990.

Lopez, R.E., A.T.Y. Lui, D.G. Sibeck, R.W. McEntire, L.J. Zanetti, T.A. Potemra, and S.M. Krimigis, The longitudinal and radial distribution of magnetic reconfigurations in the near-Earth magnetotail as observed by AMPTE/CCE, *J. Geophys. Res., 93,* 997, 1988.

Lopez, R.E., D.G. Sibeck, R.W. McEntire, and S.M. Krimigis, The energetic ion substorm injection boundary, *J. Geophys. Res., 95,* 109, 1990.

Mauk, B.H., Quantitative modeling of the "convection surge" mechanism of ion acceleration, *J. Geophys. Res., 91,* 13,423, 1986.

Mauk, B.H. and C.I. Meng, Dynamical injections as the source of near geostationary quiet time particle spatial boundaries, *J. Geophys. Res., 88,* 10,011, 1983.

Mauk, B.H. and C.I. Meng, Plasma injection during substorms, *Phys. Scripta., T18,* 128, 1987.

McPherron, R.L., Magnetospheric substorms, *Rev. Geophys. Space Phys., 17,* 657, 1979.

McPherron, R.L., C.T. Russell, and M.P. Aubry, Satellite studies of magnetospheric substorms on August 15, 1968. 9. Phenomenological model for substorms, *J. Geophys. Res., 78,* 3131, 1973.

Mitchell, D.G., D.J. Williams, C.Y. Huang, L.A. Frank, and C.T. Russell, Current carriers in the near-Earth cross-tail current sheet during substorm growth phase, *Geophys. Res. Lett., 17,* 583, 1990.

Nagai, T., Field-aligned currents associated with substorms in the vicinity of synchronous orbit, 2. GOES 2 and GOES 3 observations, *J. Geophys. Res., 92,* 2432, 1987.

Nagai, T., H.J. Singer, B.G. Ledley, and R.C. Olsen, Field-aligned currents associated with substorms in the vicinity of synchronous orbit 1. The July 5, 1979, substorm observed by SCATHA, GOES 3, and GOES 2, *J. Geophys. Res., 92,* 2425, 1987.

Nielsen, E., Ionosphere-magnetosphere mapping of dynamic auroral structures during substorms, in Auroral Physics, Cambridge Univ. Press., International Conference on Auroral Physics, Cambridge, England, p. 375, 1988.

Pu, Z.Y., A. Korth, and G. Kremser, Plasma and magnetic field parameters at substorm onsets derived from GEOS-2 observations, to be submitted, 1991.

Quinn, J.M. and C.E. McIlwain, Bouncing ion clusters in the Earth's magnetosphere, *J. Geophys. Res., 84,* 7365, 1979.

Rostoker, G., S.-I. Akasofu, W. Baumjohann, Y. Kamide, and R.L. McPherron, The roles of direct input of energy from the solar wind and unloading of stored magnetotail energy in driving magnetospheric substorms, *Space Science Rev., 46,* 93, 1987.

Roux, A., Generation of field-aligned current structures at substorm onsets, *ESA SP-235,* 151, 1985.

Roux, A., S. Perraut, P. Robert, A. Morane, A. Pedersen, A. Korth, G. Kremser, B. Aparicio, D. Rodgers, and R. Pellinen, Plasmasheet instability related to westward travelling surges, in press, *J. Geophys. Res.,* 1991a.

Roux, A., S. Perraut, A. Morane, P. Robert, A. Korth, G. Kremser, A. Pedersen, R. Pellinen, and Z.Y Pu, Role of the near Earth plasmasheet at substorms, *this issue,* 1991b.

Smits, D.P., W.J. Hughes, C.A. Cattell, and C.T. Russell, Observations of field-aligned currents, waves, and electric fields at substorm onset, *J. Geophys. Res., 91,* 121, 1986.

Spence, H.E., M.G. Kivelson, and R.J. Walker, Magnetospheric plasma pressures in the midnight meridian: observations from 2.5 to 35 R_E, *J. Geopys. Res., 94,* 5264, 1989.

Takahashi, K., L.-J. Zanetti, R.-E. Lopez, R.-W. McEntire, T.-A. Potemra, and K. Yumoto, Disruption of the magnetotail current sheet observed by AMPTE/CEE, *Geophys. Res. Lett., 14,* 1019, 1987.

7. SUBSTORM ELECTRODYNAMICS

THE CONTRIBUTION OF THE BOUNDARY LAYER EMF TO MAGNETOSPHERIC SUBSTORMS

R. LUNDIN, I. SANDAHL, J. WOCH

Swedish Institute of Space Physics, Box 812, S-981 28 Kiruna, Sweden

R. ELPHINSTONE

University of Calgary, Physics Department, Calgary, Alberta T2N 1N4, Canada

The magnetospheric substorm process has been discussed from two different standpoints; As a result of a directly driven process powered by the cross-tail electromotive force (EMF), or as a release of magnetic energy stored in the magnetospheric tail. Out of these discussions, two prevailing theories have evolved to explain the triggering of substorms, the neutral line model and the boundary layer model.

In this report we discuss the two prevalent substorm models from a slightly different point of view based on the EMF produced in the low-latitude boundary layer (LLBL) as a consequence of massive solar wind plasma injection there. Our hypothesis stems from recent observations that the dayside oval is persistently active whilst the nightside oval is only temporarily active. Improved methods for the magnetic field line mapping of the auroral oval into the boundary layer and tail region as originally introduced by *Tsyganenko [1987]* have also helped elucidating the coupling of the LLBL dynamo to the dayside and nightside oval. It will be argued here that the Region 1 current system can be divided into two partial current circuits. One circuit, entirely coupled to the LLBL, is related to the continuously active dayside oval in the cleft region. The other circuit, connected to the nightside oval and the tail current sheet, can be considered a diversion or a tail route for currents driven by the LLBL. The advantage with the current concept proposed here is that it allows for both a driven and loading/unloading mechanism for substorms. Auroral break-up is in this model a consequence of enhanced currents, induced by solar wind pressure variations within the LLBL dynamo, that has reached nightside magnetic field lines of enhanced ionospheric conductivity. The model also allows for the development of tail plasmoids, generally assumed to be the consequence of a current diversion instability within the nightside current wedge.

INTRODUCTION

The concept of substorm stems back to the mid sixties when *Akasofu [1964]* used it to explain the occasional release of magnetospheric energy manifesting itself as auroral and magnetic disturbances along the nightside oval. Akasofu also defined the various phases of a substorm: growth, break-up, expansion, and recovery - phases that still are being used in contemporary modelling attempts. Despite what appears to be a strong coupling to a southward interplanetary magnetic field (IMF) [*e.g. Akasofu, 1981*], no simple external trigger mechanism has been found [*Baumjohann, 1986*], thus suggesting that internal processes in the magnetosphere for the occurrence of substorms are also needed [*McPherron, 1979, Horwitz, 1985*]. This has lead researchers to investigate the generation of substorms from two different standpoints - as a directly driven process or as a consequence of a loading/unloading energy dissipation process.

The directly driven substorm, usually advocated by researchers seeking the energy source in the low-latitude boundary layer (LLBL) and the plasma sheet boundary layer

(PSBL), is believed to be triggered by a solar wind disturbance (IMF polarity reversal) and driven by dynamo actions and instabilities within the boundary layers [*Rostoker and Eastman, 1987*].

In the loading/unloading concept for substorms the magnetotail is assumed to store large amounts of magnetic energy that due to some internal instability can be released as substorms [*Horwitz, 1985*]. The most widely used concept for such a storage of magnetic energy is via reconnection, i.e. a transfer of dayside magnetic flux towards the nightside/tail region [*Dungey, 1961*]. However, an equally valid, but perhaps more physical, description of such an energy storage is in the form of an enhanced tail current system. This is in part the description that will be used to combine the two different substorm models in this report.

A near Earth neutral line has been used in the loading/unloading substorm model to explain the development of "plasmoids" [*Hones et al., 1984, Scholer et al, 1984*]. The existence of such plasmoids in the tail is presently a much debated topic. Arguments against the existence of a near Earth neutral line and tailward streaming plasmoids [*Huang et al., 1987*] as well as arguments for their existence [*Slavin et al., 1987, Baker et al., 1987*] have been presented. However plasmoids may also be considered from the point of view of the tail current sheet geometry, similar to the "meanders" discussed by *Heikkila and Pellinen* [1977].

DAYSIDE ACTIVITY AND THE SOURCE OF THE BOUNDARY LAYER EMF

Recent years of ground based and satellite observations have shown that the solar wind energy input leads to a persistent auroral activity along the dayside oval [*Murphree et al., 1981, Evans, 1985, Sandholt et al., 1986*], which is less dependent on the IMF B_z compared to the nightside activity. This indicates that other parameters such as the solar wind plasma pressure plays an important role for the solar wind energy transfer into the magnetosphere. An energy transfer related to solar wind plasma discontinuities may, or may not, give rise to auroral substorms along the nightside oval.

The solar wind energy transfer process is in general described as propagation of solar wind electromagnetic energy through a dayside merging line. In this (merging/ reconnection) picture the generator for magnetospheric electric fields, driving magnetospheric convection and currents, is on open field lines. The predominant region of open Terrestrial field lines is in the polar cap/lobe region. Thus, one may associate merging/reconnection with the connection of the solar wind magnetic field to polar cap magnetic field lines.

The alternative solar wind energy transfer process is related to magnetosheath plasma penetration into the magnetosphere. The momentum of plasma penetrated onto closed magnetic field lines thus represents the source of free energy that can power magnetospheric convection and currents. For historical reasons [*Axford and Hines, 1961*] it is generally referred to as "viscous" interaction, although the most profound effect is believed to be due to a real mass transfer of solar wind plasma onto closed magnetospheric field lines [*Eastman et al., 1976, Lemaire, 1977*] as in fact was first suggested by *Cole* [1961].

Although the auroral oval at first sight looks like one unit the nightside and dayside auroras are clearly separate entities. This was first proposed by *Akasofu and Kan* [1981] on basis of DMSP images. Similarly, in a study of DMSP images *Meng and Lundin* [1986] suggested that dayside aurora near local noon comprises a system completely separated from the nightside. In particular during times of low magnetic activity when the nightside oval remains in a quiescent state one may yet find active auroral forms in the dayside oval near local noon.

From a plasma injection point of view one can distinguish between the dayside and nightside plasma injections. The substorm-associated nightside plasma injection, confirmed already in the early seventies [*DeForest and McIlwain, 1971*], is an injection of plasma from the tail to the inner plasma sheet and ring current. This, more occasional, plasma injection appears to be understood relatively well [*Mauk and Meng, 1987, Lopez, 1990*]. Conversely, the dayside plasma injection of magnetosheath plasma into the magnetosphere is still the source of some controversy. Dayside plasma injection appears as a continuous process in the cusp and as a temporal but also frequent process in the cleft.

The analogy between the separation of the auroral activity into a dayside and nightside activity center and the dayside and nightside plasma injection is illustrated in Fig. 1.

The dayside oval contains a cusp centered near local noon and a cleft extending over a wider local time sector away from the noon meridian, the cleft encompassing the cusp. Both regions were early believed to be in continuous or temporal contact with the magnetosheath [*Heikkila and Winningham, 1971*]. The region may be defined as the cusp/cleft region. The polar cusp is here a region of intense structureless precipitation of magnetosheath plasma near local noon [*Burch, 1968, Heikkila and Winningham, 1971*]. The extension of the region with magnetosheath plasma precipitation to a somewhat broader region, the cleft, was discussed by *Heikkila and Winningham* [1971]. Later it was demonstrated that the penetration of magnetosheath plasma indeed extends over a wider local time sector than the "cusp proper" [*Lundin, 1988, Newell and Meng, 1988*] near local noon. This extension, denoted the cleft or the LLBL projection to the dayside oval, may be a consequence of magnetosheath plasma diffusion [*Eviatar and Wolf, 1968, Eastman et al., 1976*] or a direct plasma penetration onto closed field lines as was first suggested by *Lemaire* [1977].

The magnetopause maps theoretically to singularities near local noon - the magnetic cusps. However, a finite width of the magnetopause also means a finite width of the low-altitude cusp. The mapping of a relatively narrow region near local noon to the entire magnetopause boundary layer was first indicated by *Vasyliunas* [1979]. The advent of the improved *Tsyganenko* [1987] magnetic field model has made it possible to study in more detail the mapping geometry and to ela-

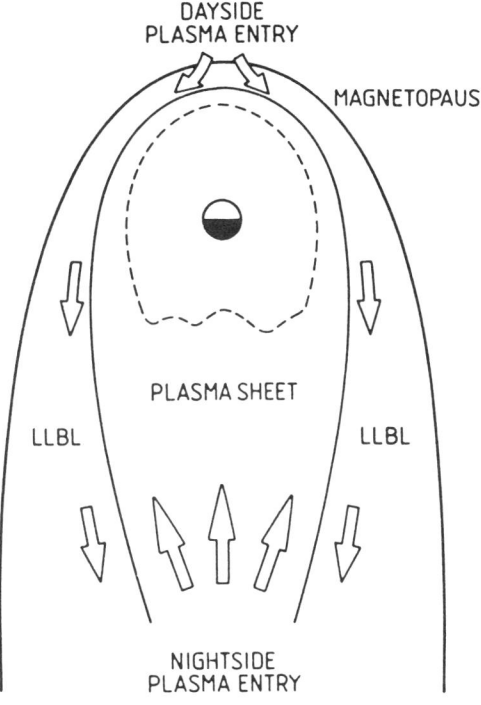 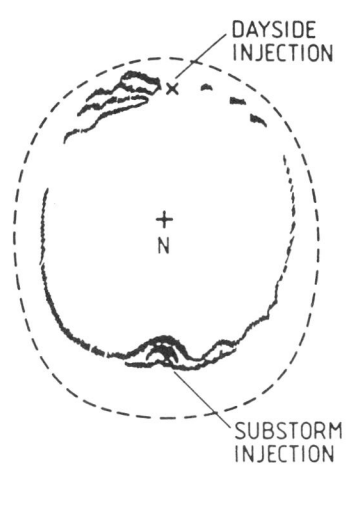

Fig. 1. Diagrammatic figure of the persistent dayside activity caused by an almost continuous plasma injection into the cusp and cleft and a more transient substorm injection into the nightside/tail plasma sheet.

borate on its consequences for large-scale magnetospheric processes. This is particularly true for a magnetospheric region that perhaps least lends itself to field line mapping, the low-latitude boundary layer (LLBL). Although one may raise doubts on the magnetic field mapping in a region where the magnetopause current sheet and local transverse currents strongly affect the local magnetic field, the Tsyganenko model appears yet to be an important tool for understanding the ionospheric footprint of the LLBL.

Fig 2 shows an example of a field line mapping of the magnetospheric boundary layer to the low-altitude dayside auroral oval, based on the *Tsyganenko* [1987] magnetic field model [*Stasiewicz, 1991*]. Although these kinds of mappings are done on the basis of an averaged magnetic field model, unable to account for time dependent features in the magnetopause current sheet, they yet demonstrate that such projections may agree well with the dayside auroral morphology. Notice for instance that relatively simple structures in the magnetopause boundary layer correspond astonishingly well to the rayed characteristics of auroral arcs observed near local noon [*Meng and Lundin, 1985*]. Furthermore, the width of the narrow ionospheric region near local noon expands when the magnetopause boundary layer widens. The cusp proper is by definition connected to the magnetopause layer and is thus in continuous contact with the magnetosheath. Thus, the polar cusps reflect the processes taking place within the dayside magnetopause. For instance, plasma crossing the magnetopause must traverse cusp field lines.

The mapping in Fig. 2 becomes quite important when comparing with statistical studies of field-aligned currents as shown in Fig. 3 [after *Ijima and Potemra, 1978*]. Notice the pronounced region 1 field aligned current maxima in the prenoon and postnoon sector, clearly within the region that could be identified by the cleft. Notice also that in these sectors the region 1 and region 2 currents are highly unbalanced, suggesting an external (magnetopause) closure of this current system besides the ionospheric closure in the cleft. Considering the mapping of Fig. 2 we may thus conclude that the LLBL dynamo/generator is related to the strongest current circuit in the magnetospheric current system.

A model for how the dayside region 1 current circuit is powered within the LLBL is shown in Fig. 4 [*after Lundin and Evans, 1985*]. Indeed, the concept of a boundary layer dynamo is not new. It was originally proposed by *Cole* [1961] and later also discussed by others [e.g. *Eastmann et al.*, 1976, *Heikkila, 1982*, and *Troschichev, 1982*]. Comparing now Fig. 4 with Figs. 2 and 3 one immediately recognizes how well such an unbalanced dynamo concept works. The polarization/dynamo voltage across the LLBL, the dynamo EMF, is driven by the boundary layer plasma flow. In the lower section of Fig.

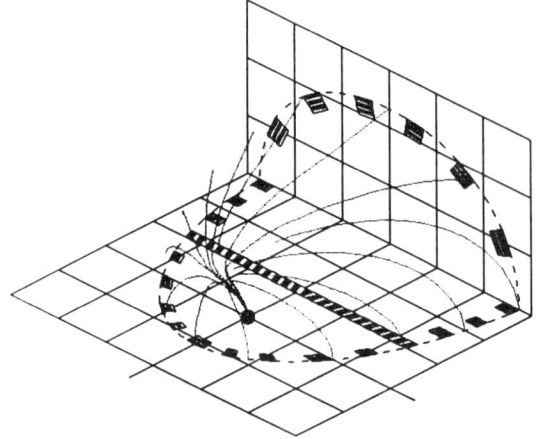

(a) MAGNETOPAUSE BOUNDARIES: 4 × 0.5 R_E

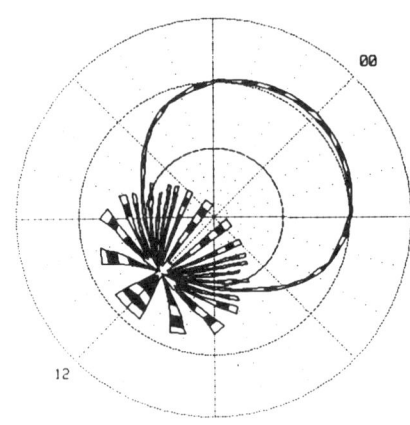

(b) IONOSPHERIC FOOTPRINTS

Fig. 2. A field line mapping (using the Tsyganenko model) of the magnetospheric boundary layers (top diagram) and its dayside ionospheric footprint (bottom diagram) centered around the polar cusp (after Stasiewicz, 1991).

plot is an illustration of the orbit track with respect to the magnetopause. Notice that the motional EMF of the dynamo plasma (here represented by solar wind protons) accumulated along the spacecraft track is in the same direction as predicted by the model in Fig. 4. The total EMF, adding dawn and dusk together, is in the range of a few tens of kV, i.e. a substantial fraction of the anticipated cross-tail potential. The line integral of the O^+ - (v × B) is substantially less than the H^+ - (v × B). This can be understood as two different plasma populations subjected to differences in e.g. partial pressure and inertia. One simple interpretation, suggested by Lundin and Evans (1985], is that the O^+ - (v × B) equals the electric drift while the H^+ - (v × B) is dominated by a kinetic pressure gradient associated with the decelerated solar wind plasma flow within the boundary layer MHD dynamo. We then consider O^+ ions, originating from the ionosphere, as passively responding to an externally applied convection electric field whilst H^+ ions are related to the cause of such an electric field. Thus, the integrated O^+ - (v × B) would correspond to the electric potential (U) over the boundary layer while the integrated H^+ - (v × B) is equivalent with the motional EMF (E). The difference (E-U) corresponds to the external loading of the boundary layer

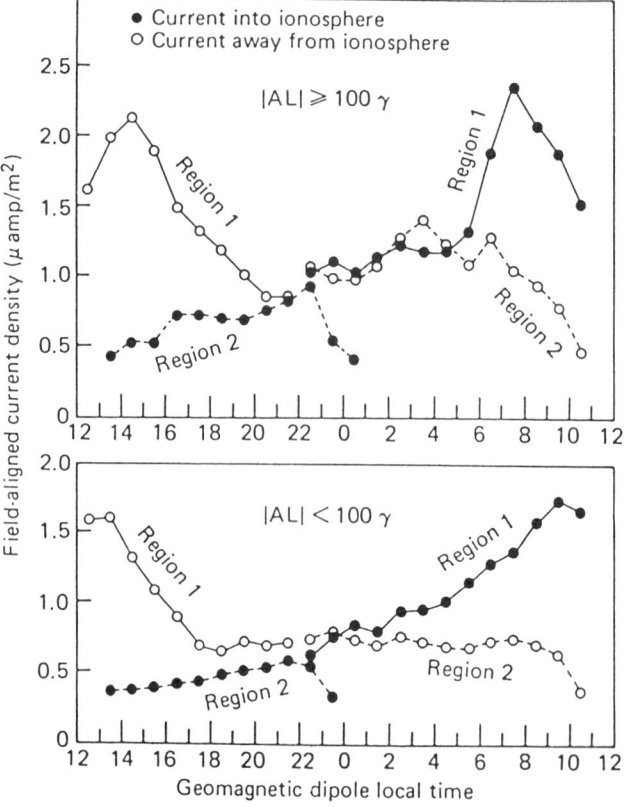

Fig. 3. Diurnal distribution of field-aligned current densities during active periods (|AL| ≥ 100 nT) (upper panel) and during weakly disturbed periods (lower panel) (after Ijima and Potemra, 1978).

4 the intrinsic charging process is described. Because the dynamo is connected to a load, i.e. currents are flowing through an ionospheric resistance, there will be a replenishment of charges in the polarization region that introduces a breaking of the dynamo plasma.

The polarization in the boundary layer is according to the LLBL model mainly radial, inward at dusk and outward at dawn (Fig. 4). This orientation has been confirmed from measurements of the plasma drift [*Eastman et al., 1976, Lundin and Evans, 1985*]. Fig. 5 gives two examples of Prognoz-7 passes in the dawn and dusk sector LLBL illustrating this. Both passes occurred during southward directed IMFs. Below each

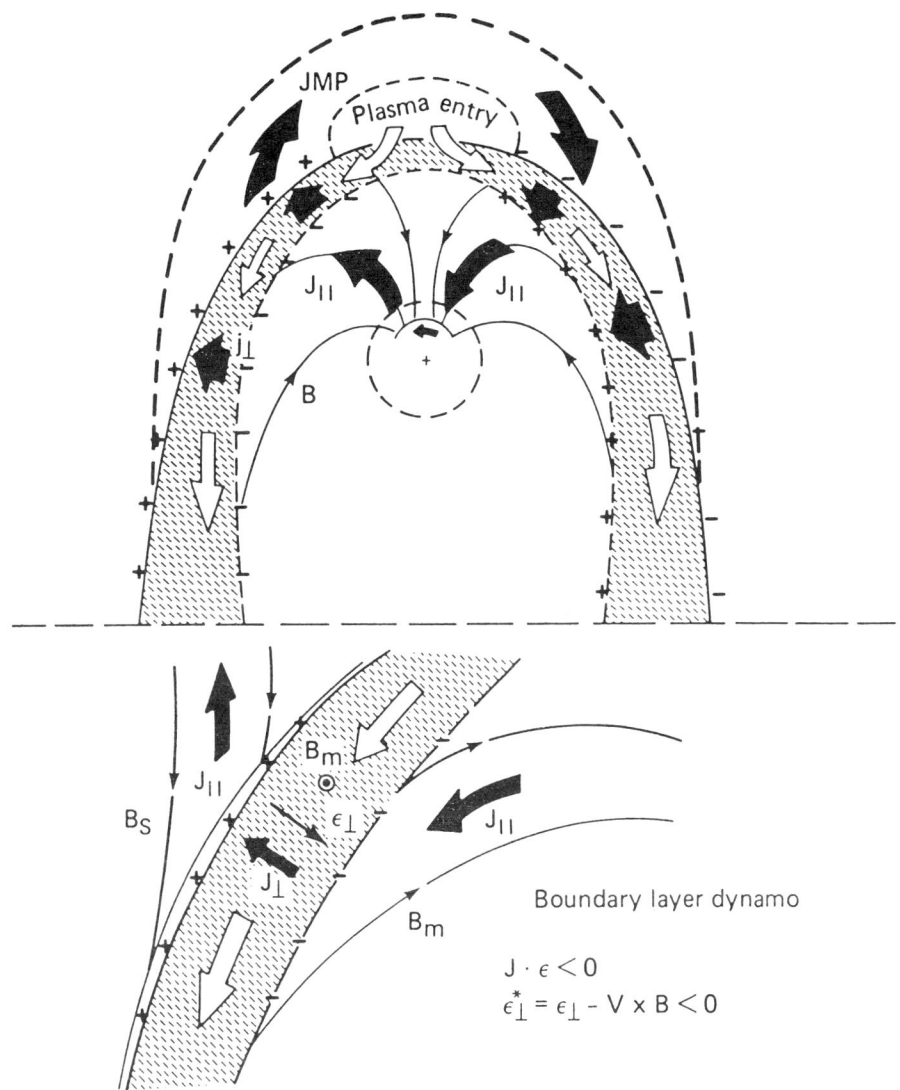

Fig. 4. A model of the magnetospheric boundary layer dynamo generated currents and fields (from Lundin and Evans, 1985). The lower panel gives a schematic model of the boundary layer polarization process. Notice that the unbalanced Birkeland current circuit designated $J_\parallel$ in the figure (region 1) closes in the ionosphere in the cusp/cleft region (Figs. 2, 3).

dynamo along the closure current, e.g. ionospheric loading and mid-altitude acceleration of plasma. Maximum loading corresponds to zero dynamo voltage ($U \approx 0$). A more detailed discussion on the intrinsic properties of the boundary layer dynamo can be found elsewhere [*Lundin, 1988*].

The polarization of the boundary layer is confirmed from electric field data as demonstrated in Fig. 6 [*adapted from Mozer, 1984*]. However, instead of the tens of kV motional EMF estimated from the crossed-field flow, *Mozer* [1984] estimated the electric potential to be of the order a few kV only. *Heikkila* [1986] disputed this very low value by Mozer on basis of the one figure of the LLBL crossing (Fig. 6) published by *Mozer* [1984]. Heikkila estimated a "radial" LLBL potential of 13 kV. However, although this represents an extreme case of boundary layer polarization there are yet a couple of conclusions that can be drawn here: First, observations indicate that the direction of the boundary layer electric field is radial, in good agreement with the boundary layer dynamo model. Secondly, a substantial deviation between the polarization voltage and motional EMF is expected in a dynamo when strong field aligned currents connected to the dynamo are driven through the conducting ionosphere (loaded dynamo). Thus, the electric field does not describe the total EMF within a dynamo under load. Yet a third conclusion is that the cases presented so far imply total (dawn-dusk) boundary layer EMF:s in the 20-30 kV range, i.e. an order of mag-

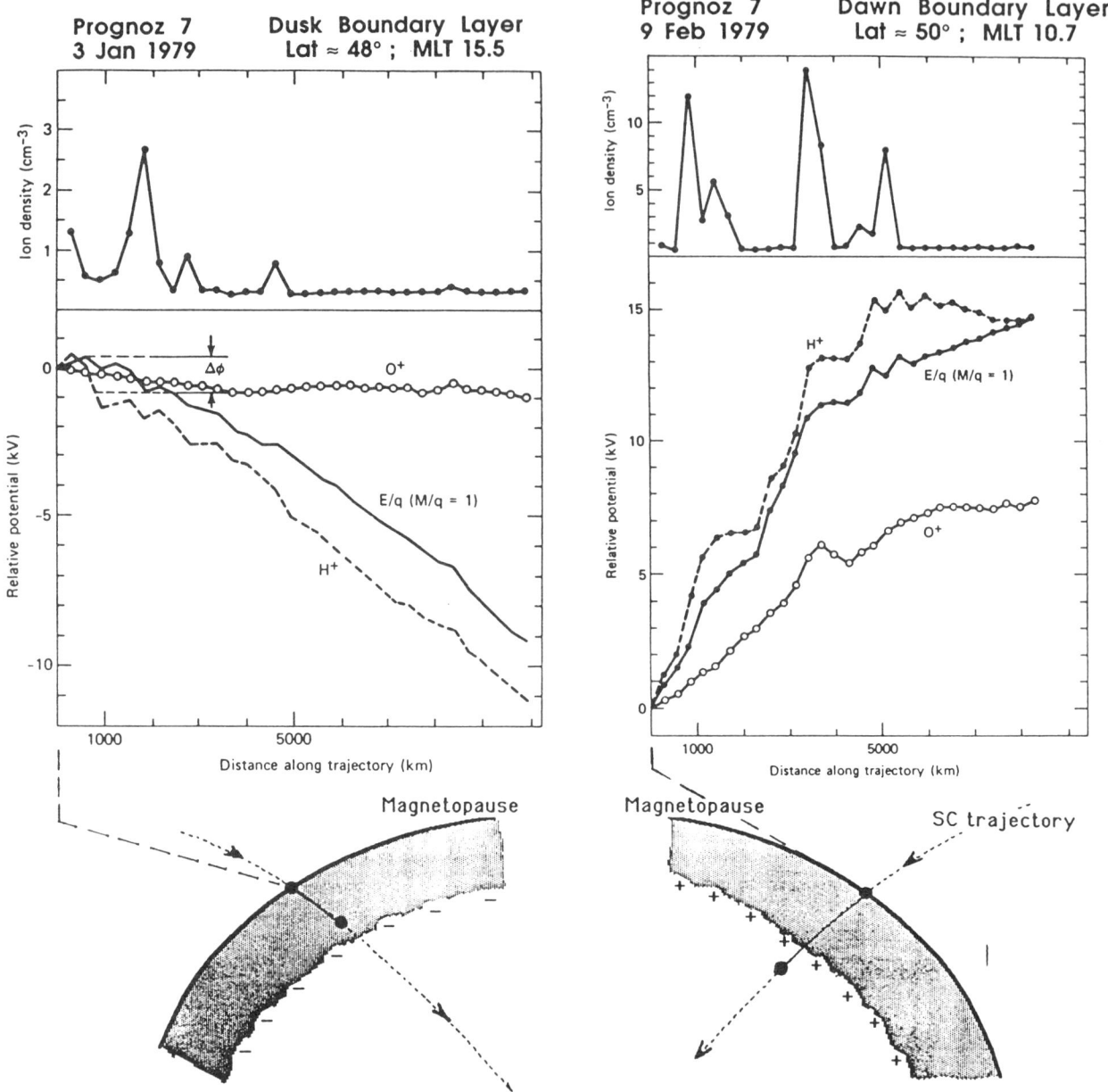

Fig. 5. Two Prognoz-7 crossings in high data rate (HDR) mode of the dawn and dusk boundary layer showing the integrated H^+, O^+ and E/q (all ions) vxB along the spacecraft trajectory. The lower part illustrating the orbit track and the associated polarization / motional emf deduced within the boundary layer. Notice that full coverage of the dusk boundary layer (left) was not obtained in the HDR-mode. (After Lundin and Evans, 1985).

nitude higher than the few kV proposed by *Sonnerup* [1980] and inferred by *Mozer* [1984]. These are substantial fractions of the dawn-dusk potentials measured over the polar cap.

SOLAR WIND PLASMA ACCESS TO THE LLBL AND THE PLASMA SHEET

The access of solar wind plasma to the LLBL and the plasma sheet is evident from numerous measurements showing the presence of solar wind ions (e.g. He^{++}) in these regions. Any controversy on this topic is related to the rate at which this entry occurs, where it occurs, and what the processes are that support such an entry.

It would lead too far to elaborate on these particular controversies here [*see e.g. Lundin, 1988 for a review*]. Thus we will merely conclude that solar wind plasma may access the magnetosphere at a rather high rate in the cusp region and the cleft/

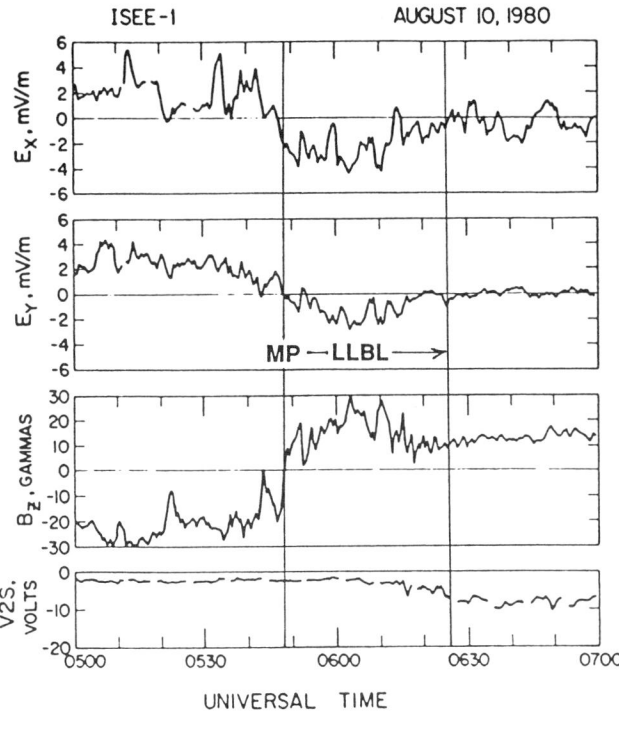

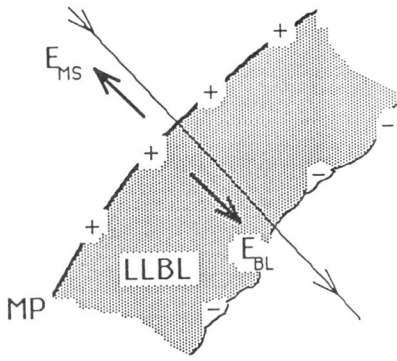

Fig. 6. ISEE-1 measurements of the electric field in the dusk LLBL (Mozer, 1984). The attached diagram illustrates the polarization across the LLBL as estimated by Heikkila (1986).

LLBL ($> 10^{27}$ ions/s) [*Eastman et al., 1985*] and that the entry process is less well understood.

One of the distinguishing differences between the cusp proper and the cleft is the apparent lack of low-altitude acceleration in the cusp and the significant energization in the cleft. Furthermore, the cusp is characterized by a continuous solar wind plasma access while plasma penetrates into the cleft/ LLBL in a more discontinuous manner.

Temporal injection of magnetosheath plasma, plasma transfer events (PTEs), have been observed with both sounding rockets and satellites in the dayside auroral oval. The first observations were presented by *Carlson and Torbert* [1980]. On basis of the ion dispersion characteristics, frequently comprising several superimposed ion dispersion events, they were able to deduce the distance to the injection point to 7 - 19 R_E away from the rocket altitude. This corresponds to somewhere between the entry layer (EL) and near equatorial region of the LLBL. Similar time dispersion features are frequently found in data from Viking passes over the dayside auroral oval. An example of this is shown in Fig. 7. The characteristic time of the ion dispersion signature suggests here an injection of plasma some 9-11 R_E upstream, i.e. into the high-latitude portion of the flankside LLBL. An analysis of the energetic electron loss cones implied that at least the three equatormost injection structures occurred on closed magnetic field lines.

Several important facts support the magnetosheath origin of these time dispersion features [*Woch and Lundin, 1991a,b*]. First of all, they are clearly temporal features because the signature is always in the "falling" sense, independent of the direction of auroral oval traversal. Secondly, the spectra of the injected ions resemble magnetosheath ion spectra and, furthermore, all events checked by the ion composition spectrometer were dominated by protons. Thirdly, the ion flow within the injection structures was essentially antisunward, i.e. in the direction of the external magnetosheath flow. Fig. 8 shows a diagrammatic representation of two PTEs and their corresponding ionospheric footprint observed by the Viking satellite [*after Woch and Lundin, 1991a*].

In a recent statistical study by *Woch and Lundin* [1991b] PTEs were found to occur preferentially on closed field lines. They also concluded that PTEs are a very common feature, with 50% frequency of occurrence on Viking orbits through the dayside oval. Fig. 9 illustrates that there is no dependence on the frequency of occurrence of PTEs on solar wind velocity (top) while there is a strong dependence on the occurrence of PTEs on solar wind dynamic pressure. Fig. 10 illustrates the unexpectedly weak dependence of PTEs on southward IMF, while the IMF B_y dependence is rather strong. Figs. 9 and 10 clearly demonstrate that the occurrence of PTEs depends on the solar wind plasma pressure and the radial component of the IMF, which makes them different to e.g. flux transfer events (FTEs).

As for the access of magnetosheath plasma into the tail plasma sheet, two main hypothesis are discussed. The prevailing hypothesis [*Pilipp and Morfill, 1974, Cowley, 1980*] has been that magnetosheath plasma gains access primarily from the plasma mantle through the lobes. The alternative hypothesis is that magnetosheath plasma is fed from the sides via the LLBL [*Heikkila, 1982, Eastman et al., 1985*]. The two rivaling ideas of the solar wind plasma access to the central plasma sheet are summarized in Fig. 11.

Experimental evidences for the plasma mantle as a source for the plasma sheet were first presented by HEOS 2 [*Rosenbauer et al., 1975*], where at least some orbits showed the expected convection pattern into the lobe region. PROGNOZ-7 data [*Lundin et al., 1981*] verified, what was also obvious from the HEOS 2 data, that the mantle plasma is too dynamical and complex to fit with a simple convection picture where mantle plasma convects through the geomagnetic lobe to the central plasma sheet. Although the flow of ionospheric plasma in the

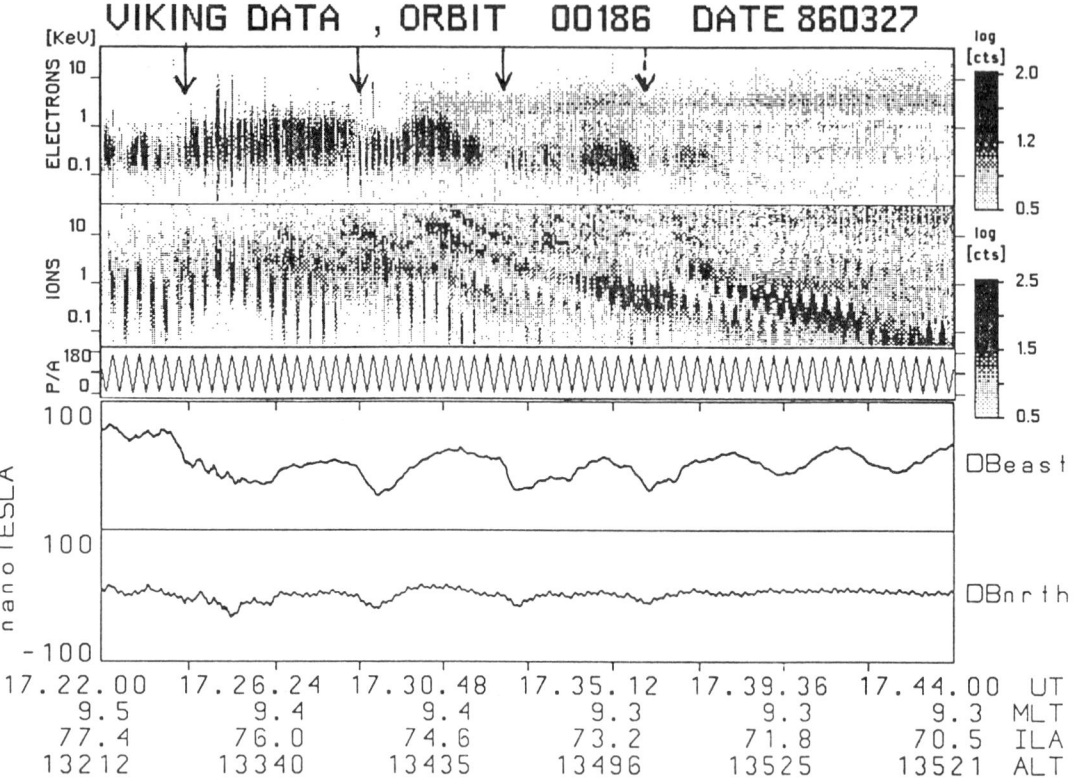

Fig. 7. Viking energy-time spectrogram illustrating several plasma transfer events (PTEs) above the dayside oval, each arrow indicating the start of a plasma injection.

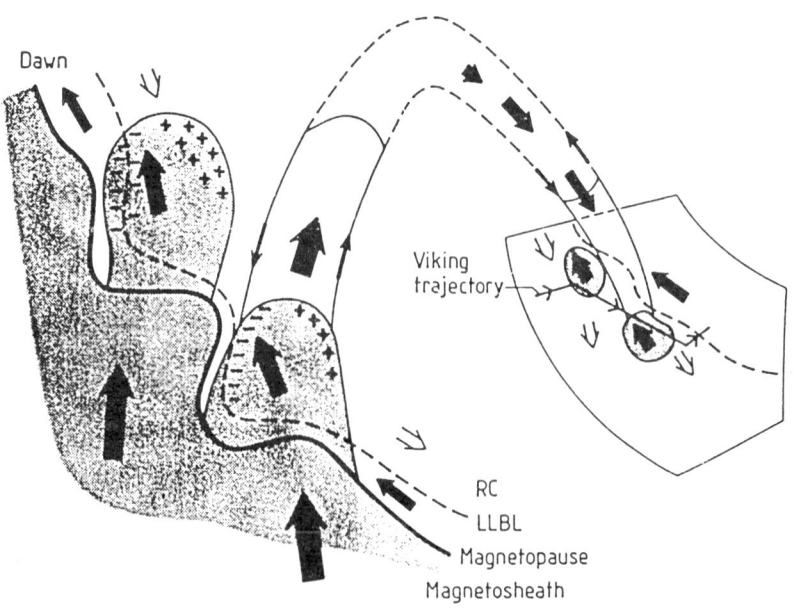

Fig. 8. Diagrammatic representation of two PTEs and the corresponding drift footprint determined from a Viking pass through the dawn oval (after Woch and Lundin, 1991a).

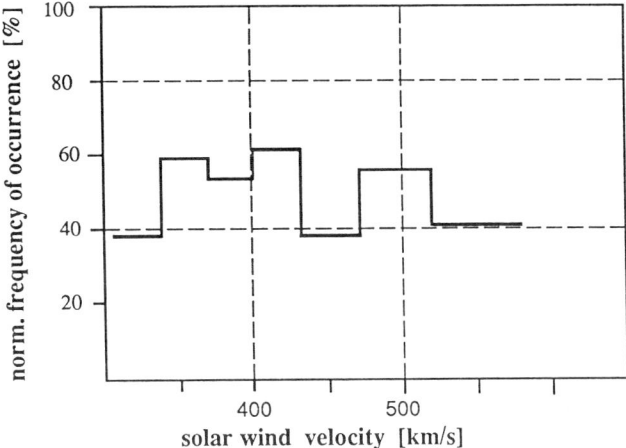

al., 1987]. Notice that the graph related to the LLBL is a qualitative estimate of the thickness based on the time ISEE 1 spent in the LLBL at various IMF B_z cooonditions. The main outcome of these two investigations is that the mantle grows in width for decreasing IMF Bz, while the LLBL shows the completely opposite behavior. If the main flow of plasma into the central plasma sheet occurred for southward Bz, the mantle would most likely be the plasma source. However, the opposite seems to be the case as is evident from Fig. 13 [*after Lennartsson, 1987*], illustrating the spatial distribution of ion density in the tail. Given the strong correlation between IMF Bz and the AE-index one can then conclude that the plasma sheet builds up primarily during low-activity periods when the LLBL is thicker, and then preferably along the flanks. Indeed, the heavy ion (e.g. O^+) contribution to the plasma sheet is highest during disturbed periods, but this is merely an expected consequence of enhanced ionospheric ion outflow. The obvious conclusion from these three investigations is that

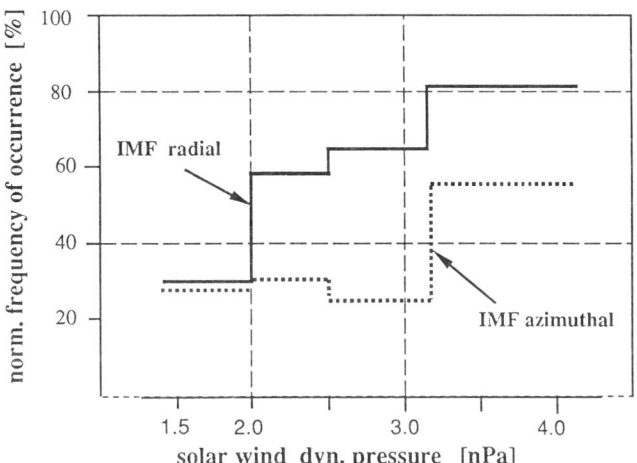

Fig. 9. Results from a statistical analysis of PTEs in the dayside, illustrating the lack of solar wind velocity influence on the occurrence of PTEs (top panel). The lower panel illustrates the strong dependence found on the solar wind dynamic pressure for PTEs, in particular for a radial IMF direction (Woch and Lundin, 1991b).

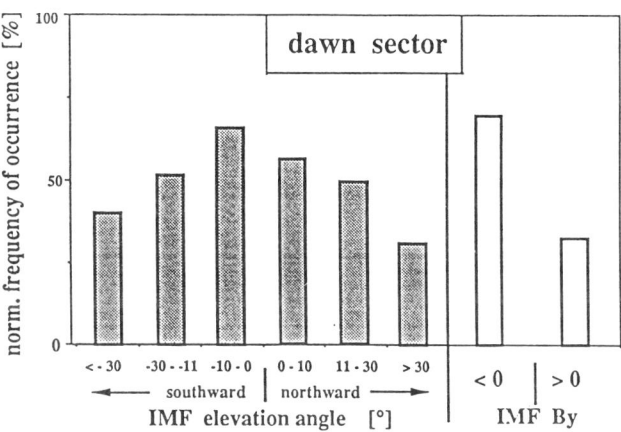

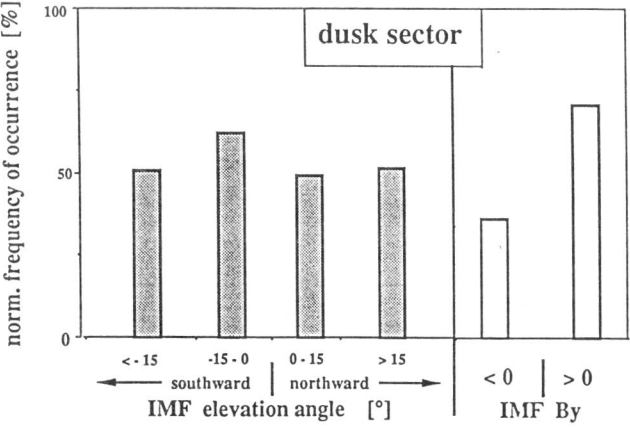

Fig. 10. Statistical results demonstrating the apparent lack of IMF Bz dependence for the occurrence of PTEs. Conversely, there is a significant dawn-dusk assymmetry of the PTE occurrence frequency versus IMF By (Woch and Lundin, 1991b).

tail appears to go in the right direction, i.e. towards the central tail [*Orsini et al., 1987*], there are yet no direct evidences presented of mantle plasma passing through the tail lobe. This is particularly true during times of enhanced convection.

Conversely, evidences for the LLBL as the prime source of solar wind plasma in the plasma sheet have been presented by *Eastman et al.* [1985] and *Lundin and Lennartsson* [1986], the latter from considerations of the ion composition in the geotail. The perhaps most convincing argument for the LLBL origin of solar wind particles in the plasma sheet originates, however, from three other independent observations.

Fig. 12 shows the magnetospheric boundary layer thickness versus IMF Bz as observed from HEOS 2 in the plasma mantle [*Sckopke et al., 1976*] and ISEE 1 in the LLBL [*Mitchell et*

PLASMA SHEET FORMATION

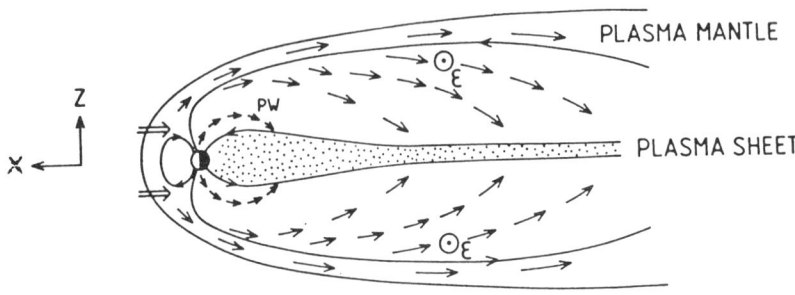

(1) Lobe/Plasma Mantle Source

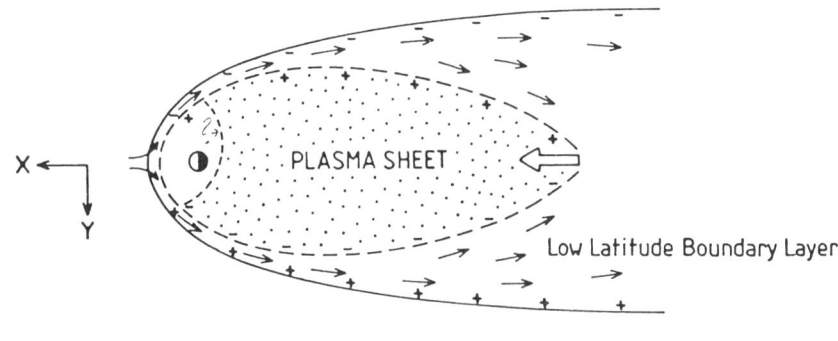

(2) Low-Latitude Boundary Layer Source

Fig. 11. Diagrammatic representation of the two processes suggested to be responsible for the plasma sheet formation.
(1) The plasma sheet is fed primarily from the "top", solar wind plasma accessing the plasma sheet through the lobe (Pilipp and Morfil, 1978, Cowley, 1980).
(2) The plasma sheet is primarily fed from the "sides" via the LLBL (Heikkila, 1982, Eastman et al., 1985).

solar wind plasma is fed into the plasma sheet primarily from the sides of the magnetosphere, during times of low magnetospheric disturbance level and high LLBL thickness.

This conclusion is consistent with a recent statistical study [*Sandahl and Lindqvist, 1990*]. Using the Tsyganenko 1987-model they performed a mapping of the quiet time oval determined from particle populations identified by Viking. Fig 14 gives a diagram of the statistical electron domains within the auroral oval (right) and its corresponding mapping to the equatorial plane (left). Comparing Fig 14 with Fig. 1 one can conclude that the region characterized by structured electrons and auroral energization processes maps to regions connected to the LLBL or regions adjacent to it.

THE LLBL AND ITS COUPLING TO SUBSTORMS

The mapping of auroral field lines into the magnetotail has so far mostly been done in a cartoon-like fashion. The advent of the Tsyganenko 1987-model, which allows to adequately map an "average" feature within the magnetosphere to the auroral oval, as discussed in connection with the boundary layer (Fig. 2), enables a highly improved magnetotail mapping. Fig 2 also shows how a simple dawn-dusk cut through the geotail at $X \approx -10\ R_E$ represents a full oval in the polar hemisphere. This demonstrates how relatively simple geometries, here a tail "slab" within the magnetosphere, may correspond to large scale features along the auroral oval. The tail

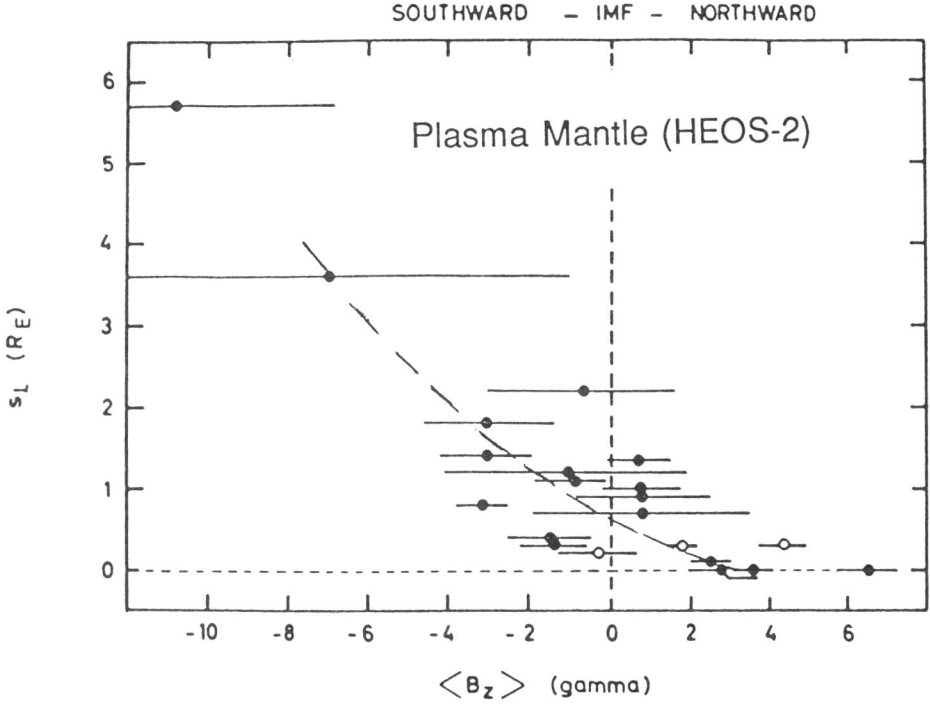

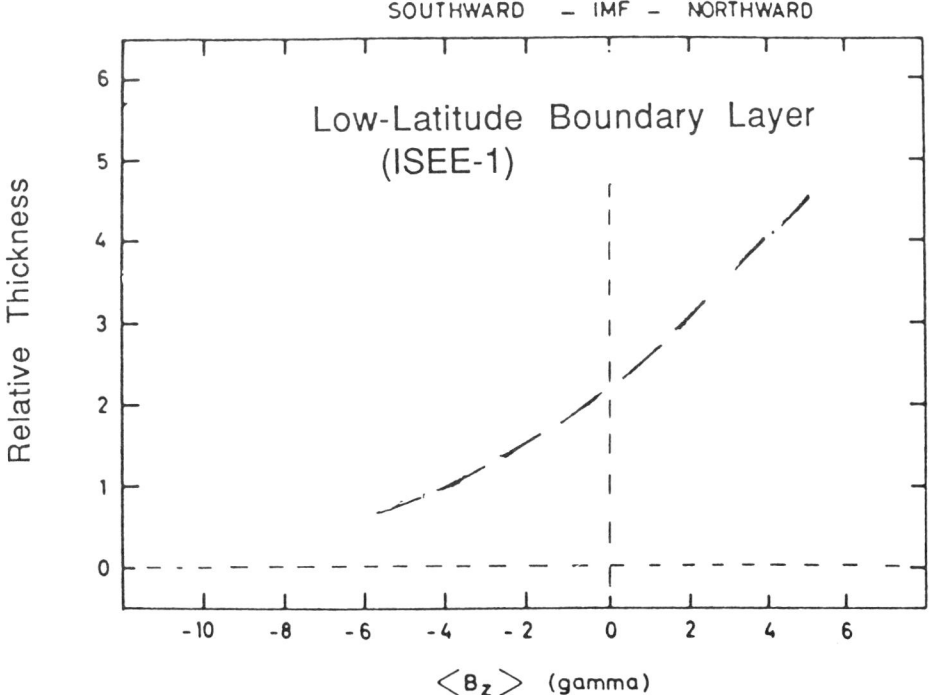

Fig 12. Magnetospheric boundary layer thickness versus the IMF Bz, the top diagram illustrating the mantle thickness (after Sckopke et al., 1976), and the bottom diagram giving a schematic picture of the LLBL thickness (as interpreted from Mitchell et al., 1987).

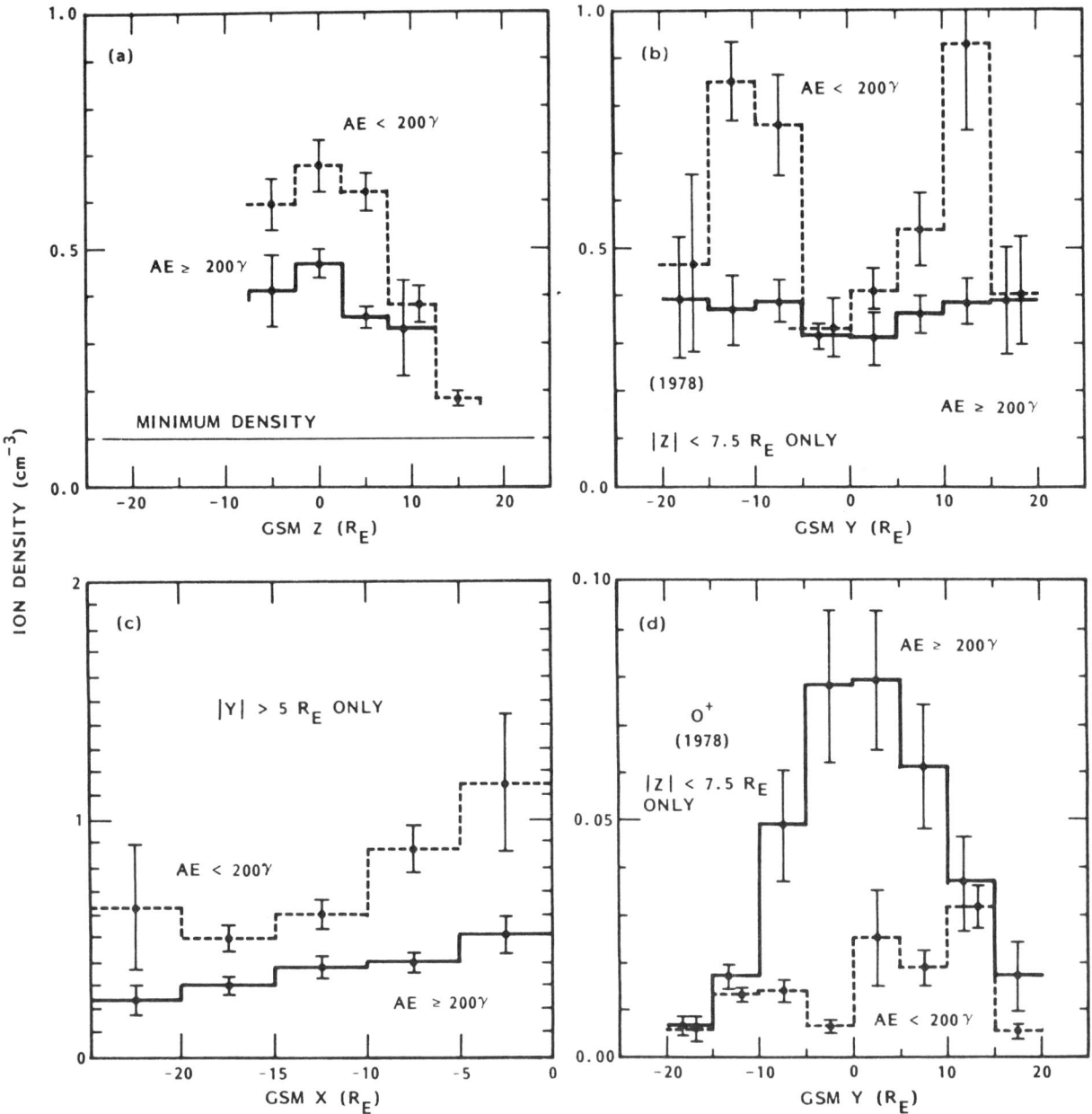

Fig. 13. Spatial distribution of H+ and O+ ion densities versus GSM X, Y, and Z illustrating the higher plasma sheet proton densities found for low magnetic activity and the quiet time flank enhancement (b) suggesting a LLBL source for the plasma sheet (after Lennartsson, 1987).

"slab" approach also leads to simple models with important consequences for the tail current sheet, the magnetospheric current system, and the substorm development. Before continuing with the model in more detail we should briefly review the triggering mechanisms proposed for substorms.

As substorms are transient phenomena expected to be related to solar wind disturbances, large efforts have been spent in unravelling the triggering mechanism within the upstream solar wind. The IMF direction has long been known to play a major role for the substorm mechanism [see Akasofu, 1977, for a review]. According to Rostoker [1983] a substorm is triggered by a northward turning of the IMF after a period of southward IMF. Similarly, triggering due to sudden solar wind plasma pressure enhancements has been proposed [Iyemori and Tsunomura, 1983]. However, in many cases the timing of substorm onset to solar wind disturbances is not possible.

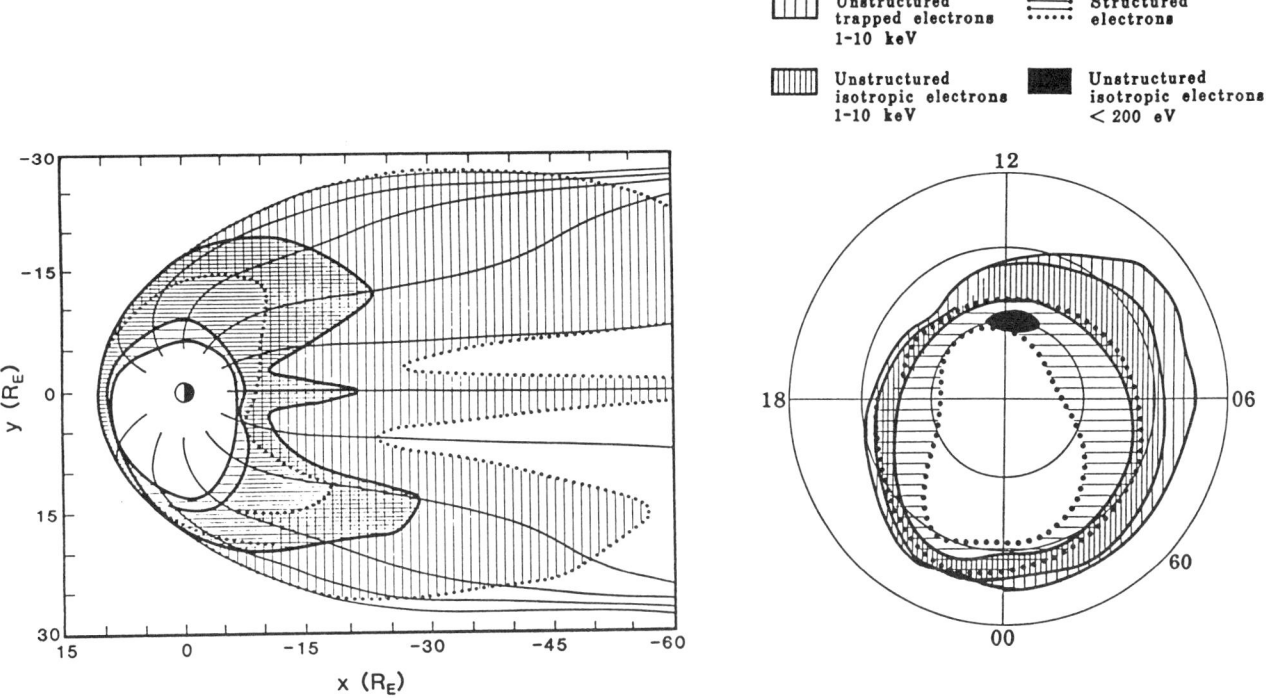

Fig. 14. Results from a statistical study of the mapping of the quiet time (Kp<2) electron populations to ionospheric heights (right) and into the equatorial plane (left) using the Tsyganenko (1987) model (after Sandahl and Lindqvist, 1990).

This has led some researchers [*Horwitz, 1985*] to believe in internal triggering processes as well. Various types of tail plasma instabilities are used to explain substorm onset [*Lui et al., 1990*]. Similarly the importance of outflowing heavy ionospheric ions for the tail current sheet thinning was pointed out [*Baker et al., 1982, 1985*] and later also found to be consistent with experimental data [*Daglis et al., 1990*]. The fact that the timing between interplanetary disturbances and substorm onset is poor should, however, not be taken as evidence against interplanetary triggering. A poor timing may very well be due to an inadequate monitoring of the upstream solar wind, as observations are generally limited to one spacecraft. Indeed, the problem is three-dimensional where gradients in the solar wind plasma may be directed almost perpendicular to the bulk flow. One can easily envisage a solar wind disturbance reaching the magnetosphere before reaching an "unsuitably" positioned upstream spacecraft. Thus, lacking an at least two-dimensional coverage of the upstream solar wind there is a great uncertainty of the correct timing of solar wind induced magnetospheric disturbances.

In our model of the substorm development we now assume that the "slab" at X ≈ -10 R_E in Fig. 2 represents the preexisting auroral oval, i.e. existing prior to the onset of the substorm [*Elphinstone et al., 1990*]. This "slab" should be considered as the magnetic mapping into the tail of a high conductance path in the ionosphere (Σ_i), the high conductance allowing magnetospheric currents to divert and close in the ionosphere. Thus, the high conductance path in the ionosphere leads to a current diversion of the cross-tail current sheet. The current diversion is most likely located on closed magnetic field lines. The important implication of this is that even regions poleward of the oval may be on closed field lines [*Feldstein and Galperin, 1985*]. The Tsyganenko 1987-model is also consistent with this, the polar cap generally being limited to a "pear" or "droplet" shaped region well poleward of the auroral oval [*Elphinstone et al., 1990*], which is more evident during quiet times [*Murphree et al., 1982, Lundin et al., 1988, Sandahl and Lindqvist, 1990*]. We shall see that this has implications also for the poleward expansion of the oval.

Fig. 15, illustrates the substorm development based on the concept of a solar wind induced pressure surge propagating in the LLBL. In essence the figure shows a continued movement of plasma penetrated into the LLBL on the dayside as already discussed in the previous section (e.g. Fig. 4). The first response of the magnetosphere to the pressure surge can be found within the dayside oval, as enhanced dayside currents (Fig. 3) and intensification of the aurora near local noon. Such "precursors" are in fact a regular behavior of the dayside oval at all disturbance levels, sometimes referred to as persistent auroral activity near noon [*Murphree et al., 1981*] or persistent 2 pm arcs [*Evans, 1985*]. The Tsyganenko magnetic field

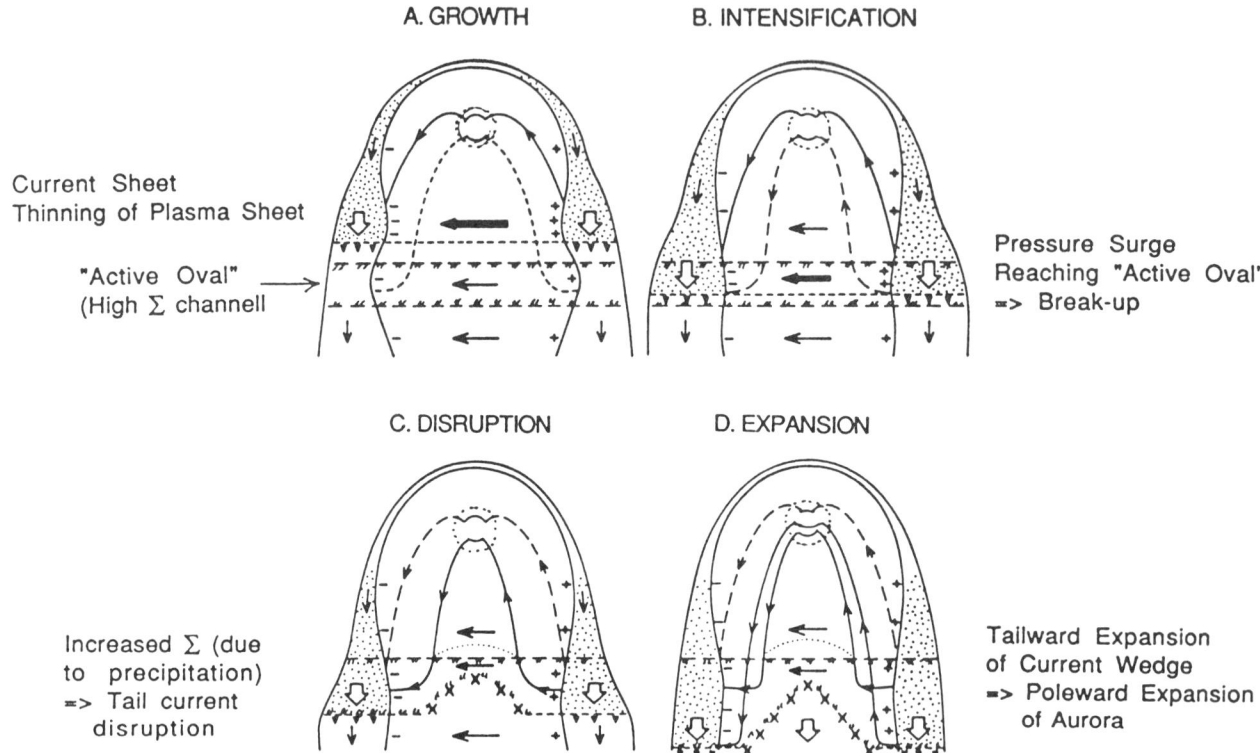

Fig. 15. Diagram illustrating the propagation of a pressure pulse in the LLBL leading to current sheet thinning (A), intensification of the aurora along the preexisting auroral oval (B), tail current diversion and the associated current sheet disruption (C), the tailward expansion of the current wedge associated with the poleward expansion of the aurora (D).

mapping suggests that even LLBL-related phenomena far downtail are associated with the dayside activity (Fig. 2). Thus, as long as the nightside current system remains at a low activity level the pressure surge continues to propel the dayside current system, affecting only the dayside oval (phase A).

When the pressure surge has reached the high conductance associated slab in the tail it can induce an enhancement of the nightside current circuit (phase B). Notice that the high conductance (Σ_i) path represents the region of main tail current diversion into the ionosphere and the associated enhanced auroral activity. The rerouting from the dayside to the nightside of the current driven by the propagating pressure surge in the LLBL may in fact start prior to the arrival at the high Σ_i path as an enhanced cross-tail current (with negligible current diversion). This would lead to the current sheet thinning which is observed during substorm growth [*McPherron, 1972*]. Thus, growth (phase A) and intensification of aurora are logical consequences of the propagating pressure surge in the LLBL. Phase A and phase B are expected to be the most common types of solar wind induced disturbances in the magnetosphere, each intensification eventually dying out after the pressure surge has passed the high Σ_i path. These types of disturbances may be considered "minisubstorms" or simply auroral enhancements because they do not lead to phases C and D, i.e. disruption of the tail current and large scale auroral expansion.

Notice that "minisubstorms" are considered from an ionospheric point of view, i.e. magnetospheric energy is transiently released to the ionosphere via field aligned currents without substantially affecting the large scale current system (no tail disruption). This means for instance that also auroral "flaring" events on the dayside [*Sandholt et al., 1986*] and the other dayside auroral features mentioned above may fall into this category. "Minisubstorms", representing mere enhancements of the current connected to a magnetospheric dynamo at one end and the ionosphere (the load) at the other, should be considered as individual features within the magnetosphere that may cause auroral intensifications simultaneously at different sites along the oval. This is also consistent with multiple surge features in the Viking data [*Rostoker et al., 1991, this issue*].

There is a difference between the dayside and nightside auroral effects by "minisubstorms" since the current circuits are different. The arc-like slots in Fig. 2 illustrate how a tail-

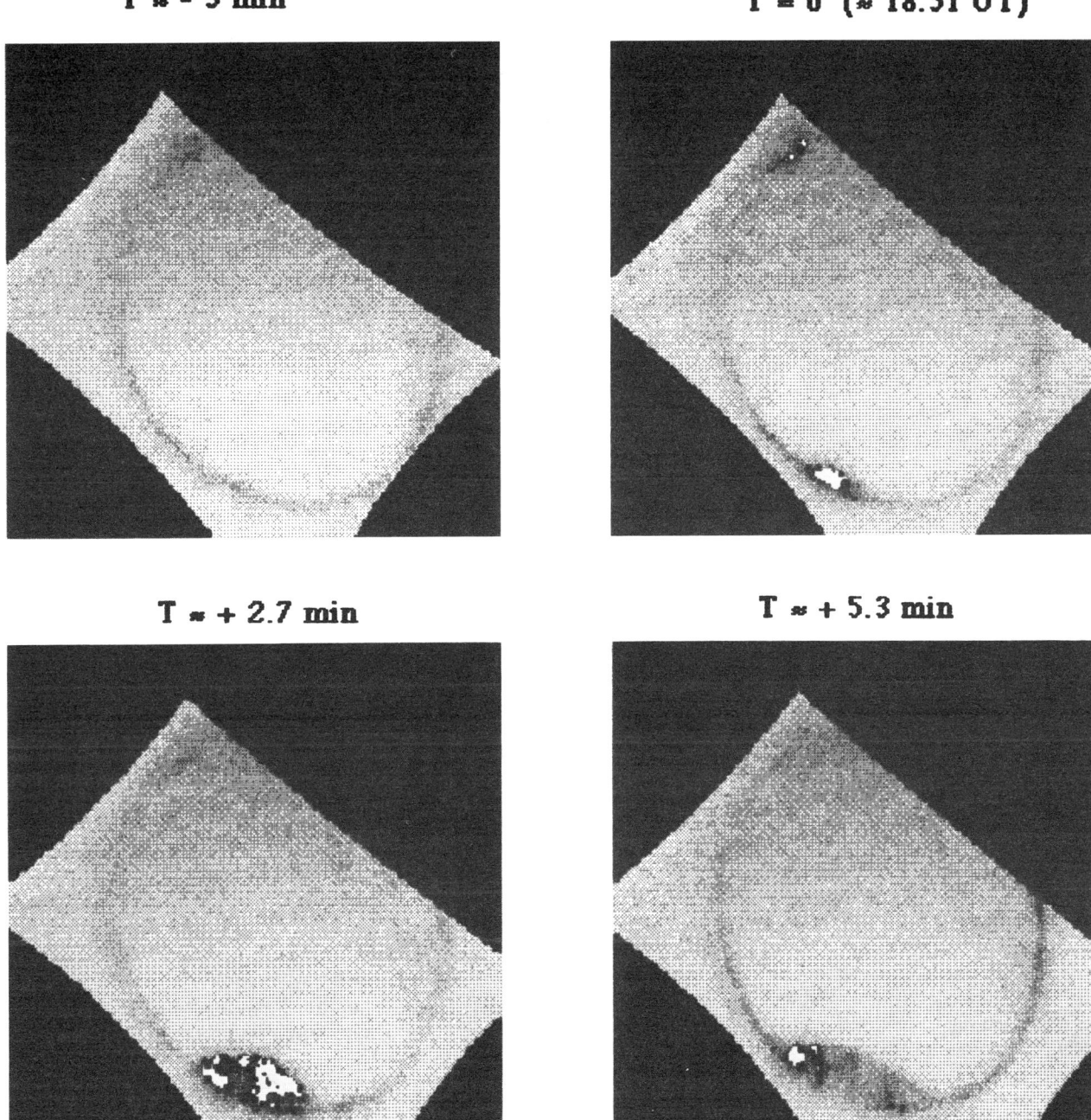

Fig. 16. Sequence of four Viking images taken before (T≈ -3 min), during (T ≈ 0 min) and after onset of a substorm during April 1, 1986. The noon-midnight meridian goes approximately from the top to the bottom on the images. Noon is on top. Notice for instance the simultaneous dayside and nightside auroral brightning at substorm onset.

ward moving pressure enhancement moves the dayside aurora in a spoke-like fashion, but retains the aurora in the narrow cleft region. On the other hand, the nightside current system and the corresponding auroral enhancement (phase B) is geometrically different. Coupled to a substantial fraction of the oval the auroral brightenings may occur within a broader local time and latitude sector and include poleward auroral expansions (increased ionospheric conductance poleward of the main

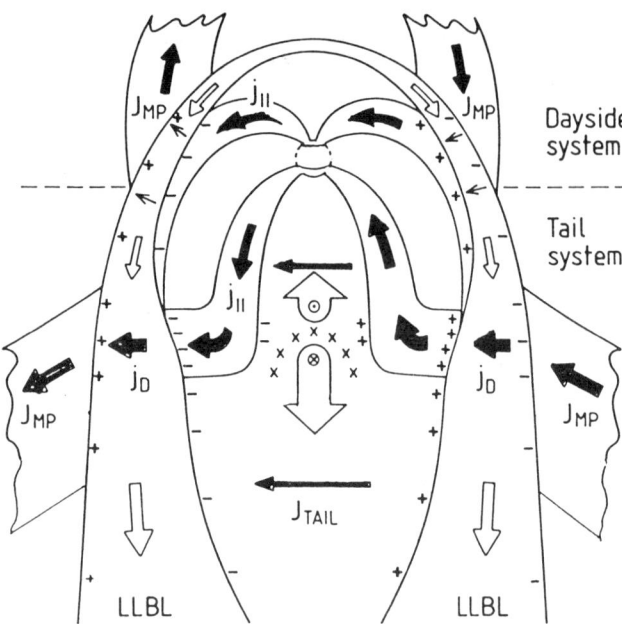

Fig. 17. Diagram showing the two main region 1 current paths, the "persistent" unbalanced dayside current circuit closing in the magnetopause together with:
- the (steady) tail current sheet system and
- the unbalanced nightside current wedge developing during substorms.

high Σ_i region due to enhanced particle precipitation poleward of the main auroral oval). Notice that the initial auroral brightenings in principle could occur anywhere along the oval.

Phase C (disruption) and D (Expansion) now represent stages when the current driven by the LLBL dynamo is almost entirely diverted to the magnetotail current circuit. Dayside auroral features may still be powered by the LLBL dynamo, but most of the power is being directed to the nightside current system. The current diversion may occur for several reasons (e.g. enhanced ionospheric conductance or tail-current instabilities), but the prime effect is a dramatic increase of the conductance along the nightside path through the ionosphere. Most of the LLBL dynamo power will then be dissipated in the nightside current circuit, specifically the nightside auroral ionosphere. Disruption of the tail current occurs as a consequence of the rerouting of the tail current through the nightside ionosphere. This leads to a local collapse of the tail [*McPherron, 1979*], a tail current sheet disruption which is usually described as reconnection in the tail. However, in the scenario presented here, the tail current sheet disruption is an instability driven by the diversion of the dynamo currents through the ionosphere.

Expansion (phase D) is now the logical consequence of the tailward propagating current sheet disruption, propagating with the speed of the plasma in the LLBL. The propagation front with newly opened field lines (marked with x) continues to "destroy" the magnetotail current sheet and the pressure surge associated current system is entirely rerouted to the ionosphere which is now the only available path besides the dayside one. Thus, the diversion of the current through the nightside ionosphere is in essence a response of the magnetosphere to maintain the geotail. Part of the tail plasma is lost as a plasmoid moving tailward while in the remaining part the cross-tail current sheet is replaced by a circuit that closes through the ionosphere. This then allows for a continued closure of the magnetopause current within the tail.

The poleward expansion of the aurora, coupled to the tailward propagation of the LLBL pressure surge, will eventually reach the preexisting, but previously "unrecognizable", polar cap. The term unrecognizable is used to illustrate that tail field lines void of significant particle precipitation is not identical with polar cap field lines. Thus the polar cap may be substantially smaller and the region of closed field lines substantially larger than what can be determined from particle and auroral imaging signatures. Notice that the main auroral activity in our scenario is believed to take place on closed field lines (earthward of the x-lines). This does not imply that all closed field lines are associated with auroral activity. Indeed, auroral activity as we envisage it here (i.e. connected to transfer of energy from an "external" energy source/dynamo) may very well be absent on closed field line. This is in fact the normal case in most parts of the magnetosphere, those parts equatorward and poleward of the high Σ_i region. There the currents are closed externally. For instance, the quiet time tail magnetopause current is closed mainly within the tail current sheet. However, there is a direct feed-back mechanism coupled to the ionospheric current circuit once the precipitation-related ionospheric conductance is locally enhanced or the external current is disrupted, thereby allowing for a current closure within the ionosphere.

Fig. 16 gives four Viking images of a substorm onset that nicely fits into the scenario described above. The first image (upper left) shows the oval 3 minutes prior to the onset. At this time (T = -3 min) only the dayside oval displayed som activity between ≈13.00 - 15.00 MLT, i.e. in the dayside cleft region. During substorm onset (T = 0) local auroral brightenings occur simultaneously in the dayside and nightside oval. This would then correspond to the arrival of the pressure surge to the "active oval" (phase B). At T ≈+2.7 min the expansion of the oval progresses (phase C and D) and the dayside activity weakens because of an increased nightside ionospheric conductance due to particle precipitation - the increased nightside ionosphereic conductance leading to an enhancement of the tail current system at the expense of the dayside current system.

Fig 17 finally summarizes the two main ionospheric current systems powered by the LLBL dynamo, the dayside current system connected to the dayside cusp/cleft region and the nightside tail current diversion system connected to the nightside high-latitude ionosphere. In Fig. 17 we show the expected currents for phase C (Fig. 16), i.e. a local tail current disruption associated with dipolarization and inward plasma injection on the earthward side and a loss of plasma (formation of plasmoid) on the tailward side of the x-line. Notice that the formation of the x-line is directly related with the tail-current geometry, i.e. it merely represents a local weakening/void of cross-tail current due to the diversion along a path with higher conductance through the nightside ionosphere. Once established (e.g. due to local plasma instabilities) it becomes self-generating because the loss of plasma on the tailward side decreases the electric conductance in the cross-tail current system. This leaves the ionosphere as the remaining path for currents to flow - further enhancing the conductance on the poleward region of the preexisting oval - thus causing a more efficient unloading of the LLBL dynamo.

CONCLUSIONS

We have discussed the generation of substorms emphasizing a directly driven process but at the same time retaining the concepts used in the unloading hypothesis for substorms. This is in fact consistent with the LLBL being the temporary source of substorm energy, and with the tail plasma processes being driven by the LLBL dynamo. That means, besides driving the dayside current system the LLBL dynamo also powers the cross-tail current system. The LLBL as the site of solar wind energy transfer [*Eastman et al., 1976, Heikkila, 1982, Lundin and Evans, 1985*] is in direct conflict with the traditional merging /reconnection hypothesis where the polar cap and plasma mantle represents the main dynamo for substorm energy.

Our main arguments in favor for the LLBL substorm dynamo are:

(1) The dayside current system, [*Vasyliunas, 1979, Stasiewicz, 1991*], represents the strongest field aligned (Region 1) currents in the magnetosphere [*Ijima and Potemra, 1976*] and maps to the dayside cleft/cusp interface. These currents are consistent with a persistent LLBL dynamo [*Eastman et al., 1976, Lundin and Evans, 1985*].

(2) Solar wind plasma penetration into the LLBL is a persistent process that seems to be little dependent on IMF-Bz, but which increases with enhanced solar wind dynamic pressure (Fig 9). Thus, the plasma pressure appears to be more important than the IMF direction per se. There is, however, a strong IMF-By coupling of the occurrence frequency of dayside plasma penetration [*Woch and Lundin, 1991b*].

(3) Observational evidences are in favor of the LLBL as the main source region for solar wind plasma in the plasma sheet, the plasma sheet predominantly building up during quiet time [*Lennartsson, 1987, Sandahl and Lindqvist, 1990*].

(4) Contemporary magnetic field mappings [*Tsyganenko et al., 1987*] combined with auroral oval images [*Elphinstone et al., 1990*] suggest that a cross section of the tail fairly well maps to the auroral oval. Using such modelling, one can easily fit an entire oval on closed field lines connected to the LLBL-dynamo, allowing for the polar cap to be located far poleward of the oval [*Feldstein and Galperin, 1985*]. This provides better geometrical agreement with a more contracted "droplet-shaped" polar cap observed during quiet times [*Murphree et al., 1982*]. According to the Tsyganenko model the "droplet-shaped" polar cap is expected to be present also during disturbed periods.

(5) The LLBL is the only magnetopause boundary layer which can connect to the dayside as well as the nightside current system. The initial brightening occurring both near local noon and local midnight at substorm onset (Fig 16) is good evidence for such a coupling.

(6) The LLBL induced currents in the non-diverted tail-current sheet can also be used to describe the loading characteristics. As the solar wind pressure increases the cross-tail current closing in the tail magnetopause also increases. This leads to plasma sheet thinning observed prior to substorm onset, a process evolving when the pressure surge progresses tailward (phase A, Fig. 15). The enhanced magnetopause currents due to the magnetospheric compression is expected to result in less significant auroral activity for the dayside as compared to the nightside. This is essentially the consequence of a higher dayside ionospheric conductance. Joule heating is expected to dominate the dayside magnetosphere-ionosphere interaction while on the nightside a substantial fraction goes into particle energization. Moreover, the dayside activity resulting from a direct coupling of the LLBL dynamo to the dayside cusp/cleft region, is also persistent while the nightside activity is more intermittent.

Finally, we here also propose a scenario whereby auroral enhancements, or "minisubstorms", are the mere consequence of the pass of solar wind-induced pressure surges in the LLBL traversing a high-conductivity channel with weak tail-current diversion. This would lead to sudden brightenings of the auroral activity along the preexisting oval, and a subsequent decrease when the pressure pulse has passed the high Σ channel.

REFERENCES

Akasofu., S.-I., The development of the auroral substorm, *Planet. Space Sci.*, 12, 273, 1964.

Akasofu., S.-I., Physics of magnetospheric substorms, D. Reidel Publ., Co., Dordrecht, Holland, 1977.

Akasofu, S.-I., and J. R. Kan, Dayside and nightside auroral arc systems, *Geophys. Res. Lett.*, 7, 753, 1980.

Akasofu, S.-I., Energy coupling between the solar wind and the magnetosphere, *Space Sci. Rev.*, 28, 121, 1981.

Axford, W. I., and C. O. Hines, A unifying theory of high latitude geophysical phenomena and geomagnetic storms, *Can. J. Phys.*, 38, 1433, 1961.

Baker, D.N., E.W. Hones, Jr., D.T. Young, and J. Birn, The possible role of ionospheric oxygen in the initiation and developement of plasma sheet instabilities, *Geophys. Res. Lett.*, 9, 1337, 1982.

Baker, D.N., T.A. Fritz, W. Lennartsson, B. Wilken, H.W. Kroehl, and J. Birn, The role of heavy ionospheric ions in the localization of substorm disturbances on March 22, 1979: CDAW 6, *J. Geophys. Res.*, 90, 1273, 1985.

Baker, D.N., S.J. Bame, W.C. Feldman, J.T. Gosling, R.D. Zwickl, J.A. Slavin, E.J. Smith, Plasma and magnetic field variAtions in the distant magnetotail associated with near-Earth substorm effects, in *Magnetotail Physics*, A.T. Lui ed., The Johns Hopkins Press, 137, 1987.

Baumjohann, W., Some recent progress in substorm studies, *J. Geomag. Geoelectr.*, 38, 633, 1986.

Burch, J.L., Low-energy electron fluxes at latitudes above the auroral zone, *J. Geophys. Res.*, 73, 3585, 1968.

Carlson C. W. and R. B. Torbert, Solar wind ion injections in the morning auroral oval, *J. Geophys. Res.*, 85, 2903, 1980.

Cole, K. D., On solar wind generation of polar magnetic disturbances, *J. Astron. Soc.*, 4, 103, 1961

Cowley, S. W. H., Plasma populations in a simple open magnetic field model magnetosphere, *Space Sci. Rev.*, 26, 217, 1980.

Daglis I.A., E.T. Sarris and G. Kremser, Indications for ionospheric participation in the substorm process from AMPTE/CCE observations, *Geophys. Res. Lett.*, 17, 57, 1990.

DeForest, S.E., and C.E. McIlwain, Plasma clouds in the magnetosphere, *J. Geophys. Res.*, 76, 3587, 1971.

Dungey, J. W., Interplanetary fields and the auroral zone, *Phys. Rev. Lett.*, 6, 47, 1961.

Eastman, T. E., E. W. Hones, Jr., S. J. Bame and J. R. Asbridge, The magnetospheric boundary layer: Site of plasma, momentum and energy transfer from the magnetosheath into the magnetosphere, *Geophys. Res. Lett.*, 3, 685, 1976.

Eastman, T.E., L.A. Frank, and C.Y. Huang, The boundary layer as the primary transport regions of the Earth's magnetotail, *J. Geophys. Res.*, 90, 9541, 1985.

Elphinstone, R.D., D. Hearn, J.S. Murphree, and L.L. Cogger, Mapping using the Tsyganenko magnetospheric model and its relationship to Viking auroral images, submitted to *J. Geophys. Res.*, 1990.

Evans, D. S., The characteristics of a persistent auroral arc at high latitude in the 1400 MLT sector, in *The Polar Cusp*, Jan A. Holtet and Alv Egeland (eds.) D. Reidel Publ. Comp., 99, 1985.

Eviatar, A. and R.A. Wolf, Transfer processes in the magnetopause, *J. Geophys. Res.*, 1968.

Feldstein, Y.I. and Yu. I. Galperin, The auroral luminosity structure in the high-latitude upper atmosphere: its dynamics and relationship to the large-scale structure of the Earth's magnetosphere, *Rev. Geophys.*, 23 217, 1985.

Heikkila, W. J., Impulsive plasma transport through the magnetopause, *Geophys. Res. Lett.*, 9, 877, 1982.

Heikkila, W. J., Comment on electric field evidence on the viscous interaction at the magnetopause, by F.S. Mozer, *Geophys. Res. Lett.*, 13, 233, 1986.

Heikkila W. J. and J. D. Winningham, Penetration of magnetosheath plasma to low altitudes through the dayside magnetic cusps, *J. Geophys. Res.*, 76, 883, 1971.

Heikkila W.J. and R.J. Pellinen, Localized induced electric field within the magnetotail, *J. Geophys. Res.*, 82, 1610, 1977.

Horwitz, J.L., The substorm as an internal magnetospheric instability: Substorms and their characteristic time scales during intervals of steady interplanetary magnetic field, *J. Geophys. Res.*, 90, 4164, 1985.

Hones, E.W., Jr., D.N. Baker, S.J. Bame, W.C. Feldtman, J.T. Gosling, D.J. McComas, R.D. Zwickl, J.A. Slavin, E.J. Smith, and B.T. Tsurutani, Structure of the magnetotail at 220 RE and its respons to geomagnetic activity, *Geophys. Res. Lett.*, 11, 5, 1984.

Huang, C.Y., L.A. Frank, T.E. Eastman, Plasma flows near the neutral sheet in the magnetotail, in *Magnetotail Physics*, A.T. Lui ed., The Johns Hopkins Press, 127, 1987.

Ijima, T., and T. A. Potemra, *J. Geophys. Res.*, 83, 599, 1978.

Iyemori, T., and S. Tsunomura, Characteristics of the association between an SC and a substorm onset, *Mem. Natl. Inst. Polar Res.*, Spec. Issue, 26, 139, 1983.

Kremser, G., and R. Lundin, Average spatial distribution of energetic particles in the midaltitude cusp/cleft region observed by Viking, *J. Geophys. Res.*, 95, 5753, 1990.

Lemaire, J., Impulsive penetration of filamentary plasma elements into the magnetospheres of the Earth and Jupiter, *Planet Space Sci.*, 25, 887, 1977.

Lennartsson, W., Dynamical features of the plasma sheet ion composition, density and energy, in *Magnetotail Physics*, A.T. Lui ed., The Johns Hopkins Press, 35, 1987

Lopez, R.E., D.G. Sibeck, R.W. McEntire, and S.M. Krimigis, The energetic ion substorm injection boundary, *J. Geophys. Res.*, 95, 109, 1990.

Lundin, R., On the magnetospheric boundary layer and solar wind energy transfer into the magnetosphere, *Space Sci. Rev.*, 48, 263, 1988.

Lundin, R., B. Hultqvist, N. Pissarenko, and A. Zakharov, The plasma mantle: composition and other characteristics observed by means of the Prognoz-7 satellite, *Space Sci. Rev.*, 31, 247, 1981.

Lundin, R., and D.S. Evans, Boundary layer plasmas as a source for high-latitude, early afternoon, auroral arcs, *Planet. Space Sci.*, 33, 1389, 1985.

Lundin R., and W. Lennartsson, On boundary layer heavy ions and the formation of the plasma sheet and magnetotail, paper presented at the XXVI COSPAR conference in Toulouse, 1986

Lundin, R., L. Eliasson and J.S. Murphree, The quiet time aurora and the magnetosphere configuration, Proceedings from the Auroral Physics International Conference, Cambridge, 1988.

Lui, A.T.Y., A. Mankovsky, C.-L., Chang, K. Papadopoulos and C.S. Wu, A current disruption mechanism in the neutral sheet: A possible trigger for substorm expansions, *Geophys. Res. Lett.*, 17, 745, 1990.

Mauk, B.H. and C.-I. Meng, Plasma injection during substorms, *Physica Scripta*, Vol T 18, 128, 1987.

McPherron, R.L., Substorm related changes in the geomagnetic tail: The growth phase, *Planet Space Sci.*, 20, 1521, 1972.

McPherron, R.L., Magnetospheric substorms, *Rev. Geophys. Space Phys.*, 17, 657, 1979.

Meng, C.-I., and R. Lundin, Auroral morphology of the midday oval, *J. Geophys. Res.*, 91, 1572, 1986.

Mitchell, D.G., F. Kutchko, D.J. Williams, T.E. Eastman, L.A. Frank, and C.T. Russell, An extended study of the low-latitude boundary layer on the dawn and dusk flanks of the magnetosphere, *J. Geophys. Res.*, 92, 7394, 1987.

Mozer, F. S., Electric field evidence on the viscous interaction at the magnetopause, *Geophys. Res. Lett.*, 11, 135, 1984.

Murphree, J.S., L.L. Cogger, and C.D. Anger, Characteristics of the instantaneous auroral oval in the 1200-1800 MLT sector, *J. Geophys. Res.*, 86, 7657, 1981.

Murphree, J.S., C.D. Anger, and L.L. Cogger, The instantaneous relationship between the polar cap and oval auroras at times of northward interplanetary magnetic field, *Can. J. Phys.*, 60, 349, 1982.

Newell, P.T., and C.-I. Meng, The cusp and the cleft boundary layer: Low-altitude identification and statistical local time variations, *J. Geophys. Res.*, 93, 14549, 1988.

Orsini S., M. Candidi, and H. Balsiger, Composition and velocity of ions streaming in the plasma mantle and in the lobe, in *Magnetotail Physics*, A.T. Lui ed., The Johns Hopkins Press, 239, 1987.

Pilipp W., and G. Morfill, The plasma mantle as the origin of the plasma sheet, in B.M. McCormac (ed.), *Magnetospheric Particles and Fields*, D. Reidel Publ. Co., Dordrecht, Holland, P.55, 1974.

Rosenbauer H., H. Grünwaldt, M.D. Montgomery, G. Paschmann, and N. Sckopke, HEOS 2 plasma observations in the distant polar magnetosphere: The plasma mantle, *J. Geophys. Res.*, 80, 2723, 1975.

Rostoker, G., Triggering of expansive phase intensifications of magnetospheric substorms by northward turnings of the interplanetary magnetic field, *J. Geophys. Res.*, 88, 6981, 1983.

Rostoker, G, and T.E. Eastman, A boundary layer model for magnetospheric substorms, *J. Geophys. Res.*, 92, 12187, 1987.

Sandahl, I. and P.-A. Lindqvist, Electron populations above the nightside auroral oval during magnetic quiet times, *Planet. Space Sci.*, 38, 1031, 1990.

Sandholt, P.E., C.S. Deehr, A. Egeland, B. Lybekk, R. Viereck, and G.J. Romick, Signatures in the dayside aurora of plasma transfer from the magnetosheath, *J. Geophys. Res.*, 91, 10063, 1986.

Scholer, M., G. Gloeckler, B. Klecker, F.M. Ipavich, D. Hovestadt, and E.J. Smith, Fast moving plasma structures in the distantmagnetotail, *J. Geophys. Res.*, 89, 6717, 1984.

Sckopke, N., G. Paschmann, H. Rosenbauer, and D.H. Fairfield, Influence of the interplanetary magnetic field on the occurrence and thickness of the plasma mantle, *J. Geophys. Res.*, 81, 2687, 1976.

Slavin et al., P.W. Daly,E.J. Smith, T.R. Sanderson, K.-P., Wenzel, R.P. Lepping, H.W. Kroehl, A.T. Lui ed., in *Magnetotail Physics,* The Johns Hopkins Press, 59, 1987.

Sonnerup, B. U. Ö., Theory of the low-latitude boundary layer, *J. Geophys. Res.*, 85, 2017, 1980.

Stasiewicz, K., A global model of gyroviscous field line merging at the magnetopause, *J. Geophys. Res.* 96, 77, 1991.

Troshichev, O. A., Polar magnetic disturbances and field-aligned currents, *Space Sci. Rev.*, 32, 275, 1982.

Tsyganenko N.A, Global quantitative models of the geomagnetic field in the cislunar magnetosphere for different disturbance levels, *Planet Space Sci.*, 35, 1347, 1987.

Vasyliunas, V. M., Interaction between the magnetospheric boundary layers and the ionosphere, in Magnetospheric Boundary Layers (edited by B. Battrick), pp 387, ESA, Paris, SP-148, 1979.

Woch, J. and R. Lundin, Temporal magnetosheath plasma injection observed with Viking: A case study, to appear in *Annales Geophysicae,*9, 133, 1991a.

Woch, J. and R. Lundin, Signatures of transient boundary layer processes observed with Viking, submitted to *J. Geophys. Res.*, 1991b.

Polar Cap Convection: Steady State and Dynamic Effects

JULIE J. MOSES AND PATRICIA H. REIFF

Department of Space Physics and Astronomy, Rice University, Houston Texas

Over the past few years, a good deal of work has been expended in qualitatively and quantitatively determining the patterns of steady state polar cap convection and their dependence on interplanetary conditions. A consensus is emerging as to the general characteristics of the patterns, at least during southward Interplanetary Magnetic Field (IMF) conditions. During times of northward IMF, however, there is still a substantial controversy as to how many distinct convection cells exist. Regions of sunward flow well poleward of the apparent location of the last closed field line are frequently observed during northward IMF (and at least sometimes during southward IMF). These regions are variously explained as either part of a separate convection cell completely confined to the polar cap, or as just a sunward meander in an otherwise "normal" two-cell convection pattern. Time variations (including the passage of a traveling convection vortex) may explain some of those events, but apparently not all of them.

Quantitative parameterizations of polar cap convection are progressing rapidly. The first such work merely related the magnitude of the cross-polar cap electrostatic potential drop versus interplanetary parameters, and this kind of study continues. Recent developments include correlating the dawnside plasmasheet potential drop with AL and the duskside with AU; and examining effects of substorm phase on the potential. In addition, attempting to develop a model which takes into account the time history of the IMF is critical. Empirical models to determine the convection pattern by making statistical averages of flow components at a given place under given IMF conditions generally yield patterns which are considerably less structured than what is typically observed. This kind of approach is severely hampered by the fact that polar cap boundary motion means that invariably sunward and antisunward flow components are averaged together, yielding a mean flow speed which is considerably smaller than the median speed. Another complementary approach is that of the "pattern recognition" type, where one uses one's experience in observing hundreds of polar cap crossings to create a typical pattern, and then creating an analytical representation of that pattern using adjustable parameters. The goal of that approach is to find a simple dependence of those parameters on interplanetary conditions.

Finally, a critical issue is substorm effects on the nightside convection pattern. The EISCAT POLAR experiment has yielded very exciting results on the local two-dimensional convection pattern and its dependence on, for example, westward traveling surges. A key topic to address in the next few years is how that convection pattern maps out to the nightside plasma sheet. This is by no means easy to answer, since the existence of parallel electric fields make mapping of ionospheric electric fields out to the magnetotail quite tricky, especially since the magnetic field structure during those times is quite dynamic. This question has serious consequences, however, in the fraction of the plasma sheet plasma which becomes injected into the inner magnetosphere.

Thus the topic of polar cap convection is far from being a stagnant field, and much interesting work is still in store to solve these fascinating controversies.

INTRODUCTION

There are two main drivers of ionospheric convection, the "viscous" interaction between the solar wind and magnetosphere and magnetic merging between the solar magnetic field and that of the Earth. *Axford and Hines* [1961] presented a scenario where the viscous interaction drives a two-cell convection pattern in the ionosphere (Figure 1a). Friction between the solar wind and the magnetopause drags plasma anti-sunward down the magnetosphere's flanks. Sunward flow in the plasmasheet results. If the ability of the ionosphere to short out the cross-tail electric field created by the viscous interaction is weak or localized, the two-cell magnetospheric convection pattern will map to the ionosphere as shown in Figure 1a.

Dungey [1961] also proposed a two-cell convection pattern but a different mechanism drives the convection. Instead, he suggested that the solar wind magnetic field merges with the Earth's field on the dayside magnetosphere and drags open field lines back across the ionosphere into the tail, creating anti-sunward convection in the polar cap (Figure 1b). Closed flux returns to the dayside through the reconnection of terrestrial field lines in the magnetotail.

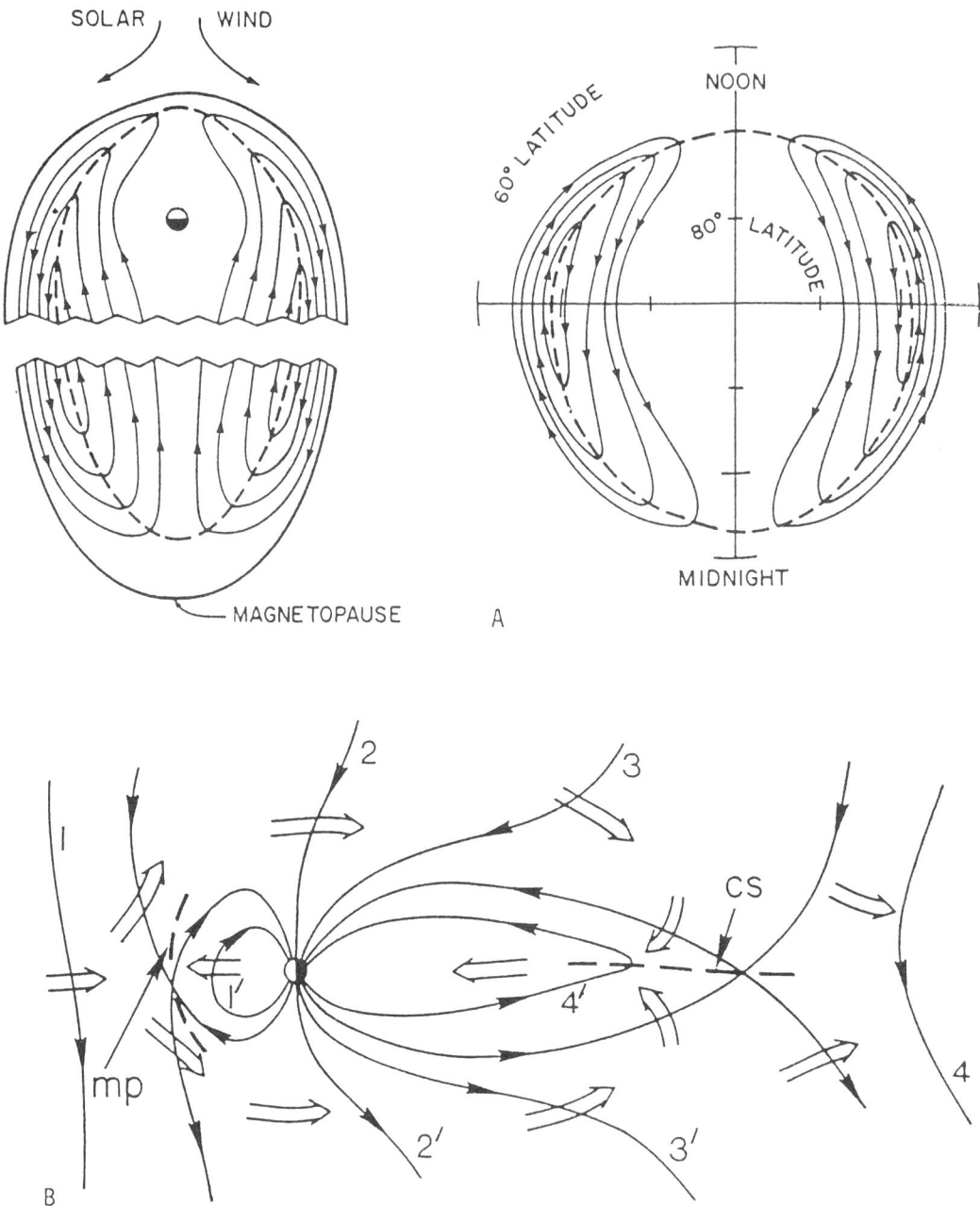

Fig. 1. (a) Magnetospheric convection and two-cell ionospheric convection as driven by the viscous interaction between the solar wind and magnetosphere. (b) A schematic of the two-cell convection pattern as driven by magnetic merging on the dayside magnetopause. (Both figures from Hill [1983].)

There is basic agreement in the field that both of these processes are responsible for driving the convection. However, it can be shown that a collisional viscous process, even operating at the Bohm diffusion rate, can only account for about 10 kV of cross-tail potential [Hill, 1983]. This amount of potential is consistent with measurements made in the low-latitude boundary layer [Mozer, 1984]. It is possible for the boundary layer driver to develop a larger potential difference if one allows momentum transfer by large amplitude (breaking) Kelvin-Helmholtz waves [Zhu and Kivelson, 1988]. Such waves may be responsible for traveling convection vortices seen deep in the magnetospheric tail [Birn et al., 1985, and references therein]. Alternatively, small-scale merging may be the driver for such low-latitude boundary layer convection [Stasiewicz, 1989; Nishida, 1989].

Two basic types of models are currently available for wide use, pattern recognition models where the authors have drawn schematic diagrams of the convection after looking at lots of data and statistical models where data for different conditions are binned and averaged together to create a pattern. Both of these types of models have been very useful. Thermospheric models such as the TIGCM [*Roble et al.*, 1988; *Sojka and Schunk*, 1987] use statistical ionospheric flow models such as *Foster et al.* [1986] to drive the thermospheric convection. Also, *Richmond et al.* [1988] used the above mentioned polar cap convection model and also the *Heppner and Maynard* [1987] pattern recognition model as a base convection pattern in the AMIE scheme. The above mentioned polar cap convection models and others are described in *Bilitza* [1990].

Another use for ionospheric convection models is as a tool for interpreting the magnetosphere-ionosphere coupling. The pattern recognition model of *Reiff and Burch* [1985] identifies different convection cells driven by separate magnetospheric processes such as merging and viscous interaction. *Heelis* [1984] and *Heelis et al.* [1976] used pattern recognition to determine convection patterns for the dayside and nightside ionosphere also.

Even though these models have been quite useful, they do have limitations. The statistical convection models smooth over the outstanding features of the convection pattern such as the convection reversal. Pattern recognition models tend to emphasize unusual convection features, such as the Harang discontinuity. More recently, now that basic features of the convection pattern are known, researchers have been attempting to rise above these problems and produce more quantitative, time-dependent, convection models.

REVIEW OF RECENT PROGRESS

Convection Patterns for Northward and Southward IMF

A lot of progress has been made on understanding the convection pattern during periods of southward IMF. All empirical and other models show an increase in the cross polar cap electric field and consequently the flow velocities with increasing negative IMF z-component [*Foster et al.*, 1986; *Heppner*, 1977; or *Lockwood et al.*, 1986, for example]. *Reiff et al.* [1981] and *Reiff and Luhmann* [1986] presented a relationship between the polar cap potential drop and the IMF B_z component. As the IMF B_z component becomes more negative, the polar cap potential drop increases. The effects of conductivity gradients across the polar cap have been duly noted by *Atkinson and Hutchison* [1978], *Yasuhara et al.* [1983] and *Moses et al.* [1987]. The effect of a day-night conductivity gradient on the polar cap convection is to concentrate the electric field toward the dawn side of the polar cap. *Hill* [1976] pointed out that the polar cap conductivity on Earth will not short out the solar wind electric field, but that a maximum potential is determined by the conductivity.

The IMF B_y component has a profound effect on convection in the polar cap also. *Svalgaard* [1968] and *Mansurov* [1969] both observed an ionospheric Hall current asymmetry related to the IMF B_y from the analysis of ground magnetometer data. Using magnetometer data from times when there was an ortho-garden hose IMF orientation, *Friis-Christensen* [1972] determined that it is the IMF y-component alone which controls the current asymmetry and consequently the electric field asymmetry (Figure 2).

So, at this juncture, it is clear that following a southward turning of the IMF, convection velocities will increase as the polar cap potential drop increases. Although the convection pattern is basically a two-cell pattern, the presence of conductivity gradients within the polar cap and an IMF B_y component will cause the pattern to concentrate to one side of the polar cap or another.

During periods of northward IMF, however, there is currently much controversy about the ionospheric convection pattern. Two different models have been proposed. *Heppner and Maynard* [1987] have proposed that the convection pattern during northward IMF consists of a distorted two-cell pattern (Figure 3a). They derived these patterns from looking at many satellite passes of DE-2 electric fields. The other model has been proposed by *Reiff and Burch* [1985] and suggests that there are three or more cells in the convection pattern during northward IMF (Figure 3b).

Since the publication of these two opposing views, numerous papers have appeared which show observations which seem to agree with one or the other proposed model. *Rasmussen and Schunk* [1988] used the pattern of Birkeland currents to infer convection patterns for positive and negative IMF B_z. They obtained a two cell pattern for southward IMF and a three-cell pattern for northward IMF, in agreement with the multiple-cell model. From Viking satellite measurements, *Zanetti et al.* [1990] used measured Birkeland and polar cap currents and auroral images to derive the convection for northward IMF. Their results did not agree with the distorted two-cell pattern. A four-cell convection pattern best modelled their observations. *Neilson et al.* [1990] showed that the convection associated with an observed theta aurora agreed with the multiple-cell model. *Reiff and Heelis* [1991] presented convection data from the AE-C satellite which they compared to their three-cell pattern and to the *Heppner and Maynard* [1987] distorted two-cell pattern. They found that the distorted two-cell pattern could not adequately explain the data without requiring additional current sheets and flow distortions (Figure 4). *Hoffman et al.* [1988] used electric field data from DE-2 during extremely quiet periods and found that the electric fields agree with the distorted two-cell model. *Clauer and Friis-Christensen* [1988] constructed a model based on Sondrestrom incoherent scatter radar observations for the northward IMF convection. They found that for positive IMF B_z and large B_y that the convection pattern agreed with the distorted two-cell model. But when the IMF B_y component was very small or 0, the convection pattern appeared to be a four-cell pattern.

In a recent diplomatic work to mesh all the above observations with the models, *Heelis and Reiff* [1991] present a scenario in which they depict how the convection pattern evolves from a two-cell pattern during southward IMF to a multiple cell or

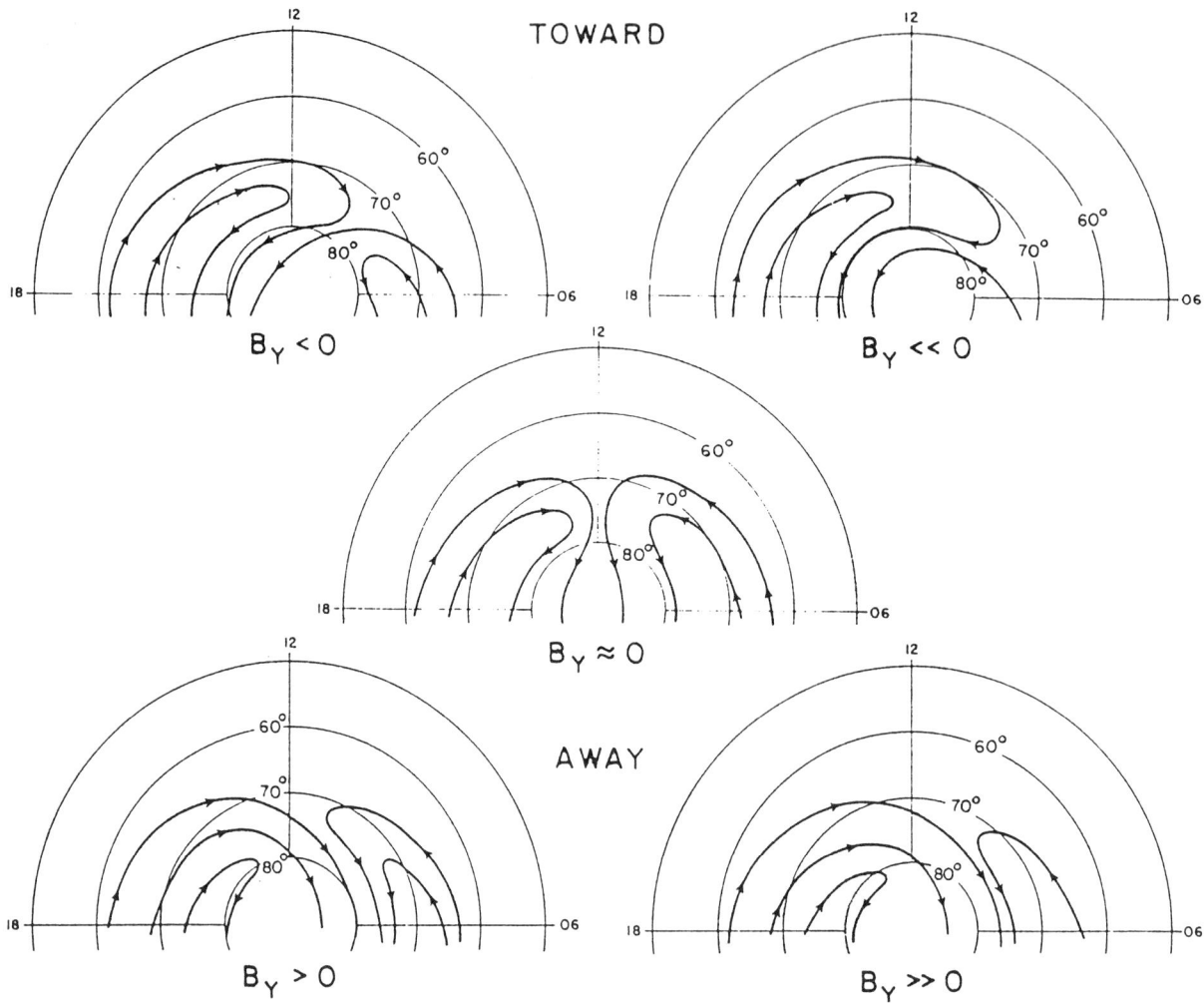

Fig. 2. Schematic of the dayside convection showing the effects of the IMF B_y [from *Heelis*, 1984].

distorted two-cell pattern during northward IMF. Fundamentally, a distorted two-cell pattern arises because the dayside merging region moves towards the nightside but the flow in the merging cell must not cross the viscous cell and instead must surround it. They conclude that it would be more useful to determine which pattern occurs under what conditions rather than to treat the two models as rivals.

Improved Quantitative Parameterizations

In addition to previous work relating the IMF B_z to the polar cap potential drop [*Reiff et al.*, 1981], new work has appeared relating the polar cap potential drop to the AE index and determining the variation of the potential around the polar cap boundary. *Lu et al.* [1989] analyzed data from AE-C and DE-2 to determine the distribution of potential around the polar cap boundary. They compared sine and arctangent fits to the polar cap boundary potentials. For some selections of IMF conditions,

both functions fit the data equally well. This finding leads the authors to conclude that the "throat" is wide or moves randomly in time. For other IMF conditions, the arctangent fit was significantly better, implying that the flow velocity shows a statistically significant constriction. They were able to check the potential patterns response to the IMF y-component and garden and ortho-garden hose orientations by locating the zero potential contour (the contour dividing the two convection cells). When the IMF goes from a garden hose to an ortho-garden hose orientation, the dayside zero potential line shifts duskward. Another result of their analysis is that the potential across the dusk cell is larger than the potential across the dawn cell. This effect is greatest for a northward IMF.

A recent statistical model by *Hairston and Heelis* [1990] has the advantage of determining the vertical component of the ionospheric flow as well as the two horizontal components. A followon study, using DMSP data, will provide for the first time

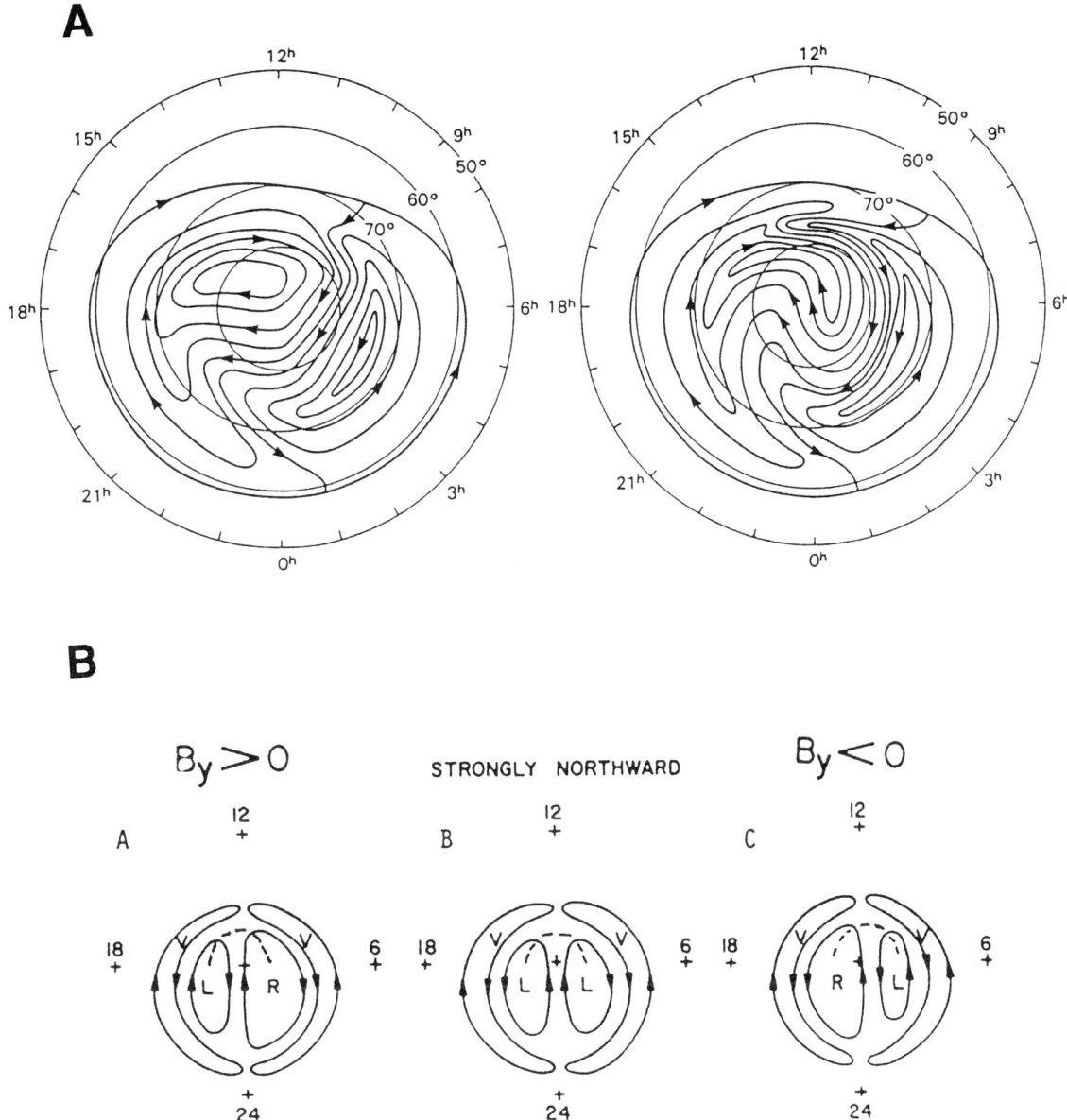

Fig. 3. (a) An example of a two-cell distorted pattern from *Heppner and Maynard* [1987] for $B_z > 0$ and $B_y > 0$ (right) and and $B_y < 0$ (left). (b) A multiple cell pattern for $B_z > 0$ and $B_y > 0$ and $B_y < 0$ [from *Reiff and Burch*, 1985].

a good statistical model of convection in the southern polar cap. Preliminary results indicate that the convection is not simply the mirror image of the northern hemisphere convection, as is frequently assumed. Part of the lack of antisymmetry is the rotation towards dawn from the noon-midnight meridian, and the increased potential drop across the dusk side plasma sheet, both effects being the same direction in each polar cap.

Weimer et al. [1990a; 1990b] related the polar cap potential drop as measured by DE-2 to the AE index and its components AL and AU to obtain the following relationship:

$$\Phi(kV) = 26.8 + 0.152 \, AE(nT) \quad \text{in summer}$$

and (1)

$$\Phi(kV) = 19.2 + 0.116 \, AE(nT) \quad \text{in winter}$$

The polar cap potential increases linearly with the AE index, but the ratio is dependent on the polar cap conductivity. They found that substorms tend to occur when the polar cap potential exceeds 60 kV. Also, the AE index saturates for IMF $B_z < -15$ nT and begins to increase again when the IMF $B_z > 20$ nT.

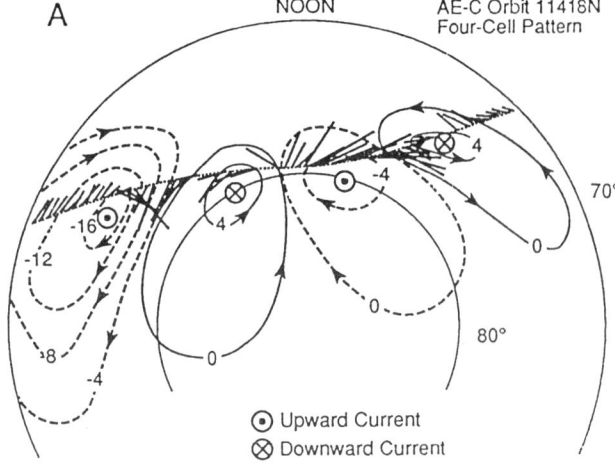

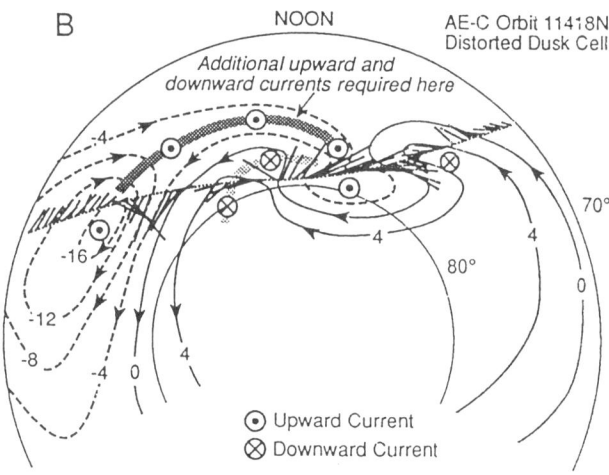

Fig. 4. (a) An example of a four cell pattern for an AE-C pass. (b) The same data with a distorted two-cell pattern overdrawn. Notice the more complicated current pattern required for the distorted two-cell pattern and also the dayside auroral arc, neither of which is observed [*Reiff and Heelis*, 1991].

Reiff et al. [1991] assembled and analyzed all 40 suitable DE-2 polar-cap passes (dawn-dusk passes with IMF of steady orientation for more than two hours) in order to study the functional dependence of the polar-cap potential drop on interplanetary parameters during the DE epoch (solar maximum) as was done previously for the AE epoch (solar minimum) [*Reiff and Luhmann*, 1986]. The objectives of this study were (1) by comparing the two datasets, to determine whether the limiting value of the polar-cap potential depends on the polar-cap conductivity (which is presumably larger in solar maximum conditions), as anticipated on theoretical grounds [*Hill*, 1976; *Hill*, 1984], and (2) by combining the two datasets, to improve the statistical determination of the functional dependence. The comparison of solar minimum versus maximum results (Figure 5 a,b) reveals no obvious difference either in the functional dependence or in the limiting value of the potential. If the average polar-cap/auroral-zone conductivity is indeed significantly larger near solar maximum than near solar minimum, then this result would appear to contradict both the simple theory cited above and the global MHD simulation results of *Fedder and Lyon* [1987] (which tend to confirm the simple theory). There are, however, some subtleties to this comparison that bear further study.

The combined datasets provide the best presently-available empirical determination of the dependence of the polar-cap potential on solar-wind parameters, namely

$$\Phi[kV] = (3.2)\,\tilde{v}^2 + (83)\,\tilde{v}^{2/3}\,\tilde{n}^{-1/6}\,\tilde{B}\,\sin^2(\theta/2) \qquad (2)$$

where $\tilde{v}$ is the solar-wind speed in units of 400 km/s, $\tilde{n}$ is the solar-wind concentration in units of $5/cm^3$, $\tilde{B}$ is the IMF strength (with an imposed ceiling of 10.8 nT – see *Reiff et al.*, [1981] or *Reiff and Luhmann* [1986]) in units of 5 nT, and θ is the polar angle of the IMF ($\cos^{-1}(B_z/|B|)$). The analysis is done in the dimensionless formulation suggested by *Vasyliunas* [1972]; thus, the IMF-independent part of the potential (first term on the rhs above) is normalized to the solar-wind proton kinetic energy per charge, while the IMF-dependent part (second term on the rhs) is normalized to the solar-wind motional emf across a distance equal to twice the Chapman-Ferraro lengthscale. This formula is only valid for times of reasonably steady IMF; a study of its time variability is in progress. It is interesting to note that estimates of the IMF-independent part, presumably attributable to processes other than magnetic merging, have declined steadily over the years, from 30-40 kV in our original study [*Reiff et al.*, 1981], to 10-20 kV [*Wygant et al.*, 1983], to 6-13 kV [*Reiff and Luhmann*, 1986], to 0-4 kV in our most recent study. This last value is consistent with theoretical estimates [*Hill*, 1983] of the potential that can be developed by microdiffusion within a low-latitude boundary layer. A paper describing these results is in preparation [*Reiff et al.*, 1991].

Dynamic Modelling

Several new techniques have been developed to depict snapshots of the convection pattern. The first is AMIE (Assimilated Mapping of Ionospheric Electrodynamics) [*Richmond and Kamide*, 1988]. The AMIE technique inverts ground magnetometer data to obtain the electric fields and currents that produced the geomagnetic variations in the first place. The method has been further upgraded to include other types of data besides ground magnetometer data. Incoherent scatter radar ion drifts [*Richmond et al.*, 1988], DMSP X-ray images [*Ahn et al.*, 1989], DE-2 ion drifts and auroral images [*Kamide et al.*, 1989], and precipitating energetic particles [*Knipp et al.*, 1989] can all be input to the model to get more accurate snapshots of the convection.

Another technique by *Marklund et al.* [1988] uses results from the Viking and DMSP F7 satellite to calculate snapshots of the ionospheric convection pattern. They made their own analytical

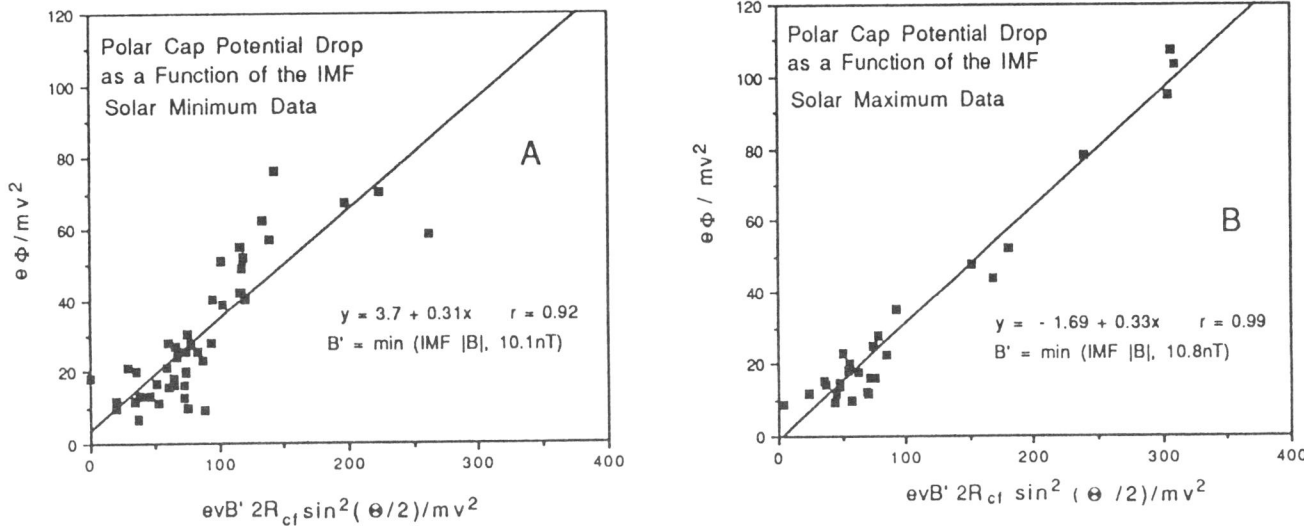

Fig. 5. Measured versus predicted polar-cap potential drop for (a) solar-minimum conditions (AE dataset) and (b) solar-maximum conditions (DE dataset). The functional fit is identical within statistical error; the best-fit joint equation is shown in Eq. 2.

conductivity model and modify it with Viking auroral images and precipitating particles. Then with the use of measured electric fields, they solve for current continuity in the ionosphere to obtain the convection pattern. They are capable of mapping electric fields to the equatorial plane through the auroral zone field-aligned potential drops also.

In response to a southward turning of the IMF, the polar cap and auroral oval are observed to expand and contract. Thus, a third technique by *Siscoe and Huang* [1986] used this observation to calculate the convection in the ionosphere caused by the expanding motion of the polar cap boundary. *Moses et al.* [1987] further developed this model by adding a day-night

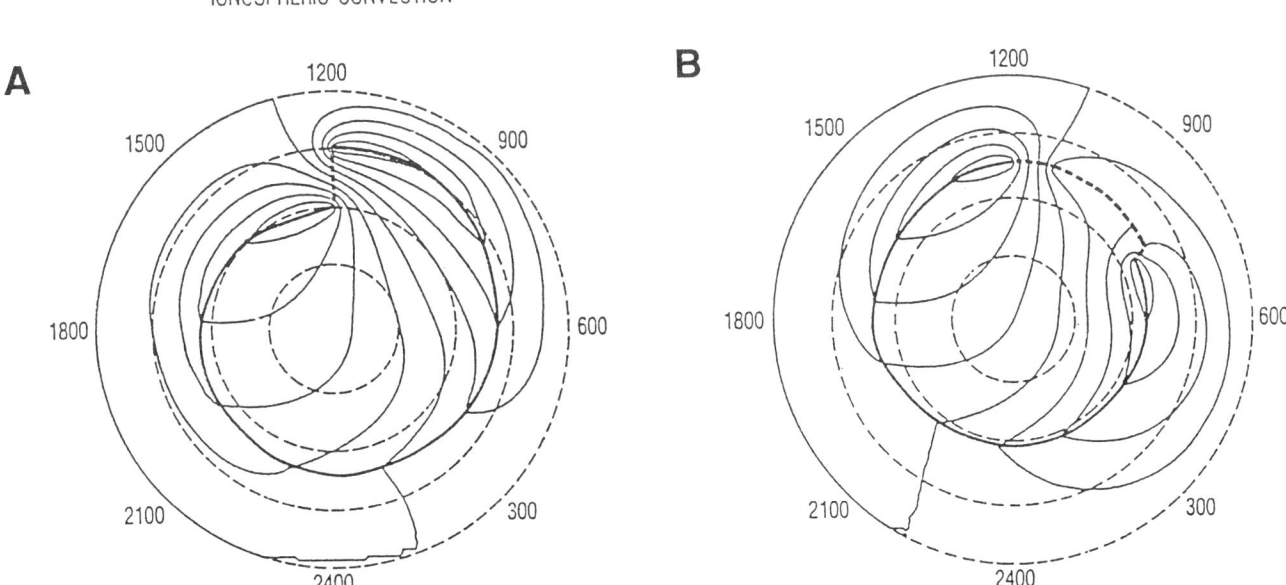

Fig. 6. (a) The expanding polar cap model of *Moses et al.* [1988]. In this example, $B_y > 0$ and $B_z < 0$. The polar cap forms a spiral which opens toward local afternoon. Flow enters the polar cap in the proper direction to give the correct $B_y > 0$ effect. The clockwise rotation of the nightside polar cap flow is due to the day-night conductivity gradient. (b) Another example from *Moses et al.* [1988]. Here, the dayside gap is extended in local time from 0830 to 1200 MLT. The flow concentrates into "throat"-like structures at 0830 and 1130 MLT, creating multiple throats.

conductivity gradient and B_y effects (Figure 6a). In this model, the B_y effect is localized to the dayside polar cap. Sunward flow on open field lines at dusk is caused by the day-night conductivity gradient. *Moses et al.* [1988] found that the only way to fit DE-2 ion drifts on the dayside was with an extended dayside gap (Figure 6b). The effect of the extended dayside gap is to produce multiple throat-like structures in the dayside convection. They suggest that the regions of enhanced dayside flows map to patchy regions of enhanced merging on the dayside magnetosphere.

Substorm-Related Convection

In addition to being able to produce snapshots of the convection during average or quiet times, researchers have made some discoveries about the nightside convection pattern during substorms. *Moses et al.* [1989] allowed the polar cap in their Expanding Polar Cap model to contract to simulate a period of tail reconnection. Flow moves from the open to closed field line region through the nightside gap, an opening created in the nightside polar cap boundary which maps to the tail reconnection region. The orientation of their nightside gap is pictured in Figure 7. They found this north-south orientation of the nightside gap always fit data with strong nightside electric fields during substorms. They searched for the nightside gap during various substorm phases and found that the nightside gap is most often present during the recovery phase of substorms. This finding implies that tail reconnection occurs through the recovery phase.

In an experiment with the Sondrestrom radar, *Robinson and Vondrak* [1990] observed the passage of a westward traveling

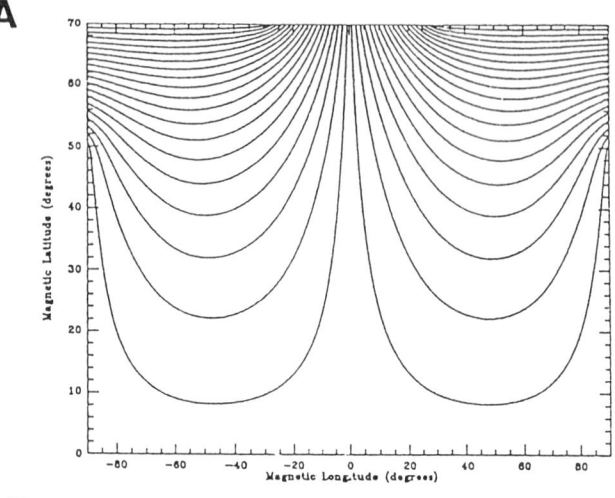

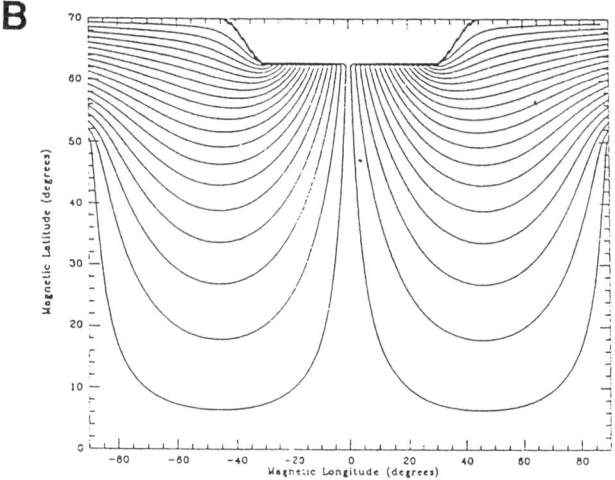

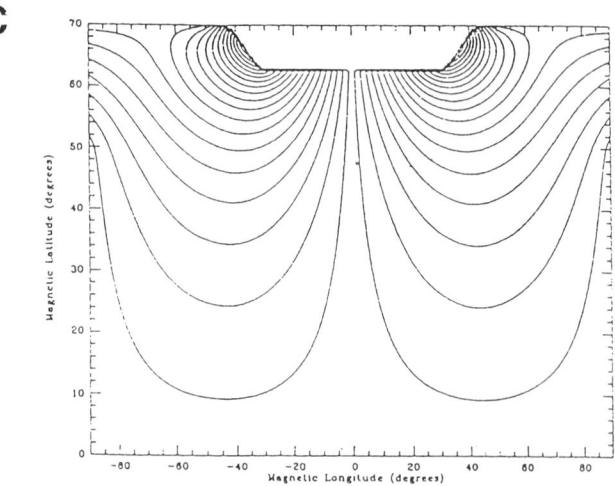

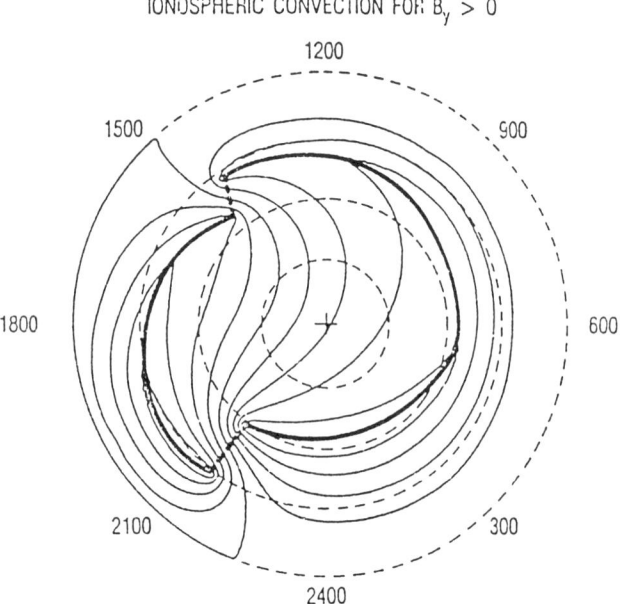

Fig. 7. An example from *Moses et al.* [1989] depicting the convection pattern from the recovery phase of a substorm. The nightside gap which maps to the reconnection region in the magnetotail is shown by the dotted line at 2100 MLT.

Fig. 8. Representation of the dayside convection local to the polar cap boundary as suggested in *Freeman and Southwood* [1988]. (a) The original two-cell dayside convection pattern. (b) The dayside convection resulting from a merging region appearing on the magnetopause. (c) The polar cap boundary returns to its original position after merging ceases.

Other EISCAT-Inspired Results

Besides the results related to substorms, EISCAT data has inspired results related to the time-dependent nature of the convection. *Etemadi et al.* [1988] tested the response of the convection to a southward turning of the IMF B_z component between 0930 and 1830 MLT. The dayside convection velocities responded in under 5 minutes to a southward turning of the IMF. *Freeman and Southwood* [1988] pointed out difficulties in interpreting radar data near the dayside polar cap boundary. They pointed out that what appears to be a convection reversal may actually be the polar cap snapping back after a period of increased dayside merging (Figure 8). As suggested in the figure, when the polar cap boundary snaps back, return flow from that event will be antisunward on closed field lines. *Lockwood et al.* [1990] gives examples of convection

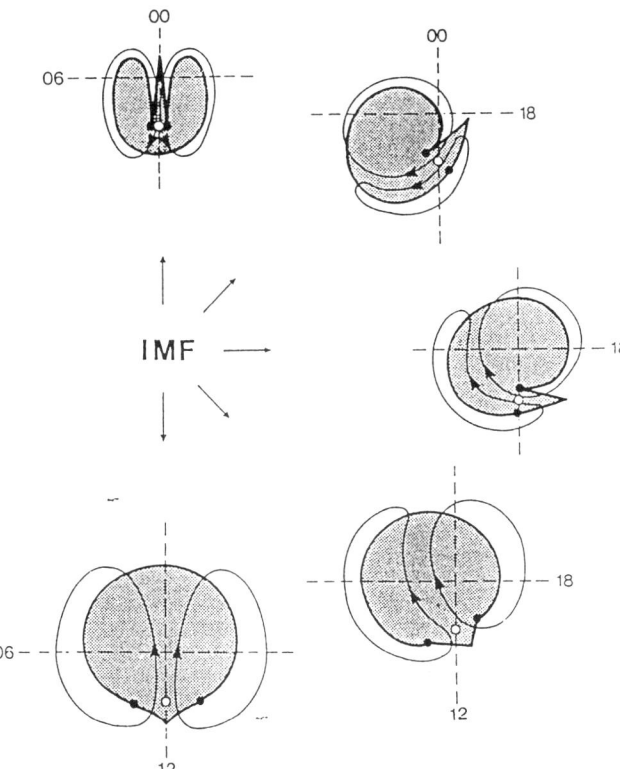

Fig. 9. The mapping of the anti-parallel dayside merging line to the dayside ionosphere for different IMF values [from *Crooker*, 1988].

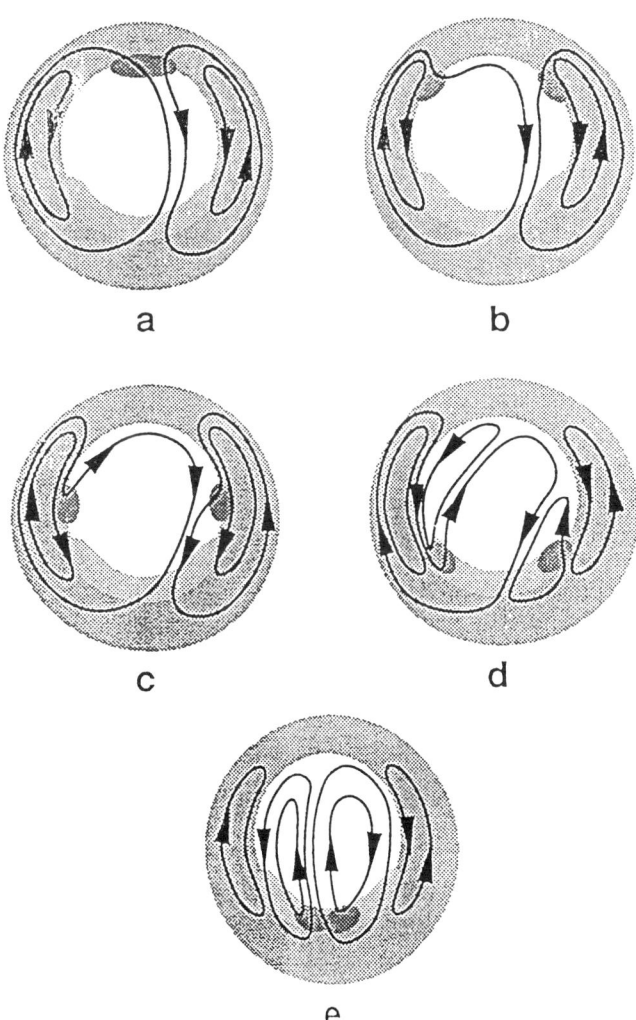

Figure 10. Convection patterns expected from motion of the dayside merging regions towards the nightside reconnection region as B_z changes from southward to northward through B_y positive. [From *Heelis and Reiff*, 1991].

surge over the radar site. They compare electric fields in the surge to local electric fields prior to the surges passing and find that the surge passage reduces the meridional electric field and moves the convection reversal to lower latitudes.

The POLAR experiment with EISCAT has yielded a number of substorm-related results. *Lester et al.* [1990] and *Lockwood et al.* [1988] noted the contraction or poleward motion of the polar cap boundary during a substorm due to tail reconnection. *Lockwood et al.* [1988] also discussed a case where the convection reversal remains on closed field lines while the polar cap boundary moved poleward. In this case, they believe they were seeing evidence of the viscous interaction on the nightside of the polar cap.

Weimer et al. [1990a] found that the functional dependence of the polar cap potential drop on the AE index significantly depended on substorm phase. This can be understood as a consequence of the conductivity enhancement which occurs in the expansion phase of substorms. Since in our work [*Reiff et al.*, 1991], we can show that the dependence of the polar cap potential drop on IMF conditions is not dependent on substorm phase, then together these results show conclusively that the solar wind-magnetosphere interaction is basically a voltage generator and not a current generator; i.e., the potential applied across the ionosphere is nearly independent of polar cap conductivity but the current in the ionosphere (as measured by the AE index) is not.

which strongly indicate that the dayside and nightside convection regions are driven by separate processes, namely merging on the dayside and reconnection in the tail on the nightside.

Mapping Potential Fields

As we continue to refine models of ionospheric convection, mapping these potential patterns out into the magnetosphere has become very attractive. What can the ionospheric convection tell us about magnetospheric processes? Perhaps we'll be able to refine our pictures of tail reconnection during substorms. Also, as a consequence of correctly mapping these fields, our understanding of magnetosphere-ionosphere coupling will increase. Recently, several papers have appeared in which the authors have mapped magnetospheric fields and structures down to the ionosphere with very interesting results. *Song et al.* [1988] took electric fields measured by the GEOS-2 satellite and mapped the fields to auroral and lower altitudes to modify the *Volland* [1978] potential pattern. *Kan and Cao* [1988] mapped a typical equatorial convection pattern to the ionosphere through a field-aligned potential drop under two different circumstances. In one case, they kept the parallel currents constant and let the parallel potential drop adjust accordingly. In the other case, the parallel potential drop was kept constant and the currents adjusted. In the constant current case, convection potentials were lowered and the evening magnetospheric convection reversal was sharper than in the constant voltage case.

Crooker [1988] mapped the dayside antiparallel merging line down to the ionosphere. The results are pictured in Figure 9. The merging line in the ionosphere appears as a "beak" shape. As the IMF turns northward, the beak moves toward dusk for negative B_y and points tailward. *Heelis and Reiff* [1991] base their evolutionary convection model on *Crooker* [1988] mappings for northward IMF, but with the additional restriction that the merging flow cell may not cross the viscous flow cell (Figure 10).

CONCLUSIONS

The study of polar cap convection is by no means finished. The complexities of magnetosphere-ionosphere coupling leave us with many more problems to solve. We have just begun to understand the evolution of the convection pattern with time, particularly changes in the convection pattern during substorms. Only with a finer-spaced grid of observing stations (spacecraft, ground-based, or combination of the above) can we hope to get a more complete understanding of the complexities of this fascinating and important subject.

ACKNOWLEDGEMENTS

This paper is supported in part by the National Science Foundation under grant ATM91-03440, and by the National Aeronautics and Space Administration under grants NAGW-1655 and NAG5-775.

REFERENCES

Ahn, B.-H., H. W. Kroehl, Y. Kamide and D. J. Gorney, Estimation of ionospheric electrodynamic parameters using ionospheric conductance deduced from bremsstrahlung x ray image data, *J. Geophys. Res., 94*, 2565-2586, 1989.

Atkinson, G. and D. Hutchison, Effect of the day-night conductivity gradient on polar cap convective flow, *J. Geophys. Res., 83*, 725-729, 1978.

Axford, W. I. and C. O. Hines, A unifying theory of high-latitude geophysical phenomena and geomagnetic storms, *Can. J. Phys., 39*, 1433-1464, 1961.

Bilitza, D., *Solar Terrestrial Models and Application Software*, National Space Science Data Center, Washington, D. C., 1990.

Birn, J., E. W. Hones Jr., S. J. Bame and C. T. Russell, Analysis of 16 plasma vortex events in the geomagnetic tail, *J. Geophys. Res., 90*, 7449-7456, 1985.

Clauer, C. R. and E. Friis-Christensen, High-latitude dayside electric fields and currents during strong northward interplanetary magnetic field: Observations and model simulation, *J. Geophys. Res., 93*, 2749-2757, 1988.

Crooker, N. U., Mapping the merging potential from the magnetopause to the ionosphere through the dayside cusp, *J. Geophys. Res., 93*, 7338-7344, 1988.

Dungey, J. W., Interplanetary magnetic field and the auroral zones, *Phys. Rev. Lett., 6*, 47, 1961.

Etemadi, A., S. W. H. Cowley, M. Lockwood, B. J. I. Bromage, D. M. Willis and H. Lühr, The dependence of high-latitude dayside ionospheric flows on the north-south component of the IMF: A high time resolution correlation analysis using EISCAT "POLAR", and AMPTE UKS and IRM data, *Planet. Space Sci., 36*, 471-498, 1988.

Fedder, J. A. and J. G. Lyon, The solar wind-magnetosphere-ionosphere current-voltage relationship, *Geophys. Res. Lett., 14*, 880-883, 1987.

Foster, J. C., J. M. Holt, R. G. Musgrove and D. S. Evans, Ionospheric convection associated with discrete levels of particle precipitation, *Geophys. Res. Lett., 13*, 656-659, 1986.

Freeman, M. P. and D. J. Southwood, The effect of magnetospheric erosion on mid- and high-latitude ionospheric flows, *Planet. Space Sci., 36*, 509-522, 1988.

Friis-Christensen, E., K. Lassen, J. Wiljelm, J. M. Wilcox, W. Gonzalez, and D.S. Colburn, Critical component of the Interplanetary Magnetic Field responsible for large geomagnetic effects in the polar cap, *J. Geophys. Res., 77*, 3371-3376, 1972.

Heelis, R. A., The effects of interplanetary magnetic field orientation on dayside high-latitude ionospheric convection, *J. Geophys. Res., 89*, 2873, 1984.

Heelis, R. A., W. B. Hanson and J. L. Burch, Ion convection velocity reversals in the dayside cleft, *J. Geophys. Res., 81*, 3803-3809, 1976.

Heelis, R. A. and P. H. Reiff, The ionospheric convection response to changes in IMF orientation, *to be submitted to J. Geophys. Res.*, 1991.

Heppner, J. P., Empirical models of high-latitude electric fields, *J. Geophys. Res., 82*, 1115, 1977.

Heppner, J. P. and N. C. Maynard, Empirical high-latitude electric field models, *J. Geophys. Res., 92*, 4467-4489, 1987.

Hill, T. W., A. J. Dessler, and R. A. Wolf, Mercury and Mars: The role of ionospheric conductivity in the acceleration of magnetospheric particles, *Geophys. Res. Lett., 3*, 429-432, 1976.

Hill, T. W., Solar-wind magnetosphere coupling, *Solar-Terrestrial Physics*, edited by R. C. Carovillano and J. M. Forbes, 261-302, Dordrecht, Netherlands, 1983.

Hill, T. W., Magnetic coupling between solar wind and magnetosphere: regulated by ionospheric conductance?, *EOS Trans. AGU., 65*, 1047, 1984.

Hoffman, R. A., M. Sugiura, N. C. Maynard, R. M. Candey, J. D. Craven and L. A. Frank, Electrodynamic patterns in the polar region during periods of extreme magnetic quiescence, *J. Geophys. Res., 93*, 14515-14541, 1988.

Kamide, Y., Y. Ishihara, T. L. Killeen, J. D. Craven, L. A. Frank and R. A.

Heelis, Combining electric field and auroral observations from DE 1 and 2 with ground magnetometer records to estimate ionospheric electromagnetic quantities, *J. Geophys. Res.*, *94*, 6723-6738, 1989.

Kan, J. R. and F. Cao, Effect of field-aligned potential drop in a global magnetosphere-ionosphere coupling model, *J. Geophys. Res.*, *93*, 7571-7577, 1988.

Knipp, D. J., A. D. Richmond, G. Crowley, O. d. l. Beaujardiere, E. Friis-Christensen, D. S. Evans, J. C. Foster, I. W. McCrea, F. J. Rich and J. A. Waldock, Electrodynamic patterns for September 19, 1984, *J. Geophys. Res.*, *94*, 16913-16923, 1989.

Lester, M., M. P. Freeman, D. J. Southwood, J. A. Waldock and H. J. Singer, A study of the relationship between interplanetary parameters and large displacements of the nightside polar cap boundary, *J. Geophys. Res.*, *95*, 21133-21145, 1990.

Lockwood, M., S. W. H. Cowley and M. P. Freeman, The excitation of plasma convection in the high-latitude ionosphere, *J. Geophys. Res.*, *95*, 7961-7972, 1990.

Lockwood, M., S. W. H. Cowley, H. Todd, D. M. Willis and C. R. Clauer, Ion flows and heating at a contracting polar-cap boundary, *Planet. Space Sci.*, *36*, 1229-1253, 1988.

Lockwood, M., A. P. v. Eyken, B. J. I. Bromage, D. M. Willis and S. W. H. Cowley, Eastward propagation of a plasma convection enhancement following a southward turning of the Interplanetary Magnetic Field, *Geophys. Res. Lett.*, *13*, 72-75, 1986.

Lu, G., P. H. Reiff, R. A. Heelis, M. R. Hairston and J. L. Karty, Distribution of convection potential around the polar cap boundary as a function of the Interplanetary Magnetic Field, *J. Geophys. Res.*, *94*, 13447-13461, 1989.

Mansurov, S. M., New evidence of a relationship between magnetic fields in space and on Earth, *Geomagn. Aeron. SSSR.*, *9*, 622-623, 1969.

Marklund, G. T., L. G. Blomberg, K. Stasiewicz, J. S. Murphree, R. Pottelette, L. J. Zanetti, T. A. Potemra, D. A. Hardy and F. J. Rich, Snapshots of high-latitude electrodynamics using Viking and DMSP F7 observations, *J. Geophys. Res.*, *93*, 14479-14492, 1988.

Moses, J. J., G. L. Siscoe, N.U. Crooker, and D.J. Gorney, IMF B_y and day-night conductivity effects in the expanding polar cap convection model, *J. Geophys. Res.*, *92*, 1193-1198, 1987.

Moses, J. J., G. L. Siscoe, R. A. Heelis and J. D. Winningham, A model for multiple throat structures in the polar cap flow entry region, *J. Geophys. Res.*, *93*, 9785-9790, 1988.

Moses, J. J., G. L. Siscoe, R. A. Heelis and J. D. Winningham, Polar cap deflation during a magnetospheric substorm, *J. Geophys. Res.*, *94*, 3785-3789, 1989.

Mozer, F. S., Electric field evidence of the viscous interaction at the magnetopause, *Geophys. Res. Lett.*, *11*, 135, 1984.

Neilson, E., J. D. Craven, L. A. Frank and R. A. Heelis, Ionospheric flows associated with a transpolar arc, *J. Geophys. Res.*, *95*, 21169-21178, 1990.

Nishida, A., Can random reconnection on the magnetopause produce the low latitude boundary layer?, *Geophys. Res. Lett.*, *16*, 227-230, 1989.

Rasmussen, C. E. and R. W. Schunk, Ionospheric convection inferred from interplanetary magnetic field-dependent Birkeland currents, *J. Geophys. Res.*, *93*, 1909-1921, 1988.

Reiff, P. H. and R. A. Heelis, Four cells or two? Are four convection cells really necessary?, *Submitted to J. Geophys. Res.*, 1991.

Reiff, P. H., D. C. Alexander and J. J. Moses, Independence of the polar cap potential on substorm phase, *To be submitted to J. Geophys. Res.*, 1991.

Reiff, P. H. and J. L. Burch, IMF B_y-dependent plasma flow and Birkeland currents in the dayside magnetosphere, 2. A global model for northward and southward IMF, *J. Geophys. Res.*, *90*, 1595-1609, 1985.

Reiff, P. H. and J. G. Luhmann, Solar wind control of the polar cap voltage, in *Solar Wind-Magnetosphere Coupling*, edited by Y. Kamide and J. A. Slavin, pp. 453-476, Terra Sci. Publ. Co., Tokyo, 1986.

Reiff, P. H., R. W. Spiro and T. W. Hill, Dependence of polar cap potential drop on interplanetary parameters, *J. Geophys. Res.*, *86*, 7639, 1981.

Richmond, A. D. and Y. Kamide, Mapping electrodynamic features of the high-latitude ionosphere from localized observations: Technique, *J. Geophys. Res.*, *93*, 5741-5759, 1988.

Richmond, A. D., Y. Kamide, B.-H. Ahn, S.-I. Akasofu, D. Alcaydé, M. Blanc, O. d. l. Beaujardière, D. S. Evans, J. C. Foster, E. Friis-Christensen, T. J. Fuller-Rowell, J. M. Holt, D. Knipp, H. W. Kroehl, R. P. Lepping, R. J. Pellinen, C. Senior and A. N. Zaitsev, Mapping electrodynamic features of the high-latitude ionosphere from localized observations: combined incoherent-scatter radar and magnetometer measurements for January 18-19, 1984, *J. Geophys. Res.*, *93*, 5760-5776, 1988.

Roble, R. G., E. C. Ridley, A. D. Richmond and R. E. Dickinson, A coupled thermosphere/ionosphere general circulation model, *Geophys. Res. Lett.*, *15*, 1325-1328, 1988.

Siscoe, G. L., and T. S. Huang, Polar cap inflation and deflation, *J. Geophy. Res.*, *90*, 543-547, 1986.

Sojka, J. J. and R. W. Schunk, Theoretical study of the high-latitude ionosphere's response to multicell convection patterns, *J. Geophys. Res.*, *92*, 8733-8744, 1987.

Song, X. T., R. Gendrin and M. Blanc, Determination of the Volland convection electric field parameters and computation of the associated field-aligned current distribution, *Planet. Space Sci.*, *36*, 631-639, 1988.

Stasiewicz, K., A fluid finite ion Larmor radius model of the magnetopause layer, *J. Geophys. Res.*, *16*, 8827-8834, 1989.

Svalgaard, L., Sector structure of the interplanetary magnetic field and daily variation of the geomagnetic field at high latitudes, *Geophys. Pap. R-6, Danish Meteorol. Inst.*, Charlottenlund, Denmark,1968.

Vasyliunas, V. M., The interrelationship of magnetospheric processes, in *Earth's Magnetospheric Processes*, edited by B. M. McCormac, pp. 29, D. Reidel Publ. Co., Dordrecht-Holland, 1972.

Volland, H., A model of the magnetospheric electric convection field, *J. Geophys. Res.*, *83*, 2695-2699, 1978.

Weimer, D. R., L.A. Reinleitner, J.R. Kan, L. Zhu, and S.-I. Akasofu, Saturation of the auroral electrojet current and the polar cap potential, *J. Geophys. Res.*, *95*, 18981-18987, 1990.

Weimer, D. R., N. C. Maynard, W. J. Burke and C. Liebrecht, Polar cap potentials and the auroral electrojet indices, *Planet. Space Sci.*, *38*, 1207-1222, 1990.

Wygant, J. R., R. B. Torbert and F. S. Mozer, Comparison of S3-3 polar cap potential drops with the Interplanetary Magnetic Field and models of magnetopause reconnection, *J. Geophys. Res.*, *88*, 5727-5735, 1983.

Yasuhara, F., R. Greenwald and S.-I. Akasofu, On the rotation of the polar cap potential pattern and associated polar phenomena, *J. Geophys. Res.*, *88*, 5773, 1983.

Zanetti, L. J., T. A. Potemra, R. E. Erlandson, P. F. Bythrow, B. J. Anderson, J. S. Murphree and G. T. Marklund, Polar region Birkeland current, convection, and aurora for northward Interplanetary Magnetic Field, *J. Geophys. Res.*, *95*, 5825-5834, 1990.

Zhu, X. and M. G. Kivelson, Analytic formulation and quantitative solutions of the coupled ULF wave problem, *J. Geophys. Res.*, *93*, 8602-8612, 1988.

POLAR HISS OBSERVED BY ISIS SATELLITES

Tadanori Ondoh

*Laboratory for Space Science, Communications Research Laboratory
Koganei, Tokyo, 184, Japan*

Abstract. The polar occurrence map for polar hiss (auroral hiss called hitherto) obtained from ISIS-VLF Syowa data is qualitatively similar to that for the inverted-V electron precipitations obtained from Atmospheric Explorer-D [Hoffman and Lin,1981], especially, concerning the low latitude boundary around invariant latitude 70° and the axial symmetry of the 10-22 hour geomagnetic local time (MLT) meridian. A statistical distribution of polar hiss in geomagnetic quiet conditions (Kp= 0-1) also shows the axial symmetry of the 10-22 hour MLT meridian and the low latitude boundary which is higher than that in various geomagnetic conditions. Thus, the occurrence map of polar hiss is different from the auroral zone which has the axial symmetry of the noon-midnight meridian. So, the auroral hiss called hitherto should be hereafter called by name of the polar hiss. The frequency range of the polar hiss is discussed in terms of whistler mode Cherenkov radiation generated from inverted-V electrons with energy below about 40 keV using realistic electron density distribution of the polar magnetosphere. The whistler mode Cherenkov radiation has an upper limit frequency given by the electron plasma frequency, electron gyrofrequency and electron energy. The frequencies of this radiation are higher when generated at lower altitudes than when generated at higher altitudes on the same field line in the polar magnetosphere. The frequency range of the downgoing polar hiss seems to be explained by the whistler-mode Cherenkov radiation generated from inverted-V electrons at geocentric distances below about 2 Re (earth's radius) along geomagnetic field lines between invariant latitudes of 70° and 77°.

Introduction

VLF radio bursts were often observed in association with auroral disturbances at high latitudes ground stations [Ellis,1957; Duncan and Ellis, 1959; Martin et al.,1960; Jorgensen and Ungstrup,1962; Morozumi,1963; Harang and Larsen,1965]. Helliwell [1965] classified these VLF emissions as auroral hiss. Ondoh [1963] calculated ionospheric absorption of auroral hiss in various polar ionospheric conditions to estimate intensity of auroral hiss above the ionosphere. Gurnett [1966] first reported satellite observations of auroral hiss associated with precipitating electrons below 10 keV. Correlative analyses of satellite data has shown that auroral hiss is generated by intense fluxes of precipitating auroral electrons with energies ranging from about 100 eV to about 40 keV [Gurnett and Frank, 1972; Hoffman and Laaspere, 1972]. Ondoh [1988] made a polar map of the occurrence rate of auroral hiss by using VLF electric field data (50 Hz - 30 kHz) for 347 ISIS VLF passes received at Syowa station (geomagnetic latitude 69.7°S, longitude 77.7°E), Antarctica, from June 1976 to January 1983.

In this paper, the ISIS polar occurrence map for polar hiss (auroral hiss called hitherto) was compared with the Atmospheric Explorer (AE)-D polar occurrence map for 280 inverted-V electron precipitation events [Hoffman and Lin, 1981]. Then, latitudinal distributions of the occurrence rate of polar hiss in geomagnetic quiet conditions (Kp= 0-1) were statistically obtained from narrow band VLF intensity data of 143 ISIS VLF passes received at Syowa station, Antarctica from January 1980 to January 1983, and they were also compared with the quiet-time auroral zone.

The Cherenkov radiation generated by auroral electrons was discussed as the origin of auroral hiss [Ellis,1957; Dowden,1960; Ondoh,1960; Jorgensen,1968]. Coherent Cherenkov radiation of the resonant Cherenkov instability was also discussed to explain auroral hiss [Gurnett and Frank,1972; Maggs, 1976,1978; James,1976]. Finally, the frequency range of polar hiss (auroral hiss) was discussed in terms of whistler mode Cherenkov radiation generated from inverted-V electrons moving along polar geomagnetic field lines in the polar magnetosphere.

Comparison of Polar Occurrence Map for Polar Hiss with Auroral Zone

Fig. 1 shows an example of dayside polar hiss observed by ISIS-2 from 1626 UT (1840-1329 MLT) in a geomagnetic quiet condition (Kp= 1+) on May 24, 1982. In Fig.1, simultaneous intensity increases of about 10 dB are seen at 5, 8, 16 and 20 kHz bands at invariant latitudes from about 68° to about 78°, and a weak ELF hiss is seen at 300 Hz band in the same latitudes. Since the latitudinal extent of auroral hiss called hitherto is clearly wider than that of the quiet auroral zone (Fig. 4 right), we call hereafter the auroral hiss by name of the polar hiss. Fig. 1 shows the narrow-band intensity data processed from DR outputs of ISIS VLF wide-band electric field (50 Hz - 30 kHz) tapes by narrow-band DC amplifiers with a minimum reading circuit similar to a usual VLF hiss receiver used at high latitude ground stations. Charging and discharging time constants of the minimum reading circuits are 10 sec and 10 millisec, respectively, and output signals from the minimum reading circuit are compressed by a logarithmic amplifier for chart recording. Dynamic range for the relative intensity variation of the narrow-band ISIS intensity data is about 30 dB in arbitrary scale, while that of the f-t spectrum film data is about 10 dB, and also the narrow-band ISIS intensity data is convenient to see a latitudinal variation of VLF hiss activity.

Fig. 2 shows an another example of morningtime polar hiss observed by ISIS-2 at invariant latitudes above 75° on December 13, 1980 (Kp= 1+). Strong daytime ELF hiss was continuously observed at 300 Hz and 1.5 kHz bands at invariant latitudes below 78°. Between 75° and 78°, the polar hiss with relative intensity of −40 dB at frequencies above 5 kHz was simultaneously observed with the ELF hiss with 300 Hz relative intensity of −20 dB. So, the weak polar hiss at frequencies above 5 kHz was observed together with the strong ELF hiss, although the ISIS VLF receiver adopts the automatic gain control (AGC) circuit. Thus, the polar hiss is mainly observed at frequency range from 2 kHz to 30 kHz or more, and the ELF hiss and chorus are usually observed at frequencies below 3 kHz. The ELF hiss is mainly observed in the plasmasphere, and it often extends up to the auroral zone. The high latitude limit of ELF hiss, which shows a rapid decrease of the 300 Hz intensity as shown by Fig. 2, approximately follows the auroral zone [Ondoh et al., 1983]. The chorus frequency decreases with invariant latitude and its latitudinal variation roughly agrees with a latitudinal variation of one half of the equatorial gyrofrequency. So, the chorus appears at frequencies below about 3 kHz in the auroral zone or at invariant latitudes above 62°[Ondoh et al., 1982]. Ondoh [1988] defined the polar hiss by simultaneous intensity increases above 5 dB, as compared with the quiet-time level at 5, 8, 16 and 20 kHz bands at high invariant latitudes above 60° as shown by Figs. 1 and 2, and also defined the occurrence rate of polar hiss over a unit area (1°interval in invariant latitude and 1 hour interval in MLT) by the ratio of the number of ISIS passes for which polar hiss was received to the number of ISIS passes over each area. The narrow-band VLF data are compiled in

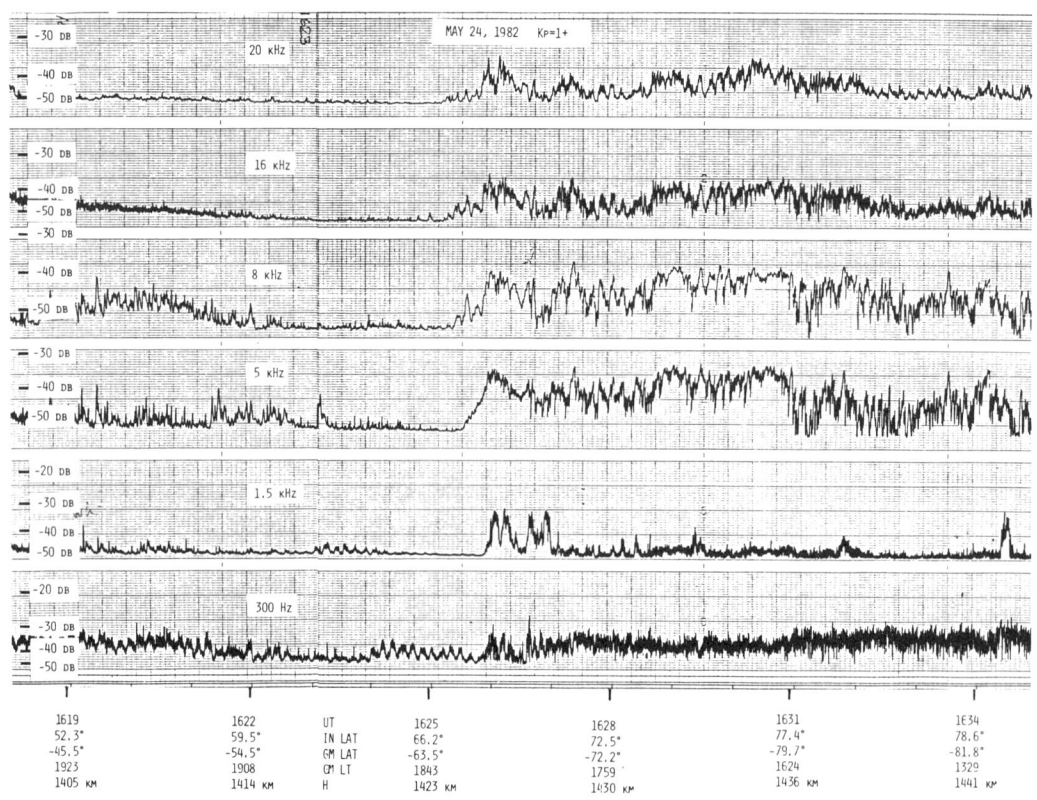

Fig. 1 Daytime polar hiss observed by ISIS-2 at invariant latitudes above 68°on May 24, 1982, Kp = 1+.

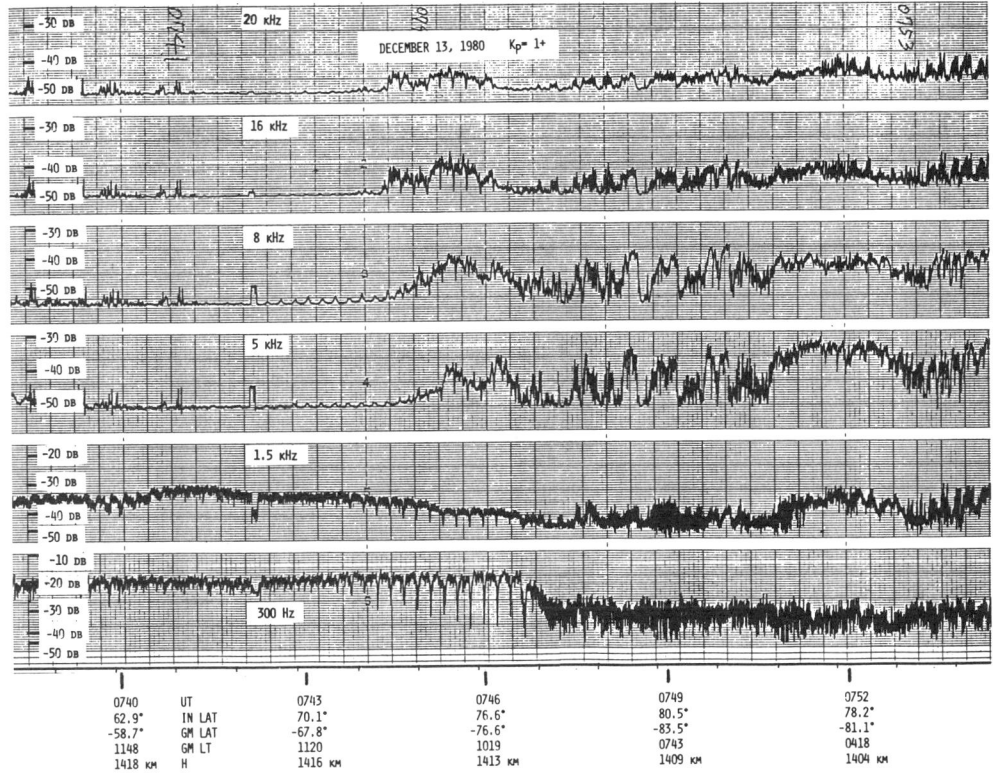

Fig. 2 Morningtime polar hiss observed by ISIS-2 at invariant latitudes above 75° on December 13, 1980, Kp = 1+.

"ISIS VLF data received at Syowa station, Antarctica", Radio and Space Data, No. 9 (1981), No. 13 (1983), and No. 15 (1984) published by Radio Research Laboratories, Tokyo.

Fig. 3 shows a polar map of the occurrence rate of polar hiss for various geomagnetic activities, Kp= 0 to 7 obtained from 347 VLF passes of ISIS-1 and -2 received at Syowa station, Antarctica, from June 1976 to January 1983, where contours of occurrence rates of 0.3, 0.4 and 0.5 are illustrated. Geomagnetic activities for most ISIS VLF passes were below Kp= 4. So, Fig. 3 seems to represent the average occurrence condition of polar hiss observed above the polar ionosphere (altitude 100 - 500 km). The high latitude contour of occurrence rate 0.3 lies at an invariant latitude of about 82° at all local times. However, the low latitude contour of 0.3 lies at an invariant latitude of about 74° at 10 hour MLT, and it extends down to 67° at 22 hour MLT. Thus, the low latitude contour of occurrence rate 0.3 is symmetric with respect to the geomagnetic meridian of 10 - 22 hour MLT. In 09 - 11 hour MLT, polar cusp hiss occurs mainly and f-t spectra of the polar cusp hiss usually are very irregular in frequency, time and intensity, while f-t spectra of the polar hiss observed in other local time intervals are steady and stable in frequency, time and intensity [Ondoh et al., 1980]. So, the occurrence rate of polar hiss is the lowest of all local time intervals at invariant latitudes from 65° to 82°, since polar cusp hiss is not included in the occurrence map of the polar hiss (Fig. 3). It should be noted that

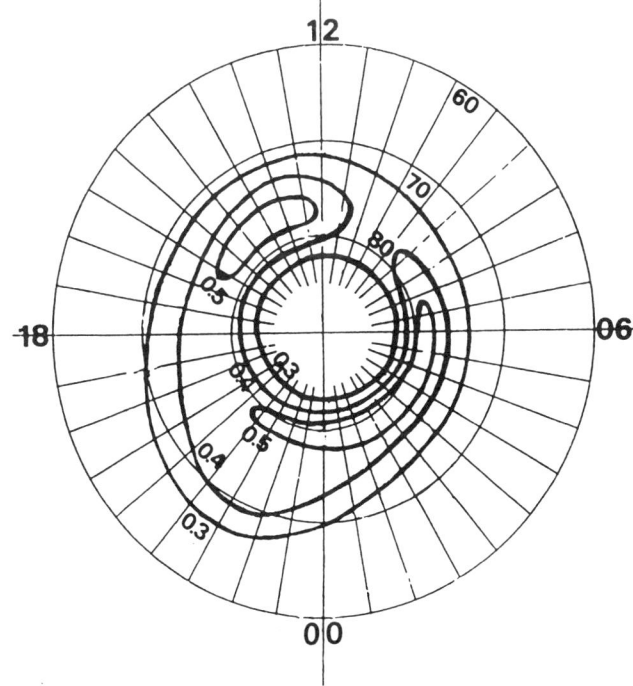

Fig. 3 Spatial map of occurrence rate of polar hiss obtained from 347 ISIS VLF pass data (Kp = 0 - 7) in invariant latitude and geomagnetic local time coordinates. Contours of occurrence rate of 0.3, 0.4 and 0.5 are shown.

the DP-2 current system associated with magnetospheric plasma convection also has parallel current flows at the 10 - 22 hour MLT meridian in the polar cap region [Nishida, 1968]. Active regions of high occurrence rate above 0.5 lie between 76° and 78° in the afternoon sector, and 78° and 81° in the late night-morning sector.

Vasyliunas [1970] projected the inner edge of the plasma sheet along the geomagnetic field lines on a polar map for the auroral zone during geomagnetically quiet (Fig. 4 right) and disturbed (Fig. 4 left) conditions. Polar maps for these quiet and disturbed auroral zones show an approximately symmetrical axis for the noon-midnight meridian. The latitudinal extent of the quiet (Q = 0) auroral zone is from invariant latitude 76° to 78° at the geomagnetic noon, and from 70° to 73° at the geomagnetic midnight, respectively. Also, that of the disturbed (Q = 4) auroral zone is from 74° to 76° at the geomagnetic noon, and from 65° to 72° at geomagnetic midnight, respectively. On the one hand, the latitudinal extent of the occurrence rate for polar hiss (Fig. 3) is from 74° to 82° at 10 hour MLT and from 67° to 82° at 22 hour MLT. The polar hiss region is wider than both of these zones in latitude, though the former includes many ISIS passes at various geomagnetic activities (Kp = 0 - 7). Thus, the polar map for polar hiss differs noticeably from the auroral zone. This is the reason why the auroral hiss used hitherto should be called hereafter by name of the polar hiss. This difference suggests that the energy range of the electrons producing the polar hiss above the ionosphere is different from that causing the aurora in the ionosphere.

Particle precipitation in the high latitude is traditionally broken into three regions, auroral oval, dayside cusp and polar cap. Precipitation of a few keV electrons from the central plasma sheet varies smoothly and produces diffuse auroral patterns. The diffused auroral region in the high latitude ionosphere lies alongside of the low-latitude half-side of the auroral oval [Akasofu, 1981]. Precipitating electrons are the free energy sources of certain plasma instabilities in the polar magnetosphere. Correlative studies have clearly indicated that strong auroral VLF hiss occurs simultaneously with inverted-V electron events in the auroral oval and the low-latitude side of the polar cap [Gurnett and Frank, 1972; Hoffman and Laaspere, 1972].

Fig. 5 shows a correlation example of auroral VLF hiss with inverted-V electron precipitations observed first by Injun-5 in the polar topside ionosphere [Gurnett and Frank, 1972]. This was observed at an invariant latitude of about 72° around 19 hour MLT. The electron energy seems to have been observed up to a few tens of keV. The region of inverted-V events projected along geomagnetic field lines onto the equatorial plane corresponds to the plasma sheet and magnetotail. Many inverted-V events in the dusk-midnight hemisphere have been found to have a maximum peak energy above 10 keV. In the polar cap, the inverted-V's usually had much lower peak energies, ranging from several hundred eV to about 1 keV [Hoffman and Lin, 1981; Lin and Hoffman, 1982]. Precipitation near the high latitude boundary of the auroral oval, referred to as a boundary plasma sheet precipitation, is characterized by discrete structures. The average energy of boundary plasma sheet electrons is typically a few hundred eV in quiet periods, but increases dramatically during substorm periods [Ossakow et al., 1984]. According to AE-D observations of inverted-V events [Hoff-

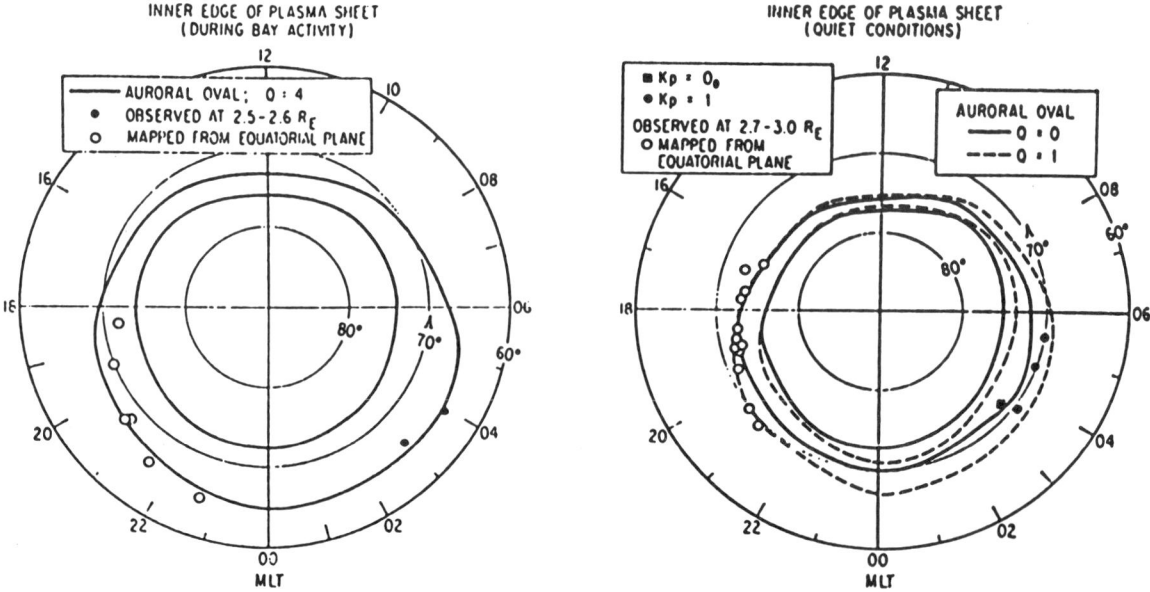

Fig. 4 Polar maps of auroral zones during geomagnetic quiet (Q = 0, 1, right) and disturbed (Q = 4, left) periods (after Vasyliunas, 1970).

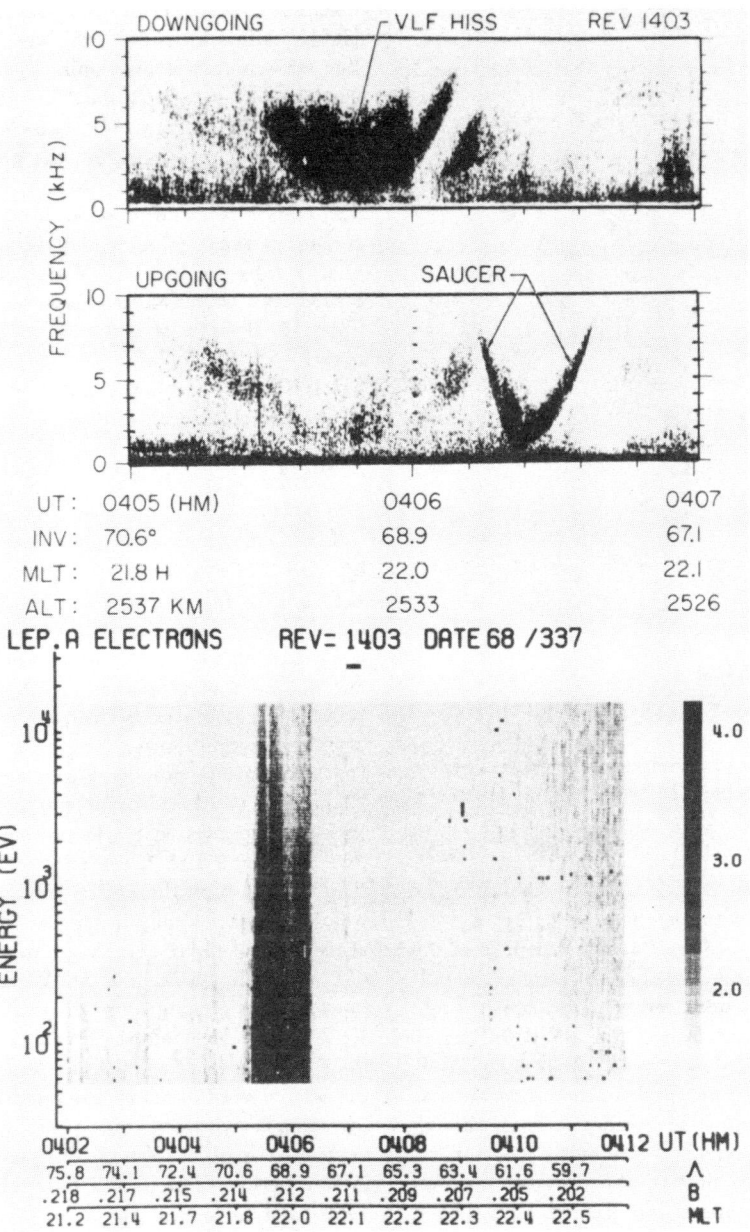

Fig. 5 Frequency-time spectrograms of electric (upper panel) and magnetic (middle panel) fields and energy vs time spectrogram (lower panel) of inverted-V electron precipitation events observed by Injun-5 (after Gurnett and Frank, 1972).

man and Lin, 1981; Lin and Hoffman, 1982], the inverted-V electrons precipitate mainly from the boundary plasma sheet, and they occur essentially all the time, independently of geomagnetic activity, especially in the pre-midnight region.

Fig. 6 shows a spatial occurrence map of 280 inverted-V events observed by AE-D in invariant latitude and geomagnetic local time coordinates [Hoffman and Lin, 1981]. Two thick lines showing the low-latitude and high -latitude boundaries have been drawn by the author for comparison with the polar map of polar hiss observed on 347 ISIS polar passes (Fig. 3). The high latitude boundary of inverted-V events may be represented better by a dotted curve elongated toward 22 hour MLT [Hoffman, 1988, Private communication]. The inverted-V event map (Fig. 6) shows the distribution of occurrence location of inverted-V events observed at various geomagnetic activities (Kp= 0 - 6), and the polar hiss map (Fig. 3) shows the distribution of occurrence rate of polar hiss at various geomagnetic activities (Kp= 0 - 7). Both polar maps have the 10-22 hour MLT meridian

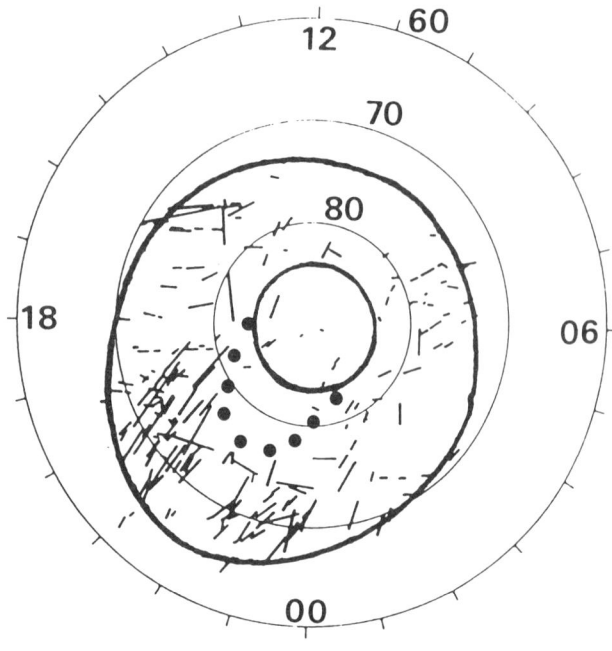

**AE-D
"INVERTED V" EVENT OCCURRENCE MAP**

Fig. 6 Spatial occurrence map of 280 inverted-V events observed by AE-D in invariant latitude and geomagnetic local time coordinates (after Hoffman and Lin, 1981).

as an axis of symmetry, with rare occurrence regions in the 09-11 hour MLT sector and in the polar cap region. Especially, the low-latitude contour of occurrence rate of 0.3 for the polar hiss is very similar to the low-latitude boundary of the inverted-V event occurrence. Both low-latitude boundaries lie at an invariant latitude of about 74° at 10 hour MLT, and they extend down to 66° at 22 hour MLT. Thus, in general features, the polar occurrence map of polar hiss corresponds well to that of inverted-V events, though the former indicates the occurrence rate and the latter the occurrence location. The distribution of occurrence location (or spatial extent) of polar hiss is represented by a spatial distribution of occurrence rate of polar hiss when the occurrence rate approaches to zero. The low-latitude boundary of polar hiss when the occurrence rate approaches to zero is about 4° to 7° lower than that of occurrence rate 0.3 (Fig. 3) in latitude depending on local time, and the high-latitude boundary of polar hiss in this case is about 2° to 4° higher than that of occurrence rate 0.3 (Fig. 3) in latitude depending on local time.

Magnetospheric electrons precipitate toward the ionosphere along geomagnetic field lines, while polar hiss waves generated from the electrons do not always propagate along geomagnetic field lines. Polar hiss waves propagate at large wave normal angles in the non-ducted whistler mode since the waves are emitted at various wave normal angles to the field lines. As a result, the polar hiss waves deviate appreciably at low altitudes from the original geomagnetic field line on which the polar hiss is generated by precipitating electrons at high altitudes. This effect produces the spatial extent of polar hiss wider than that of causative inverted-V electron precipitations. So, the polar map of occurrence location of inverted-V events may be compared with that of occurrence rate 0.3 of polar hiss. In other words, a broad frequency range of polar hiss seems to correspond to a broad energy range of inverted-V electrons precipitated from the boundary plasma sheet. However, since there are no available AE-D data of inverted-V electrons corresponding to individual ISIS polar hiss data, we do not discuss further correlation between the both phenomena and dependence of the correlation on altitude and latitude.

POLAR DISTRIBUTION OF POLAR HISS IN GEOMAGNETIC QUIET CONDITIONS

Fig. 7 shows an example of darkside polar hiss observed by ISIS-2 in a geomagnetic quiet time (Kp= 0) from 0814 UT to 0823 UT (07 - 02 hour MLT) on February 13, 1980. Simultaneous intensity increases of polar hiss are seen at 5, 8, 16 and 20 kHz bands at invariant latitudes between 73° and 83° even at geomagnetic activity of Kp= 0. In this section, we examine statistically whether a polar distribution of polar hisses observed for geomagnetic quiet times has also the low-latitude boundary and axial symmetry of the 10 - 22 hour MLT meridian or not as discussed in Fig. 3. So, ISIS polar passes with a latitudinal extent greater than 20°, like Figs. 1, 2 and 7, are selected to take a uniform statistical weight over the entire polar region. For this purpose, we obtain statistically latitudinal distributions of the occurrence rate of polar hiss in geomagnetic local time intervals of 8 hours for 06-14 hour MLT, 14-22 hour MLT and 22-06 hour MLT, by using narrow-band intensity data of 143 ISIS VLF passes received at Syowa station, Antarctic for Kp= 0 - 1+ between January 1980 and January 1983. Since the period between January 1980 and January 1983 is a declining solar activity one, many ISIS VLF data were received in geomagnetic quiet (Kp= 0 - 1+) conditions at Syowa station in this period. The period between June 1976 and January 1983 used for Fig. 3 contains both geomagnetic quiet and disturbed conditions, since there were relatively many disturbed days in the increasing solar activity period from June 1976 to December 1979.

Figs. 8a-8c show latitudinal distributions of the occurrence rate for polar hiss (or auroral hiss called hitherto) in 06-14 hour MLT, 14-22 hour MLT and 22-06 hour MLT intervals for geomagnetic quiet times (Kp= 0 - 1+) obtained from narrow band ISIS VLF data observed at Syowa station during 143 passes from January 1980 to January 1983. Regions of the occurrence rate above 0.3 lie from invariant latitude 74° to 84° in 06-14 hour MLT, from 70° to 82° in 14-22 hour MLT and from 70° to 83° in 22-06 hour MLT, respectively. The low latitude boundary of polar hiss in 06-14 hour MLT is clearly higher than that in 14-22 hour

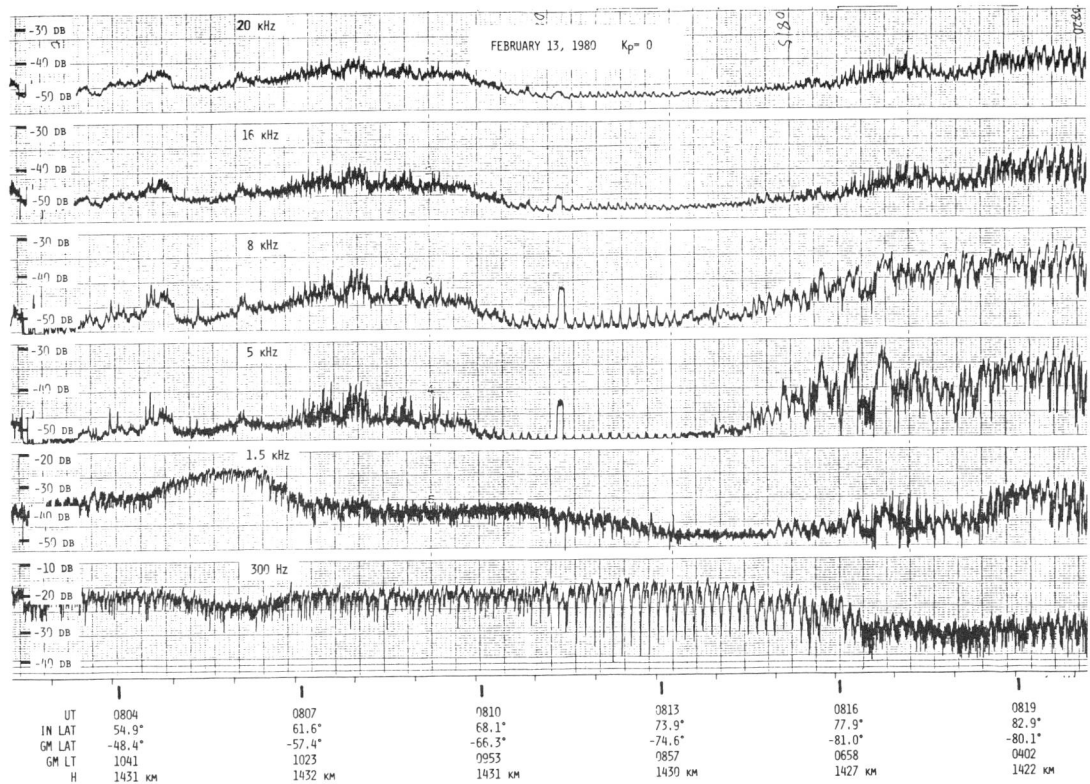

Fig. 7 Darkside polar hiss observed by ISIS-2 at Kp = 0 on February 13, 1980.

MLT and 22-06 hour MLT in latitude, but the high latitude boundary lies at around 83° in the three intervals. The maximum occurrence rate of polar hiss lies at invariant latitude 79° - 80° in 06-14 hour MLT and 14-22 hour MLT and at 77° -78° in 22-06 hour MLT, respectively. In lower latitude side of the maximum occurrence rate, the occurrence rate of polar hiss in 06-14 hour MLT decreases faster with latitude than those in 14-22 hour MLT and 22-06 hour MLT. In the general features, the latitudinal distributions of the occurrence rate for polar hiss in 14-22 hour MLT and 22-06 hour MLT are similar to each other, but are considerably different from that in 06-14 hour MLT. Thus, the polar distribution of polar hiss (or auroral hiss called hitherto) in geomagnetic quiet conditions (Kp= 0 - 1+) has also the low latitude boundary elongating toward 22 hour MLT and the axial symmetry of the 10-22 hour MLT meridian, since the middle time of 06-14 hour MLT interval is 10 hour MLT as shown by the polar occurrence map of polar hiss at various geomagnetic activities (Kp = 0 - 7) in Fig. 3. The latitudinal extent of the hiss occurrence rate above 0.3 for geomagnetic quiet times (Kp - 0 - 1+) in Fig. 8 is appreciably narrower than that of the occurrence map of polar hiss at various geomagnetic activities (Kp = 0 - 7) in Fig. 3, but it is clearly wider than the quiet (Q = 0 - 1) auroral zone in Fig. 4, which has the noon-midnight symmetry. Therefore, it is concluded that the occurrence map for quiet time polar hiss is also qualitatively similar to that for the inverted-V electron precipitation events (Fig. 6) and that the former is clearly different from the quiet-time auroral zone.

WHISTLER-MODE CHERENKOV RADIATION GENERATED FROM INVERTED-V ELECTRON AS ORIGIN OF POLAR HISS

Polar hiss is downgoing whistler-mode waves in a frequency range from 2 kHz to above 30 kHz, and it is continuously observed in a wide latitudinal range of polar region by ISIS satellites [Ondoh et al., 1980]. Gurnett and Frank [1972] demonstrated that auroral VLF hiss occurred in direct association with inverted-V electrons. The polar hiss is continuously observed for more than 1000 km along an ISIS orbit at invariant latitudes above 70° as shown by 5 kHz intensity increases in Figs. 1 and 2. The latitudinal widths of regions of inverted-V electron precipitations and of polar hiss are much larger than the auroral arc width of less than 10 km. Latitudinal width of most inverted-V events are from 0.1° to 2° in invariant latitude, with a mean width of about 0.5°, and the average longitudinal width appears to be much larger than the typical auroral arc length of 0.1°[Lin and Hoffman, 1982]. Thus, it is no longer necessary to discuss a relationship between the auroral VLF hiss and aurora as observed from ground observations. Polar hiss is believed to be produced by the Landau resonance [Maggs,

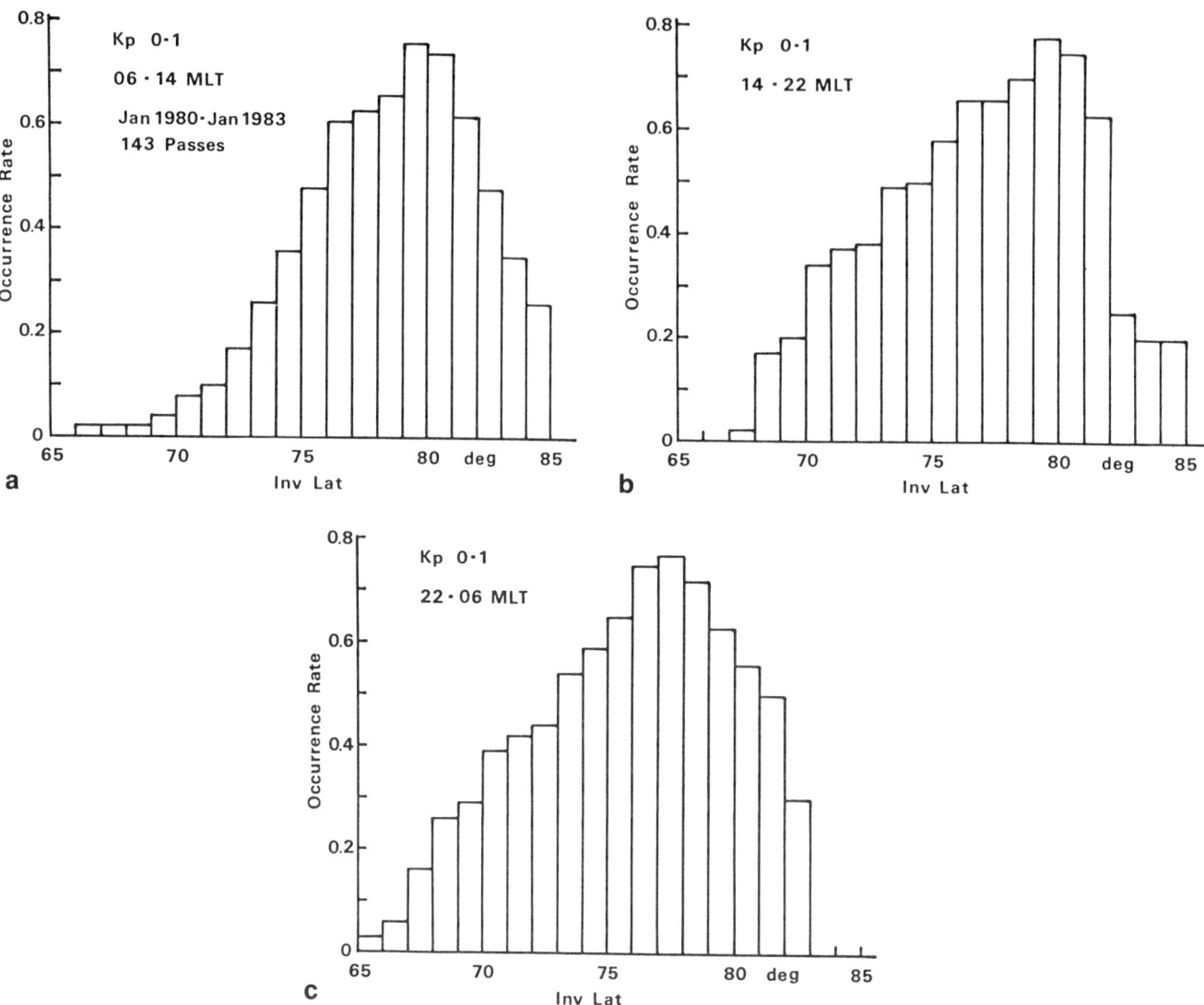

Figs. 8a - 8c Latitudinal distributions of the occurrence rate for polar hiss (or auroral hiss) at Kp = 0 - 1 in 06-14 hour MLT (a), 14-22 hour MLT (b) and 22-06 hour MLT (c) intervals obtained from 143 ISIS polar passes between January, 1980 and January, 1983.

1976], in which the inverted-V electron beam moves in the same direction as the downgoing hiss wave, i.e. $\omega/k = V$, where V is the beam velocity parallel to the field line, ω is the angular wave frequency and k is the parallel wave number. Next, we discuss the frequency range of whistler-mode Cherenkov radiation generated by precipitating inverted-V electrons in the polar magnetosphere as the origin of the downgoing polar hiss. The whistler-mode refractive index, n is given by

$$n^2 = \frac{f_N^2}{f f_H \cos \theta}$$

for conditions of $f_H, f_N \gg f$ and $f_H \cos \theta \gg f$ where f denotes the whistler mode frequency, f_H the electron gyrofrequency, f_N the electron plasma frequency and θ the wave normal angle to geomagnetic field line [Storey,1953]. For simplicity, we consider the Cherenkov radiation from electrons with velocity, V, parallel to the geomagnetic field lines. The frequency of whistler mode Cherenkov radiation,

$$f = \frac{f_N^2 \cos \theta}{f_H} \frac{V^2}{c^2}$$

is derived from the whistler mode refractive index and the Cherenkov resonant condition, $c/n = V \cos \theta$, where c is the light speed, n the whistler mode refractive index and θ the wave normal angle to field line. The frequency of whistler mode Cherenkov radiation is rewritten as

$$f = \frac{f_N^2 \cos\theta \cdot E}{250 f_H}$$

where $E = 250(V/c)^2$ keV is the energy of an electron with parallel velocity, V. So, the whistler mode Cherenkov radiation has an upper limit frequency given by

$$f \leq \frac{f_N^2 E}{250 f_H} \quad \text{kHz}$$

If the polar hiss is the whistler mode Cherenkov radiation, the upper limit frequency of polar hiss should be observed at each altitude along the geomagnetic field line, and the higher frequency components of the radiation should be generated at the lower altitudes. The upper limit frequency of whistler mode Cherenkov radiation is calculated as a function of electron energy, E, at various altitudes along two dipolar geomagnetic field lines of invariant latitude 70° and 77°, using an electron gyrofrequency of

$$f_H = 896 \left(\frac{R_e}{R}\right)^3 (1 + 3\sin^2\theta_m)^{1/2} \quad \text{kHz}$$

for a geomagnetic field of $H_o = 0.32$ gausses at the geomagnetic equator and the polar altitude profile of electron density (Fig. 9) obtained from the Alouette/ISIS data and the DE-1 plasma wave data [Persoon et al., 1983], where R is the geocentric distance, R_e the earth's radius and θ_m the geomagnetic latitude. The two geomagnetic field lines of 70° and 77° represent, respectively, the low latitude occurrence boundary and the active region of polar hiss as shown by Fig. 3. A thick curve between data of the Alouette/ISIS density and the DE-1 plasma wave density (crosses) in Fig. 9 was interpolated by the author. The electron densities at high altitudes [Persoon et al., 1983] and at low altitudes in Fig. 9 were obtained poleward of the auroral zone, inside the polar cap. Uncertainties of the electron densities in the nightside auroral zone are as high as ±40 % [Persoon et al., 1988]. The upper limit frequency of the whistler mode Cherenkov radiation was calculated at four geocentric distances ; 1.32 Re (altitude about 2000 km), 1.5 Re (about 3200 km), 2.0 Re (about 6400 km) and 3.0 Re (about 13000 km), along the two geomagnetic field lines of 70° and 77°. Electron gyrofrequency and mean electron plasma frequency at geocentric distance, 3.0 Re on the geomagnetic field line, 77°are 62.5 kHz and 20 kHz (N = 5/cm^3), respectively. The maximum and minimum electron plasma frequencies corresponding to electron density uncertainties of 40 % (N = 5±2/cm^3) are 24 kHz and 16 kHz respectively. So, the whistler mode condition (f_H, $f_N \gg f$) is not satisfied at 3.0 Re on the two geomagnetic field lines at frequencies above 1 kHz, which is actually the frequency range of the polar hiss.

Fig. 10 shows the relationship of the calculated upper limit frequency of the whistler mode Cherenkov radiation to the electron energy at four geocentric distances on geomagnetic field lines of invariant latitude 70° and 77°, respectively. Vertical rods in Fig. 10 show the error ranges corresponding to the electron density uncertainties of ± 40%. The error range of the upper limit frequency increases with

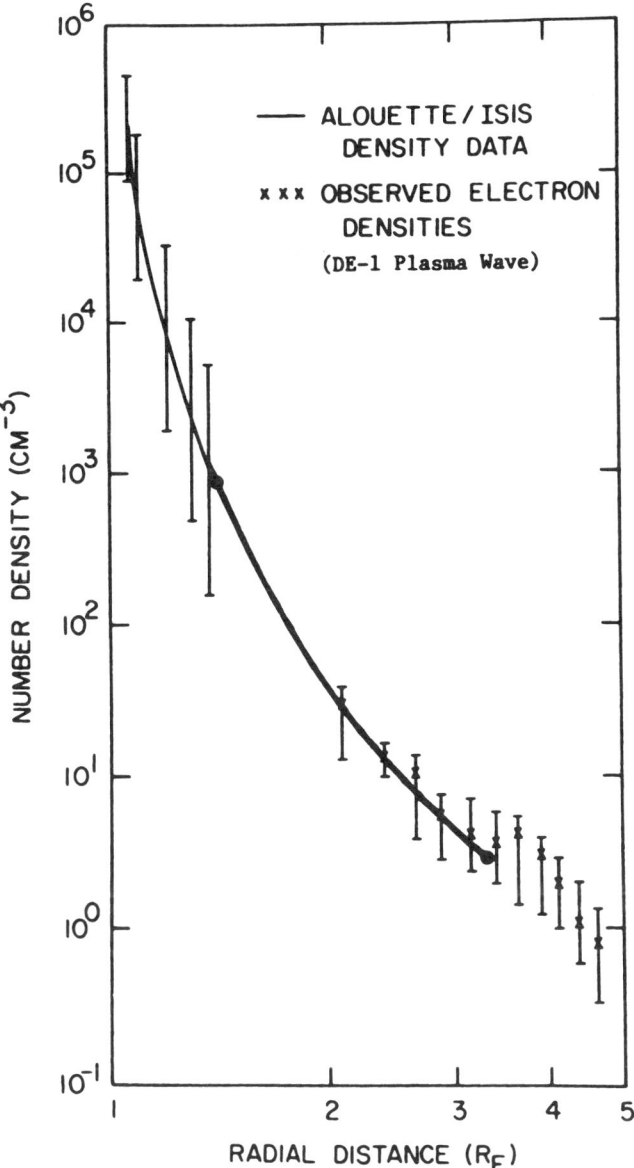

Fig. 9 Altitude profile of electron density in the polar magnetosphere obtained from Alouette/ISIS radio sounding data and DE-1 plasma wave data (after Persoon et al., 1983).

increasing electron energy at the same geocentric distance, and it also increases with decreasing geocentric distance at the same electron energy. The frequency range of the whistler mode Cherenkov radiation at each geocentric distance lies in a region below each line. The energy range of the inverted-V electrons associated with polar hiss is usually below about 40 keV [Hoffman and Lin, 1981]. The discussion of the altitude range of the generation region is not directly derived from the discussion on the upper limit frequency of the whistler mode Cherenkov radiation at each altitude. Fre-

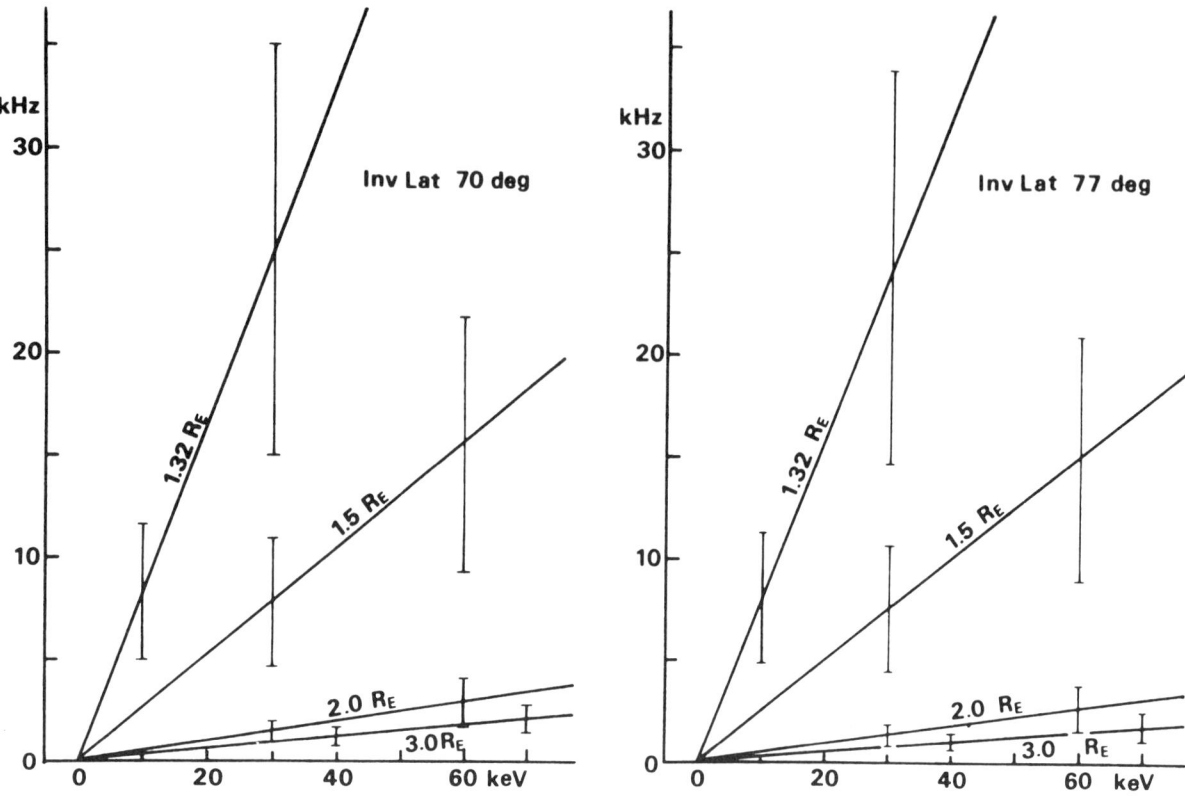

Fig. 10 Relation of electron energy to the upper limit frequency of whistler mode Cherenkov radiation from energetic electrons at geocentric distances of 1.32, 1.5, 2.0 and 3.0 Re on two geomagnetic field lines with invariant latitudes of 70° and 77°. Vertical rods represent the error ranges corresponding to electron density uncertainties of ±40 %.

quency ranges for whistler mode Cherenkov radiation from electrons with energies below 40 keV are below 34 kHz at a geocentric distance of 1.32 Re (altitude about 2000 km), below 12 kHz at 1.5 Re (altitude 3200 km), below 2 kHz at 2.0 Re (altitude 6400 km) and below 1 kHz at 3.0 Re (altitude 13000 km) on both geomagnetic field lines. In other words, the lower frequency components below 2 kHz and 12 kHz of the whistler mode Cherenkov radiation are respectively generated at altitudes below 6400 km and 3200 km in the polar magnetosphere. The high frequency components below 34 kHz are generated at altitudes below 2000 km in the polar topside ionosphere. This result is considerably different from Shawhan's model [1979] in that V-shaped downgoing polar hiss (2 - 30 kHz) is generated by inverted-V electrons (1 - 10 keV) at altitudes of around 4000 km on auroral field lines. It is difficult to obtain experimentally altitude profiles for the upper limit frequency of observed downgoing polar hiss for examining whistler mode Cherenkov radiation. The frequency of the whistler mode Cherenkov radiation means that the frequency of whistler mode Cherenkov radiation depends on the wave normal angle to the geomagnetic field line, that is, its frequency range at each altitude lies from the upper limit frequency at wave normal angle $\theta = 0°$ to zero at $\theta = 90°$. However, as a whole, the good agreement

between the regions of polar hiss occurrence and inverted-V electron precipitation suggests that polar hiss does not deviate largely from the geomagnetic field line. Consequently, the direction of the wave group velocity of polar hiss may be nearly parallel to the field line and the wave normal angle θ of polar hiss may be large when taking into account of the shape of the refractive index surface. The wave normal angle of polar hiss is actually large [Ondoh et al., 1980]. So, the effective generation frequency of whistler mode Cherenkov radiation at each altitude must be lower than the upper limit frequency by the factor $\cos\theta$.

Gurnett et al.[1983] have often observed polar hiss with a characteristic "funnel-shaped" f-t spectrum over the auroral zone. This is very similar to the saucer emission observed in the auroral zone by low-altitude polar orbiting ISIS satellites [James, 1976]. However, the funnel-shaped hiss was observed on a time scale less than 20 minutes (less than 10° in geomagnetic latitude) at altitudes from 15000 km to 20000 km by DE-1 and the VLF saucer was observed on a time scale of less than 10 sec (less than 1° in geomagnetic latitude) at 1000 - 3000 km by ISIS satellites. Both emissions have been explained by a propagation effect which occurs for whistler mode waves with wave vectors near the resonance cone [James, 1976 ; Gurnett et al., 1983], but with different temporal or spatial scales. At the ISIS altitudes, the VLF saucer normally occurs during a short period between curtain-shaped polar hisses with an irregular low frequency cutoff around 2 kHz, but, at the DE-1 altitudes, the funnel-shaped polar hiss normally occurs over the auroral zone. Gurnett et al.[1983] have compared ray path boundaries computed by the whistler mode propagation model with the observed f-t profile of a funnel-shaped polar hiss event. The comparison shows that the funnel-shaped polar hiss is propagating upward from a localized source below the DE-1, and that the best fit radial distance of the low altitude boundary of the upgoing polar hiss source is about 1.7 - 1.8 Re. Lin et al.[1984] found several events in which upward-moving electron beams were associated with funnel-shaped polar hiss. However, studies of attenuation bands in VLF saucers show that their source regions are located above or below the ISIS satellite [Watanabe et al.,1979; Horita and James,1982; Ondoh et al.,1984]. So, we can not rule out the possibility that the funnel-shaped polar hiss is produced by a source generating downgoing waves. The broad band frequency range of the downgoing polar hiss with low frequency cutoff of about 2 kHz seems to be explained by whistler mode Cherenkov radiation from inverted-V electrons with energies up to 50 keV at geocentric distances below 2 Re along polar field lines, since the whistler mode condition for frequencies above 1 kHz is not satisfied at 3.0 Re on the polar field lines. The coherent generation mechanism for polar hiss proposed by Maggs [1976, 1978] requires the electron plasma frequency larger than the electron gyrofrequency to produce strong polar hiss. However, in the auroral plasma cavity above the auroral ionosphere, the electron gyrofrequency is usually larger than the electron plasma frequency. As a result, the proposed coherent generation mechanism for polar hiss is insufficient to explain the strong polar hiss observed in the auroral plasma cavity.

CONCLUSION

A polar map of the occurrence rate for the polar hiss at various geomagnetic activities (Kp = 0 - 7) made from 347 ISIS polar passes is qualitatively similar to that of the occurrence locations at various geomagnetic activities (Kp = 0 - 6) obtained from 280 inverted-V electron precipitation events by AE-D [Hoffman and Lin, 1981], especially, concerning the low-latitude boundary and axial symmetry of the 10-22 hour MLT meridian.

Latitudinal distributions of the occurrence rate for quiet time (Kp = 0 - 1+) polar hiss in 06-14 hour MLT, 14-22 hour MLT and 22-06 hour MLT intervals are statistically obtained from 143 ISIS VLF passes received at Syowa station, Antarctica during January 1980 to January 1983. Regions of the occurrence rate above 0.3 lie from invariant latitude 74° to 84° in 06-14 hour MLT, from 70° to 82° in 14-22 hour MLT and from 70° to 83° in 22-06 hour MLT, respectively, and they are wider than the quiet auroral zone in latitude. In lower latitude side of the peak occurrence rate, the occurrence rate of polar hiss in 06-14 hour MLT decreases faster with latitude than those in 14-22 hour MLT and 22-06 hour MLT. The latitudinal distributions of polar hiss occurrence in 14-22 hour MLT and 22-06 hour MLT are similar to each other, but are considerably different from that in 06-14 hour MLT. Therefore, the polar occurrence distribution of quiet time polar hiss has also the axial symmetry of the 10-22 hour MLT meridian and the low latitude boundary which elongates toward 22 hour MLT. Thus, it is statistically clear that the polar hiss observed in a quiet or disturbed time is closely related to the inverted-V electrons precipitated from the boundary plasma sheet as shown by the simultaneous observation of polar hiss and inverted-V electron precipitation [Gurnett and Frank,1972]. Frequencies of whistler mode Cherenkov radiation generated from inverted-V electrons in the polar magnetosphere are calculated to explain downgoing polar hiss, at geocentric distances of 1.32, 1.5, 2.0 and 3.0 Re for geomagnetic field lines at invariant latitude 70° and 77°, which respectively represent the low latitude boundary and the active occurrence latitude for polar hiss. Low frequency components of the whistler mode Cherenkov radiation below 2 kHz are generated over a wider region at altitudes below 6400 km, and high frequency components above 12 kHz are generated at altitudes below 3200 km. Downgoing polar hiss with frequencies above 2 kHz seems to be explained by whistler mode Cherenkov radiation generated from inverted-V electrons with energies below 40 keV at geocentric distances below about 2 Re in the polar magnetosphere.

Acknowledgments.

The author wishes to express his sincere thanks to many Japanese wintering party members at Syowa station, Antarctica, Space Physics Section and Satellite Data Analysis Section of Radio Research Laboratories, National Institute of Polar Research, Dr. R. E. Barrington of ISIS VLF experimenter, and ISIS Working Group for their collaboration in ISIS spacecraft operation and data acquisition at Syowa station. He is also grateful to French CNES members at Kergulen Island and Terre Adelie Island for their ISIS spacecraft command operations and to Canadian Communications Research Centre and NASA Goddard Space Flight Center for sending him ISIS orbital element data over a long period.

REFERENCES

Akasofu, S. -I., Auroral arcs and auroral potential structures, *Physics of Auroral Arc Formation*, edited by Akasofu, S. -I. and J. R. Kan, American Geophys. Union, 1 - 14, Geophys. Monogr. 25, 1981.

Dowden, R. L., Geomagnetic noise at 230 kc/s, *Nature, 187,* 677, 1960.

Duncan, R. A. and G. R. Ellis, Simultaneous occurrence of subvisual aurorae and radio noise bursts on 4.6 kc/s, *Nature, 183,* 1618, 1959.

Ellis, G. R., Low-frequency radio emission from aurorae, *J. Atmos. Terr. Phys., 10,* 302, 1957.

Gurnett, D. A., A satellite study of VLF hiss, *J. Geophys. Res., 71,* 5599, 1966.

Gurnett, D. A. and L. A. Frank, VLF hiss and related plasma observations in the polar magnetosphere, *J. Geophys. Res., 77,* 172, 1972.

Gurnett, D. A., S. D. Shawhan and R. R. Shaw, Auroral hiss, z mode radiation, and auroral kilometric radiation in the polar magnetosphere : DE-1 observations, *J. Geophys. Res.*, 88, 329, 1983.

Harang, L. and R. Larsen, Radio wave emissions in the v. l. f. band observed near the auroral zone - I ; Occurrence of emissions during disturbances, *J. Atmos. Terr. Phys.*, 27, 481, 1965.

Helliwell, R. A., *Whistlers and Related Ionospheric Phenomena*, Stanford University Press, P. 205, 1965.

Hoffman, R. A. and T. Laaspere, Comparison of very-low-frequency auroral hiss with precipitating low-energy electrons by the use of simultaneous data from two Ogo 4 experiments, *J. Geophys. Res.*, 77, 640, 1972.

Hoffman, R. A. and C. S. Lin, Study of inverted-V auroral precipitation events, *Physics of Auroral Arc Formation*, edited by Akasofu, S. -I. and J. R. Kan, American Geophys. Union, P. 80, Geophys. Monogr., 25, 1981.

Horita, R. E. and H. G. James, Source regions deduced from attenuation bands in VLF saucers, *J. Geophys. Res.*, 87, 9147, 1982.

James, H. G., VLF saucers, *J. Geophys. Res.*, 81, 501, 1976.

Jorgensen, T. S. and E. Ungstrup, Direct observation of correlation between aurorae and hiss in Greenland, *Nature*, 194, 462, 1962.

Lin, C. S. and R. A. Hoffman, Observations of inverted-V electron precipitation, *Space Sci. Rev.*, 33, 415, 1982.

Lin, C. S., J. L. Burch, S. D. Shawhan and D. A. Gurnett, Correlation of auroral hiss and upward electron beams near the polar cusp, *J. Geophys. Res.*, 89, 925, 1984.

Maggs, J. E., Coherent generation of VLF hiss, *J. Geophys. Res.*, 81, 1707, 1976.

Maggs, J. E., Electrostatic noise generated by the auroral electron beam, *J. Geophys. Res.*, 83, 3173, 1978.

Martin, L. H., R. A. Helliwell and K. R. Marks, Association between aurorae and very low frequency hiss observed at Byrd station, Antarctica, *Nature*, 187, 751, 1960.

Morozumi, H. M., Semi-diurnal auroral peak and VLF emissions observed at the south pole, *Trans. American Geophys. Union*, 44, 798, 1963.

Nishida, A., Geomagnetic DP-2 fluctuations and associated magnetospheric phenomena, *J. Geophys. Res.*, 73, 1795, 1968.

Ondoh, T., On the origin of VLF noise in the earth's exosphere, *J. Geomag. Geoelectr.*, 12, 77, 1961.

Ondoh, T., The ionospheric absorption of the VLF emissions at the auroral zone, *J. Geomag. Geoelectr.*, 15, 90, 1963.

Ondoh, T., Y. Nakamura and T. Murakami, Characteristics of auroral VLF emissions observed by ISIS, *J. Radio Res. Labs., Japan*, 27, 141, 1980.

Ondoh, T., Y. Nakamura, S. Watanabe and T. Murakami, Latitudinal variation of chorus frequency observed in the topside ionosphere, *J. Radio Res. Labs., Japan*, 29, 1, 1982.

Ondoh, T., Y. Nakamura, S. Watanabe, K. Aikyo and T. Murakami, Plasmaspheric hiss observed in the topside ionosphere at mid- and low-latitudes, *Planet. Space Sci.*, 31, 411, 1983.

Ondoh, T., S. Watanabe and Y. Nakamura, VLF emissions observed by ISIS satellites during the IMS, *Proc. Conf. Achievements of the IMS*, ESA SP-217, P. 525, 1984.

Ondoh, T., Polar occurrence map of broad-band auroral hiss observed by ISIS satellites, *J. Radio Res. Labs., Japan*, 35, 1, 1988.

Ossakow, S., W. Burke, H. C. Carlson, P. Gary, R. Heelis, M. Keskinen, N. Maynard, C. Meng, E. Szuszczewicz and J. Vickrey, High latitude ionospheric structure, *NASA Reference Publication*, 1120, P. 12. 1, 1984.

Persoon, A. M., D. A. Gurnett and S. D. Shawhan, Polar cap electron densities from DE-1 plasma wave observations, *J. Geophys. Res.*, 88, 10123, 1983.

Persoon, A. M., D. A. Gurnett, W. K. Peterson, J. H. Jr. Waite, J. L. Burch and J. L. Green, Electron density depletions in the nightside auroral zone, *J. Geophys. Res.*, 93, 1871, 1988.

Shawhan, S. D., Magnetospheric Plasma Waves, *Solar System Plasma Physics*, edited by Lanzerotti, L. J., C. F. Kennel and E. H. Parker, North-Holland, P. 211, 1979.

Storey, L. R. O., An investigation of whistling atmospherics, *Phil. Trans. Roy. Soc., London*, Ser. A, 246, 113, 1953.

Vasyliunas, V. M., *The Polar Ionosphere and Magnetospheric Processes*, edited by Skovlii, G., P. 26, Gordon and Breach, 1970.

Watanabe, S., T. Ondoh, Y. Nakamura and T. Murakami, Attenuation band and electric field of VLF saucers, *Antarctic Record*, 64, 159, (in Japanese), 1979.

Driven and Unloading Electrojets During the Main Phase of a Magnetic Storm

K. LASSEN AND E. FRIIS-CHRISTENSEN

Danish Meteorological Institute
Lyngbyvej 100, DK 2100 Copenhagen, Denmark

During the main phase of an ssc type magnetospheric storm the auroral oval is continuously active and displaced towards equator, accompanied by a strong ionospheric electrojet, the intensity of which varies smoothly, roughly in time with the variation of B_Z. A case study of a particular ssc-storm is shown, in which several substorms are observed during the main phase. Data from the meridian chain of magnetometers in Greenland show that substorm activity during this storm was accompanied by auroral electrojets separated from and situated well poleward of the storm electrojet, which continued to exist in the equatorward displaced auroral oval during the substorm activity. Whereas the substorm precipitation is assumed to be connected to the plasma sheet it is proposed that the auroral oval precipitation continuously observed during a magnetic storm originates mainly in the region between the inner boundary of the plasma sheet and the ring current.

INTRODUCTION

During the main phase of a great magnetic storm the auroral oval is continuously occupied by diffuse as well as active discrete auroral forms. The oval is displaced toward the equator. A relationship between the polar distance of the auroras and the intensity of the ring current, represented by the index Dst, has been demonstrated by Feldstein and Starkov [1968]. Magnetic records from observatories at oval latitude show the existence of strong persistent ionospheric electrojets along the oval. It has been shown [Pytte et al., 1978] that such enduring magnetic activity may be due to strong magnetospheric activity driven by a continuously southward interplanetary magnetic field, in contrast to substorm magnetic activity, which is believed to be caused by unloading of energy from the magnetosphere through the so-called 'substorm wedge', a combination of field-aligned currents and ionospheric electrojets [McPherron et al., 1973]. The driven activity is generally identified with the so-called growth phase preceding a substorm break-up and poleward expansion. Wiens and Rostoker [1975] have suggested that the westward expansion of the substorm westward electrojet takes place as a series of discrete steps. Each jump is accompanied by signatures of growth in the adjacent sector to the west.

Clauer and Kamide [1985] have used the KRM computer algorithm to analyze ground magnetic data during a selected CDAW 6 substorm interval and found that the ground magnetic effects could be numerically separated into effects from two simultaneous components of the westward electrojet, one thought to be related to enhanced convection associated with the DP2 current system and another component apparently related to the expansion phase westward electrojet, the DP1 current system. It is the purpose of this paper to demonstrate the validity of this distinction during the prolonged activity of a magnetic storm and further that the DP1 component consists of several, apparently independent electrojets, not necessarily behaving according to the scheme presented by Wiens and Rostoker.

THE MAGNETIC STORM OF JAN. 9-10, 1983

A great magnetospheric storm was initiated by a sudden commencement at 1545 UT on January 9, 1983 (Figure 1).

During the following nine hours the variations of Dst indicate injection of particles into the ring current which are associated with variations in B_Z according to Burton et al., 1975. Finally, near 01 UT on Jan. 10 a sudden large negative change of B_Z initiated a strong injection of particles resulting in a great increase of the ring current through the following seven hours.

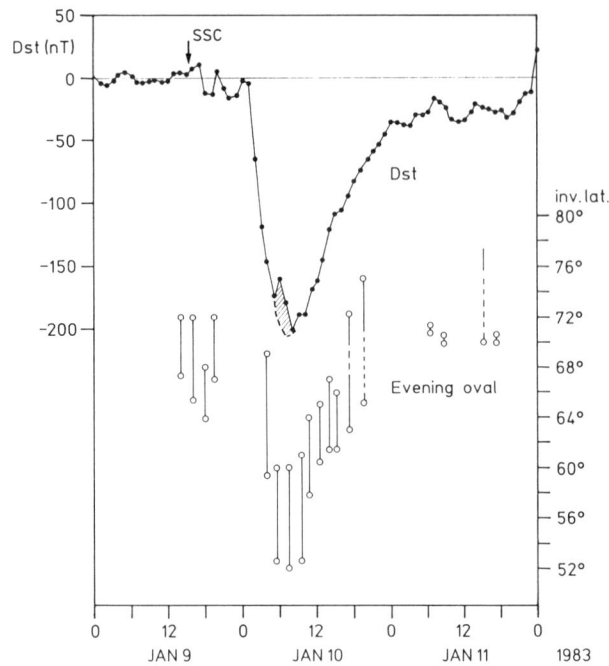

Fig. 1. Variation of Dst and the position of the evening auroral oval near 18 MLT as observed by DMSP during the magnetic storm Jan. 9-11, 1983.

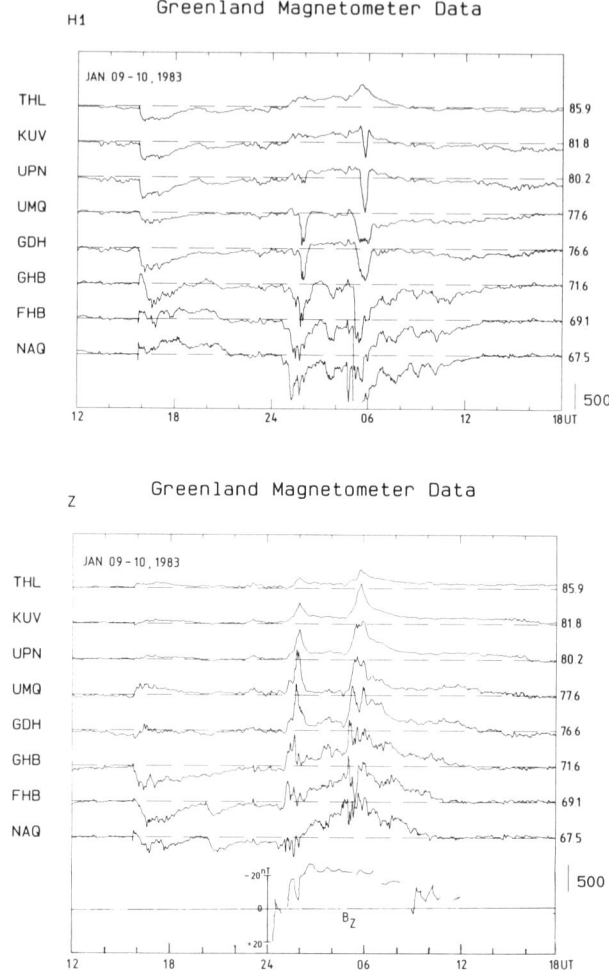

Fig. 2 Stacked magnetograms from the Greenland magnetometer chain Jan. 9-10, 1983. Also shown is the variation of B_Z as observed by IMP-8.

The maximum was reached at 06 UT on Jan. 10 with Dst -200 nT. B_Y was strongly negative; the maximum Kp value was 8+ at 03-06 UT. With a gradual decrease of the negative B_Z after 09 UT and a subsequent shift to positive B_Z values near 13 UT the injection seems to have become insufficient to maintain the ring current, and the recovery phase of the storm became the dominant feature.

From the onset of the storm the auroral oval was filled with active auroral forms, as judged from the imager on the DMSP F-6 satellite. In Figure 1 it is demonstrated how the (evening side) of the oval shifted toward lower latitude in the course of the storm. Whereas the position of the oval in general changes with Dst [Akasofu and Chapman, 1963; Feldstein and Starkov, 1968] it is interesting to notice that the oval reaches its equatormost position at a time when the Dst-variation undergoes a positive inflection. This positive excursion of Dst, shown hatched between 4 and 7 UT in Figure 1, is caused by a large substorm, which gave rise to an increase in H at low-latitude stations and hence in the value of Dst. It appears that the latitude of the oval in this case would be a better signature of the strength of the ring current than Dst.

THE QUASI-PERMANENT STORM ELECTROJET

Co-located with the auroral oval was a strong westward electrojet. In the magnetograms from the Greenland magnetometer chain shown in Figure 2 the effect of this electrojet is seen as a strong perturbation, negative in H and positive in Z, persisting during the hours when B_Z was negative. This perturbation is particularly evident in the Z-component at the southernmost stations of the magnetometer chain (e.g. NAQ and FBH). The variation in H is obvious too, but less clear due to the greater influence from superposed substorm activity. Presumably, the reason for this is that the substorm electrojets are located near the magnetometer stations, while the longer-durating oval electrojet is co-located with the visible auroras equatorward of the stations (Figure 1). The persistent storm electrojet is part of a pattern of increased convection. Based on magnetic records from a number of high-latitude observatories (included in Table 1) this pattern has been constructed for a situation in which substorm perturbation was absent. In Figure 3 the convection is represented by the equivalent current vectors at the nearly substorm-free hour 0415 UT. As might be expected from the large negative values of both B_Z and B_Y the morning vortex dominates, and the presence of an intense westward auroral electrojet is indicated at relatively low

TABLE 1. Location of Stations

Station	Sym	Geog N	Geog E	INVL
Thule	THL	77.5	291	86.1
Kullorsuaq	KUV	74.6	303	81.8
Upernavik	UPN	72.8	304	80.2
Umanaq	UMQ	70.7	308	77.6
Godhavn	GDH	69.2	306	76.6
Godthåb	GHB	64.2	308	71.6
Frederikshåb	FHB	62.0	310	69.1
Narsarsuaq	NAQ	61.2	315	67.5
Nord	NRD	81.6	343	80.8
Danmarkshavn	DMH	76.8	341	77.4
Daneborg	DNB	74.3	340	75.4
Scoresbysund	SCO	70.5	338	72.2
Leirvogur	LRV	64.2	338	65.9
New Ålesund	NAL	78.9	12	75.9
Bear Island	BJN	74.5	19	71.3
Arkhangelsk	ARK	64.6	40	60.5
Dixon Isl.	DIK	73.6	80	67.9
C. Chelyuskin	CCS	77.7	104	71.3
Tixie Bay	TIK	71.6	129	65.2
C. Wellen	CWE	66.2	190	62.5
College	COL	64.9	212	64.9
Mould Bay	MBC	76.2	241	80.8
Yellowknife	YKC	62.4	246	69.8
Meanook	MEA	54.6	247	62.5
Cambridge Bay	CBB	69.1	255	77.6
Baker Lake	BLC	64.3	264	74.6
Resolute Bay	RES	74.7	265	83.9
Fort Churchill	FCC	58.8	266	69.7
Great Whale R.	GWC	55.3	282	66.9
Ottawa	OTT	45.4	284	57.5
Brorfelde	BRF	55.6	12	52.8
Askhabad	ASH	38.0	58	32.2
Tucson	TUC	32.2	249	40.1
Fredericksburg	FRD	38.2	283	50.8
San Juan	SJG	18.1	294	33.0

The position of the very active auroral oval as observed by DMSP at about 04 UT is shown shaded. Unfortunately, the brilliancy and activity of the aurora do not allow a clear distinction between discrete and diffuse forms, although the discrete forms seem to be concentrated near the poleward edge of the oval. The change of sign of the vertical component of the magnetic perturbation indicating the latitude where the electrojet is overhead, appears to be located in the northern part of the oval.

The convection pattern in Figure 3 is assumed to be characteristic for the smoothly varying part of the magnetic perturbation. Since this perturbation appears to follow the variation of B_Z (Figure 2) we assume that it is the signature of a process which is driven by the solar wind (cross-tail) electric field [Pytte et al., 1978]. In the course of the perturbation the pattern is supposed to expand and contract concurrently with the latitudinal shift of the oval (Figure 1), while the intensity of the electrojet is assumed to increase and decrease according to the B_Z-variation.

THE SUBSTORM ELECTROJETS

Superposed on the smooth magnetic perturbation are a number of magnetic substorms expansions of the type usually associated

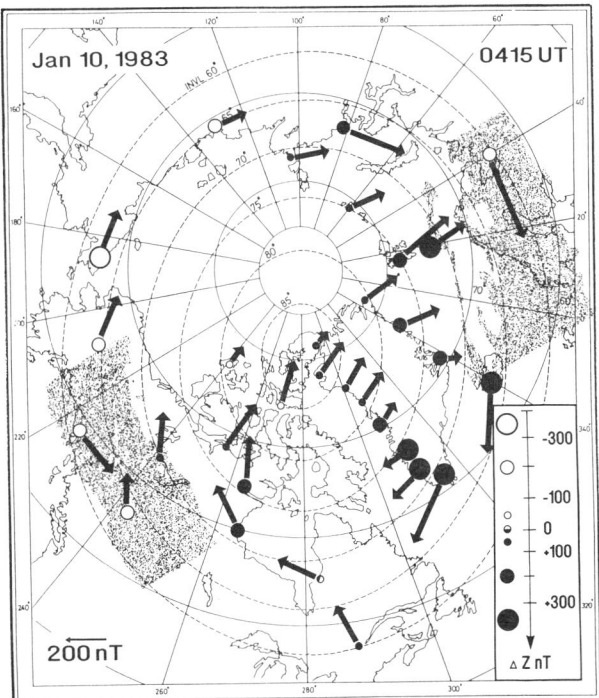

Fig. 3 Convection pattern 0415 UT, Jan. 10, 1983. The convection is represented by the equivalent current vectors derived from magnetometer observations. The magnitude of the vertical component of the magnetic perturbation is indicated by the radius of the open (for negative Z) and filled (for positive Z) circles. Dark areas indicate position of the very active auroral oval as observed by DMSP at about 04 UT.

latitude across Scandinavia and Canada, well to the south of Iceland and Greenland.

The unusual extension of the morning vortex as well as of the westward electrojet is confirmed by the fact that the auroral electrojet index AU, which is expected to describe the most positive horizontal disturbance of the magnetic field at selected auroral zone stations caused by the eastward electrojet, was predominantly negative between 03 and 08 UT. During these hours the disturbance in H at the AE-chain of stations was dominated by the longitudinally extended westward electrojet, while the eastward electrojet is supposed to have been situated far to the south of these stations.

with poleward expansion of brilliant auroras. The most active interval was between 0430 and 0700 UT.

It has been demonstrated [Akasofu and Meng, 1969; McPherron, 1973] that the expansion phase of a magnetic substorm is accompanied by a positive bay in H at low-latitude stations situated in the midnight sector. During this particular substorm San Juan was located close to magnetic midnight when the disturbance set in. The variation of H and D at this locality is shown in Figure 4. The overall deviation during the interval from 04 to 08 UT is positive in H as well as in D which indicates that this station was not located in the central location of the substorm wedge. The large substorm is preceded by several minor disturbances. We have chosen the beginning of the large increase near 0440 UT as the onset of the substorm.

The disturbance is initiated by a rapid increase in H which from about 0458 continues more rapidly to a maximum near 0530. Then a gradual decrease takes place during the following two hours. However, a closer inspection reveals that the disturbance may be subdivided into several events. There is a regular increase in H between the onset at 0440 and 0458, when a more abrupt increase sets in. This increase continues to a maximum near 0515. Shortly after the passage of this maximum a new increase occurs, reaching a maximum near 0530, which is also the maximum of the whole event. During the declining phase secondary maxima are observed near 0540, 0610, and 0630.

The variation at San Juan is assumed to have been caused by the combined effect of distant field-aligned currents and the associated ionospheric electrojet(s), which together constitute the substorm wedge [McPherron et al., 1973]. In this case, it is expected to depict the variation of the total current intensity in the wedge. High-latitude observatories located near the ionospheric part of the substorm wedge show a more irregular variation. The magnetic field at such stations is influenced not only by the varying intensity, but also, especially in the Z-component, by the position of the substorm electrojet as well as by irregular changes of the number of parallel conductors forming part of the instantaneous current flow.

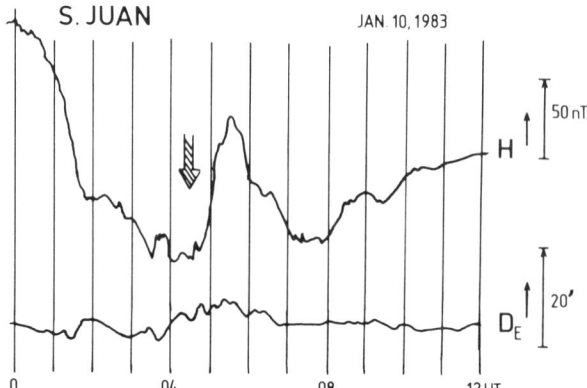

Fig. 4 Variation of the geomagnetic H and D components at San Juan 00-12 UT, Jan. 10, 1983. Mn indicates local magnetic midnight.

The complexity of the magnetic substorm variations at high-latitude stations may be seen in Figure 2. Part of this figure, covering the interval 04 - 08 UT, is reproduced in Figure 5. At the southernmost station of the chain (NAQ) the variation of H reflects similar occurrence of events as San Juan. Actually, the shape of the H-trace indicates intensifications of the current system, each of which may be directly associated with characteristic features of the San Juan record. But whereas the absolute maximum of the disturbance in H is observed near 0530 at the low-latitude station, the maximum at NAQ occurs already in connection with the 0515 event. With increasing latitude of the chain stations the maximum becomes progressively delayed, thus indicating a poleward shift of the ionospheric part of the substorm wedge.

With the assumption that the magnetic disturbances are caused by infinite oval-aligned westward horizontal electrojets we can use the Z-variations in Figure 5 to estimate the position of such electrojets that may have caused the recorded perturbation. Ideally, the vertical component of the disturbance should be negative (positive) at a station to the south (north) of the electrojet and zero at a station situated beneath the center of the current. Using this criterion we have shown schematically in Figure 5 how the position of the currents may have varied through the substorm. A horizontal bar between two magnetograms indicates that the current is located in the interval between the corresponding stations, though not necessarily at a fixed latitude. Vertical lines connecting the bars indicate passage of the current from one interval to the next following one. In the schematic a current was initiated around 0440 UT to the south of NAQ. Shortly after 0500 UT it passed NAQ in a poleward motion, which may have continued, monotonicly or in steps corresponding to the subdivision into single events at San Juan, till 76° or possibly right up to 80°. Behind the leading current other filaments appear to have lit up intermittently, so that the pattern may include 3-4 parallel conductors in addition to the persistent electrojet.

The schematic in Figure 5 is in qualitative agreement with the observations in the sense that it may account for the sign of the perturbations. To present a model which is also in quantitative agreement with the observations, we have modeled the horizontal west-east directed currents using the latitude profiles of the H- and Z-components along the West Greenland coast. The latitude profiles were calculated every two minutes and infinitely long west-east directed line current filaments at an altitude of 110 km were assumed to account for the magnetic perturbations. Although the southernmost station of Greenland (NAQ) is located at 67°.5 the use of latitude profiles for both the H- and Z-components allows some estimation of the east-west directed currents also south of the line of stations. The modeled current intensities as a function of UT and invariant latitude have been plotted as a contour plot in Figure 6.

The contour plot confirms the interpretation based on the collection of magnetograms regarding the position and motion of the ionospheric current filaments. A strong westward current is located well south of the stations in Greenland as indicated by the strong positive vertical perturbation throughout the interval. At the first break-up around 0440 to the South of NAQ a westward

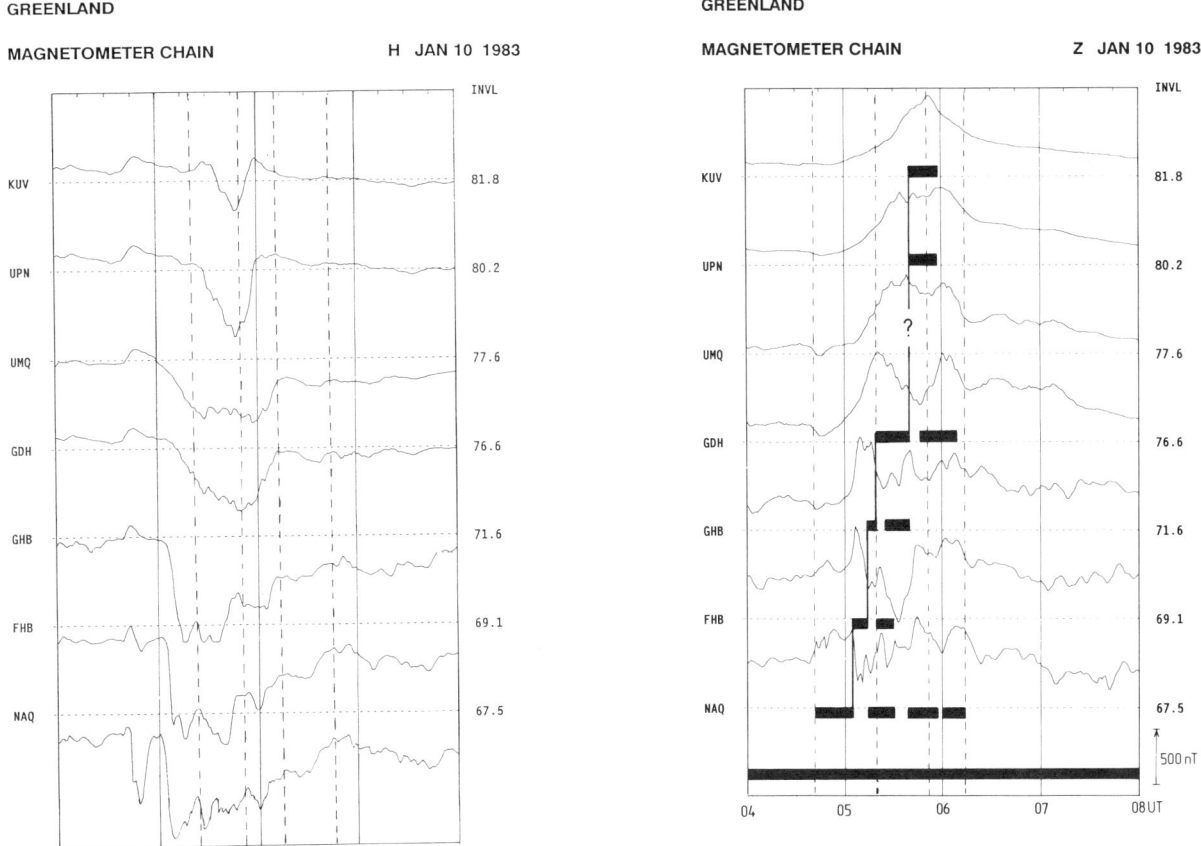

Fig. 5 Stacked magnetograms from the West Greenland magnetometer chain 04-08 UT, Jan. 10, 1983.

current filament is created which lasts for about 10 minutes. This expansion is followed by a larger substorm expansion starting a little South of NAQ at 0500 UT. This expansion is associated with an electrojet moving poleward to about 70° where it seems to settle and remain at the same latitude till around 0540 UT. At the same time, however, a second electrojet filament has started near 76° at 0520 UT. Also this current is seen to exist till after 0600 UT. Finally a local minimum in the west-east directed current intensity indicates the existence of a short-lived (10-15 minute) line current at such high latitudes as 79°.

The contour plot illustrates that both the poleward expansion and the subsequent retreat is composed by a series of intermittently activated current filaments. A different interpretation of the poleward shift could be that the jumps in the plot are artificial, resulting from incomplete coverage of magnetometer stations, in particular between 72° and 76°. Whether the jumps are real or not, the model in Figure 6 confirms our conclusion that several parallel conductors may be present simultaneously in the widened auroral bulge during the expansive phase of a substorm.

The intrusion of substorm poleward expansions over Greenland was observed directly with all-sky cameras at Danmarkshavn (77° inv.lat.) and Thule (86° inv.lat.) several times during the storm. During the particular substorm 04 - 07 UT the poleward boundary was seen from Danmarkshavn over latitude 76°.0 at 0526 and over 77°.0 at 0532 UT. The poleward movement continued several degrees beyond 77° From Thule the poleward boundary was sighted a few degrees above the southeastern horizon corresponding to a maximum intrusion to about 79°, or about twenty degrees poleward of the instantaneous oval. The auroral activity at Thule was highest at 0530 to 0600 UT. A faint arc which may have been related to the expanding bulge was visible in a direction parallel to the front as far north as 86° between 0548 and 0624 UT. The poleward shift of the leading electrojet appears to have followed the front of the visual auroral bulge, either continuously or in steps.

Until now the discussion of the configuration of the ionospheric electrojets during the substorm has been confined to the meridian of West Greenland, which was the area best covered by magnetometer stations. However, it would be desirable to extend the mapping to other meridians by including records from other relevant high-latitude observatories, although their number and distribution is insufficient to give but an approximate position of the currents. The observatories used in this connection have been listed in Table 1, together with the stations of the Greenland magnetometer chain.

The position of the electrojet was sketched qualitatively in Figure 7a-d at selected times from the direction of the horizontal and the sign of the vertical components of the disturbance field.

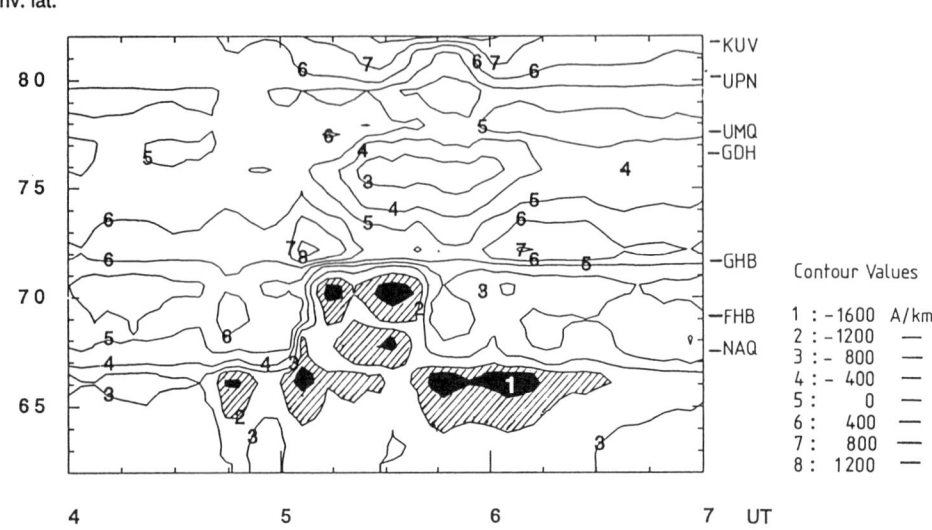

Fig. 6 Contour plot showing the current intensity of modeled west-east current filaments along the West Greenland coast 04-07 UT, Jan. 10, 1983. The contour line labelled '5' represents zero current intensity. Lower numbers represent westward current.

The onset of the disturbance near 0440 was recorded simultaneously, i.e. within a few minutes, over a large range of longitudes, at least between the meridian of Scandinavia (New Aalesund, M'Bour) and Western Canada (Yellowknife). The small differences in onset time are unsystematic and easily explained as uncertainty in timing from the available record. As shown in Figure 7a the electrojet derived for the first disturbance 0440 - 0458 was observed along the whole night side from dawn to dusk. Simultaneously with the occurrence of the electrojet the all-sky camera at Danmarkshavn recorded increased intensity of an auroral arc over the south-eastern horizon. Assuming a height of the arc of 100 km its calculated position (indicated by a dotted line in the figure) was close to the estimated position of the electrojet. A comparison with Figures 1 and 3 indicates that the onset took place less than a few degrees poleward of the quasi-permanent driven electrojet.

Figure 7b shows the estimated position of the electrojet together with the horizontal disturbance vectors at 0515 when the activity according to Figures 5 and 6 had shifted polewards to 70°. In agreement with the model of Kisabeth [Wiens and Rostoker, 1975, Figure 7] the vectors near the eastern end of the electrojet show a tendency to a clockwise rotation pattern, corresponding to the effect of the downward current in the substorm wedge. The associated upward current is apparently situated beyond Yellowknife, somewhere to the north of Alaska.

At 0530 (Figure 7c) the southernmost current in Figure 7b is shifted a few degrees poleward. The filament near 70° remains in its position, while a new current has lit up near 76°.

The northernmost current appears to be situated just behind the front of the expanding auroral bulge, part of which is seen in the DMSP-image from 0533. A sketch of the satellite image as well as of the position of the front of the aurora as observed from Danmarkshavn have been included in Figure 7c.

Figure 7d shows the distribution at 0545, near the last of the major maxima prior to the recovery. The system of electrojets has now reached its most poleward position near 79°, which is also the latitude of the auroral front estimated from all-sky camera observations at Thule. In accordance with Figure 6 we have added to the currents in Figure 7c a parallel current filament near 79° poleward of the 0530 pattern. As in Figure 7b horizontal magnetic disturbance vectors indicate occurrence of field-aligned current over the dawn- respectively dusk ends of the main electrojet.

The horizontal vectors poleward of the electrojet in Figure 7b indicate a shift between clockwise and anticlockwise direction near 290° geographic longitude (Baffin Land). Hence, this longitude, which corresponds to 1 hr Local Magnetic Time, appears to mark the center of the substorm wedge. The additional current filaments seem to be situated at later magnetic time, roughly from magnetic midnight to dawn. Although this impression may in part be due to the scarcity of stations in northwestern Canada, the observations do not support the existence of electrojets of considerable intensity in the dusk-midnight sector.

Simultaneously with the eastward shift of the center of the ionospheric currents there appears to be a shift in the same direction of the footprints of the field-aligned currents. This is concluded from the change of direction of the horizontal vectors at five low-latitude stations (Figure 8). Whereas San Juan, which is located close to the central line of the wedge and far from the electrojets, maintains a horizontal disturbance in the direction of H, the vectors at TUC and FBD, both to the west of the center, turn clockwise as would be expected, if the upward current

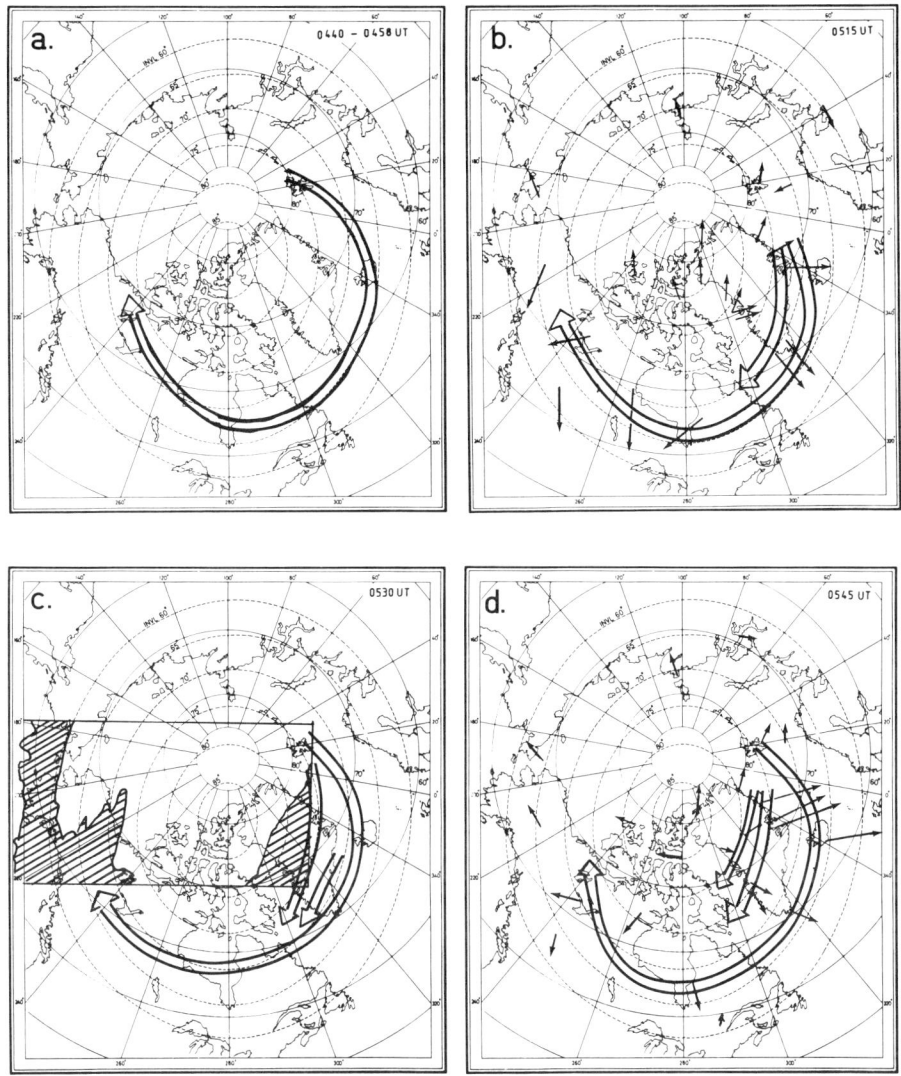

Fig. 7 Distribution of electrojet filaments on Jan. 10, 1983 at 0440-58, 0515, 0530, and 0545 UT.

approached the stations from west. Similarly, BRF and ASH turn clockwise, corresponding to a passage of the downward current toward east.

SUMMARY AND DISCUSSION

The magnetospheric substorm is believed to consist of two basic processes, the 'driven' component and the 'loading-unloading' part [Rostoker et al., 1987]. The 'driven' component is related to the global magnetospheric convection directly correlated with the solar wind energy input. During this process convection currents are created in the auroral ionosphere and particle energy is deposited in the ionosphere and in the ring current. In case of enhanced solar wind energy input due to a southward B_Z component a considerable amount of the energy is stored in the Earth's magnetotail. This surplus of energy is then later released during an explosive process, the so-called 'unloading' process, also called the expansion phase of the substorm. One of the major questions regarding the magnetospheric substorm is the relative importance of these two energy dissipation processes. The current systems related to the magnetospheric substorm contain two basic types, the DP1 system related to the substorm current wedge and therefore to the 'unloading' process, and the DP2 system which is believed to be associated with the 'driven' process. Clauer and Kamide [1985] showed that these two current systems may coexist during periods of continued energy input from the solar wind. The substorm presented in this paper takes place during unusually active conditions, caused by a prolonged period of a large southward component of the IMF. This condition is favourable for both processes to occur and the

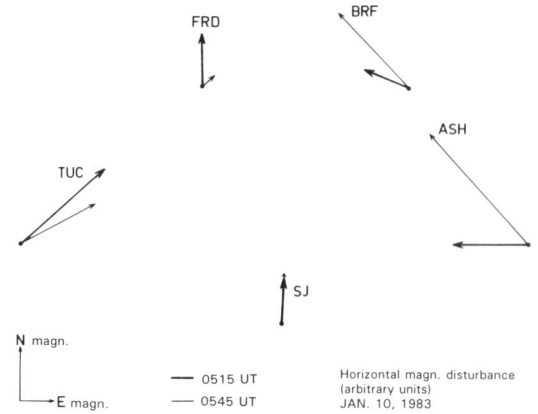

Fig. 8 Horizontal magnetic disturbance at selected observatories at low and medium latitude on Jan. 10, 1983, 0515 and 0545 UT.

observations in fact show that it was possible to clearly distinguish between the effects from the two basic concurrent substorm processes.

Beginning with the SSC there was persistent great activity in the auroral oval accompanied by increased two-cell convection over the polar cap. The main phase of the storm, which is characterized by increasing ring current intensity (Dst) and equatorward shift of the position of the auroral oval, set in shortly after midnight as B_Z went negative to below -20 nT and continued through the whole 10-12 hr interval where B_Z was negative. During this phase an intense westward electrojet was maintained in all local time sectors of the oval as a result of intensified two-cell convection and increased precipitation of auroral particles (i.e. increased conductivity). With the return of B_Z to low numerical values this 'driven' electrojet current faded out, and the recovery phase of the storm began. During the first 12 hr of this phase both the ring current and a broad belt of mainly diffuse auroral precipitation were still observed. Concurrently with the gradual decay of the ring current the oval slowly contracted towards its normal undisturbed position.

The persistence of a broad diffuse precipitation of auroral particles at low latitude several hours after the termination of the 'driven' processes seem to indicate that the precipitating particles originate from a population of particles trapped in connection with the injection into the inner magnetosphere during the early hours of the storm. Obviously, this interpretation would be consistent with the conclusion by Galperin and Feldstein [1990] that the diffuse auroral maps to the region between the inner edge of the plasma sheet and the ring current, which they call the 'remnant layer'.

Several substorm expansions occurred during the storm. During the poleward expansions the substorm electrojets moved with the auroral bulge from oval latitude toward the pole, while the strong persistent electrojet and active oval auroras remained in their position in lower latitude.

A particular series of substorm expansions has been analyzed in detail using data from the meridian magnetometer chain along the West Greenland coast together with other polar cap and auroral zone magnetic observations. From the magnetometer observations it is concluded that the first substorm expansion phase starts around 0440 UT. This onset takes place at latitudes close to the center of the persistent electrojet. According to Galperin and Feldstein [1990] a 'typical' substorm starts in the tail at radial distances less than 10 R_E and just near the trapping boundary. At auroral altitudes this corresponds to the border between the diffuse and the discrete aurora near the midnight sector. The magnetic and optical observations in the present case are consistent with this view.

Due to lack of observations in the Canadian sector poleward of Great Whale River, the first expansion can not be observed in details. In the southern part of Greenland it does not expand further north than NAQ. A second substorm expansion, however, takes place very close to Greenland at 0500 UT. From the contour plot of the current intensity it looks as if the expansion moves poleward associated with intermittent intensifications of the electrojet. The results indicate the continued existence of a westward electrojet at the original latitude concurrent with a superposed expansion electrojet which is moving poleward, associated with intermittent intensifications. The expansion is thus not a mere broadening of the quasi-permanent westward electrojet region caused by enhanced precipitation and hence enhanced ionospheric conductivity. Rather the results indicate that the poleward motion of singular current filaments takes place in conjunction with the creation of different current filaments at higher latitudes.

New current filaments are created immediately behind the poleward boundary of the expanding optical auroral bulge. This is particularly marked near the maximum of the expansion, where two new filaments lit up several degrees poleward of the original ones and stayed in this position during the withdrawal of these. Almost simultaneously an isolated faint auroral arc which could be a signature of similar precipitation as, but of lower energy and density than the precipitation associated with the current filaments, appeared near 86°, directed parallel to the front of the auroral bulge.

Apparently, the appearance of new electrojets poleward of the original ones at the transition from the expansion to the recovery phase would tend to support the idea of a poleward leap [Hones et al., 1973]. On the other hand, we find it difficult to see, how the persistence of substorm electrojets at lower latitude after the poleward leap could fit into the model of Hones et al..

Our interpretation of the magnetograms in terms of stepwise penetration of current filaments toward higher latitude is consistent with the results of Wiens and Rostoker [1975], although the intermittent occurrence of filaments equatorward of the leading ones was not reported by them. However, our observation of a shift of the whole substorm wedge in an easterly direction is in apparent disagreement with their result. One explanation of the shift could be that the appearance of additional currents mainly after midnight influence the position of the center of the integrated disturbance. Another possibility is that the westward jumps reported by Wiens and Rostoker are restricted to the area in the evening sector where the westward travelling surge is located. This possibility seems to be supported by observations by Craven et al. [1988].

References

Akasofu, S.-I, and Chapman, S. The lower limit of latitude (US sector) of northern quiet auroral arcs, and its relation to Dst (H), *J. Atmos. Terr. Phys.*, 25, 9-12, 1963.

Akasofu, S.-I and C.-I. Meng, A study of polar magnetic substorms, *J. Geophys. Res.*, 74, 293-313, 1969.

Burton, R.K., R.L. McPherron, and C.T. Russell, An empirical relationship between interplanetary conditions and Dst, *J. Geophys. Res.*, 80, 4204-4212, 1975.

Clauer, C. R., and Y. Kamide, DP1 and DP2 current systems for the March 22, 1979 substorms, *J. Geophys. Res.*, 90, 1343, 1985.

Craven, J.D., L.A. Frank, and S.-I. Akasofu, Propagation of a westward travelling surge and the development of persistent auroral features, *J. Geophys. Res.*, 94, 6961-6967, 1989.

Feldstein, Y.I. and G.V. Starkov, Auroral oval in the IGY and IQSY period and a ring current in the magnetosphere, *Planet. Space Sci.*, 16, 129-133, 1968.

Galperin, Yu. I. and Ya. I. Feldstein, Auroral luminosity and its relationship to magnetospheric plasma domains, in *Auroral Physics*, edited by C.I. Meng, M.J. Rycroft, L.A.Frank, pp. 207-219, Cambridge University Press, Cambridge, England, 1991.

Hones, E.W. Jr., J.R. Asbridge, S.J. Bame, and S. Singer, Substorm variations of the magnetotail plasma sheet from X_{SM} -6R_E to X_{SM} -60R_E, *J. Geophys. Res.*, 78, 109-132, 1973.

McPherron, R.L., Satellite studies of magnetospheric substorms on August 15, 1968, *J. Geophys. Res.*, 78, 3044-3053, 1973.

McPherron, R.L., C.T. Russell, and M.P. Aubry, Satellite studies of magnetospheric substorms on August 15, 1968, *J. Geophys. Res.*, 78, 3131-3149, 1973.

Pytte, T., R.L. McPherron, E.W. Hones, and H.I. West, Multiple-satellite studies of magnetospheric substorms: distinction between polar magnetic substorms and convection-driven negative bays, *J. Geophys. Res.*, 83, 663-679, 1978.

Rostoker, G., S.-I. Akasofu, W. Baumjohann, Y. Kamide, and R.L. McPherron, The Roles of direct input of energy from the solar wind and unloading of stored magnetotail energy in driving magnetospheric substorms, *Space Sci. Rev.*, 46, 93-111, 1987.

Wiens, R.G. and G. Rostoker, Characteristics of the development of the westward electrojet during the expansive phase of magnetospheric substorms, *J. Geophys. Res.*, 80, 2109-2128, 1975.

OCCURRENCE OF MAGNETOSPHERIC FLUX TRANSFER EVENTS DURING SUBSTORM

Hideaki Kawano and Susumu Kokubun

Department of Earth and Planetary Physics, Faculty of Science, University of Tokyo, Bunkyo-ku, Tokyo 113, Japan

Abstract. Transient variations in the magnetic field and plasma parameters, called flux transfer events (FTEs), have been detected in the magnetosheath and the magnetosphere near the magnetopause. FTEs have been interpreted to be generated by transient and patchy reconnection.

In this report, four series of recurring transient magnetic field variations identifiable as FTEs, observed with AMPTE/CCE in the outer magnetosphere, are studied in relation to substorm activities. Since AMPTE/CCE resides in the region of 8<L<9 for about 6 hours on each orbit, the spacecraft can continuously monitor the time-sequence of the events occurring near the magnetopause.

It is found that the transient variations are observed during the substorm growth phase and continue to occur during the substorm expansion phase until the spacecraft enters the magnetosheath. This time-sequence can be interpreted as follows: the southward turning of the interplanetary magnetic field starts the dayside merging, which generates FTEs near the dayside magnetosphere and initiates the growth phase of substorm in the magnetosphere.

Introduction

Flux transfer events (FTEs) are phenomena occurring in the magnetosheath and the magnetosphere near the dayside magnetopause. FTEs, which were first identified in the magnetic field data of ISEE spacecraft by Russell and Elphic [1978], are characterized by their bipolar perturbation in the magnetic field component normal to the magnetopause. Russell and Elphic [1978] interpreted that the FTEs were observed when the spacecraft encountered open flux tubes generated by patchy and transient reconnection. The bipolar variation of the boundary normal component of the magnetic field is explained in terms of the passage of the flux tube structure which includes "draping" of the external field and the field aligned current inside the flux tube [e.g., Russell and Elphic., 1978; Paschmann et al., 1982; Cowley, 1982; Saunders et al., 1984]. The patchy and transient reconnection model also explains many of plasma properties of FTEs in terms of the mixing of magnetosheath plasma and magnetospheric plasma in the reconnected tubes [Daly et al., 1981; Daly and Keppler, 1982; Paschmann et al., 1982; Scholer et al., 1982]. Moreover, the results of occurrence statistics for FTEs [Berchem et al., 1984; Rijnbeek et al., 1984; Russell et al., 1985; Southwood et al., 1986] are reported to be consistent with the idea of transient and patchy reconnection. On the basis of the concept of transient reconnection, FTEs are proposed to play a significant role in eroding the dayside magnetic flux [Rijnbeek et al., 1984].

In spite of general success in explaining the observed feature of FTEs, the transient and patchy reconnection process may not be the sole possibility for the cause of the phenomenon. Sibeck et al. [1989] and Sibeck [1990] have suggested that the signatures identifiable as FTEs can be generated by solar wind pressure pulses. They interpreted the bipolar magnetic field perturbation in terms of the passage of the troughs (in the magnetosphere) or the bulges (in the magnetosheath) generated by compressional waves. Sibeck [1990] also interpreted the plasma properties characteristic of FTEs as a result of the spacecraft entry into the quasi-static mixing layers of different plasmas, which can be identified as the low latitude boundary layer and the plasma depletion layer. In one sequence of pressure pulses studied by Sibeck et al. [1989], transient magnetic field variations in the magnetosphere were well correlated with pressure variations in the solar wind, although not all of the magnetic field variations showed clear solitary bipolar perturbation identifiable as FTEs. Then the identification of the generation mechanism is an important issue in studying the transient magnetic field variations.

The purpose of this paper is to examine the recurring transient events which exhibit clear magnetic field signatures identifiable as FTEs (i.e., isolated bipolar perturbation in the component normal to the magnetopause) in the AMPTE/CCE magnetic field data. We will call them transient magnetic field events (TMFEs) here, in order to avoid their a priori association with a specific physical process. In this paper we present four cases of such recurring TMFEs and show that they are well related to the development of substorms. Since the CCE spacecraft resides at 8<L<9 for about 6 hours on each orbit, the spacecraft can be considered as a stationary observatory for these events occurring near the magnetopause.

Data

We mainly use AMPTE/CCE magnetic field data [Potemra et al., 1985] in this study. The inclination of the orbit of AMPTE/CCE was 4.8° in geographic coordinate system when it was launched in August, 1984. The magnetometer on board AMPTE/CCE measures vector magnetic field every 0.124-sec. 6.2-sec median samples are used in order to identify the occurrence of TMFEs. In this paper the CCE magnetometer data are expressed in the V-D-H coordinate system, where H is parallel to the geomagnetic dipole axis and points northward, D is perpendicular to the dipole meridian plane of the spacecraft and points eastward, and V completes the (V,D,H) triad. We identify TMFEs according to the following features: (1) A bimodal pulse in the V component of the magnetic field and an one-sided pulse in the H component are clearly observed. (2) The peak of the H component pulse appears roughly at the center between the maximum and minimum of the bipolar pulse in the V component. Note that we make no limitation for amplitude of the fluctuation; events with very small amplitudes are also identified as TMFEs as long as they fill the above criteria.

We also utilize simultaneous observations of the magnetic field at GOES 5 and 6 geostationary satellites. The geographical longitudes of GOES 5 and 6 are roughly $-75°$ and $-108°$, respectively. The geomagnetic latitudes of GOES 5 and 6 are roughly $11°$ and $9°$, respectively. The magnetometers on board GOES sample the magnetic field every 0.75-sec, and 3.06-sec averages are used in this study.

We also use magnetic field and plasma data from IMP 8 to monitor the interplanetary medium. The spacecraft had an apogee of $\sim 40 R_E$ for the period examined. The magnetic field vector is expressed in GSM coordinate system.

In order to monitor the time-sequence of substorms, 1-min values of the AE(12) indices (AU and AL) are mainly used. Low-latitude ground magnetogram data are also used to identify the onset time of the expansion phase.

Observation

In the survey of two year (1984−1986) data from AMPTE/CCE we found four series of TMFEs which were observed to recur with recurrence period of less than 20-min for 2−4 hours. After selection of these events we noted that they were associated with the enhancement of ground magnetic activities. In this section we present the four cases of recurring TMFEs respectively. We discuss two of the four cases (the first and last case below) in detail, and the others are briefly discussed because they have essentially the same features as the first case. In former three cases the magnetopause-crossings took place, and their relation to the time-sequence of substorm will form an interesting topic. However in this paper we concentrate on the occurrence of the recurrent TMFEs in the magnetosphere.

August 27, 1984

At first we briefly exhibit the existence of the recurrent TMFEs. Figure 1 shows the magnetic field data from AMPTE/CCE for the first case of recurrent TMFEs. AU and AL indices are also shown in the figure. Clear TMFEs were observed in the period of 1038∼1106 UT, marked by vertical lines and letters A-H in the figure. We will examine this figure in detail later.

Figure 2 shows the plasma and magnetic field data simultaneously observed with IMP 8. In the period shown in the figure the IMP 8 was situated in the interplanetary space. In the interval from ~ 0900 to ~ 1040 UT, the solar wind flowed nearly along the sun-earth line (see the graph labeled AGL). It is also noted that density, speed and pressure of the solar wind plasma were nearly constant in this period ($\sim 17/cm^3$, $\sim 390 km/sec$ and $\sim 5.0 \times 10^{-9} N/m^2$, respectively). On the other hand, the IMF condition was not constant in this period. That is, the IMF turned southward at ~ 1010 UT (see the graph labeled BZ and a vertical line in the figure). At this time IMP 8 was situated around (4.7, −35.8, 8.8) (R_E) in GSE coordinate, and the magnetic field vector was (−4.3, 8.8, 0.0) in GSM coordinates and (−4.3, 7.4, 4.8) in GSE coordinates.

In order to determine the arrival time of this southward turning at the subsolar bow shock, we utilize the method presented by Lockwood et al. [1989]. Their equation (5) is reproduced here:

$$T_{sb} \approx \{X_s - X_b - (Y_s \cdot B_x / B_y)\} / [\{\cos \delta + (\sin \delta \cdot B_x / B_y)\} \cdot V_{sw}]$$

where T_{sb} is a time between the change reaching the spacecraft and the subsolar bow shock, X_s and Y_s are the positions of spacecraft, X_b is the subsolar distance of the bow shock, B_x and B_y are the components of the IMF, V_{sw} is the speed of the solar wind, and δ is the solar wind angle to the sun-earth line. The position of the spacecraft and the magnetic field vector are expressed in GSE coordinates (see the original paper for detail). Unlike Lockwood et al. [1989], we simply assume $13 R_E$ for X_b. By substituting the values at 1010 UT (given above) into the above equation, we obtain -8 min for T_{sr}. Thus, it is estimated that the southward turning of the IMF arrived at the subsolar bow shock at ~ 1002 UT.

In Figure 2 we can also get a perspective of the geomagnetic activities from AE indices. A series of enhancement in the geomagnetic activity took place from ~ 1000 to ~ 1400 UT. The preceding substorm activity ended near 0700 UT, and a geomagnetically quiet condition lasted for ~ 3 hours after ~ 0700 UT. It is likely that the enhancement of geomagnetic activity after ~ 1000 UT was initiated by the southward turning of the IMF.

Let us now return to Figure 1 and discuss the time-sequence of substorm in detail. AU began to increase at ~ 0950 UT, followed by a continuous increase. This would correspond to the growth phase. Following the gradual increase in AL, sudden enhancements of AL occurred at ~ 1113 UT and at ~ 1144 UT (marked by arrows with letters "EX" in the figure). These are likely to correspond to substorm expansion onsets. We have confirmed these onsets by using low-latitude ground magnetograms. The geomagnetic activity reached a maximum level of AU = 500nT and AL = -1000nT at ~ 1215 UT, and then started returning to the quiet-time level. We thus estimate that the period after ~ 1215 UT corresponds to the recovery phase of the substorm.

Finally we discuss the properties of the magnetic field variation observed with AMPTE/CCE (Figure 1). The first TMFE was observed at ~ 1038 UT (event A). Irregular transient magnetic field perturbations were observed from ~ 1010 UT, but they did not show the feature characteristic of the TMFEs and their amplitude were small (~ 7nT) compared with the following TMFEs. The first TMFE was followed by 7 recurring TMFEs with an average recurrence interval of 4-min. The maximum amplitude (difference between maximum and minimum of the perturbation) of the V component perturbation was ~ 59 nT (event G). These TMFEs occurred at southern latitudes in the afternoon sector, and they were all "reverse" type events as defined by Rijnbeek et al. [1984]. Although the fine structure of the TMFEs are not prime concern of this paper, it is noteworthy that the last 3 TMFEs (events F,G,H) had a "crater" signature with a reduction in the field magnitude in the central part, as reported by LaBelle et al. [1987] and Farrugia et al. [1988]. After ~ 1106 UT (event H), the time scale of the magnetic field perturbation became smaller, as is evident in the V and D components. Thus they are not identified as TMFEs although they may be the same phenomena as the previous TMFEs. The

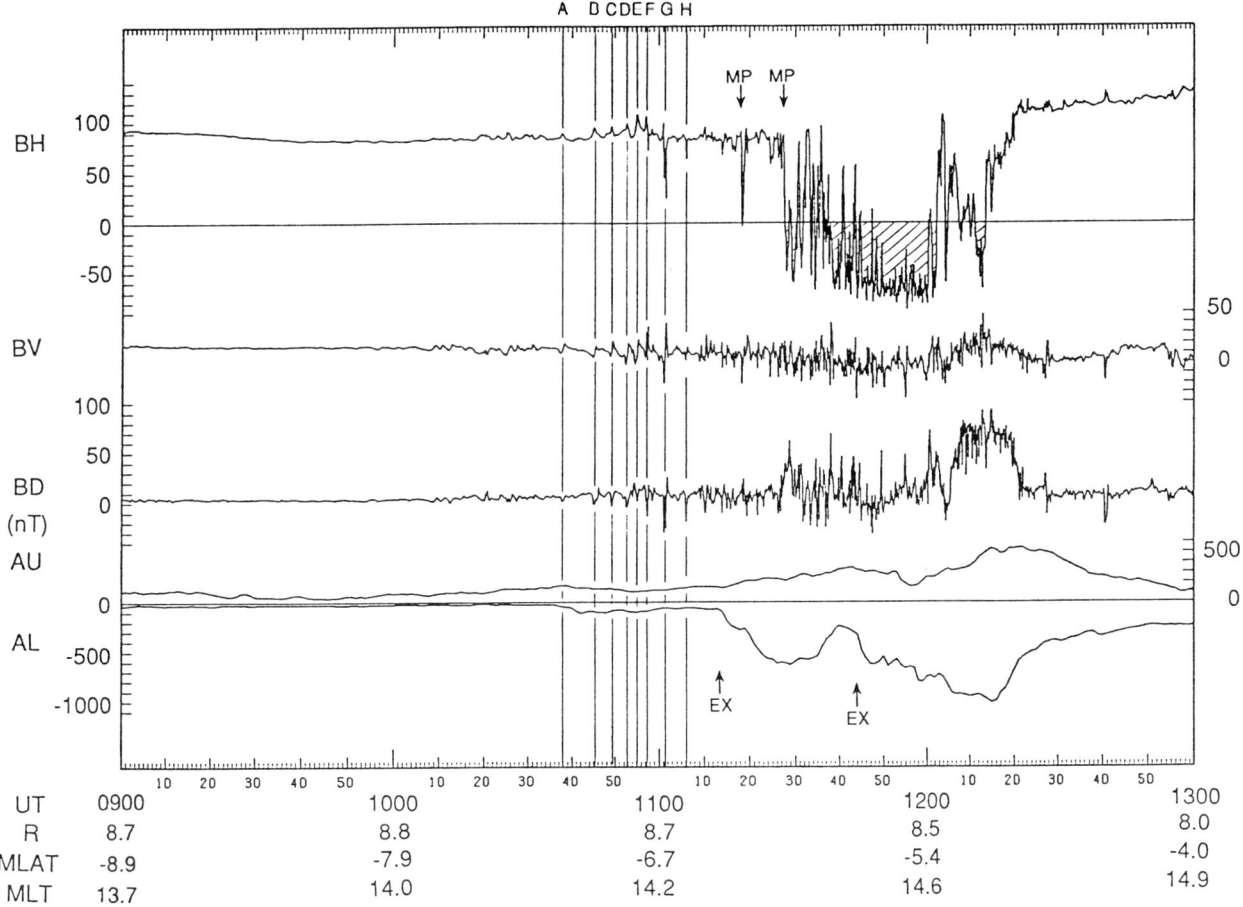

Fig. 1. The magnetic field data of AMPTE/CCE and AU and AL indices for the recurrent transient magnetic field events (TMFEs) on August 27, 1984. The figure presents, from top to bottom, H, V and D components of the magnetic field in the V-D-H coordinates, AU and AL indices, universal time, position (radial distance, magnetic latitude and magnetic local time) of AMPTE/CCE. Vertical lines and letters A-H in the period 1038~1106 UT mark the occurrence of TMFEs. Arrows with letters 'MP' mark the magnetopause-crossings. Arrows with letters 'EX' mark the expansion onsets of substorm. Shaded is the period when the spacecraft observed the southward field in the magnetosheath.

first magnetopause-crossing occurred at ∼ 1118 UT (marked by an arrow with letters "MP" in the figure), followed by an immediate reentry into the magnetosphere at ∼ 1120 UT. The second and major magnetopause-crossing occurred at ∼ 1127 UT. The spacecraft stayed in the magnetosheath for subsequent ∼ 50-min. The magnetosheath field was clearly directed southward (shaded), which suggests that the IMF was also directed southward in the ∼ 50-min period centered at ∼ 1200 UT.

March 13, 1986

Figure 3 shows the magnetic field data from AMPTE/CCE and AU and AL indices for the second series of recurrent TMFEs. The format of this figure is the same as Figure 1, where vertical lines and letters A-K mark the occurrence of TMFEs.

For this case we have no interplanetary data because IMP 8 was not in the interplanetary space. However CCE detected the southward magnetic field in the magnetosheath (0721 ∼ 0726 UT, 0733 ∼ 0738 UT and 0816 ∼ 0853 UT; shaded), suggesting the southward IMF.

The time-sequence of substorm activity is as the following. The previous substorm enhancement lasted until ∼ 0500 UT, followed by an hour of quiet condition. At ∼ 0600 UT AL started to decrease, and after a slight increase near 0645 UT it suddenly decreased by ∼ 150 nT at ∼ 0707 UT. In order to define the expansion phase onset more definitely we examined the low-latitude ground

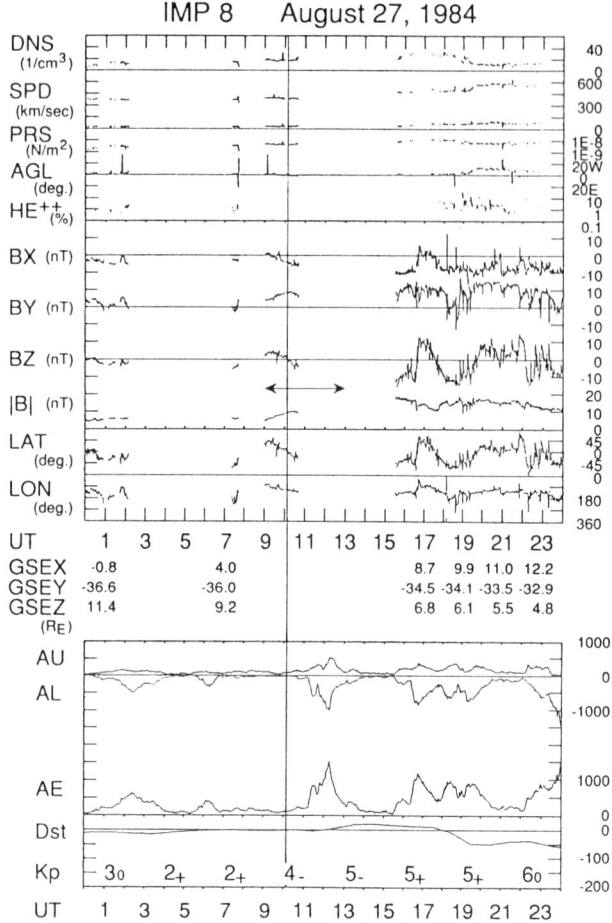

Fig. 2. The plasma and magnetic field data of IMP 8 for the TMFEs on August 27, 1984. The plasma data are of 1-2 minute resolution, and the magnetic field data are of 15 second resolution. The figure presents, from top to bottom, the proton number density (DNS; cm^{-3}), the plasma bulk speed (SPD; km/sec), the thermal speed (SPD; km/sec), the solar wind dynamic pressure (PRS; Newton/m^2), flow angle (AGL; degree; east-west direction from which the plasma flow comes), ratio of alpha current to proton peak current (HE^{++}; %), X, Y, and Z components (BX,BY,BZ; nT) of the magnetic field in the GSM coordinate system, total magnitude of the magnetic field ($|B|$; nT), the latitudinal and longitudinal angles (LAT,LON; degree) of the magnetic field vector in GSM, universal time (UT), X, Y, and Z components (GSEX,GSEY,GSEZ; R_E) of the satellite position in the GSE coordinate system, the 1 minute resolution AE(12) indices (AU,AL,AE; nT), the hourly Dst index (nT), and the three hourly Kp indices. Horizontal arrow between plots of the BZ and $|B|$ shows the time range shown in Fig. 1. In this time range the IMP 8 was in the interplanetary space. Vertical line at $\sim$ 1010 UT marks the southward-turning of the interplanetary magnetic field (IMF).

magnetograms. We found that the onset was at $\sim$ 0650 UT. The onset time is marked by an arrow in the figure. AL decreased to reach its minimum value around 0740 UT, and then gradually returned toward the quiet-time level until $\sim$ 0900 UT. We thus estimate that the period from $\sim$ 0650 to $\sim$ 0740 UT corresponds to the expansion phase of substorm.

The occurrence of TMFEs at the position of AMPTE/CCE is as the following. Before $\sim$ 0550 UT we found no TMFE. The magnetic field data showed small and irregular variations of amplitudes less than 4 nT in this period. From $\sim$ 0550 UT the amplitude became larger ($\sim$ 8nT) but still no clear TMFE was observed. The first clear TMFE (event A) occurred at $\sim$ 0628 UT (in the growth phase of the substorm) and was followed by 7 TMFEs until $\sim$ 0711 UT (in the expansion phase of the substorm) with recurrence interval of $\sim$ 7-min. The maximum amplitude of the V component magnetic field perturbation was $\sim$ 95 nT (event H). These TMFEs occurred near the geomagnetic equator in the morning sector (near 9 hour MLT), and they were all "standard" type events [Rijnbeek et al., 1984]. Some of the TMFEs (events B,C,I,J,K) had "crater" signatures [LaBelle et al., 1987]. An interesting feature of this case is that the TMFEs continued to occur after the substorm expansion onset time ($\sim$ 0650 UT).

January 21, 1986

Figure 4 shows the magnetic field data from AMPTE/CCE and AU and AL indices for the third case of recurrent TMFEs. The format of this figure is the same as Figure 1, where vertical lines and letters A-I mark the occurrence of TMFEs.

Also in this case we have no interplanetary data because of a data gap in the IMP 8 observation. However, as in the previous case, CCE detected the southward magnetic field when it entered into the magnetosheath (1321 $\sim$ 1344 UT; shaded), suggesting the southward IMF.

The time-sequence of geomagnetic activity for this case is different from the previous cases with clear onsets of the expansion phase, as stated below. The enhancement of the AE activity after $\sim$ 1100 UT was preceded by $\sim$ 8 hours of low activity. From $\sim$ 1100 UT to $\sim$ 1320 UT AU and $|AL|$ (the absolute value of AL) made gradual increase without typical expansion phase onset signature. Low-latitude ground magnetograms also indicated the lack of positive bay onset signature. Although there was a spiky decrease of fairly large amplitude ($\sim$ 250 nT) in AL at $\sim$ 1210 UT, the duration of this spike was less than 20-min, and AL returned to its original decreasing gentle slope after the spiky event. Moreover, only one out of 12 AE stations (Barrow, at 0140 LT) recorded this perturbation. This spiky decrease may therefore be very much localized phenomenon. Thus, the geomagnetic disturbance during the period from $\sim$ 1100 to $\sim$ 1320 UT may be identical to so called "convection bay" (i.e., S_q^p-like or DP2-like convection enhancement accompanied by no expansion phase bay disturbances) [Kokubun et al., 1977; Pytte et al., 1978].

The time-sequence of this case observed with AMPTE/CCE is as the following. We found very small amplitude of fluctuations and no TMFEs before $\sim$ 0910 UT. From $\sim$ 0910 UT to $\sim$ 1126 UT irregular transient fluctuations were observed but none of them was identified as TMFE. After the first clear TMFE at $\sim$ 1126 UT (event A), 7 TMFEs repeatedly occurred until $\sim$ 1314 UT with average recurrence interval of 15-min. The maximum amplitude of the V component magnetic field perturbation of the TMFEs was $\sim$ 36 nT (event H). The TMFEs occurred at northern latitudes west of the noon meridian, and they were all "standard" type events.

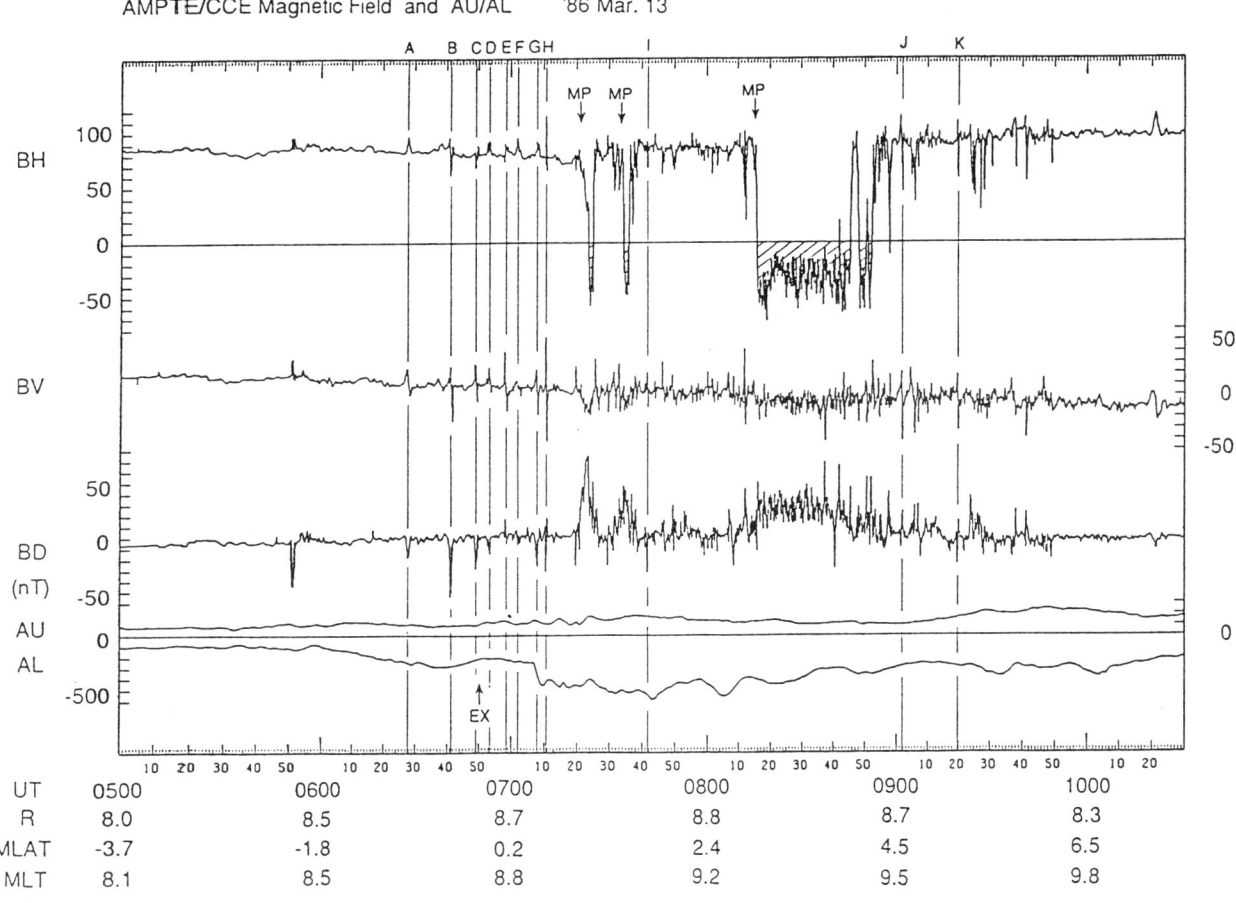

Fig. 3. The magnetic field data of AMPTE/CCE and AU and AL indices for the TMFEs on March 13, 1986. The format of the figure is the same as Fig. 1.

December 25, 1985

Figure 5 shows the magnetic field data from AMPTE/CCE and AU and AL indices for the fourth case. AU and AL in the figure show an isolated substorm activity. The interplanetary data were unavailable for this case.

The geomagnetic activity, as manifest in AU and AL indices, is as the following. The geomagnetic condition was quiet before $\sim$ 1600 UT. At $\sim$ 1736 UT, AL sharply decreased to reach its minimum of $\sim$ -500 nT within 3-min, followed by a rapid recovery with similar time scale. This corresponds to the expansion onset of substorm. It is noted that the only one AE station (Dixon Island, at 2230 LT) recorded this spiky perturbation. Thus, this spiky enhancement in AL is likely to be associated with an localized enhancement of geomagnetic activity. AL reached its second minimum of $\sim$ -430 nT at $\sim$ 1746 UT. This minimum was recorded at several AE stations on the night side. We further confirmed that the corresponding bay disturbance was observed at low-latitude stations and that the positive bay onset time was $\sim$ 1736 UT.

The magnetic field data of AMPTE/CCE shows the existence of TMFEs, as marked by vertical lines and letters A-O in Figure 5.

TMFEs occurred in average recurrence interval of 15-min. The maximum amplitude of the V-component magnetic field perturbation of the TMFEs was $\sim$ 12 nT (events F and I). The TMFEs were observed at northern latitudes and they were all "standard" type events. This case has a feature different from previous cases; TMFEs were observed more than two hours before the onset time of the substorm expansion phase ($\sim$ 1736 UT). However, the following feature is the same as that of previous cases; TMFEs continued occurring across the expansion phase onset time. An another prominent feature of this case is that the TMFEs recurred in irregular fashion before 1700 UT while in regular fashion after 1700 UT.

For this case, we can utilize the magnetic field data from GOES 5 and 6. The orbits of the three spacecraft are depicted in Figure 6. Local times of AMPTE/CCE, GOES 5 and 6 at $\sim$ 1736 UT were 13.3, 12.6 and 10.4 hour, respectively. AMPTE/CCE and GOES 5 were at the same meridian around 1830 UT, and AMPTE/CCE and GOES 6 were at the same meridian around 2200 UT. Figure7 shows the comparison of the total magnetic field observed with the three spacecraft. In this figure, vertical lines and letters A-O mark the times when TMFEs were observed at AMPTE/CCE. Arrows in the figure indicate the events at GOES 5 and 6, which

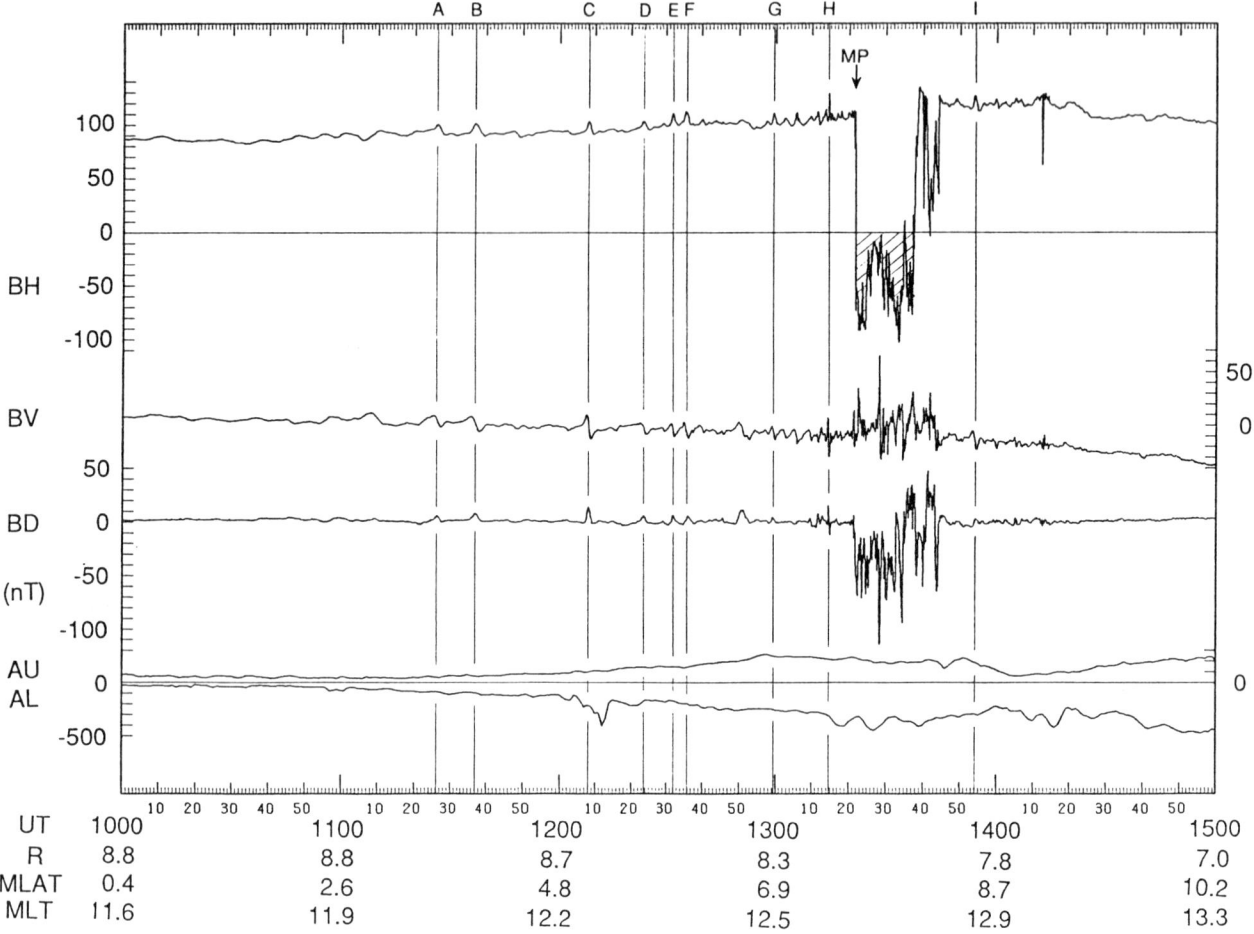

Fig. 4. The magnetic field data of AMPTE/CCE and AU and AL indices for TMFEs on January 21, 1986. The format of the figure is the same as Fig. 1.

are visually identified as disturbances corresponding to TMFEs at AMPTE/CCE. The peaks corresponding to the TMFEs are shaded.

As evident in Figure 7, a step-like increase in the field magnitude around 1700 UT was simultaneously observed with all three spacecraft. This step-like increase probably corresponds to a global compression of the magnetosphere with inward motion of the magnetopause, caused by a sudden enhancement in the solar wind pressure. It is also noted in Figure 7 that the observability of the events from the three spacecraft changed at this time (1700 UT), as is discussed below. Before 1700 UT, the most of the TMFEs observed at AMPTE/CCE accompanied clear corresponding perturbation at both GOES 5 and 6. Amplitudes of perturbation at GOES 5 and 6 were also comparative to those at AMPTE/CCE. The average ratios of amplitude at GOES 5 and 6 to that at AMPTE/CCE for the events before 1700 UT were both ~ 0.3. On the other hand, the TMFEs after 1700 UT were harder to identify at GOES 5 and 6 because of much smaller amplitude than at AMPTE/CCE. We could identify corresponding variations for 5 out of 9 TMFEs at AMPTE/CCE after 1700 UT (marked by arrows; events G,I,K,M,O). The average ratios of amplitude at GOES 5 and 6 to that at AMPTE/CCE, for the 5 events, were both ~ 0.1. This change in observability from the three spacecraft at 1700 UT may reflect the difference in generation mechanism, as will be discussed in the next section.

Discussion

The four series of recurrent TMFEs, which were found in the survey of two year (1984–1986) data from AMPTE/CCE, had the following common properties. (1) Recurrent TMFEs were observed in association with the occurrence of substorm. (2) TMFEs started to occur during the growth phase of substorm, after the southward turning of the IMF. (3) TMFEs continued to occur during the expansion phase.

The item (1) simply suggests that the observability of the TMFEs is related to the substorm activity. It is an established notion that

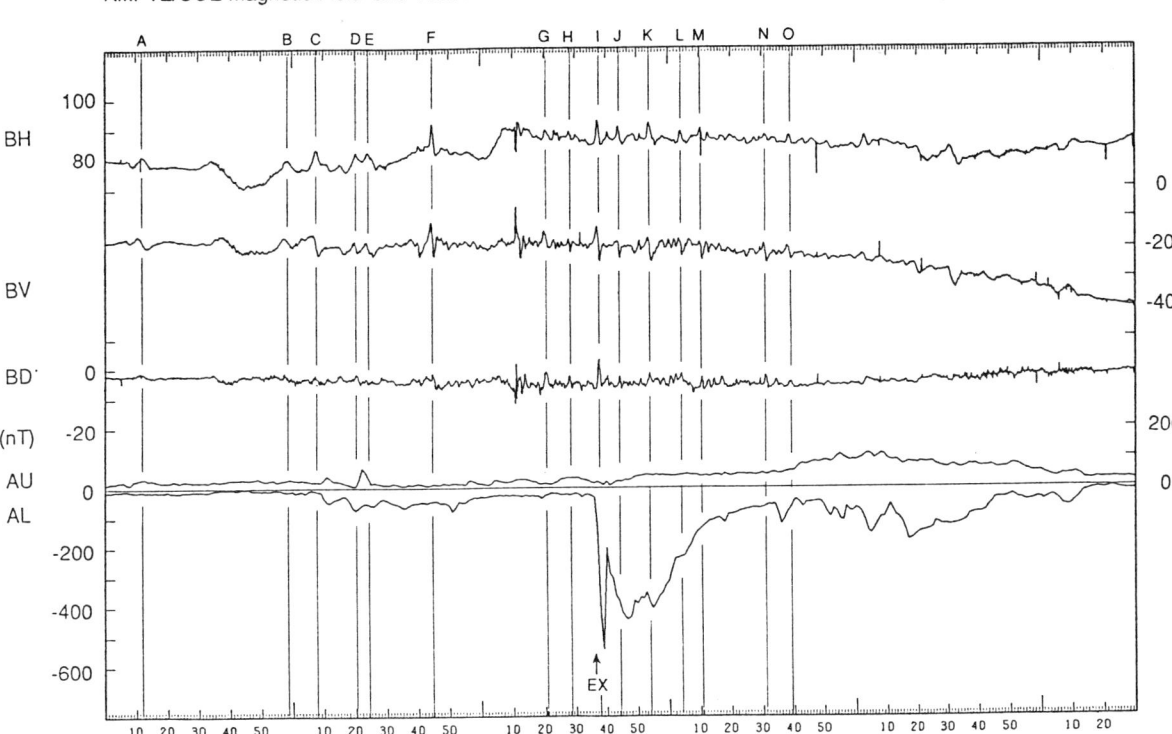

Fig. 5. The magnetic field data of AMPTE/CCE and AU and AL indices for TMFEs on December 25, 1985. The format of the figure is the same as Fig. 1. Vertical lines and letters A-O mark the occurrence of TMFEs.

the substorm activity is enhanced when the IMF has southward component [Fairfield and Cahill, 1966; Schatten and Wilcox, 1967; Rostoker and Fälthammer, 1968; Hirshberg and Colburn, 1969; Kokubun, 1971, 1972]. It has also been suggested that when the IMF is southward the dayside magnetopause is more inward than when the IMF is northward, both from direct observations of the magnetopause-crossings [Aubry et al., 1970, 1971; Fairfield, 1971; Maezawa, 1974; Holzer and Slavin, 1978, 1979; Rufenach et al., 1989; Fairfield, Solar wind control of the size and shape of the magnetosphere, submitted to J. Geomag. Geoelectr., 1990] and from the IMF Bz-dependence of the cusp latitude [Burch, 1972; Yasuhara et al, 1973; Kamide et al, 1976; Carbary and Meng, 1986]. This inward displacement has been observed to occur in wide spatial range of the magnetopause [e.g., Maezawa, 1974; Fairfield, submitted to J. Geomag. Geoelectr., 1990]. As a cause of this inward displacement of the magnetopause, the dayside flux erosion due to large-scale dayside reconnection has been proposed in the above papers. The dayside flux erosion has also been predicted in the theoretical papers [Coroniti and Kennel, 1973; Kan and Akasofu, 1974; Holzer and Reid, 1975; Reid and Holzer, 1975]. Thus it is very likely that, for all of the cases discussed in the previous section, the flux erosion took place owing to large-scale dayside reconnection and the magnetopause moved inward.

As a cause of the TMFEs we take into account of two possibilities; one is the transient and patchy reconnection and the other is the solar wind pressure pulse, as discussed in introduction. For simplicity we assume here that the two effects are mutually exclusive, and then we examine which process is more proper for the interpretation of the cases under study. Note that we assume, as a background, an existence of large-scale reconnection, as discussed above.

First we consider the possibility that the observed TMFEs are generated by transient and patchy reconnection. Then we find that the characteristics (1)–(3) are deduced concisely, as the following. The transient and patchy reconnection is activated when the IMF is directed southward. The substorm activity is also enhanced after the southward turning of the IMF. It is therefore reasonable that the period in which the TMFEs are frequently observed overlaps the period of the substorm (item 1). Moreover the first generation of TMFEs are expected to be close to the beginning of the substorm growth phase (item 2). Furthermore, since the IMF does not always return northward after the substorm expansion phase onsets,

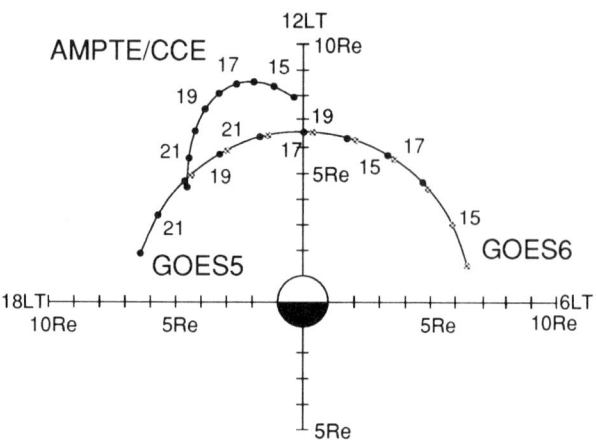

Fig. 6. Trajectories of AMPTE/CCE, GOES 5 and 6 during the period 1400−2200 UT on December 25, 1985, projected into the geographic equatorial plane.

the TMFEs can continue to be generated beyond the time of the expansion onset (item 3).

Second we consider another possibility that the observed TMFEs are generated by solar wind pressure pulses. In this case, the generation of the TMFEs can be independent of the IMF condition and/or the substorm activity. Item (3) is therefore easily explained (simply independent), but items (1) and (2) look contrary to the expected independence from the substorm activity. The idea of the solar wind pressure pulses therefore looks improper in explanation of the observed TMFEs. However, one may still claim that the idea of solar wind pressure pulses explains the items (1) and (2); the two items may be explained in terms of the change in the distance from the magnetopause to the spacecraft. That is, after the IMF have turned southward, the large-scale dayside reconnection takes place, the magnetopause moves inward, and the substorm activity (growth phase) starts to enhance. As a result of the inward motion of the magnetopause, the effect of the solar wind pressure pulse becomes more detectable from the spacecraft because of closer magnetopause (i.e., source region). Then, if it is assumed that the effect of the solar wind pressure pulse is too small to be detected before an inward displacement of the magnetopause (thus before the substorm enhancement due to the southward turning of the IMF), items (1) and (2) may be expected. However it is questionable whether this assumption is a realistic one. Then, let us below estimate the amplitude before and after the inward displacement of the magnetopause by using simple equations.

In order to obtain equations for estimation of the perturbation amplitude, we assume that the signal of solar wind pressure pulse travels in the magnetosphere as the fast mode MHD wave. The uniform background field and the low-β condition are also assumed for simplicity. Then the magnitude of the perturbed magnetic field approximately satisfies the following wave equation:

$$(V_A^2 \nabla^2 - \frac{\partial^2}{\partial t^2})\delta B = 0 \quad (1)$$

where δB is the magnitude of the perturbation field ($|\delta \vec{B}|$) and V_A is the Alfvén velocity. We take the coordinate system as follows: the x axis is normal to the magnetopause and points outward, and the y axis is tangential to the magnetopause and parallel to the velocity of the solar wind pressure pulse ($\vec{u}_s$). Then we get

$$k_y = \omega/|\vec{u}_s| \quad (2)$$

where ω and $\vec{k}$ are the angular frequency and the wave vector of the Fourier component of the δB. With (1) and (2) we obtain the following relation:

$$k_x^2 = -(1 - u_s^2/V_A^2)k_y^2 \quad (3)$$

If we assume $|\vec{u}_s|/V_A \ll 1$, equation (3) reduces to

$$k_x = \pm i k_y$$

Thus the perturbation of the total magnetic field at the position x is expressed as

$$\delta B = b_0 e^{i(k_y y - \omega t)} e^{k_y(x - x_{mp})} \quad (4)$$

where x_{mp} is the position of the magnetopause, b_0 is the amplitude of the perturbation at x_{mp}. In this way the amplitude decreases in an exponential fashion as the distance from the magnetopause increases. In order to see the ratio of amplitudes before and after the inward motion of the magnetopause, we fix b_0, k_y (source wave condition), fix x and y (the position of an observer), and compare two values of δB for two values of x_{mp}, which satisfy equation (4). That is, pairs (δB_1, x_{mp1}) and (δB_2, x_{mp2}) are set to satisfy equation (4) with the same values of b_0, k_y, x and y. Here x_{mp1} corresponds to the position of the magnetopause before the inward motion, x_{mp2} to that after the motion. Then we have

$$|\delta B_1/\delta B_2| = e^{-k_y(x_{mp1} - x_{mp2})} \quad (5)$$

Now we can estimate the ratio of amplitude. We assume that $|\vec{u}_s| = 150$km/sec (typical speed of the solar wind along the dayside magnetopause) and that $2\pi/\omega = 2$min (typical duration of the observed events). Then we get k_y of $2.2 R_E^{-1}$. x_{mp1} is set to be the nominal magnetopause distance of $10.0 R_E$ [Fairfield, submitted to J. Geomag. Geoelectr., 1990]. x_{mp2} is set to be $8.8 R_E$ (the apogee of AMPTE/CCE). Using above values, we obtain for $|\delta B_1/\delta B_2|$ of 0.07. Since the observed TMFEs had the maximum amplitude of ~ 95 nT (event H in the March 13, 1986 case; see Figure 3), the maximum expected amplitude when $x_{mp} = x_{mp1}$ is ~ 7 nT, which would be observable with its typical bipolar pattern. Although the calculation is on a tentative basis, we can expect that at least the solar wind pressure pulses with large amplitudes are identifiable as TMFEs even before the inward motion of the magnetopause (before the substorm enhancement). On the other hand, we found no TMFEs before the substorm enhancement for the cases of August

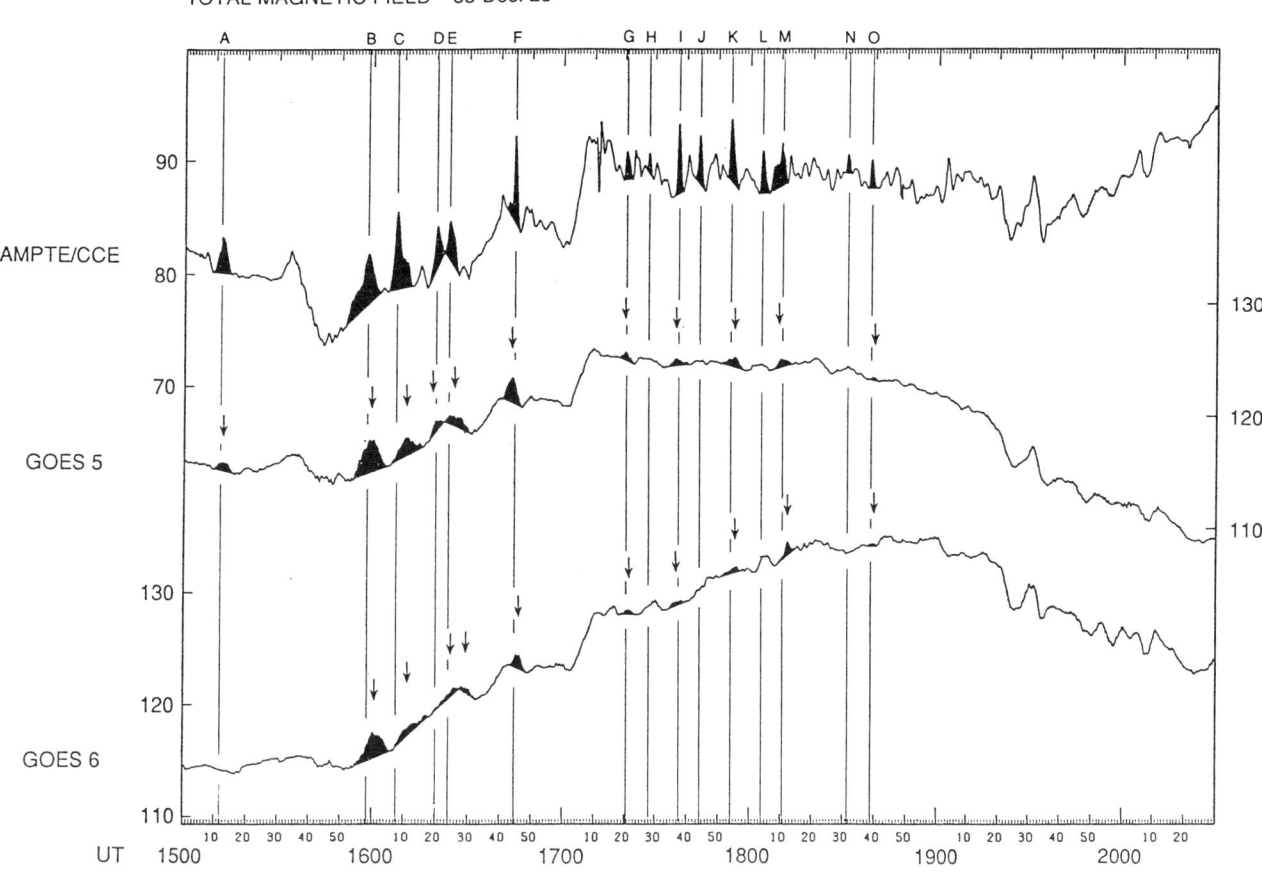

Fig. 7. Total magnetic field data of AMPTE/CCE, GOES 5 and 6 for the same period as Fig. 5. Vertical lines and letters A-O mark the occurrence of TMFEs at AMPTE/CCE. Arrows attached to the graphs of GOES 5 and 6 mark the occurrence of the corresponding perturbation at each spacecraft. The peaks corresponding to the TMFEs are shaded.

27, 1984, March 13, 1986 and January 21, 1986. Hence these cases are more likely to be explained by transient and patchy reconnection. For the December 25, 1985 case, the TMFEs before $\sim$ 1700 UT does not satisfy the condition (2) summarized before; they occurred prior to the enhancement of the substorm activity. These events were observed with AMPTE/CCE, GOES 5 and 6, suggesting they were caused by solar wind pressure pulses. On the other hand, some of the TMFEs after $\sim$ 1700 UT were not identified at GOES 5 and 6. Thus the TMFEs after $\sim$ 1700 UT may have been generated by transient and patchy reconnection.

Finally we present another possible observational disadvantage for the idea of solar wind pressure pulse. Under a condition of continuously strong northward IMF and solar wind dynamic pressure, we can expect that TMFEs are clearly observed at AMPTE/CCE during the period of low substorm activity. However we have found no such cases in spite of search for two years data.

In summary, we have found that all of the recurrent TMFEs found from two years (1984-1986) magnetic field data of AMPTE/CCE were associated with substorm activity. From this association we suggest that the observed recurrent TMFEs were caused by transient and patchy reconnection and thus our TMFEs can be called FTEs. Then, it will be worthwhile to study the efficiency of energy transfer achieved by transient and patchy reconnection during substorm.

Acknowledgments. We are most grateful to K. Takahashi at the Johns Hopkins University Applied Physics Laboratory for his great help in obtaining the CCE magnetometer data which were kindly made available by the investigators of the magnetometer experiment: T. A. Potemra and L. J. Zanetti at the Johns Hopkins University Applied Physics Laboratory, and M. H. Acūna at the NASA Goddard Space Flight Center. We also thank K. Takahashi and T. K. Nakamura for their useful comments for improving the manuscript. Plots of IMP 8 magnetic field data and the 1-min digital data of AE(12) indices (AU and AL) were supplied by the Data Analysis Center for Geomagnetism and Space Magnetism, Kyoto University. Ground magnetogram data were supplied by the Data Analysis Center for Aurora, National Institute of Polar Research. The work at Geophysics Research Laboratory, University of Tokyo, was supported by the Grant-in-Aid for Scientific Research, project 62420013, Ministry of Education, Science and Culture, Japan.

References

Aubry, M. P., C. T. Russell, and M. G. Kivelson, Inward motion of the magnetopause before a substorm, *J. Geophys. Res.*, 75, 7018, 1970.

Aubry, M. P., M. G. Kivelson, and C. T. Russell, Motion and structure of the magnetopause, *J. Geophys. Res.*, 76, 1673, 1971.

Berchem, J., and C. T. Russell, Flux transfer events on the magnetopause: Spatial distribution and controlling factors, *J. Geophys. Res.*, 89, 6689, 1984.

Burch, J. L., Precipitation of low-energy electrons at high latitudes: Effects of interplanetary magnetic field and dipole tilt angle, *J. Geophys. Res.*, 77, 6696, 1972.

Carbary, J. F., and C. I. Meng, Correlation of cusp latitude with Bz and AE(12) using nearly one year's data, *J. Geophys. Res.*, 91, 10047, 1986.

Coroniti, F. V., and C. F. Kennel, Can the ionosphere regulate magnetospheric convection? *J. Geophys. Res.*, 78, 2837, 1973.

Cowley, S. W. H., The causes of convection in the earth's magnetosphere: A review of developments during the IMS, *Rev. Geophys. Space Phys.*, 20, 531, 1982.

Daly, P. W., D. J. Williams, C. T. Russell, and E. Keppler, Particle signature of magnetic flux transfer events at the magnetopause, *J. Geophys. Res.*, 86, 1628, 1981.

Daly, P. W., and E. Keppler, Observation of a flux transfer event on the earthward side of the magnetopause, *Planet. and Space Sci.*, 30, 331, 1982.

Fairfield, D. H., Average and unusual locations of the earth's magnetopause and bow shock, *J. Geophys. Res.*, 76, 6700, 1971.

Fairfield, D. H., and L. Cahill, Jr., Transition region magnetic field and polar magnetic disturbance, *J. Geophys. Res.*, 71, 155, 1966.

Farrugia, C. J., P. Rijnbeek, M. A. Saunders, D. J. Southwood, D. J. Rodgers, M. F. Smith, C. P. Chaloner, D. S. Hall, P. J. Christiansen, and L. J. C, Wooliscroft, A multi-instrument study of flux transfer event structure, *J. Geophys. Res.*, 93, 14465, 1988.

Hirshberg, J., and D. S. Colburn, Interplanetary field and geomagnetic variations - A unified view, *Planet. and Space Sci.*, 17, 1183, 1969.

Holzer, T. E., and G. C. Reid, The response of the day side magnetosphere-ionosphere system to time-varying field line reconnection at the magnetopause 1. Theoretical model, *J. Geophys. Res.*, 80, 2041, 1975.

Reid, G. C., and T. E. Holzer, The response of the day side magnetosphere-ionosphere system to time-varying field line reconnection at the magnetopause 2. Erosion event of March 27, 1968, *J. Geophys. Res.*, 80, 2050, 1975.

Holzer, R. E., and J. A. Slavin, Magnetic flux transfer associated with expansions and contractions of the dayside magnetosphere, *J. Geophys. Res.*, 83, 3831, 1978.

Holzer, R. E., and J. A. Slavin, A correlative study of magnetic flux transfer in the magnetosphere, *J. Geophys. Res.*, 84, 2573, 1979.

Kamide, Y., J. L. Burch, J. D. Winningham, and S.-I Akasofu, Dependence of the latitude of the cleft on the interplanetary magnetic field and substorm activity. *J. Geophys. Res.*, 81, 698, 1976.

Kan, J. R., and S.-I. Akasofu, A model of the open magnetosphere, *J. Geophys. Res.*, 79, 1379, 1974.

Kokubun, S., Polar substorm and interplanetary magnetic field, *Planet. and Space Sci.*, 19, 697, 1971.

Kokubun, S., Relationship of interplanetary magnetic field structure with development of substorm and storm main phase, *Planet. and Space Sci.*, 20, 1033, 1972.

Kokubun, S., and R. L. McPherron, and C. T. Russell, Triggering of substorms by solar wind discontinuities, *J. Geophys. Res.*, 82, 74, 1977.

LaBelle, J., R. A. Treumann, G. Haerendel, O. H. Bauer, G. Paschmann, W. Baumjohann, H. Lühr, R. R. Anderson, H. C. Koons, and R. H. Holzworth, AMPTE IRM observations of waves associated with flux transfer events in the magnetosphere, *J. Geophys. Res.*, 92, 5827, 1987.

Lockwood, M., P. E. Sandholt, S. W. H. Cowley, and T. Oguti, Interplanetary magnetic field control of dayside auroral activity and the transfer of momentum across the dayside magnetosphere, *Planet. and Space Sci.*, 37, 1347, 1989.

Maezawa, K., Dependence of the magnetopause position on the southward interplanetary magnetic field, *Planet. and Space Sci.*, 22, 1443, 1974.

Paschmann, G., G. Haerendel, I. Papamastorakis, N. Sckopke, S. J. Bame, J. T. Gosling, and C. T. Russell, Plasma and Magnetic field Characteristics of magnetic flux transfer events, *J. Geophys. Res.*, 87, 2159, 1982.

Potemra, T. A., L. J. Zanetti, and M. H. Acuna, The AMPTE CCE magnetic field experiment, *IEEE Trans. Geosci. Remote Sens.*, GE-23, 246, 1985.

Pytte, T., R. L. McPherron, E. W. Hones, Jr., and H. I. West, Jr., Multiple-satellite studies of magnetospheric substorms : Distribution between polar magnetic substorms and convection-driven negative bays, *J. Geophys. Res.*, 83, 663, 1978.

Rijnbeek, R. P., S. W. H. Cowley, D. J. Southwood, and C. T. Russell, A survey of dayside flux transfer events observed by ISEE 1 and 2 magnetometers, *J. Geophys. Res.*, 89, 786, 1984.

Rostoker, G., and C. G. Fälthammer, Relationship between changes in the interplanetary magnetic field and variations in the magnetic field at the earth's surface, *J. Geophys. Res.*, 72, 5853, 1967.

Rufenach, C. L., R. F. Martin, and H. H. Sauer, A study of geosynchronous magnetopause crossings, *J. Geophys. Res.*, 94, 15125, 1989.

Russell, C., T., and R. C. Elphic, Initial ISEE magnetometer results; Magnetopause observations, *Space Sci. Rev.*, 22, 681, 1978.

Russell, C. T., J. Berchem, and J. G. Luhmann, On the source region of flux transfer events, *Adv. Space Res.*, 5, 363, 1985.

Saunders, M. A., C. T. Russell, and N. Sckopke, Flux transfer events; Scale size and interior structure, *Geophys. Res. Lett.*, 11, 131, 1984.

Schatten, K. H., and J. M. Wilcox, Response of the geomagnetic activity K_p to the interplanetary magnetic field, *J. Geophys. Res.*, 72, 5185, 1967.

Scholer, M., D. Hovestadt, F. M. Ipavich, and G. Gloeckler, Energetic protons, alpha particles, and electrons in magnetic flux transfer events, *J. Geophys. Res.*, 87, 2169, 1982.

Sibeck, D. G., W. Baumjohann, R. C. Elphic, D. H. Fairfield, J. F. Fennell, W. B. Gail, L. J. Lanzerotti, R. E. Lopez, H. Luehr, A.

T. Y. Lui, C. G. Maclenann, R. W. McEntire, T. A. Potemra, T. J. Rosenberg, and K. Takahashi, The magnetospheric response to 8 minute-period strong-amplitude upstream pressure variations, *J. Geophys. Res.*, *94*, 2505, 1989.

Sibeck, D. G., A model for the transient magnetospheric response to sudden solar wind dynamic pressure variations, *J. Geophys. Res.*, *95*, 3755, 1990.

Southwood, D. J., M. A. Saunders, M. W. Dunlop, W. A. C. Mier-Jedrzjowicz, R. P. Rijnbeek, A survey of flux transfer events recorded by the UKS spacecraft magnetometer, *Planet. Space Sci.*, *34*, 1349, 1986.

Yasuhara, F., S.-I. Akasofu, J. D. Winningham, and W. J. Heikkila, Equatorward shift of cleft during magnetospheric substorms as observed by Isis I, *J. Geophys. Res.*, *78*, 7286, 1973.

SUBSTORM ELECTRODYNAMICS

DAVID P. STERN

Laboratory for Extraterrestrial Physics,
Code 695, Goddard Space Flight Center, Greenbelt MD 20771, USA

What happens in a substorm? About half a dozen competing explanations exist, suggesting that we really do not know. Substorms are mainly noted for their rapid energy release, but their nature may also be studied by examining other features, such as their current wedge and their poleward motion. Here some new ideas about those features are outlined: a more detailed presentation also exists [Stern, 1990].

It will be assumed that the substorm arises from magnetic reconnection (or "magnetic merging") in the Earth's magnetotail, the oldest explanation and perhaps the one most widely held. The arguments do not depend critically on the distance to the merging site, often claimed to be around 15-20 R_E.

To make the substorm model mathematically tractable, several assumptions must be made:

(1) Merging occurs in a neutral *sheet* where $B_z=0$. The shape of the merging region is furthermore assumed to be roughly rectangular (Figure 1).

(2) *Inside* the merging region the electric field E due to merging is constant and directed along y. This part of E is assumed to be negligibly small *outside* that region.

Above and below the sheet E is related to the plasma inflow v (from the lobes towards the middle of the sheet) by $E = -v \times B$, and it may also be viewed as the inductive field caused by the reduction of accumulated magnetic flux in the tail lobes. The flux taken from the lobes flows earthwards and tailwards, in the sequence 1-4 of Figure 5 (discussed further below).

(3) The ions determine the plasma's large-scale behavior, while the density of electrons adjusts itself to maintain charge neutrality.

Magnetospheric Substorms
Geophysical Monograph 64
This paper is not subject to U.S. copyright. Published in 1991 by the American Geophysical Union.

(4) The energization process acts primarily on ions *from the plasma sheet*. When a substorm occurs, it is assumed that a section of the plasma sheet thins down and both magnetic flux and charged particles flow into it from the tail lobes. The flux arriving from the lobes supplies the magnetic energy which fuels the substorm, while the ions arriving from the lobes are always greatly outnumbered by the original plasma sheet population.

Alfvén [1968] studied the reconnection process in a sheet geometry. He assumed antiparallel fields in the $\pm x$ direction separated by a neutral sheet, and a y-directed electric field E which gave the plasma an $E \times B$ drift towards the middle (Figure 2). Applied to the present case his theory would mean that all the accelerated plasma originates in the lobes and that none comes from the plasma sheet.

After the ions have reached the middle their motion becomes nonadiabatic and they undergo acceleration along E; such motion was studied by Sonnerup [1971], who identified its adiabatic invariant.

In a steady state, inflow and outflow in this configuration must be equal, and Ampére's law $I = \Delta B/\mu_0$ must relate the sheet current density with the jump in B across the sheet, equal to $\Delta B = 2B_{lobe}$. From these Alfvén estimated E_y. He showed that the energy which E imparts to particles from a unit volume of the lobe equals the magnetic energy $B^2/2\mu_0$ contained in that volume, plus an equal amount associated with that volume being "squeezed out". Thus magnetic energy is converted into kinetic.

The plasma of the tail lobes is very rarefied: if its magnetic energy is converted in this way, each of its particles gains about 100 keV. If on the other hand ions and electrons from the lobes share their energy gain with a much larger number of plasma sheet particles, the final energy per ion is expected to be much smaller, as observed. Alfvén's calculation can be modified to

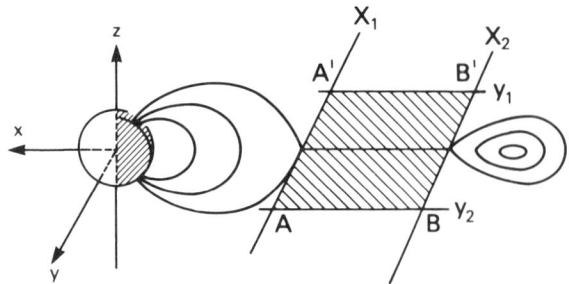

Fig. 1. Simple model of the region of magnetic merging (reconnection).

allow such "energy sharing" and this confirms the expected result.

Alfvén's calculations assume the *continuity* of particle flow--what goes in must come out again. *But what about charge neutrality?* If ions in a neutral sheet are accelerated along y from dawn to dusk, and if the x-dependence of the flow is weak and the small number of ions entering from the lobes is ignored, the equation of continuity for ions requires $N_i v_y$ to be constant; here N_i is the ion density integrated over the thickness of the plasma sheet, which remains continuous no matter how the thickness of the layer changes. This means

$$N_i \, dv_y/dy = -v_y \, dN_i/dy$$

As y increases and v_y grows, the ion density N_i *decreases*. Electrons are accelerated in the opposite direction, hence if the ion density N_i *decreases* in the direction of the electric field, the density N_e *increases* in that direction and neutrality cannot exist.

A similar problem arises if an electric field **E** is imposed in the laboratory on a plasma with **B**=0 (note that in the central layer of a neutral sheet geometry, the average of **B** is also close to zero). Nature solves the problem by compressing **E** into a narrow layer near the plasma boundary, known as a *sheath*. The sheath is not neutral, but its charge density is consistent with its strong **E**, while in the rest of the plasma **E**=0 and neutrality problems do not arise.

Cowley [1973] actually did calculate a sheath-like solution for the electric field associated with magnetic reconnection in a neutral sheet. If this happened, **E** would be severely distorted and particle motion would acquire a conspicuous y-component *before* reaching the neutral sheet.

It is proposed here instead that neutrality is maintained, not by a rearrangement of **E** but by the ionosphere to which all tail lines are connected.

Let ions be accelerated along y. Then in the acceleration region, from y_1 to y_2 (Figure 3), their density N_i decreases as one advances from dawn to dusk. If one goes far enough in that direction, however, at some point y_3 one will find ions from before the substorm, whose density N_i has remained at its undisturbed level.

Electrons arrive from the other side and until they reach y_3 their density N_e matches that of undisturbed ions. As they advance past y_3, however, the ambient ion density drops and an excess of electrons develops. We assume this excess is dumped into the ionosphere, so that even the minimal ion density at y_2 is matched by the density of surrounding electrons.

Beyond this minimum point, N_i rises again and neutrality demands additional electrons: they are therefore pulled out of the ionosphere. The downward

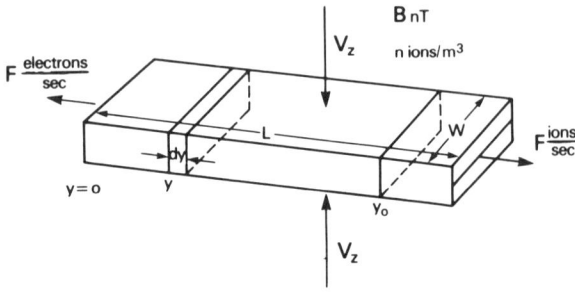

Fig. 2. Geometry used in Alfvén's calculation: lobe plasma flows into a slab-like neutral sheet configuration and comes out at the ends.

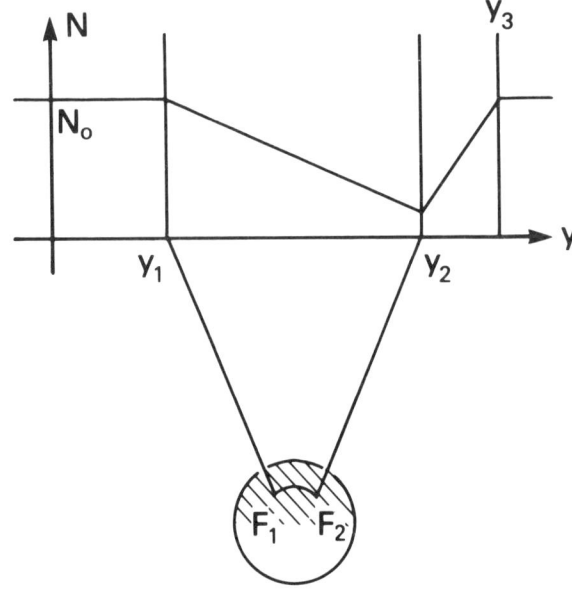

Fig. 3. Schematic view of the variation of the integrated ion density N_i across the merging region $y_1 y_2$ and its fringes. The current diverted to the ionospheric segment $F_1 F_2$ constitutes the current wedge.

and upward currents carried by the electrons, as well as the ionospheric current which completes the circuit, have the same polarities as those observed in the substorm wedge. It is proposed that this is how the wedge current actually arises.

In fact B_z in the tail may be not be zero but just small, even in most of the reconnection region, and this modifies the motion into one that is 3-dimensional and much harder to analyze [Büchner and Zelenyi, 1989; Karimabadi et al., 1990]. In one class of such motions studied by Speiser [1965], the ions enter the weak-field region, weave back and forth across the reversal while advancing from dawn towards dusk in a semicircle, and then are ejected again (Figure 4). At least for this class, ions increase their energy and one therefore expects N_i to decrease in the direction of E as before. Neutrality problems like those described before can then be expected, though their quantitative evaluation will be difficult.

We next look at the global properties of the electric field E, best studied by considering the plasma velocity v as the signature of E.

Consider a set of field lines associated with the reconnection process, labeled at some time t=0 with numbers 1 to 4 (Figure 5). If a y-directed electric field E exists across the reconnection region, the plasma and field lines flow towards the middle. At a later time t_1 one can visualize two possible extreme situations:

(1) The magnetic configuration remains exactly the same, but the particles from line (1) have moved to (2), those from (2) are now on (3), and so on.

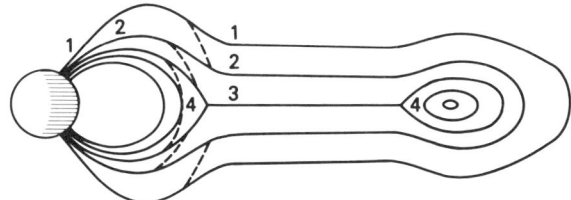

Fig. 5. Labeled field lines in the midnight plane, to illustrate two basic types of electric fields possibly associated with magnetic merging.

Since B is not changed, $\partial B/\partial t = 0$, which means that no magnetic energy is released and also that $E = -\nabla V$, i.e. that E can be expressed as the gradient of some scalar potential V. Some external energy source must maintain V, since magnetic energy is not available.

For instance, V might be associated with the dawn-to-dusk potential across the tail, of the order of 50,000 V, which is attributed to open magnetic field lines and is inferred from observations by polar near-earth satellites [e.g., Stern, 1977].

The plasma will flow with a finite velocity v which satisfies $E = -v \times B$, but this will be a "relabeling velocity" [Stern, 1966], permuting the (α, β) labeling of field lines without changing B.

Because the magnetic field is fixed, the ionospheric footprint of the reconnection region always stays in the same place. Furthermore, because V is constant along each field line, the electric field from the reconnection region spreads along field lines down to the ionosphere. Processes responsible for field-aligned electrical fields are expected to further modify V, but are ignored here.

(2) A different situation exists if the motion of charged particles from (1) to (2) etc. exists mainly *inside the merging region*. Outside that region field lines change and establish new connections, as shown by the broken lines.

Here $\partial B/\partial t \neq 0$ and magnetic energy *is* released. The electric field again satisfies $E = -v \times B$, but because the labeling near Earth stays fixed, v contains no "relabeling velocity" and hence no potential component. Near Earth v=0 and therefore E=0, so that E is mainly confined to the reconnection region and does *not* spread along field lines. Note that $\partial E_y/\partial x \neq 0$, because $E_y \neq 0$ in the merging region ($x < -10\ R_E$) but $E_y = 0$ at Earth ($x < 1$); since E_y is the leading field component, that suggests $\nabla \times E \neq 0$. We may call this the "purely inductive" scenario.

Which of the two alternatives describes the substorm better? In the inductive scenario, at t=0 the field line labeled "3" is the one leading to the reconnection region. However, at the later time $t=t_1$ the ionospheric

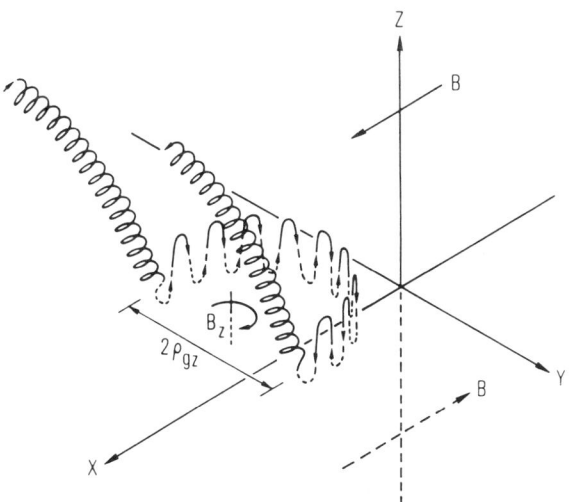

Fig. 4. Schematic view of the particle motion studied by Speiser in a neutral sheet configuration with finite B_z [Speiser, 1965]

line leading to the reconnection region is the one originally labeled at Earth as "2". Auroral activity in substorms is observed to expand poleward. If such activity marks field lines leading to the reconnection region (and those that have already passed through it), this suggests that the substorm has a large "inductive" component. In addition to this this component, of course, there may also exist an external dawn-to-dusk cross-tail potential, consistent with the first scenario.

It has been proposed that the poleward motion of auroral activity occurs because the merging region moves down the tail and away from Earth. Figure 6 gives a map of the polar cap, with footpoints of tail lines starting on the equator at x = -15, -25, -35 and -45 R_E, derived using the 1987 Tsyganenko model [Tsyganenko, 1987]. At midnight a shift of the accelerating region from -25 to -45 R_E advances the footpoints by only one degree in latitude; the observed motion of the substorm aurora is much larger and extends to lines which before the substorm were clearly in the lobe, suggesting that tailward motion only gives part of the effect.

One may now ask: if the electric field **E** in a substorm is purely inductive, how can it affect the ionosphere? The answer proposed here is: only through the wedge current.

If the ionosphere were a perfect conductor (and field lines also conducted perfectly), the wedge current could detour through it without requiring any appreciable electric field **E**. The fact that the ionosphere has a finite resistivity means that when the wedge current passes through it, a secondary electric field $\mathbf{E}' = -\nabla V'$ arises.

This **E**´ is a potential field and it propagates along magnetic field lines back to the reconnection region, where it modifies the primary accelerating field **E**. The magnitude of V´ depends on the resistivity of the ionosphere, but in principle it could reach a few kilovolts or even tens of kilovolts. Details of this are are given elsewhere [Stern, 1990].

In summary, although the model proposed here is crude and simplified, it does give a surprisingly detailed picture and it explains the substorm wedge without resorting to some unspecified "current interruption." It would be interesting if new ways could be devised for testing and extending it.

References

Alfvén, H., Some properties of the magnetospheric neutral surfaces, J. Geophys. Res., 73, 4379-4381, 1968.

Büchner, J. and L.M. Zelenyi, Regular and chaotic charged particle motion in magnetotaillike field reversals, 1. Basic theory of trapped motion, J. Geophys. Res., 94, 11,821-11,842, 1989.

Cowley, S.W.H., Self-consistent model of a simple magnetic neutral sheet system surrounded by a cold, collisionless plasma, Cosm. Electrodyn., 3, 448-501, 1973.

Karimabadi, H., P.L. Pritchett and F.V. Coroniti, Particle orbits in two-dimensional equilibrium models for the magnetotail, J. Geophys. Res., 95, 17,153-17,166, 1990.

Sonnerup, B.U.Ö., Adiabatic particle orbits in a magnetic null sheet, J. Geophys. Res., 76, 8211-8222, 1971.

Speiser, T.W., Particle trajectories in model neutral sheets, 1. Analytical solutions, J. Geophys. Res., 70., 4219-4226, 1965.

Stern, D.P., The motion of magnetic field lines, Space Sci. Rev., 4, 147-173, 1966.

Stern, D.P., Large-scale electric fields in the earth's magnetosphere, Rev. Geophys., 14, 199-214.

Stern, D.P., Substorm electrodynamics, J. Geophys. Res., 95, 12,057-12,067, 1990.

Tsyganenko, N.A., Global Quantitative models of geomagnetic field in the cislunar magnetosphere for different disturbance levels, Planet. Space Sci., 35, 1347-58, 1987.

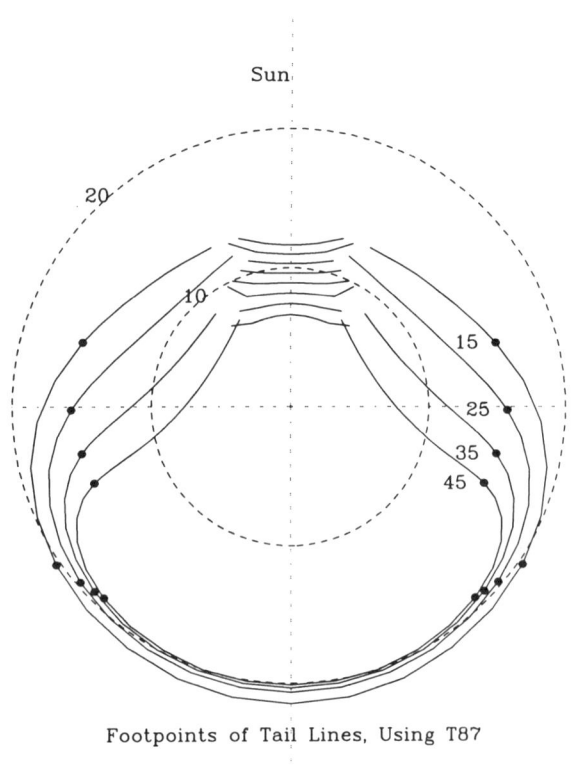

Fig. 6. Ionospheric footpoints of equatorial cross-sections of the tail at x=-15, -25, -35 and -45 R_E. Dark circles mark footpoints of lines from y = ±10, 20 R_E.

Characteristics of the Fields and Particle Acceleration During Rapidly Induced Tail Thinning and Reconnection

R. M. Winglee[1]

Department of Astrophysical, Planetary and Atmospheric Sciences, University of Colorado, Boulder

Two-dimensional (three-velocity) electromagnetic particle simulations are used to investigate the thinning and reconnection of the current sheet under substorm conditions. It is shown that the evolution of the tail is strongly dependent on the strength of the normal magnetic field component and the dawn-to-dusk electric field. A criterion for tearing which relates these fields is derived and verified by simulations. If tearing occurs, high-energy tails are produced during the coalescence of magnetic islands. If tearing is suppressed, the current sheet can be subject to a thermal disruption due to excess particle heating and cancellation of some of the ion current by an unbalanced electron convective drift. In both cases, strong electrostatic fields and broadband electrostatic noise are generated by the differential motion between the electrons and ions through the current sheet. This motion also produces an x current and B_y which change sign during the growth and expansion phases, similar to observations in the tail.

1. Introduction

During the growth phase of substorms, the current sheet that supports the magnetotail is known to thin. This thinning is followed by substorm onset which is associated with brightening and poleward expansion of the auroral zone. In addition, intense field-aligned currents develop along auroral field lines in conjunction with the rapid injection of energetic particles and the expansion of the current sheet [for a review see Kan 1990 and references therein]. At some point along the tail, magnetic reconnection occurs leading to the formation of a plasmoid and the conversion of magnetic field energy into particle energy [*Hones*, 1984].

Much of the above morphology for the evolution of the tail current sheet has been reproduced in MHD simulations [e.g., *Birn and Hones*, 1980; *Birn*, 1984]. However, these simulations have been unable to clearly identify the location of the neutral point where reconnection and plasmoid formation occurs. Estimates range from 10–15 R_E [e.g., *McPherron et al.*, 1968; *Hones*, 1984] to several tens of R_E [e.g., *Baumjohann*, 1988; *Lyons and Nishida*, 1988]. Part of the reason for the uncertainty arises because the particle dynamics is approximated by global transport coefficients (including anomalous resistivity). Differences in the implementation of these transport coefficients can lead to different results.

The key to resolving this uncertainty is understanding the particle dynamics under realistic magnetotail conditions. Initial theory by *Lembége and Pellat* [1982] predicted that tearing is stabilized in the presence of even very weak, normal magnetic field components. This suppression arises because a normal magnetic field component causes the electrons to become magnetized in the center of the current sheet, making them incompressible.

Swift [1983] and *Swift and Allen* [1987] used simulations of a Harris current sheet with a fixed, uniform, normal magnetic field component to investigate thinning of the plasma sheet via an imposed dawn-to dusk electric field. In their work, tearing of the current sheet was suppressed due to the magnetization of the electrons via the relatively strong normal magnetic field assumed. Despite the lack of tearing, strong particle acceleration can be produced via the dawn-to-dusk electric field. Recent results by *Zwingmann et al.* [1990] confirm that unless the current sheet is very thin and the normal magnetic field is small then tearing is likely to be suppressed.

Wave perturbations can possibly lead to enhanced tearing [e.g., *Pritchett et al.*, 1989] by locally reducing the normal magnetic field strength, thereby allowing the electrons to become unmagnetized and tear. An alternate

---

[1]Now at Department of Space Sciences, Southwest Research Institute, San Antonio.

way, which is the subject of this paper, is to compress the current sheet via the dawn-to dusk electric field, thereby boosting the growth rate of tearing above threshold. A recent study by *Pritchett and Coroniti* [1990] also investigated tearing in the presence of a dawn-dusk electric field and a normal (but non-uniform) magnetic field component. Their results predict that tearing can occur even in the presence of small dawn-to-dusk electric fields and that the tail current sheet may always be unstable to tearing, even during quiet times. Because of this lack of threshold it is difficult to relate their work to the development of substorms.

There are two possible reasons for the lack of threshold in the work of *Pritchett and Coroniti* [1990]. First, they do not include the electron dynamics so that suppression from electron magnetization is not included. Second, they use reflecting particle boundary conditions which can lead to the pile up of current sheet particles about the near-earth boundary, thereby artificially producing the tearing of the current sheet (section 2). Indeed, tearing in these simulations tends to first appear near the near-earth boundary.

In this paper, a simpler but more comprehensive study of tearing in the presence of a finite normal field is made. The model (section 2) uses a Harris current sheet (similar to *Swift* [1983]) to represent the initial tail configuration. Both the electrons and ion dynamics are included self-consistently. A uniform normal magnetic field and dawn-to-dusk electric field are imposed, but their values are varied in a series of simulations to derive a threshold for tearing. This threshold, which is derived analytically in section 3, provides a means for estimating the position where the near-earth neutral point should develop given the initial conditions. The simulation results are presented in section 4 and confirm the derived threshold. A summary of results is given in Section 5.

While the above model reduces effects from boundary conditions, it is limited in that there is an unbalanced $j_y \times B_z$ force which causes the plasma to flow in the earthward direction. This acceleration takes place on the order of the period of an ion gyrating about the normal magnetic field (B_z). This period for typical magnetotail parameters takes place on time scales of the order of a few minutes. For the cases considered here as well as in many of the existing particle simulations of tearing (e.g., *Pritchett et al.*, 1989; *Pritchett and Coroniti*, 1990; *Zwingmann et al.*, 1990] the tearing occurs on shorter time scales. These shorter time scales are particularly relevant to investigations of tearing stabilization via the magnetization of the electrons. On these time scales, which is the focus of this paper, the acceleration from the unbalanced pressure is not significant. On longer time scales, the acceleration becomes more important and tends to suppress tearing. For such cases, the threshold dervied gives an overestimate of the conditions required for tearing onset.

2. SIMULATION MODEL

The development of tearing and associated particle acceleration is examined using a two-dimensional (three velocity) particle simulation code. The code has been used previously in applications involving the active injection of electron beams from spacecraft [*Winglee and Kellogg*, 1990; *Winglee*, 1990]. The code incorporates both electromagnetic and electrostatic fields and utilizes a system of multiple scale lengths to optimize computer memory while treating both the electromagnetic and electrostatic fields on large system sizes.

In the simulations, the particles initially have a Harris current sheet distribution [*Harris*, 1962] as described below. Added to this are (i) a uniform normal magnetic field component and (ii) a dawn-to dusk electric field. Because the system is uniform in x, periodic boundary conditions in x are used and particles are reinjected at the opposite boundary. This models an extended magnetotail with the particles continually flowing along the same field line. Reflecting x boundary conditions are not used in order to avoid pile-up of particles on the closed field lines near the earth. Indeed, test simulations with reflecting boundary conditions and a normal magnetic field had faster growth rates than the Harris current sheet with no normal magnetic field, which is unphysical.

In the z direction, an inverted-mirror image is added and field solutions are then derived from this enlarged system. With these boundary conditions, the magnetic field and particle trajectories across the z boundaries are continuous and the magnetic field in the lobes is allowed to float. This boundary condition is not very important since there are very few particles in the lobe regions.

The electron and ion density in a Harris current sheet has the form

$$n(z) = n_e / cosh^2(z/L_c) \qquad (1)$$

where n_e is the density in the center of the current sheet. The half width of the current sheet L_c is determined from Maxwell's equation and pressure equilibrium and is given by

$$L_c = 2 \frac{c}{v_c} \lambda_D \qquad (2)$$

where $\lambda_D = v_{Te}/\omega_{pe}$ is the electron Debye length in the center of the current sheet, v_{Te} and ω_{pe} are the electron thermal speed and plasma frequency, respectively, and v_c is the drift velocity in the y direction of the ions. The electrons drift with opposite velocity in y to the ions but have equal temperatures. The factor of 2 in (2) arises from the addition of the inverted-mirror image.

In the following, v_{Te} is taken to be $0.1\ c$. The relatively high value of v_{Te}/c is chosen to maximize the effective scale lengths simulated as well as the time step ($\Delta t = 0.15/\omega_{pe}$) without significantly modifying the relevant physics. In addition, an artificially light ion mass ($m_i = 10 - 50 m_e$) is used. This assumption further compresses the scale lengths, allowing the examination of relatively thick current sheets (i.e., sheets many ion-gyroradii wide), similar to actual conditions in the tail. Simulations with the different ion mass show that the initial growth rate of tearing is independent of the ion mass. However, the saturation time increases with increasing ion-to-electron mass. Thus, the predictions for tearing onset are not strongly modified by the use of artificially small ion masses.

The initial drift velocity v_c is set at 0.15 v_{Te} ($= 0.5v_{Ti}$) so that the electron cyclotron frequency in the lobe Ω_e is equal to $0.2\omega_{pe}$ and $L_c \simeq 8\rho_i$ where ρ_i is the gyro-radius of an ion in the lobe field. For an initial ion temperature of a few hundred eV, L_c is about 1000 km. The simulation system size is $512\Delta \times 256\Delta$ where the grid spacing is $\Delta = 2\lambda_D = 0.125\rho_i$. A total of about 200,000 particles are used in the simulations.

The growth rate for tearing in a Harris current sheet is [e.g., Galeev, 1982]

$$\gamma_t \simeq \pi^{-1/2}\omega_e(\frac{\rho_e}{L_c})^{5/2}(1 + \frac{T_i}{T_e}) \qquad (3)$$

which for the present parameters reduces to about $3.5 \times 10^{-4}\ \omega_{pe}$. This growth rate was verified in test simulations.

A simple criteria for suppression of tearing in the presence of a normal magnetic field is that the electrons become magnetized before tearing occurs, i.e.,

$$\gamma_t \lesssim \Omega_N \qquad (4)$$

where Ω_N is the electron cyclotron frequency arising from the normal magnetic field. For the above parameters, a normal magnetic field of 0.2% of the lobe fields is sufficient to suppress tearing. This field is much smaller than typical, normal fields in the tail.

3. Conditions for Compression and Tearing

Not only does the presence of a normal magnetic field lead to suppression of tearing, but it also allows the plasma to flow out along the field lines to higher altitudes via their own thermal motions (Figure 1a). This flow can lead to the depletion of the current sheet, further suppressing tearing. This outward flux is in part balanced by flows from the outer plasma sheet and lobe. However, due to the relatively low density in these regions, this inflow is unable to balance the outflow.

In particular, consider the flow of particles along a field line that stretches the entire length of the simulation system which starts near the lobe at z_l and then enters the central current sheet at z_c. The outward flux from the center of the current sheet is given approximately by

$$\text{Flux}_{\text{out}} \simeq <\frac{B_z}{B_x}> v_{Te} n_e(z_c) \qquad (5)$$

where $n_e(z_c)$ denotes the density near the inner part of the current sheet and $<\frac{B_z}{B_x}> v_{Te}$ is the average value of the particle speed in z (assuming $B_x^2 \gg B_z^2$). The inward

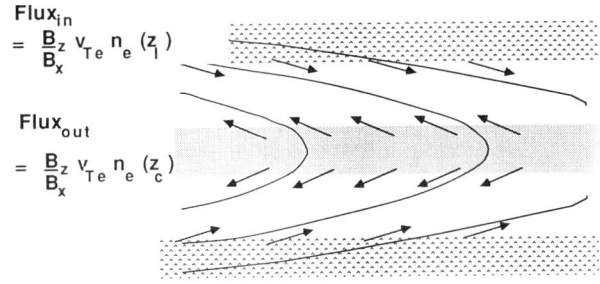

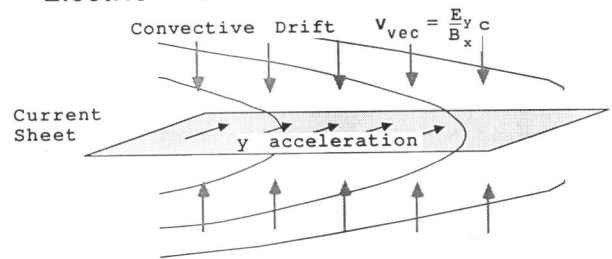

Fig. 1. Schematic of (a) thermal outflow of plasma in the presence of a normal magnetic field, and (b) the convective drifts and y acceleration produced by the dawn-to-dusk electric field.

flux moving along the same field line is approximately

$$\text{Flux}_{\text{in}} \simeq <\frac{B_z}{B_x}> v_{Te} n_e(z_l) \qquad (6)$$

and hence, the net outward flux is

$$\text{Flux}_{\text{net}} \simeq v_{Te}(n_e(z_c) - n_e(z_l)) \qquad (7a)$$

$$\simeq <\frac{B_z}{B_x}> v_{Te}\frac{\Delta z}{L_c} <n_e> \qquad (7b)$$

where $<n_e>$ is the average density between z_1 and z_2 and $\Delta z = \text{abs}(z_c - z_l) \simeq <\frac{B_z}{B_x}> L_x$. Hence, (7b) reduces to

$$\text{Flux}_{\text{net}} \simeq <\frac{B_z}{B_x}>^2 \frac{L_x}{L_c} v_{Te} <n_e>. \qquad (8)$$

This net outflow can be impeded in two ways. First, because the electrons with their faster speeds tend to initially propagate outward in front of the ions, a space-charge field (essentially an electrostatic E_z) develops and slows further propagation to a speed approximately equal to the maximum of the ion thermal speed v_{Ti} and the ion sound speed $v_s = (m_e/m_i)^{1/2}v_{Te}$ so that (8) becomes

$$\text{Flux}_{\text{net}} \simeq <\frac{B_z}{B_x}>^2 \frac{L_x}{L_c} v_0 <n_e> \qquad (9a)$$

with

$$v_o = \max(v_{Ti}, v_s). \qquad (9b)$$

The second way is through the convection of plasma into the center of the current sheet produced by the dawn-to-dusk electric field (Figure 1b). The drift speed of this convection is approximately

$$v_{vec} \simeq \frac{E_y}{<B_x>} c \qquad (10)$$

where $<B_x>$ is average value of B_x along the field line. Thus, from (9) and (10) the required electric field to stop the outflow is

$$\frac{E_y}{<B_x>} \simeq <\frac{B_z}{B_x}>^2 \frac{L_x}{L_c} \frac{v_o}{c} . \qquad (11)$$

It should be noted that the value of E_y needed to compress or convect the outer edges of the current sheet is smaller than that for the inner current sheet because of the difference in B_x. Thus, for the outer current sheet

$$\frac{E_{y,outer}}{B_{xL}} \simeq (\frac{B_z}{B_{xL}})^2 \frac{L_x}{L_c} \frac{v_o}{c} \qquad (12a)$$

where B_{xL} in the x component of the lobe magnetic field. For the parameters considered here (12a) reduces to

$$\frac{E_{y,outer}}{B_{xL}} \gtrsim 0.25 (\frac{B_z}{B_{xL}})^2 . \qquad (12b)$$

To compress the inner part of the current sheet, (10) should be evaluated where the electrons become unmagnetized, i.e, $d = \sqrt{\rho_e L_c}$ [cf., Galeev, 1982] so that

$$\frac{E_{y,inner}}{B_{xL}} \gtrsim (\frac{L_c}{\rho_e})^{1/2} (\frac{B_z}{B_{xL}})^2 \frac{L_x}{L_c} \frac{v_o}{c} \qquad (13a)$$

$$\simeq (\frac{L_c}{\rho_i})^{1/2} (\frac{B_z}{B_{xL}})^2 \frac{L_x}{L_c} \frac{v_{Te}}{c} \qquad (13b)$$

$$\gtrsim 2 (\frac{B_z}{B_{xL}})^2 . \qquad (13c)$$

In the present study, B_z/B_{xL} is varied from 1.5% to 3% and E_y/B_{xL} from 0 to 0.005 (i.e., a few mV/m). A reduced value of v_{Te}/c (which is relatively high in the simulations) would require smaller values of the electric field to produce the compression.

In the following numerical results, the dawn-to-dusk electric field is linearly ramped up over a period of $1000/\omega_{pe}$ to the quoted value of E_y. After this initial period E_y is held constant. This period is long with respect to both the electron and ion cyclotron periods (for the assumed mass ratios) as determined by the lobe fields but is short compared with the ion cyclotron period in the center of the sheet. This initial ramping allows the motion of the electrons and ions to develop self-consistently in the absence of unphysical jumps. The results are not strongly effected by the duration of the ramp provided it is short compared with the ion cyclotron period in the center of the current sheet.

Although the dawn-to-dusk electric field is varied in time, it is uniformly applied across the system. This is only an approximation to the real magnetotail where the variations in the solar wind conditions change the dawn-to-dusk electric field and strong non-uniformities may be present. Some of these non-uniformities develop in the simulation system since it is allowed to evolve self-consistently under the influence of the applied fields. In particular, it is shown in section 4.1 that the current sheet is inductive, i.e., the currents cannot be changed rapidly, so that the actual electric field in the center of the current sheet remains small. As a result of this inductance, the actual electric fields experienced by the particles is smaller than the applied electric field. Thus, the applied electric field has to be slightly higher than in (12) and (13) before tearing threshold is attained.

4. NUMERICAL RESULTS

4.1. Tearing as a Function of E_y

In this section, results from a series of simulations are presented to show the evolution of the tail magnetic field as a function of the dawn-to-dusk electric field. Time histories of the total energy in the three magnetic field components are shown in Figure 2 for five different values of E_y. In all cases, the normal magnetic field B_z is 1.5% of the lobe B_{xL} field, and the energies ϵ are normalized to the initial particle energy ϵ_0. For no applied electric field, the plasma sheet diffuses out along the field lines through their thermal motion (cf. Section 3) and the energy of the tail magnetic field (i.e., ϵ_x in Figure 2a) decreases. As this expansion occurs, currents in the x direction develop due to the differential flow between the ions and electrons along the field lines towards the earth and generate the y magnetic fields in Figure 2b. These B_y fields are most intense on the edges of the current sheet where their local magnitude can exceed that of the normal magnetic field component.

At an applied electric field of $E_y/B_{xL} = 0.0005$, this outward flow is stopped (Figure 2a). This value is about twice that for $E_{y,outer}$ given by (7b). As E_y is increased above this value, there is compression of the plasma sheet which increases the current density and associated tail-magnetic field energy. For $E_y/B_{xL} = 0.00125$, the current sheet is marginally unstable to tearing with small increases in the energy of B_z evident in Fig 2c. This value for $E_{y,inner}$ is about 2 – 3 times the value given by (8c), similar to the simulation value for $E_{y,outer}$.

There are also intense B_y fields generated during the compression, similar to the case for $E_y = 0$. However, their origin is different, being produced by the enhanced convection of electrons into the center of the sheet and their net earthward motion. This difference in the origin of the current produces a change in sign in B_y during compressive and expansive phases.

At still larger E_y, the initial increase in B_x is followed by a short period of magnetic field dissipation. This dissipation is associated with tearing of the current sheet and the creation of plasmoids with large increases in energy B_z (Figure 2c).

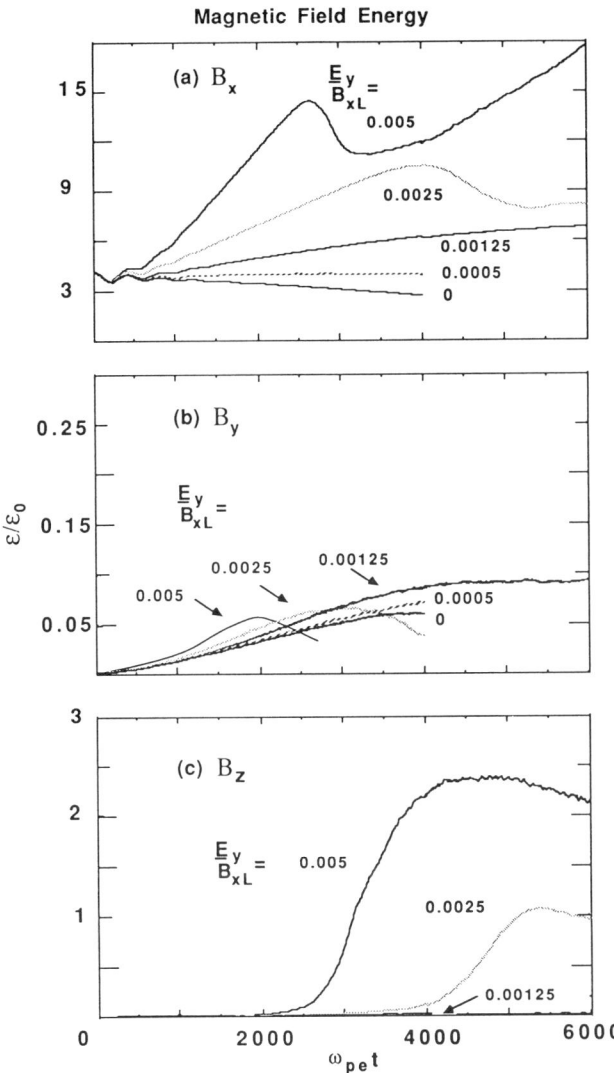

Fig. 2. Time histories of the energy ε in the three magnetic field components normalized to the initial energy in the particles ε_0 as a function of the applied E_y field. Initially B_z/B_{xL} is equal to 1.5%. For $E_y = 0$ the current sheet disperses, causing the decrease in the energy in B_x. At higher values, this dispersion is stopped and tearing can occur with a resultant large increase in B_z. In all cases strong B_y fields are generated due to the differential motion between electrons and ions.

The development of tearing of the current sheet is illustrated in Figure 3 which shows the magnetic field lines at four different times for the simulation with $B_z/B_{xL} = 0.015$ and $E_y/B_{xL} = 0.005$. At early times (Figure 3a), there is little change in the magnetic field lines but by Figure 3b tearing has begun with the formation of two magnetic islands. These islands eventually coalesce, to form the single large magnetic island or plasmoid in Figure 3c. This coalescence produces the temporary decrease in tail magnetic field

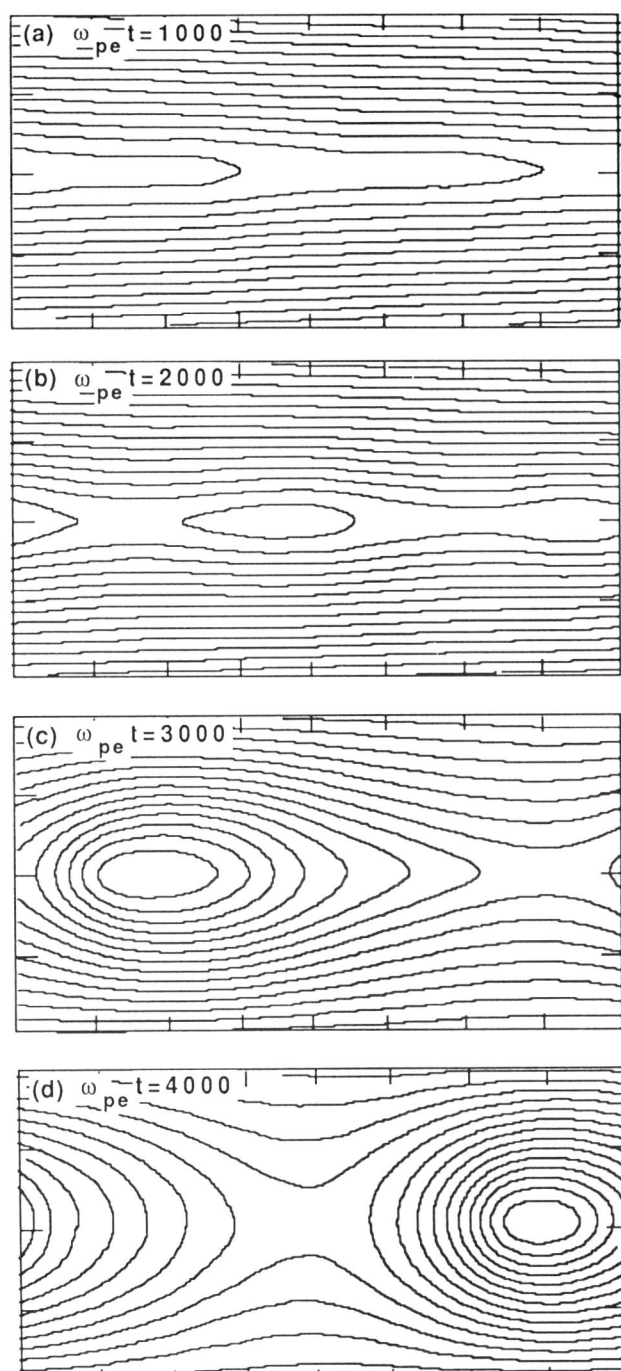

Fig. 3. Evolution of the magnetic field for $B_z/B_{xL} = 0.15$ and $E_y/B_{xL} = 0.005$. The initial tearing is accompanied by the formation of two magnetic islands which then coalesce to form one large plasmoid.

energy in Figure 2a. However, because of the continued presence of E_y, there is continued y acceleration and the field strength again grows and the plasmoid becomes more compact.

The particle acceleration and heating are illustrated in Figure 4 which shows the evolution of the ion $v_y - x$ phase space for the same times as in Figure 3. It is seen that the initial tearing (Figures 3b and 4c) is accompanied by some clumping of the ions in the center of the magnetic islands (as evidenced by the slightly enhanced phase space density in these regions) but little acceleration is evident (except for some weak bulk acceleration). The strongest acceleration appears during the coalescence of the islands in Figure 4d, with some ions being accelerated to velocities nearly 7 times their initial thermal velocity, i.e., an increase of nearly 50 in energy. Thus, while the applied electric field is responsible for some bulk acceleration it is the fields associated with the tearing and coalescence of islands that produces the most energetic particles.

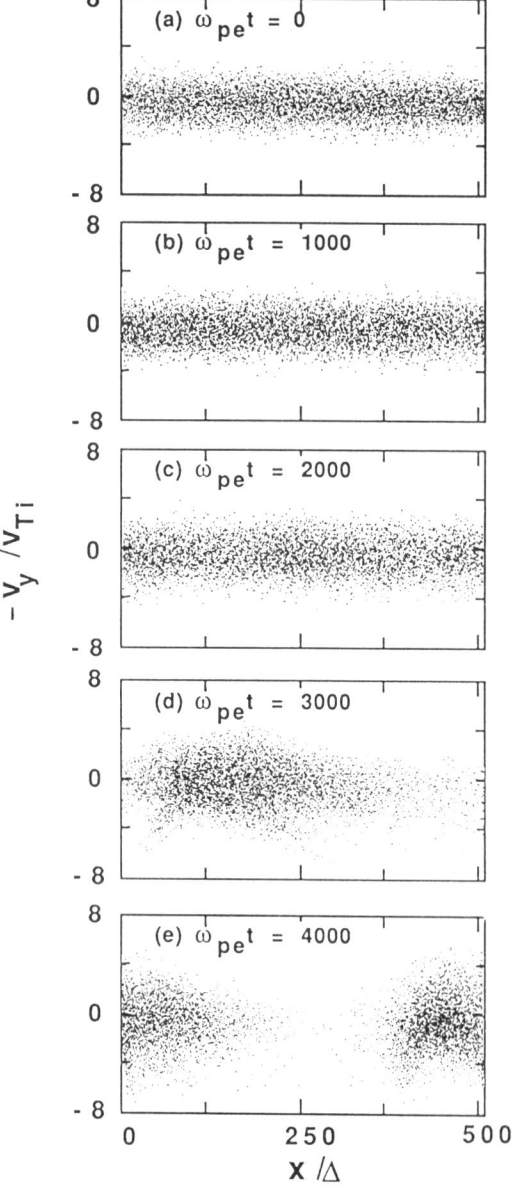

Fig. 4. The ion $v_y - x$ phase space corresponding to Figure 3. The initial tearing is accompanied by only local increases in density in phase space at the center of the magnetic islands with little acceleration. When the magnetic islands coalesce, high-energy tails are produced.

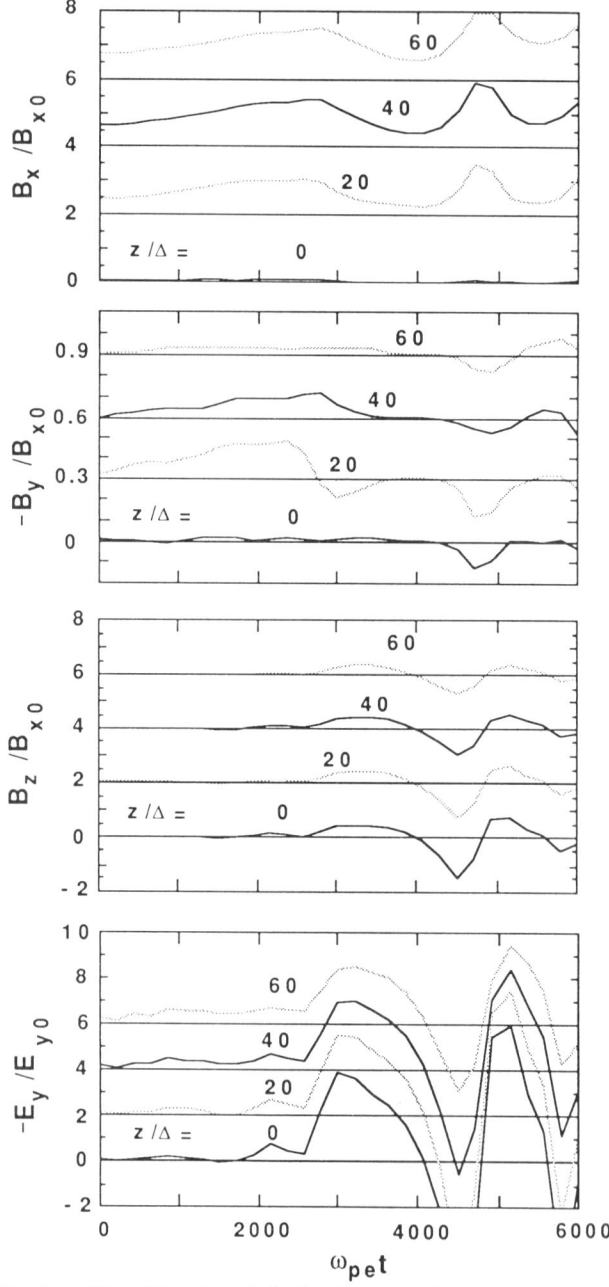

Fig. 5. Time histories of the local magnetic field components and E_y along the current sheet at $x/\Delta = 128$ and at the given z displacements from the center of the current sheet. Time averages of $500/\omega_{pe}$ have been made to reduce any high-frequency fluctuations. Before tearing occurs, B_y is anti-correlated with B_z while after tearing it is correlated. Both B_z and E_y are 90° out of phase. Note that the E_y field that the particles actually experience is much smaller than the applied field, particularly in the center of the current sheet.

In order to compare the predicted evolution of the fields with observations, Figure 5 shows the time histories of the three magnetic field components and E_y at fixed position along the magnetic field (at $x/\Delta = 128$) and at four different z displacements from the center of the current sheet. The time histories for the different points have been shifted linearly to aid clarity and the fields have been averaged over a period of $500/\omega_{pe}$ to remove high-frequency fluctuations. During the initial compression of the plasma sheet (i.e., $\omega_{pe}t \lesssim 2000$), the local value of B_x increases by nearly a factor of two (Figure 5a). This increase occurs in association with large increases in the magnitude of B_y but its sign is opposite to that of B_x, i.e., changes in B_y are anti-correlated with B_x. This B_y, as discussed earlier, is due to the differential motion between the electrons and ions and is strongest at about $d = \sqrt{\rho_i L_c} = 22\Delta$ where the ions become unmagnetized while the electrons are still magnetized. There is essentially no significant change in B_z at this stage.

With the formation of the plasmoid, the average B_y flips sign so that changes in B_y become correlated (rather than anti-correlated) with B_x. This change in sign is due to the fact that as the plasmoid forms, the ions with their larger gyro-radius are able to tear more readily than the electrons so that fractionally more ions are trapped in the plasmoid (this is also seen in plots of the electrostatic potential, not shown). As a result, the x current and corresponding B_y reverse sign. With the tearing, large changes in B_z are generated (as previously discussed) but they are 90° out of phase with those seen in B_x and B_y.

This phase lag between B_z and B_x is essentially that predicted by MHD theory. However, the generation of B_y and its correlation or anti-correlation with B_x is strictly a particle effect. These relative variations in all three components appear to be consistent with the tail observations presented by *Cattell and Mozer* [1984].

The time histories of the actual E_y that develops across the system is shown in the bottom panel of Figure 5 normalized to the applied field E_{y0}. It is seen that, at all points prior to plasmoid formation, the actual field is smaller than the applied field. It is largest near the lobe where the plasma is tenuous, being nearly 50% of E_{y0}. However, in the center of the current sheet where the plasma density is high and the particles are essentially unmagnetized, E_y remains small, typically being only about 20% of the applied field. Due to this inductance, higher electric fields than given in the criteria in (12) and (13) have to be applied before compression and tearing can occur.

With the development of the plasmoid, large E_y fields are seen. These large perturbations are anti-correlated with B_z and are due to the induced electric fields arising from a time rate of change in B_z as the plasmoid propagates past the observing point. This electric field is as predicted by MHD theory.

4.2. *Tearing as a Function of the Normal Magnetic Field*

As discussed in section 3, the thresholds for compression and tearing are also dependent on the strength of the normal magnetic field. This dependence is now investigated through a series of simulations in which the dawn-to-dusk electric field is held fixed at $E_{y0}/B_{xL} = 0.005$ and B_z is varied. From Figure 2, for $B_z/B_{xL} = 0.015$ this value of the applied electric field is about a factor of 4 above threshold for tearing. Thus, according to (12), increasing B_z by a factor of 2 should lead to the suppression of tearing. This suppression is illustrated in Figure 6 which shows the time histories of the energy in the three magnetic field components for the given values of the normal magnetic field component. For the two lower values of B_z, the initial increase in

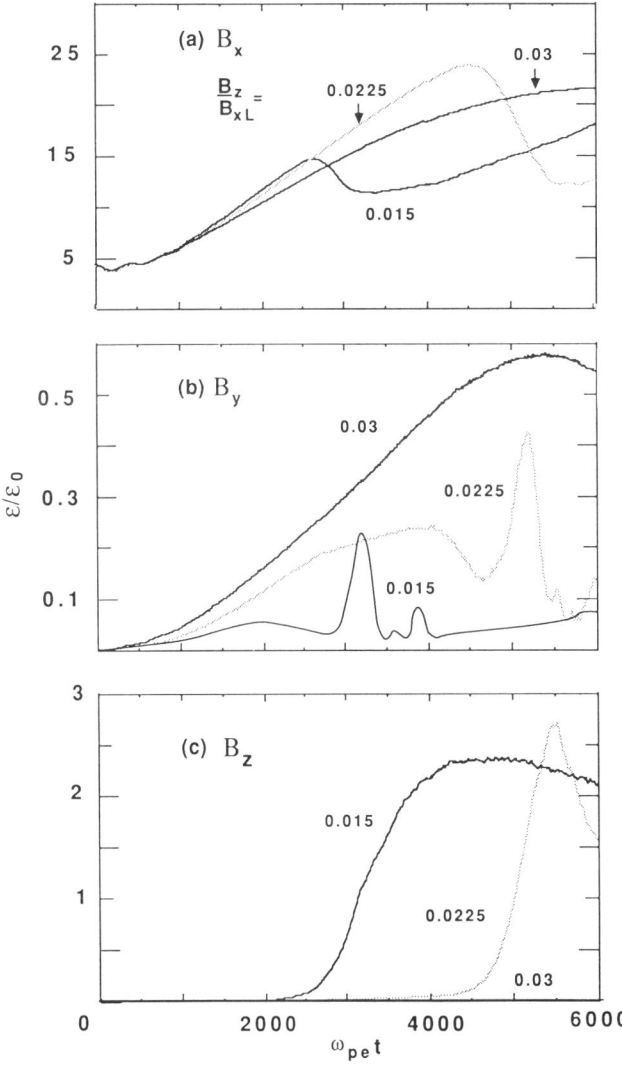

Fig. 6. As in Fig. 2, except $E_y/B_{xL} = 0.005$ and B_z/B_{xL} is varied. Tearing is suppressed for the highest value of normal magnetic field.

B_x is followed by a period of rapid dissipation of B_x and a large increase in the energy in B_z (Figures 6a and 6c). These variations, as discussed in the previous subsection, are associated with tearing and the formation of a plasmoid.

For the largest value of B_z, this tearing is not seen. Instead, the energy in B_x increases monotonically reaching its maximum at about $\omega_{pe}t = 6000$ (extension to longer times scales does not show tearing). At the same time, the energy in B_y essentially tracks that of B_x except that it reaches its maximum at about $\omega_{pe}t = 5500$ and there is no significant increase in the energy of B_z. This suppression is consistent with criterion (12).

The evolution of the magnetic profile for the highest value of B_z in Figure 6 is shown in Figure 7. The dotted lines indicate the spacing in x between adjacent field lines. During the compression of the current sheet (i.e., for $\omega_{pe}t \lesssim 6000$), the x displacement is seen to increase and the field lines become more horizontal. There is no evidence of any island formation. This stage is essentially equivalent to the growth phase of a substorm.

At later times, the current sheet is seen to expand or dissipate (this dissipation can also be produced by a reduction in E_y as in Figures 7e and 7f). Onset of this expansion phase is partially due to the development of intense electrostatic E_z fields during the compression or growth phase. The development of these fields is illustrated in Figure 8 which shows the electrostatic potential for the same times as in Figure 7. The shaded areas (solid lines) indicate regions of positive potential. As noted in subsection 4.1, due to the fast response of the electrons and their small gyro-radius, the electrons are more easily able to be convected into the center of the current sheet via the dawn-to-dusk electric field. This enhanced convection leads to the development of a net negative potential in the center of the current sheet (Figures 8b–8d). The associated E_z can be nearly an order of magnitude greater than the applied E_y at the height of compression. This electric field produces a convective y motion in the electrons through an $E_z \times B_x$ drift which actually overcomes the y acceleration produced by E_y. As a result, the electron current (not shown) reverses sign at about $\omega_{pe}t \simeq 6000$. This reversal leads to partial cancellation of the ion current, causing the annihilation of some of the tail magnetic field.

Another important factor causing the expansion is the particle heating during the compression phase. This heating is aided by waves induced by the electrostatic

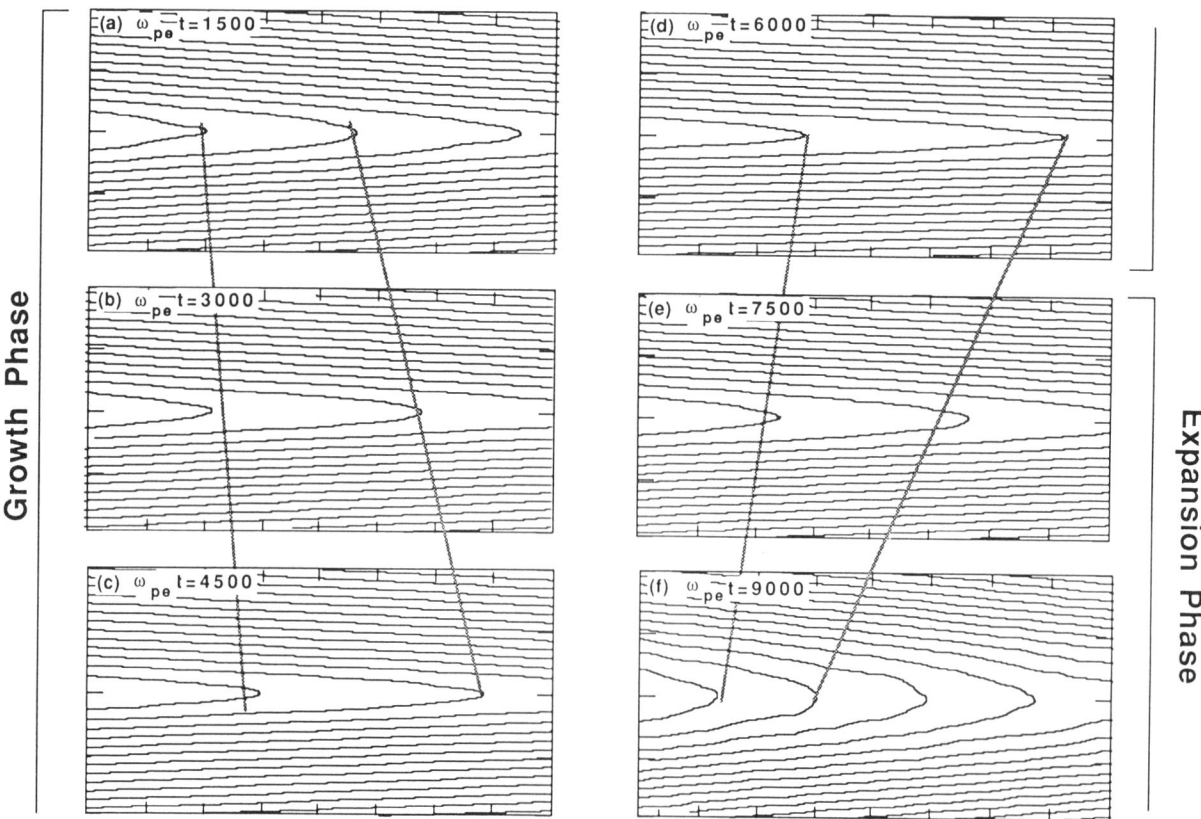

Fig. 7. Evolution of the magnetic field in the absence of tearing. Due to the heating of the plasma and the development of the quasi-static E_y, the current sheet is subject to a thermal dissipation at late times.

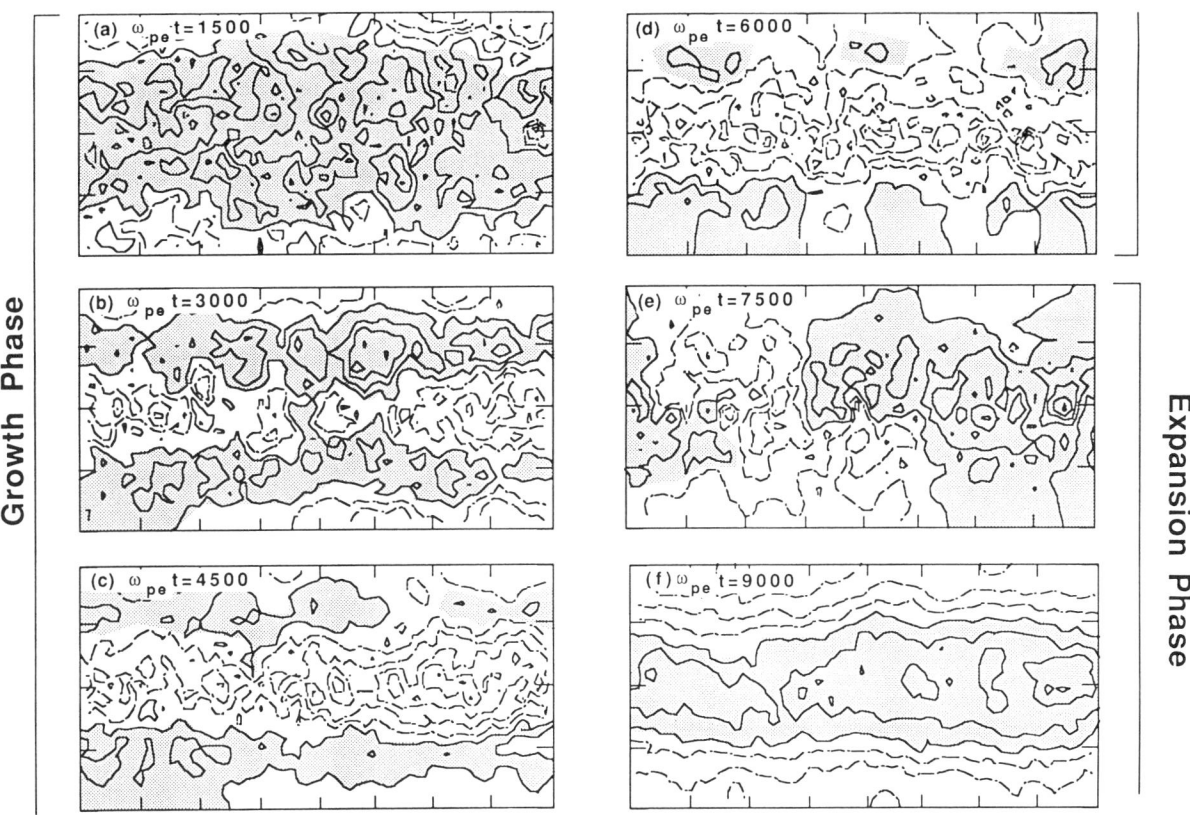

Fig. 8. The electrostatic potential for the same times as in Fig. 7. The shaded areas indicate positive potentials. During the growth phase, the inner current sheet has a net negative potential while during the expansion phase it becomes positive.

fields. Spectra of these waves across the current sheet during the growth and expansion phases are shown in Figure 9. The waves are broadband with the upper frequency cutoff decreasing with distance from the center of the current sheet. On the outer edges of the current sheet, the spectrum decreases approximately monotonically with frequency, with some enhancement at low frequencies during the expansion phase. Near the center of the current sheet, there is a strong enhancement of wave intensity near the electron plasma frequency as the electron motion changes from magnetized to unmagnetized. These spectra have similar properties to broadband electrostatic noise associated with the plasma sheet boundary layer [Scarf et al., 1974; Gurnett et al., 1976; Grabbe and Eastman, 1984].

Due to the heating produced by these waves, the thermal velocity of the particles can be raised so that (12) and (13) are no longer satisfied. At this stage, if the applied electric field does not increase (as in the present work), the hot particles can no longer be confined and the current sheet expands or dissipates, producing the more dipole-like fields in Figures 7e and 7f. This expansion is accompanied by reversals in E_z (Figures 8e and 8f) and B_y (not shown), as the hot electrons propagate out in front of the ions.

Finally it should be noted that while tearing does not occur, the induced earthward drifts give the ions an average energy which is fractionally larger than for the cases where tearing occurs. However, the energy stored in the magnetic field is smaller.

5. SUMMARY

The evolution of the tail magnetic field and associated particle acceleration depend strongly on the strength of (i) the normal magnetic field relative to the tail field and (ii) the dawn-to-dusk electric field. The normal magnetic field can lead to suppression of the tearing by forcing the electrons in the center of the current to follow magnetized orbits rather than unmagnetized orbits. Except for very thin normal magnetic fields, small normal fields (less than a percent) can suppress tearing. However, the dawn-to-dusk electric field can accelerate the plasma, boosting the growth rate of tearing. The competition between these two effects is examined in this paper, both analytically and via particle simulations.

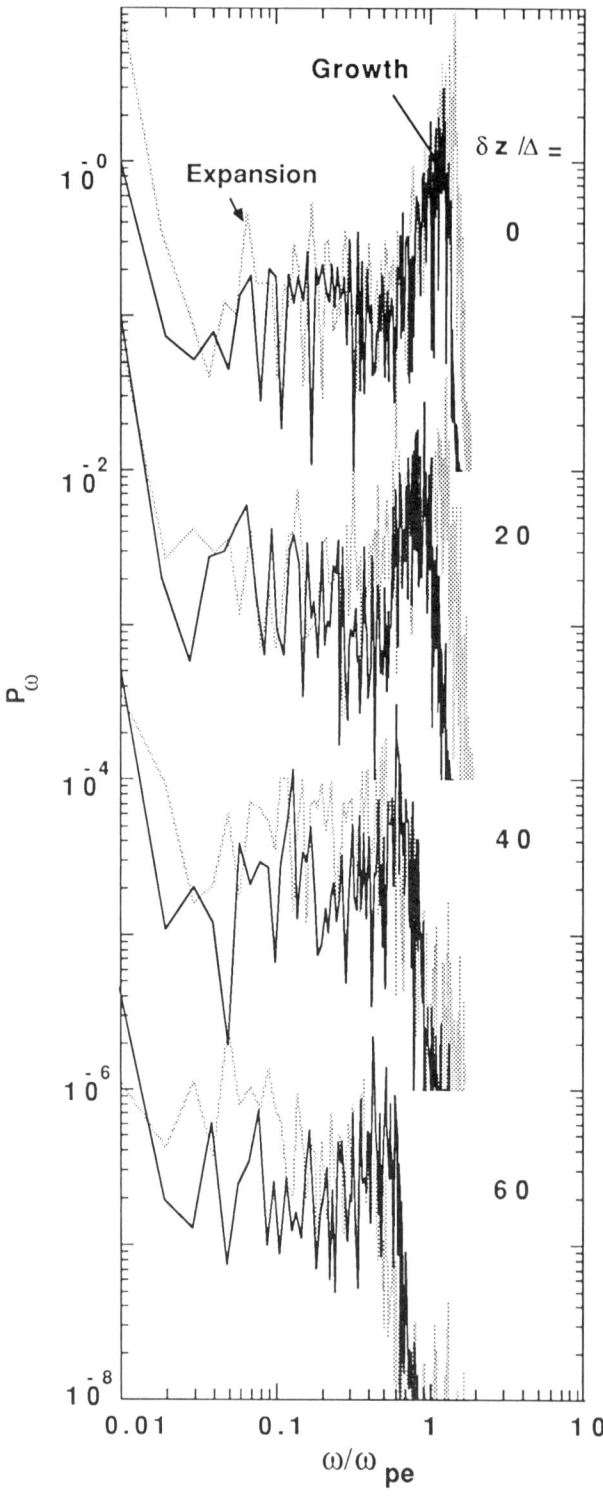

Fig. 9. The spectra of the electrostatic waves for the example in Figures 7 and 8 taken at the same points as in Figure 6. The spectra are characterized by a broad banded emissions. Near the outer edge, there is an enhancement at low frequencies particularly during the expansion phase. Near the center where the electron motion changes from magnetized to unmagnetized, there is an enhancement near the local plasma frequency.

It is shown that, in addition to this suppression of tearing, the presence of a normal magnetic field allows the plasma to diffuse outward due to thermal motions. This flow can lead to the depletion of the current sheet and the tail magnetic field. However, this outflow can be inhibited by two factors: (i) the development of quasi-static E_z arising from space-charge separation between the ions and electrons and (ii) convection into the plasma sheet driven by the dawn-to-dusk electric field. A criterion for tearing which relates the dawn-to-dusk electric field to the normal magnetic field is derived in section 3.

A series of simulations were performed verifying, semi-quantitatively, this criterion. However, these simulations show that the current sheet is inductive, with the induced fields being generated in opposition to the applied electric field thereby inhibiting rapid changes in the tail current. As a result, fractionally larger electric fields need to be applied before the actual electric field that is experienced by the particles is sufficient to satisfy the criterion. It should be noted that the simulation system used is similar to that of *Swift* [1983] and *Swift and Allen* [1987]. However, for their parameters, their system is below threshold and tearing is not observed in their simulations. In the present work, the normal magnetic field and/or dawn-to-dusk electric field are varied to drive the system from states that are tearing stable to states that are tearing unstable.

During the growth phase, it was shown that the compression of the current sheet was associated with the development of intense quasi-static E_z and B_y fields. These fields are due to particle effects and have not been previously reported in MHD simulations. They arise because the electrons with their smaller gyroradius are able to convect more easily into the center of the current sheet, producing a net negative potential (or E_z) and an x current. This E_z leads to the generation of broadband electrostatic noise with similar properties to that observed in the plasma sheet boundary layer [*Grabbe and Eastman*, 1984]. The magnitude of B_y tracks that of B_x but the sign depends on whether the current sheet is undergoing compression or expansion. Similar E_z and B_y fields have also been observed in particle simulations of *Swift* [1983] and *Swift and Allen* [1987].

After this compressive phase, the tail evolves in one of two ways. If the dawn-to-dusk electric field is sufficiently strong, tearing can occur. With the tearing, magnetic islands or plasmoids develop. The coalescence of the magnetic islands can lead to the production of high-energy tails with energies nearly 50 times their initial thermal energy in the present simulations. In addition, because the ions have larger gyro-radii than the electrons, they are more susceptible to tearing than the electrons, and the plasmoids tend to be positively charged. As a result, E_z and B_y flip sign so that changes in B_y become correlated (rather than anti-correlated) with B_x. The B_z field associated with the plasmoid are $90°$ out of phase with B_x as predicted by MHD theory. These correlations in all three components appear to be consistent with observations of plasmoids in the tail [*Cattell and Mozer*, 1984].

If the dawn-to-dusk electric field is insufficient to produce tearing, then the current sheet can be subject to thermal dissipation. This disruption is due to two processes. The first is due to the E_z field which can produce a bulk y current in the electrons via an $E_z \times B_x$ drift, thereby cancelling some of the ion current and annihilating some of the tail field. Second, strong particle heating via the broadband electrostatic noise associated with the quasi-static E_z fields can raise the temperature above the value that can be contained by the dawn-to-dusk field and the current sheet expands or disrupts. Even though tearing does not occur, there is strong particle energization through induced earthward drifts with the average energy of the ions being fractionally greater than when tearing occurs.

This, dissipation is in many ways analogous to the "thermal catastrophe" proposed by *Goertz and Smith* [1989]. The main difference is that the particle heating is not produced by Alfvén waves but rather via electrostatic processes This dissipation may be inhibited if effects arising from magnetospheric-ionospheric coupling are included. In particular, development of the perpendicular electric field E_z can lead to the development of strong field-aligned currents which couple into the ionosphere. These currents and the associated particle acceleration are discussed in a companion paper [*Dusenbery et al.*, 1991]. The current can lead to outflow of the heated electrons from the current sheet and their replacement by cold return-current ionospheric electrons. Through this coupling the temperature of the electrons in the current sheet may be sustained at lower values, thereby allowing better compression of the plasma sheet than predicted here.

Acknowledgments. This work was supported by NASA's Ionospheric Physics Program and under grant NAGW-1587, and National Science Foundation grants ATM-8719371 and ATM-9020577 to the University of Colorado. The particle simulations were performed on the CRAY Y-MP at the San Diego Supercomputer Center which is supported by the National Science Foundation.

REFERENCES

Baumjohann, W., The plasma sheet boundary layer and magnetospheric substorms, *J. Geomagn. Geoelectr.*, 40, 157, 1988.

Birn, J., Three-dimensional computer modelling of dynamic reconnection in the magnetotail: Plasmoid signatures in the near and distant tail, in *Magnetic Reconnection in Space and Laboratory Plasmas*, Geophys. Monogr., vol. 30, edited by E. W. Hones, Jr., AGU, Washington, D.C., 1984.

Birn, J., and E. W. Hones, Jr., Three-dimensional computer modelling of dynamic reconnection in the geomagnetic tail, *J. Geophys. Res.*, 85, 1214, 1980.

Cattell, C. A., F. S. Mozer, Substorm electric fields in the earth's magnetotail, in *Magnetic Reconnection in Space and Laboratory Plasmas*, Geophys. Monogr., vol. 30, edited by E. W. Hones, Jr., AGU, Washington, D.C., 1984.

Dusenbery, P. B., R. M. Winglee, and G. A. Dulk, Development of field-aligned currents and auroral particle acceleration during substorms, *Proc. of the Chapman Conf. on Substorms*, this issue, 1991.

Galeev, A. A., Magnetospheric tail dynamics, in *Magnetospheric Plasma Physics*, A. Nishida, ed., Developments in Earth and Planetary Sciences No. 04, D. Reidel, Boston, 1982.

Goertz, C. K., and R. A. Smith, Thermal catastrophe model of substorms, *J. Geophys. Res.*, 94, 6581, 1989.

Grabbe, C. L., T. E. Eastman, Generation of broadband electrostatic noise by ion beam instabilities in the magnetotail, *J. Geophys. Res.*, 89, 3865, 1984.

Gurnett, D., L. Frank, and R. Lepping, A region of intense plasma wave turbulence on auroral field lines, *J. Geophys. Res.*, 81, 6059, 1976.

Harris, E. G., On a plasma sheath separating regions of l directed magnetic field, *Nuovo Cimento*, 23, 115, 1962.

Hones, E. W., Jr., Plasma sheet behavior during substorms, in *Magnetic Reconnection in Space and Laboratory Plasmas*, Geophys. Monogr., vol. 30, edited by E. W. Hones, Jr., AGU, Washington, D.C., 1984.

Kan, J. R., Developing a global model of magnetospheric substorms, *EOS*, 71, 1083, 1990.

Lembége, B., and R. Pellat, Stability of a thick two-dimensional quasineutral sheet, *Phys. Fluids*, 25, 1995, 1982.

Lyons, L. R., and A. Nishida, Description of substorms in the tail incorporating boundary layer and neutral line effects, *Geophys. Rev. Lett.*, 15, 1337, 1988

McPherron, C. T. Russell, and M. P. Aubry, Satellite studies of Magnetospheric substorms on August 15, *J. Geophys. Res*, 78, 3131, 1968.

Pritchett, P. L., and F. V. Coroniti, Plasma sheet convection and the stability of the magnetotail, *Geophys. Res. Lett.*, 17, 2233, 1990.

Pritchett, P. L., F. V. Coroniti, R. Pellat, and H. Karimabadi, Collisionless reconnection in a quasi-neutral sheet near marginal stability, *Geophys. Res. Lett.*, 16, 1269, 1989.

Scarf, F. L., L. A. Frank, K. L. Ackerson, and R. P. Lepping, Plasma wave turbulence at distant crossings of the plasma sheet boundaries and neutral sheet, *Geophys. Res. Lett.*, 1, 189, 1974.

Swift, D. W., A two-dimensional simulation of the interaction of the plasma sheet with the lobes of the earth's magnetotail, *J. Geophys. Res*, 88, 125, 1983.

Swift, D.W., and C. Allen, Interaction of the plasma sheet with the lobes of the earth's magnetotail, *J. Geophys. Res*, 92, 10015, 1987.

Winglee, R. M., and P. J. Kellogg, Electron beam injection during active experiments. 1. Electromagnetic wave emissions, *J. Geophys. Res.*, 95, 6167, 1990.

Winglee, R. M., Electron beam injection during active experiments. 2. Collisional Effects, *J. Geophys. Res.*, 95, 6190, 1990.

Zwingmann W., J. Wallace, K. Schindler, and J. Birn, Particle simulation of magnetic reconnection in the magnetotail configuration, *J. Geophys. Res*, 95, 20877, 1990.

The Development of Field-Aligned Currents and Auroral Particle Acceleration During Active Times

P. B. DUSENBERY, R. M. WINGLEE[1], AND G. A. DULK

Department of Astrophysical, Planetary and Atmospheric Sciences
University of Colorado, Boulder

The evolution of the auroral current system, particle acceleration and wave emissions during active times are examined through two-dimensional (three-velocity) electrostatic particle simulations which include a collisional ionosphere and convergent field lines. It is shown that cross-field currents associated with the compression and expansion of the tail current sheet during active times can induce intense field-aligned currents down into the ionosphere. The associated particle acceleration is determined via coupling between the auroral potential, wave-particle interactions and the mirror force. The coupling of these processes lead to the development of a variety of distinct features in the particle distributions both along and across the field lines including (a) electron beams, loss-cones, trapped components and conics in association with Langmuir, electron acoustic and upper hybrid waves and (b) ion beams, conics and the preferentially accelerated heavy ions in association with lower-hybrid and ion cyclotron waves.

1. INTRODUCTION

During the expansion phase of a substorm, intense field-aligned currents develop in association with the brightening and poleward expansion of the auroral zone. In a companion paper [*Winglee*, 1991], large quasi-static potentials across the field lines are predicted to develop during this phase in the magnetotail due to the differential motion of the electrons and ions through the current sheet. In this paper, the response of the ionosphere to these cross-field potentials and currents is investigated, particularly, the characteristics of the particle acceleration and wave emissions in the auroral region.

The auroral acceleration region is a highly structured and inhomogenous system. In-situ electric field observations show the presence of quasi-static fields (or electrostatic shocks) and double layers localized to very small spatial scale lengths *across* the field lines, from a few hundred kilometers down to less than ten kilometers [e.g., *Mozer et al.*, 1980; *Redsun et al.*, 1985; *Weimer et al.*, 1985]. There are also strong non-uniformities *along*

[1]Now at Department of Space Sciences, Southwest Research Institute, San Antonio.

Magnetospheric Substorms
Geophysical Monograph 64
Copyright 1991 American Geophysical Union

the field lines. For example, ion beams and associated electrostatic shocks are most commonly observed above about 3000 km [*Collin et al.*, 1981; *Gorney et al.*, 1981; *Redsun et al.*, 1985] with the occurrence probability increasing with altitude. On the other hand, the occurrence probability of ion conics (i.e., perpendicularly accelerated or heated ions) tends to be more uniformly distributed along the field lines. Attempts to model the observed electron distributions also suggests a non-uniform distribution of electric fields [*Gurgiolo and Burch*, 1988].

There is also observational evidence that the particle acceleration is not solely determined by acceleration via a simple quasi-static electric field. *Redsun et al.* [1985] report that in a statistical study of the properties of 1073 electrostatic shocks, the average peak energy of the downflowing electrons tends to be higher than the peak ion energy. *Reiff et al.* [1988] using data from near-conjugate observations from DE 1 and DE2, have shown that the peak energy of the ions is typically 30–50% smaller than the electric potential inferred from the electron distribution.

Wave-particle interactions may also be important in determining the particle dynamics. An important example is the generation of ion conics and the preferential acceleration of heavy ionospheric ions [e.g., *Dusenbery and Lyons*, 1981; *Chang and Coppi*, 1981; *Ashour-Abdalla and Okuda*, 1983; *Bergmann et al*, 1988; *Dusenbery et al.*, 1988; *Winglee et al.*, 1989]. These models, while predicting some of the observed characteristics, neglected the effects of the auroral potential and the mirror force. As a result, a direct comparison with observations is difficult and an unambiguous signature in observations for some of these processes has yet to be made [e.g., Kintner and Gorney, 1984].

The purpose of this paper is to present initial results from a model for the development of auroral particle acceleration during active times as the tail current varies. It characterizes the magnetospheric conditions and the induced auroral particle acceleration and wave emissions. The model differs from existing models by including self-consistently the effects of (a) quasi-static electric fields, (b) the mirror force associated with convergent field lines and (c) modification of the particle flow and energy transport produced by wave-particle interactions in a multi-ion component plasma.

Such a comprehensive study has not been attempted before. For example, the above references that investigate the wave-particle interactions neglect the effects of the auroral potential and propagation of particles along the field lines. Conversely studies that investigate the evolution of the particle distribution in the auroral potential [e.g., Croley et al., 1978; Gurgiolo and Burch, 1988] neglect the effects of wave-particle interactions on the particle dynamics. The simulation model presented in this paper provides one of the first comprehensive investigations of the coupling of these processes on long time scales.

It should be noted though, that because the simulation model tracks the full particle dynamics it is unable to model realistic scale lengths along the field lines. Instead, the simulation model uses compressed scale lengths but maintains the ordering of important scale lengths like the ion gyroradii relative to the scale lengths of the auroral potential along and across the field lines. In particular, a particle's interaction with the macroscopic fields (i.e. auroral potential and mirror force) does not depend on their scale lengths provided they are long relative to the plasma scale lengths. Since this is true in the simulation model presented, the the particles can experience similar potential drops and wave-particle interactions as actual particles moving along the auroral field lines. The validity of the modelling can also be tested against its ability to predict the features observed in the wave and particle data.

The model is described in section 2. The free energy (section 3) used to drive the auroral potential and associated particle acceleration is provided by a cross-field current in the magnetotail [cf. Winglee et al., 1988]. These cross-field currents are shown in a companion paper [Winglee, 1991] to be a natural consequence of the differential acceleration of the ions and electrons in the magnetotail current sheet. Two important applications of the model are considered.

The first application (section 4) is the generation of electron beams and conics, their associated wave emissions and spatial distribution along and across the field lines. The second application (section 5) is the generation of ion conics and relative heating of H and heavy ionospheric ion species. Ion-ion streaming instabilities are important in this latter application, leading to the preferential acceleration of the heavy ions. The mirror force plays an important role in the perpendicular heating of the electrons and the generation of upflowing ion conics in the return-current regions. A summary of results is given in section 6.

2. SIMULATION MODEL

The simulation model used is similar to that of Winglee et al. [1988] except that three additional features have been included : (i) a two-ion component plasma, (ii) convergent field lines and (iii) a collisional ionosphere. The presence of two-ion components allows a comparison of the acceleration and heating of the dominant H ions relative to minority heavy ionospheric ions such as He^+ and O^+. In the following, the plasma is assumed to comprise 80% H^+ and the remaining 20% comprising He^+.

The presence of convergent field lines plays two important roles. First, the mirror force effectively creates a resistance for downward propagating particles. Due to this resistance, electric fields are set up along the field lines in an effort to maintain the auroral current system which modifies both the electron and ion dynamics. Second, the outward acceleration of ionospheric ions is aided via the conversion of perpendicular energy to parallel energy.

The simulation model also includes a collisional ionosphere with elastic scattering collisions of low-energy electrons and ions, and ionizing and energy-degrading collisions between high-energy electrons and neutrals [cf. Winglee, 1990]. These collisions provide (i) Pederson conductivity which allows the closing of currents in the ionosphere, (ii) additional source of plasma and (iii) a sink for the energy of the precipitating energetic electrons. This more realistic modelling, particularly the dissipation of the energy of the high-energy precipitating electrons in a collisional ionosphere, allows the simulations to be run more than an order of magnitude longer than previously possible by removing problems with boundary conditions.

With these three effects included, the particle dynamics is fairly complicated due to the coupling between wave-particle interactions and the macroscopic fields. To allow some simplification, only electrostatic fields are considered in the present work. This approximation is justified to first order since in low beta plasmas much of the particle dynamics is dominated by electrostatic processes rather than by electromagnetic fields. The presence of intense double layers, perpendicular electrostatic shocks, electrostatic ion-cyclotron and lower-hybrid waves are just a few examples of important electrostatic processes. Further, fully electromagnetic simulations by Winglee and Kellogg [1990] of the propagation of electron beams in the ionosphere indicate that the electrostatic fields are the dominant ones.

Computer memory and time are further saved by using an artificially small ion-to-electron mass ratio with $m_H = 50 m_e$ which compresses both the time and length scales without significantly changing the physics. In the following the simulations are run over several thousand electron plasma periods, or equivalently several hundred ion cyclotron periods. This duration while short to a substorm is long compared with the transit time of ions out from the ionosphere through the auroral potential, i.e., it is sufficiently long to model the characteristics of induced ionospheric flows for given conditions in the magnetosphere.

A system size of $L_x \times L_y = 1024\Delta \times 128\Delta$ is used where Δ is equal to a Debye length of the magnetospheric electrons or equivalently to $1/\sqrt{m_H}$ of a gyro-radius of the magnetospheric H ions. The system when scaled to the latter length scale corresponds to several tens of kilometers.

While this distance is only a few percent of the actual auroral acceleration region, we believe that this is not a strong limitation. In particular, the macroscopic properties of the distributions are determined by only the amount of electric field and mirror force experienced by the particles and not by their scale lengths, provided the macroscopic scale lengths remain large compared with intrinsic plasma scale lengths (particularly in the absence of wave-particle interactions). This ordering of macroscopic and plasma scale lengths is maintained in the present model.

The presence of intense wave-particle interactions can lead to the scattering of some of the particles and thereby modify their subsequent flow along and across the field lines. Some of these wave-interactions may not be able to reach complete saturation because of variations over several wavelengths of the plasma conditions arising from the compressed scale lengths. Thus, while the simulations are able to excite many of the important or dominant wave modes, their effect on the particle distributions may be underestimated. Nevertheless, at least some of the effects of these wave-particle interactions will appear in the particle distributions so that their relative importance in the particle dynamics can be investigated. A close comparison between the modelling and the observed characteristics of the variety of waves and particle acceleration and heating seen on auroral field lines provides stringent testing of the modelling and allows the relative importance of various processes to be measured.

The initial density profile assumed in the simulations is shown in Figure 1a. The left-hand side represents high altitudes and has relatively low density. The density increases by a factor of four, approximately linearly with x (i.e., decreasing altitude), until the collisional ionosphere is reached at $x/\Delta = 825$. The plasma density and frequency at this point are hereafter

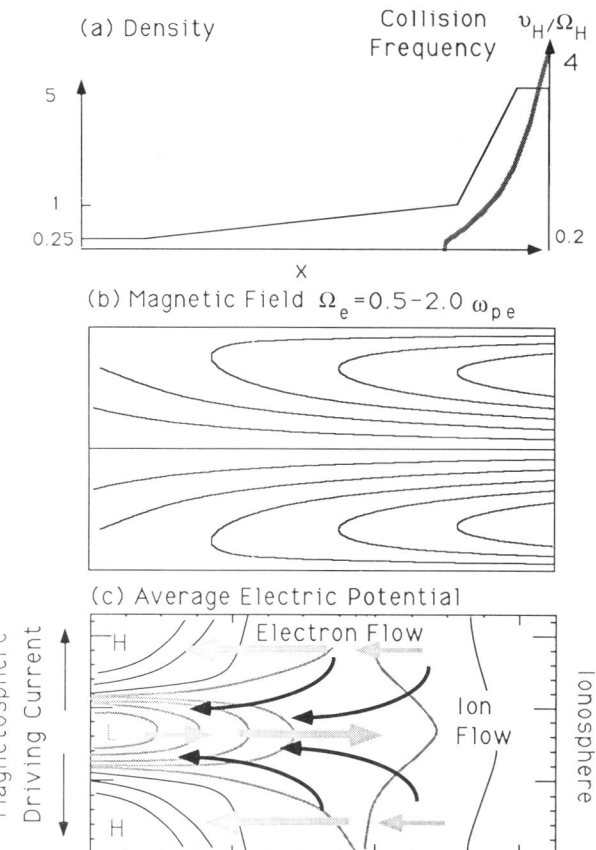

Fig. 1. Schematic of the simulation model showing (a) the density profile and the variation of the collision frequency across the ionosphere, (b) the assumed magnetic field and (c) the induced currents and electrostatic potential driven by the magnetospheric cross-field current.

denoted by n_e and ω_{pe}, respectively. Between the region $825 < x/\Delta \leq 975$, the density increases by an additional factor of five, after which the density is assumed to be constant. Thus, there is a total change of a factor of 20 in density across the system.

The initial temperature of the plasma is assumed to vary inversely with density so that the system is in approximate pressure equilibrium. The thermal velocity of the electrons at the magnetospheric boundary v_{Te} is assumed to be equivalent to 100 eV and that at the ionospheric boundary equal to 5 eV. The ion temperatures are assumed to be equal to the local electron temperature which is approximately valid for initially quiet auroral conditions. This assumption is not very important to the present work since the particles are accelerated and heated to energies much higher than their initial energies.

Collisions are first introduced just prior to the start of the large jump associated with the ionosphere (i.e., at $x/\Delta = 775$) with an ion-neutral collision frequency smaller than both the local ion plasma and cyclotron frequencies, specifically being 4% of the ion cyclotron frequency. The collisional frequency is increased by a factor of 40 across the remaining length of the simulation system, so that the ion-neutral collisional frequency fractionally exceeds the ion-cyclotron frequency at lowest altitudes. This variation in collision frequency is similar to ionospheric conditions and thereby allows the plasma near this region to evolve self-consistently without the need for the inclusion of any artifical boundary conditions. At the other end of the system, any particles crossing the magnetospheric boundary are taken out from the system and are not replaced.

The magnetic field assumed in the simulations is shown in Figure 1b. The overall magnetic field is given by

$$B_x = B_{xo}(1 + \beta x/L_x) \quad (1)$$

$$B_y = -B_{xo}\beta(y - L_y/2)/L_x . \quad (2)$$

In the following, B_{xo} is taken to be such that the electron cyclotron frequency at $x/\Delta = 0$ is equal to the local plasma frequency and β is equal to 3 so that $B_{min}/B_{max} = 0.25$. At the y boundaries, the condition $B_y = 0$ is imposed so that there are no jump discontinuities at the y boundaries. This is achieved by placing an additional y dependence on B_y and modifying B_x so that Maxwell's equations remain satisfied. The exact form of the magnetic field in this region is not important since most of the action occurs in the center of the simulation system.

3. DEVELOPMENT OF THE AURORAL CURRENT SYSTEM

As discussed in section 1, the auroral currents and particle acceleration are driven by a cross-field current, similar to that in *Winglee et al.* [1988]. These cross-field currents can be generated by the differential acceleration of ions and electrons in the current sheet in the magnetotail [e.g., *Swift and Allen*, 1987; *Winglee*, 1991]. The present model only seeks to model the effects of such a current on the auroral zone; a self-consistent model with a true magnetotail included is beyond the scope of this work. Instead, we model this cross-field current as it would appear mapped down to the top of the auroral zone by placing a cross-field current at the magnetospheric boundary of the simulation system. This is achieved by continuously placing ions and electrons at the left-hand boundary in the center of the system over a width $L_b = 32\Delta$. This width when scaled to an ion gyro-radius corresponds to tens of kilometers which is typical of discrete arcs [e.g., *Mozer et al.*, 1980; *Redsun et al.*, 1985; *Weimer et al.*, 1985]. These stuctures would map to features of several hundred to thousands of kilometers in the magnetotail. The v_y velocity of the ions is made sufficiently large that they propagate across the field line setting up the cross-field current. They are also given a slightly upward velocity so that by the time they stream across the field lines and are neutralized by return-current electrons, they leave the system. This

assumption isolates any effects these energetic ions might have on the auroral plasma.

The induced flows and potential are shown schematically in Figure 1c. Electrons in the middle of the system see a negative charge excess near the magnetospheric boundary which accelerates the electrons down into the ionosphere. This region is hereafter called the beam region. Along adjacent field lines, there is a positive charge excess which accelerates the ionospheric electrons upward to produce a return current. Current closure is produced by the acceleration of ions into the beam region as well as scattering of electrons in the collisional ionosphere from the beam region out into the return-current region.

The characteristics of the driving current can be related to the auroral currents via current continuity as follows. Suppose that the total driving current has the form

$$I_D = 2\alpha(t)en_e v_{TH} L_x \quad (3)$$

where L_x is the system length and α is a time-dependent proportionality constant and is a measure of the strength of the driving current In addition, suppose that the integrated parallel current induced in the plasma is given by

$$I_{\|b} \simeq en_e(x)v_{be}L_b \quad (4)$$

where v_{be} is the bulk velocity of the plasma electrons and $n_e(x)$ is the local plasma density. (The factor of two in (3) takes into account that there are outward currents from both sides of the beam region.) Then, current continuity requires that the time average of the two currents match, i.e., $I_{\|b} \simeq I_D$, and hence the bulk velocity expected to be induced in the plasma electrons is

$$v_b \simeq 2\alpha(t)\frac{L_x}{L_b}\frac{n_e}{n(x)}\left(\frac{m_e}{m_H}\right)^{1/2}v_{Te} \quad (5)$$

where $v_{TH} = (m_e/m_H)^{1/2}v_{Te}$ has been used. If the ions fall through the same potential as the electrons then they should attain a bulk velocity

$$v_{bi} = -\left(\frac{m_e}{m_i}\right)^{1/2}v_{be} \quad (6)$$

where the subscript i denotes either H or He.

In the following, the $\alpha(t)$ is essentially zero initially and then ramped up to its maximum value of 0.3 over a period of $240/\omega_{pe}$. After this period, α is kept constant. The corresponding bulk electron speed given by (6) is about 5 v_{Te} (i.e., 3 keV), taking $n(x)$ as the average density in the collisionless part of the plasma, i.e., $n(x) = 0.5n_e$.

4. ELECTRON ACCELERATION AND HIGH-FREQUENCY WAVES

4.1 The Beam Region

The influence of the quasi-static fields, mirror force and wave-particle interactions in producing the electron acceleration is illustrated in Figure 2 which shows the early evolution of the $v_x - x$ phase space of the beam electrons. Initially (Figures 2a–2c), the beam electrons are only mildly energetic with a mean parallel velocity of about 4–5 v_{Te} and a maximum perpendicular velocity of about v_{Te} (not shown). This parallel speed is consistent with (6). At this stage the beam has not travelled sufficiently far into the convergent magnetic field to produce strong mirroring. However, the beam is unstable to a variety of waves, including Langmuir waves, which produces the vortices and scatters some of the electrons to lower pitch angles. The mirror force acting on these electrons causes them to mirror at higher altitudes than would be expected from single particle trajectories. These two processes produce the large back-scattered component indicated by the arrows.

Due to this reflection the parallel current is no longer able to maintain current continuity. As a result, the parallel electric field increases so that later electrons are accelerated to higher parallel energies and hence inhibits their mirroring. For example, in Figures 2d and 2e electrons in the region $x/\Delta \lesssim 250$ are accelerated to speeds of the order of 10 v_{Te}.

The evolution of the total electron distribution in the beam region at high altitudes ($x/\Delta = 150$, $y/\Delta = L_y/2$ and averaged over a region $100\Delta \times 20\Delta$) is shown in Figure 3. At this altitude, the beam density is relatively high compared with the ambient density so that there is always a well-defined beam component with $v_\| \gtrsim 4 v_{Te}$ and an ambient backscattered component with $v_\| \simeq -v_{Te}$. During the period shown on the left-hand side, there is continual increase in the bulk velocity of

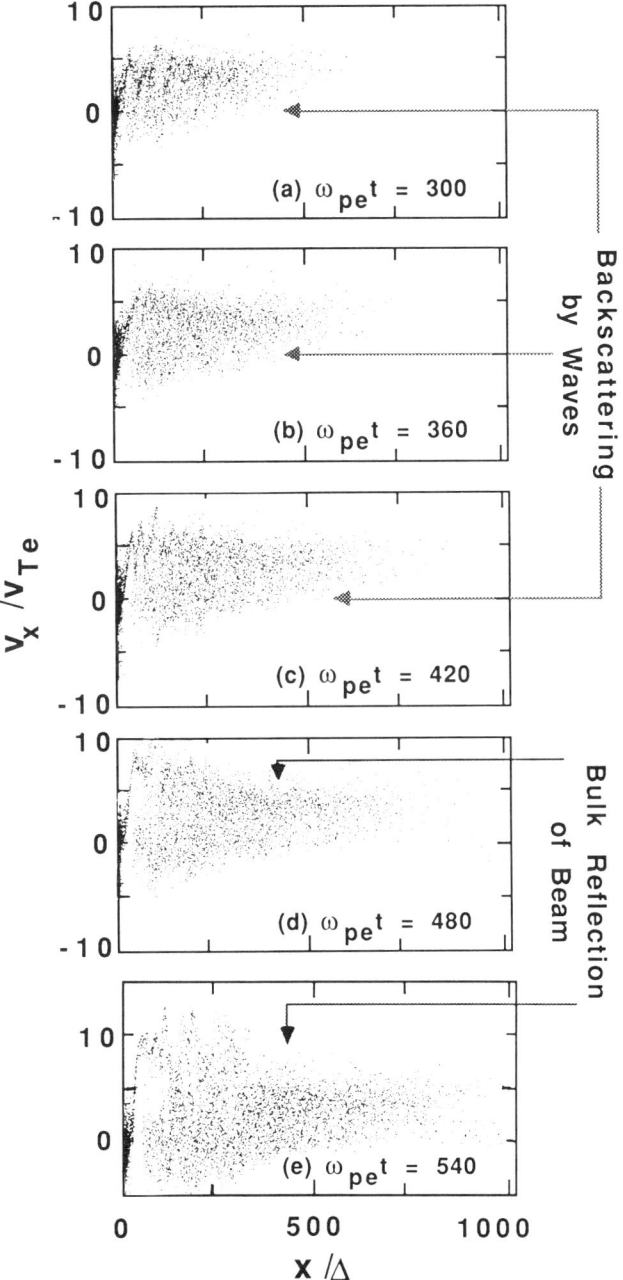

Fig. 2. The early evolution of parallel velocity of the magnetospheric electrons as they are accelerated downwards by the induced quasi-static potential. A portion of the downflowing electrons are back-scattered by a combination of wave-particle interactions and the mirror force.

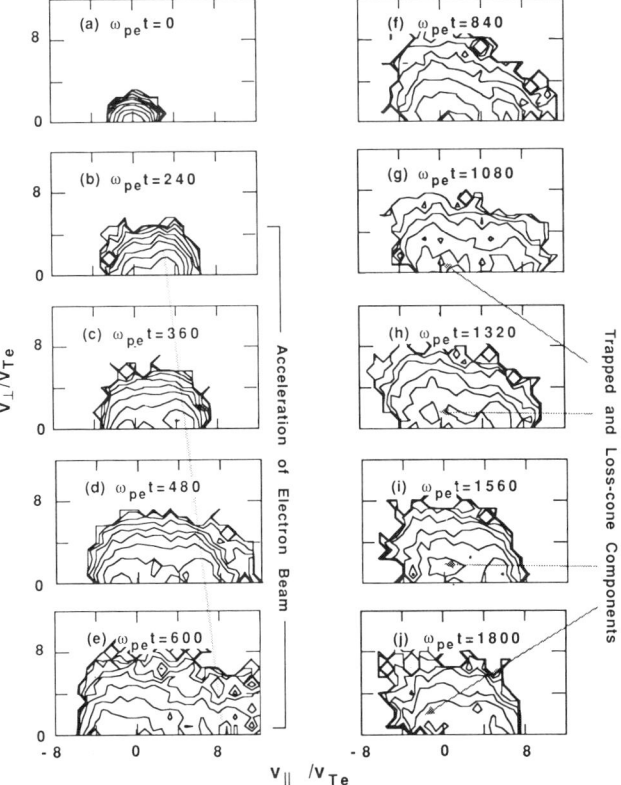

Fig. 3. The evolution of the total electron distribution at high altitudes ($x/\Delta = 150$) in the beam region showing the rapid generation of a beam, followed by the development of loss-cone and trapped components.

the downflowing electrons (as indicated by the dashed line) as the induced currents try to match the driving current. This increase in parallel velocity is accompanied by an increase in the perpendicular velocity from about 3 v_{Te} in Figure 3a to about 10 v_{Te} in Figure 3e. This increase is due to the scattering of particles by wave-particle interactions and/or conversion of parallel energy to perpendicular energy by the mirror force.

The evolution of the distribution at later times (right-hand side of Figure 3) has three distinct features. First, the average velocity of the downflowing energetic electrons is seen to decrease from its peak value of about 9 v_{Te} to its equilibrium value of about 4–5 v_{Te}. This decrease is primarily due to the increase in the conductivity of the ionosphere (and hence, reduced electric fields needed to support the current system) as the ionosphere becomes heated by the precipitating energetic electrons. The second is a loss-cone feature (e.g., Figure 3g). This feature is slow to develop because the electrons have to propagate down the field lines, mirror and propagate back up to high altitudes. The third feature which is only short lived is the development of a trapped component (Figures 3g–3j). This feature develops during periods of the intense electric fields where ambient electrons along the field lines can be confined between mirroring point at low altitudes and the parallel electric field at high altitudes. The presence of the beam, loss-cone and trapped features in the electron distribution are all common features of observed electron distributions in the auroral zone [e.g., Croley et al., 1978]

The evolution of the electron distribution at lower altitudes is somewhat different as illustrated in Figure 4 which shows the distribution at $x/\Delta = 550$. This point is near or below the mirror point of most of the back-scattered electrons. At this lower altitude, the beam does not appear in the distribution until about $360/\Omega_{pe}$. Because of the high ambient density at low altitudes, the beam particles primarily appear as a tail, with only a small local maximum appearing in the distribution at late times. In addition, most of the mirroring particles only have small parallel velocities so that the effective loss-cone angle is nearly 90° degrees.

With this steep loss-cone angle the distribution at small parallel velocities can be represented by a two-component plasma (dashed circles in Figure 4i). The first component represents isotropically heated ambient electrons with a thermal velocity of approximately v_{Te}. The second component represents primarily mirroring electrons with a relative density of about 10% and a temperature anisotropy of $T_\perp/T_\| \sim 2$. This temperature anisotropy can also be seen by comparing the ratio of $(\partial f/\partial v_\perp)/(\partial f/\partial v_\|)$ near the left-hand edges of the dashed circles in Figure 4i.

4.2 Induced Wave Emissions

The above features in the distributions in Figures 3 and 4 provide a variety of free energy sources for driving high-frequency waves. Spectra for the induced waves are shown in Figure 5. At high altitudes, the spectrum is characterized by broadband emissions associated with the Langmuir turbulence driven by the beam particles with the upper cutoff extending well above the local plasma frequency [cf. Winglee et al., 1988]. In addition, there is a local spectral peak between the local hybrid and plasma frequencies. This peak is attributed to an electron-acoustic instability driven by the motion of the beam electrons relative to the trapped component [Winglee and Pritchett, 1986]. This emission is not seen at the lower altitude (bottom panels) since there the beam component only forms a tail in the distribution.

Another important difference in the spectra is the generation of upper-hybrid waves. These waves appear very intense at low altitudes and only at moderate intensities at high altitudes. There are three possible sources of free energy for the upper-hybrid waves: loss-cones, trapped distributions and temperature anisotropies [e.g., Lin et al., 1990 and references therein]. The present results show that the most intense waves are associated with the appearance of a temperature anisotropy at

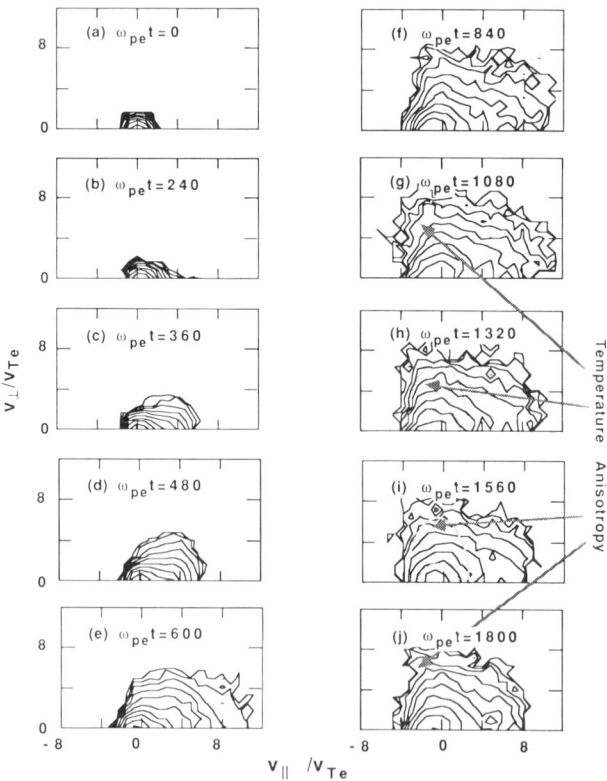

Fig. 4. As in Figure 3 except at low altitudes. The beam electrons appear only as a high-energy tail due to the relatively high ambient density at these altitudes. Because the relatively low altitude, most of the mirroring electrons have large perpendicular energies and small parallel energies, giving the upflowing part of the distribution a large temperature anisotropy.

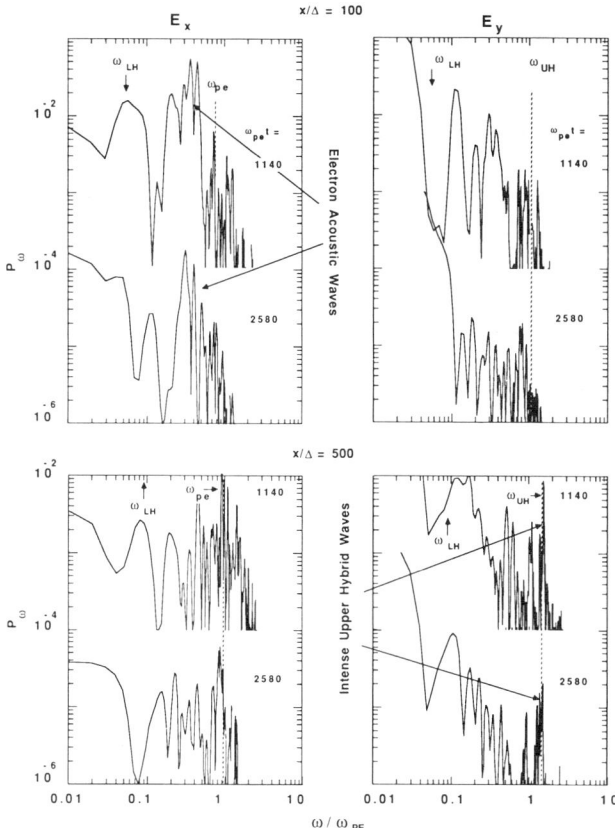

Fig. 5. Power spectra of the parallel electric field (left-hand side) and perpendicular electric field (right-hand side) at high altitudes and low altitudes. The spectra are taken over a period of $720/\omega_{pe}$ about the given times. Spectra are broadbanded at both altitudes. They differ in that the induced waves at high altitudes has an intense peak between the lower-hybrid and electron plasma frequencies associated with electron-acoustic waves while at low altitudes intense upper-hybrid waves are seen.

low altitudes. This altitude dependence of the induced waves may account for the intense upper hybrid waves reported by EXOS-D but not by Viking [Roux, private communication, 1990].

4.3 The Return-Current Region

Because the return-current region is 2–3 times wider than the beam region, the bulk velocity of the return-current electrons should be smaller by this factor. This reduced acceleration is illustrated in Figure 6 which shows the evolution of the electron distributions in the return-current region for the same x as in Figures 3 and 4. At low altitudes (right-hand side) where the density is relatively high, the bulk upflow velocity is in general less than about v_{Te}. At high altitudes (left-hand side) the velocity is seen to increase to about $2\ v_{Te}$ which is about a third to a half of the bulk velocity seen in the beam region.

The bulk of the electrons in Figure 6 show little increase in their perpendicular energy since the mirror is actually converting perpendicular energy into parallel energy as they move up the field lines. However, high-energy tails are generated, producing the conic-like features in the electron distribution (e.g., Figures 6b, 6c and 6e). These tails or conics appear to be generated by the absorption of upper-hybrid waves, excited in the beam region and which propagate into the return-current region. There is little free energy available for wave generation in the return-current region and indeed wave spectra (not shown) indicate more than an order of magnitude drop in power of the high frequency waves in the return-current region.

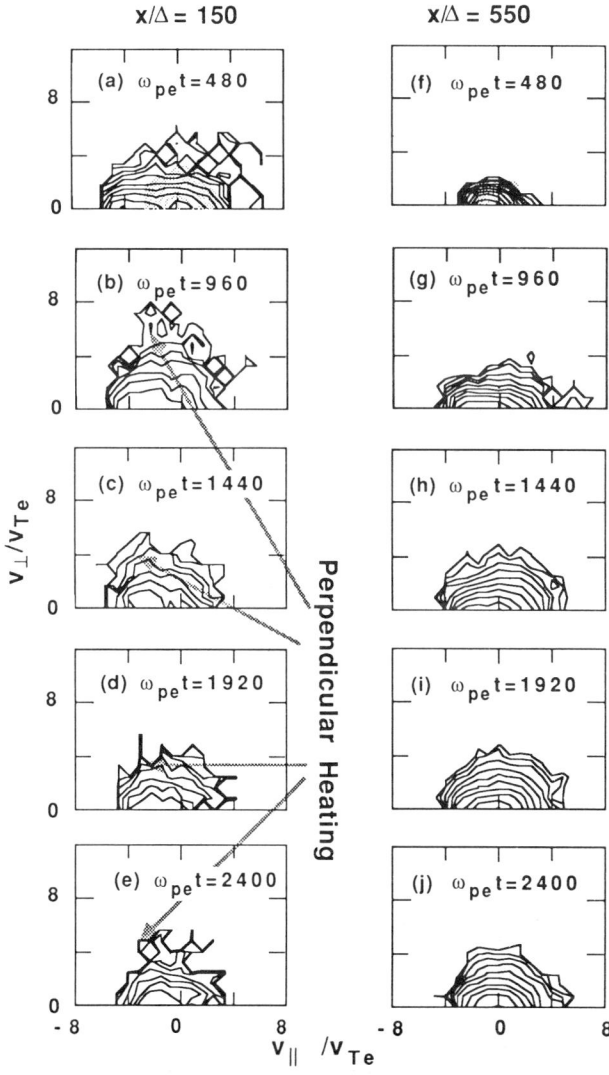

Fig. 6. Evolution of the electron distribution in the return-current region. Perpendicular heating is primarily driven by upper-hybrid waves, leading the formation of electron conics.

5. Ion Acceleration and Low-Frequency Waves

While the electron acceleration occurs relatively quickly, the ion acceleration *along* the field lines is relatively slow to occur. However, because of the relatively small spatial scales *across* the field lines, there is rapid perpendicular acceleration. The difference in time scales for the parallel and perpendicular acceleration is illustrated in Figure 7 which shows the time histories of the average parallel and perpendicular energies of the H and He ions at three different locations in the beam region. The values are derived for spatial averages of $100\Delta \times 20\Delta$ about the central position. Sample distributions at the highest altitude are shown in Figure 8.

Within $1000/\omega_{pe}$, both ion species experience strong perpendicular acceleration, with the average energy of the ions being nearly 15 times their initial energy. At this time, the electrons attain their fastest parallel velocities (viz. Figure 3). However, because of the faster response of the electrons, the average energy of the electrons is higher than that of the ions, similar to the observations of *Redsun et al.* [1985] and *Reiff et al.* [1986].

After this period, many of the energetic ions propagate across and out of the beam region, producing the apparent cooling (e.g., compare Figures 8b and 8h with Figures 8c and 8i). The H ions at this period have fractionally more perpendicular energy than the heavy ions (e.g., Figures 7b, 7d and 8d, 8j). This preferential heating of the light ions is due to the rapid rise time of the driving current which the heavy ions have problems tracking. Simulations with slower rise times show that the heavy ions can actually gain more energy than the H ions.

The parallel energy shows a local maximum coincident with the peak in the perpendicular energy. This local

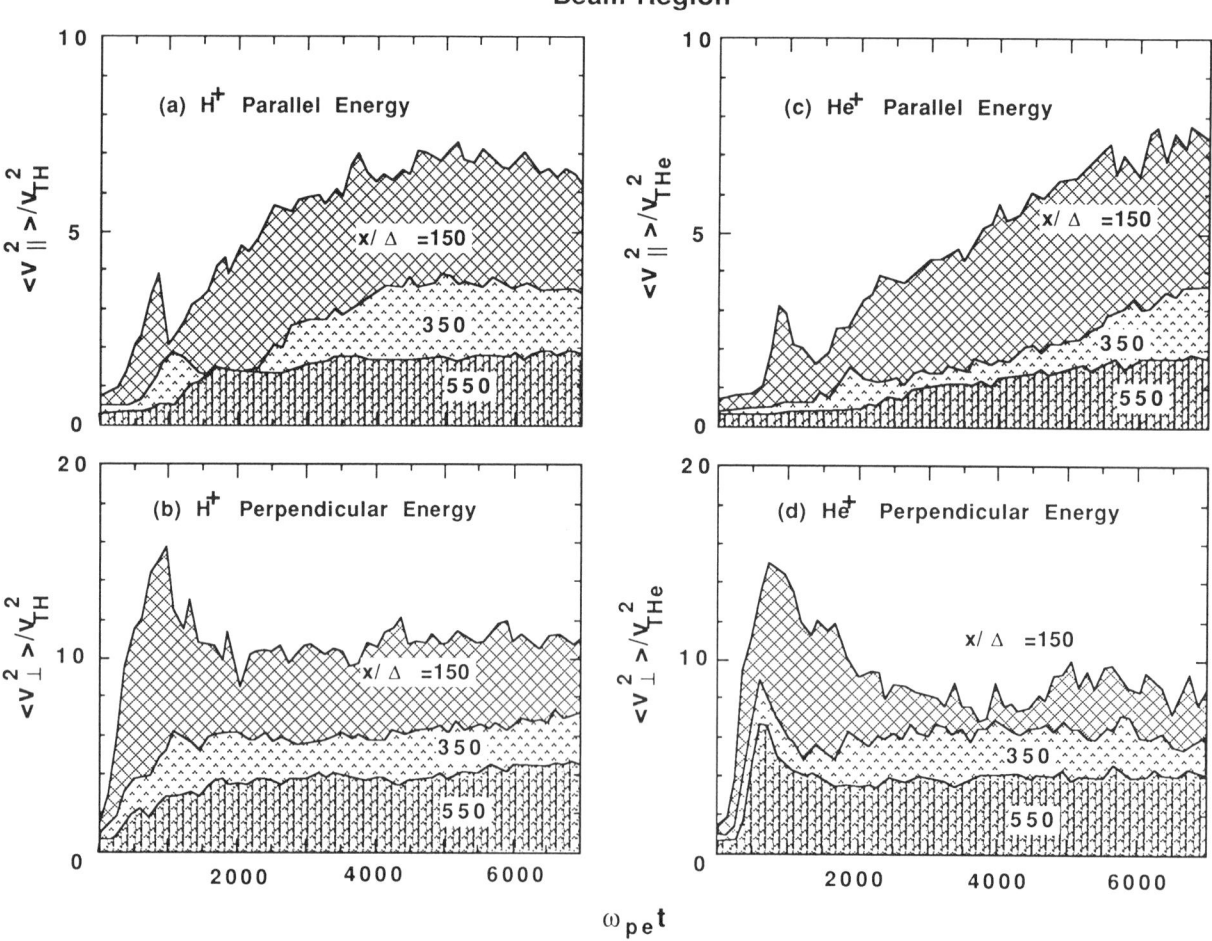

Fig. 7. Time histories of the average parallel and perpendicular energies of the H and He ions in the beam region at three different locations along the field lines. The averages are taken over a region $100\Delta \times 20\Delta$. The perpendicular energy is seen to reach saturation very quickly and then cools. The parallel energy is much slower to increase. At the end of the simulations the He ions, particularly at high altitudes, have higher parallel energy than the H ions.

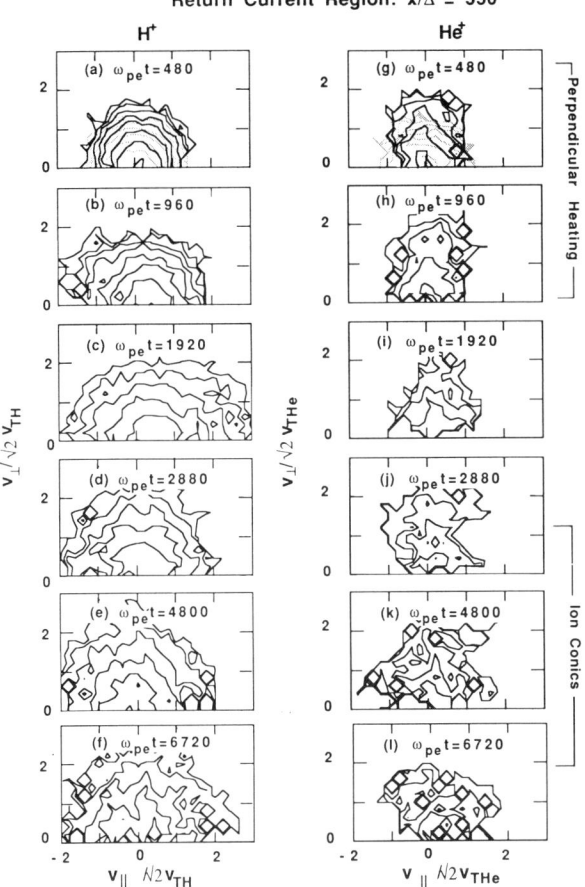

Fig. 8. The evolution of the H and He ions distributions in the beam region. The dotted lines indicate the bulk speed of the upflowing ions. These beams at late times are subject to an ion-ion instability which leads to the generation of low-energy tail in the H ions and a high-energy tail in the He ions.

to the transfer of energy from the light ions to the heavy ions. At the end of the simulations the He ions have on average about 25% more parallel energy than the light ions. Thus, wave-particle interactions can make significant contributions to the energization of upflowing ionospheric ions.

The mirror force also plays an important role in the energetization, particularly in the return-current regions. This is illustrated in Figure 9 which shows the evolution of the H and He distributions in the return-current region at $x/\Delta = 550$. Both ion species experience strong perpendicular heating at early times similar to the ions in the beam region (Figures 9a, 9b, 9g and 9h). However, the parallel electric field of the return current tends to accelerate the ions downwards. This downward acceleration is particularly evident in the H distributions in Figures 9b and 9c where the center of the distribution has been shifted to a positive velocity approximately equal to $0.5 \times \sqrt{2}v_{TH}$. However, the mirror force acts on the perpendicular energy in opposition to the electric

maximum is due primarily to heating rather than bulk acceleration with the distributions in Figures 8b and 8h showing little bulk upward flow. The bulk acceleration, as indicated by the dashed lines, develops at latter times (Figs 8d–8f). The H ions reach their maximum speed and energy at about $\omega_{pe}t \simeq 5000$ (Figures 7a and 8e), which is approximately the expected transit time of the ions through the potential drop.

After this time, a low velocity tail develops in the H ions (e.g., compare Figures 8e and 8f). This is accompanied by a net decrease in their average parallel energy (Figure 8a). At the same time, the He develops as a high speed tail (Figures 8k and 8j) with the fast He ions attaining speeds approximately equal to the bulk velocity of the H ions.

The development of the low-energy tail in the H and the high-energy tail in He is characteristic of the ion-ion streaming instability [Winglee et al., 1989] and leads

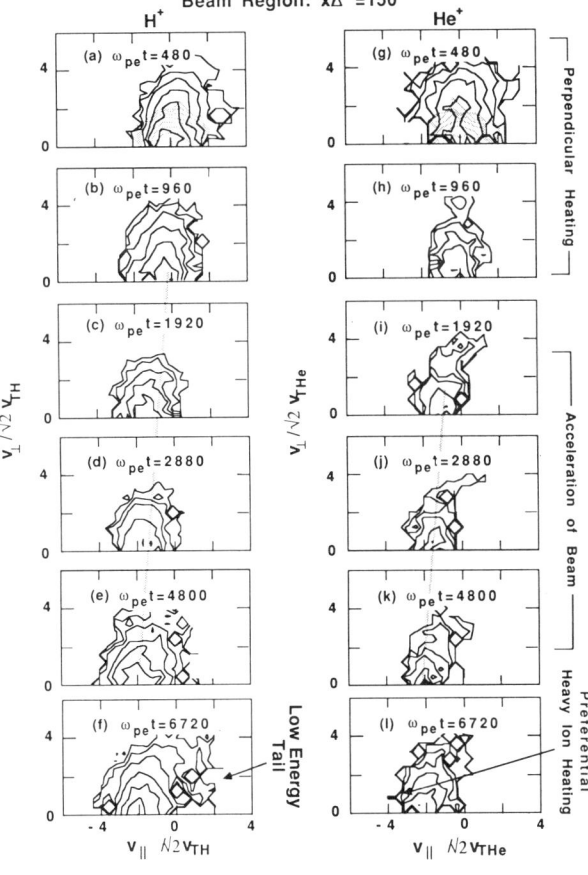

Fig. 9. The evolution of the ions distributions in the return-current region. Due to influence of the mirror force on the perpendicularly heated ions, upflowing ion conics develop against the downward acceleration of the electric field which supports the return current.

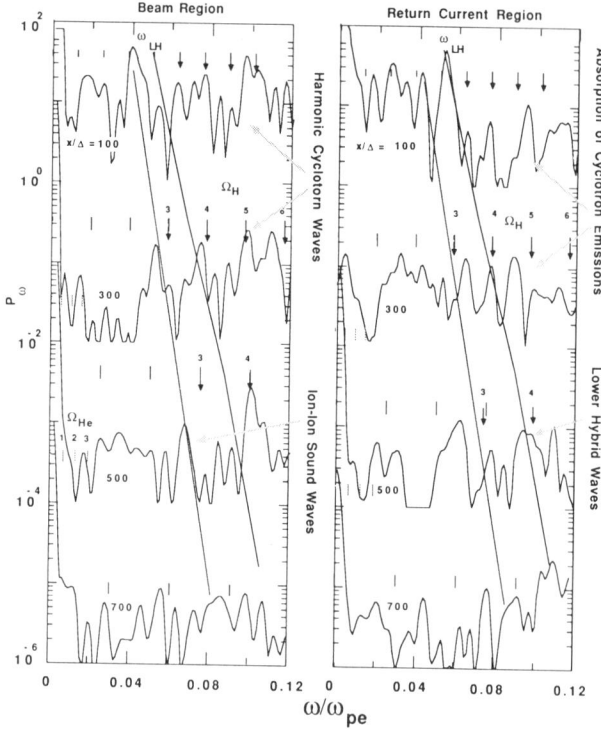

Fig. 10. Spectra of the low-requency waves generated in association with the ion acceleration. Strong cyclotron emissions are seen in the beam region and absorption features in the return-current region (except when the cyclotron harmonic lies near the lower-hybrid frequency). There is also a peak near the local H plasma frequency in the beam region associated with the ion-ion streaming instability.

field and can even dominate because of the preferential perpendicular heating of the ions. As s result, upflowing ion conics are produced in regions in which upflowing electron beams are also present.

The generation of these conics is most notable in the heavy ion distributions in Figure 9 for two reasons. First, the heavy ions, particularly at low altitudes, experience stronger perpendicular heating so that, on average, they have higher pitch angles and more energy is available to overcome the downward electric field. Second, because the H ions tend to have smaller pitch angles, many more of them are accelerated by the electric field from high altitudes to low altitudes so that there is stronger mixing of plasma which can mask the development of conics.

The spectra of the low-frequency waves associated with the above ion beams and conics is shown in Figure 10. Spectra for the beam region are shown on the left-hand side and those for the return-current region on the right-hand side for four different positions along the magnetic field. Consecutive spectra have been displaced by a factor of 100. The local values of the H plasma frequency ω_{pH}, lower-hybrid frequency ω_{LH}, and H and He cyclotron frequencies and harmonics are indicated.

The spectrum is broadband because of the number of free energy sources.

The more intense waves include lower-hybrid waves and H cyclotron waves in the vicinity of the beam region. They are driven by quasi-static perpendicular electric fields (associated with the auroral potential) and the sloshing of the ions through these perpendicular fields. These waves are most intense in the beam region because the relatively high electron temperature favors the growth of electrostatic waves while in the return-current region where the electron temperature is relatively low, strong absorption of the cyclotron waves can be seen (except where the lower-hybrid frequency lies near a harmonic). This absorption of wave energy can lead to transfer of energy from the beam region to the return-current region and the enhanced perpendicular heating of ions. A particularly clear example is at $x/\Delta = 300$ where harmonic emissions in association with ion beams are present in the beam region while absorption features in association with conics appear in the return-current region.

In addition, the spectra in the beam region (but not in the return-current region) have a strong maximum near the local H plasma frequency, at essentially all altitudes. This maxima occurs at the frequency expected for the ion-ion instability that produces the preferential parallel acceleration of the He discussed above [e.g., *Dusenbery et al.*, [1988]; *Winglee et al.*, 1989]. At even lower frequencies there is some evidence of He cyclotron waves. However, the spectrum tends to be masked by the quasi-static fields and a definitive identification of the characteristics of these waves is not possible.

6. SUMMARY

In this paper, a model for the development of the auroral current system during active times has been presented. The model is important in two areas. First, it provides some insight into the coupling between the ionosphere and magnetosphere during substorms by prescribing the conditions in the magnetotail needed to establish the auroral current system and the response of the ionosphere to these currents. Second, it provides a detailed analysis of many of the important processes responsible for some of the unique features associated with particle acceleration in the auroral zone.

The model utilizes two-dimensional (three-velocity) particle simulations with a multi-ion component plasma, a collision ionosphere and convergent magnetic field lines. The auroral current system is driven by a cross-field current in the magnetotail that can develop due to the differential motion of the electrons and ions through the magnetotail current sheet [*Winglee*, 1991]. As a result of this differential flow, there is charge separation of the magnetotail ions and electrons which can be shorted out by field-align flows into and out of the ionosphere (section 3).

As these field-aligned flows develop, the quasi-static fields, mirror forces and wave-particle interactions drive the particle acceleration and heating. For the electrons, distinct beam features at high altitudes are generated in association with intense Langmuir and electron-acoustic waves. At lower altitudes, the accelerated electrons appear more as a high-energy tail in the distribution and the electron-acoustic waves are no longer present. The scattering produced by these waves enhances the influence of the mirror force producing a moderate upflowing component in the beam region which leads to the appearance of loss-cone and trapped distributions at high altitudes. Intense upper-hybrid waves are generated by these features but they are most intense at low altitudes where mirror force also leads to a local temperature anisotropy. These waves can propagate across the field lines to produce electron conics in the return-current regions.

The ion dynamics is just as intricate. They initially experience rapid perpendicular heating due to the relatively short scale lengths across the field lines compared to the parallel scale lengths. These energetic ions can propagate across the field lines to produce conics in both the beam and return-current regions. On longer time scales, upflowing ion beams are generated in the primary current region. The parallel energy of the light ions tends to exceed that of the heavy ionospheric ions during the initial stage of beam formation, primarily due to their shorter transit time through the potential drop. However, ion beams eventually become unstable to an ion-ion streaming instability which transfers energy from the light ions to the heavy ions. In the present work, the heavy ions have an average parallel energy about 20% larger than the light ions. This result is similar to the observations of Collin et al. [1987].

At the same time, the mirror force acts on the perpendicularly heated ions. This is particularly important for ions in the return-current regions where the mirror force produces an upward push against the electric field which drives the return current. Because of the preferential perpendicular heating of the ions, the mirror force dominates and upflowing ion conics can be generated in the same region as upflowing return-current electrons. Ion cyclotron waves are seen in association with the beams and cyclotron absorption is seen in association with the conics, similar to the observations of Gorney et al. [1981].

Acknowledgments. This work was supported by NASA's Ionospheric Physics Program under grants, NAGW-1587 and NAGW-1593 and National Science Foundation grants ATM 87-19371 and ATM-9020577 to the University of Colorado. The particle simulations were performed on the CRAY Y-MP at the San Diego Supercomputer Center which is supported by the National Science Foundation.

REFERENCES

Ashour-Abdalla, M., and H. Okuda, Turbulent heating of heavy ions on auroral field lines, *J. Geophys. Res., 89*, 2235, 1984.

Bergmann, R., I. Roth, and M. K. Hudson, Linear stability of the $H^+ - He^+$ two-stream interaction in a magnetized plasma, *J. Geophys. Res., 93*, 4005, 1988.

Chang, T. T. S., and B. Coppi, Lower Hybrid acceleration and ion evolution in the supra-auroral region, *Geophys. Res. Lett., 8*, 1253, 1981.

Collin, H. L., R. D. Sharp, E. G. Shelley and R. G. Johnson, Some general characteristics of upflowing ion beams over the auroral zone and their relationship to auroral electrons, *J. Geophys. Res., 86*, 6820, 1981.

Collin, H. L., W. K. Peterson and E. G. Shelley, Solar cycle variation of some mass dependent characteristics of upflowing beams of terrestrial ions, *J. Geophys. Res., 92*, 4757, 1987.

Croley, D. R., Jr., P. F. Mizera, and J. F. Fennell, Signature of a parallel electric field in ion and electron distribution, *J. Geophys. Res., 83*, 629, 1978.

Dusenbery, P. B., and L. R. Lyons, Generation of ion-conic distributions by upgoing ionospheric electrons *J. Geophys. Res., 86*, 7627, 1981.

Dusenbery, P. B., R. F. Martin, Jr., and R. M. Winglee, Ion-ion waves in the auroral region: wave excitation and ion heating, *J. Geophys. Res., 93*, 5655, 1988.

Gorney, D. J., A. Clarke, D. Croley, J. Fennell, J. Luhmann, and P. Mizera, The distribution of ion beams and conics below 8000 km, *J. Geophys. Res., 86*, 83, 1981.

Gurgiolo, C., and J. L. Burch, Simulation of electron distributions within auroral acceleration regions, *J. Geophys. Res., 93*, 3989, 1988.

Kintner, P. M., and D. J. Gorney, A search for the plasma processes associated with perpendicular ion heating, *J. Geophys. Res., 89*, 937, 1984.

Lin, C. S., J. D. Menietti and H. K. Wong, Perpendicular heating of electrons by upper hybrid waves generated by a ring distribution, *J. Geophys. Res., 95*, 12295, 1990.

Mozer, F. S., C. A. Cattell, M. K. Hudson, R. L. Lysak, M. Temerin and R. B. Torbert, Satellite measurements and theories of low altitude auroral particle precipitation, *Space Sci. Rev., 27*, 255, 1980.

Redsun, M. S., M. Temerin, and F. S. Mozer, Classification of auroral electrostatic shocks by their ion and electron associations, *J. Geophys. Res., 90*, 9615, 1985.

Reiff, P. H., H. L. Collin, E. G. Shelley, J. L. Burch, and J. D. Winningham, Heating of upflowing ionospheric ions on auroral field lines, in Ion Acceleration in the Magnetosphere and Ionosphere, edited by T. Chang, Geophysical Monograph 38, p.83, 1986.

Reiff, P. H., H. L. Collin, J. D. Craven, J. L. Burch, J. D. Winningham, E. G. Shelley, L. A. Frank and M. A. Friedman, Determination of auroral electrostatic potentials using high- and low-altitude particle distributions, *J. Geophys. Res., 93*, 7441, 1988.

Swift, D.W., and C. Allen, Interaction of the plasma sheet with the lobes of the earth's magnetotail, *J. Geophys. Res, 92*, 10015, 1987.

Weimer, D. R., C. K. Geortz, D. A. Gurnett, N. C. Maynard, and J. L. Burch, Auroral zone electric fields from DE 1 and DE 2 at magnetic conjunction, *J. Geohys. Res., 90*, 7479, 1985.

Winglee, R. M., Characteristics of the fields and particle acceleration during tail thinning and reconnection, *Proc. of the Chapman Conf. on Substorms*, this issue, 1991.

Winglee, R. M., Electron beam injection during active experiments 2. Collisional Effects, *J. Geophys. Res., 95*, 6190, 1990.

Winglee, R. M., and P. J. Kellogg, Electron beam injection during active experiments 1. Electromagnetic wave emissions, *J. Geophys. Res., 95* 6167, 1990.

Winglee, R. M. and P. L. Pritchett, "The Generation of Low

Frequency Electrostatic Waves in Association with Auroral Kilometric Radiation," *J. Geophys. Res.*, *91*, 13,531, 1986.

Winglee, R. M., P. L. Pritchett, P. B. Dusenbery, A. M. Persoon, J. H. Waite, Jr., T. E. Moore, J. L. Burch, H. L. Collin, J. A. Slavin, and M. Sugiura, "Particle Acceleration and Wave Emissions Associated with the Formation of Auroral Cavities and Enhancements," *J. Geophys. Res.*, *93*, 14,567, 1988.

Winglee, R. M., P. B. Dusenbery, H. L. Collin, C. S. Lin and A. M. Persoon, "Simulations and Observations of Heating of Auroral Ion Beams," *J. Geophys. Res.*, *94*, 8663, 1989.

A Nonlinear Dynamic Analogue Model of Substorms

A. J. Klimas, D. N. Baker, D. A. Roberts, and D. H. Fairfield

NASA/GSFC, Laboratory for Extraterrestrial Physics, Code 690
Greenbelt, Maryland

J. Büchner

Central Institute for Astrophysics
Potsdam, Germany

Linear prediction filter studies have shown that the magnetospheric response to energy transfer from the solar wind contains both directly driven and unloading components. These studies have also shown that the magnetospheric response is significantly nonlinear and, thus, the linear prediction filtering technique and other correlative techniques which assume a linear magnetospheric response cannot give a complete description of that response. Here, the solar wind-magnetosphere interaction is discussed within the framework of deterministic nonlinear dynamics. An earlier dripping faucet mechanical analogue to the magnetosphere is first reviewed and then the plasma physical counterpart to the mechanical model is constructed. A Faraday loop in the magnetotail is considered and the relationship of electric potentials on the loop to changes in the magnetic flux threading the loop is developed. This approach leads to a model of geomagnetic activity which is similar to the earlier mechanical model but described in terms of the geometry and plasma contents of the magnetotail. This Faraday loop response model contains analogues to both the directly driven and the storage-release magnetospheric responses and it includes, in a fundamental way, the inherent nonlinearity of the solar wind-magnetosphere system. It can be characterized as a nonlinear, damped harmonic oscillator that is driven by the loading-unloading substorm cycle. The model is able to explain many of the features of the linear prediction filter results. In particular, at low geomagnetic activity levels the model exhibits the "regular dripping" response which provides an explanation for the unloading component at 1 hour lag in the linear prediction filters. Further, the model suggests that the disappearance of the unloading component in the linear prediction filters at high geomagnetic activity levels is due to a chaotic transition beyond which the loading-unloading mechanism becomes aperiodic. The model predicts the existence of a global plasma sheet oscillation mode with a 10 to 20 minute period. The conjecture is made that the peak in the linear prediction filters at 20 minutes lag is related to this global oscillation mode. Generally, the facets of the mechanical dripping faucet analogue model which were considered encouraging are retained in the new Faraday loop model when realistic physical estimates of the parameters which enter the model are used.

Introduction

The goal that motivates this work, a predictive capability for geomagnetic activity, has remained an elusive one despite considerable effort. Statistical or correlative studies [*Baker*, 1986; *Baker, et al.*, 1986], culminating in the linear prediction filter studies [*Iyemori et al.*, 1979; *Iyemori and Maeda*, 1980; *Clauer et al.*, 1981, 1983; *Bargatze et al.*, 1985; *McPherron et al.*, 1988], relating solar wind properties to indices of geomagnetic activity have proven only partly successful. Numerical simulation of the solar wind-magnetosphere plasma system for this purpose remains far in the future. Prior to several phase space reconstruction studies which have been reported recently and are reviewed below [*Vassiliadis et al.*, 1990; *Roberts et al.*, 1991; *Shan et al.*, 1991], possibilities for further significant progress seemed limited. Those reconstruction studies, however, have opened up an alternate path that does appear promising. They have shown that a simple analogue model of the substorm phenomenon does exist. Unfortunately, however, they do not reveal what that model is. For that, we must reason from basic physical principles. This paper reviews some early developments in the search for that analogue model and presents a new approach which has evolved from that early work. We do not intend to suggest that the new model presented here is the model; we doubt that. However, we would argue on the basis of the phase space reconstruction studies that the analogue model

Magnetospheric Substorms
Geophysical Monograph 64
Copyright 1991 American Geophysical Union

approach is a valid one and, further, we would argue that the particular approach taken here is also a valid one. Beyond that, it is possible to discuss the detailed assumptions that are involved in the Faraday loop model construction. Indeed, further model development is anticipated. At this point, however, we feel that the present model accounts for several macrophysical aspects of geomagnetic activity and that it includes, in a fundamental way, the inherent nonlinearity of the solar wind-magnetosphere system. Thus, the present model represents a significant step forward toward the goal of geomagnetic activity prediction.

Linear Prediction Filters

It has been recognized in recent years that the solar wind-magnetosphere interaction is a dynamic nonlinear phenomenon [*Bargatze et al.*, 1985]. Correlative and linear prediction filter methods that have been used to analyze the magnetospheric response to the solar wind are strictly applicable to only linear processes [e.g., *Baker et al.*, 1986 and references therein]. To the extent that the magnetospheric response is, indeed, nonlinear these traditional methods cannot give a complete description of that response. While these linear techniques have been used with some success, we believe the limit of their applicability has been reached.

In the linear prediction filter technique it is assumed that a filter $H(t)$ exists which provides the most general linear relationship between an input time series $I(t)$ and an output time series $O(t)$ through the convolution

$$O(t) = \int_0^\infty H(s) I(t-s) ds . \qquad (1)$$

Within this construct it is possible to relate a particular solar wind input time series that is a measure of solar wind energy loading to another magnetospheric output time series that is an indicator of geomagnetic activity through the properties of the solar wind-magnetosphere coupling as expressed in the filter $H(t)$. If $H(t)$ were known, then this relationship provides a predictive capability for geomagnetic activity given measurements in the solar wind that determine the energy loading. $H(t)$ might be deduced through physical analysis applied to the solar wind-magnetosphere coupling mechanism. In applications of this technique [*Iyemori et al.*, 1979; *Iyemori and Maeda*, 1980; *Clauer et al.*, 1981, 1983; *Bargatze et al.*, 1985; *McPherron et al.*, 1988] an alternate approach has been taken. From satellite and ground based measurements, data indicating solar wind loading and data indicating geomagnetic activity have been inserted into (1) to deduce the properties the filter must have in order to relate the observed input and output time series. This process has usually been repeated using data from many time intervals to produce average filters. Then these average filters have been used for geomagnetic activity prediction during any particular time interval.

Bargatze et al. [1985] have carried out an extensive study using the linear prediction filter technique to deduce the properties of the solar wind-magnetosphere coupling mechanism as expressed in linear filters. They have produced the set of 30 averaged filters shown in Figure 1 which relate the input VB_s time series as measured in the upstream solar wind and the simultaneous output AL index time series. They classified each of the intervals within which a particular filter was constructed by the average level of geomagnetic activity during that interval. The averaged filters were constructed by grouping together particular filters according to average geomagnetic activity level and then averaging over each group. The result was that the evolution of average filter with geomagnetic activity level could be displayed. At low activity levels they found that the solar wind-magnetosphere coupling, as expressed in the linear prediction filters, is bimodal. They found two well defined peaks in the low activity filters at approximately 20 minutes and 1 hour. They further found that the averaged filters evolve as the level of average activity increases. The peak at 20 minutes grows with increasing activity but remains at a relatively constant lag time. The 1 hour peak also remains at a relatively constant lag and grows somewhat with increasing activity until, at a moderate activity level, it vanishes. Bargatze et al. [1985] and McPherron et al. [1988] have ascribed the separate peaks in the linear prediction filters to two different magnetospheric response mechanisms to solar wind loading. The 20 minute peak, in their view, is produced by the directly driven response mechanism [*Perreault and Akasofu*, 1978; *Akasofu*, 1979, 1980] and the 1

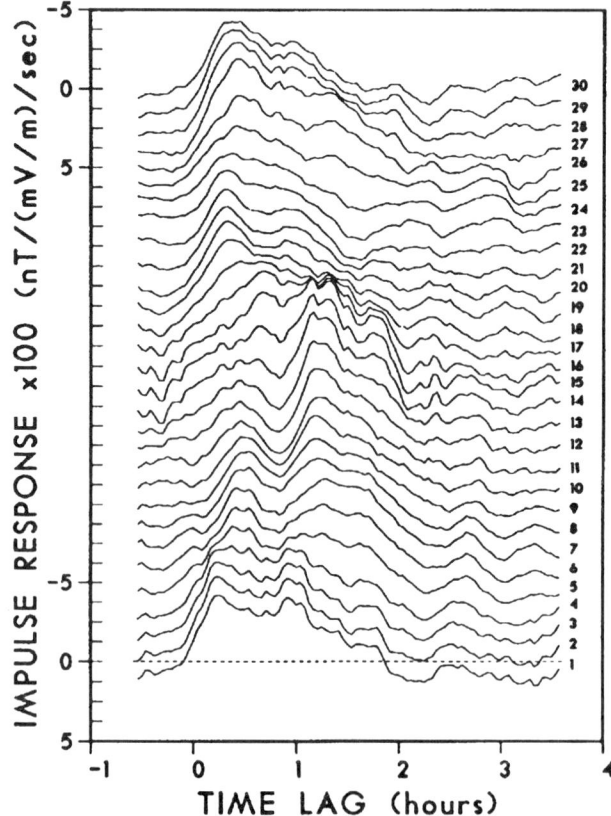

Fig. 1. A stack plot of linear prediction filters for all levels of geomagnetic activity. The geomagnetic activity level increases unevenly from filter 1 through 30. (Taken from Bargatze et al. [1985])

hour peak by the storage-release mechanism [McPherron, 1970; McPherron et al., 1973 Hones, 1979; Baker et al., 1979, 1981a, 1981b]. Then, does the storage-release mechanism stop operating at high activity levels? Baker et al. [1990] have argued that it is more likely that this mechanism continues, but aperiodically due to a transition into chaotic magnetospheric response at high activity levels.

The linear prediction filter results have shown that the magnetospheric response to solar wind loading operates through two distinct physical mechanisms, each with its own well defined average time scale. This result is the primary success of this technique applied to the solar wind-magnetosphere coupling problem. However, when viewed as a technique for geomagnetic activity prediction it is only partially successful. The reason for this partial success was a primary motivator for the modeling effort of Baker et al. [1990] and continues to be so for the modeling effort to be discussed here: On applying the deduced linear prediction filters to specific intervals of data for which both the appropriate input and output time series were available, Clauer et al. [1983], Bargatze et al. [1985], and McPherron et al. [1988] found that approximately 40 to 45 percent of the AL index variance could be predicted. McPherron et al. [1988] studied this issue further and found that individual substorm events could, in fact, be predicted quite accurately with approximately 90 percent efficiency but only by varying the parameters that enter the linear prediction filter (lag times, peak heights) for each event. They argued that the magnetospheric response depended, therefore, on both the details of the solar wind loading time series and the prior state of the magnetosphere when the loading begins. Thus, the magnetospheric response is, to a significant degree, a nonlinear response that lies outside the realm of the linear prediction filter technique. Baker et al. [1990] responded to these results by adopting the methods of nonlinear deterministic dynamics to the study of solar wind-magnetosphere coupling. In view of an earlier discussion by Hones [1979], in which the storage-release mechanism was compared to the formation and release of a liquid drop from a faucet, Baker et al. [1990] attempted an analysis in terms of an earlier analogue model of a laboratory dripping faucet.

Dripping Faucet Analogue Model of Geomagnetic Activity

Baker et al. [1990] have studied the solar wind-magnetosphere coupling problem using a nonlinear dripping faucet analogy of the system. Their approach was motivated by the dripping faucet description of plasmoid formation and release discussed by Hones [1979] (see Figure 2) and, also, by the laboratory study of a dripping faucet done by Shaw [1984]. Their dripping faucet analogue model is a generalization of the dripping faucet analogue of Shaw [1984] which consists of a variable mass on a spring that increases steadily until a piece of it falls away according to a nonlinear algorithm that leads to dripping faucet behavior. Analysis of the Baker et al. [1990] model has shown that the analogue approach is a promising one for extending our present understanding of the solar wind-magnetosphere interaction.

The Shaw [1984] model is illustrated in Figure 3; it consists of a variable mass hanging on a spring. The displacement downward

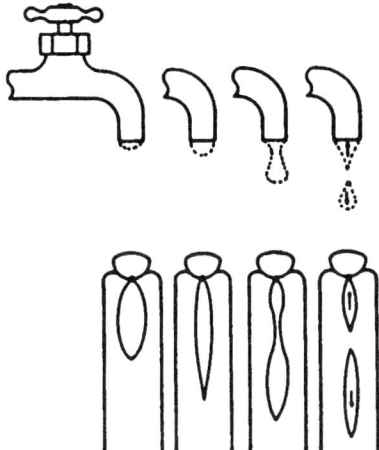

Fig. 2. Analogy of plasma sheet severance and plasmoid formation to the behavior of water dripping from a leaky faucet. (Taken from Hones, [1979])

from the unstressed spring position is measured by the variable D. To model drop formation, the mass m is generally increased linearly with time at a constant rate, $\dot{m}_L$. The exception to this constant mass loading occurs if D reaches a critical displacement Dc. Then a piece of the mass Δm is dropped so that m decreases discontinuously by the same amount. The equation that $D(t)$ obeys is

$$\frac{d}{dt}\left(m\frac{dD}{dt}\right) + KD = Gm \qquad (2)$$

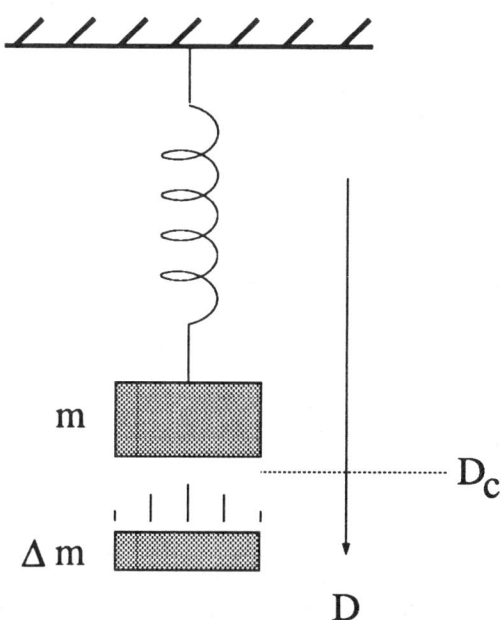

Fig. 3. Dripping faucet analogue of Shaw [1984].

There are two natural oscillation periods in the model when variations in the mass are small; the first is the obvious oscillation period of the mass on the spring determined by the ratio of the spring constant to the mass. The second oscillation period is a consequence of the algorithm chosen by Shaw [1984] for determining the magnitude of Δm each time $D = D_c$. Shaw chose $\Delta m = \beta v_c$ where β is a proportionality constant and where v_c is the downward velocity of the mass when $D = D_c$. This assumption leads to periodic mass loss (steady dripping) with a period that is relatively constant for a large range of low $\dot{m}_L$. To understand how this steady dripping period follows from Shaw's algorithm, notice that (2) has a non-oscillating solution for $D < D_c$ given by

$$K\dot{D} = G\dot{m}_L \quad (3)$$

with $\ddot{D} = 0$. The growing mass moves downward at a constant rate while maintaining its instantaneous equilibrium position $KD = Gm$. Thus, in the absence of spring oscillations we have $\Delta m = \beta v_c = \beta(G/K)\dot{m}_L$. A second dripping period emerges given by

$$T_{drip} = \frac{\Delta m}{\dot{m}_L} = \beta\left(\frac{G}{K}\right) \quad (4)$$

Notice that if we vary $\dot{m}_L$ then Δm varies with it while maintaining T_{drip} fixed; larger loading rates lead to larger drops while the dripping period remains fixed. Generally, for low $\dot{m}_L$, such that variations in the mass remain small compared to the mass, these two oscillation periods coexist without interfering with each other. In this low mass loading (low faucet flow rate) regime the drops fall periodically even though they may oscillate up and down several times as they are filling. For larger loading rates the situation becomes considerably more complex with, ultimately, a transition into chaotic dripping with no definable dripping period.

Baker et al. [1990] modified the Shaw [1984] dripping faucet analogue by adding friction and by changing the mode of mass loss. The friction was added to model dissipative processes in the magnetosphere. The mass loss was made continuous to model the finite time of magnetotail field line merging during plasmoid formation. In their model mass is dumped continuously at a rate $\dot{m}_D$ for as long as $D > D_c$ although the transition from loading to dumping, or back again, remains discontinuous. Further, they assumed $\dot{m}_D = \beta v_c$ and recovered the regular dripping mode discussed above when β is appropriately large.

The Baker et al. [1990] dripping faucet response model exhibits several properties in its dynamical evolution which are encouraging when compared to the linear prediction filter results of Bargatze et al. [1985]. For low loading rates the model exhibits two well defined time scales in its evolution that are a factor of three apart in magnitude; both time scales are insensitive to loading rate over a large range of values. (The filter results show two response peaks that remain at relatively constant lags, one at twenty minutes and the other at one hour, for low to moderate levels of average activity.) As the loading rate increases within the low loading rate regime, the size of the mass fluctuations also increases; the insensitivity of the loading-unloading cycle period to loading rate magnitude is a consequence of this behavior. (The one hour response peak increases in magnitude with increasing average activity indicating that larger plasmoids are released for higher magnetospheric loading rates.) For large loading rates, the loading-unloading cycle becomes chaotic with no definable period but the oscillations at the shorter time scale remain ordered and well defined. (For high levels of average activity, the one hour response peak in the filter results vanishes while the twenty minute peak remains, still at a relatively constant lag.)

Phase Space Reconstruction

Phase space reconstruction studies [*Vassiliadis et al.*, 1990; *Roberts et al.*, 1990; *Shan et al.*, 1990] of geomagnetic activity indicators, such as the AL and AE indices, have shown that a low-dimensional analogue model of the type proposed by Baker et al. [1990] may be sufficient for a study of the magnetospheric response to loading. We review these methods below and summarize the results of our recent study [*Roberts et al.*, 1990] of the AL index time series for the intervals studied by Bargatze et al. [1985] in their linear prediction filter analysis.

The number of variables involved in a description of the magnetosphere is in principle very large, involving many current and flow systems coupled in complex ways. As a driven, damped system, however, the behavior of the magnetosphere may settle down to a much lower dimensional subset of the whole phase space available. The attractor (the set the system settles onto) for complex systems may often be recovered from the time series of one variable alone, a fact that was used by Grassberger and Procaccia [1983] to develop an embedding method that will be used here. Suppose we make N measurements of a variable x at equal time intervals. To embed these measurements in an m dimensional space, we may form the vectors $\mathbf{X}_i = (x(t_i)\,x(t_i + \tau)\,....\,x(t_i + (m-1)\tau))$ using a lag τ to make m-tuples from a single time series. There will be almost N vectors $\mathbf{X}_i$ as the starting point t_i for the m-tuple is allowed to move uniformly through the set. A simple way of finding the dimension is to count the points in progressively larger m-spheres about each point. For a linear structure embedded in a space of arbitrary dimension, the number of such points $C(r)$ will increase linearly with the distance r. More generally we define, after Grassberger and Procaccia [1983],

$$C(r) = \frac{1}{N^2}\sum_{i,j=1}^{N}\Theta(r - |\mathbf{X}_i - \mathbf{X}_j|)\,, \quad (5)$$

with the sum being over all pairs of distinct points, and where Θ is the theta function, equal to zero for negative arguments and 1 for positive, and $|\,|$ indicates the usual Euclidean distance. Then for small r,

$$C(r) \propto r^d \,, \quad (6)$$

can be used to determine the dimension of the attractor by finding the slope of the $\log(C) - \log(r)$ plot. Note that this method works whether d is an integer or not, and in the latter case the attractor is a fractal, and thus strange. However, it is important to realize that a low dimension does not immediately imply that an attractor exists, and other tests are needed (see e.g. Roberts et al. [1991] and Roberts [1991]). Of course the embedding dimension m must be higher than the dimension of the attractor for this to work, and what is done in practice is to vary m until the slope settles down to a value independent of m. We have implemented this procedure and checked that it works with known fractal and nonfractal attractors. The method works more poorly when d becomes as high as 5 or 6, but d is smaller than this in the present case.

In Figure 4 we present the results of an embedding dimension analysis of one interval from the data set compiled by Bargatze et al. [1985]. This data set consists of a number of intervals of 2.5 min. averaged data, each about two days long, ordered such that the overall level of activity is increasing throughout the data set. The endpoints of the intervals were near zero to avoid spurious jumps in the composite data set. The interval for Figure 4 is at the beginning of the data set and consists of 5000 points, a number that is perhaps too low, but has been found to be adequate for at least a rough embedding dimension analysis. The figure shows the slope of the $\log(C(r))$ vs $\log(r)$ curves for small r. The slope levels off at 3.5 (with a relatively large uncertainty indicated by the error bar which is typical for all points) for larger m, and thus we have good evidence that the attractor for the underlying process can be described by a model with four degrees of freedom. An examination of cases with different activity levels yields a dimension of 3.5 to 3.9 for the AL index, in agreement with the AE analysis of Vassiliadis et al. [1990]. Subsequent analysis with 40,000 points has yielded a dimension of 4.0 ± 0.2 [Roberts et al., 1991]. Not all the cases are as simply convergent as this one, sometimes exhibiting increasing slopes at the smallest r's, and it remains to be seen whether this is significant or not. Nonetheless, the present evidence makes it seem reasonable to seek a low-dimensional description of the physics underlying substorms.

Faraday Loop Response Model

We are actively developing a new analogue model of the solar wind-magnetosphere interaction that remains closely related to the dripping faucet model but is constructed from first principles and is expressed in terms of the shape and plasma contents of the magnetotail. The present state of this model is outlined below.

Consider, at approximately 20 R_E tailward of Earth, a closed path in the magnetotail that passes through the tail in the current sheet, from the dawn to dusk sides, and then closes on itself in the magnetopause as illustrated in Figure 5. The loop encircles one of the magnetotail lobes. If the lobe expands or contracts, then the loop must also expand or contract to remain on the magnetopause. The total magnetic flux, Φ, passing through the loop changes in time according to Faraday's law

$$\frac{1}{c}\frac{d\Phi}{dt} = -\int_{loop} d\mathbf{l} \cdot \mathbf{E} = (E_0 - E_y)d \qquad (7)$$

with, in this case, the total potential drop around the loop divided into two parts, the first due to the cross-tail electric field in the current sheet E_y and the second E_0 due to the electric field imposed on the loop at the magnetopause.

Flux Estimate

By construction the magnetic flux through the loop is given by

$$\Phi = \int_{surface} d\mathbf{A} \cdot \mathbf{B} = \int_{surface} dA\, B_x \qquad (8)$$

We estimate this flux by assuming a thin, low flux, current sheet. Then

$$\Phi \simeq \frac{4\pi}{c} A\, h\, j_y \qquad (9)$$

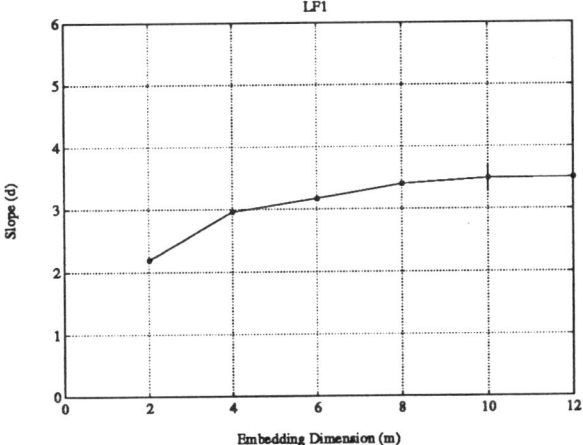

Fig. 4. The slope d of the $\log(C(r))$ vs $\log(r)$ curves as a function of the embedding dimension m.

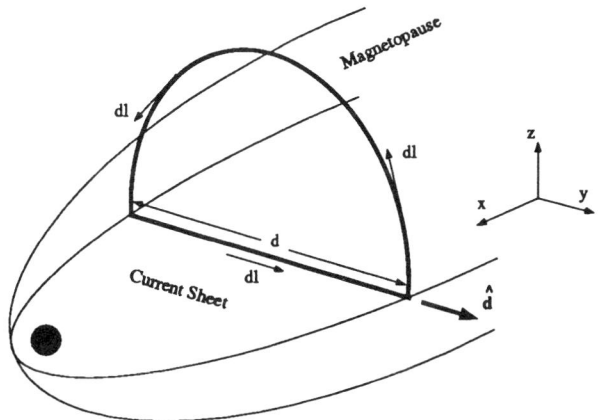

Fig. 5. A Faraday loop in the magnetotail.

in which h is the current sheet half-thickness, A is the cross-sectional area of the lobe, and j_y is the cross-tail current density in the current sheet.

Current Contributions

Combining (9) with (7), we obtain

$$\frac{1}{c}\frac{d\Phi}{dt} = \frac{4\pi}{c^2}\frac{d}{dt}(hAj_y) = (E_0 - E_y)d \quad (10)$$

To proceed, it is necessary to express j_y in terms of E_y and d. At present, we assume that j_y is dominated by three contributions: $j_y = j_0 + j_R + j_p$. We first assume the magnetotail has an average shape consistent with a current sheet contribution, j_0. We estimate the magnitude of this current by assuming that it must provide the lobe field pressure capable of withstanding the ram pressure of the solar wind at the magnetopause. This is the dominant current sheet contribution in the tail and in the Faraday loop model; it varies slowly and by only a small amount over the course of a substorm. With the addition of solar wind loading and/or internal tail dynamics, additional currents that may be driven by a cross-tail electric field must be included in the model. In the central current sheet where magnetic field line curvature is maximum, it is possible for the guiding center approximation for ion motion to fail so that the ions can be accelerated in the direction of the cross-tail electric field. The resulting transfer of energy from field to particles can be expressed in terms of a "resistive" current, j_R, which is supported by the Speiser conductivity [*Speiser*, 1970; *Lyons and Speiser*, 1985]. Finally, if the cross-tail electric field is time dependent, then a polarization current, j_p, is induced [*Longmire*, 1963]. Thus, $j_y = j_0 + j_R + j_p$ with

$$\rho V_{sw}^2 \sin^2\theta = \frac{2\pi}{c^2}(hj_0)^2 \quad (11)$$

$$hj_R = \frac{E_y D}{R} \quad (12)$$

and

$$j_p = \frac{1}{4\pi c}\left(\frac{c}{V_A}\right)^2 \frac{\partial E_y}{\partial t} \quad (13)$$

The quantity V_A is the Alfvén speed in the current sheet and D is a constant measure of the average tail diameter.

Model Equations

Substituting $j_y = j_0 + j_R + j_p$ into (10), and writing the result in a dimensionless form, we obtain

$$\frac{d}{d\tau}\left[a\left(\alpha\sin\theta(a) + \nu E + \frac{dE}{d\tau}\right)\right] = (\mathcal{E}_0 - E)\sqrt{a} \quad (14)$$

In this equation, the electric fields have been normalized to the strength of the $\mathbf{v}\times\mathbf{B}$ electric field in the solar wind. Thus

$$E = \frac{E_y}{\left(\frac{V_{SW}}{c}\right)B_{SW}} \quad (15)$$

and

$$\mathcal{E}_0 = \frac{E_0}{\left(\frac{V_{SW}}{c}\right)B_{SW}} \quad (16)$$

Time is measured in units of the natural oscillation frequency of the model; i.e. $\tau = \omega t$ with

$$\omega^2 = \frac{8}{\pi}\left(\frac{D}{h}\right)\left(\frac{V_A}{D}\right)^2 \quad (17)$$

The cross sectional area of the loop has been put in the dimensionless form

$$a = (d/D)^2 \quad (18)$$

The parameter ν that determines the damping rate in the model can written as

$$\nu = \frac{\pi}{8}\left(\frac{h}{D}\right)\left(\frac{D\omega_{p_i}}{c}\right)^2 \omega\tau_{ac} \quad (19)$$

and

$$\alpha = \frac{\sqrt{\pi}}{2}\sqrt{\frac{D}{h}}\left(\frac{V_A}{V_{ASW}}\right) \quad (20)$$

with V_{ASW} the Alvén speed in the solar wind and $\tau_{ac} = 1/\Omega_z$ where Ω_z is the proton gyro frequency in the z component of the central current sheet magnetic field. Notice that we have indicated the possible dependence of the tail flaring angle on tail diameter through the symbol $\theta(a)$. At present this quantity and all others that appear in (14) except the variables E and a are held constant.

Plasma Sheet Oscillations

The model predicts the existence of a global plasma sheet oscillation mode with a 10 to 20 minute period. Because of its integrated nature, the model does not give any detailed information concerning the nature of this oscillation mode. However, at minimum the mode must execute the axial displacements (along the earth-sun line) illustrated in Figure 6 or it would not lead to fluctuations in the flux passing though the Faraday loop. The conjecture is made that the peak in the linear prediction filters at 20 minutes lag is related to this global oscillation mode. Evidence has been found through spectral analysis of AL index data that supports the existence of this oscillation mode.

Assuming for the moment that a is also constant, then (14) reduces to an equation for a damped harmonic oscillator. E oscillates about $\mathcal{E}_0$ and goes asymptotically to $\mathcal{E}_0$ as the oscillations decay. When $E = \mathcal{E}_0$, then from (7), the flux Φ reaches a new steady state value with the total potential drop about the Faraday loop equal to zero. The period of these oscillations is given by

$$\tau = \pi\left(\frac{d}{V_A}\right)\sqrt{\frac{\pi}{2}}\sqrt{\frac{h}{d}} \simeq 100\,\text{min.}\times\sqrt{\frac{h}{d}} \quad (21)$$

in which we have used $d = 40\,R_E$ and $V_A = 160$ km/sec. Our estimate of V_A is based on the results of Fairfield [1979,1986], from which we conclude that a reasonable estimate of B_z, aver-

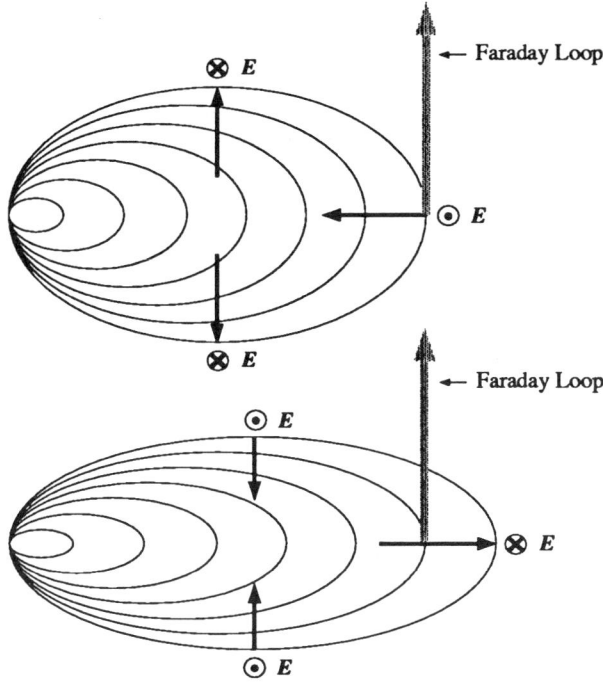

Fig. 6. Plasma sheet oscillations lead to variations in cross-tail electric field and Faraday loop flux. The variations illustrated here should be viewed as perturbations on a background convective equilibrium with possible associated field and convection contributions.

aged over the current sheet both in z and in y, is $\langle B_z \rangle \simeq 4$ nT, and on the results of Fairfield [1987], from which we conclude that a reasonable estimate of the average proton number density in the same region is $\langle n_p \rangle \simeq .25 \, cm^{-3}$. The value for $\langle B_Z \rangle$ used here is a simple average of all of the values given by Fairfield [1986] in the 10 to 20 R_E range of tailward distances; it is raised somewhat above typical central current sheet values (1–2 nT) by the larger values found in the flanks of the tail. If we assume h is somewhere in the range 1/2 to 2 R_E, then τ is in the range 11 to 23 minutes. We expect these global plasma sheet oscillations to be reflected in oscillations in the ionospheric current systems where they might be detected. Roberts, et al. [1990], through spectral analysis of intervals of AL data selected from the Bargatze et al. [1985] data set, have discovered a persistent peak at approximately fifteen minutes period which supports the existence of this global plasma sheet oscillation mode. The Roberts et al. [1990] results are illustrated in Figure 7. We presume that the twenty minute peak in the linear prediction filter results [Bargatze et al., 1985; McPherron et al., 1988] is a consequence of this response mode in the magnetotail. However, to confirm this presumption an analysis of our Faraday loop model response to varying solar wind conditions will have to be done.

The Loading-Unloading Response

We introduce the loading-unloading cycle into the model by *imposing* a time dependence on a with the intention of maintaining the Faraday loop at the magnetopause over the course of a substorm. Although it is known that the magnetotail expands during the substorm growth phase and collapses during the expansion phase [Maezawa, 1975; Baker et al., 1984; Fairfield, 1985] when the tail is unloaded, further details concerning the expansion and collapse are not available. Accordingly, in the model we impose either growth or decay of the cross sectional area that is simply linear in time. During the growth phase we let $a(\tau)$ grow linearly in time at a rate that is proportional to $\mathcal{E}_0$; the proportionality constant λ is a free parameter in the model. In this way, the tail diameter grows only when there is flux transfer into the loop through the magnetopause and it grows at a rate that increases with the rate of flux transfer. The rate at which $a(\tau)$ decreases during unloading in the model depends on the dynamical evolution of the model.

A transition in the model from loading to unloading must be introduced. We take the point of view that field line reconnection in

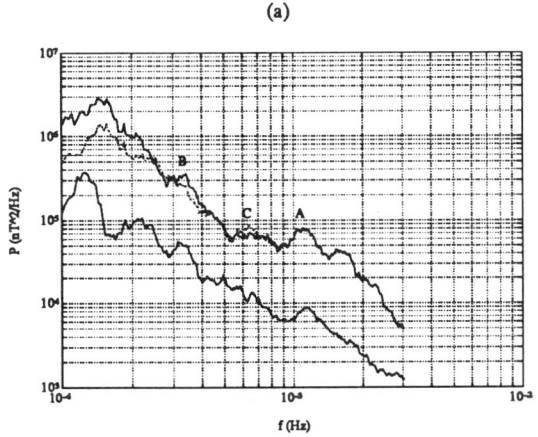
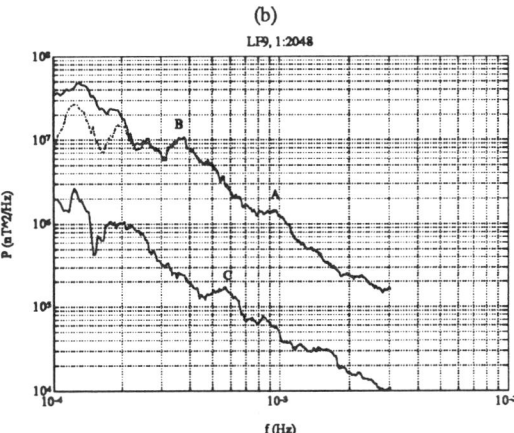

Fig. 7. Spectra of AL (upper curve), AL without "jumps", and low values of AL (lowest curve) for (a) low activity level, and (b) high activity level. Peak labeled "A" indicates a persistent oscillation at 15 min. period. (Taken from Roberts et al. [1990])

the magnetotail begins with the onset of the tearing mode instability [e.g., *Schindler*, 1974] when the plasma sheet has been thinned sufficiently [*Büchner and Zelenyi*, 1987]. The half thickness of the current sheet h does enter the loop model. However, at present h is not treated as a dynamic quantity. For now it is assumed that there is a critical flux value ϕ_c at which the plasma sheet has been thinned sufficiently to establish the onset of the tearing mode instability. If the flux reaches this value during the loading stage the unloading stage is entered by setting $\dot{a} = \dot{a}_D = $ const. < 0 and if the flux reaches this value during the unloading stage the loading stage is entered by setting $\dot{a}$ back to $\dot{a}_L$. In the Baker et al. [1990] dripping faucet analogue model the transition to dumping was made when the displacement of the mass reached a critical value and the rate of dumping was set proportional to the rate of change of the displacement at that instant. In this loop model the transition to unloading is made when $\phi = \phi_c$ and, in analogy to the Baker et al. [1990] model, the unloading rate is set proportional to the rate of change of ϕ at that instant. Thus,

$$\dot{a}_D = \beta \frac{(\mathcal{E}_0 - E_c)\sqrt{a_c}}{\nu} \tag{22}$$

in which E_c and a_c are the values of E and a when the critical flux is reached and β is a proportionality constant which is a free parameter in the model.

Analysis of the Faraday Loop Response Model

The linear prediction filter results of Bargatze et al. [1985] show us that there are certain natural time scales in the response of the magnetosphere to loading by flux transfer. The filter results also show the evolution of the magnetospheric response as the overall level of activity changes. In the following some properties of the model behavior are presented that show the model is capable of explaining the natural time scales and the evolution with activity found in the linear filter results.

The regular dripping mode at low loading rates

As illustrated in Figure 8, the loop model retains the "regular dripping" mode. That is, the period of the loading-unloading cycle remains constant over a large range in loading rate while the size of the plasmoids released scales with the loading rate. In this figure, the evolution with time of the flux passing through the Faraday loop is shown for a broad range of low loading rates. With the constant solar wind conditions in use, and for these low loading rates, substorms occur periodically in our model. The slow growth of the flux in each of the eight substorms pictured corresponds to the growth phase of the substorms. Superimposed on this slow growth are damped plasma sheet oscillations whose period is approximately 15 minutes. The rapid decreases in flux

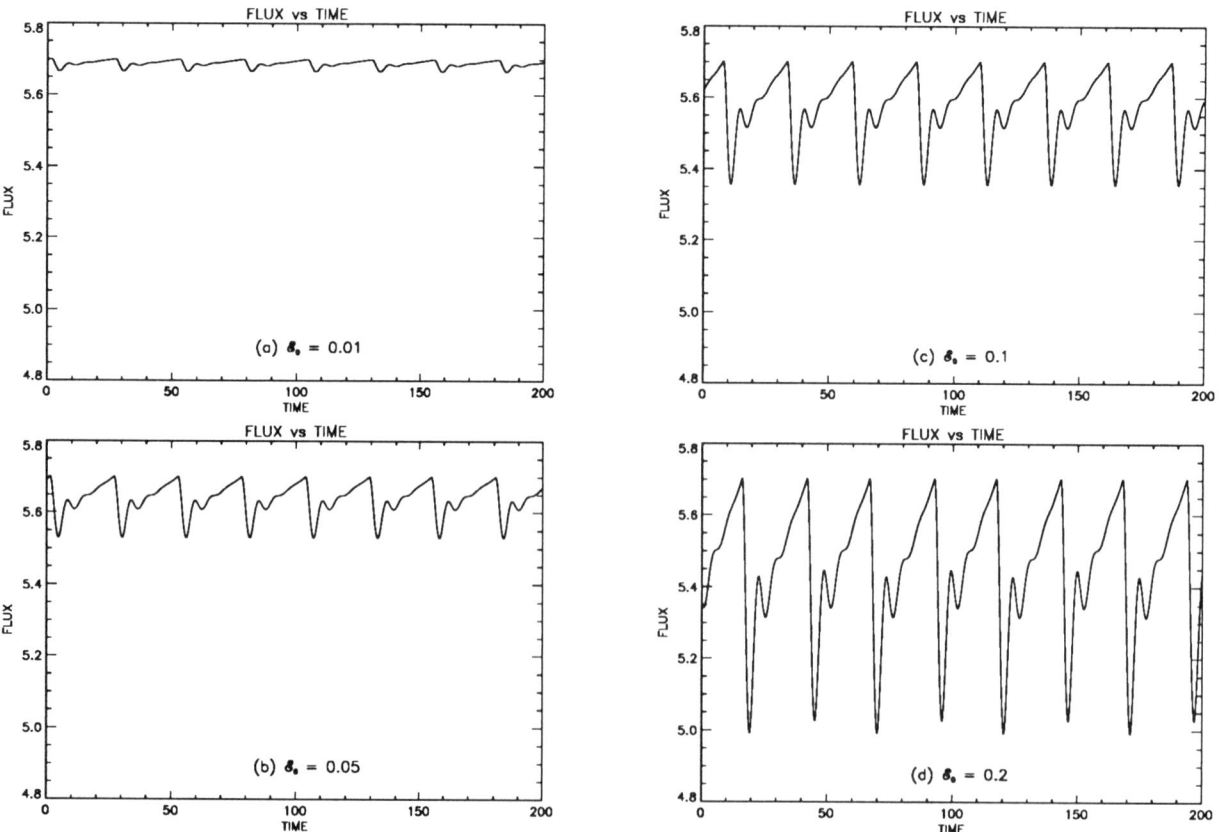

Fig. 8. The regular dripping mode over a large range of loading rates in the loop model.

are due to plasmoid releases that occur, for the parameters in use, over a period of several minutes. The overall period of the loading-unloading cycle is approximately one hour. In the linear prediction filter results, the peak ascribed to the loading-unloading cycle is at an approximately constant one hour lag and, over the range of activity for which it exists, the amplitude of the peak increases with activity. In our view, the regular dripping mode in the model behavior provides a possible explanation of the constant lag and increasing amplitude of the one hour peak in the filter prediction filter results.

Bifurcation sequence into chaos with increasing loading rate

The Faraday loop model exhibits a 1 hour loading-unloading time scale over a large range of low loading rates. As the loading rate is increased to high values, however, the model makes a transition into chaotic behavior and, then, the loading-unloading times become widely distributed. We propose this transition into chaos as an explanation for the vanishing 1 hour response peak in the linear prediction filter results.

From the model results, the period of a loading-unloading cycle versus the period of the previous cycle, for many pairs, are plotted in Figure 9 for loading rates ranging from low to high. For the lowest loading rate, the model behavior is periodic. Thus, all of the periods overlap and a single point appears at approximately 1 hour. The transition into chaos is through a bifurcation sequence during which all of the periods remain close to 1 hour. In the chaotic regime, the loading-unloading cycle period is distributed randomly as shown.

Persistence of the 15 minute oscillations

In the chaotic high loading rate regime, the loading-unloading period becomes unpredictable. However, as shown in Figures 10 and 11, little else in a model substorm changes significantly as the transition to chaos is made. Although the maximum cross tail electric field and the plasmoid flux content may vary somewhat more in the chaotic regime, the basic physical process is unchanged. In particular, the 15 minute oscillations evident both in the electric field and the lobe flux remain a persistent feature of the substorm for all loading rates. If the 15 minute oscillations lead to the peak at 20 minutes lag in the linear prediction filter results [*Bargatze et al.*, 1985], then the persistence of that peak over all levels of average activity can be understood on this basis.

Geomagnetic Activity Prediction

Upon successful completion of our analogue model construction, we anticipate a study of the feasibility of predicting the mag-

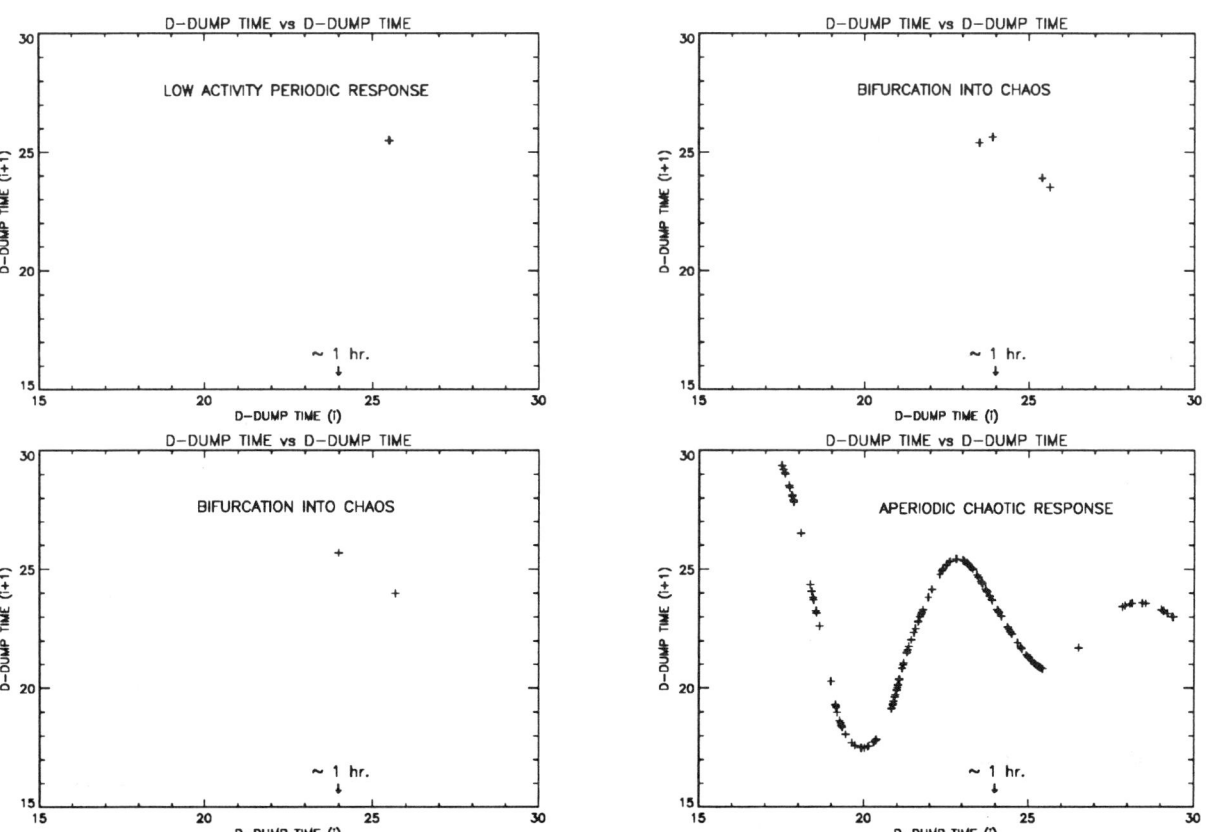

Fig. 9. The loading-unloading time scale goes from predictable to unpredictable through a bifurcation sequence.

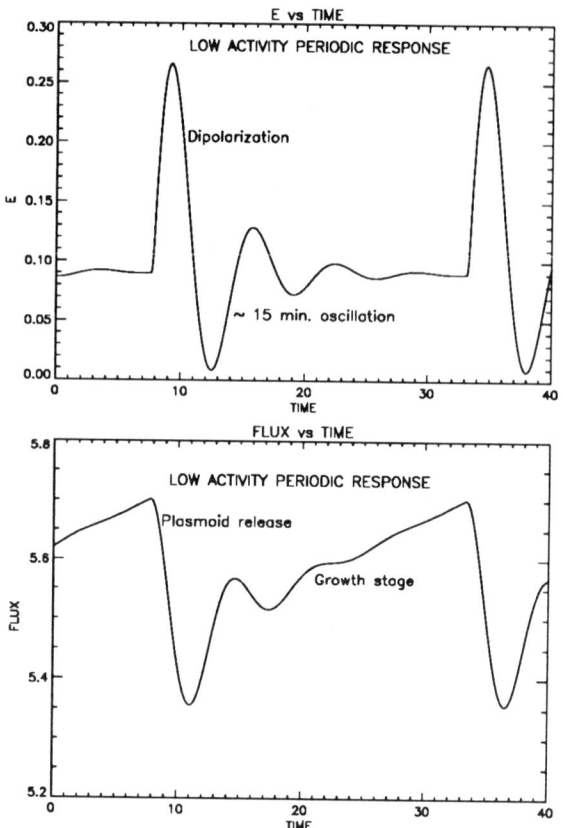

Fig. 10. Evolution of cross tail electric field and lobe flux over the course of a model substorm for low loading rate (periodic evolution).

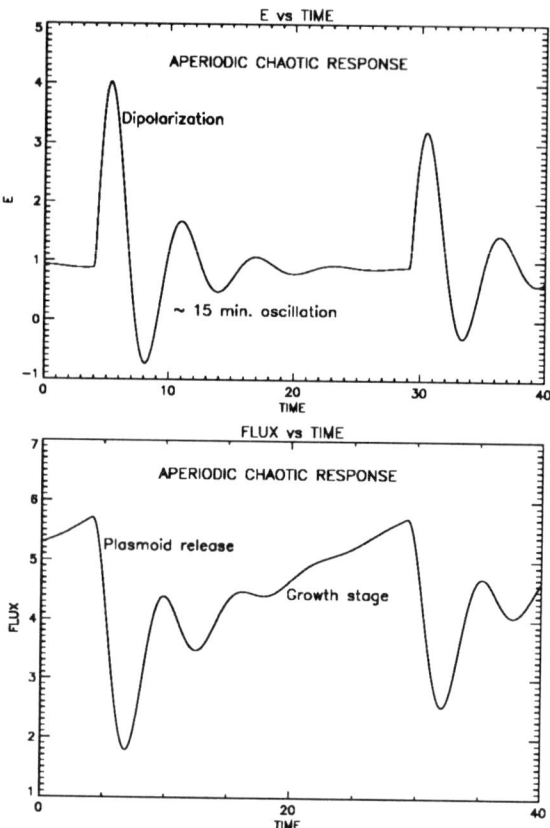

Fig. 11. Evolution of cross tail electric field and lobe flux over the course of a model substorm for high loading rate (chaotic evolution).

netospheric response to the solar wind on a case by case basis. As with the dripping faucet analogue model, our present analogue model exhibits a transition into chaotic behavior for model parameters consistent with high levels of geomagnetic activity. In this regime, the model exhibits extreme sensitivity to variations in state variables and model parameters, indicating the possibility of a fundamental limitation in the predictive power of any realistic magnetospheric model. The existence of a sound model of this activity, however, will help us to understand where the limits of prediction are. Further, in its present form, the model exhibits predictable behavior at low levels of geomagnetic activity. In this case, we have a renewed basis for making practical predictions concerning variations of the geospace environment.

Summary

Our Faraday loop analogue to geomagnetic activity and the related methods of nonlinear dynamics provide powerful new tools for proceeding beyond the limitations of the linear solar wind-magnetosphere coupling studies done in the past. We are actively involved in a program of research designed to take advantage of these new tools to better understand the nonlinear magnetospheric response to the solar wind. We are studying both solar wind-magnetosphere coupling based on the analogue model approach plus the related methods of deterministic nonlinear dynamics. We anticipate an ongoing analogue model development program in the future. Ultimately, we wish to provide a predictive capability for geomagnetic activity plus a firm understanding of whatever fundamental limits to this capability may exist.

Acknowledgments. We wish to acknowledge the significant contributions that have been made by L. Bargatze, R. McPherron, J. Scudder, and D. Stern as we have constructed our response model.

References

Akasofu, S.-I., Interplanetary energy flux associated with magnetospheric substorms, *Planet. Space Sci.*, 27, 425, 1979.

Akasofu, S.-I., The solar wind-magnetosphere energy coupling and magnetospheric disturbances, *Planet. Space Sci.*, 28, 495, 1980.

Baker, D. N., Statistical analyses in the study of solar wind-magnetosphere coupling, In *Solar Wind-Magnetosphere Coupling*, Kamide, Y., and J. A. Slavin, editors, pages 17–38, Tokyo, Terra Scientific Publishing, 1986.

Baker, D. N., S. J. Bame, R. D. Belian, W. C. Feldman, J. T. Gosling, P. R. Higbie, E. W. Hones, Jr., D. J. McComas, and R. D. Zwickl, Correlated dynamical changes in the near-earth and distant magnetotail regions: ISEE 3, *J. Geophys. Res.*, 89, 3855, 1984.

Baker, D. N., L. F. Bargatze, and R. D. Zwickl, Magnetospheric response to the IMF: Substorms, *J. Geomag. Geoelect.*, 38, 1047, 1986.

Baker, D. N., R. D. Belian, P. R. Higbie, and E. W. Hones, Jr., High-energy magnetospheric protons and their dependence on geomagnetic and interplanetary conditions, *J. Geophys. Res.*, *84*, 7138, 1979.

Baker, D. N., E. W. Hones, Jr., P. R. Higbie, R. D. Belian, and P. Stauning, Global properties of the magnetosphere during a substorm growth phase: A case study, *J. Geophys. Res.*, *86*, 8941, 1981a.

Baker, D. N., E. W. Hones, Jr., J. B. Payne, and W. C. Feldman, A high time resolution study of the interplanetary parameter correlations with AE, *Geophys. Res. Lett.*, *8*, 179, 1981b.

Baker, D. N., A. J. Klimas, R. L. McPherron, and J. Büchner, The evolution from weak to strong geomagnetic activity: An interpretation in terms of deterministic chaos, *Geophys. Res. Lett.*, *17*, 41, 1990.

Bargatze, L. F., D. N. Baker, R. L. McPherron, and E. W. Hones, Magnetospheric impulse response for many levels of geomagnetic activity, *J. Geophys. Res.*, *90*, 6387, 1985.

Büchner, J., and L. M. Zelenyi, Chaotization of the electron motion as the cause of an internal magnetotail instability and substorm onset, *J. Geophys. Res.*, *92*, 13,456, 1987.

Clauer, C. R., R. L. McPherron, and C. Searls, Solar wind control of the low-latitude asymmetric magnetic disturbance field, *J. Geophys. Res.*, *88*, 2123, 1983.

Clauer, C. R., R. L. McPherron, C. Searls, and M. G. Kivelson, Solar wind control of auroral zone geomagnetic activity, *Geophys. Res. Lett.*, *8*, 915, 1981.

Fairfield, D. H., On the average configuration of the geomagnetic tail, *J. Geophys. Res.*, *84*, 1950, 1979.

Fairfield, D. H., Solar wind control of magnetospheric pressure (CDAW 6), *J. Geophys. Res.*, *90*, 1201, 1985.

Fairfield, D. H., The magnetic field of the equatorial magnetotail from 10 to 40 R_E, *J. Geophys. Res.*, *91*, 4238, 1986.

Fairfield, D. H., Structure of the geomagnetic tail, In *Magnetotail Physics*, Lui, A. T. Y., editor, p. 23. Baltimore: Johns Hopkins University Press, 1987.

Grassberger, P., and I. Procaccia, Measuring the strangeness of strange attractors, *Physica*, *9D*, 189, 1983.

Hones, Jr., E. W., Transient phenomena in the magnetotail and their relation to substorms, *Space Science Reviews*, *23*, 393, 1979.

Iyemori, T., and H. Maeda, Prediction of geomagnetic activity from solar wind parameters based on the linear prediction theory, In *Solar-Terrestrial Predictions Proceedings*, volume 4, National Oceanic and Atmospheric Administration, Boulder, Colo., 1980.

Iyemori, T., H. Maeda, and T. Kamei, Impulse response of geomagnetic indices to interplanetary magnetic field, *J. Geomagn. Geoelectr.*, *6*, 577, 1979.

Longmire, C. L., *Elementary Plasma Physics*, Interscience Publishers, a division of John Wiley and Sons, New York, London, 1963.

Lyons, L. R., and T. W. Speiser, Ohm's law for a current sheet, *J. Geophys. Res.*, *90*, 8543, 1985.

Maezawa, K., Magnetotail boundary motion associated with geomagnetic substorms, *J. Geophys. Res.*, *80*, 3543, 1975.

McPherron, R. L., Growth phase of magnetospheric substorms, *J. Geophys. Res.*, *28*, 5592, 1970.

McPherron, R. L., D. N. Baker, L. F. Bargatze, C. R. Clauer, and R. E. Holzer, IMF control of geomagnetic activity, *Adv. Space Res.*, *8*, 71, 1988.

McPherron, R. L., C. T. Russell, and M. P. Aubry, Satellite studies of magnetospheric substorms on August 15, 1968, 9, Phenomenological model for substorms, *J. Geophys. Res.*, *78*, 3131, 1973.

Perreault, P., and S.-I. Akasofu, A study of geomagnetic storms, *Geophys. J. R. Astron. Soc.*, *54*, 547, 1978.

Roberts, D. A., Is there a strange attractor in the magnetosphere?, *J. Geophys. Res.*, In press, 1991.

Roberts, D. A., D. N. Baker, and A. J. Klimas, Indicators of low-dimensionality in magnetospheric dynamics, *Geophys. Res. Lett.*, *18*, 151, 1991.

Schindler, K., A theory of the substorm mechanism, *J. Geophys. Res.*, *79*, 2803, 1974.

Shan, L. H., P. Hansen, C. K. Goertz, and R. A. Smith, Chaotic appearance of the AE index, *Geophys. Res. Lett.*, *18*, 147, 1991.

Shaw, R., *The dripping faucet as a model chaotic system*, The Science Frontier Express Series. Aerial Press, Santa Cruz, CA, 1984.

Speiser, T. W., Conductivity without collisions or noise, *Planet. Space Sci.*, *18*, 613, 1970.

Vassiliadis, D. V., A. S. Sharma, T. E. Eastman, and D. Papadopoulos, Low-dimensional chaos in magnetospheric activity from AE time series, *Geophys. Res. Lett.*, *17*, 1841, 1990.

Copyright 1990 by the American Geophysical Union.

Synergetic Approach to Substorm Phenomenon

ZOLTAN VÖRÖS

Geophysical Institute, Slovak Academy of Sciences, 947 01 Hurbanovo, CSFR

In order to study the nonlinear physical processes connected with substorm activity we analyse time series of geomagnetic field variations obtained at the Hurbanovo geomagnetic observatory. The concept of deterministic chaos and magnetospheric strange attractors are examined in the paper. On this basis we propose a new qualitative, synergetic model of global substorm activity. The basic notion of suggested model is the perception that, in the process of energy redistribution or substorm breakup, some kind of informational-energetical magnetosphere-ionosphere coupling plays the decisive role.

INTRODUCTION

The Earth's magnetosphere is a macroscopic, nonlinear, open system with a characteristic permanent state that is far from thermal equilibrium. The non-equilibrium state manifests itself by the distorted comet-shaped configuration of the magnetosphere and by a system of large scale currents that flow in the tail, into and away from the auroral region, around the equator, etc. The presence of the current system indicates that, through energy dissipation combined with mass and momentum exchange between magnetosphere-ionosphere subsystems the whole system proceeds toward equilibrium. The non-equilibrium state is insured by the permanent solar wind magnetosphere interaction. As a consequence of the external forcing of the solar wind and of the internal magnetosphere-ionosphere dynamics, a large variety of spatio-temporal and functional structures are present in the external geospace. Fortunately, there exists a unified theory for spatio-temporal pattern formation in non-equilibrium systems like pattern formation in lasers, hydrodynamics, thermodynamics or chemical reactions [Haken, 1978; Nicolis and Prigogine, 1977]. This theory of self-organized macroscopic structures was called "synergetics" by Haken. As stated in his monographs on this field [Haken, 1983; Haken, 1988]: "Synergetics is an interdisciplinary field of research dealing with the emergence of spatial, temporal, spatio-temporal as well as functional structures in complex, nonlinear, dissipative and open systems".

In this paper we briefly describe the essential ideas of synergetic theory and on the basis of nonlinear time-series analysis and presumed universality present a new qualitative picture of magnetosphere-ionosphere coupling.

SELF-ORGANIZATION IN COMPLEX SYSTEMS

The basic idea of synergetic theory is a qualitative change of considered macroscopic states. A new macroscopic state as a consequence of qualitative changes occurs due to the loss of linear stability of a given system's basic state. A hierarchy of instabilities may exist when one deals with stationary, time-periodic or quasi-periodic basic states depending on special external conditions and fluctuating forces. So, the hierarchy of new macroscopic states may exist, too.

The unified nonlinear treatment of the transition:

$$\begin{array}{c} \text{basic macroscopic state (stationary,} \\ \text{time-periodic, quasi-periodic)} \\ \downarrow \\ \text{instability} \\ \downarrow \\ \text{new macroscopic state (stationary, time-} \\ \text{periodic, quasi-periodic)} \end{array} \quad (1)$$

is incorporated in the "slaving" principle of synergetics. This principle can be described as follows: close to the transition/instability point it becomes possible to adiabatically eliminate/enslave many physical variables (stable modes of motion). Then the macroscopic behavior (structure formation) of the complex system is governed only by very few degrees of freedom called order parameters.

In many cases of macroscopic systems the evolving pattern (spatial, temporal, functional structure) is determined by a very small number of order parameters and its nonlinear equations called generalized Ginzburg-Landau equations [Haken, 1988]. The physical meaning of mathematically formulated slaving principle is based on nonlinear competition of stable and unstable modes of motion and on strong dissipation of energy of enslaved modes.

Using the synergetic slaving principle, there is a high probability that we may understand the transition between old and new magnetosphere-ionosphere structures forced by the change in external (solar wind) conditions (Vörös, manuscript in preparation, 1991). Particularly interesting is the case when, after a hierarchy of instabilities, a new state corresponding to a low-dimensional turbulence, called deterministic chaos is developed. A special sequence of instabilities and transitions leading to the chaos realized by a specific system is often called a "route" or "scenario" [Eckmann and Ruelle, 1985; Feigenbaum, 1978].

So, the modified transition scheme similar to (1) is valid:

basic macroscopic state (stationary,
time-periodic, quasi-periodic)
↓
hierarchy of instabilities   "scenarios"   (2)
↓
deterministic chaos (soft turbulence).

Various theoretically predicted "routes" to chaos have also been experimentally ascertained [Yahata, 1984].

Naturally, the first question which arises after such results is: to what extent do real geophysical, astrophysical and other complex systems undergo the transition to chaos in the low dimensional manner? In our opinion, deterministic chaos is a possible regime of "erratic" motion (turbulent temporal or spatio-temporal patterns) in real complex systems when a truncated system of differential equations corresponds to a few evolution equations of unstable modes deduced by the synergetic slaving principle. This means that, as a consequence of nonlinear mode competition, a very high number of degrees of freedom become enslaved and the macroscopic dynamics and structures will be determined only by a few degrees of freedom.

This assertion is not evident for the case of deterministic chaos because, in a chaotic regime the primarily stable and adiabatically eliminated modes lose their stability so that the slaving principle could become inoperative. Such an example is given by Haken [1978]. However, as has been shown theoretically (e.g., for the baroclinic instability of oceanic circulation) only a few of the unstable modes become significant in a chaotic regime while others remain enslaved [Seidov, 1989]. Then an assembly of the slaving principle by a chaotic regime of motion is basically possible.

We should physically interpret these facts as follows: Deterministic chaos and ordered, self-organized structures (even in chaos) arise from the same sort of nonlinear laws. Chaotic and ordered structures coexist in nature.

In the magnetosphere-ionosphere system one can observe many macroscopic structures, turbulence, depending on selected spatial and time scales. We mention here the aurora [Frank and Craven, 1988], magnetopause field structures [Dubinin at al., 1980; Galeev et al, 1986], or auroral convention patterns on turbulent background [Ahn et al., 1989; Heelis at al., 1986]. More detailed discussion is to be found in (Vörös, manuscript in preparation, 1991). In our opinion, the formation and the evolution of magnetosphere-ionosphere macroscopic structures and their qualitative changes (phase transitions) play a key role in energy redistribution and coupling processes during geomagnetic storms.

In the following two sections our main interest will be concentrated on nonlinear time series analysis. We will show below that the apparently irregular, stochastic field variations during geomagnetic storms correspond to deterministic chaos on the time scale of substorms.

NONLINEAR TIME-SERIES ANALYSIS

As usual, we assume that the time evolution of a dynamical system can be described by a differentiable dynamical system in a phase space of possible infinite dimensions.

In many cases, due to strong dissipation, there corresponds to a turbulent temporal pattern (chaotic state) a finite dimensional chaotic or strange attractor in phase space [Eckmann and Ruelle, 1985]. The closed set A is called an attractor if, for $t \to \infty$, all nearby trajectories in phase space tend to A. Strange attractors are typically characterized by fractal dimensionality while the term "chaotic" refers to Lyapunov instability of phase trajectories. For the purpose of quantitative characterization of strange attractors the dimension spectrum of attractors has been defined [Atmanspacher et al., 1988] as

$$D^{(\gamma)} = \frac{1}{\gamma - 1} \lim_{\varepsilon \to 0} \frac{\ln\left[\sum_{i=1}^{m} P_i^{\gamma}\right]}{\ln \varepsilon} \quad (3)$$

where P_i is the probability that a point on a trajectory falls into the i-th box of size ε in phase space partitioned to m hypercubes. The most commonly used dimension is the so-called correlation dimension $D^{(2)}$ which gives a lower estimate of the minimum number of variables (degrees of freedom) which are at least required to describe the system.

The details of the temporal evolution of phase trajectories on a chaotic attractor can be quantitatively characterized by the spectrum of Lyapunov exponents. In general, the Lyapunov exponents, L_i, may be calculated by linearizing the relevant equations and studying the evolution of small perturbations w [Haken, 1983]. Then

$$L_i = \lim_{t \to 0} \sup \left\{ \frac{1}{t} \ln |w_i(t)| \right\} \quad (4)$$

If $L_i > 0$; ($i=1, 2, ..., j$, $j \leq N$) then the motion along the trajectories is unstable and small perturbations grow exponentially with time t. This phenomenon of an error's exponential growth is called a sensitive dependence on the initial conditions. The cases when all L_i's are nonpositive correspond to stationary or periodic states.

Unfortunately, for geomagnetic storm activity, neither the relevant variables nor even their total number or evolution are known, so that the attractor of the system is not a priori accessible. In such a case, the attractor can be reconstructed in an artificial phase space if a time series of one single variable is measured. The embedding theorems of Takens [1981] ensure that the attractor is reliably reconstructed in an artificial phase space of sufficiently large embedding dimension m. Usually $m \geq 2D^{(2)} + 1$. Then the attractor in a phase space of m dimensions may be obtained from scalar time series $\{X_0(t_i)\}$ by the time delayed method [Simm et al., 1987].

The set of N points on the attractor A embedded in m dimensional Euclidean phase space is reconstructed as follows:

$$X_i = \{X_0(t_i), ..., X_0(t_i + (m-1)\tau)\}$$
$$i = 1, 2, ..., N; \quad X_i \in A \subset R^m \quad (5)$$

τ is the time delay.

To compute the dimension and to estimate the sum of positive Lyapunov exponents of a reconstructed magnetospheric attractor - phase space portrait - we use the method of Grassberger and Procaccia [Grassberger and Procaccia, 1983a; Grassberger and Procaccia, 1983b].

Following the procedure proposed by Grassberger and Procaccia, the dimension D_m of an attractor behaves as a power of some prescribed distance r such that

$$C_m(r) = \lim_{k \to \infty} \frac{1}{k^2} \sum_{\substack{i,j=1 \\ i \neq j}}^{k} \theta(r - |X_i^{(m)} - X_j^{(m)}|) \doteq r^{D_m} \quad (6)$$

where θ - is the Heaviside step function,
$C_m(r)$ is the correlation function in m dimensional phase space.

If we subsequently compute D_m in a phase space of increasing dimension m, then for deterministic time series, D_m reaches its saturation limit

$$\lim_{m \to \infty} D_m = D \sim D^{(2)}$$

This limit will be regarded as the correlation dimension for an attractor as it is defined by Eq. (3). A stochastic signal requires that $D_m \to \infty$ for $m \to \infty$.

To estimate the sum of all positive Lyapunov exponents we use the quantity called 2nd order Renyi entropy [Grassberger and Procaccia, 1983b]

$$K_2 = \lim_{m \to \infty} \lim_{r \to 0} \frac{1}{t} \ln \frac{C_m(r)}{C_{m+1}(r)} \quad (7)$$

The K_2 entropy has the following properties

i. $K_2 \geq 0$,
ii. $K_2 \leq K \leq \Sigma L_i^+$ (L_i^+ - positive Lyapunov exponents),
iii. K_2 is infinite for random systems,
iv. $K_2 \neq 0$ for chaotic systems.

So, $0 < K_2 << \infty$ is a sufficient condition for deterministic chaos.

DATA ANALYSIS

We assume that, during intense geomagnetic storms, the global magnetospheric dynamics is well represented in local magnetic field variations on the time scale of substorms. Such an assumption is founded on the world-wide character of geomagnetic storms. On this basis the time series of geomagnetic field H component's variations obtained by digital magnetometers at the Hurbanovo geomagnetic observatory (CSFR) in the form of a sequence of 1 min mean values are used in the analysis [Podsklan and Brunner, unpublished data, 1989].

The geomagnetic storm of march 13, 1989 was selected for the analysis, because of its relatively long duration and world-wide intensity. The duration of this storm is approximately 32 h. It is expected that the time series of geomagnetic storm variations will necessarily be noisy. For this reason, we decided to filter the data. To eliminate a high frequency variability in data set, we use a smoothing procedure with a 17 min running mean. The filtered and unfiltered data sets are compared in Fig. 1.

Chennaoui et al. [1990] have shown for one-dimensional maps that a weak filter can induce an increase of an attractor dimension. However, it is necessary to underline the fact that still there exists a big uncertainty surrounding dimension estimates of filtered data.

The data set given by Eq. (5) contains the free parameters N, $t_s = t_{i+1} - t_i$ and τ which are still to be determined. The number of data points N and the sampling interval t_s should be chosen so as to yield sufficient resolution over the desired frequency range of the Fourier spectrum of the scalar variable. Clearly t_s is related to the time scale (substorm time scale) on which the process of interest (geomagnetic storm) appears to take place. There should be at least a few dozens of points within this time scale, so the temporal resolution of 1 min is adequate for the analysis performed on the substorm

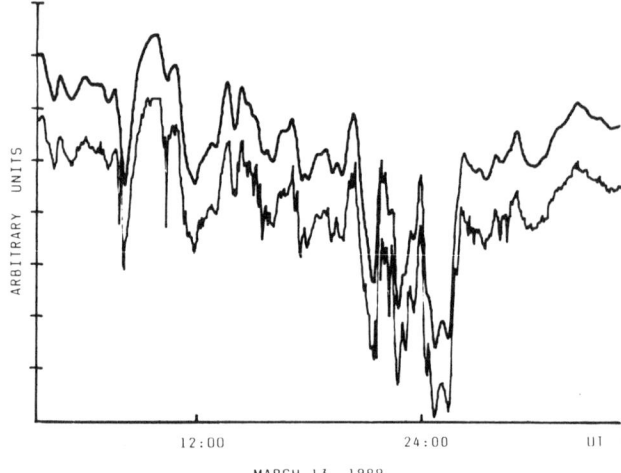

Fig. 1. Unfiltered and filtered time series of geomagnetic storm of march 13, 1989.

time scale (~30 ÷ 90 min). At the same time, too small a value of t_s often provides a noisy signal.

The total number of data points N is determined by the length of a time series (geomagnetic storm) and by the temporal resolution. As shown by Wolf et al. [1985] the minimum number of data points required for the reliable reconstruction of an attractor of dimension D is $\sim 10^D$. Although, there are good indications that the structure of a low dimensional attractor can be analysed even if the available data length are rather short [Kurths and Herzel, 1987]. Our algorithm was tested for known, low dimensional model systems (periodic function, Henon map, Lorenz attractor [Vörös, manuscrip in preparation, 1991]).

Using a few hundred of data points the dimension computation's error is ~5%. In the case of entropy it is about 20%. This means that the value of computed entropy is rather an estimation. Nevertheless, our main question, whether there exist positive Lyapunov exponents which correspond to chaotic variations of geomagnetic field can be answered.

The choice of time delay τ is much more problematical. For an ideal data set (e.g., obtained as a solution of a system of nonlinear equations) the size of time delay τ is essentially arbitrary. For a realistic data set, Fraser and Swinney [1986] have developed an idea to determine an optimum value of τ based on the concept of mutual information. However, this method has the drawback of being too complicated. Moreover, to obtain utilizable results one, usually needs a larger data set then the one corresponding to a geomagnetic storm. For analysed geomagnetic data set we use the simple criterion $T/m < \tau < T$, where T is defined as the quarter period of a pseudo-periodic signal [Simm et al., 1987]. For geomagnetic storms, the value of τ relevant to substorm time scales is approximately $3 < \tau < 30$ min. The lower limit approximately corresponds to the time scale of Alfvén waves. The upper limit in order is equal to the time delay of 30 - 50 min between a southward turning of the IMF and the substorm onset [Kan and Akasofu, 1989], which is roughly the autocorrelation time scale. It is a time period over which the magnetosphere adjust itself to sudden changes.

In addition, after Atmanspacher et al. [1988], we use an approximate test for a suitable choice of τ by a careful examination of two dimensional phase space portraits defined by $X(t_i)$ versus $X(t_i + \tau)$. In Fig. 2 we show such a phase-space portrait reconstructed for various time delays τ in the case of the geomagnetic storm of march 13, 1989.

The plotted points do not show a strong compression around the line $X(t_i) = X(t_i + \tau)$; hence there is no significant linear dependence between them. Also, the two-dimensional phase portraits are very similar for the various values of τ which is also a necessary condition for trustworty computation of the correlation dimension.

Using this procedure the correlation dimension (Eq. 6) and the K_2 entropy (Eq.7) have been calculated for the geomagnetic storm of march 13, 1989. Fig.3 and Fig.4 show examples of logarithmic plots of $C_m(r)$ versus r and the dependence (saturation) of the correlation dimension, D_m, on the embedding dimension, m, for the considered geomagnetic storm. Curves corresponding to the time delay of $\tau = 20\ t_s$ and to the total number of data points N = 1900 are shown.

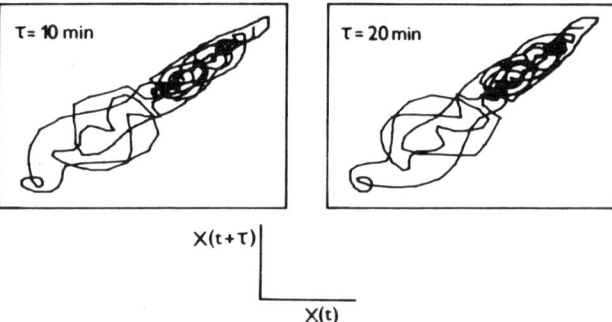

Fig. 2. Phase-space portraits of the investigated geomagnetic time-series for the various values of time delay τ.

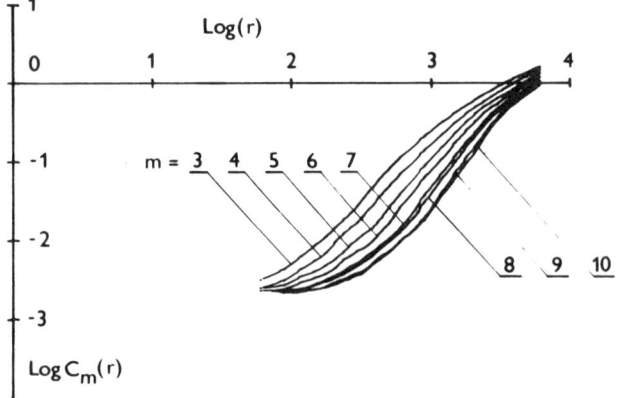

Fig. 3. Logarithmic plots of the correlation integral $C_m(r)$ versus r.

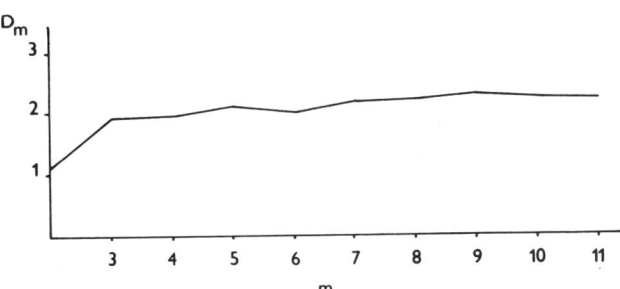

Fig. 4. Correlation dimension D_m versus embedding dimension m.

By careful inspection of curves in Fig.3 we determined the "meaningful range of r" in which the statistics for the power law (Eq. 6) is reliable. We denote the characteristic size of the magnetospheric attractor as R. Then, the slopes of $C_m(r)$ become unreliable due to a lack of sufficient attractor coverage with the finite number of data points available to us, when r < 0.03R. On the other hand, for too large r, approximately all the points become correlated within the hypersphere of radius r, and C(r) tends to constant value. This yield r > 0.4R. Then the meaningful range of r is: r∈(0.03R; 0.4R). This range is somewhat shifted for largest embedding dimensions. The correlation dimension, D=2.2±0.04, obtained in such a way give a good estimate for $D^{(2)}$ as defined in Sect.3 by Eq.(3). After the embedding theorems of Takens [1981] for reliable reconstruction of an attractor is required m ≥ 2D + 1. In our case it is: m ≥ 5.4. Indeed, the knee in Fig.4 occurs at small embedding dimension.

We have also computed the K_2 entropy according to Eq.(7). Here K_2 is simply obtained from the vertical spacing between the lines of the log $C_m(r)$ versus log r plots, within the meaningful range of r. The obtained value of K_2 entropy for the analyzed geomagnetic storm is $K_2 = 0.02 \pm 0.01$.

We mention yet, that there exist other methods for dimension and entropy calculations. On the basis of nonlinear prediction techniques it is also possible to estimate the dimensions and entropies of chaotic attractors [Farmer and Sidorowich, 1987].

DISCUSSION

The finite attractor dimension and positive K_2 entropy estimated from geomagnetic data give some indications of deterministic chaos on substorm time scale. This suggests that observed geophysical phenomena should be described by nonlinear system in low dimensional phase space which are capable of deterministic chaos. These results do not anticipate the validity of any particular model of substorm activity. Rather, they set a number of constraints that should be satisfied by a model.

There are other indications of low dimensional deterministic dynamics of chaotic substorm activity. We refer only to some examples. Using the phase space reconstruction technique, the AE index time series has been examined by Vassiliadis et al. [1991], Roberts et al. (1991) and by Shan et al. (1991). They obtained a correlation dimension which is between 2 and 4 depending on the time delay τ. The K_2 entropy for the AE index data is finite and positive as well as in our case of local field variations. The dimension of the reconstructed magnetospheric attractor, as expressed through the AE index or local field variation time series, has somewhat different values, nevertheless, the agreement is surprising. We mention that, using a very large number of data points (better resolution, longer time series), the phase space reconstruction technique may reveal a multifractal structure of magnetospheric dynamics.

Baker et al. [1990] have developed a simple, but dynamically elegant, chaotic model of loading-unloading cycle based on a generalization of the dripping faucet model [Shaw, 1985]. Beyond a critical energy input rate from the solar wind, during a geomagnetic storm, a chaotic loading-unloading cycle develops in their substorm model which is represented by a chaotic attractor in three dimensional phase space.

The results obtained on the basis of fractal analysis of several geomagnetic storms also suggest that to time patterns of local field's variations correspond a self-similar dynamics [Vörös, 1990].

These results strongly suggest that models involving 3 or 4 variables already should provide a description of the salient features of chaotic substorm activity during geomagnetic storms.

We suggest that the excited degrees of freedom should correspond to unstable modes of motion as there are defined by slaving principle of synergetics. It is very challenging that, on the time scale of daily variations the reconstructed magnetospheric attractor is also low dimensional. In this case D~4.6 [Vörös, 1989].

As an approximation, we suggest that soft-turbulent regimes can be studied in terms of corresponding chaotic attractors within the magnetopause boundary layer, auroral and magnetotail regions, too [Vörös, manuscript in preparation, 1991]. The sensitive dependence on the initial conditions plays the central role in our qualitative model of chaotic, global magnetospheric activity on the substorm time scale.

One can imagine, for example, that the auroral ionosphere (or another aforementioned region) is represented by a box. Fig.5 shows, inside this box, a schematic phase picture - an assumed chaotic attractor - with unstable trajectories in the sense of Lyapunov. The phase trajectories may correspond, for example, to the evolution of ionospheric Hall or Pedersen conductances, to the changes of ion-neutral collision frequency, to the configuration of the ionospheric current system, etc. Past work has also demonstrated that substorm activity results from nonlinear coupling of the magnetosphere and auroral ionosphere. In their magnetosphere-ionosphere (M-I) coupling model of substorms, Kan and Akasofu (1989) showed that the substorm onset is controlled to a large extent

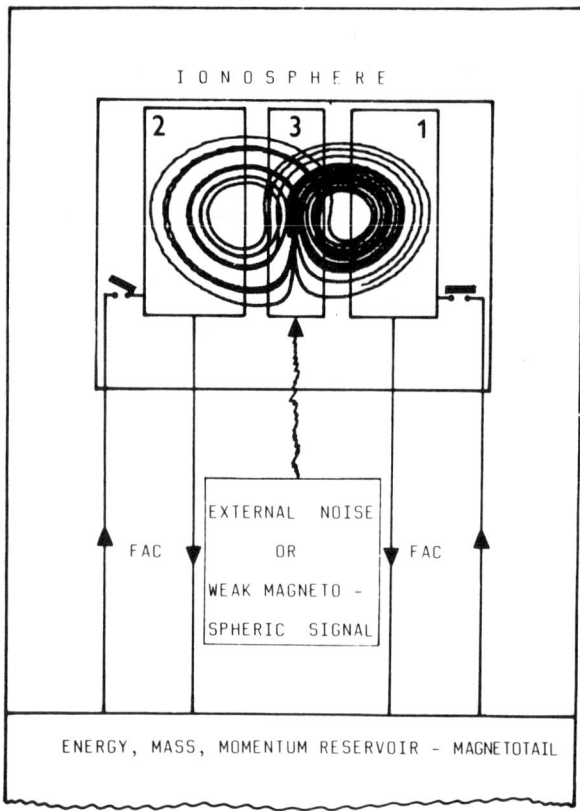

Fig. 5. Schematic view of magnetosphere-ionosphere informational-energetical coupling; FAC - field-aligned currents.

by the configuration and evolution of Hall and Pedersen currents. Due to the co-location of divergences of the Hall and Pedersen currents the upward field-aligned currents become stronger and the substorm onset should starts. Taking into account both a possible ionospheric chaos and the energetic coupling depicted by the M-I coupling model, we propose a new qualitative picture of M-I coupling based on the sensitive dependence on initial conditions of chaotic attractors.

We suppose that an energetic connection due to the field-aligned currents is possible only when the phase trajectories wander say, in Region 1 of the phase space (a substorm onset starts). On the other hand, coupling is impossible when the phase trajectories are in Region 2 as shown in Fig.5. Within Region 3 of the phase space a small external noise due to Lyapunov instability or magnetospheric signal (by itself insufficient for coupling) may regulate the connection between ionospheric and magnetospheric energetic channels (e.g. current systems) through sudden and chaotic overswitching between trajectories. Chaotic trajectories continually reveal or "create" new information about the system. The mean rate of this "information creation" is measured by the K_2 entropy of the system [Nicolis, 1986]. Naturally, our capability to make predictions, decreases.

Our results correspond to global magnetosphere-ionosphere chaos, however, it should originate in complex dynamics of magnetosphere-ionosphere subsystems. Thus, instead of a single energetical coupling we have to deal with informational-energetical coupling. It is necessary to emphasize, however, that because of the structural stability of the chaotic attractors [Eckmann and Ruelle, 1985] no sudden energy explosions - substorms are possible without energetic channels (energy, mass, momentum reservoirs). This means that, in addition to a chaotic internal dynamics some energy reservoirs (solar wind, loading, magnetotail storage) are necessary for an informational-energetical coupling. Thus, some kind of competition between distinct parts of the magnetosphere-ionosphere system with chaotic internal dynamics and informational-energetical coupling, as well as fulfilment of geometrical conditions (e.g. co-location of divergences of the Hall and Pedersen currents) may lead to chaotic appearance of substorms as we have observed.

The most similar model to our synergetic one is the substorm model of Zeithamer [1987, 1988]. In his model, during a substorm breakup, the unstable state of the magnetosphere-ionosphere system is represented by the point cusp-catastrophe manifold whose perpendicular projection to the plane of control parameters belongs to a bifurcation set [Zeithamer, 1990].

Perhaps, our qualitative model seems to be speculative, however, chaotic behavior, self-organized macroscopic structures, information creation are typical features of macroscopic, nonlinear systems. The magnetosphere-ionosphere system is evidently nonlinear and, in a sense of deterministic, low dimensional dynamics, perhaps, chaotic. Taking into account an informational coupling, we only recognize the fact that the macroscopic system is nonlinear.

We would like to make it perfectly clear that Fig.5 is only a cartoon which corresponds to the approximation level of our nonlinear approach. Moreover, the problems of hard turbulence or interactions with unionized gases, etc., which evidently exist within the magnetosphere, are not involved in the outlined qualitative model. The large scale nature and the complexity of the substorm breakup clearly indicates that the complete process can not be included in one model. There is no contradiction between our qualitative model and the existing substorm models as [Goertz and Smith, 1989; Kan et al., 1988; Liu et al., 1988; Coroniti, 1985]. The synergetic approach rather guides the way to the direction of global, nonlinear description of magnetosphere-ionosphere activity, and helps to find a common language on the basis of existing substorm models.

We would like to refer yet to pathways of spontaneous self-organization outlined in Haken [1983]. One of the possible pathways of self-organization is pattern formation through change of the number of coupled components of a system. During substorm activity, self-organization (e.g., auroral pattern formation or phase transition) may origin in enhanced intensity of informational-energetical M-I coupling. Other ways as self-organization through transients are also possible.

Finally, we would like to emphasize that the description of global substorm activity under this point of view is still in its infancy and should be continued by further carefull analysis.

Acknowledgements: The author would like to thank M. Dryer for making numerous helpful corrections and suggestions. Discussions with N. Manaenkova, A. A. Ruzmaikin, J. Brunner, Z. Kolláth, and R. Szlizs as well as the reading of the manuscript by J. Verö and A. Prigancová are also gratefully acknowledged. I thank J. Podsklan for his help with the graphical work.

REFERENCES

Ahn, B. H., H. W. Kroehl, Y. Kamide, D. J. Gorney, S. I. Akasofu, and J. R. Kan, The auroral energy deposition over the polar ionosphere during substorms, *Planet. Space Sci.*, 37, 239-252, 1989

Atmanspacher, H., H. Scheigraber, and W. Voges, Global scaling properties of a chaotic attractor reconstructed from experimental data, *Phys. Rev.*, 37, 1314-1322, 1988

Baker, D. N., A. J. Klimas, R. L. McPherron, and J. Büchner, The evolution from weak to strong geomagnetic activity: an interpretation in terms of deterministic chaos, *Geophys. Res. Lett.*, 17, 41-44, 1990

Chennaoui, A., J. Liebler, and H. G. Schuster, The mechanism of the increase of the generalized dimension of a filtered time series, *J. Statist. Phys.*, 59, 1311-1327, 199

Coroniti, F., Explosive tail reconnection: the growth and expansion phases of magnetospheric substorms, *J. Geophys. Res.*, 90, 7427-7447, 1985

Dubinin, E. M., I. M. Podgorny, and Yu. N. Potanin, On magnetic field structure on the magnetosphere boundary, (in Russian), *Kosm. Issled.*, 18, 99-111, 1980

Eckmann, J. P., and D. Ruelle, Ergodic theory of chaos and strange attractors, *Rev. Mod. Phys.*, 57, 617-656, 1985

Farmer, J. D., and J. J. Sidorowich, Predicting chaotic time series, *Phys. Rev. Lett.*, 59, 845-848, 198

Feigenbaum, M. J., Quantitative universality for a class of nonlinear transformations, *J. Stat. Phys.*, 19, 25-52, 197

Frank, L. A., and J. D. Craven, Imaging results from Dynamics Explorer 1, *Rev. Geophys.*, 26, 249-283, 1988

Fraser, A. M., and H. L. Swinney, Independent coordinates for strange attractors from mutual information, *Phys. Rev. A*, 33, 1134-1140, 1986

Galeev, A. A., M. M. Kuznetsova, and L. M. Zeleny, Magnetopause stability threshold for patchy reconnection, *Space Sci. Rev.*, 44, 11-41, 1985

Goertz, C. K., and R. A. Smith, The thermal catastrophe model of substorms, *J. Geophys. Res.*, 94, 6581-6596, 1989

Grassberger, P., and I. Procaccia, Characterization of strange attractors, *Phys. Rev. Lett.*, 50, 346-349, 1983a

Grassberger, P., and I. Procaccia, Estimation of the Kolmogorov entropy from a chaotic signal, *Phys. Rev. A*, 28, 2591-2593, 1983b

Haken, H., Synergetics, Springer Verlag, Berlin-Heidelberg-New York, pp. 371, 1978

Haken, H., Advanced Synergetics, Springer Verlag, Berlin-Heidelberg-New York, pp. 356, 1983

Haken, H., Information and self-organization, Springer Verlag, Berlin-Heidelberg-New York, pp. 196, 1988

Heelis, R. A., P. H. Reiff, J. D. Winningham, and W. B. Hanson, Ionospheric convection signatures observed by DE 2 during northward interplanetary magnetic field, *Geophys. Res.*, 91, 5817-5830, 1986

Kan, J. R., and S. I. Akasofu, Electrodynamics of solar wind-magnetosphere-ionosphere interactions, *Plasma Sci.*, 17, 83-108, 1989

Kan, J. R., L. Zhu, and S. I. Akasofu, A theory of substorms: onset and subsidence, *J. Geophys. Res.*, 93, 5624-5640, 1988

Kurths, J., and H. Herzel, An attractor in a solar time series, *Physica*, 25D, 165-172, 1987

Lin, Z. X., L. C. Lee, C. Q. Wei, and S. I. Akasofu, Magnetospheric substorms: an equivalent circuit approach, *J. Geophys. Res.*, 93, 7366-7375, 1988

Nicolis, J. S., Dynamics of hierarchical systems: an evolutionary approach, Springer Verlag, Berlin-Heidelberg-New York-Tokyo, pp. 397, 1986

Nicolis, G., and I. Prigogine, Self-organization in nonequilibrium systems, Wiley-Interscience, New York, 1977

Roberts, D. A., D. N. Baker, and A. J. Klimas, Indicators of low-dimensionality in magnetospheric dynamics, *Geophys. Res. Lett.*, 18, 151, 1991

Seidov, D. G., Synergetics of the ocean processes, (in Russian), Gidrometeoizdat, Leningrad, pp. 287, 1989

Shan, L., P. Hansen, C. K. Goertz, and R. A. Smith, Chaotic appearance of the AE index, *Geophys. Res. Lett.*, 18, 147, 1991

Shaw, R. S., The dripping faucet as a model chaotic system, Aerial Press, Santa Cruz, 1985

Simm, C. W., M. L. Sawley, F. Skiff, and A. Pochelon, On the analysis of experimental signals for evidence of deterministic chaos, *Helv. Phys. Acta*, 60, 510-596, 1987

Takens, F., Detecting strange attractors in turbulence, in Lect. Notes in Math., edited by A. Dold, and B. Eckmann, Springer Verlag, Berlin-Heidelberg-New York, 366-381, 1981

Vassiliadis, D. V., A. S. Sharma, T. E. Eastman, and K. Papandopoulos, Low-dimensional chaos in magnetospheric activity from AE time series, *Geophys. Res. Lett.*, 17, 1841-1844, 1990

Vörös, Z., Deterministic chaos in solar-terrestrial physics, (in Hungarian), in *Ionoszféra magnetoszféra fizika XVI*, edited by P. Bencze, 99-109, 1989

Vörös, Z., Fractal analysis applied to some geomagnetic storms observed at the Hurbanovo geomagnetic observatory, *Ann. Geophys.*, 8, 191-194, 1990

Wolf, A., J. B. Swift, H. L. Swinney, and J. A. Vastano, Determining Lyapunov exponents from a time series, *Physica*, 16D, 285-317, 1985

Yahata, H., Onset of chaos in the Rayleigh-Benard convection, *Prog. Theor. Phys. Suppl.*, 79, 26-74, 1984

Zeithamer, T., Topological stability of the Earth's magnetosphere and catastrophic features of its behaviour, in *Proc. XIX-th IUGG Gen. Ass., Vancouver*, 2, 686, 1987

Zeithamer, T., Dynamics of synergetic systems in solar terrestrial relationships, in *Proc. Symp. Human Biometeorology*, Strbské Pleso, edited by J. Zvonár, 452-456, 1988

Zeithamer, T., Structural stability of the Earth's magnetosphere, in *Geo-cosmic relations; the earth and its macro-environment*, edited by G. J. M. Tomassen et al., PUDOC, Wageningen, 321-326, 1990

RING CURRENT ION INTERACTION WITH MICROPULSATIONS

Xinlin Li, Anthony Chan and Mary Hudson

Department of Physics and Astronomy, Dartmouth College, Hanover New Hampshire 03755-1895

Ilan Roth

Space Science Laboratory, University of California, Berkeley, California 94720

The interaction between ring current ions and Pc 5 micropulsations has been studied numerically using a test particle code. Due to the convection potential, which increases during geomagnetic storms, the low energy($\lesssim 10$ keV) ions injected at midnight from the magnetotail cannot complete a drift path around the earth and are lost to the dayside magnetopause. The intermediate energy O^+ ions (tens of keV) interact with Pc5 micropulsations via a drift-bounce resonance. The resonant interaction leads to energy loss or gain for the particles, depending on their guiding center position with respect to the phase of the micropulsations. Once the energy drops below that required to overcome the convection potential, the particle will be lost to the magnetopause. H^+ ions satisfy the drift-bounce resonance condition at higher energies, where the convection potential is negligible, and at lower energies where the convection potential dominates. The mechanism reported here appears to contribute to O^+ ion loss observed *[Cladis and Lennartsson, 1986]* during the recovery phase of geomagnetic storms.

INTRODUCTION

Measurements of H^+ and O^+ phase-space densities by the ISEE-1 satellite show significant loss of 5—17 keV ring current O^+. The densities of the ions were measured while the satellite was inbound at 0900 local time and then while it was outbound around 0600 local time during the recovery phase of the storm of December 11, 1977 [*Cladis and Lennartsson, 1986*]. Fig. 1 shows the trajectory of ISEE-1 during that time.

Using the inbound measurements as initial conditions, *Cladis and Lennartsson, [1986]* calculated phase-space densities on the outbound path by following ion drift orbits in a dipole magnetic field with corotation and convection electric fields included. It was found that, for a convection electric field appropriate for $K_p = 2$, the measured and calculated H^+ phase-space densities on the outbound path were in agreement, whereas the measured O^+ phase-space densities were much lower than the calculated values. Fig. 2 shows the phase space density measurements. Apparently there is a rapid loss mechanism which acts selectively on the O^+ ions along their drift paths. We note that loss by charge exchange [*Smith and Bewtra, 1978*] is too slow to account for the observations.

Magnetospheric Substorms
Geophysical Monograph 64
Copyright 1991 American Geophysical Union

At the same time as the ISEE-1 measurements were made, Pc 5 magnetic pulsations were measured in the pre-noon local time sector, near synchronous orbit, by a number of satellites[*Cladis and Lennartsson,1986*]. Pc 5 magnetic pulsations have also been observed during quiet-to-moderate geomagnetic activity, and it has been suggested that the occurrence of Pc 5s may be correlated with substorm onsets [*Pangia et al, 1990*]. Briefly, Pc 5 pulsations have the following properties: (1) ultra low frequency (2—5 mHz), (2) westward propagation with $V_\phi \approx$ 10km/s, (3) antisymmetric standing wave structure along the field line, (4) azimuthal wavenumber $m = 40$–140. Most of the pulsations are measured on the dayside, at an equatorial radial distance $r = 5$–$8R_e$, where R_e is an earth radius, within about $20°$ of the magnetic equator [*Takahashi et al, 1985, 1987; Higbie et al, 1982*].

Since the Pc 5 wave frequency is comparable to the O^+ bounce frequency, *Cladis and Lennartsson [1986]* suggested that the loss of O^+ was due to a bounce interaction with the Pc 5 waves. In this work, we show that a combination of a drift-bounce resonant interaction and the $\mathbf{E} \times \mathbf{B}$ drift in the convection and corotation electric fields can lead to loss of O^+ from the dayside ring current. We investigate this interaction using a test particle code. In the following, we first investigate which ring current ions resonantly interact with the Pc 5 wave, then we describe an analytical model for the waves, based on spacecraft observations. Next we

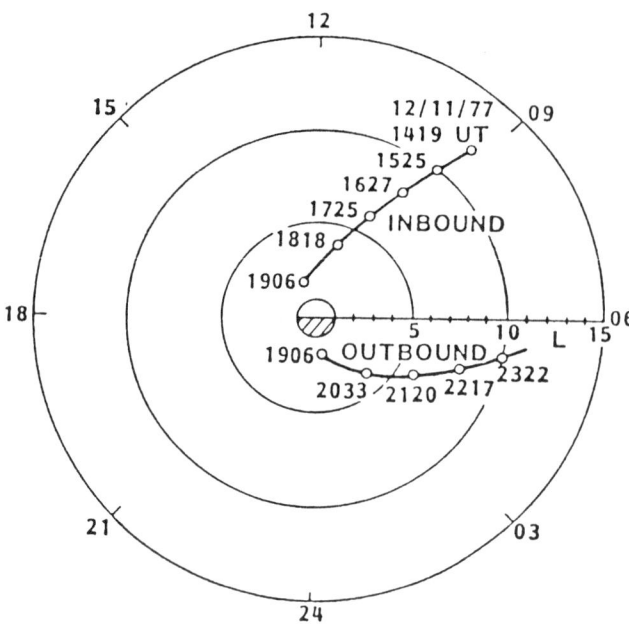

Fig. 1. ISEE-1 satellite trajectory (L versus magnetic local time) through the magnetosphere during recovery phase of storm of December 11, 1977 [*Cladis and Lennartsson, 1986*].

present our simulation results of the test particle motion in different field conditions, followed by a brief discussion.

ANALYSIS

The wave-particle interaction will be strongest for particles which satisfy the following drift-bounce resonance condition [*Southwood et al., 1969*]:

$$\omega_b = \pm(m\bar{\omega}_d - \omega)$$

where ω is the wave frequency, ω_b is particle's bounce frequency and $\bar{\omega}_d$ is the particle's bounce-averaged drift frequency. Here $\omega_b \propto W^{\frac{1}{2}}/L$ and $\bar{\omega}_d \propto WL$, where L is the shell number of the earth's dipole magnetic field and W is particle kinetic energy; the frequencies ω_b and $\bar{\omega}_d$ also depend weakly on pitch angle [*Hamlin et al., 1961*]. Physically, viewed in a frame moving with the wave, resonant ions drift in one bounce period by a perpendicular wavelength: $\lambda_\perp = 2\pi L R_e/m$. To investigate which ring current ions will be resonant, we have calculated and plotted in Fig. 3 the resonant energy for a range of azimuthal wave numbers. There is a high and low energy branch corresponding to the $\pm$ sign above. The low energy resonance is satisfied for H^+ and for O^+ with $W \leq 10$ keV. These ions will be lost to the dayside magnetopause by $\mathbf{E} \times \mathbf{B}$ drift under the effect of the convection electric field. The high energy resonance is satisfied by H^+ with $W \sim 100$ keV and O^+ with $W \sim$ (tens of keV). The high energy resonant H^+ have been studied by Chan et al [1989]; here we focus on the high-energy resonant O^+.

TEST PARTICLE MODEL

A test particle model is used to study the particle behavior in a Pc 5 pulsation. The electromagnetic field is assumed to consist of a time-independent part, which is the dipole magnetic field superimposed with corotation and convection electric fields, and a time-dependent Pc 5 wave.

The corotation and convection electric field are defined as follows:

$$\mathbf{E}_{cor} = -(\mathbf{\Omega}_e \times \mathbf{r}) \times \mathbf{B}, \quad (1a)$$
$$\mathbf{E}_{con} = -\nabla \Psi. \quad (1b)$$

Here $\mathbf{B}$ is the earth's dipole magnetic field, $\mathbf{\Omega}_e$ is the angular frequency of the earth, $\mathbf{r}$ is the position vector from the center of the earth to the field point and Ψ is a K_p-dependent potential [*Maynard and Chen, 1975*] of the form:

$$\Psi = 0.045(1 - 0.159 K_p + 0.0093 K_p^2)^{-3} L^2 \sin(\phi_s). \quad (2)$$

Here Ψ is in kV and ϕ_s is azimuthal angle.

Based on spacecraft observations, the compressional Pc 5 is assumed to be a standing wave traveling in the azimuthal direction. The wave fields are specified by the time varying components $b_\parallel$, b_α and e_ϕ which are the compressional and 'radial' perturbed magnetic field components and the azimuthal perturbed electric field component, collectively satisfying Faraday's Law and $\nabla \cdot \mathbf{b} = 0$. Dipole coordinates used in the calculation are

$$\alpha = \frac{R_e}{r}\cos^2\lambda, \quad (3a)$$

$$\delta = \left(\frac{R_e}{r}\right)^2 \sin\lambda, \quad (3b)$$

and longitude ϕ (positive westward), where $\hat{\alpha}, \hat{\phi}, \hat{\delta}$ form a right hand coordinate system. At the magnetic equator $\hat{\alpha} = -\hat{r}$, $\hat{\delta} = -\hat{\theta}$ and $\hat{\phi} = -\hat{\phi}_s$, where (r, θ, ϕ_s) are the usual spherical polar coordinates. Here λ is magnetic latitude. The fields are determined from a vector potential which has only an azimuthal component $\mathbf{A}_\phi$ such that

$$\mathbf{b} = \nabla \times \mathbf{A}_\phi, \quad (4a)$$

$$\mathbf{e}_\phi = -\frac{1}{c}\frac{\partial \mathbf{A}_\phi}{\partial t}, \quad (4b)$$

where $\mathbf{A}_\phi = A_0 \alpha H(\alpha) K(\lambda) e^{i(m\phi - \omega t)}$. Here $H(\alpha)$ and $K(\lambda)$ are arbitrary functions of α and λ, respectively, and $\alpha = 1/L$ in our dipole coordinates. If we choose : $K(\lambda) = \sin(k\lambda)$,

$$H(\alpha) = \frac{1}{L^{n-1}} exp[-\frac{(L-L_0)^2}{4a^2}]$$

and $A_0 = R_e B_0 \beta_\parallel$, one obtains the following components of the wave fields:

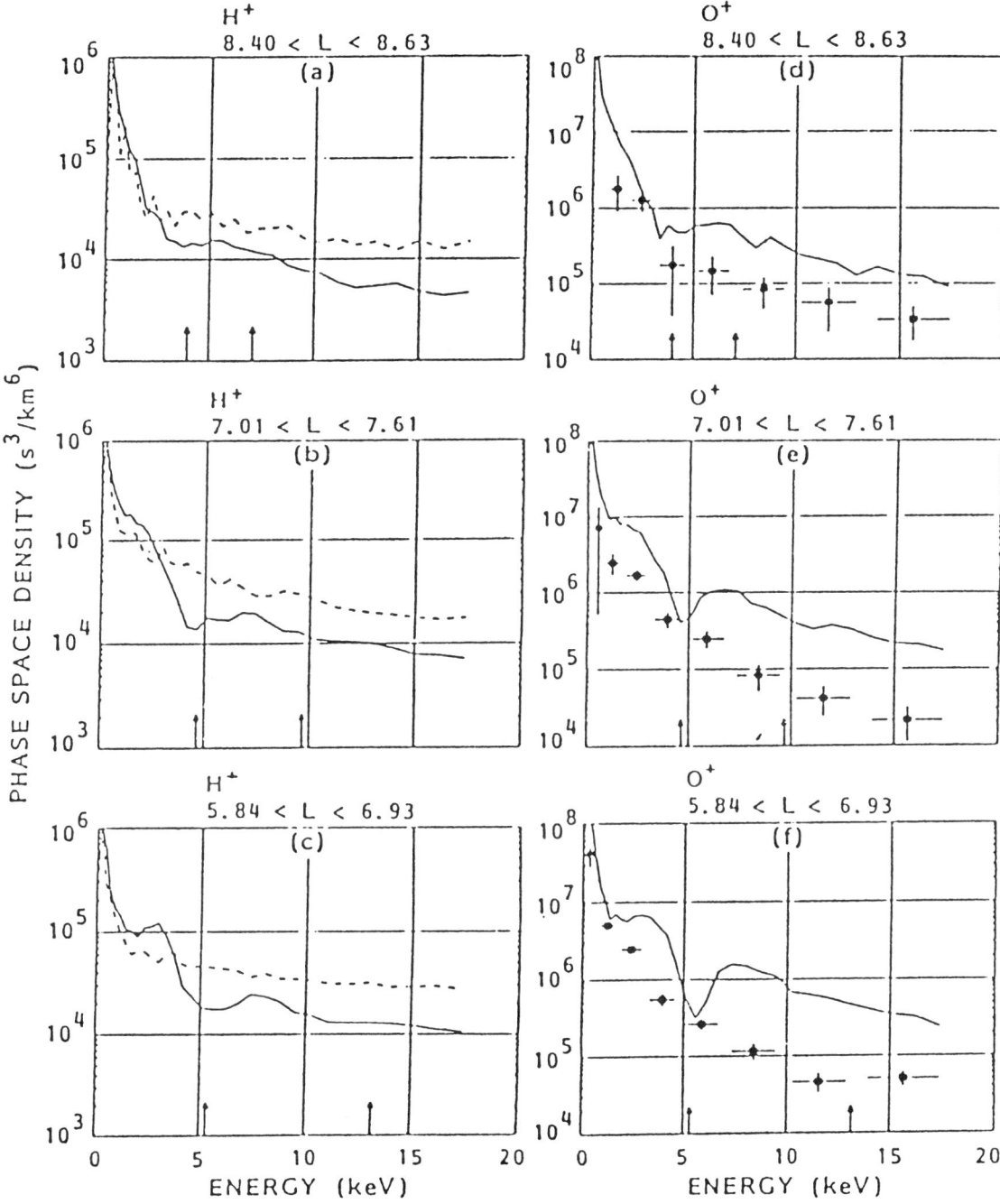

Fig. 2. Phase densities of H^+ (graphs a, b, c) and O^+ (graphs d, e, f) measured while the ISEE-1 satellite was inbound (solid lines) and outbound (broken lines in a, b, c; data points in d, e, f). The L value intervals are given at the tops of the graphs[Cladis and Lennartsson, 1986].

$$\frac{b_\|}{B} = -\frac{\beta_\|}{L^{n-2}} exp[-\frac{(L-L_0)^2}{4a^2}] \times$$
$$\frac{\cos^3\lambda}{1+3\sin^2\lambda}\{-[(n-1)+\frac{L(L-L_0)}{2a^2}](1+3\sin^2\lambda)\sin k\lambda$$
$$+2\sin\lambda[-3\sin\lambda\sin k\lambda + k\cos\lambda\cos k\lambda]\}e^{i(m\phi-\omega t)}, \quad (5a)$$

$$\frac{b_\alpha}{B} = -\frac{\beta_\|}{L^{n-2}} exp[-\frac{(L-L_0)^2}{4a^2}]$$
$$\frac{\cos^4\lambda}{1+3\sin^2\lambda}[-3\sin\lambda\sin k\lambda + k\cos\lambda\cos k\lambda]e^{i(m\phi-\omega t)}, \quad (5b)$$

$$e_\phi = i\frac{\omega R_e B_0}{c}\frac{\beta_\|}{L^n} exp[-\frac{(L-L_0)^2}{4a^2}]\sin k\lambda e^{i(m\phi-\omega t)}. \quad (5c)$$

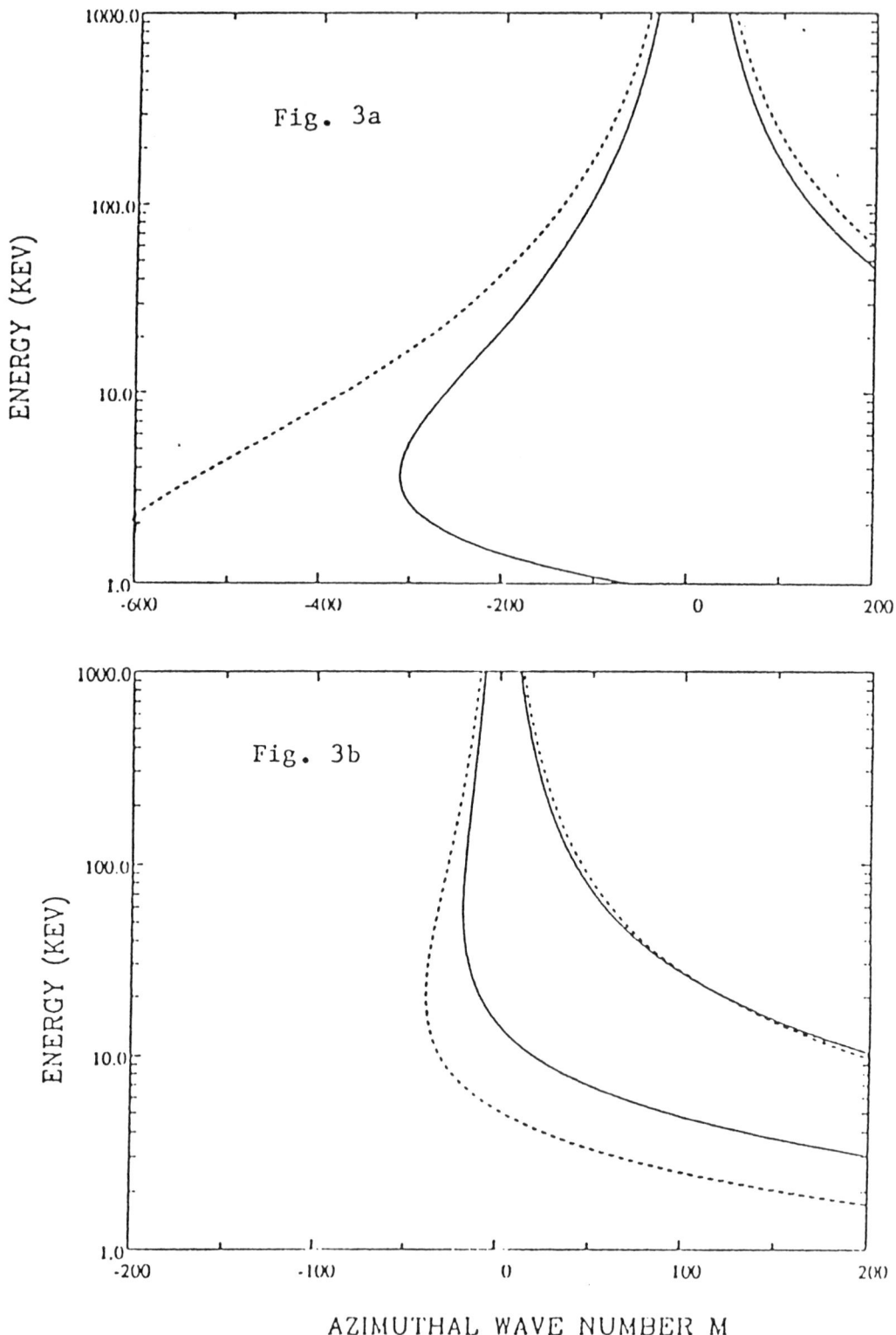

Fig. 3. Drift-bounce resonant energy versus azimuthal wavenumber for H^+ (upper graph) and O^+ (lower graph), calculated from: $\omega_b = \pm(m\bar{\omega}_d - \omega)$, where $f = 2 mHz$, L=6.5. + (−) sign corresponds to higher (lower) branch. Solid (broken) lines correspond to a pitch angle of 5° (80°).

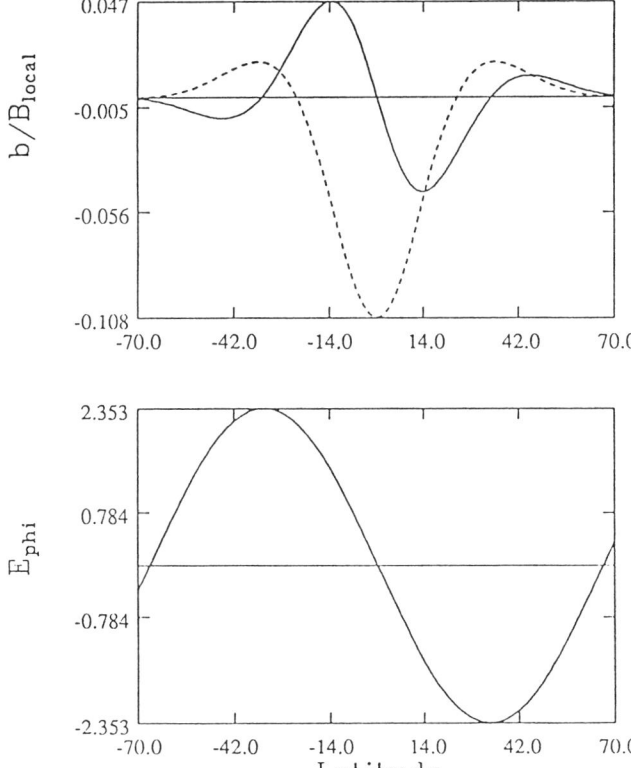

Fig. 4. Amplitudes of the wave field of components $b_\|$ (solid line), b_α (dashed line) and e_ϕ in mV/m in our Pc 5 wave model. $f = 2$ mHz, $L=6.5$, $L_0=6.5$ and $a=1$ [Hudson et al., 1991].

Here $\beta_\|$ is an amplitude, chosen to give wave amplitudes comparable to the measured amplitudes of $b_\alpha/B \approx 10\%$, and n is an integer, which gives the monotonic radial dependence of the wave magnetic fields. The Gaussian factor confines the wave fields to a radial region consistent with observations. L_0 is the center of the Gaussian distribution and a is the half-width in units of Earth radii. For $n=0$, $f=2$mHz, parallel wavenumber $k=2.7$ in units of inverse radians, $L_0 = 6.5$ and $a=1$, we plot $b_\|, b_\alpha$ and e_ϕ in Fig. 4. The model fields in Fig. 4 are consistent with the spacecraft measurements, and are in good agreement with the eigenfunction fields computed by Chan and Chen, [1990].

Since $k_\perp \rho$ is $\sim O(1)$ in our case, where $k_\perp = 2\pi/\lambda_\perp$ and ρ is the Larmor radius of the O^+, the particle orbit is obtained by numerically integrating the Newton-Lorentz equations of motion (rather than the guiding center equations)

$$m\frac{d\mathbf{V}}{dt} = q(\mathbf{E} + \frac{\mathbf{V} \times \mathbf{B}}{c}), \quad (6a)$$

$$\frac{d\mathbf{X}}{dt} = \mathbf{V}, \quad (6b)$$

where $\mathbf{E}$ and $\mathbf{B}$ represent the total electric and magnetic fields (i.e., background and wave fields).

We use the Richardson extrapolation with modified midpoint method[Press et al., 1986] to push the particles in phase space. At each step of the extrapolation procedure, we decompose the perturbed fields as well as the background fields into Cartesian components, where **x** points toward midnight, **y** points dawnward and **z** points northward. The origin is at the center of the earth. After evaluation of the new positions and velocities of the particle, we calculate the fields at the new location as functions of the dipole coordinates, Eqs. (3) and (5).

RESULTS OF SIMULATIONS

Fig. 5 shows the orbit of a 25.5 keV O^+ ion injected with 60 degree pitch angle at 1800 LT into the earth's magnetic field with convection and corotation electric fields but no Pc 5 wave. The orbit displayed is a projection onto the equatorial plane. The orbit appears irregular because we plot roughly every ten cyclotron periods. The inner extent of the particle's orbit is the mirror point and the outer extent is the equatorial crossing point. The particle does not have enough initial energy to overcome the convection field and corotation field in order to make a complete drift path across the dayside. Near dawn it follows the equipotential lines and is lost to the dayside magnetopause. Fig. 6 shows the orbit of a 27 keV O^+ ion (other parameters the same as in the case of Fig. 5). This particle has enough energy to complete a drift orbit.

In Fig. 7, we show an orbit of a 27 keV O^+ ion, now assuming there are no convection and corotation electric

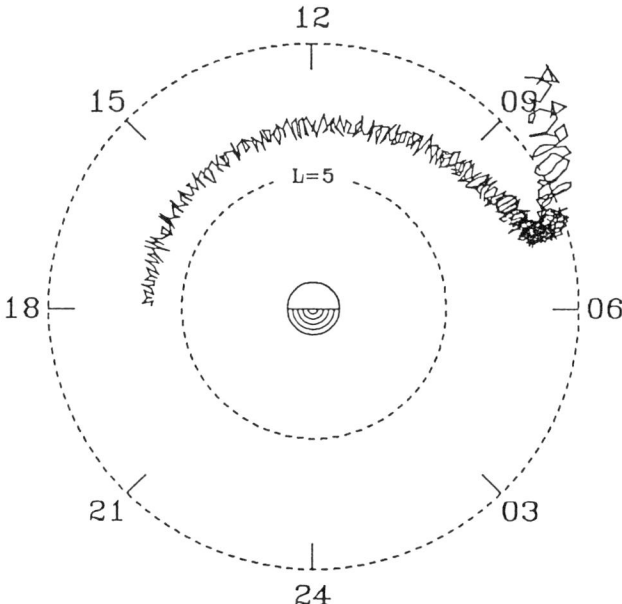

Fig. 5. The orbit of a 25.5 keV O^+ ion injected with 60° pitch angle at 1800 LT and at L=6.5, projected onto the equatorial plane. Background fields only, $K_p = 1.5$.

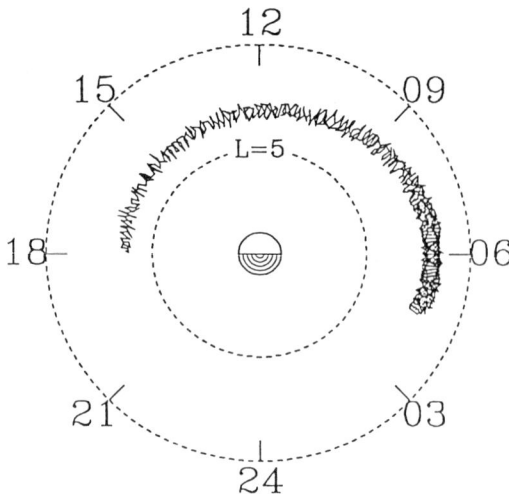

Fig. 6. Same as Fig. 5, except for a 27 keV O^+ ion.

we have plotted equatorial trajectories of 36 O^+ ions having the same initial energy, pitch angle and L shell, but different gyrophase angle, $\arctan(V_y/V_x)$, where V_x and V_y are the x and y components of the particle's velocity, respectively. Without the Pc 5 wave, all 36 particles can make a complete path around the earth. If we turn on the wave (the same wave described before, e.g., in Fig. 7), as shown in Fig. 10, 18 of them are lost due to combined effects of the wave and the convection and corotation fields. Here the different gyrophase corresponds to a different guiding center location with respect to the wave phase which deter-

fields; but including the Pc 5 wave fields, with mode number $m=100$; other parameters are the same as in Fig. 4. The particle is trapped in the Pc 5 wave and undergoes a long period oscillation with a large L variation, a characteristic phenomenon of resonance with the wave.

In Fig. 8a, we turn on both the convection and corotation electric fields and the Pc 5 wave. The particle interacts resonantly with the Pc 5 wave, it loses energy, and is subsequently lost to the magnetopause when the convection electric field $\mathbf{E} \times \mathbf{B}$ drift begins to dominate. Fig. 8b shows the corresponding energy versus time history when the particle crosses the equator.

To further demonstrate the significance of Pc 5 waves for causing O^+ to be lost to the dayside magnetopause, in Fig. 9

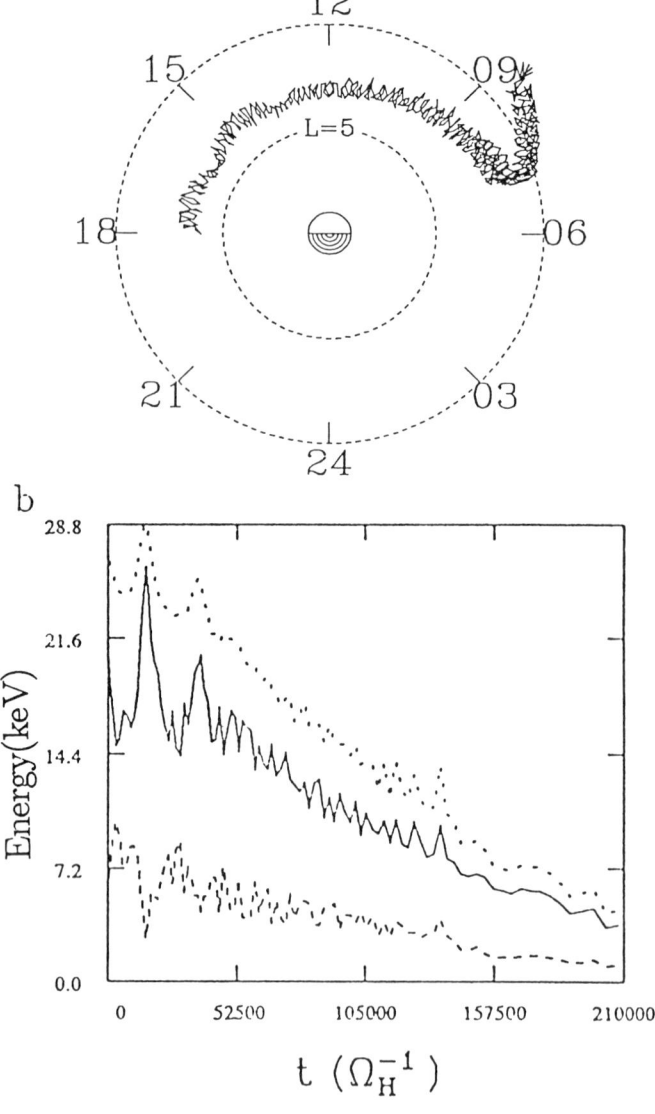

Fig. 8. (a) Same as Fig. 7, except the convection and corotation fields are turned on. (b) The corresponding energy versus time history when the particle passes the equator. Dotted line is for the total energy, solid line is for perpendicular and dashed line for parallel.

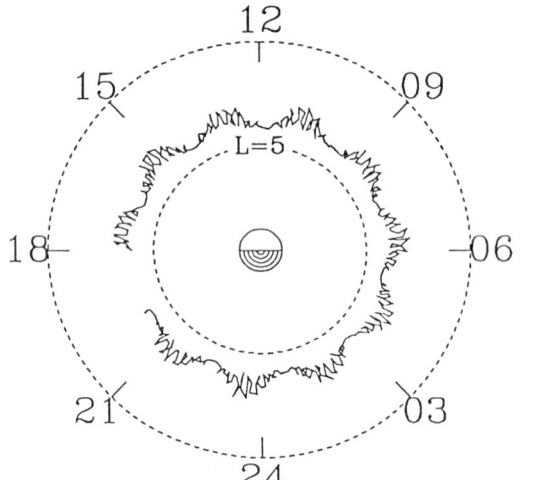

Fig. 7. Same as Fig. 6, except no convection and corotation electric fields; Pc 5 wave plotted in Fig. 4 is present.

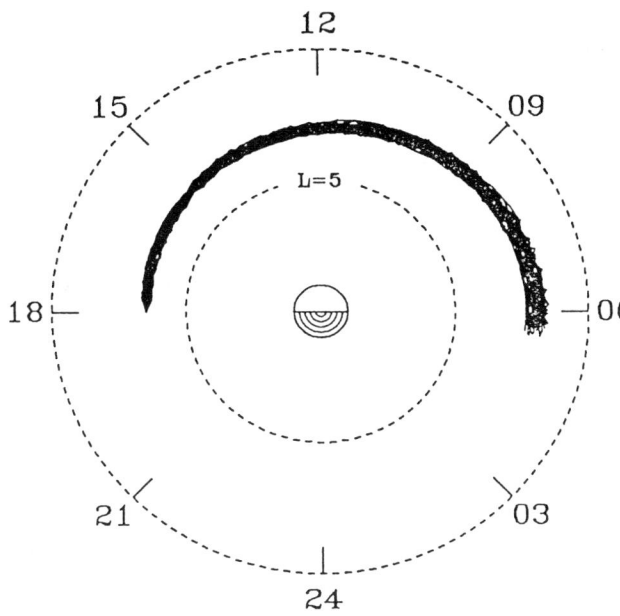

Fig. 9. The equatorial trajectories of 36 O^+ ions with the same initial energy of 28 keV, pitch angle of 60° and $L=6.5$, but different gyrophase(=arctan(V_y/V_x)), i.e., gyrophases between 0° and 350° in steps of 10°.

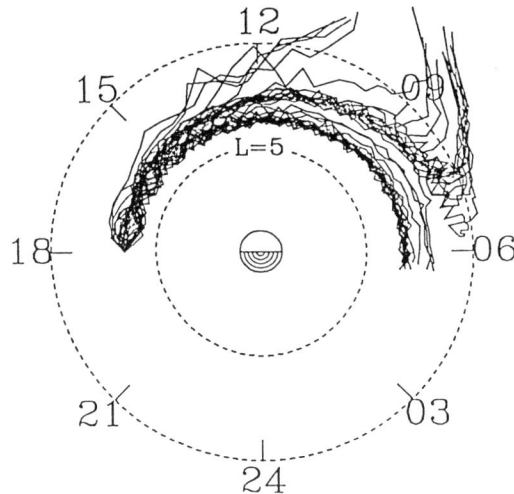

Fig. 10. Same as in Fig. 9, except the Pc 5 wave, described in Fig. 4, is turned on.

mines the interaction. In other words, these 36 O^+ ions are initially exposed to different wave phases for a given wave. Some of them (at least 18 of them) lose energy due to the presence of the Pc 5 wave. Others gain energy and move radially inward. Those which move inward gain a maximum of a few keV from the Pc 5 wave; in addition, their equatorial pitch angles are increased to about 70°.

Pc 5 waves have a wide range of azimuthal wave number (m=40–140). Our preliminary study of a wave consisting of a sum of waves, each with a different m value, shows further enhancement of O^+ loss. In Fig. 11, the Pc 5 wave has ten different azimuthal wave numbers, ranging from $m=95$ to $m=105$, each with a random initial azimuthal phase but the same amplitude, which now is reduced by $1/\sqrt{M}$, where $M=10$ is the number of waves in the sum, to keep the total wave power at the same level as in a single mode wave. There are 30 O^+ ions with the same initial gyrophase, L shell and pitch angle, but different energies. In the presence of this wave, 21 of them are lost. If there are no Pc 5 waves present, then only 11 of them will be lost, the ones with lowest initial energies. Counting from 0900LT to dawn, there are 11 O^+ ions lost out of 20, so the loss rate is about 55%. The observed number density loss rate is around 75% from Fig. 2. Other loss mechanisms may also contribute, for example, the charge-exchange loss is expected to be approximately 5% [Smith and Bewtra, 1978]. Further runs are needed to make a better quantitative comparison with observations and with other loss mechanisms.

We have analytically calculated several aspects of the wave-particle interaction between ring current ions, time-independent fields and wave fields. For example, the threshold energy which an O^+ ion must have when injected at 1800 LT in order to traverse the dayside in the presence of time-independent fields only is about 25 keV for our simulation parameters (Hudson et al., 1991). The addition of model

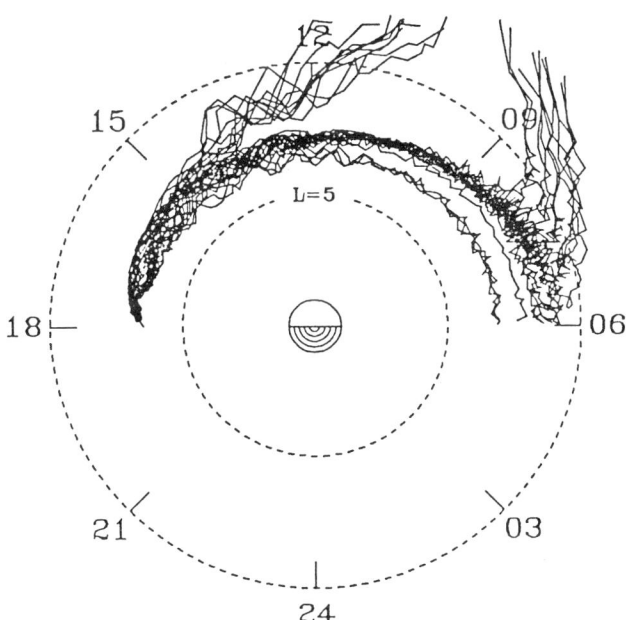

Fig. 11. The equatorial trajectories of 30 O^+ ions with the same initial gyrophase of 0°, pitch angle of 60° and $L=6.5$, but different initial energy, i.e., in the range of 24.00 keV to 29.80 keV, in steps of 0.20 keV. Here the Pc 5 wave has ten different azimuthal wave numbers, ranging from $m=95$ to $m=105$, each with a random initial azimuthal phase and the same amplitude, which now is reduced by $1/\sqrt{10}$.

Pc 5 wave fields produces an oscillation in total particle energy at the same frequency as the radial oscillation period evident in Fig. 8a, see Fig. 8b. This energy oscillation has been calculated by Qian et al, (1991) for ions not too close to resonance, and the resulting energy oscillation scales as

$$\frac{\delta W}{W_\perp} \propto \beta_\parallel \frac{\omega}{\Delta\omega}(e^{i\Delta\omega} - 1),$$

where $\Delta\omega = \omega_b \pm (m\bar{\omega}_d + \omega)$. The amplitude of the energy oscillations can be large enough to cause ions to drop below the threshold energy required to traverse the dayside in the presence of the corotation and convection electric fields. We plan to present further detailed calculations of the energy exchange in a future publication.

Discussion

As the particles travel across the dayside sector, they are de-energized by the convection electric field. The total energy loss is determined by the potential drop from dawn to dusk, which depends on the geomagnetic activity. The net loss rate in the presence of Pc 5 waves is not expected to vary significantly for different convection field models as long as they represent the same total potential drop. For all the runs, we took $K_p = 1.5$ corresponding to the recovery phase of a magnetic storm, which gives a dawn-dusk potential drop of 12 keV at $L = 8$. Under such a circumstance, the convection and corotation electric fields remove particles with lower energies ($\lesssim 20$ keV at dusk) by $\mathbf{E} \times \mathbf{B}$ drift to the dayside magnetopause. Additionally, they can tune and detune the resonance of the particles with the Pc 5 wave. O^+ with high energies ($\gtrsim 40$ keV at dusk) will drift around the earth, regardless of the presence of the Pc 5 wave. Those O^+ ions with intermediate energies can be lost to the magnetopause due to combined effects of the Pc 5 wave and the background fields. As ions drift from dusk to dawn, they lose energy to the background electric fields, and they may gain or lose energy due to interaction with the Pc 5 wave, depending on their guiding center positions with respect to the azimuthal phase of the wave.

In the case of the ISEE-1 measurements in Fig. 2, 5-17 keV ions starting at 0900 LT fall in the lower and intermediate energy range at dusk. The lower energy ions in the range of the ion composition experiment on ISEE-1 will be swept away by the convection potential, or else corotate inside the plasmapause, in the absence of Pc 5 wave interaction. The more energetic 17 keV O^+ ions measured at 0900 LT and $L=7.5$ correspond to 27 keV ions at 1800 LT in the corotation and convection fields given by (1) and (2) for the parameters of our simulations. Thus the combined effects of background fields and Pc 5s are important for these ions. In future work, we plan to use a more realistic particle distribution function, obtained from observed particle data, and to compare quantitatively with other loss mechanisms.

Acknowledgements. We gratefully acknowledge helpful discussions with Richard Denton and Perry Gray; and a travel grant from the American Geophysical Union to attend the Chapman Conference on Magnetospheric Substorms (X. Li). This work was supported by NASA Grants NAG 5-1098, NAGW-1652, NAGW-1626, and Air Force Contract F19628-87-K-0038. Computations were performed on the Convex C1 at Dartmouth, and the SDSC Cray.

References

Chan, A. A. and L. Chen, MHD ballooning mode calculations for dipole magnetic field, *EOS Trans. Am. Geophys. Union, 71*, 612, 1990.

Chan, A. A., L. Chen and R. B. White, Nonlinear interaction of energetic ring current protons with magnetospheric hydromagnetic waves, *J. Geophys. Res. Lett., 16*, 1133, 1989.

Cladis, J. B. and O. W. Lennartsson, On the loss of O^+ ions (< 17keV/e) in the ring current during the recovery phase of a storm, in *Ion Acceleration in the Magnetosphere and Ionosphere*, Tom Chang, ed. AGU, Washington, D.C., pp. 153–157, 1986.

Hamlin, D. A., R. K. Karplus, R. C. Vik, and K. M. Watson, Mirror and azimuthal drift requencies for geomagnetically trapped particles, *J. Geophys. Res., 66*, 1, 1961.

Higbie, P. R., D. N. Baker, R. D. Zwickl, R. D. Belian, J. R. Asbridge, J. F. Fennel, B. Wilken, and C. W. Arthur, The global Pc 5 event of November 14 -15, 1979, *J. Geophys. Res., 87*, 2337 1982.

Hudson, M. K., A. A. Chan, X. Li and I. Roth, Ring current ion interaction with Pc 5 micropulsations, in *Physics of Space Plasmas, SPI Conf. Proc. and Rep. Series, 10*, Scientific Publishers, Inc. pp. 263–275, 1991.

Maynard, N. C. and J. J. Chen. Isolated cold plasma region: observations and their relation to possible production mechanisms, *J. Geophys. Res., 80*, 1009, 1975.

Pangia, M. J., C. S. Lin and J. N. Barfield, A correlative study of Pc 5 magnetic pulsations with substorm onsets, *J. Geophys. Res., 95*, 10699, 1990.

Press, W. H., B. P. Flannery, S. A. Teukolsky, and W. T. Vetterling, *Numerical Recipes*, chap. 5, Cambridge University Press 1986.

Qian, S., M. K. Hudson and I. Roth, Ring current O^+ interaction with Pc 5 micropulsations, *Geophysical Monograph Series*, G. Wilson and M. Chandler, eds., American Geophysical Union, in press, 1991.

Smith, P. H. and N. K. Bewtra, Charge exchange lifetimes for ring ions, *Space Sci. Rev. 22*, 301, 1978.

Southwood, D. J., J. W. Dungey, and R. L. Etherington, Bounce resonant interaction between pulsations and trapped particles, *Planet. Space Sci. 17*, 349, 1969.

Takahashi, K., P. R. Higbie, and D. N. Baker, Azimuthal propagation and frequency characteristic of compressional Pc 5 waves observed at geostationary orbit, *J. Geophys. Res., 90*, 1473, 1985.